# Handbook of Hydrocarbon and Lipid Microbiology

**Series Editors**

Kenneth N. Timmis (Editor-in-Chief)
Emeritus Professor
Institute of Microbiology
Technical University Braunschweig
Braunschweig, Germany

Matthias Boll
Institute of Biology/Microbiology
University of Freiburg
Freiburg, Germany

Otto Geiger
Universidad Nacional Autónoma de
México
Centro de Ciencias Genómicas
Cuernavaca, Morelos, México

Howard Goldfine
Department of Microbiology
University of Pennsylvania Perelman
School of Medicine
Philadelphia, PA, USA

Tino Krell
Department of Environmental
Protection
Estación Experimental del Zaidín,
Consejo Superior de Investigaciones
Científicas
Granada, Spain

Sang Yup Lee
Department of Chemical and
Biomolecular Engineering
Korea Advanced Institute of Science
and Technology (KAIST)
Daejeon, Republic of Korea

Terry J. McGenity
School of Biological Sciences
University of Essex
Wivenhoe Park, Colchester, UK

Fernando Rojo
Centro Nacional de
Biotecnología, CSIC
Madrid, Spain

Diana Z. Sousa
Laboratory of Microbiology
Wageningen University
Wageningen, The Netherlands

Alfons J. M. Stams
Laboratory of Microbiology
Wageningen University
Wageningen, The Netherlands

Robert Steffan
Cape Coral
FL, USA

Heinz Wilkes
Organic Geochemistry, Institute for
Chemistry and Biology of the Marine
Environment (ICBM)
Carl von Ossietzky University
Oldenburg
Oldenburg, Germany

This handbook is the unique and definitive resource of current knowledge on the diverse and multifaceted aspects of microbial interactions with hydrocarbons and lipids, the microbial players, the physiological mechanisms and adaptive strategies underlying microbial life and activities at hydrophobic material: aqueous liquid interfaces, and the multitude of health, environmental and biotechnological consequences of these activities.

**Scientific Advisory Board**
Victor de Lorenzo, Eduardo Diaz, Otto Geiger, Ian Head, Sang Yup Lee, Terry McGenity, Colin Murrell, Balbina Nogales, Roger Prince, Juan Luis Ramos,Wilfred Röling, Eliora Ron, Burkhard Tümmler, Jan Roelof van der Meer, Willy Verstraete, Friedrich Widdel, Heinz Wilkes and Michail Yakimov.

More information about this series at http://www.springer.com/series/13884

Tino Krell
Editor

# Cellular Ecophysiology of Microbe: Hydrocarbon and Lipid Interactions

With 80 Figures and 18 Tables

 Springer

*Editor*
Tino Krell
Department of Environmental Protection
Estación Experimental del Zaidín
Consejo Superior de Investigaciones Científicas
Granada, Spain

ISBN 978-3-319-50540-4      ISBN 978-3-319-50542-8 (eBook)
ISBN 978-3-319-50541-1 (print and electronic bundle)
https://doi.org/10.1007/978-3-319-50542-8

Library of Congress Control Number: 2017963780

This Springer imprint is published by the registered company Springer International Publishing AG, part
of Springer Nature.
The registered company address is: Gewerbestrasse 11, 6330 Cham, Switzerland

# Preface

Hydrocarbons have played a key role in bacterial evolution. Few other compounds cause physiological effects as varied as those elicited by hydrocarbons; many are toxic and cause bacterial cell death, while others serve as important growth substrates. This diversity makes the field of research focused on hydrocarbons and bacteria complex and exciting. Here, we will review different mechanisms by which bacteria sense the presence of hydrocarbons and how these signals translate into bacterial responses. Bacteria rely on two basic evolutionary strategies to cope with the toxic effects of hydrocarbons: fight or flee. Currently, data suggests that the "fight" strategy is predominant because a multitude of different adaptive mechanisms have been identified that enable bacteria to cope with toxic hydrocarbons, while there are relatively few examples of directed cell movement away from hydrocarbons. Of the different strategies that bacteria use to cope with toxic hydrocarbons, active expulsion of the toxic compound from the cell has been proven to be a particularly successful evolutionary approach. As well, there is a wealth of data available on the different pathways that mediate degradation of hydrocarbons. However, there is relatively little information available on how hydrocarbons enter the cell; thus, exploration of processes related to hydrocarbon uptake represents a major research need. Hydrocarbons also play regulatory roles in the cell by acting as signal molecules. Furthermore, biological molecules are often chemically modified via the attachment of a hydrocarbon – a process that can mediate important effects. Accordingly, the mechanisms and functional consequences of DNA and protein methylation, which is of central importance to the cell, are also summarized in this book. Other signaling hydrocarbons – many of which contain oxygen and nitrogen in addition to hydrogen and carbon – have been shown to play key roles in quorum sensing processes, and they are also covered here.

Biodegradation and biotransformation are the two primary biotechnological applications of hydrocarbon-related research. However, these biotechnological processes often lack the efficiency required to compete with existing processes. Deciphering the ecophysiological consequences of bacterial exposure to hydrocarbons has the potential to provide new and valuable insights – insights that may ultimately enable us to better optimize such processes.

Granada, Spain                                                                 Tino Krell

# Contents

**Part I   Problems of Hydrophobicity, Bioavailability** . . . . . . . . . . . .   **1**

1   **Problems of Hydrophobicity/Bioavailability: An Introduction** . . .   3
Hauke Harms, Kilian E. C. Smith, and Lukas Y. Wick

2   **Microorganism-Hydrophobic Compound Interactions** . . . . . . . .   17
Lukas Y. Wick, Hauke Harms, and Kilian E. C. Smith

3   **Matrix - Hydrophobic Compound Interactions** . . . . . . . . . . . . . .   33
Hauke Harms, Lukas Y. Wick, and Kilian E. C. Smith

4   **Assimilation of Hydrocarbons and Lipids by Means of Biofilm
Formation** . . . . . . . . . . . . . . . . . . . . . . . . . . . . . . . . . . . . . . . . . . .   47
Pierre Sivadon and Régis Grimaud

5   **Uptake and Assimilation of Hydrophobic Substrates by
the Oleaginous Yeast *Yarrowia lipolytica*** . . . . . . . . . . . . . . . . . . .   59
France Thevenieau, Athanasios Beopoulos, Thomas Desfougeres,
Julia Sabirova, Koos Albertin, Smita Zinjarde, and Jean Marc Nicaud

6   **Biodiversity of Biosurfactants and Roles in Enhancing
the (Bio)availability of Hydrophobic Substrates** . . . . . . . . . . . . . .   75
Amedea Perfumo, Michelle Rudden, Roger Marchant, and
Ibrahim M. Banat

7   **Biofilm Stress Responses Associated to Aromatic
Hydrocarbons** . . . . . . . . . . . . . . . . . . . . . . . . . . . . . . . . . . . . . . . . .   105
Laura Barrientos-Moreno and Manuel Espinosa-Urgel

**Part II   Sensing, Signaling and Uptake** . . . . . . . . . . . . . . . . . . . . . .   **117**

8   **Sensing, Signaling, and Uptake: An Introduction** . . . . . . . . . . . .   119
Tino Krell

9   Bioinformatic, Molecular, and Genetic Tools for Exploring
    Genome-Wide Responses to Hydrocarbons ................. 127
    O. N. Reva, R. E. Pierneef, and B. Tümmler

10  One-Component Systems that Regulate the Expression
    of Degradation Pathways for Aromatic Compounds .......... 137
    G. Durante-Rodríguez, H. Gómez-Álvarez, J. Nogales,
    M. Carmona, and E. Díaz

11  Transcriptional Regulation of Hydrocarbon Efflux Pump
    Expression in Bacteria ................................. 177
    Cauã Antunes Westmann, Luana de Fátima Alves, Tiago Cabral
    Borelli, Rafael Silva-Rocha, and María-Eugenia Guazzaroni

12  The Family of Two-Component Systems That Regulate
    Hydrocarbon Degradation Pathways ...................... 201
    Andreas Busch, Noel Mesa-Torres, and Tino Krell

13  Chemotaxis to Hydrocarbons ............................ 221
    Rebecca E. Parales and Jayna L. Ditty

14  The Potential of Hydrocarbon Chemotaxis to Increase
    Bioavailability and Biodegradation Efficiency ................ 241
    Jesús Lacal

15  Amphiphilic Lipids, Signaling Molecules, and
    Quorum Sensing ...................................... 255
    M. Dow and L. M. Naughton

16  Fatty Acids as Mediators of Intercellular Signaling ........... 273
    Manuel Espinosa-Urgel

17  Substrate Transport .................................... 287
    Rebecca E. Parales and Jayna L. Ditty

18  Strategies to Increase Bioavailability and Uptake of
    Hydrocarbons ......................................... 303
    J. J. Ortega-Calvo

19  The Mycosphere as a Hotspot for the Biotransformation
    of Contaminants in Soil ................................. 315
    Lukas Y. Wick and Hauke Harms

Part III   Problems of Solventogenicity, Solvent Tolerance ....... 325

20  Problems of Solventogenicity, Solvent Tolerance:
    An Introduction ....................................... 327
    Miguel A. Matilla

21  Toxicity of Hydrocarbons to Microorganisms  ................  335
    Hermann J. Heipieper and P. M. Martínez

22  Genetics of Sensing, Accessing, and Exploiting Hydrocarbons  ...  345
    Miguel A. Matilla, Craig Daniels, Teresa del Castillo,
    Andreas Busch, Jesús Lacal, Ana Segura, Juan Luis Ramos, and
    Tino Krell

23  Extrusion Pumps for Hydrocarbons: An Efficient
    Evolutionary Strategy to Confer Resistance to Hydrocarbons  ...  361
    Matilde Fernández, Craig Daniels, Vanina García,
    Bilge Hilal Cadirci, Ana Segura, Juan Luis Ramos, and Tino Krell

24  Membrane Composition and Modifications in Response to
    Aromatic Hydrocarbons in Gram-Negative Bacteria  ...........  373
    Álvaro Ortega, Ana Segura, Patricia Bernal, Cecilia Pini,
    Craig Daniels, Juan Luis Ramos, Tino Krell, and
    Miguel A. Matilla

25  Cis–Trans Isomerase of Unsaturated Fatty Acids:
    An Immediate Bacterial Adaptive Mechanism to Cope
    with Emerging Membrane Perturbation Caused by Toxic
    Hydrocarbons  .........................................  385
    Hermann J. Heipieper, J. Fischer, and F. Meinhardt

26  Surface Properties and Cellular Energetics of Bacteria in
    Response to the Presence of Hydrocarbons  .................  397
    Hermann J. Heipieper, Milva Pepi, Thomas Baumgarten, and
    Christian Eberlein

27  Ultrastructural Insights into Microbial Life at the
    Hydrocarbon: Aqueous Environment Interface  ..............  409
    Nassim Ataii, Tyne McHugh, Junha Song, Armaity Nasarabadi, and
    Manfred Auer

28  Microbiology of Oil Fly Larvae  ..........................  419
    K. W. Nickerson and B. Plantz

Part IV  Problems of Feast or Famine  ......................  429

29  Nitrogen Fixation and Hydrocarbon-Oxidizing Bacteria  ........  431
    J. Foght

30  Kinetics and Physiology at Vanishingly Small Substrate
    Concentrations  .......................................  449
    D. K. Button

31    Feast: Choking on Acetyl-CoA, the Glyoxylate Shunt, and
      Acetyl-CoA-Driven Metabolism ......................... 463
      M. Peña Mattozzi, Yisheng Kang, and Jay D. Keasling

Part V    Hydrophobic Modifications of Biomolecules  ...........    475

32    Hydrophobic Modifications of Biomolecules: An Introduction  ...  477
      Álvaro Ortega

33    DNA Methylation in Prokaryotes: Regulation and Function  .....  487
      Saswat S. Mohapatra and Emanuele G. Biondi

34    DNA Methylation in Eukaryotes: Regulation and Function  .....  509
      Hans Helmut Niller, Anett Demcsák, and Janos Minarovits

35    Methylation of Proteins: Biochemistry and Functional
      Consequences  ........................................  571
      Álvaro Ortega

Index  ....................................................  585

# About the Series Editor-in-Chief

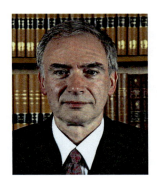

**Kenneth N. Timmis**
Emeritus Professor
Technical University Braunschweig
Institute of Microbiology
Braunschweig
Germany

Kenneth Timmis studied microbiology and obtained his Ph.D. at Bristol University. He undertook postdoctoral training at the Ruhr-University Bochum, Yale and Stanford, at the latter two as a Fellow of the Helen Hay Whitney Foundation. He was then appointed Head of an independent research group at the Max Planck Institute for Molecular Genetics in Berlin and subsequently Professor of Biochemistry in the University of Geneva, Faculty of Medicine. Thereafter, for almost 20 years, he was Director of the Division of Microbiology at the National Research Centre for Biotechnology (GBF)/now the Helmholtz Centre for Infection Research (HZI), and concomitantly Professor of Microbiology in the Institute of Microbiology of the Technical University Braunschweig. He is currently Emeritus Professor in this institute.

The Editor-in-Chief has worked for more than 30 years in the area of environmental microbiology and biotechnology, has published over 400 papers in international journals, and is an ISI Highly Cited Microbiology-100 researcher. His group has worked for many years, *inter alia*, on the biodegradation of oil hydrocarbons, especially the genetics and regulation of toluene degradation, and on the ecology of hydrocarbon-degrading microbial communities, discovered the new group of marine oil-degrading hydrocarbonoclastic bacteria, initiated genome sequencing projects on bacteria that are paradigms of microbes that degrade organic compounds (*Pseudomonas putida* and *Alcanivorax borkumensis*), and pioneered the topic of experimental evolution of novel catabolic activities.

He is Fellow of the Royal Society, Member of the European Molecular Biology Organisation, Fellow of the American Academy of Microbiology, Member of the European Academy of Microbiology, and Recipient of the Erwin Schrödinger Prize. He is the founder and Editor-in-Chief of the journals *Environmental Microbiology*, *Environmental Microbiology Reports*, and *Microbial Biotechnology*.

# About the Volume Editor

**Tino Krell**
Department of Environmental Protection
Estación Experimental del Zaidín
Consejo Superior de Investigaciones Científicas
Granada
Spain

Tino Krell began his studies in biochemistry at the universities of Leipzig (Germany) and Glasgow (UK). He completed his Ph.D. at the University of Glasgow, and his postdoctoral studies in the laboratory of Dr. J. R. Coggins, where he studied the different enzymes of the shikimate pathway. In Lyon, France, he studied the different aspects of multiple drug resistance at the Institut de Biologie et Chimie des Protéines. Also in Lyon, he served as leader of a research unit at the pharmaceutical company Sanofi Pasteur, which was focused on protein vaccines. In 2004, he moved to Granada, Spain, where he served initially as a tenure track scientist in the group of Dr. Juan Luis Ramos at the Estación Experimental del Zaidín (EEZ). In 2007, Dr. Krell received a permanent position at the EEZ and initiated his own laboratory, which is focused on bacterial sensing and signal transduction (http://krell-laboratory.com/). Specifically, his team studies the receptors and molecular mechanisms by which bacteria sense environmental signals, including various one- and two-component systems, and more recently, chemoreceptor-based signaling mechanisms. He has made several key contributions to our understanding of hydrocarbon sensing proteins that induce efflux pumps and degradation pathways, and has also shed light on receptors involved in hydrocarbon chemotaxis. He has authored 129 articles in international peer-reviewed journals such as *Nucleic Acids Research*, *Science Signaling*, and *PNAS*.

# Contributors

**Koos Albertin** Department of Microbial, Biochemical and Food Biotechnology, University of the Free State, Bloemfontein, South Africa

**Luana de Fátima Alves** Departamento de Biologia, FFCLRP – University of São Paulo, Ribeirão Preto, SP, Brazil

Departamento de Bioquímica e Imunologia, FMRP – University of São Paulo, Ribeirão Preto, SP, Brazil

**Cauã Antunes Westmann** Departamento de Biologia Celular, FMRP – University of São Paulo, Ribeirão Preto, SP, Brazil

**Nassim Ataii** Molecular Biophysics and Integrated Bioimaging Division, Lawrence Berkeley National Laboratory, Berkeley, CA, USA

**Manfred Auer** Molecular Biophysics and Integrated Bioimaging Division, Lawrence Berkeley National Laboratory, Berkeley, CA, USA

**Ibrahim M. Banat** School of Biomedical Sciences, Ulster University, Coleraine, Northern Ireland, UK

**Laura Barrientos-Moreno** Department of Environmental Protection, Estación Experimental del Zaidín, CSIC, Granada, Spain

**Thomas Baumgarten** Center for Biomembrane Research, Department of Biochemistry and Biophysics, Stockholm University, Stockholm, Sweden

**Athanasios Beopoulos** Laboratoire de Microbiologie et Génétique Moléculaire, AgroParisTech, Centre de Biotechnologie Agro-Industrielle, Thiverval-Grignon, France

**Patricia Bernal** Imperial College London, London, UK

**Emanuele G. Biondi** Aix Marseille University, CNRS, IMM, LCB, Marseille, France

**Andreas Busch** Confo Therapeutics, VIB Campus Technologiepark, Zwijnaarde, Belgium

**D. K. Button** Institute of Marine Science, University of Alaska, Fairbanks, AK, USA

**Tiago Cabral Borelli** Departamento de Biologia, FFCLRP – University of São Paulo, Ribeirão Preto, SP, Brazil

**Bilge Hilal Cadirci** Department of Bioengineering, Gaziosmanpasa University, Tokat, Turkey

**M. Carmona** Environmental Biology Department, Centro de Investigaciones Biológicas-CSIC, Madrid, Spain

**Craig Daniels** Developmental and Stem Cell Biology Program, Brain Tumour Research Centre, The Hospital for Sick Children, Toronto, ON, Canada

**Teresa del Castillo** Group of Physics of Fluids, Interfaces and Colloidal Systems, Department of Applied Physics, Faculty of Science, University of Granada, Granada, Spain

**Anett Demcsák** Department of Oral Biology and Experimental Dental Research, Faculty of Dentistry, University of Szeged, Szeged, Hungary

**Thomas Desfougeres** Laboratoire de Microbiologie et Génétique Moléculaire, AgroParisTech, Centre de Biotechnologie Agro-Industrielle, Thiverval-Grignon, France

**E. Díaz** Environmental Biology Department, Centro de Investigaciones Biológicas-CSIC, Madrid, Spain

**Jayna L. Ditty** Department of Biology, College of Arts and Sciences, University of St. Thomas, St. Paul, MN, USA

**M. Dow** School of Microbiology, University College Cork, Cork, Ireland

**G. Durante-Rodríguez** Environmental Biology Department, Centro de Investigaciones Biológicas-CSIC, Madrid, Spain

**Christian Eberlein** Department Environmental Biotechnology, Helmholtz Centre for Environmental Research – UFZ, Leipzig, Germany

**Manuel Espinosa-Urgel** Department of Environmental Protection, Estación Experimental del Zaidín, CSIC, Granada, Spain

**Matilde Fernández** Department of Environmental Protection, Estación Experimental del Zaidín, Consejo Superior de Investigaciones Científicas, Granada, Spain

**J. Fischer** Department of Environmental Biotechnology, Helmholtz Centre for Environmental Research, Leipzig, Germany

**J. Foght** Department of Biological Sciences, University of Alberta, Edmonton, AB, Canada

**Vanina García** The University of Nottingham, Nottingham, Nottinghamshire, UK

**H. Gómez-Álvarez** Environmental Biology Department, Centro de Investigaciones Biológicas-CSIC, Madrid, Spain

**Régis Grimaud** CNRS/ UNIV PAU and PAYS ADOUR, Institut des Sciences Analytiques et de Physico-Chimie pour l'Environnement et les Matériaux – MIRA, UMR5254, PAU, France

**María-Eugenia Guazzaroni** Departamento de Biologia, FFCLRP – University of São Paulo, Ribeirão Preto, SP, Brazil

Faculdade de Filosofia, Ciências e Letras de Ribeirão Preto, Universidade de São Paulo, Ribeirão Preto, São Paulo, Brazil

**Hauke Harms** Department of Environmental Microbiology, Helmholtz Centre for Environmental Research – UFZ, Leipzig, Germany

**Hermann J. Heipieper** Department Environmental Biotechnology, Helmholtz Centre for Environmental Research – UFZ, Leipzig, Germany

**Yisheng Kang** Department of Chemical Engineering, Washington University, St. Louis, MO, USA

**Jay D. Keasling** Department of Chemical Engineering, Washington University, St. Louis, MO, USA

Department of Bioengineering, University of California, Berkeley, CA, USA

Physical Biosciences Division, Lawrence Berkeley National Laboratory, Berkeley, CA, USA

Joint BioEnergy Institute, Emeryville, CA, USA

**Tino Krell** Department of Environmental Protection, Estación Experimental del Zaidín, Consejo Superior de Investigaciones Científicas, Granada, Spain

**Jesús Lacal** Department of Microbiology and Genetics, University of Salamanca, Salamanca, Spain

**Roger Marchant** School of Biomedical Sciences, Ulster University, Coleraine, Northern Ireland, UK

**P. M. Martínez** Department of Bioremediation, Helmholtz Centre for Environmental Research—UFZ, Leipzig, Germany

**Miguel A. Matilla** Department of Environmental Protection, Estación Experimental del Zaidín, Consejo Superior de Investigaciones Científicas, Granada, Spain

**Tyne McHugh** Molecular Biophysics and Integrated Bioimaging Division, Lawrence Berkeley National Laboratory, Berkeley, CA, USA

**F. Meinhardt** Institut für Molekulare Mikrobiologie und Biotechnologie, Westfälische Wilhelms-Universität Münster, Münster, Germany

**Noel Mesa-Torres** Department of Environmental Protection, Estación Experimental del Zaidín – CSIC, Granada, Spain

**Janos Minarovits**  Department of Oral Biology and Experimental Dental Research, Faculty of Dentistry, University of Szeged, Szeged, Hungary

**Saswat S. Mohapatra**  Department of Genetic Engineering, School of Bioengineering, SRM University, Kattankulathur, TN, India

**Armaity Nasarabadi**  Molecular Biophysics and Integrated Bioimaging Division, Lawrence Berkeley National Laboratory, Berkeley, CA, USA

**L. M. Naughton**  School of Microbiology, University College Cork, Cork, Ireland

**Jean Marc Nicaud**  Laboratoire de Microbiologie et Génétique Moléculaire, AgroParisTech, Centre de Biotechnologie Agro-Industrielle, Thiverval-Grignon, France

**K. W. Nickerson**  School of Biological Sciences, University of Nebraska, Lincoln, NE, USA

**Hans Helmut Niller**  Institute of Medical Microbiology and Hygiene, University of Regensburg, Regensburg, Germany

**J. Nogales**  Environmental Biology Department, Centro de Investigaciones Biológicas-CSIC, Madrid, Spain

**Álvaro Ortega**  Department of Environmental Protection, Estación Experimental del Zaidín, Consejo Superior de Investigaciones Científicas, Granada, Spain

**J. J. Ortega-Calvo**  Instituto de Recursos Naturales y Agrobiologia de Sevilla, CSIC, Sevilla, Spain

**Rebecca E. Parales**  Department of Microbiology and Molecular Genetics, College of Biological Sciences, University of California, Davis, CA, USA

**M. Peña Mattozzi**  Department of Plant and Microbial Biology, University of California, Berkeley, CA, USA

Center for Life Sciences Boston, Harvard Wyss Institute for Biologically Inspired Engineering, Boston, MA, USA

**Milva Pepi**  Stazione Zoologica Anton Dohrn, Villa Comunale, Naples, Italy

**Amedea Perfumo**  Helmholtz Centre Potsdam, GFZ German Research Centre for Geosciences, Potsdam, Germany

**R. E. Pierneef**  Centre for Bioinformatics and Computational Biology, Department of Biochemistry, University of Pretoria, Hillcrest, Pretoria, South Africa

**Cecilia Pini**  Shionogi Limited, London, UK

**B. Plantz**  School of Biological Sciences, University of Nebraska, Lincoln, NE, USA

**Juan Luis Ramos**  Department of Environmental Protection, Estación Experimental del Zaidín, Consejo Superior de Investigaciones Científicas, Granada, Spain

**O. N. Reva** Centre for Bioinformatics and Computational Biology, Department of Biochemistry, University of Pretoria, Hillcrest, Pretoria, South Africa

**Michelle Rudden** Department of Biology, University of York, York, UK

**Julia Sabirova** Laboratory of Industrial Microbiology and Biocatalysis, Department of Biochemical and Microbial Technology, Faculty of Bioscience Engineering, Ghent University, Ghent, Belgium

**Ana Segura** Department of Environmental Protection, Estación Experimental del Zaidín, Consejo Superior de Investigaciones Científicas, Granada, Spain

**Rafael Silva-Rocha** Departamento de Biologia Celular, FMRP – University of São Paulo, Ribeirão Preto, SP, Brazil

**Pierre Sivadon** CNRS/ UNIV PAU and PAYS ADOUR, Institut des Sciences Analytiques et de Physico-Chimie pour l'Environnement et les Matériaux – MIRA, UMR5254, PAU, France

**Kilian E. C. Smith** Institute for Environmental Research (Biology 5), RWTH Aachen University, Aachen, Germany

**Junha Song** Molecular Biophysics and Integrated Bioimaging Division, Lawrence Berkeley National Laboratory, Berkeley, CA, USA

**F. Thevenieau** Laboratoire de Microbiologie et Génétique Moléculaire, AgroParisTech, Centre de Biotechnologie Agro-Industrielle, Thiverval-Grignon, France

Oxyrane UK Limited, Greenheys House, Manchester Science Park, Manchester, UK

**B. Tümmler** Klinische Forschergruppe, Medizinische Hochschule Hannover, Hannover, Germany

**Lukas Y. Wick** Department of Environmental Microbiology, Helmholtz Centre for Environmental Research – UFZ, Leipzig, Germany

**Smita Zinjarde** Institute of Bioinformatics and Biotechnology, University of Pune, Pune, India

# Part I
# Problems of Hydrophobicity, Bioavailability

# Problems of Hydrophobicity/ Bioavailability: An Introduction

1

Hauke Harms, Kilian E. C. Smith, and Lukas Y. Wick

## Contents

1  Introduction ................................................................ 4
2  The Bioavailability of Hydrocarbons ....................................... 6
3  Living with Poorly Bioavailable Substrates ................................ 8
4  The Bioaccessibility of Hydrocarbons ...................................... 10
5  Influence of Chemical Hydrophobicity on Bioavailability and Bioaccessibility ............ 11
6  Research Needs ............................................................ 13
References .................................................................... 14

**Abstract**

This chapter discusses how the hydrophobicity and other properties of oil hydrocarbons influence their availability for toxic exposure, microbial degradation and growth. It also describes how the hydrocarbon bioavailability can control the maximum population size of a degrading microbial community in a given habitat (carrying capacity). Bioavailability is operationalized and presented as a process at the interface between microbial dynamics and physicochemical constraints.

H. Harms · L. Y. Wick (✉)
Department of Environmental Microbiology, Helmholtz Centre for Environmental Research – UFZ, Leipzig, Germany
e-mail: hauke.harms@ufz.de; lukas.wick@ufz.de

K. E. C. Smith
Institute for Environmental Research (Biology 5), RWTH Aachen University, Aachen, Germany
e-mail: kilian.smith@bio5.rwth-aachen.de

© Springer International Publishing AG, part of Springer Nature 2018
T. Krell (ed.), *Cellular Ecophysiology of Microbe: Hydrocarbon and Lipid Interactions*, Handbook of Hydrocarbon and Lipid Microbiology, https://doi.org/10.1007/978-3-319-50542-8_38

## 1    Introduction

Hydrocarbons have in common that they are repelled or expulsed from water. This characteristic of chemicals is referred to as *hydrophobicity* (from the Greek *hydro* for water and *phobos* for fear). Hydrophobicity has its physical origin in the polar nature of water and the tendency of water molecules to form hydrogen bonds with each other. Hydrogen bond formation between water molecules is energetically more favourable than the interaction of water molecules with non-polar (i.e., non-hydrogen bond forming) molecules or phases (Schwarzenbach et al. 2017). Water thus repels or expels hydrophobic chemicals in favour of bonding with itself. From a thermo-dynamic point of view this can be ascribed to the high entropic cost of forming a cavity inside a water mass around nonpolar molecules. The term hydrophobicity is thus misleading, as the force giving rise to the so-called hydrophobic effect in fact arises from the hydrophilic interaction partner. Hydrophobicity, most frequently, but not always, goes along with lipophilicity, i.e., the tendency of a molecule to partition into and accumulate in lipids or other non-polar organic phases.

Hydrophobicity as such is difficult to measure and to describe in quantitative terms. A convenient proxy for a compound's hydrophobicity is its tendency to partition between water and a liquid organic phase. To simplify comparison between compounds and sorbing phases, octanol is often chosen as the reference organic solvent to quantify the hydrophobicity of many kinds of chemicals. The octanol-water partition coefficient $K_{OW}$ (units: $L\,L^{-1}$) is the ratio of the concentration of a compound in an octanol phase to its concentration in an adjacent water phase at equilibrium. The $K_{OW}$ is typically determined in partition experiments, but there are now reliable methods for the calculation of unknown $K_{OW}$ from the chemical structure available as free internet resources (e.g., http://www.ufz.de/lserd that allows the calculation of partition coefficients of chemicals between water and octanol as well as other phases). Observed $K_{OW}$ values span such a wide range that conventionally $\log K_{OW}$ values are reported. Chemicals with $K_{OW}$ values higher than 1 can be regarded as being hydrophobic since they prefer the organic phase to the aqueous phase. Oil hydrocarbons generally have $\log K_{OW}$ values above 3. In Fig. 1 octanol-water partition coefficients of a series of environmentally relevant oil hydrocarbons are presented together with two other properties that additionally influence microbial bioavailability, namely the tendency to partition from water into air (i.e., become expelled from water into an air phase) and aqueous solubility (which sets the maximum dissolved bioavailable concentrations of a compound).

From Fig. 1, it becomes obvious there are some general rules for the degree of the hydrophobicity of hydrocarbons as a function of their structures. The hydrophobic-ities of hydrocarbons increase with the molecular mass and the degree of saturation of the carbon-carbon bonds of a molecule. Long alkane chains are thus more hydrophobic than short ones, saturated aliphatics more hydrophobic than unsatu-rated aliphatics and aromatics of the same molecular mass, and polycyclic aromatic hydrocarbons consisting of a higher number of rings are more hydrophobic than low molecular mass PAHs. In the environment, hydrocarbons can be present as individ-ual molecules in the gaseous, water-dissolved, surface-adsorbed or matrix-absorbed

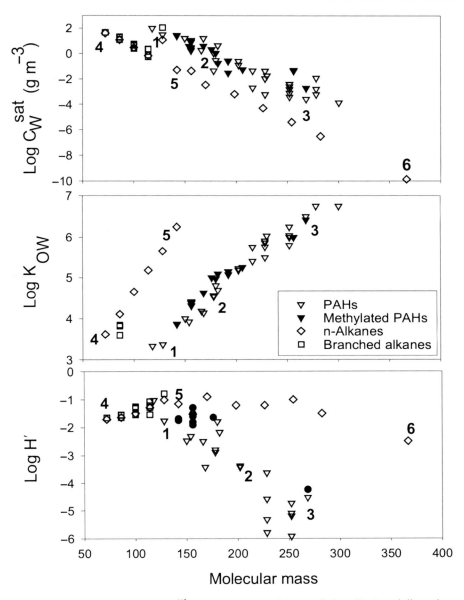

**Fig. 1** Plots of water solubility ($C_w^{sat}$), octanol-water partition coefficient ($K_{ow}$), and dimensionless Henry coefficient (H′) versus molecular mass for four typical groups of hydrocarbons: PAHs, methylated PAHs, n-alkanes and branched alkanes. Typical low, middle and high mass compound from the PAHs and n-alkanes have been marked on the plots using numbers (PAHs: 1 naphthalene; 2 phenanthrene; 3 benzo(ghi)perylene. n-Alkanes: 4 n-pentane; 5 n-decane; 6 hexacosane). Data from Schwarzenbach et al. (2017); Eastcott et al. (1988), and http://www.lec.lancs.ac.uk/ccm/research/database/index.htm

form or as separate bulk phase (liquid or solid) that floats on water, forms blobs or aggregates in the pores of sediments and soil or spreads as thin layers on biota.

## 2    The Bioavailability of Hydrocarbons

Active microbes are typically surrounded by water and their interaction with oil hydrocarbons is thus mediated by the water phase. Therefore, whenever microorganisms interact with hydrocarbons, the hydrophobic effect plays a central role since it largely determines how the oil hydrocarbons partition to various phases and are removed from the water.

The hydrophobicity of a chemical is thus often simplistically taken as an indicator for its *bioavailability*, in particular for microorganisms, which are believed to take up chemicals predominantly as water-dissolved molecules. The term bioavailability is generally used to refer to *the degree of interaction of chemicals with living organisms*. Unfortunately none of the existing refinements of this all-encompassing definition are universally accepted (NRC 2003; Harmsen 2007). This despite the need for practical definitions and approaches (cf. ISO 17402 (2008)) for specific uses, e.g., in risk assessment and remediation (Ortega-Calvo et al. 2015; Naidu et al. 2015). It appears useful to start out from the above-mentioned definition and develop an operational definition of bioavailability in the context of hydrocarbon microbiology before beginning to scrutinize the various bioavailability processes that govern the interactions of microbes with hydrocarbons. Obviously, quantification requires that bioavailability has a physical dimension. As there is no general agreement on a dimension or unit of bioavailability either, the issue of a suitable operational dimension of bioavailability remains open.

Before we address this issue, it is important to distinguish between bioavailability for microbial degradation on the one hand and for bioaccumulation on the other hand (Fig. 2).

(i) Microbial hydrocarbon degradation is a consumptive process that tends to strive towards an at least transient steady state between biological uptake and environmental re-provision of the hydrocarbons (cell 1 in Fig. 2). It has been shown that the rate of substrate diffusion to an organism and its rate of uptake by the organism tend to become equal (Koch 1990). The hydrocarbon concentration to which the microbe is exposed to and the actual rate at which the hydrocarbon enters the organism are controlled by both biological kinetics and environmental mass transfer. The steady-state mass transfer rate is determined by the specific affinity (i.e., the slope of the first order part of the whole-cell Michaelis-Menten curve) of the cell for the chemical in question and the water-dissolved concentration in contact with the cell. Microorganisms having a high specific affinity for their substrate can reduce the exposure concentration to extremely low values, because they take up their substrate so rapidly (drive such a fast substrate flux) that they create a steep concentration gradient between their surface and the bulk water at some distance.

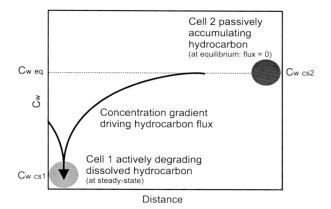

**Fig. 2** Illustration of the difference between chemical bioavailability to a cell actively degrading a dissolved hydrocarbon (Cell 1) and to a cell passively accumulating hydrocarbons (Cell 2). Note that cell 1 may behave like cell 2 with respect to other hydrocarbons in a mixture. The water-dissolved cell surface exposure concentrations Cw cs, the hydrocarbon transfer flux (only present in cell 1 as indicated by the arrow) and the hydrocarbon concentration in the cells (visualized as different shades of grey) differ substantially. Cw and Cw eq are the water-dissolved and the equilibrium water-dissolved concentrations at equilibrium, respectively

(ii) The situation is different for a microbe that is subject to the toxic effects of a bioaccumulating, but non-degradable compound (cell 2 in Fig. 2). Here, an equilibrium situation will be approached. The catabolically inactive microorganism will be subject to passive inflow and bioaccumulation of the chemical. With time, equilibrium will be reached that is characterized by, firstly, a hydrocarbon content of the organism that is mainly controlled by the microbe's lipid content and the lipophilicity of the hydrocarbon, secondly, exposure of the cell surface to the aqueous equilibrium concentration, and, thirdly, the absence of a net flux of hydrocarbon into the organism. Unlike the steady state observed for biodegradation (characterized by a low exposure concentration and high mass transfer), the endpoint of bioaccumulation is thus characterized by a relatively high aqueous equilibrium concentration and zero mass transfer. Bioavailability for passive bioaccumulation can therefore be quantified by approaches addressing the chemical activity of a compound, such as equilibrium extraction using solid phase microextraction (SPME) fibers (Reichenberg and Mayer 2006; Booij et al. 2016).

In the following we will see that it is more difficult to describe bioavailability for biodegradation. This is due to the dynamics of biodegradation and the frequently long duration over which biodegradation needs to be sustained by the bioavailable compound. From a practical perspective (e.g., in bioremediation or industrial biotransformation) we are interested in reaction rates and reaction endpoints, e.g., because remediation goals need to be matched in reasonable time scales. The reaction rate appears thus to be an immediate measure and appropriate dimension of bioavailability. However, there are some problems with this view since the measured rate may be limited by the capacity of the organisms to degrade or transform the chemical. One might say that the bioavailability is even higher than is apparent from the degradation rate since the mass transfer capacity remains partly unexploited. Bosma et al. (1997) have therefore defined bioavailability as the ratio of the rate at which a volumetric unit of the environment can theoretically provide a chemical to organisms to the rate at which the microbes can theoretically degrade the compound. As these rates (or capacities) of consumption and re-provision have equal units, a dimensionless bioavailability Ba number was proposed. Values of Ba above 1 indicate degradation rates predominantly controlled by the degradation capacity of the organisms (i.e., high bioavailability), whereas Ba values below 1 indicate degradation rates predominantly controlled by the mass transfer capacity of the environment (i.e., low bioavailability). Thullner et al. (2008) have simplified the above concept by defining bioavailability as the ratio of the actual biodegradation rate of an extant microbial community to its degradation capacity. Both concepts account for the fact that exposure concentrations are inappropriate descriptors of bioavailability since they can be very low even at high degradation rates (cell 1 in Fig. 2). Bioavailability according to both definitions depends on how the habitat can compensate for substrate degradation. This capacity of the habitat will depend on the physical state of the chemical (for example dissolved, sorbed, separate phase, occurrence as individual substance vs. mixture), its physical character-istics (such as hydrophobicity, effective diffusivity etc.) and its spatial distribution relative to the catabolically active biota (Johnsen et al. 2005).

## 3  Living with Poorly Bioavailable Substrates

Microbial growth depends on the degradation of suitable substrates by catabolically active microorganisms. Despite this obvious causal link, the relationship between biodegradation and growth is more complicated than often thought. In particular, the Michaelis-Menten relationship (i.e., an equation describing the rate of enzymatic reactions) and the Monod relationship (i.e., an equation describing the growth of microorganisms) are often mixed up or taken as equivalent. This is especially tempting when the former is applied to the kinetics of entire cells. It appears thus necessary to explain some important differences before using extended Monod kinetics to explain the constraints of life with poorly bioavailable substrates. Both relationships are presented in the form of hyperbolic plots of activity or growth against the substrate concentration. Both models are based on an equation of the general form '$a = b * c/(d + c)$', where $a$ is the actual rate, $b$ the maximum rate (the biological capacity), c the concentration and d the concentration resulting in half maximal rate. The similarity hides that the Michaelis-Menten equation is a mathematically derived relationship whereas the Monod relationship is an empirical model that has been found to fit growth data. It also hides that the experimental verification of Michaelis-Menten is performed with constant biological materials (pure enzyme, a raw protein extract or resting cells), whereas Monod data are obtained in monoclonal chemostat cultures that are slowly shifted from one steady state substrate concentration to the next one, i.e., shifts that allow adaptation and regulation events to occur (Kovarova-Kovar and Egli 1998). Even more

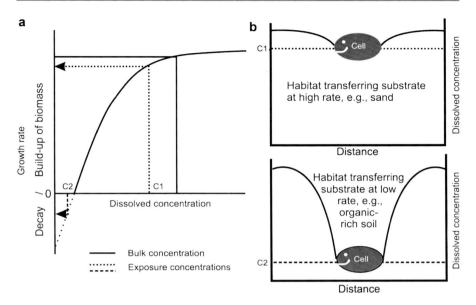

**Fig. 3** Monod-plot accounting for the maintenance energy requirements of bacteria (**a**) and visualization of different substrate consumption-dependent concentration gradients around bacteria in habitats permitting different rates of substrate transfer (**b**). Low substrate transfer in organic-rich soil (bottom graph in b) reduces the cell exposure concentration (c2 in a and b) to below the threshold for population maintenance, whereas high substrate transfer in sand (top graph in b) sustains a cell exposure concentration (c1 in a and b) permitting population growth

important for our discussion of bioavailability is another difference between both concepts. Whereas Michaelis-Menten enzyme kinetics applies down to minute substrate concentrations, since even the last substrate molecule has the chance to meet an enzyme (or a transporter on a cell surface), the classical Monod concept disregards the fact that minute substrate concentrations corresponding to minute substrate fluxes do not provide enough energy to organisms to allow them to grow. A certain critical substrate flux will correspond to an exposure concentration that will only be sufficient for the maintenance of the existing cells and even lower concentrations will lead to the cell death of existing cells. This offset of the hyperbolic relationship has been accounted for by a maintenance rate coefficient extending the classical Monod-kinetics (van Uden 1967), which is equivalent to the rate of culture decay (also termed negative growth) at zero substrate concentration (Fig. 3). A consequence of this offset is that in many environments, populations will grow as long as the substrate flux allows them to do so, thereby approaching a situation where the substrate flux only allows for maintenance of the extant population. This has been shown in the past for cultures in chemostats with biomass retention (Tros et al. 1996), in biofilms growing on poorly soluble anthracene crystals (Wick et al. 2001, 2002) and for bacteria relying on the diffusion of naphthalene from spatially separate sources (Harms 1996).

Under conditions of reduced bioavailability (i.e., high bias between the cell exposure concentration and the total concentration in the system), even high total

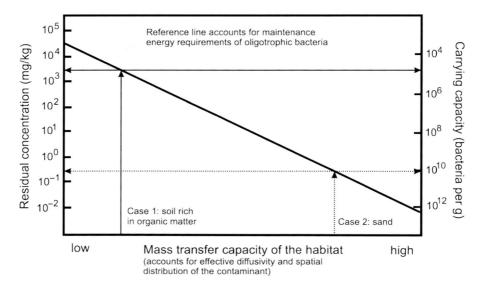

**Fig. 4** Graph illustrating the relationship between the mass transfer capacity of a microbial habitat, the carrying capacity of this habitat and the observed residual concentration of a hydrophobic chemical. Low substrate transfer in organic-rich soil (upward arrow of case 1) limits the carrying capacity of this habitat to low cell numbers (upper arrow pointing to the right), which are insufficient for observable contaminant degradation although residual concentrations are still relatively high (upper arrow pointing to the left). High mass transfer in sand (case 2) increases the carrying capacity thereby facilitating the reduction of residual chemical

concentrations may only allow for the maintenance of relatively small populations and, hence, for only limited degradation progress (Rein et al. 2016) (Fig. 4). In ecological terms, the substrate bioavailability controls the carrying capacity of microbial habitats, i.e., their capacity to maintain a population given there are no other substrates available for the degrader communities. The maintenance requirements have consequently been proposed as one reason for the observation of remediation thresholds particularly in environments that permit only slow substrate mass transfer (Bosma et al. 1997).

## 4 The Bioaccessibility of Hydrocarbons

The previous discussion focused largely on the degradation rate. However, reaction endpoints cannot be predicted from bioavailability that is defined in a way that it is linked to actual, ephemeral degradation rates. The capacity of an environment to provide a chemical at a constant rate may become exhausted because certain labile pools of the chemical may become emptied. This possibility has inspired the differentiation of an immediate bioavailability (compound at hand actually being taken up) from a potential bioavailability named *bioaccessibility* (compound that may become bioavailable by dissolution, desorption, diffusion etc.) (Semple et al. 2004). In this concept, the bioaccessible compound has the dimension of a fraction of the total compound and is thus an appropriate descriptor of the possible degradation endpoint. The bioaccessible fraction comprises the chemical mass that can reach the biota (or *vice versa*) within a predefined time frame. The distinction between bioavailability and bioaccessibility is to some extent equivalent to that between a compound's chemical activity as a quantitative descriptor of spontaneously mobilizable compound and an accessible quantity that has to be operationally defined from case to case as introduced by Reichenberg and Mayer (2006). These authors propose equilibrium sampling devices for the quantification of the chemical activity aspect of bioavailability and mild extraction methods for the quantification of the fraction of bioaccessible compound.

## 5    Influence of Chemical Hydrophobicity on Bioavailability and Bioaccessibility

Various phenomena of relevance for microbe-hydrocarbon interactions that arise from a compound's hydrophobicity are presented in Table 1. Effects at the molecular level such as low water solubility and the tendency of hydrophobic compounds to sorb to surfaces of differing sorption strengths or partition into animate or inanimate materials can be distinguished from bulk phase-related phenomena such as immiscibility of phases and low wettability of hydrophobic phases. Only short descriptions will be given here since many of these phenomena will be treated in more detail in subsequent chapters.

Low water solubility controls the maximum substrate concentration or energy density in the aqueous environment of the microbes. Unless microorganisms take up hydrocarbons directly from a non-aqueous phase (a possibility that will be discussed in another chapter of this volume), or co-solvents increase the water solubility (Mao et al. 2015), the water solubility will cut off a population's Michaelis-Menten curve thus setting an upper limit to the specific degradation rate. On the positive side low solubility partly limits a chemical's toxicity to microorganisms.

Sorption to solid surfaces and accumulation at the air-water interface (Hoff et al. 1993) will reduce the aqueous concentration with the consequences discussed above. One can distinguish a first case where the sorbing interfaces are exposed to bacteria and a second case where the sorbent is present in pores too small to be accessed by bacteria. Increased substrate and energy densities are found at these interfaces. Transfer from the bioaccessible to the bioavailable state, however, will require desorption. At equilibrium, the chemical activities of the sorbed compound and its dissolved counterpart will be the same, meaning that microorganisms will not benefit from higher concentrations at interfaces (because they represent the bioaccessible and bioavailable fraction of the compound) but only from the short transport distance of desorbing chemicals (van Loosdrecht et al. 1990). Except for the second case that molecules need to diffuse out of pores, the overall influence of adsorption on bioavailability will hence be relatively low.

**Table 1** Environmentally relevant phenomena associated with the hydrophobicity of hydrocarbons

| Phenomenon | Associated phenomena and environmental expression | Relevance for microbe-hydrocarbon interaction |
|---|---|---|
| Low water solubility | | Low substrate/energy density of the aqueous habitat<br>Low toxicity |
| Sorption | Adsorption to solid surfaces | Reduced bioavailability in the bulk water<br>Lowered toxicity<br>High substrate density at the solid surface<br>Heterogeneous substrate density at differently sorbing surfaces |
| | Pore condensation | Removal into pores inaccessible to the microbial cells |
| | Accumulation at air-water interfaces | Reduced bioavailability in the bulk water<br>High substrate density at the air-water interface |
| Partitioning | Absorption in particulate and dissolved organic matter | Reduced bioavailability due to lower aqueous concentration<br>Increased bioaccessibility when the organic matter is small and thus mobile |
| | Bioaccumulation | Increased substrate density in organisms<br>Increased toxicity |
| | Dissolution in micelles and NAPLs | Reduced bioavailability in the bulk water<br>Increased bioaccessibility due to micelle mobility |
| Formation of immiscible separate phases | Minimized interfacial area | Small area for settlement and microbial attack<br>Slow dissolution limiting the re-supply of consumed substrate in the aqueous phase |
| | Floating of light non-aqueous phase liquids (lNAPLs) | Shielding of oxygen<br>Lowered bioaccessibility due to heterogeneous substrate distribution<br>Limited provision of e-acceptors and nutrients |
| | Sinking of dense non-aqueous phase liquids (dNAPLs) | Heterogeneity of substrate distribution<br>Limited provision of e-acceptors and nutrients |
| | Low bulk mobility due to viscosity and sequestration in pores | Low bioaccessibility |
| | Instability of emulsions | Reduced area for settlement and microbial attack |
| Low wettability | Beading of water on hydrocarbon phase in unsaturated systems | Reduced contact of hydrocarbon with the aqueous habitat of microbes |

Partitioning into inanimate bulk phases, also referred to as absorption, has more drastic consequences for bioavailability, since it spatially separates the chemical from the aqueous phase that mediates the microbial uptake. Molecules partitioning into organic matter (OM) diffuse towards the centre of the OM until partition equilibrium is achieved. Their subsequent release may take a long time, since the same distance will have to be passed again prior to their diffusion back into the surrounding water. This is compounded by chemical changes to the absorbing matrix that can occur and that strongly retain the molecules (Luthy et al. 1997). Partitioning into small entities of OM such as dissolved organic carbon can have the effect that the chemical is absorbed, but in a highly mobile form of OM that may act as a vector and facilitate its transfer to biota. The same holds true for surfactant micelles. Partitioning into the microbes themselves obviously concentrates the chemical at the point where it can be consumed or act as a toxicant.

The formation of separate bulk hydrocarbon phases that are immiscible with water has consequences both at the microscopic and macroscopic scales. At the microscopic scale it reduces the interfacial area through which the hydrocarbon molecules are released into the water phase. This limits the dissolution rate thereby affecting bioavailability via the reduced substrate re-provision rate. The minimized interface also limits the area available for settlement of microorganisms. As only few bacteria will be located in the proximity of the hydrocarbon phase, immiscibility increases the mean distance between the major part of the hydrocarbon mass and the biota in the system. Alternatively, one can say that it increases the heterogeneity of the hydrocarbon distribution. Measures such as emulsification help to overcome this effect. In soils and sediments the formation of viscous non-aqueous phases also has consequences for the bulk movement of hydrocarbons, which tend to stick inside pores, a phenomenon known from the difficulty in extracting residual oil from oil reservoirs. At the macroscopic scale non-aqueous phase liquids, depending on their gravimetric density, may form layers on the groundwater surface (or the ocean) or sink to the confinement layers (or ocean sediment). In both cases the ratio of the NAPL-water interfacial area to the NAPL mass is very low and considerably restricts the transfer of bioaccessible into the bioavailable form. In the case of light NAPLs it may in addition shield oxygen or nutrients from the biota sitting below the NAPL, thereby affecting the bioavailability of these important cofactors.

Finally, low wettability plays a role where an air phase is present that competes (and succeeds in the competition) for the formation of an interface with hydrocarbons. Examples can be found in the beading of raindrops on the waxy surfaces of leaves or in the vadose zone of terrestrial environments. Here, surface-attached microcolonies and biofilms harbour most of the microbes that will be exposed to and degrade highly bioavailable air-borne hydrocarbons (Hanzel et al. 2012).

## 6    Research Needs

On-going research into hydrocarbon bioavailability should lead to the development of reliable (i.e., validated and preferably standardized chemical and biological) methods for the prediction of bioavailability and bioaccessiblity of hydrocarbons and mixtures thereof for application in both retrospective and prospective risk assessment approaches. This will require an improved understanding of physico-chemical bioavailability processes, particularly those prevailing in complex hydro-carbon mixtures as well as the elucidation of microbial strategies that improve bioavailability under multiphase environmental conditions. The latter should also address possible roles of microbial community members not directly involved in hydrocarbon degradation. An example is the function of fungal mycelia as facilita-tors of bacterial movement treated in another chapter of this volume. Another largely neglected aspect is the capability of hydrocarbon degrading populations and com-munities to decouple their biomass build-up and survival from the hydrocarbon provision by either co-metabolism or oligotrophic multi-substrate utilisation. Such approaches make clear that bioavailability of a hydrocarbon should be perceived as an emergent ecosystem property rather than just a chemical- or organism-specific trait.

# References

Booij K, Robinson CD, Burgess RM, Mayer P, Roberts CA, Ahrens L et al (2016) Passive sampling in regulatory chemical monitoring of nonpolar organic compounds in the aquatic environment. Environ Sci Technol 50:3–17

Bosma TNP, Middeldorp PJM, Schraa G, Zehnder AJB (1997) Mass transfer limitation of bio-transformation: quantifying bioavailability. Environ Sci Technol 31:248–252

Eastcott L, Shiu YS, Mackay D (1988) Environmentally relevant physical-chemical properties of hydrocarbons: a review of data and development of simple correlations. Oil Chem Pollut 4:191–216

Hanzel J, Thullner M, Harms H, Wick LY (2012) Walking the tightrope of bioavailability: growth dynamics of PAH degraders on vapour-phase PAH. Microb Biotechnol 5:79–86

Harms H (1996) Bacterial growth on distant naphthalene diffusing through water, air, water-saturated and nonsaturated porous media. Appl Environ Microbiol 62:2286–2293

Harmsen J (2007) Measuring bioavailability: from a scientific approach to standard methods. J Environ Qual 36:1420–1428

Hoff JT, Mackay D, Gillham R, Shiu WY (1993) Partitioning of organic-chemicals at the air water interface in environmental systems. Environ Sci Technol 27:2174–2180

Johnsen AR, Wick LY, Harms H (2005) Principles of microbial PAH-degradation in soil. Environ Pollut 133:71–84

Koch AL (1990) Diffusion – the crucial process in many aspects of the biology of bacteria. Adv Microb Ecol 11:37–70

Kovarova-Kovar K, Egli T (1998) Growth kinetics of suspended microbial cells: from single-substrate-controlled growth to mixed-substrate kinetics. Microbiol Mol Biol Rev 62:646–666

Luthy RG, Aiken GR, Brusseau ML, Cunningham SD, Gschwend PM, Pignatello JJ, Reinhard M, Traina SJ, Weber WJ Jr, Westall JC (1997) Sequestration of hydrophobic organic contaminants by geosorbents. Environ Sci Technol 31:3341–3347

Mao XH, Jiang R, Xiao W, Yu JG (2015) Use of surfactants for the remediation of contaminated soils: a review. J Hazard Mater 285:419–435

Naidu R, Channey R, McConnell S, Johnston N, Semple KT, McGrath S, Dries V, Nathanail P, Harmsen J, Pruszinski A, MacMillan J, Palanisami T (2015) Towards bioavailability-based soil criteria: past, present and future perspectives. Environ Sci Pollut Res 22:8779–8785

NRC Committee (2003) NRC Committee on Bioavailability of Contaminants in Soils and Sediments. Bioavailability of contaminants in soils and sediments: processes, tools and applications. The National Academic Press, Washington, DC

Ortega-Calvo JJ, Harmsen J, Parsons JR, Semple KT, Aitken MD, Ajao C, Eadsforth C, Galay-Burgos M, Naidu R, Oliver R, Peijnenburg W, Rombke J, Streck G, Versonnen B (2015) From bioavailability science to regulation of organic chemicals. Environ Sci Technol 49:10255–10264

Reichenberg F, Mayer P (2006) Two complementary sides of bioavailability: accessibility and chemical activity of organic contaminants in sediments and soils. Environ Toxicol Chem 25:1239–1245

Rein A, Adam IKU, Miltner A, Brumme K, Kastner M, Trapp S (2016) Impact of bacterial activity on turnover of insoluble hydrophobic substrates (phenanthrene and pyrene)-model simulations for prediction of bioremediation success. J Hazard Mater 306:105–114

Schwarzenbach RP, Gschwend PM, Imboden DM (2017) Environmental organic chemistry. Wiley, New York

Semple KT, Doick KJ, Jones KC, Burauel P, Craven A, Harms H (2004) Defining bioavailability and bioaccessibility of contaminated soil and sediment is complicated. Environ Sci Technol 38:228A–231A

Soil Quality Requirements and Guidance for the Selection and Application of Methods for the Assessment of Bioavailability of Contaminants in Soil and Soil Materials (2008) ISO No. 17402; International Organization for Standardization: Geneva, Switzerland, http://www.iso.org/iso/catalogue_detail.htm?csnumber=38349

Thullner M, Kampara M, Harms H, Wick LY (2008) Impact of bioavailability restrictions on microbially induced stable isotope fractionation: 1. Theoretical calculation. Environ Sci Technol 42:6544–6551

Tros ME, Bosma TNP, Schraa G, Zehnder AJB (1996) Measurement of minimum substrate concentration (S-min) in a recycling fermenter and its prediction from the kinetic parameters of *Pseudomonas* sp. strain B13 from batch and chemostat cultures. Appl Environ Microbiol 62:3655–3661

van Loosdrecht MCM, Lyklema J, Norde W, Zehnder AJB (1990) Influences of interfaces on microbial activity. Microbiol Rev 54:75–87

van Uden N (1967) Transport-limited fermentation in the chemostat and its competitive inhibition: a theoretical treatment. Arch Mikrobiol 58:145–154

Wick LY, Colangelo T, Harms H (2001) Kinetics of mass-transfer-limited growth on solid PAHs. Environ Sci Technol 35:354–361

Wick LY, de Munain AR, Springael D, Harms H (2002) Responses of *Mycobacterium* sp. LB501T to the low bioavailability of solid anthracene. Appl Biotechnol Microbiol 58:378–385

# Microorganism-Hydrophobic Compound Interactions

**2**

Lukas Y. Wick, Hauke Harms, and Kilian E. C. Smith

## Contents

1   Introduction ............................................................................  17
2   Adhesion to Hydrocarbons as a Microbial Adaptive Response ..........................  21
3   Forces and Microbial Characteristics Mediating Attachment to Hydrocarbons ...........  21
4   Accumulation of Hydrophobic Compounds in Bacterial Membranes and Cell Walls ......  22
5   Is Dissolution of Hydrophobic Compounds in Water Required for Microbial Uptake? ....  23
6   Bacterial Interactions with Hydrocarbons in Mobile Sorbents ..........................  24
7   Research Needs ........................................................................  28
References ................................................................................  28

**Abstract**

The low solubility and high hydrophobicity of hydrocarbons means that they sorb to various solids and nonaqueous-phase liquids (NAPLs), obliging hydrocarbon-degrading microorganisms to physically interact with these phases. This has various implications for the physicochemical characteristics of these microbes, their modes of hydrocarbon uptake, and their behavioral and physiological strategies aimed at promoting such interactions.

L. Y. Wick (✉) · H. Harms
Department of Environmental Microbiology, Helmholtz Centre for Environmental Research – UFZ, Leipzig, Germany
e-mail: lukas.wick@ufz.de; hauke.harms@ufz.de

K. E. C. Smith
Institute for Environmental Research (Biology 5), RWTH Aachen University, Aachen, Germany
e-mail: kilian.smith@bio5.rwth-aachen.de

© Springer International Publishing AG, part of Springer Nature 2018                    17
T. Krell (ed.), *Cellular Ecophysiology of Microbe: Hydrocarbon and Lipid Interactions*,
Handbook of Hydrocarbon and Lipid Microbiology, https://doi.org/10.1007/978-3-319-50542-8_40

# 1    Introduction

The consensus in the literature is that water-dissolved chemicals are available to microbes (Harms and Wick 2004). Hydrocarbons in the environment, however, often occur as pure water-immiscible liquids or solids, dissolved in nonaqueous-phase liquids (NAPLs), volatilized as gases, absorbed in solids, or adsorbed to interfaces. This often means that only tiny fractions of the total hydrocarbon mass are actually dissolved in the aqueous phase for supporting microbial uptake and growth. This has resulted in the development of different strategies by the degrading microorganisms to maximize their access to this limited source of substrate (Table 1). The questions then arise as to (i) what degree mono-disperse dissolution of hydro-phobic compounds in water is required for microbial uptake and (ii) whether direct contact with substrate in other forms could allow for additional uptake?

In this context, it is important to realize that bacteria:hydrophobic compound interactions take place at an extended interphase rather than at a sharp interface (Köster and van Leeuwen 2004). This concept of an inter*pha*se considers that the microorganisms are surrounded by so-called boundary layers characterized by a gradual transition from the biological to the environmental phase. From the perspective of a microbial cell, the actual process of chemical uptake typically includes transport in the aqueous medium (i.e., across the water boundary layer around the bacterial cell), adsorption to and absorption into extracellular hydrogel-like cell wall components, uptake by transfer across the cell membrane, and finally biodegradation or (nonconsumptive) bioaccumulation (Köster and van Leeuwen 2004). The availability of a chemical for degradation is thus controlled by the rate of mass transfer from the surrounding environment into the microbial cells relative to their intrinsic catabolic activity (Bosma et al. 1997). As this chemical transport occurs by concentration-gradient driven diffusion, the relative locations of hydrocarbon source (e.g., sorbed to NAPLS or solids) and sink (i.e., the degrading cells) are of primary importance. This means that regardless of the uptake mechanism, bacterial attachment to solid, liquid, or sorbed substrates immediately becomes a powerful way to maximize substrate mass transfer.

According to Fick's law, $J = D_{eff}{}^{*}A^{*}\partial c/\partial x$, the diffusive mass flux of a substrate towards the cell surface $J$ (µg/s) is strongly affected by the space coordinate in direction of the transport $\partial x$(m), the concentration difference between the location of bioconversion and the substrate source $\partial c$(µg/m$^3$), the area $A$(m$^2$) through which the compound is diffusing and the effective diffusion coefficient of the substrate molecules $D_{eff}$ (m$^2$/s) (Fig. 1a). Therefore, any reduction in the distance between a substrate and the microorganism enhances the diffusive mass flux, thereby increasing the availability of the hydrocarbon. For hydrophobic compounds associated with either a solid or NAPL, the so-called aqueous boundary layer diffusion is the limiting step in the dissolution (Bird et al. 1960) and controls the rate of appearance of molecules in the bulk water phase. A surrounding bulk water phase containing a subequilibrium concentration of the solute will thus drive desorption across this

**Table 1**  Overview of the different microbial strategies employed to enhance the bioavailability of nonaqueous-phase hydrocarbons to microorganisms. The literature references document where the strategies have been shown to alter the hydrocarbon bioavailability.

| Category | Strategy | Consequence | Reference |
|---|---|---|---|
| **Passive** | Degradation in the bulk aqueous phase | High gradient in chemical activity maintained | Harms and Zehnder 1994 |
| **Positional Interaction with the NAPL** | Adhesion to interface/biofilm formation | Positioning in region of highest chemical activity and reduced diffusion distance | Wick et al. 2001 |
| | | Additional diffusive barrier | Wick et al. 2001 |
| | Preferential dispersal and chemotaxis | Positioning in region of highest chemical activity and reduced diffusion distance | Wick et al. 2007; Otto et al. 2016; Marx and Aitken 2000 |
| | | Positioning in region of nontoxic chemical activity and reduced diffusion distance | Hanzel et al. 2011. |
| | Surfactant production | Break up of NAPL phase increasing surface area for mass transfer and colonization | Barkay et al. 1999 |
| | | Increased micellar concentrations in the aqueous phase | Volkering et al. 1995 |
| | | Direct bioavailability of micellar hydrocarbons | Guha and Jaffé 1996a |
| | | Change in bacterial surface hydrophobicity leading to altered attachment to interfaces | de Carvalho et al. 2009 |
| | | Inhibition of cell attachment | Neu 1996 |

distance resulting in a concentration gradient between the bulk phase dissolved concentration and the interfacial aqueous pseudo-equilibrium concentration (Fig. 1a). Furthermore, even under conditions of percolation or mild stirring, particles (including bacteria) are typically surrounded by a diffusive boundary layer of the thickness 10–100 μm (Levich 1962; Köster and van Leeuwen 2004). Therefore, when attached to a solid, liquid or sorbed hydrocarbon, bacteria drive desorption via substrate uptake from inside of these boundary layers, which may result in an up to roughly 100-fold steeper concentration gradient and mass transfer to the cell (Fig. 1b). It has thus been concluded that the lower desorption rates determined in the absence of bacteria as compared to degradation rates can lead to wrong interpretations regarding the mechanisms of uptake, unless distance effects (i.e., the possibility that bacteria invade diffusion boundary layers) are accounted for (Harms and Zehnder 1995). Hence, generalizations about the bioavailability of hydrocarbons are very much dependent on the habitat and the organisms living in them.

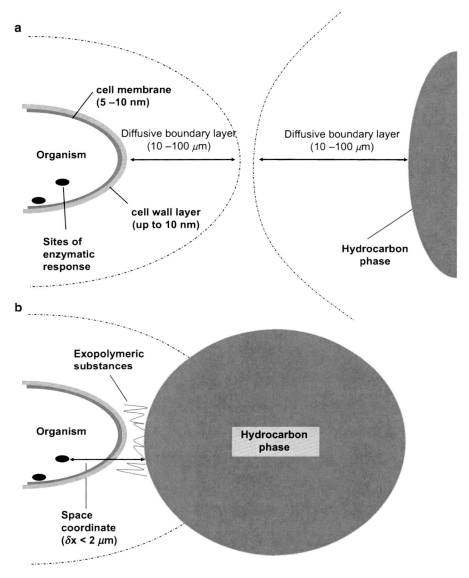

**Fig. 1** Conceptualized outline of the physically relevant layers at the microorganism/medium (**a**) and the microorganism/hydrocarbon (**b**) interphase. Diffusive hydrocarbon flux towards the cell surface is strongly affected by the space coordinate in the direction of the transport $\partial x$ between the location of compound transformation in a cell and the substrate source. When attached to a solid, dissolved, liquid, or sorbed hydrocarbon, bacteria drive desorption by substrate uptake from inside of the boundary layer, which may result in an up to roughly 100-fold steeper concentration gradient and mass transfer to the cell (**b**)

## 2 Adhesion to Hydrocarbons as a Microbial Adaptive Response

Given its potential effects, it is thus not surprising that bacterial adhesion to and the formation of biofilms on hydrocarbon phases have been frequently reported for transfer-limited conditions in the presence of solid (e.g., Mulder et al. 1998; Wick et al. 2002) and liquid hydrocarbons (e.g., McLee and Davies 1972; Rosenberg and Rosenberg 1981; Mounier et al. 2014). Further evidence for the importance of the close contact to the substrate comes from the observed drastic loss of microbial activity when association to the substrate source (Otto et al. 2016) was suppressed by nontoxic surfactants or when adhesion-hindered mutants were used (e.g., Rosenberg and Rosenberg 1981; Efroymson and Alexander 1991). Bacteria may also adapt to the presence of hydrophobic (or from their perspective lipophilic) hydrocarbons by changing their membrane and outer cell wall properties, e.g., in order to better attach to hydrophobic surfaces (van Loosdrecht et al. 1990; Wick et al. 2002; Johnsen et al. 2005; de Carvalho et al. 2009; Naether et al. 2013).

## 3 Forces and Microbial Characteristics Mediating Attachment to Hydrocarbons

The adhesion of bacteria to interfaces is believed to be mediated by long-ranging colloidal interactions, which attract bacteria and hold them at a close proximity of the interface, thereby facilitating short-ranging van der Waal (hydrophobic), acid-base, and electrostatic interactions to become effective and enabling the subsequent adsorption of extracellular polymeric substances, e.g., mediated by hydrogen bonds (van Loosdrecht et al. 1990). It is important to note that any electrostatic forces will be shielded by high ion concentrations in sea water, whereas they may play an important role in less saline groundwater (Busscher et al. 1995). In the case of adhesion to liquids, which do not possess a rigid surface, these interaction forces will induce changes in the interface topography to optimize the interaction. It is thus to be expected that more hydrophobic bacteria will establish a larger interfacial area with a liquid hydrocarbon. Besides the establishment of more interfacial area for substrate transfer, the effect of separation distance on mass transfer (as discussed above) may also explain the higher hydrophobicity of organisms degrading hydrophobic compounds. It has been observed for instance that more hydrophobic bacteria were enriched when hydrophobic membranes were used to isolate degraders of polycyclic aromatic hydrocarbons (PAH) from soil as compared to conventional water extraction protocols (Bastiaens et al. 2000). The physiological role of the hydrophobicity of these organisms was supported by the observation that the PAH-degrading mycobacteria modulated their surface hydrophobicity depending on the hydrophobicity of their growth substrate by varying the length of their mycolic acids (Wick et al. 2002). It should be noted, however, that the cell surface hydrophobicity is not the only characteristic controlling the contact with hydrocarbon phases. For instance, a study described the strong dependency of the surface charge (zeta potential, $\zeta$) of suspended *R. erythropolis* cells during growth on *n*-alkanes (with extremes of varying $\zeta = -35.8$ mV (*n*-heptane) and $\zeta = +4.7$ mV (*n*-hexadecane)) that went along with preferential association to *n*-hexadecane in microbial adhesion to hydrocarbons (MATH) tests (de Carvalho et al. 2009). This may be explained by the observation that the zeta potentials of hydrocarbon surfaces in water were found to be highly negative. The origin of the measured charge is not entirely clear, but preferred adsorption of OH⁻ as opposed to $H_3O^+$ at hydrophobic liquids and solids (e.g., Teflon and polystyrene surfaces) has been suggested as the cause.

## 4    Accumulation of Hydrophobic Compounds in Bacterial Membranes and Cell Walls

Various physiological adaptations of bacteria to the lipophilicity of hydrocarbons and the concomitant potential toxicity have been described and will be discussed in other chapters of this volume (for an excellent review see also Sikkema et al. 1995). Adaptations include (i) changes in the membrane fluidity by modifications of the membrane fatty acid composition, (ii) changes in the membrane's protein content and composition, (iii) active excretion of toxic hydrocarbons by energy-consuming transport systems (e.g., multidrug resistance system), (iv) increase of the cross-linking between the cell-wall constituents and changes in the cell wall hydrophobicity and charge, (v) modifications of the lipopolysaccharides (LPS) of the outer membrane of Gram-negative bacteria, and/or (vi) biotransformation of the compound. Changes in the cell wall properties have also been suggested to increase the accumulation of hydrophobic compounds (e.g., Klein et al. 2008). Sikkema et al. correlated the *n*-octanol-water partition constants $K_{OW}$ of hydrophobic organic contaminants to their membrane-buffer partitioning coefficient $K_B$ as $\log K_B = 0.97$ * $\log K_{OW} - 0.64$ (Sikkema et al. 1994). More recent studies and freely accessible on-line tools based on polyparameter linear free energy relationships (pp-LFERs) (e.g., www.ufz.de/lserd) also predict the capacities of membrane lipids (Endo et al. 2011) and structural proteins (Endo et al. 2012) for accumulating neutral organic chemicals. This underlines the fugacity-driven bioaccumulation of hydrophobic substances in cell membranes, cell walls, and the cytoplasm. Bioaccumulation of hydrophobic compounds appears to be a passive process and, thermodynamically speaking, is driven by the high fugacity of hydrophobic compounds in the water-phase that promotes their expulsion from water. It is thus likely that hydrophobic cell envelopes act as transient reservoirs (biosolvents) of hydrophobic substances. The resulting bioaccumulation of lipophilic compounds in the lipid bilayer or in cell wall proteins may thus enhance their availability for biotransformation or potentially cell-toxic effects. Accumulation in biofilms as a possible sink for hydrocarbon solutes has also been described. Despite the hydrogel-like character (96%-98% of water), extracellular polymeric substances of biofilms were found to contain about 60%-70% of the total monoaromatic hydrocarbons isolated from the total biomass (Späth et al. 1998).

Despite the normally passive character of hydrocarbon uptake, different microbial adaptations have evolved to augment their uptake rates. For instance, it has been postulated that outer-membrane related lipopolysaccharides are released to encapsulate hydrocarbon droplets and to increase the efficiency of alkane mass transfer (Witholt et al. 1990). Lipopolysaccharides may influence the mass transfer via several mechanisms including the improvement of the dissolution process, dispersion of the soil matrix, dispersion of nonaqueous-liquid phases and crystalline contaminants (Neu 1996), the solubilization in micelles that may act as kinetic contaminant carriers through the diffuse boundary layer (Garcia et al. 2001) or as an additional dispersed phase or "biosolvent" (Noordman et al. 1998). Release of outer membrane vesicles (MV) in Gram-negative bacteria has also been discussed as a stress response to toxic levels of environmental chemicals and stresses (Baumgarten et al. 2012; Orench-Rivera and Kuehn 2016).

## 5   Is Dissolution of Hydrophobic Compounds in Water Required for Microbial Uptake?

This brings us to the long-standing question whether monodisperse dissolution of hydrophobic compounds in water is a prerequisite for microbial uptake. Although it is generally accepted that substrates which occur in pure or matrix-sorbed form have to dissolve or to desorb in order to become available for microbial uptake (Ogram et al. 1985; van Loosdrecht et al. 1990), over the past 20 years several observations have been interpreted as evidence for endocytotic uptake of "undissolved" substrate. This, however, is experimentally difficult to verify (Taylor and Simkiss 2004). Three microbial uptake mechanisms of nonaqueous-phase liquids have been postulated. Firstly, entire droplets of these liquids may be channeled through pores in the cell envelope into the cytoplasm. Secondly, entire droplets could enter the cell surrounded by parts of the cell membrane or the cell envelope (i.e., pinocytosis) from where the hydrocarbons can be slowly released into the cytoplasm or be degraded by membrane-bound enzymes (Taylor and Simkiss 2004). Thirdly, it is imaginable that droplets may fuse with and flow into the lipid bilayer upon contacting a cell membrane. Microbial uptake of solid or sorbed substrates is somewhat different. Similar to the process of pinocytosis, solid substrates could directly be taken up by phagocytosis (Taylor and Simkiss 2004). To our knowledge, however, this has never been observed and seems unlikely for bacteria, which are covered by rigid cell walls. On the other hand, solid-state molecules might directly dissolve in the membranes of contiguous bacteria upon physical contact excluding the water phase (e.g., Southam et al. 2001). The conceptual difference between adsorbed molecules and solid substrates is that the latter are adsorbed to molecules of their own kind, and, with respect to their bioavailability and possibility of uptake, may behave like a solid substrate. Biophysically speaking, the problem reduces to the questions whether bacteria are able (i) to enzymatically attack molecules that are still sorbed while the enzyme-substrate-complex is formed, (ii) to take up molecules that are still sorbed while the transporter-substrate-complex is formed, and/or (iii) to passively take up molecules without a desorption step into the aqueous phase involved (Harms and Wick 2004). In case of absorbed molecules, pore-sequestered molecules have to move to the sorbent-water interface and desorb (i.e., dissolve) into the water phase, before film diffusion into the bulk water, and subsequent microbial uptake can take place. Intrasorbent transfer to the sorbent-water interface has often been found to control the overall desorption rate of absorbed or pore-sequestered compound. Direct contact of bacteria with absorbed hydrocarbons is thus unlikely before they appear at the sorbent-water interface and uptake of molecules by microorganisms in other than the water-dissolved state may be the exception rather than the rule (for a review cf. Harms and Wick 2004). At present, there seems to be insufficient evidence for the direct uptake of hydrocarbon droplets by prokaryotes. More evidence appears to exist for the direct uptake of hydrocarbons from mobile sorbents, which will be addressed in the following section.

## 6    Bacterial Interactions with Hydrocarbons in Mobile Sorbents

The interaction of cells with hydrocarbons sorbed to small and highly mobile sorbents has been shown to increase the mass flux of hydrophobic substrates to degrading cells and is given specific attention in this section. With respect to hydrocarbon bioavailability, such mobile sorbent phases (e.g., surfactants or dissolved organic carbon (DOC)) possess the following relevant properties: (i) propensity to sorb hydrocarbons, (ii) small size and large surface area to volume ratio, (iii) mobility with, for example, the flow of water, and (iv) the ability to interact with biological membranes. In general, it is thought that the diffusive transfer of hydrocarbons from a bulk dissolved pool supplies the degrading microorganism (van Loosdrecht et al. 1990). Therefore, sorption to mobile sorptive phases generally leads to a reduction in the dissolved concentrations and is thus expected to lead to a reduction in the biodegradation rates. This does not imply that this sorption removes a hydrocarbon from the biodegradable pool because of a permanent association, but that the reduction in the dissolved concentrations driving the diffusive uptake lowers the uptake rates so that the total biodegradation is extended over a longer period of time.

   An increasing number of studies point to other transfer pathways in addition to diffusive uptake from the dissolved phase (also called enhanced or facilitated diffusion). Under certain conditions, it appears that a fraction of hydrocarbon sorbed to such small and mobile sorbing phases is "directly bioavailable" to the degrading microorganisms, i.e., it is transferred directly from the sorbed state to the degrading cell without having to first desorb into the bulk medium followed by diffusive uptake. Primarily, evidence for this has come from studies investigating the degradation kinetics of PAHs in the presence of artificial and natural surfactants (Guha and Jaffé 1996a, b; Guha et al. 1998; Brown 2007).

The role of surfactants in the biodegradation of hydrocarbons is affected by many factors including surfactant toxicity or preferential degradation, enhanced hydrocarbon solubilization or emulsification, and altered cell surface characteristics leading to changes in attachment behavior. Consequently, the results from studies using different surfactants appear at first to be contradictory, either showing an enhancement, inhibition, or no effect on biodegradation. Nevertheless, a series of studies using different nontoxic surfactants have shown that a fraction of the micellar-associated PAHs is directly bioavailable to degrading bacterial cells, without previous desorption into the dissolved phase (Guha and Jaffé 1996a; Guha et al. 1998). This led to the development and refinement of a mechanistic model detailing this micellar transport pathway (Guha and Jaffé 1996b; Brown 2007). This model envisages that the PAHs associated with the micelles are transported to, and subsequently fuse with, a hemicellar surfactant layer covering the bacterial surface, thus directly releasing the intercalated PAHs into the cell. This is shown schematically in Fig. 2a.

Of course, such a micelle-mediated PAH delivery pathway depends on the micelle forming properties of surfactants. Nevertheless, studies with other types of sorbents, which are also small and mobile, show enhanced PAH degradation kinetics, pointing to the existence of transfer pathways additional to diffusive uptake from

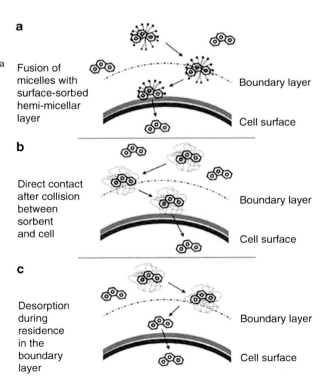

**Fig. 2** Suggested mechanisms by which a mobile sorbent phase might enhance the mass transfer of a hydrocarbon to degrading bacteria. The mechanism described in (**a**) is based on that described in Guha and Jaffé 1996b

**a** Fusion of micelles with surface-sorbed hemi-micellar layer

Boundary layer

Cell surface

**b** Direct contact after collision between sorbent and cell

Boundary layer

Cell surface

**c** Desorption during residence in the boundary layer

Boundary layer

Cell surface

the dissolved phase. Barkay et al. (1999) investigated the effect of the bioemulsifier alasan on the biodegradation kinetics of phenanthrene and fluoranthene. Mineralisation rates of both PAHs were enhanced by the presence of increasing amounts of alasan, indicating an enhanced mass flux to the degrading cells. Alasan is a tightly bound complex of anionic polysaccharides and proteins, which forms multimolecular aggregates rather than micelles. Several studies with dissolved organic carbon (DOC) also point to a possible carrier role. Phenanthrene degradation experiments using *Sphingomonas* sp. LH162 in the presence of increasing DOC concentrations indicated increased degradation rates by up to a factor of five in a DOC concentration-dependent manner (Smith et al., 2009). This was interpreted by a DOC-mediated transport of phenanthrene to the cells, supplementing diffusive uptake from the freely dissolved phase. Ortega-Calvo and Saiz-Jimenez (1998) looked at the mineralization of phenanthrene in batch cultures and found that high DOC loadings increased the mineralization rates, despite a reduction in the dissolved aqueous concentrations due to sorption. Laor et al. (1999) analyzed the effect of sorption to DOC on the biodegradation of phenanthrene. In their studies, the reduction in dissolved concentrations with increasing amounts of DOC did not result in appreciable differences in the rate or extent of mineralisation. This would suggest a mass transport pathway in addition to the diffusive transfer of phenanthrene in the dissolved phase. Holman et al. (2002) studied in-situ the influence of DOC on the degradation of a thin film of pyrene deposited on a magnetite surface. Here, humic acids shortened the onset of biodegradation. As discussed in another chapter of this volume, DOC can also increase mass transfer rates from the sorbed state and into the aqueous phase. For example, NAPL to water mass fluxes of PAHs increased by up to a factor of four in the presence of DOC, with the greatest increases observed for more hydrophobic compounds and highest DOC concentrations (Smith et al. 2011). Therefore, enhanced diffusion might play a dual role of increasing the delivery hydrocarbons *from* the source but also *to* the cells. However, under certain conditions, DOC has also been suggested to limit the adhesion of the cells to the hydrocarbon source thus decreasing their bioavailability (Tejeda-Agredano et al. 2014). Clearly, additional work is required to understand under which conditions hydrocarbon bioavailabiltiy is increased or decreased by DOC.

Recently, it has been shown that small and motile organisms such as protozoa can also enhance the diffusion of PAHs across an aqueous boundary layer (Gilbert et al. 2014). Typically, after an oil spill, a succession of microbial degraders sequentially degrade the different classes of oil components (Head et al. 2006). This raises the question as to whether the initial growth of one group of degraders can increase the availability of other hydrocarbons via a similar mechanism of enhanced diffusion, either due to the cells themselves or their secretions entering the unstirred boundary layer. This of course would be offset by the formation of biofilms on the oil surface.

A very different example of enhanced diffusion is given by the mass transfer of PAHs across an aqueous boundary layer from a source to a sink polymer (Mayer et al. 2005). In the presence of humic acids, mass transfer was enhanced above that resulting from molecular aqueous diffusion. A follow-up study confirmed this to be the case for wide range of different aqueous media including surfactant solutions, humic acid solutions, aqueous soil and horse manure extracts, digestive fluid of a deposit feeding worm, and root exudates from willow plants (Mayer et al. 2007). Again, it remains to be seen exactly which types of small and mobile hydrocarbon-sorbing phases can play a role in enhancing mass fluxes and in which circumstances.

The above studies indicate that the association of a hydrocarbon to certain types of dissolved organic matter or similar phases can in some cases lead to an enhancement in the biodegradation rates. These sorbents all have the common property that they are mobile and/or can intimately associate with the microbial surface. This appears to lead to an efficient transfer of hydrocarbon from the sorbed state to the cell. However, despite this being a widespread phenomenon (different types of immotile and mobile sorbing phases, microorganisms, etc.), the nature of the mechanisms behind these observations are not always clear. The very lack of a unifying framework to explain the above observations makes this a good point to speculate about the mechanisms that might explain such an enhanced flux. In the case of surfactants, the fusion of hydrocarbon-filled micelles with a cell surface hemimicellar layer and direct transfer (Fig. 2a) is well backed up by experimental evidence (Guha and Jaffé 1996b; Brown 2007). However, such a mechanism is also plausible for hydrocarbons sorbed to dissolved organic matter. Dissolved organic matter is composed of low-molecular-weight components which form dynamic associations via hydrophobic interactions and hydrogen bonds (Buffle and Leppard 1995; Sutton and Sposito 2005). Furthermore, dissolved organic matter associates with living membranes (Maurice et al. 2004), which under certain conditions leads to changes in the membrane permeability (Vigneault et al. 2000). Therefore, for a transfer pathway analogous to that of surfactants to operate, the necessary prerequisites exist. These include the surfactant-like aggregation of smaller units and adsorption to the biological membrane, although this has not been experimentally verified. The fact that dissolved organic matter can intimately associate with biological membranes indicates that close contact is indeed possible. In the case of silicone polymers, the direct contact between a PAH-loaded donor and empty-receiver compartment resulted in a very effective transfer, exceeding the transfer across an aqueous boundary layer by several orders of magnitude (Mayer et al. 2005, 2007). Therefore, it is feasible that when collisions between bacteria and DOC occur, this leads to a direct albeit temporary contact and a transfer of hydrocarbon from the sorbent to the cell (Fig. 2b). Finally, for many hydrocarbons the main resistance to diffusive transfer to or from an organic-type matrix lies in the aqueous boundary layer (Schwarzenbach et al. 2003). This implies that the chemical activity in this layer is below that in the bulk solution. Therefore, if a mobile sorbent phase, at equilibrium with the bulk solution, enters this depleted boundary layer, desorption could occur. The extent would depend on the residence time in this layer and the kinetics of hydrocarbon release (Fig. 2c).

# 7    Research Needs

One challenge is to identify the nature of the mechanisms behind the often observed increase in direct bioavailability to degrading microorganisms of carrier-sorbed hydrocarbons. This knowledge would contribute to understanding those situations where direct hydrocarbon transfer plays a dominating role in bioavailability. Application of the newest generation of powerful analytical and microscopic techniques, including time of flight and nanoscale secondary ion mass spectrometry (ToF- and nanoSIMS) and stable isotope labeling approaches, may offer ways to obtain more mechanistic insight into cellular uptake processes and/or to elucidate to which degree microbial interactions take place for obtaining preferential access to hydrocarbons. Furthermore, it still remains unclear what role the sorption to other mobile phases such as bacteria has for the availability of contaminants. The cotransport of hydrocarbons by motile microorganisms (Gilbert et al. 2014), or the cytoplasmic transport by fungal mycelia (Schamfuss et al. 2013), may lead to enhanced mass transfer under diffusion-limited conditions and enhanced bioavailability to degrader organisms. Such studies exemplify that the bioavailability of a hydrocarbon goes beyond the purely physical interactions and is a function of both biological activity and the physical availability of a hydrocarbon. The bioavailability of a hydrocarbon is thus the result of highly dynamic physical, chemical, and biological interactions that shape the spatio-temporal exposure of individual organisms to chemicals in an environment (Hanzel et al. 2012). Many bacteria have developed evolutionary adaptations that help them to cope with an unfavorable (too high or limited) availability of chemicals. Next to the cell wall changes discussed above, cells have developed chemotactic behavior that allows them to move along compound gradients (Krell et al. 2012) or specific mutualistic interactions with other organisms such as fungi that attack hydrophobic hydrocarbons exo-enzymatically and allow the transfer of hydrophilic metabolites to catabolically interacting commensals. Another future challenge will be to describe and predict the bioavailability of hydrocarbons as a spatio-temporal ecosystem property.

**Acknowledgements** This work contributes to the research topic Chemicals in the Environment (CITE) within the research program Terrestrial Environment of the Helmholtz Association.

# References

Barkay T, Navon-Venezia S, Ron EZ, Rosenberg E (1999) Enhancement of solubilization and biodegradation of polyaromatic hydrocarbons by the bioemulsifier alasan. Appl Environ Microbiol 65:2967–2702

Baumgarten T, Sperling S, Seifert J, von Bergen M, Steiniger F, Wick LY, Heipieper HJ (2012) Membrane vesicle formation as multiple stress response mechanism enhances cell surface hydrophobicity and biofilm formation of *Pseudomonas putida* DOT-T1E. Appl Environ Microbiol 78:6217–6224

Bastiaens L et al (2000) Isolation of adherent polycyclic aromatic hydrocarbon (PAH) degrading bacteria using PAH sorbing carriers. Appl Environ Microbiol 66:1834–1843

Bird RB, Stewart WE, Lightfoot EN (1960) Transport phenomena, 1st edn. Wiley, New York

Bosma TNP, Middeldorp PJM, Schraa G, Zehnder AJB (1997) Mass transfer limitation of bio-transformation: Quantifying bioavailability. Environ Sci Technol 31:248–252

Brown DG (2007) Relationship between micellar and hemi-micellar processes and the bioavailability of surfactant-solubilized hydrophobic organic compounds. Environ Sci Technol 41:1194–1199

Buffle J, Leppard GG (1995) Characterization of aquatic colloids and macromolecules. 1. Structure and behavior of colloidal material. Environ Sci Technol 29:2169–2175

Busscher HJ, van de Beltgritter B, van derMei HC (1995) Implications of microbial adhesion to hydrocarbons for evaluating cell-surface hydrophobicity: 1. Zeta potentials of hydrocarbon droplets. Colloids Surf B Biointerfaces 5:111–116

de Carvalho CCR, Wick LY, Heipieper HJ (2009) Cell wall adaptations of planktonic and biofilm *Rhodococcus erythropolis* cells to growth on C5 to C16 *n*-alkane hydrocarbons. Appl Microbiol Biotechnol 82:311–320

Efroymson RA, Alexander M (1991) Biodegradation by an Arthrobacter species of hydrocarbon partitioned into an organic solvent. Appl Environ Microbiol 57:1441–1447

Endo S, Bauerfeind J, Goss KU (2012) Partitioning of neutral organic compounds to structural proteins. Environ Sci Technol 46:12697–12703

Endo S, Escher BI, Goss KU (2011) Capacities of membrane lipids to accumulate neutral organic chemicals. Environ Sci Technol 45:5912–5921

Garcia JM, Wick LY, Harms H (2001) Influence of the nonionic surfactant Brij 35 on the bioavailability of solid and sorbed dibenzofuran. Environ Sci Technol 35:2033–2039

Gilbert D, Jakobsen HH, Winding A, Mayer P (2014) Co-Transport of polycyclic aromatic hydrocarbons by motile microorganisms leads to enhanced mass transfer under diffusive conditions. Environ Sci Technol 48:4368–4375

Guha S, Jaffé PR (1996a) Biodegradation kinetics of phenanthrene partitioned into the micellar phase of nonionic surfactants. Environ Sci Technol 30:605–611

Guha S, Jaffé PR (1996b) Bioavailability of hydrophobic compounds partitioned into the micellar phase of nonionic surfactants. Environ Sci Technol 30:1382–1391

Guha S, Jaffé PR, Peters CA (1998) Bioavailability of mixtures of PAHs partitioned into the micellar phase of a nonionic surfactant. Environ Sci Technol 32:2317–2324

Hanzel J, Thullner M, Harms H, Wick LY (2011) Microbial growth with vapor-phase substrate. Environ Poll 159:858–864

Harms H, Wick LY (2004) Mobilization of organic compounds and iron by microorganisms. In: van Leeuwen HP, Köster W (eds) Physicochemical kinetics and transport at biointerfaces. Wiley, Chichester, pp 401–444.

Harms H, Zehnder AJB (1994) Influence of substrate diffusion on degradation of dibenzofuran and 3-chlorodibenzofuran by attached and suspended bacteria. Appl Environ Microbiol 60:2736–2745.

Harms H, Zehnder AJB (1995) Bioavailability of Sorbed 3-Chlorodibenzofuran. Appl Environ Microbiol 61:27–33

Head IM, Jones DM, Röling WFM (2006) Marine microorganisms make a meal of oil. Nature Rev Microbiol 4:173–182

Holman HN, Nieman K, Sorensen DL, Miller CD, Martin MC, Borch T, McKinney WR, Sims RC (2002) Catalysis of PAH biodegradation by humic acid shown in synchrotron infrared studies. Environ Sci Technol 36:1276–1280

Johnsen AR, Wick LY, Harms H (2005) Principles of microbial PAH degradation. Environ Pollut 133:71–84

Klein B, Grossi V, Bouriat P, Goulas P, Grimaud R (2008) Cytoplasmic wax ester accumulation during biofilm-driven substrate assimilation at the alkane – water interface by *Marinobacter hydrocarbonoclasticus* SP17. Res Microbiol 159:137–144

Krell T, Lacal J, Reyes-Darias JA, Jimenez-Sanchez C, Sungthong R, Ortega-Calvo JJ (2012) Bioavailability of pollutants and chemotaxis. Curr Opin Biotechnol 24:451–456

Köster W, van Leeuwen HP (2004) Physicochemical kinetics and transport at the biointerface: setting the stage. In: van Leeuwen HP, Köster W (eds) Physicochemical kinetics and transport at biointerfaces. Wiley, Chichester, pp 2–14.

Laor Y, Strom PF, Farmer WJ (1999) Bioavailability of phenanthrene sorbed to mineral-associated humic acid. Water Res 33:1719–1729

Levich V (1962) Physicochemical hydrodynamics. Prentice Hall, Englewood Cliffs

Maurice PA, Manecki M, Fein JB, Schaefer J (2004) Fractionation of an aquatic fulvic acid upon adsorption to the bacterium, *Bacillus subtilis*. Geomicrobiol J 21:69–78

Mayer P, Fernqvist MM, Christensen PS, Karlson U, Trapp S (2007) Enhanced diffusion of polycyclic aromatic hydrocarbons in artificial and natural aqueous solutions. Environ Sci Technol 41:6148–6155

Mayer P, Karlson U, Christensen PS, Johnsen AR, Trapp S (2005) Quantifying the effect of medium composition on the diffusive mass transfer of hydrophobic organic chemicals through unstirred boundary layers. Environ Sci Technol 39:6123–6129

McLee AG, Davies SL (1972) Linear growth of a *Torulopsis* sp. on n-alkanes. Canad J Microbiol 18:315–319

Mounier J, Camus A, Mitteau I, Vaysse PJ, Goulas P, Grimaud R, Sivadon P (2014) The marine bacterium *Marinobacter hydrocarbonoclasticus* SP17 degrades a wide range of lipids and hydrocarbons through the formation of oleolytic biofilms with distinct gene expression profiles. Fems Microbiol Ecol 90:816–831

Mulder H, Breure AM, van Honschooten D, Grotenhuis JT, Andel JGV, Rulkens WH (1998) Effect of biofilm formation by Pseudomonas 8909N on the bioavailability of solid naphthalene. Appl Microbiol Biotechnol 50:277–283

Naether DJ, Slawtschew S, Stasik S, Engel M, Olzog M, Wick LY, Timmis KN, Heipieper HJ (2013) Adaptation of hydrocarbonoclastic *Alcanivorax borkumensis* SK2 to alkanes and toxic organic compounds – a physiological and transcriptomic approach. Appl Environ Microbiol 79:4282–4293

Neu TR (1996) Significance of bacterial surface-active compounds in interaction of bacteria with interfaces. Microbiol Rev 60:151–166

Noordman WH, Ji W, Brusseau ML, Janssen DB (1998) Effects of rhamnolipid biosurfactants on removal of phenanthrene from soil. Environ Sci Technol 32:1806–1812

Ogram AV, Jessup RE, Ou LT, Rao PS (1985) Effects of sorption on biological degradation rates of (2,4-dichlorophenoxy)acetic acid in soils. Appl Environ Microbiol 49:582–587

Orench-Rivera N, Kuehn MJ (2016) Environmentally controlled bacterial vesicle-mediated export. Cellular Microbiology 18:1525–1536

Ortega-Calvo JJ, Saiz-Jimenez C (1998) Effect of humic fractions and clay on biodegradation of phenanthrene by a Pseudomonas fluorescens strain isolated from soil. Appl Environ Microbiol 64:3123–3126

Otto S, Banitz T, Thullner M, Harms H, Wick LY (2016) Effects of facilitated bacterial dispersal on the degradation and emission of a desorbing contaminant. Environ Sci Technol 50:6320–6326

Rosenberg M, Rosenberg E (1981) Role of adherence in growth of *Acinetobacter cacoaceticus* RAG-1 on hexadecane. J Bacteriol 148:51–57

Schamfuss S, Neu TR, van der Meer JR, Tecon R, Harms H, Wick LY (2013) Impact of mycelia on the accessibility of fluorene to PAH-degrading bacteria. Environ Sci Technol 47:6908–6915

Schwarzenbach RP, Gschwend PM, Imboden DM (2017) Environmental organic chemistry, Wiley, New York

Sikkema J, de Bont JAM, Poolman B (1994) Interactions of Cyclic Hydrocarbons With Biological-Membranes. J Biol Chem 269:8022–8028

Sikkema J, de Bont JAM, Poolman B (1995) Mechanism of membrane toxicity of hydrocarbons. Microbiol Rev 59:201–222

Smith KEC, Thullner M, Wick LY, Harms H (2009) Sorption to humic acids enhances polycyclic aromatic hydrocarbon biodegradation. Environ Sci Technol 43:7205–7211

Smith KEC, Thullner M, Wick LY, Harms H (2011) Dissolved organic carbon enhances the mass transfer of hydrophobic organic compounds from nonaqueous phase liquids (NAPLs) into the aqueous phase. Environ Sci Technol 45:8741–8747

Southam G, Whitney M, Knickerbocker C (2001) Structural characterization of the hydrocarbon degrading bacteria-oil interface: implications for bioremediation. International Biodeter Biodeg 47:197–201

Späth R, Flemming HC, Wuertz S (1998) Sorption properties of biofilms. Wat Sci Technol 37:207–210

Sutton R, Sposito G (2005) Molecular structure in soil humic substances: the new view. Environ Sci Technol 39:9009–9015

Taylor MG, Simkiss K (2004) Transport of colloids and particles across biological memnbranes. In: van Leuven HP, Köster W (eds) Physicochemical kinetics and transport at chemical-biological interphases. Wiley, Chichester, pp 358–400

Tejeda-Agredano M-C, Mayer P, Ortega-Calvo J-J (2014) The effect of humic acids on biodegradation of polycyclic aromatic hydrocarbons depends on the exposure regime. Environ Pollut 184:435–442

van Loosdrecht MCM, Lyklema J, Norde W, Schraa G, Zehnder AJB (1990) Influence of interfaces on microbial activity. Microb Rev 54:75–87

Vigneault B, Percot A, Lafleur M, Campbell PGC (2000) Permeability changes in model and phytoplankton membranes in the presence of aquatic humic substances. Environ Sci Technol 3:3907–3913

Volkering F, Breure AM. van Andel JG, Rulkens WH (1995) Influence of nonionic surfactants on bioavailability and biodegradation of polycyclic aromatic hydrocarbons. Appl Environ Microbiol 61:1699–1705

Wick LY, Colangelo T, Harms H (2001) Kinetics of mass-transfer-limited growth on solid PAHs. Environ Sci Technol 35:354–361

Wick LY, deMunain AR, Springael D, Harms H (2002) Responses of *Mycobacterium* sp. LB501T to the low bioavailability of solid anthracene. Appl Microbiol Biotechnol 58:378–385

Wick LY et al (2007) Effect of fungal hyphae on the access of bacteria to phenanthrene in soil. Environ Sci Technol 41:500–505

Witholt B et al (1990) Bioconversions of Aliphatic-Compounds by *Pseudomonas-Oleovorans* in Multiphase Bioreactors – Background and Economic-Potential. Trends Biotechnol. 8:46–52

# Matrix - Hydrophobic Compound Interactions

3

Hauke Harms, Lukas Y. Wick, and Kilian E. C. Smith

## Contents

1   Introduction .................................................................... 34
2   Phenomena of Molecular Sorption to Solid Matrices ..................................... 34
3   Transfer of Hydrocarbons Between the NAPL and Aqueous Phases ...................... 36
4   Sorption of HOCs to Mobile-Sorbing Phases .............................................. 39
5   Contaminant Aging and Release Kinetics ................................................. 41
6   Research Needs ................................................................... 43
References ........................................................................... 43

### Abstract

The fate and transport of hydrophobic organic compounds (HOCs) such as oil hydrocarbons are strongly influenced by their interactions with environmental matrices including soils and sediments. These interactions can be grouped into those of nonaqueous phase liquids (NAPLs), e.g., the spreading of oil on solid surfaces and its movement in porous media, and those of water-dissolved HOC molecules which sorb onto solid surfaces or partition into organic matter or NAPL phases. Generally, these different types of sequestration phenomena lead to reduced contact between organisms and the bioavailable HOC molecules dissolved in the surrounding water phase, and thus to lower uptake and biodeg-

H. Harms · L. Y. Wick (✉)
Department of Environmental Microbiology, Helmholtz Centre for Environmental Research – UFZ, Leipzig, Germany
e-mail: hauke.harms@ufz.de; lukas.wick@ufz.de

K. E. C. Smith
Institute for Environmental Research (Biology 5), RWTH Aachen University, Aachen, Germany
e-mail: kilian.smith@bio5.rwth-aachen.de

© Springer International Publishing AG, part of Springer Nature 2018
T. Krell (ed.), *Cellular Ecophysiology of Microbe: Hydrocarbon and Lipid Interactions*,
Handbook of Hydrocarbon and Lipid Microbiology, https://doi.org/10.1007/978-3-319-50542-8_39

radation. However, in certain situations, sorption of the HOCs to small and highly mobile HOC-sorbing phases such as dissolved organic carbon or surfactants may mobilize the HOCs and increase their bioavailability and/or toxicological risk.

# 1    Introduction

Interactions of NAPLs with solids are of importance wherever oil is present in porous media, for instance as is the case in oil reservoirs. Here the interest in extracting the oil has motivated much research into the possibility of pumping the oil out or, if pumping fails, to push residual oil out of the porous matrix by injecting gases or aqueous solutions (e.g., brine). The latter often occurs in combination with selective blocking of alternate flow paths, thermal treatment (e.g., steam injection), or physicochemical enhancement of oil movement using detergents (Banat 1995). A second field of environmental concern is the behavior of spilled fuels, coal tar, or other oily masses in soil, aquifer sediment, fractured rock, on beaches, or in the sediments of rivers, lakes, and oceans. A detailed description of the physics of the residence behavior and movement of oil in these phases goes beyond the purpose of this chapter, but some factors of influence shall be mentioned here. Oil masses of lower viscosity move more readily in porous media than more viscous oil. This viscosity depends on the chemical composition of the oil and is reduced by an increase in temperature. The injection of gases during enhanced oil recovery also reduces the viscosity of oil as some of the gas dissolves in the oil, a phenomenon called oil swelling (McInerney et al. 2007). Emulsifiers are surface-active substances capable of stabilizing emulsions by accumulating at water-oil interfaces. Emulsification of oils with water can influence the viscosity in both ways depending on the oil-water ratio, i.e., if the oil is the continuous phase or the dispersed phase in the emulsion, but also depending on other factors such as the size of the droplets of the dispersed phase. In the environment, quasisolid emulsions of hydrocarbons and water have been observed as viscous interfacial films around aged tar globules (Nelson et al. 1996). Finally, the chemical composition of the solid material and the pore diameter of the solid matrix influence the oil movement via capillary forces. If the oil has a tendency to spread on the solid or, in other words, to form a contact angle (defined as the angle between the oil droplet surface and the underlying solid, and which provides a measure of the surface wettability) with the solid surface below 90°, it will be retained better in pores of smaller diameters. The opposite will be the case if the oil forms a contact angle above 90° with the solid surface. In this case capillary forces will retain the oil better in larger pores. This behavior is described by the Young-Laplace equation (Mozes et al. 1991). Capillary forces can be influenced by surfactants of chemical or biological origin, leading to improved movement of the oil in the porous medium.

# 2    Phenomena of Molecular Sorption to Solid Matrices

The total amount of HOC in an environmental compartment can be conceptually divided into three pools: the irreversibly bound, the reversibly bound, and the freely dissolved pool (Fig. 1) (Ortega-Calvo et al. 2015). The mechanistic interpretation of

**Total amount of the compound**

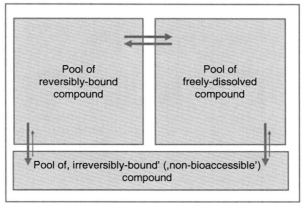

**Fig. 1** The schematic shows how the total amount of a compound in an environmental system can be conceptually divided into the freely dissolved, the reversibly and the irreversibly bound pools. From the perspective of microbial degradation, the kinetics of the release of hydrophobic organic compounds (HOCs) into the water phase, i.e., into the freely dissolved pool, is of primary importance. Soft organic matter and NAPL-absorbed HOC as well as surface-adsorbed HOC appear to be more readily bioaccessible than quasi irreversibly bound HOC "stuck" in hard organic matter and micropores

the macroscopically observed sorption behavior of HOCs in soils and sediments has been an issue of much debate in the last decades. It is therefore highly appreciable that in 1997 leading experts in the field, among them proponents of contrasting views, jointly published an article categorizing the various mechanisms of HOC sorption to geosorbents (Luthy et al. 1997). These authors distinguish five microscopic sorption mechanisms of HOC in geosorbents: (A) **ab**sorption **into** amorphous or "soft" natural organic matter or NAPL; (B) **ab**sorption **into** condensed or "hard" organic polymeric matter or combustion residues such as soot; (C) **ad**sorption **onto** water-wet organic surfaces such as soot; (D) **ad**sorption **onto** exposed water-wet mineral surfaces; and (E) **ad**sorption **into** microvoids or microporous minerals. An important conclusion was that none of these sorption mechanisms is likely to occur exclusively in natural geomaterials, and that the complex sorption and desorption equilibria and kinetics that are often observed can be explained as overall effects of varying contributions of these different mechanisms. These mechanisms were then examined for their impact on the behavior of HOCs in terms of linearity of the sorption isotherms, competition between sorbates for sorption sites, sorption kinetics, and, in the case of desorption hysteresis (i.e., when the sorption and desorption curves cannot be superimposed), selectivity for steric features of the sorbents as well as the ease at which the HOCs can be extracted.

Absorption into soft organic matter and NAPLs (A in the above list), as well as both types of adsorption onto exposed surfaces (C, D), were identified to be fast and readily reverted by solvent extraction. In contrast, absorption into hard organic matter (B) and adsorption in microvoids (E) were characterized as being slower, difficult to revert by solvent extraction and often showing desorption hysteresis

(Greenberg et al. 2005). From the perspective of microbial degradation, the release kinetics of sorbed HOCs into the water phase is of primary importance since uptake of dissolved compound is generally required for biodegradation (Volkering et al. 1992). Soft organic matter and NAPL-absorbed HOCs, as well as surface-adsorbed HOCs, appear to be more readily bioaccessible than HOCs "stuck" in hard organic matter and micropores (Cornelissen et al. 1997). The case of pore-obstructed HOC illustrates that there is also a geometric aspect to bioaccessibility (in this case the exclusion of microbes from micropores that spatially separates the HOC source from its biological sink), further to the influence of chemical interactions. This is due to the dynamic nature of microbial degradation and the fact that rates of mass transfer depend largely on the distances that need to be bridged (Bosma et al. 1997, Harms and Wick 2004).

## 3    Transfer of Hydrocarbons Between the NAPL and Aqueous Phases

NAPLs are varied with respect to both the environmental compartment in which they are found (e.g., the open sea, sediment, or soil), but also their physical and chemical characteristics. Typical examples of environmentally relevant NAPLs include crude oil, its various refinement products, and anthropogenic wastes such as the aromatic rich coal-tars contaminating groundwater at industrial sites. Natural oil seeps in marine and terrestrial environments mean that the interplay between microorganisms and certain types of NAPLs is not a recent phenomenon, and there has been sufficient time for microbial populations to evolve strategies for an increased exploitation of this rich hydrocarbon resource (Head et al. 2006). Furthermore, NAPLs are typically composed of many classes of hydrocarbons in combination with other nonhydrocarbon compounds that are often unresolved, e.g., crude oil (Marshall and Rodgers 2004). It is therefore not usually the case that a single microorganism has the metabolic ability to degrade the full range of components present in oil, and a range of microorganisms are involved in the biodegradation process of NAPL hydrocarbons (Head et al. 2006). Therefore, biodegradation in the field becomes a complex interplay between HOC bioavailability, toxicity, and microbial ecology.

NAPL hydrocarbons represent a potential source of carbon and energy in a form that is difficult to exploit (Volkering et al. 1992). Lighter oil hydrocarbons rapidly dissipate via volatilization or dissolution, and the following focusses on hydrophobic NAPL compounds and groups them under the general term HOCs. Thermodynamic considerations indicate that even at equilibrium (i.e., the maximum dissolved aqueous concentrations that can be attained when partitioning is involved), HOCs preferentially remain in the NAPL with only low concentrations being reached in the aqueous phase (Efroymonson and Alexander 1995). Furthermore, in non-equilibrium situations their hydrophobicity means that the mass transfer between the NAPL and the aqueous phase is usually slow (Schluep et al. 2002). Therefore, it is normally the case that despite the high levels of HOCs in the NAPL phase, these have a low bioavailability and biodegradation is mass transfer limited (e.g., Ramaswami et al. 1997). Although this low bioavailability is not ideal for degrading microorganisms, it does mean exposure to potentially toxic NAPL phase HOCs is reduced for nondegraders and other organisms.

Adjacent NAPL and aqueous compartments, each with homogenous bulk HOC concentrations, are separated by the NAPL:water interface. On each side of this interface, thin unstirred boundary layers (BLs) exist as depicted in Fig. 2. Within these BLs, transport of the HOCs occurs by the relatively slow process of molecular diffusion, and transfer across the BLs has the determining role in the overall mass flux from the NAPL into the bioavailable dissolved phase. Therefore, the mass transfer pathway between a NAPL and the aqueous phase can be considered as being composed of two barriers to mass transfer aligned in series (Schwarzenbach et al. 2017). For HOCs, it is particularly the BL on the aqueous side that limits the overall mass flux (Ghoshal and Luthy 1996). Special cases where the main resistance to mass transfer of HOCS occurs in the NAPL phase are discussed at the end of this section.

The mass flux can be defined as the HOC mass from the NAPL phase entering the aqueous phase (where the degrading microorganisms are present) over a given period of time. This can be described in general terms as follows

$$\text{Mass flux} = \text{Area} \times \text{Transfer velocity} \times \text{Driving force} \quad (1)$$

The role of the interface area is obvious, the larger this is the bigger the surface for mass exchange. The transfer velocity can be further rationalized as being composed of the ratio between the aqueous diffusivity of the HOC molecules and the thickness of the BL (Schwarzenbach et al. 2017). This is also intuitively obvious, a higher aqueous diffusivity of a molecule is indicative of more rapid motion and a thinner BL will be more quickly traversed. The driving force is determined by the gradient in chemical activity of the HOC between the bulk NAPL and aqueous compartments. Molecular diffusion occurs from regions of high to low chemical activity (Reichenberg and Mayer 2006), and the greater this difference the greater the mass flux. Chemical activity is a function of the HOC concentration, its physicochemical properties as well as those of the environmental matrix in which it is found (Schwarzenbach et al. 2017). This means a compound can have the same chemical activity in two phases (i.e., be at equibrium) but at very different concentrations. A well-known illustration of this is the equilibrium octanol:water partition coefficient, where a HOC has the same chemical activity in the water and octanol phases but very different concentrations. The usual practice is to measure mass concentrations in the NAPL and aqueous phases; the driving force is then computed from these mass concentrations in conjunction with experimentally determined partition coefficients (e.g., Schluep et al. 2002). Note that equilibrium sampling techniques such as passive sampling allow direct determination of chemical activity gradients between adjacent phases (Mayer et al. 2003).

Equation 1 indicates that the magnitude of the mass flux is influenced by the NAPL:aqueous interface area, the speed of transfer across the BLs, and the driving force between the NAPL and aqueous compartments. Therefore, a change in any of the above parameters can result in an increase or decrease in the mass flux of HOCs into the aqueous phase, and thus have a knock-on effect on HOC bioavailability and biodegradation. This is shown schematically in Fig. 2, where different scenarios have been depicted in terms of the above three factors. Figures 2a–d illustrate that a range of abiotic processes can potentially impact the mass transfer of HOCs into the bioavailable aqueous phase. These processes can occur simultaneously and may vary with respect to one another over time and have been summarized in Table 1. Some of these changes can have other beneficial effects with respect to microbial growth, in addition to any enhancement in the HOC mass transfer. For example, increased mixing might also lead to improved aeration and prevent the formation of nutrient-depleted patches or changes in temperature might lead to an increase (or decrease) in growth depending on the degrader.

In specific situations, the resistance to diffusive mass transfer on the NAPL side can become significant (Fig. 2e). Such scenarios include the mass transfer of more water-soluble hydrocarbons (Schluep et al. 2002), highly viscous NAPLs (Ortiz et al. 1999), or situations where there is a weathering of the surface layers of multicomponent NAPL mixtures leading to the formation of more impermeable surface skins. The latter is believed to be the result of viscous NAPL:water emulsions forming at the surface rather than because of changes in the composition due to preferential dissolution of the more soluble components (Nelson et al. 1996). In terms of microbial degradation of NAPL hydrocarbons such cases are significant for

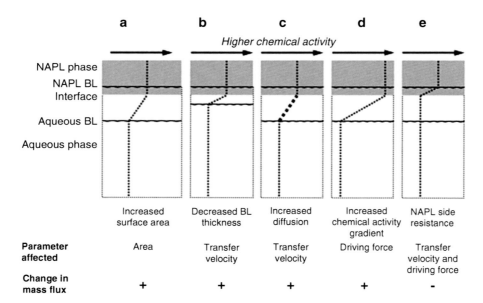

**Fig. 2** Chemical activity profile (·····) of a hypothetical hydrophobic organic compound (HOC) across the NAPL:aqueous interface under various environmental conditions. The resultant effect on the HOC mass flux is shown at the bottom by a positive or negative symbol

**Table 1** Overview of the physical, environmental, and chemical factors influencing the mass transfer of a hydrocarbon from the NAPL into the aqueous phase. The letters in brackets correspond to the scenarios depicted in Fig. 2

| Category | Description | Parameter affected |
|---|---|---|
| **Physical** | NAPL architecture and amount, e.g., film, droplets, micropores | Surface area [A] |
| **Environmental** | Hydrodynamic mixing | Break-up of NAPL [A] Boundary layer thickness [B] |
| | Temperature | Molecular diffusion [C] Chemical activity gradient [D] |
| | Sorption in the aqueous phase | Chemical activity gradient [D] |
| **Chemical** | Hydrocarbon concentration | Chemical activity gradient [D] |
| | Hydrocarbon properties | Diffusivity [C] Chemical activity gradient [D] |
| | NAPL properties | Diffusivity [C] Chemical activity gradient [D] Wetting properties [A] |
| | Weathering of multicomponent NAPLs | Formation of impermeable skins [E] |
| | NAPL side resistance | Diffusivity and chemical activity gradient [E] |

two reasons. Firstly, there is a reduction in HOC mass transfer into the aqueous phase (Luthy et al. 1993). Secondly, those environmental processes listed in Table 1 that affect the mass transfer through the aqueous BL no longer play the same role. For example, hydrodynamic mixing will not increase or decrease the mass transfer via altering the aqueous BL thickness as this no longer controls the mass flux.

From the point of view of the degrading microorganism, there is little that can be done to influence such abiotic processes and thus they can only react passively to any resulting changes in the mass flux and bioavailability. An interesting example is that of sorption, which had the consequence of lowering the dissolved phase activity and thus maintaining a high diffusive gradient between the NAPL and aqueous phases as illustrated in Fig. 2d. This figure shows that there is potentially a feedback between biotic growth of the microorganism due to HOC degradation and abiotic sorption. A bacterial population actively degrading a certain NAPL component(s) will increase in biomass, which in turn leads to an increase in the sorption capacity of the aqueous phase, thus potentially enhancing the mass transfer of other NAPL HOCs.

# 4   Sorption of HOCs to Mobile-Sorbing Phases

Various types of hydrocarbon interactions with the geosorbents present in soils and sediments have been considered above. These sorbents can in the main be considered as relatively immobile, with only small fractions being slowly displaced by processes such as bioturbation or erosion by wind and water. Therefore, for micro-organisms to colonize such sorbents, they need to initially be brought into contact with and subsequently attach to the surface.

Some sorbing phases, though, are smaller and much more mobile and even move with the aqueous phase. This is a diverse category, encompassing everything from suspended inorganic minerals, particulate and dissolved organic matter, living bacteria- and phytoplankton to biosurfactants. Of course, there is no defined cut-off between what constitutes a mobile- and immobile-sorbing phase, and furthermore, the significance these have in terms of the total HOC sorption depends on the environment in which they are found. For example, in soils and sediment, most of the HOCs are primarily associated with larger sized and stationary particulate material (e.g., Hawthorne et al. 2005). However, in the fresh and marine water column, mobile sorbents such as plankton and particulate organic carbon can comprise the dominant-sorbing phase (e.g., Schulz-Bull et al. 1998). Some of these small and mobile sorbing phases have been shown to play an important role with respect to HOC bioavailability and biodegradation.

HOCs associate with these matrices via the same set of sorption mechanisms discussed above (Luthy et al. 1997). Therefore, their role in terms of reducing the bioavailability of HOCs can be understood using the same terms of reference. However, their small size, and thus high surface area to volume ratio, imply that the exchange kinetics are rapid (Poerschmann et al. 1997). Therefore, processes such as retarded diffusion play less of a bioavailability-limiting role. Their small size also enables them to readily move with, for example, the advective flow of water. This, together with their propensity to associate with cellular membranes, confers on them a particular role in the bioavailability and biodegradation of HOCs.

Exactly what role do such mobile HOC-sorbing phases play with respect to bioavailability? They can alter both the kinetics of abiotic mass transfer and the equilibrium distribution of HOCs between the (stationary) sorbed and aqueous compartments. In many environments, bioavailability and thus biodegradation is limited by the slow HOC mass transfer from the sorbed state and into the aqueous phase (Harms and Bosma 1997). Here, the stationary sorbing material can be anything from a geosorbent such as particulate matter to a nonaqueous phase liquid (NAPL). The presence of an additional sorbing phase in the surrounding aqueous medium can enhance the rate of dissolution. For example, both natural and synthetic surfactants increase the dissolution rate of HOCs such as PAHs from their pure solid state (Grimberg et al. 1995) or when present in NAPLs (Garcia-Junco et al. 2001). Dissolved organic carbon (DOC) (Smith et al. 2011) or DOC associated with mineral surfaces (Garcia-Junco et al. 2003) enhance the mass transfer of HOCs from NAPLs into the water phase. Finally, particularly relevant in terms of the microbial blooms developing after an oil spill is the observation that small motile organisms such as protozoa increase the mass transfer of HOCs from the sorbed to the dissolved state (Gilbert et al. 2014). In part, the increased mass transfer observed in the above studies can be explained by HOC sorption to the surfactant micelles, DOC, or organisms in the aqueous phase reducing the dissolved phase concentrations and maintaining the high chemical activity gradients driving the dissolution process.

However, it also appears that in parallel these small and mobile sorbing phases function as carriers, enhancing the transport of the HOCs across the aqueous BL and into the bulk solution (Grimberg et al. 1995).

An enhancement in the dissolution rates has two implications for bioavailability. Firstly, the rate of mass transfer out of the nonaccessible phase is increased which can be particularly important for those HOCs where dissolution is very low to start with. Secondly, the total amount of HOCs in the aqueous phase can be increased above solubility, albeit partly sorbed. Should a fraction of this sorbed aqueous amount be accessible by the degrading microorganisms then this could have a positive overall effect on bioavailability.

When considering natural environments, the distribution of HOCs is generally heterogeneous. In soils and sediments, such microscale inhomogeneities are particularly important in lowering the overall bioavailability. Since microbial colonies are spatially distributed and mainly exist attached to various surfaces rather than suspended in the interstitial solution, they "see" a relatively small volume of the environment (Postma and Vanveen 1990). Should this volume become depleted of HOCs via consumption, then it is rapidly the case that the distance to a replenishing source becomes too large for a sufficient resupply via solely aqueous diffusion (Bosma et al. 1997). Therefore, HOC association with an advectively transported sorbing phase such as surfactants or DOC could also have implications for redistribution, functioning at the microscale as "carriers" from a site of contamination to that of biodegradation. When considering the aquatic ecosystem, the more thorough mixing of the water column by turbulence probably means that such microscale inhomogeneities are less of an issue. Nevertheless, in some cases heterogeneity in compound distribution might also play a role but over a larger scale. An oil spill at the water surface is an example. Oil hydrocarbon sorption to suspended mineral and organic matter in the water column initially might enhance the dissolution process, and then be transported away via the water currents to more distant locations, forming a plume of bioavailable HOC. Although the relevance of such roles remains to be demonstrated in the field, these are the general principles behind bioremediation of soils using surfactant washing solutions.

## 5      Contaminant Aging and Release Kinetics

Contaminants in geomaterials may undergo changes that have been summarized as processes of contaminant aging or weathering. It has been observed that the efficiency of chemical extraction and biodegradation of contaminants is lower when the contact time between contaminant and the geomaterials before these interventions was longer (Cornellissen et al. 1997). In many cases, recent contaminations may thus be treated with higher efficiency than historical ones (Hatzinger and Alexander 1995). For instance, it has been frequently seen that when the biodegradation of a historical contamination in soil has come to an end despite a still relatively high residual concentration, a contaminant of the same kind that is freshly spiked to the same soil is rapidly degraded (Valo and Salkinoja-Salonen 1986). From this experiment and similar real-world observations, it can be inferred that the different portions of the contaminant show different degrees of bioaccessibility (Ortega-Calvo et al. 2015). This differential behavior of contaminant fractions is not only a problem for the remediation process but also limits the value of spiking soil with radioactively labelled contaminant as an indicator of the biodegradation potential.

*Which kinds of mechanisms can lead to reduced contaminant bioaccessibility and extractability?* Mechanisms include chemical changes, changes in the soil or sediment structure, and shifts in the spatial distribution of the chemical due to diffusive transfer in combination with the exclusion of microorganisms (or extractants) from certain parts of the geomaterial. An example for chemical changes would be the successive occupation of high-energy sorption sites by contaminant molecules. The probability that individual contaminant molecules that are initially absorbed in "soft" organic matter or NAPL come into contact with either "hard" organic matter, high-energy sorption sites (e.g., on the surface of soot) or enter the swollen interlayers of clay minerals increases with time. A declining reversibility of sorption would also arise from the metamorphosis of organic matter into forms that retain absorbed molecules more efficiently. Structural changes in the soil matrix could lead to contaminant aging by burying formerly labile contaminant pools under poorly permeable phases. The encapsulation of "soft" organic matter by "hard" organic matter as exemplified by Luthy et al. (1997), or by mineral soil constituents, would be examples. One can easily imagine that bioturbation or processing of soil materials inside the digestive tracts of soil-dwelling organisms could lead to the obstruction of diffusion pathways. The successive formation of poorly permeable interphases around tar globules would also fall into the category of structural changes. Aging can occur even in the absence of chemical or structural changes due to diffusive mass transfer. Soil and sediment contaminants typically enter via the larger pores. From thereon diffusive processes carry part of the contaminant in regions and size classes of pores that are increasingly difficult to access by microbes. Before an equilibrium distribution of the contaminant is achieved, there will be a continuous inward-bound diffusive flux of contaminant into soil aggregates that carries contaminant fractions further away from the biota. Using experimental model polymers that exclude the possibility of chemical or structural changes, the effects of diffusion distances due to the pore-size exclusion of microbes was shown (Harms and Zehnder 1995).

The observable effect of all types of contaminant aging is an apparent desorption hysteresis, i.e., much longer time scales are needed to complete desorption compared to adsorption. Biodegradation curves as well as release curves obtained by continuous mild extraction (e.g., flushing with water) show pronounced tailing that may be interpreted as successive emptying of sorption sites of increasing sorption energy or as growing diffusion distances of the contaminant molecules within a matrix composed of particles that empty from the outside to the center (also referred to as "shrinking core" desorption). Mathematically, the observed desorption progress can be easily fitted by applying the first-order two-compartment models that distinguish a rapidly and a slowly desorbing contaminant fraction with largely different release rate constants (Cornelissen et al. 1998). Further distinction of a third, very slowly desorbing fraction, may give even better fits (e.g., Greenberg et al. 2005). However, given the wealth of possible mechanisms that may bring about the earlier release of one contaminant molecule than another one, the existence of a continuum of sorption strengths and travel distances appears much more likely than the existence of two or three distinct states of sorption. The rapidly desorbing fraction has nevertheless been found to be a relatively good descriptor of the bioaccessible fraction (Cornelissen et al. 1998), whereas the extremely slowly desorbing chemicals have been conceptually defined as "nonbioaccessible" (Semple et al. 2004) or "irreversibly bound" (Reichenberg and Mayer 2006).

# 6    Research Needs

To date, the influence of small and mobile sorbents and their potential to increase HOC mobility at the microscale has been largely neglected. Therefore, the common opinion that sorption generally reduces both bioavailability and risk is a simplification requiring knowledge-based specification, particularly using systems that more accurately mimic the complexity in the field. Another area requiring further research is the interaction of the components found in complex contaminant mixtures such as oils. For instance, biologically inactive components may well influence the partitioning behavior of bioactive components either directly or via stimulating the growth of specific degraders. Finally, directly applying chemical activity-based measurements to understand the dynamics and fate of HOCs may in some cases be more appropriate than common water-solubility and concentration-based approaches.

# References

Banat IM (1995) Biosurfactants production and possible uses in microbial enhanced oil-recovery and oil pollution remediation - a review. Bioresour Technol 51:1–12

Bosma TNP, Middeldorp PJM, Schraa G, Zehnder AJB (1997) Mass transfer limitation of biotransformation: quantifying bioavailability. Environ Sci Technol 31:248–252

Cornelissen G, van Noort PCM, Govers HAJ (1997) Desorption kinetics of chlorobenzenes, polycyclic aromatic hydrocarbons and polychlorinated biphenyls: sediment extraction with Tenax® and effects of contact time and solute hydrophobicity. Environ Toxicol Chem 16:1351–1357

Cornelissen G, Rigterink H, Ferdinandy MMA, van Noort PCM (1998) Rapidly desorbing fractions of PAHs in contaminated sediments as a predictor of the extent of bioremediation. Environ Sci Technol 32:966–970

Efroymonson RA, Alexander M (1995) Reduced mineralization of low concentrations of phenanthrene because of sequestering in nonaqueous-phase liquids. Environ Sci Technol 29:515–521

Garcia-Junco M, De Olmeda E, Ortega-Calvo JJ (2001) Bioavailability of solid and non-aqueous phase liquid (NAPL)-dissolved phenanthrene to the biosurfactant-producing bacterium *Pseudomonas aeruginosa* 19SJ. Environ Microbiol 3:561–569

Garcia-Junco M, Gomez-Lahoz C, Niqui-Arroyo JL, Ortega-Calvo JJ (2003) Biosurfactant- and biodegradation-enhanced partitioning of polycyclic aromatic hydrocarbons from nonaqueous-phase liquids. Environ Sci Technol 37:2988–2996

Ghoshal S, Luthy RG (1996) Bioavailability of hydrophobic organic compounds from nonaqueous phase liquids: the biodegradation of naphthalene from coal tar. Environ Toxicol Chem 15:1894–1900

Gilbert D, Jakobsen HH, Winding A, Mayer P (2014) Co-transport of polycyclic aromatic hydrocarbons by motile microorganisms leads to enhanced mass transfer under diffusive conditions. Environ Sci Technol 48:4368–4375

Greenberg MS, Burton AB Jr, Landrum PF, Leppänen MT, Kukkonen JVK (2005) Desorption kinetics of fluoranthene and trifluralin from Lake Huron and Lake Erie, USA, sediments. Environ Toxicol Chem 24:31–39

Grimberg SJ, Nagel J, Aitken MD (1995) Kinetics of phenanthrene dissolution into water in the presence of nonionic surfactants. Environ Sci Technol 29:1480–1487

Harms H, Bosma TNP (1997) Mass transfer limitation of microbial growth and pollutant degradation. J Ind Microbiol Biotechnol 18:97–105

Harms H, Wick LY (2004) Mobilization of organic compounds and iron by microorganisms. In: van Leuven HP, Köster W (eds) Physicochemical kinetics and transport at chemical-biological interphases. Wiley, Chichester, pp 401–444

Harms H, Zehnder AJB (1995) Bioavailability of sorbed 3-chlorodibenzofuran. Appl Environ Microbiol 61:27–33

Hatzinger P, Alexander M (1995) Effect of aging of chemicals in soil on their biodegradability and extractability. Environ Sci Technol 29:537–545

Hawthorne SB, Grabanski CB, Miller DJ, Kreitinger JP (2005) Solid-phase microextraction measurement of parent and alkyl polycyclic aromatic hydrocarbons in milliliter sediment pore water samples and determination of K-DOC values. Environ Sci Technol 39:2795–2803

Head IM, Jones DM, Röling WFM (2006) Marine microorganisms make a meal of oil. Nat Rev Microbiol 4:173–182

Luthy RG, Ramaswami A, Ghoshal S, Merkel W (1993) Interfacial films in coal tar nonaqueous-phase liquid-water systems. Environ Sci Technol 27:2914–2918

Luthy RG, Aiken GR, Brusseau ML, Cunningham SD, Geschwend PM, Pignatello JJ, Reinhard M, Traina SJ, Weber WJ, Westall JC (1997) Sequestration of hydrophobic organic contaminants by geosorbents. Environ Sci Technol 31:3341–3347

Marshall AG, Rodgers RP (2004) Petroleomics: the next grand challenge for chemical analysis. Acc Chem Res 37:53–59

Mayer P, Tolls J, Hermens JLM, Mackay D (2003) Equilibrium sampling devices. Environ Sci Technol 37:184A–191A

McInerney MJ, Voordouw GE, Jenneman GE, Sublette KL (2007) Oil field microbiology. In: Hurst CJ, Crawford RL, Lipson DA, Mills AL, Stetzenbach LD (eds) Manual of environmental microbiology. ASM Press, Washington, DC, pp 898–911

Mozes N, Handley PS, Busscher HJ, Rouxhet PG (1991) Microbial cell surface analysis: structural and physicochemical methods. Verlag Chemie, New York, Weinheim, Cambridge

Nelson EC, Ghoshal S, Edwards JC, Marsh GX, Luthy RG (1996) Chemical characterization of coal tar-water interfacial films. Environ Sci Technol 30:1014–1022

Ortega-Calvo JJ, Harmsen J, Parsons JR, Semple KT, Aitken MD, Ajao C, Eadsforth E, Galay-Burgos M, Naidu R, Oliver R, Peijnenburg WJGM, Römbke J, Streck G, Versonnen B (2015) From bioavailability science to regulation of organic chemicals. Environ Sci Technol 49:10255–10264

Ortiz E, Kraatz M, Luthy RG (1999) Organic phase resistance to dissolution of polycyclic aromatic hydrocarbon compounds. Environ Sci Technol 33:235–242

Poerschmann J, Zhang ZY, Kopinke FD, Pawliszyn J (1997) Solid phase microextraction for determining the distribution of chemicals in aqueous matrices. Anal Chem 69:597–600

Postma J, Vanveen JA (1990) Habitable pore-space and survival of *Rhizobium-Leguminosarum Biovar Trifolii* introduced into soil. Microb Ecol 19:149–161

Ramaswami A, Ghoshal S, Luthy RG (1997) Mass transfer and bioavailability of PAH compounds in coal tar NAPL-slurry systems. 2. Experimental evaluations. Environ Sci Technol 31:2268–2276

Reichenberg F, Mayer P (2006) Two complementary sides of bioavailability: accessibility and chemical activity of organic contaminants in sediments and soils. Environ Toxicol Chem 25:1239–1245

Schluep M, Gälli R, Imboden DM, Zeyer J (2002) Dynamic equilibrium dissolution of complex nonaqueous phase liquid mixtures into the aqueous phase. Environ Toxicol Chem 21:1350–1358

Schulz-Bull DE, Petrick G, Bruhn R, Duinker JC (1998) Chlorobiphenyls (PCB) and PAHs in water masses of the northern North Atlantic. Mar Chem 61:101–114

Schwarzenbach RP, Gschwend PM, Imboden DM (2017) Environmental organic chemistry. Wiley, New York

Semple K, Doick K, Jones K, Burauel P, Craven A, Harms H (2004) Defining bioavailability and bioaccessibility of contaminated soil and sediment is complicated. Environ Sci Technol 15:229A–231A

Smith KEC, Thullner M, Wick LY, Harms H (2011) Dissolved organic carbon enhances mass fluxes of hydrophobic organic compounds from NAPLs into the aqueous phase. Environ Sci Technol 45:8741–8747

Valo R, Salkinoja-Salonen M (1986) Bioreclamation of chlorophenol-contaminated soil by composting. Appl Microbiol Biotechnol 25:68–75

Volkering F, Breure AM, Sterkenburg A, van Andel JG (1992) Microbial degradation of polycyclic aromatic hydrocarbons: effect of substrate availability on bacterial growth kinetics. Appl Microbiol Biotechnol 36:548–552

# Assimilation of Hydrocarbons and Lipids by Means of Biofilm Formation

**4**

Pierre Sivadon and Régis Grimaud

## Contents

1   Introduction ................................................................................. 47
2   Multispecies Biofilms on Hydrophobic Interface ........................................... 49
3   Cell Adhesion to Hydrophobic Compounds ................................................... 50
4   Regulation and Determinism of Biofilm Formation at the HOC-Water Interfaces ......... 51
5   Biofilm Formation as an Adaptive Response to Optimize Acquisition of
    Insoluble HOCs ............................................................................ 53
6   Research Needs ............................................................................ 55
References ..................................................................................... 56

**Abstract**

Hydrophobic organic compounds (HOCs) that are used as substrates by bacteria encompass a great variety of molecules, including contaminants such as hydrocarbons and natural components of the organic matter such as lipids. It is now well known that many bacterial strains use HOCs as carbon and energy sources for growth and form biofilms at the HOCs-water interface that are referred to as oleolytic biofilms. The formation of these biofilms appears to be a strategy to overcome the low accessibility of nearly water-insoluble substrates and is therefore a critical process in the biodegradation of hydrocarbons and lipids. Because oleolytic biofilms develop on a nutritive interface serving as both physical support and growth substrate, they represent an original facet of biofilm biology.

P. Sivadon · R. Grimaud (✉)
CNRS/ UNIV PAU and PAYS ADOUR, Institut des Sciences Analytiques et de Physico-Chimie pour l'Environnement et les Matériaux – MIRA, UMR5254, PAU, France
e-mail: pierre.sivadon@univ-pau.fr; regis.grimaud@univ-pau.fr

© Springer International Publishing AG, part of Springer Nature 2018
T. Krell (ed.), *Cellular Ecophysiology of Microbe: Hydrocarbon and Lipid Interactions*,
Handbook of Hydrocarbon and Lipid Microbiology, https://doi.org/10.1007/978-3-319-50542-8_41

# 1    Introduction

Early studies on hydrocarbon biodegradation led to the observation that hydrocarbon-degrading bacteria have high affinity for oil droplets. Phase contrast and electron microscopy examination of *Acinetobacter* sp. growing on *n*-hexadecane revealed hydrocarbon spheres densely covered with bacterial cells and suggested close contact between the cells and oil droplets (Kennedy et al. 1975). Since then, similar observations have been reiterated with various alkane-degrading strains, such as *Rhodococcus* sp. Q15 (Whyte et al. 1999), *Acinetobacter venetianus* RAG-1 (Baldi et al. 1999), *Oleiphilus messinensis* (Golyshin et al. 2002), *Pseudomonas* UP-2 (Zilber Kirschner et al. 1980), and *Marinobacter hydrocarbonoclasticus* SP17 (Fig. 1) (Klein et al. 2008). Bacterial attachment to polycyclic

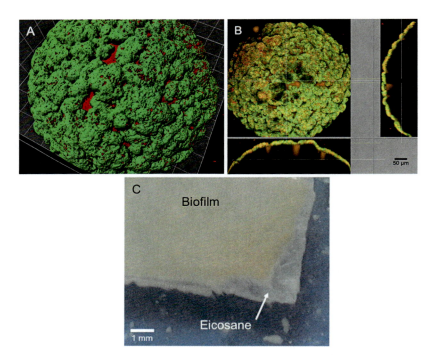

**Fig. 1** *M. hydrocarbonoclasticus* SP17 biofilms growing on alkanes. (**a** and **b**), confocal scanning laser microscopy images of a biofilm covered *n*-hexadecane droplet. Hydrophobic regions including bacteria as well as hydrocarbon were stained with red Nile (red signal), glycoconjugates were stained with PSA lectin (green signal). In (**a**), the data are presented as an isosurface projection where the two signals have been split. In (**b**), the dataset is presented as an XYZ projection. The two signals were not separated; colocalized signals of the green and red channel appear in yellow indicating the colocalization of Nile red and lectin stain (Images courtesy by Pierre-Jo Vaysse and Thomas R. Neu (Helmholtz Centre for Environmental Research – UFZ, Magdeburg, Germany)). (**c**), picture of *M. hydrocarbonoclasticus* SP17 biofilm growing at the surface of solid eicosane. A biofilm of *M. hydrocarbonoclasticus* SP17 (cream) was grown on a piece of solid eicosane (white) sticked on the bottom of a petri dish (blue) and submerged with culture medium (Régis Grimaud, unpublished)

aromatic hydrocarbons (PAHs) has also been described for *Pseudomonas* spp. (Eriksson et al. 2002; Mulder et al. 1998), *Sphingomonas* sp. CHY-1 (Willison 2004), and *Mycobacterium frederiksbergense* LB501T (Bastiaens et al. 2000). In biphasic culture medium containing a poorly water-soluble substrate and the aqueous phase, substrate-bound cells often coexist with cells floating freely in the aqueous phase. Although the presence of hydrocarbon-bound cells at the interface assumes interfacial growth, demonstration of actual substrate degradation and growth of the attached cells was provided in only a few cases (Efroymson and Alexander 1991; Wick et al. 2003; Zilber Kirschner et al. 1980).

The sessile mode of life and the multilayered structure of cell populations growing at the interface between hydrophobic organic compounds (HOCs) and water are reminiscent of biofilms. During the last two decades, biofilms have been the subject of extensive investigations. Most of our knowledge about the molecular biology of biofilms has been derived from model strains such as *Pseudomonas aeruginosa* and *Escherichia coli*. This research has revealed that biofilms are much more than the simple accretion of cells attached to an interface. Biofilms are heterogeneous, highly organized structures possessing an architecture that is essential to their functioning. Biofilm growth follows a stepwise pattern involving differentiation and collective behavior of cells (Stewart and Franklin 2008; Webb et al. 2003). Molecular studies of oleolytic biofilms growing at hydrophobic interfaces have not yet gone far enough to conclude whether they share all the characteristics of extensively studied model biofilms. However, properties characteristic of the biofilm lifestyle have been identified. CSLM (Confocal Scanning Laser Microscopy) observation of a biofilm community developing at polychlorinated biphenyl-water interface provided evidence of a stepwise development pattern of the biofilm (Macedo et al. 2005). Transcriptomic and proteomic studies indicated that cells growing at the alkane-water interface experienced a profound reshaping of their gene expression profile as compared to planktonic cells growing on acetate (Mounier et al. 2014; Vaysse et al. 2009, 2010). Extracellular polymeric substances (EPS), polysaccharides, DNA, and proteins were detected during growth at the alkane-water interface, indicating production of an extracellular matrix, which is a typical trait of biofilms (Ennouri et al. 2017; Whyte et al. 1999).

In this chapter, biofilms on HOCs refer to multilayered, matrix-embedded bacteria or bacterial communities growing at the HOCs–water interface and using these compounds as a substrate. In such biofilms, the energy and the carbon, which fuel bacterial growth, are provided by the degradation of the substrate, which thus constitutes a nutritive interface serving as both growth substrate and physical support.

## 2     Multispecies Biofilms on Hydrophobic Interface

During enrichment procedures on hydrocarbons, microbiologists have very often observed multispecies biofilms developing at the hydrocarbon–water interface. For example, Deppe et al. (2005) observed by phase contrast microscopy oil droplets covered by a biofilm during isolation of a consortium enriched on crude oil from Arctic sea ice and seawater from Spitzbergen. Unfortunately, such observations received little attention and in consequence they have not been always mentioned in the literature and have rarely been fully documented, making it difficult to measure the occurrence of these biofilms among bacterial species. Stach and Burns (2002) carried out a study devoted to the diversity of biofilm communities developing on PAHs. Biofilms developing on naphthalene- and phenanthrene-coated flow cells were isolated, and their diversity compared with planktonic cultures enriched on the same hydrocarbons. The biofilm system showed a three times higher diversity of cultivable bacteria as compared to the enrichment culture. Molecular approaches revealed that the biofilm community contained a greater diversity of active species and of PAH-degradation genes than the planktonic enrichment community. The diversity of active species found in the biofilm closely matched the diversity found in the PAH-contaminated soil used as inoculum. This study demonstrates that biofilm cultures represent a means to obtain PAH–degrading communities closely related to environmental situations suggesting that biofilm formation on hydrocarbons is a likely lifestyle in natural ecosystems. The existence of biofilms at oil-water interfaces in natural environments has been reported during the Deepwater Horizon oil spill where bacterial flocs were observed in the oil plume. Synchrotron radiation–based Fourier-transform infrared-spectra of these flocs indicated that their formation occurred on the surface of oil droplets and revealed the presence of oil-degradation products, polysaccharides, and proteins (Baelum et al. 2012; Hazen et al. 2010).

## 3    Cell Adhesion to Hydrophobic Compounds

In order to either develop a biofilm or to grow as single cell layers at the interface, bacteria must first approach and then adhere to the interface. Bacteria can reach surfaces by passive diffusion, random swimming, or taxis that are directed motility in response to gradients of chemical and physical stimuli. Chemotaxis has been observed in response to single ring aromatic hydrocarbons, naphthalene, and hexadecane. Regrettably, no experiment designed to determine whether this behavior led to biofilm formation at the interface between water and hydrocarbons has been conducted so far (Lanfranconi et al. 2003; Pandey and Jain 2002). To date, random mobility like swimming has never been shown to play a role in adhesion to HOCs. Once cells have reached the interface, the initial adhesion step is a purely physicochemical process described by the traditional and extended DLVO (Derjaguin-Landau-Verwey-Overbeek) theories of colloidal stability, which describe contact of cells to surfaces as the result of van der Waals interactions, Lewis acid–base interactions, and electrostatic interactions (Hermansson 1999). The intensity of these interactions and hence the effectiveness of the binding depend on the cell surface properties (hydrophobicity, charge, roughness, etc....) as well as interface properties. This means that only bacteria exhibiting the proper surface properties will adhere on hydrophobic surfaces (for a detailed review of bacterial adhesion to hydrocarbon, see Abbasnezhad et al. 2011). In contrast to adhesion to biotic surfaces or insoluble polysaccharides like chitin or cellulose, no specific adhesins or receptors recognizing hydrocarbons or lipids have been identified so far.

In many cases, adhered cells exhibit surface properties that are different from their soluble substrate-grown counterparts. These alterations of the cell surface are thought to reinforce adhesion after the initial interaction with the interface. For example, anthracene-grown cells of *Mycobacterium* sp. LB501T are more hydrophobic and more negatively charged than glucose-grown cells (Wick et al. 2002). Changes in cell surface can be achieved by modification, production, or removal of their surface molecules. In Gram-positive bacteria, the presence and the chain length of mycolic acids were correlated with hydrophobicity and adherence (Bendinger et al. 1993). Lipopolysaccharides are important determinants of cell surface properties in Gram-negative bacteria. Their chain length variation or their removal from the cell surface were shown to be important for interacting with hydrocarbons (Al-Tahhan et al. 2000). Capsular polysaccharides are another class of surface molecules playing a role in adhesion to hydrophobic compounds (Baldi et al. 1999).

Adhesion to hydrophobic surfaces is also mediated by extracellular appendages such as fimbriae and pili. The importance of fimbriae in adherence to *n*-hexadecane was demonstrated by the isolation of a nonadherent mutant of *Acinetobacter calcoaceticus* RAG-1, which was devoid of fimbriae and defective for growth on hydrocarbons. The reappearance of fimbriae in adherent revertants is a strong argument in favor of the involvement of fimbriae in adhesion to hexadecane (Rosenberg et al. 1982). The strains *Acinetobacter haemolyticus* AR-46 and *Acinetobacter* sp. Tol 5 produced pili at their cell surfaces when grown on *n*-hexadecane or triglycerides, respectively. Although the function of these organelles has not been elucidated, it was presumed that they play a role during the adhesion to the hydrophobic substrate (Bihari et al. 2007; Katsutoshi et al. 2011). Changes in surface properties of cells grown on hydrophobic substrates and inhibition of adhesion in the presence of soluble substrates indicate that the process of adhesion is regulated and that cells are able to respond to contact with hydrophobic interfaces.

## 4  Regulation and Determinism of Biofilm Formation at the HOC-Water Interfaces

In biofilms growing on an inert substratum, all nutrients are supplied through the aqueous phase. In biofilms on HOCs, the situation is very different. These biofilms develop in a biphasic medium where the electron donor is provided by the nonaqueous phase and the electron acceptor (e.g., oxygen) is available from the aqueous phase. It results in a geometry in which one side of the biofilm is in close association with the electron source, while the other contacts the source of the electron acceptor. Thus, cells within the biofilm experience two opposite gradients of acceptor and donor of electrons generated by their simultaneous diffusion and consumption. These micro-scale chemical gradients presumably contribute to the physiological heterogeneity in the biofilm and exert a control on its development. The experiment carried out by Joannis-Cassan et al. (2005) demonstrated that biofilm growth on hydrocarbons can be limited either by carbon or by oxygen depletion. The authors studied biofilm growth in a liquid–liquid system consisting of an emulsion obtained by stirring dodecane in mineral medium. Biofilm growth occurred at the surface of a dodecane droplet. During growth, the droplet diameter was reduced from 200 μm to 160 μm. Biofilm growth ceased when it reached a maximum thickness of about 80 μm. A series of experiments demonstrated that inhibition of growth was caused by the diffusion limitation of both dodecane and oxygen within the biofilm but not by others factors such as nutrient exhaustion or product inhibition (Joannis-Cassan et al. 2005).

Although adhesion to hydrocarbons does not seem to depend on the recognition of molecular structures, many HOC-degrading bacteria show a preference or specificity to the surface of metabolizable hydrocarbons or lipids for biofilm formation (Johnsen and Karlson 2004; Klein et al. 2008; Rodrigues et al. 2005). The substrate/ substratum specificity of biofilms on hydrocarbons is certainly one remarkable feature that distinguishes them from other biofilms. Biofilm formation tends to occur preferentially on poor-soluble substrates and seems to be regulated in function of substrate availability. Screening for biofilm formation capacity by isolated PAHs-degrading strains showed that the majority of the tested strains formed biofilm in microtiter wells coated with PAH crystals. For strains capable of growing on different PAHs, it was observed that the percentage of adhering cells decreased with the solubility of the PAHs, indicating that aqueous solubility of the substrate exerts a regulation on biofilm development (Johnsen and Karlson 2004). *Pseudomonas putida* ATCC 17514 exhibits different growth patterns depending on the PAH properties on which it is feeding. CSLM observation of a gfp-labeled derivative of this strain showed that growth on phenanthrene occurred by forming a biofilm at the crystal surface, while on fluorene, which is more soluble than phenanthrene, *P. putida* grew randomly between the crystals feeding on dissolved PAH (Rodrigues et al. 2005). Insoluble substrate preference for biofilm formation has also been observed in *M. hydrocarbonoclasticus* SP17. This bacterium forms biofilms on a variety of HOCs, including *n*-alkanes, wax esters, and triglycerides but is unable to form biofilm (in presence of acetate as substrate) on nonmetabolizable alkanes (branched alkanes and *n*-alkanes with more than 28 carbon atoms) and forms only weak biofilms on polystyrene with 10 times less biomass than on paraffin (Ennouri et al. 2017; Klein et al. 2008). The preference for insoluble substrates suggests that bacteria forming biofilms on HOCs are able to detect and recognize nutritive interfaces. It is reasonable to anticipate that control of biofilm formation by substrate/substratum is exerted through a signal transduction pathway and genetic regulatory mechanisms. Indeed, induction of genes at an interface was demonstrated for the *pra* gene encoding the PA protein, an alkane-inducible extracellular protein exhibiting an emulsifying activity involved in hexadecane assimilation, and the *rhlR* gene coding for the transcriptional activator of rhamnolipids biosynthesis. Studies with liquid cultures on *n*-hexadecane of *P. aeruginosa* harboring a *pra::gfp* or *rhlR::*

*gfp* fusion revealed specific transcriptional activity at the hexadecane–water interface (Holden et al. 2002).

## 5 Biofilm Formation as an Adaptive Response to Optimize Acquisition of Insoluble HOCs

The first intuitive indications that biofilms could favor access to poorly soluble HOCs came from the observations that biofilm formation occurs in function of substrate solubility since it was shown that the more insoluble the substrate is, the more growth occurs at the water–HOC interface. Moreover, it was observed that several strains growing at the interface between nearly insoluble hydrocarbons and water did not release emulsifier or surface-active compounds in the bulk medium (Bouchez et al. 1997; Bouchez-Naïtali et al. 1999, 2001; Klein et al. 2008; Wick et al. 2002). In these cases, cells did not access the substrate by surfactant-mediated transfer, during which cells contact emulsified, solubilized, or pseudo solubilized hydrocarbons. Access to the insoluble substrate would have rather occurred by direct contact of the cells or extracellular structures with the hydrocarbon–water interface. Rosenberg demonstrated the importance of adhesion to hydrocarbons in the growth of *Acinetobacter calcoaceticus* RAG-1 on *n*-hexadecane in absence of any emulsifier (Rosenberg and Rosenberg 1981). Thus, biofilm formation and adhesion to hydrocarbons would promote growth on hydrocarbons by facilitating interfacial access. The strongest evidence of an increase in access to HOCs by adhesion or biofilm formation arose from kinetic studies showing that growth at the interface occurred faster than the mass transfer rate of HOCs in the absence of bacteria would suggest (Bouchez-Naïtali et al. 2001; Calvillo and Alexander 1996; Harms and Zehnder 1995; Wick et al. 2002).

Mechanisms employed in biofilms for accessing HOCs are still poorly understood. On the one hand, it is not difficult to imagine that biofilms offer a way to optimize the effect of known mechanisms of acquisition of poorly soluble hydrophobic substrates. Surfactant production within a biofilm would limit surfactants dilution in the bulk phase, facilitating the formation of micelles by keeping the concentration of the surfactant close to the critical micelle concentration (CMC). Biofilms also offer the advantage of holding the cell population in the vicinity of the HOC–water interface thus stimulating the mass transfer of HOCs by shortening the diffusive pathway (Wick et al. 2002). On the other hand, biofilm lifestyle might offer possibilities of biofilm-specific mechanisms for HOCs accession. Biofilms are typically characterized by dense cell clusters embedded in extracellular polymeric substances. The formation of such structures involves profound changes in cell physiology and behavior requiring regulation of the expression of hundreds of genes. Such changes in cellular physiology have been indeed revealed by transcriptomic and proteomic studies on biofilms of *M. hydrocarbonoclasticus* growing at HOCs-water interfaces. The transition from the planktonic to the biofilm mode on

HOCs entailed change in the expression level of more than one thousand genes (Mounier et al. 2014; Vaysse et al. 2009, 2010). Although most of these genes are of unknown function, some of them are involved in cellular processes like lipid, alkanes, and central metabolisms; chemotaxis; motility; transport; and protein secretion. In view of such a reshaping of cellular functions and structural organization, the existence of biofilm-specific mechanisms for HOCs accession is conceivable.

For instance, biofilms can improve the assimilation of HOCs through their extracellular matrix. Many functions currently attributed to the biofilm matrix, e.g., adhesion to surfaces and retention of enzymes and metabolites, can have implications in the assimilation of HOCs (Flemming and Wingender 2010). The retention properties of the matrix could maintain exoproducts in the vicinity of cells and prevent their loss in the bulk medium. The action of biosurfactants, which has been demonstrated to improve assimilation of hydrocarbons in some cases (Perfumo et al. 2010), could be greatly increased within a matrix by enabling their accumulation up to the CMC and thus allow micellar transport of hydrocarbons (Guha and Jaffé 1996). In addition, EPS of the biofilm matrix may act as sorbents or emulsifiers that could stimulate the mass transfer rate of HOCs (Harms et al. 2010). Various strains of *Acinetobacter* sp. produce extracellular complexes of polysaccharides or lipopolysaccharides and proteins called bio-emulsans that have the capacity to emulsify and increase the solubility of hydrocarbons (Barkay et al. 1999; for reviews see Ron and Rosenberg 2002, 2010). Although the emulsifying activities were not localized within a biofilm matrix, these results indicate that biopolymers could increase mass transfer rates of hydrocarbons and hence stimulate their biodegradation. Direct interactions between EPS and hydrocarbons have been evidenced in *A. venetianus* VE-C cells growing on diesel fuel where nanodroplets incorporated in an extracellular matrix containing glycoconjugates were observed (Baldi et al. 1999, 2003). Similarly, oil droplets were completely covered with cells and EPS in a culture of *Rhodococcus* sp. strain Q15 on diesel fuel (Whyte et al. 1999). In both cases, EPS mediated the adhesion of cells to hydrocarbons and are thought to participate in the uptake of hydrocarbons, although the mechanisms involved remain unknown. Biofilm EPS can also serve as an adsorbent to store HOCs and allow their subsequent utilization by the biofilm community (Wolfaardt et al. 1995).

Biofilm matrices comprise extracellular proteins with various functions such as hydrolytic enzymes or adhesion (Flemming and Wingender 2010). The involvement of extracellular proteins in alkane utilization was first evidenced in *A. calcoaceticus* ADP1 as it was shown that a Type-2 Secretion System (T2SS) mutant showed reduced growth on alkanes (Parche et al. 1997). The OmpA-like AlnA protein of *A. radioresistens* KA53 and the PA protein from *P. aeruginosa* PG201 and S7B1 are two extracellular proteins that have been shown to play a role in alkane utilization in planktonic culture (Hardegger et al. 1994; Kenichi et al. 1977). These two proteins exhibit emulsifying properties that have been claimed to be the basis of their function in alkane assimilation, although no clear cause-to-effect relationship has been established. During biofilm development on alkanes or triglycerides, *M. hydrocarbonoclasticus* SP17 uses cytoplasmic proteins released by cell lysis and proteins secreted through the T2SS to form a proteinaceous matrix (Ennouri et al. 2017). It was hypothesized that the surface activity of proteins could be exploited in oleolytic biofilm matrices to form a conditioning film at the HOC-water interface that could promote cell adhesion and colonization. Matrix proteins could also participate in the mass transfer of HOCs to the biofilm cells by forming micelles or acting as mobile sorbents.

## 6  Research Needs

The most exciting aspect of biofilms on HOCs is certainly to identify the features that distinguish them from other biofilms, that is to say, their substrate/substratum specificity and their capacity to overcome the low accessibility of a hydrophobic substrate. These two properties make biofilm formation a very efficient adaptive strategy to assimilate HOCs, which can provide a serious advantage in environments where carbon sources are scarce. The processes by which biofilms stimulate interfacial accession to nearly insoluble substrates remain to be elucidated. Substrate specificity of biofilm formation for HOCs surfaces presumably involves surface sensing and signal transduction pathways, which have not been revealed yet. Biofilm development during the assimilation of HOCs most likely requires coordination of fundamental processes such as architectural biofilm organization, physicochemical interactions between biofilm and substrate, and the control of gene expression. Investigation of these processes will require multidisciplinary approaches aimed at (1) identifying the genes/proteins involved in biofilm formation, (2) deciphering the architecture of biofilms, and (3) characterizing at the physicochemical level the interactions between biofilm components (cells and extracellular matrix) with hydrophobic substrates. So far, investigations on biofilms on HOCs have been conducted on different strains growing on various substrates in diverse experimental setups. It was therefore not possible to correlate these results in order to draw a picture of the physiology of these biofilms. The study of additional model bacteria, chosen for their ability to form readily and reproducibly biofilms on HOCs, their genetic amenability, and the availability of their genome sequence would ensure the complementarity of the data obtained from multidisciplinary approaches. In addition to studies examining model single-species biofilms on HOCs, investigations of the activities and biodiversity of multispecies biofilms isolated from samples collected from various environments are critical to gain full understanding of the ecological significance of these biofilms.

Due to their wide distribution in the environment, their recalcitrance, and their deleterious effect on human health, hydrocarbons have been the main molecules used in studies of biofilm formation on HOCs. However, other classes of HOCs should also be taken into consideration. In the natural environment, lipids represent a very abundant class of HOCs. For example, in sea water they represent up to about 15% of the organic carbon and its biodegradation is relevant to the global carbon cycle (Lee et al. 2004). Consistent with this, some bacterial strains isolated for their hydrocarbon-degrading capacities also form biofilms on a larger panel of HOCs. This suggests that HOCs-degrading bacteria may have the ability to form oleolytic biofilms that can cope with several types of hydrophobic organic carbon they may encounter in the environment by adapting their physiology according to the chemical nature of the HOC.

# References

Abbasnezhad H, Gray M, Foght JM (2011) Influence of adhesion on aerobic biodegradation and bioremediation of liquid hydrocarbons. Appl Microbiol Biotechnol 92:653–675

Al-Tahhan RA, Sandrin TR, Bodour AA, Maier RM (2000) Rhamnolipid-induced removal of lipopolysaccharide from *Pseudomonas aeruginosa*: effect on cell surface properties and interaction with hydrophobic substrates. Appl Environ Microbiol 66:3262–3268

Baelum J, Borglin S, Chakraborty R, Fortney JL, Lamendella R, Mason OU, Auer M, Zemla M, Bill M, Conrad ME, Malfatti SA, Tringe SG, Holman HY, Hazen TC, Jansson JK (2012) Deep-sea bacteria enriched by oil and dispersant from the Deepwater Horizon spill. Environ Microbiol 14:2405–2416

Baldi F, Ivoševic N, Minacci A, Pepi M, Fani R, Svetličic V, Žutic V (1999) Adhesion of *Acinetobacter venetianus* to diesel fuel droplets studied with *in situ* electrochemical and molecular probes. Appl Environ Microbiol 65:2041–2048

Baldi F, Pepi M, Capone A, della Giovampaola C, Milanesi C, Fani R, Focarelli R (2003) Envelope glycosylation determined by lectins in microscopy sections of *Acinetobacter Venetianus* induced by diesel fuel. Res Microbiol 154:417–424

Barkay T, Navon-Venezia S, Ron EZ, Rosenberg E (1999) Enhancement of solubilization and biodegradation of polyaromatic hydrocarbons by the bioemulsifier alasan. Appl Environ Microbiol 65:2697–2702

Bastiaens L, Springael D, Wattiau P, Harms H, deWachter R, Verachtert H, Diels L (2000) Isolation of adherent polycyclic aromatic hydrocarbon (PAH)-degrading bacteria using PAH-sorbing carriers. Appl Environ Microbiol 66:1834–1843

Bendinger B, Rijnaarts HHM, Altendorf K, Zehnder AJB (1993) Physicochemical cell surface and adhesive properties of coryneform bacteria related to the presence and chain length of mycolic acids. Appl Environ Microbiol 59:3973–3977

Bihari Z, Pettko-Szandtner A, Csanadi G, Balazs M, Bartos P, Kesseru P, Kiss I, Mecs I (2007) Isolation and characterization of a novel *n*-alkane-degrading strain, *Acinetobacter haemolyticus* AR-46. Z Naturforsch 62:285–295

Bouchez M, Blanchet D, Vandecasteele JP (1997) An interfacial uptake mechanism for the degradation of pyrene by a *Rhodococcus* strain. Microbiology 143:1087–1093

Bouchez-Naïtali M, Rakatozafy H, Leveau JY, Marchal R, Vandecasteele JP (1999) Diversity of bacterial strains degrading hexadecane in relation to the mode of substrate uptake. J Appl Microbiol 86:421–428

Bouchez-Naïtali M, Blanchet D, Bardin V, Vandecasteele JP (2001) Evidence for interfacial uptake in hexadecane degradation by *Rhodococcus equi*: the importance of cell flocculation. Microbiology 147:2537–2543

Calvillo YM, Alexander M (1996) Mechanism of microbial utilization of biphenyl sorbed to polyacrylic beads. Appl Microbiol Biotechnol 45:383–390

Deppe U, Richnow HH, Michaelis W, Antranikian G (2005) Degradation of crude oil by an arctic microbial consortium. Extremophiles 9:461–470

Efroymson RA, Alexander M (1991) Biodegradation by an *Arthrobacter* species of hydrocarbons partitioned into an organic solvent. Appl Environ Microbiol 57:1441–1447

Ennouri H, d'Abzac P, Hakil F, Branchu P, Naïtali M, Lomenech AM, Oueslati R, Desbrières J, Sivadon P, Grimaud R (2017) The extracellular matrix of the oleolytic biofilm of *Marinobacter hydrocarbonoclasticus* comprises cytoplasmic proteins and T2SS effectors that promote growth on hydrocarbons and lipids. Environ Microbiol 19:159–173

Eriksson M, Dalhammar G, Mohn WW (2002) Bacterial growth and biofilm production on pyrene. FEMS Microbiol Ecol 40:21–27

Flemming HC, Wingender J (2010) The biofilm matrix. Nat Rev Microbiol 8:623–633

Golyshin PN, Chernikova TN, Abraham WR, Lunsdorf H, Timmis KN, Yakimov MM (2002) *Oleiphilaceae* fam. nov., to include *Oleiphilus messinensis* gen. nov., sp. nov., a novel marine bacterium that obligately utilizes hydrocarbons. Int J Syst Evol Microbiol 52:901–911

Guha S, Jaffé PR (1996) Biodegradation kinetics of phenanthrene partitioned into the micellar phase of nonionic surfactants. Environ Sci Technol 30:605–611

Hardegger M, Koch AK, Ochsner UA, Fiechter A, Reiser J (1994) Cloning and heterologous expression of a gene encoding an alkane-induced extracellular protein involved in alkane assimilation from *Pseudomonas aeruginosa*. Appl Environ Microbiol 60:3679–3687

Harms H, Zehnder AJB (1995) Bioavailability of sorbed 3-chlorodibenzofuran. Appl Environ Microbiol 61:27–33

Harms H, Smith KEC, Wick LY (2010) Microorganism-hydrophobic compound interactions. In: Timmis KN (ed) Handbook of hydrocarbon and lipid microbiology. Springer, Berlin/Heidelberg, pp 1479–1490

Hazen TC, Eric A, Dubinsky EA, DeSantis TZ, Andersen GL, Piceno YM, Singh N, Jansson JK, Probst A, Borglin SE, Fortney JL, Stringfellow WT, Bill M, Conrad ME, Tom LM, Chavarria KL, Alusi TR, Lamendella R, Joyner DC, Spier C, Baelum J, Auer M, Zemla ML, Chakraborty R, Sonnenthal EL, D'haeseleer P, Holman HYN, Osman S, Lu Z, Van Nostrand JD, Deng Y, Zhou J, Mason OU (2010) Deep-sea oil plume enriches indigenous oil-degrading bacteria. Science 330:204–208

Hermansson M (1999) The DLVO theory in microbial adhesion. Colloids Surf B: Biointerfaces 14:105–119

Holden PA, LaMontagne MG, Bruce AK, Miller WG, Lindow SE (2002) Assessing the role of *Pseudomonas aeruginosa* surface-active gene expression in hexadecane biodegradation in sand. Appl Environ Microbiol 68:2509–2518

Joannis-Cassan C, Delia ML, Riba JP (2005) Limitation phenomena induced by biofilm formation during hydrocarbon biodegradation. J Chem Technol Biotechnol 80:99–106

Johnsen AR, Karlson U (2004) Evaluation of bacterial strategies to promote the bioavailability of polycyclic aromatic hydrocarbons. Appl Microbiol Biotechnol 63:452–459

Katsutoshi H, Ishikawa M, Yamada M, Higuchi A, Ishikawa Y, Hironori E (2011) Production of peritrichate bacterionanofibers and their proteinaceous components by *Acinetobacter* sp. Tol 5 cells affected by growth substrates. J Biosci Bioeng 111:31–36

Kenichi H, Nakahara T, Minoda Y, Yamada K (1977) Formation of protein-like activator for n-alkane oxidation and its properties. Agric Biol Chem 41:445–450

Kennedy RS, Finnerty WR, Sudarsanan K, Young RA (1975) Microbial assimilation of hydrocarbons. I. The fine structure of a hydrocarbon oxidizing *Acinetobacter* sp. Arch Microbiol 102:75–83

Klein B, Grossi V, Bouriat P, Goulas P, Grimaud R (2008) Cytoplasmic wax ester accumulation during biofilm-driven substrate assimilation at the alkane-water interface by *Marinobacter hydrocarbonoclasticus* SP17. Res Microbiol 159:137–144

Lanfranconi MP, Studdert CA, Alvarez HM (2003) A strain isolated from gas oil-contaminated soil displays chemotaxis towards gas oil and hexadecane. Environ Microbiol 5:1002–1008

Lee C, Wakeham S, Arnosti C (2004) Particulate organic matter in the sea: he composition conundrum. Ambio 33:565–575

Macedo AJ, Kuhlicke U, Neu TR, Timmis KN, Abraham WR (2005) Three stages of a biofilm community developing at the liquid-liquid interface between polychlorinated biphenyls and water. Appl Environ Microbiol 71:7301–7309

Mounier J, Camus A, Mitteau I, Vaysse PJ, Goulas P, Grimaud R, Sivadon P (2014) The marine bacterium *Marinobacter hydrocarbonoclasticus* SP17 degrades a wide range of lipids and hydrocarbons through the formation of oleolytic biofilms with distinct gene expression profiles. FEMS Microbiol Ecol 90:816–831

Mulder H, Breure AM, Van Honschooten D, Grotenhuis JTC, Van Andel JG, Rulkens WH (1998) Effect of biofilm formation by *Pseudomonas* 8909n on the bioavailability of solid naphthalene. Appl Microbiol Biotechnol 50:277–283

Pandey G, Jain RK (2002) Bacterial chemotaxis toward environmental pollutants: role in bioremediation. Appl Environ Microbiol 68:5789–5795

Parche S, Geißdöfer W, Hillen W (1997) Identification and characterization of xcpR encoding a subunit of the general secretory pathway necessary for dodecane degradation in *Acinetobacter calcoaceticus* ADP1. J Bacteriol 179:4631–4634

Perfumo A, Smyth TJP, Marchant R, Banat IM (2010) Production and roles of biosurfactants and bioemulsifiers in accessing hydrophobic substrates. In: Timmis KN (ed) Handbook of hydrocarbon and lipid microbiology. Springer, Berlin/Heidelberg, pp 1501–1512

Rodrigues AC, Brito AG, Wuertz S, Melo LF (2005) Fluorene and phenanthrene uptake by *Pseudomonas putida* ATCC 17514: kinetics and physiological aspects. Biotechnol Bioeng 90:281–289

Ron EZ, Rosenberg E (2002) Biosurfactants and oil bioremediation. Curr Opin Biotechnol 13:249–252

Ron EZ, Rosenberg E (2010) Protein emulsifiers. In: Timmis KN (ed) Handbook of hydrocarbon and lipid microbiology. Springer, Berlin/Heidelberg, pp 3031–3035

Rosenberg M, Rosenberg E (1981) Role of adherence in growth of *Acinetobacter calcoaceticus* RAG-1 on hexadecane. J Bacteriol 148:51–57

Rosenberg M, Bayer EA, Delarea J, Rosenberg E (1982) Role of thin fimbriae in adherence and growth of *Acinetobacter calcoaceticus* RAG-1 on hexadecane. Appl Environ Microbiol 44:929–937

Stach JEM, Burns RG (2002) Enrichment versus biofilm culture: a functional and phylogenetic comparison of polycyclic aromatic hydrocarbon-degrading microbial communities. Environ Microbiol 4:169–182

Stewart PS, Franklin MJ (2008) Physiological heterogeneity in biofilms. Nat Rev Microbiol 6:199–210

Vaysse PJ, Prat L, Mangenot M, Cruveiller S, Goulas P, Grimaud R (2009) Proteomic analysis of *Marinobacter hydrocarbonoclasticus* SP17 biofilm formation at the alkane-water interface reveals novel proteins and cellular processes involved in hexadecane assimilation. Res Microbiol 160:829–837

Vaysse PJ, Sivadon P, Goulas P, Grimaud R (2010) Cells dispersed from *Marinobacter hydrocarbonoclasticus* SP17 biofilm exhibit a specific protein profile associated with a higher ability to reinitiate biofilm development at the hexadecane-water interface. Environ Microbiol 13:737–746

Webb JS, Givskov M, Kjelleberg S (2003) Bacterial biofilms: prokaryotic adventures in multicellularity. Curr Opin Microbiol 6:578–585

Whyte LG, Slagman SJ, Pietrantonio F, Bourbonnière L, Koval SF, Lawrence JR, Inniss WE, Greer CW (1999) Physiological adaptations involved in alkane assimilation at a low temperature by *Rhodococcus* sp. strain Q15. Appl Environ Microbiol 65:2961–2968

Wick LY, De Munain AR, Springael D, Harms H (2002) Responses of *Mycobacterium* sp. LB501T to the low bioavailability of solid anthracene. Appl Microbiol Biotechnol 58:378–385

Wick LY, Pasche N, Bernasconi SM, Pelz O, Harms H (2003) Characterization of multiple-substrate utilization by anthracene-degrading *Mycobacterium frederiksbergense* LB501T. Appl Environ Microbiol 69:6133–6142

Willison JC (2004) Isolation and characterization of a novel sphingomonad capable of growth with chrysene as sole carbon and energy source. FEMS Microbiol Lett 241:143–150

Wolfaardt GM, Lawrence JR, Robarts RD, Caldwell DE (1995) Bioaccumulation of the herbicide diclofop in extracellular polymers and its utilization by a biofilm community during starvation. Appl Environ Microbiol 61:152–158

Zilber Kirschner I, Rosenberg E, Gutnick D (1980) Incorporation of $^{32}$P and growth of *pseudomonad* UP-2 on *n*-tetracosane. Appl Environ Microbiol 40:1086–1093

# Uptake and Assimilation of Hydrophobic Substrates by the Oleaginous Yeast *Yarrowia lipolytica*

**5**

France Thevenieau, Athanasios Beopoulos, Thomas Desfougeres, Julia Sabirova, Koos Albertin, Smita Zinjarde, and Jean Marc Nicaud

## Contents

1 Introduction ................................................................. 60
2 Solubilization and Binding of HS onto the Cell Surface ..................... 62
3 Transport/Export of HS into the Cells ..................................... 64
4 Modification of HS into the Cells ......................................... 68
5 Degradation Through the β-Oxidation Pathway ............................... 70
6 Storage as Triglyceride and Sterol Ester ................................. 71
7 Research Needs ............................................................ 72
References .................................................................. 73

F. Thevenieau (✉)
Laboratoire de Microbiologie et Génétique Moléculaire, AgroParisTech, Centre de Biotechnologie Agro-Industrielle, Thiverval-Grignon, France

Oxyrane UK Limited, Greenheys House, Manchester Science Park, Manchester, UK

A. Beopoulos · T. Desfougeres · J. M. Nicaud
Laboratoire de Microbiologie et Génétique Moléculaire, AgroParisTech, Centre de Biotechnologie Agro-Industrielle, Thiverval-Grignon, France
e-mail: jean-marc.nicaud@inra.fr

J. Sabirova
Laboratory of Industrial Microbiology and Biocatalysis, Department of Biochemical and Microbial Technology, Faculty of Bioscience Engineering, Ghent University, Ghent, Belgium

K. Albertin
Department of Microbial, Biochemical and Food Biotechnology, University of the Free State, Bloemfontein, South Africa

S. Zinjarde
Institute of Bioinformatics and Biotechnology, University of Pune, Pune, India

© Springer International Publishing AG, part of Springer Nature 2018          59
T. Krell (ed.), *Cellular Ecophysiology of Microbe: Hydrocarbon and Lipid Interactions*,
Handbook of Hydrocarbon and Lipid Microbiology, https://doi.org/10.1007/978-3-319-50542-8_42

**Abstract**
The transport of hydrophobic substrates such as fatty acids, triglycerides, and alkanes into a microbial cell has recently begun to receive interest by the scientific community, especially due to the potential biotechnological applications. Here we present an overview on how this process is likely to proceed in the oleaginous yeast *Yarrowia lipolytica*, an organism which is known to inhabit various lipid containing environments. It is, therefore, well adapted to utilizing these hydrophobic substrates, a process involving firstly their transport into the cells followed by their entry into the subsequent metabolic pathways. Among the strategies employed by *Yarrowia* in response to exposure to hydrophobic substances are surface-mediated and direct interfacial transport processes, production of biosurfactants, hydrophobisation of the cytoplasmic membrane and the formation of protrusions. Several transport systems have been found to be essential for the growth on hydrophobic compounds, these being either involved in importing the hydrophobic substrate, or in exporting cellular intermediates in order to maintain intracellular concentrations of such compounds at non-toxic levels. Finally, this review also discusses recent advances on the metabolic fate of hydrophobic compounds inside the cell: their terminal oxidation, further degradation or accumulation in the form of intracellular lipid bodies.

# 1    Introduction

In order to assimilate hydrocarbons/oils substrates, microorganisms have to elaborate sophisticated mechanisms. In particular, these carbon sources have to be modified, transported into the cells without damage and then are to be either degraded for the production of energy or stored for further use.

The yeast *Yarrowia lipolytica* is often found in environments that are rich in hydrophobic substrates such as alkanes or lipids. This organism was recently used as a model system to study the mechanisms involved in the degradation of hydrophobic substrates (HS). Among the various yeasts that are able to assimilate HS. *Y. lipolytica* presents several advantages: (1) the entire sequence of the six chromosomes has been determined (Dujon et al. 2004); (2) this species is haploid which is not the case for most HS degrading yeast species such as *Candida tropicalis*; (3) many genetic tools are nowadays available for gene manipulation in *Y. lipolytica* which has been reviewed previously (Barth and Gaillardin 1996). For example, classical *in vivo* genetics was used for the isolation of mutants affected in oleic acid utilisation and in peroxisomal assembly (Nuttley et al. 1993). The identification of the mutated genes was obtained by complementation studies and sequencing of the complementing genes. Tagged mutants affected in HS utilization and in colony morphology were isolated. The mutated genes were identified by sequencing of the insertion border using reverse PCR (Mauersberger et al. 2001). In addition, recently, microarrays have become available for this yeast.

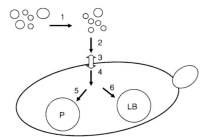

**Fig. 1** Schematic representation of the fate of hydrophobic substrates (*HS*) in *Y. lipolytica*. The emulsion of the HS is modified by liposan (a surfactant) in order to decrease the size of the HS droplets (*1*), which then could bind onto the cell surface (*2*), from where they enter into the cells *via* transport/export systems (*3*) after modification of the HS through several pathways (*4*), they are finally degraded by β-oxidation (*5*) or stored into lipid bodies as triglyceride (*6*)

This review will focus on the recent advances of the physiology and genetics of *Y. lipolytica* in relation to the assimilation of HS.

The fate of HS from the media into the cells is depicted schematically in Fig. 1. The first change that occurs is the modification of the substrates in order to improve their accessibility. *Y. lipolytica* is well known for its extracellular lipolytic and proteolytic activities (Barth et al. 2003). For this purpose *Y. lipolytica* utilizes two mechanisms: (1) the *surface-mediated transport* by producing surfactants during growth on HS. Such surfactants reduce the size of the HS droplets, thereby increasing the possible contact surface with the substrate (Mlickova et al. 2004). *Y. lipolytica*, apart from an extracellular emulsifier, called liposan, also secretes an extracellular lipase, allowing the hydrolysis of triglycerides into fatty acids and glycerol (Hadeball 1991; Barth and Gaillardin 1996). The second mechanism is the *direct interfacial transport* which allows the binding of the HS droplets onto the cell surface thus increasing the access of the HS for their transport into the cells (Mlickova et al. 2004). Hence, alkanes attached to the protrusions or hydrophobic outgrowths may migrate through the channels via the plasma membrane to the ER, the site of alkane hydroxylation by P450 monooxygenase systems (Tanaka and Fukui 1989). The exact mechanism by which hydrophobic compounds pass through a membrane is in all organisms still highly controversial. (2) The docking of the HS droplets results from the increase in the apolar properties of the cell surface and the induction of *protrusions* on the surface of the cells. (3) The substrates must be transported into the cells, while regulating the concentration and export systems seems to be involved in maintaining this non-toxic local concentration. (4) Several pathways are involved in the degradation of HS such as monoterminal alkane oxidation, β-oxidation and the crosstalk between the glyoxylate and the citrate cycles. These pathways are localised in different compartments such as the endoplasmic reticulum, the peroxisomes and mitochondria, (5) HS degradation is finalized through the β-oxidation pathway in the peroxisomes. (6) Alternatively, under conditions of HS excess, the substrates could be stored into lipid bodies.

## 2    Solubilization and Binding of HS onto the Cell Surface

Due to the poor water-miscibility of HS, these substrates form an emulsion or a thin layer on the surface of the media. In order to improve this solubility and the quality of the emulsion, *Y. lipolytica* secretes emulsifiers, called *liposan* that vary in composition. This was first observed in the work of Cirigliano and Carman (1984), where *Candida lipolytica* (later re-named in *Yarrowia lipolytica*) produced an inducible extracellular emulsification activity when it was grown with HS. Characterization came from Zinjarde and Pant (2002) who reported that the surfactant liposan contains 5% protein, 20% carbohydrate and 75% lipid. However, Vance-Harrop et al. (2003) reported that *liposan* contained approximately 50% protein, 40% carbohydrate and 10% lipid.

Such variations in liposan composition may be attributed to the differences in the purification procedures that were used. Alternatively, it may correspond to different emulsifiers being produced by different strains or on the media composition. Several groups are currently working on the optimisation of surfactant production and purification. This will result soon in the identification of the gene(s) coding for liposan. This will open a completely new field for the identification of the gene (s) encoding these surfactant(s), the analysis of their regulation and their biosynthetic pathway involved in the addition of carbohydrate and lipid.

In natural environments, it may not be possible to secrete surfactant in sufficient quantities. Therefore, some microorganisms use a second strategy in order to utilize HS. Instead of solubilizing the substrate, the microorganisms may bind to the substrate. Yeasts undergo modifications of the cell surface properties in order to bind to the HS droplets. This adhesion of the cells to large HS droplets can be easily visualised by light microscopic observation of cells in an emulsion. The cells can be observed clustered around the HS droplets. As shown in Fig. 2a, *Y. lipolytica* cells are surrounding the hexadecane droplets and covering a large surface of the droplet. On the other hand, smaller HS droplets were bound to the yeast cell surface. This adherence was demonstrated for alkanes and oils (Fickers et al. 2005a).

The first evidence of the presence of specific structures on the cell surface was reported in 1975 by Osumi and colleagues for some *Candida* spp. grown on *n*-alkanes (Osumi et al. 1975). Through cryo-scanning election microscopy (*SEM*) and transmission electron microscopy (*TEM*), the presence of *protrusions* was observed on the cell surface of *Candida tropicalis* that had been grown on alkanes. These protrusions consisted of a 50 nm high by 150 nm large structure, presenting an electron dense channel connecting the top of the protrusion with the interior. The electron dense membrane was characteristic of endoplasmic reticulum membrane. This probably constitutes a transport/export mechanism which may contain specific transporters and/or exporters taking the HS directly from the droplet to the site of first modification (e.g., the transporter ABC1p, see below). Such structures were also observed on the *Y. lipolytica* surface grown on either fatty acid (oleic acid) or alkanes (decane, hexadecane) as shown in Fig. 2b (see also Mlickova et al. 2004). These authors observed that these protrusions correspond to a docking platform for the binding of lipid droplets. They furthermore reported that the production of these

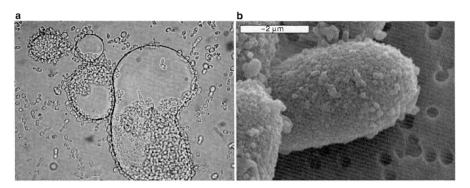

**Fig. 2** Physiological and surface modifications of *Y. lipolytica* grown on hydrophobic substrates (*HS*). (**a**) Example of an interaction of *Y. lipolytica* cells which adhere on the surface of large hexadecane droplets. Cells were observed by optical microscopy. (**b**) Example of cell surface of *Y. lipolytica* grown on hexadecane. SEM micrograph of cells grown for 18 h on YNB-hexadecane minimal medium featuring protrusions on the surfaces of yeast cells. Cells were treated for cryo-scanning electron microscopy (*SEM*). Method for electronic microscopy preparation (*SEM*) was previously described by Mlickova et al. (2004). Bar represents 1 nm

protrusions was inducible. Such protrusions were thus not detected on the cell surface of *Y. lipolytica* during exponential phase with glucose as carbon source.

In addition to these cell surface changes, morphological and physiological changes occurred. Apart from the development of compartments containing enzymes involved in the degradation of HS such as ER, peroxisomes, lipid bodies, and mitochondria, there was an increase of the periplasmic space width from 80 to 150 nm. On the contrary, the cell wall thickness decreased from 40 to 25 nm (Mlickova et al. 2004).

The binding of *Y. lipolytica* cells to the alkane phase can be easily demonstrated. As an example, we present in Fig. 3 the binding of *Y. lipolytica* strain H222 to hexadecane over time.

Glucose grown cells (YNB-glucose) in exponential phase were inoculated into alkane media (YNB-hexadecane media). Total cell number, alkane-bound cells and free cells were determined. As shown in Fig. 3a, these cells do not bind to the alkane phase; therefore 100% of the cells are present in the media as free cells. But very quickly, protrusions are induced, allowing the binding of the cells to the alkane phase and within less than 10 h nearly all the cells are bound to the alkane droplets with the free cells fraction representing less than 0.02 $OD_{600}$. After 20 h, the alkane phase is saturated and therefore the cells could not bind to the alkane phase. We observed at this time an increase in free cells. However, these free cells did not bind to the alkane droplets due to the saturation of the alkane phase. However, when additional alkane was added to the water phase containing the free cells and mixed by vortexing, the cells immediately partitioned into the alkane phase (data not shown). When cells entered stationary phase, they lost their alkane-binding capacity. In contrast, glucose grown cells entering stationary phase started to exhibit binding properties (data not shown).

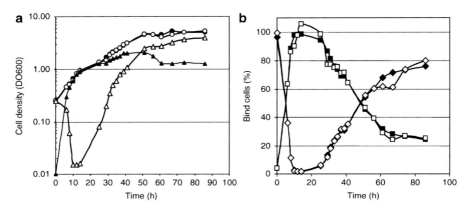

**Fig. 3** Binding of *Y. lipolytica* cells during growth on hexadecane. (**a**) Cell numbers; total cells (measured cells, *full circle*; calculated cells, *open circle*), bound cells to the alkane phase (*full triangle*) and free cells in the water phase (*open triangle*). (**b**) Percentage of cells bound to the alkane phase (*square*) and percentage of free cells in the media (*diamond*) from the total cells (measured cells, *full symbols*; calculated cells, *open symbols*). *Y. lipolytica* strain H222 was grown for 18 h on minimal medium YNB-glucose. Cells were collected in exponential phase (OD $_{600}$ lower than 1.5), collected by centrifugation and used to inoculate at OD = 0.25 a culture in hexadecane (YNB-C16). Samples were collected over time and cell concentration measured. Total cells measured (OD of cells centrifuged after NaOH treatment), free cells (OD of cells centrifuged without NaOH treatment), cells bound to hexadecane (OD of the cells in the alkane phase centrifuged after NaOH treatment); the total cells (calculated) correspond to the total of free cells and bound cells

This inducibility of the binding was not observed for all strains of *Y. lipolytica*. Indeed, some strains, especially strains isolated from marine environment (such as strain NCIM3589), presented a constitutive binding property (Zinjarde et al. 1997). In contrast to these strains, the binding capacity of strain H222 seems to be regulated in an oleate-responsive manner, i.e., inducible by alkane and fatty acids (oleic acid, ricinoleic acid) and repressed by glucose and glycerol. However, glucose had a stronger repressing effect than glycerol (Fig. 4a). No cells grown on glucose or glycerol media could bind to hexadecane (Fig. 4a, line 1 and 2 respectively). In contrast, all cells grown on hexadecane could adhere to the alkane droplets (Fig 4a, line 3). Cells grown on a combination of glucose (0.3%) and hexadecane (2%) or glycerol (0.3%) and hexadecane (2%) (Fig 4a, line 4 and 5 respectively) adhered to the alkane droplets. This adherence occurred more rapidly in the glycerol-hexadecane medium, however, not as rapidly as in the hexadecane medium.

# 3    Transport/Export of HS into the Cells

Hydrophobic substrates are products which could interfere with lipid bilayers and therefore modify the membrane structure and/or fluidity. Some of these substrates may be even toxic to the cells and there is a strong correlation between toxicity

**Fig. 4** Binding of *Y. lipolytica* cells. (**a**) Adhesion depending on media composition. (**b**) Adhesion depending on pronase treatment. (**a**) Exponentially pre-grown H222 cells on glucose were used to inoculate different media: glucose 2% (*1*), glycerol 2% (*2*), hexadecane 2% (*3*), and mixed substrates glucose-hexadecane 0.3–2% (*4*), and glycerol-hexadecane 0.3–2% (*5*), 100 µl of hexadecane were added to 1 ml of culture media, mixed, and left undisturbed for 5 min to allow phase separation. (**b**) Cells grown on decane media (1 ml) were treated with pronase for various time intervals and the binding capacity was tested by adding 1 ml of decane, mixed and allowed to undergo phase separation. The upper phase corresponds to the decane phase containing the bound cells; the lower phase corresponds to the media water-phase containing the non-bound cells

and chain length with short-chain hydrophobic substrates being the most toxic. For example, *Y. lipolytica* could not use alkanes shorter than C9 (Klug and Markovetz 1967). Similarly, some fatty acids are toxic to the cell, for example *Y. lipolytica* could utilise a large variety of fatty acids with a chain length over C9 at concentrations above 3%, while it could grow in presence of nonanoic acid (C9) only at concentration lower than 0.02% (data not shown). When shorter-chain fatty acids where used as a carbon source no growth was observed. It is therefore very important for the cells to regulate the transport/export of these substrates within the cells.

Transport of HS into *Y. lipolytica* cells is substrate type dependent (triglyceride, fatty acid and alkanes).

*Triglycerides*, first need to be hydrolysed into free fatty acid and glycerol by lipolytic enzymes. Owing to its lipophilic lifestyle, *Yarrowia* seems to be well-adapted for the efficient use of triglycerides as suggested by the presence of 16 lipase-encoding genes. In *Y. lipolytica* the number of lipases far outnumbers that of other ascomycetous yeasts which have only one or two lipases (Thevenieau et al. 2009). The lipase family vC.6174 (formerly family GLS.94) contains 16 members encoded by the genes *LIP2* (YALI0A20350g), *LIP4*, *LIP5* and *LIP7* to *LIP19*. The lipase/esterase family vC.14039 (formerly GLS.95) present four members; *LIP1*

(YALI0E10659g), *LIP3* (YALI0B08030g), *LIP6* (YALI0C00231g) and *LIP20* (YALI0E05995g) (Thevenieau et al. 2009).

Some of their corresponding gene products have been characterized, including extracellular lipases lip2p, lip7p, and lip8p. These three lipases have different substrate specificities ranging from medium- to long-chain fatty acids. The *LIP2* gene codes for a secreted lipase which preferentially hydrolyses long-chain fatty acid esters (C18). Two other genes, *LIP7* and *LIP8*, encode lipases which were shown to be partially secreted. Lip7p has substrate specificity towards C6 esters and Lip8p towards C10 esters (Fickers et al. 2005b). They also show different expression pattern; e.g., *LIP2* was shown to be induced by oleic acid, while *LIP11* was the only lipase expressed on glucose (Nicaud et al. unpublished).

Transport of *fatty acids* (FA) is not yet well-understood. Two carrier systems were proposed to be involved when fatty acid concentration was below the threshold of 10 mM (Kohlwein and Paltauf 1984). Aggelis and coll have demonstrated that fatty acids may have different incorporation rates depending on their chain lengths and unsaturation levels and thus suggested the presence of two transporters, one being specific for the short chain (i.e., C12-C14) fatty acids and the second one for C16-C18 fatty acids (Papanikolaou and Aggelis 2003). Nevertheless, the genes encoding for HS transport remain to be identified in *Y. lipolytica*. Fatty acids uptake is also not well understood in other yeast species. In *S. cerevisiae*, when fas1p (Fatty Acid Synthase) is inactivated using drugs or gene deletion, *FAA1* and *FAA4* are essential to activate extracellular fatty acid import from medium (Johnson et al. 1994; Knoll et al. 1995). These two genes encode fatty acid activators which catalyse acyl-CoA synthesis from free fatty acids and CoA-SH. In addition, Faa1p and Faa4p are lipid droplet- or mitochondrion-associated proteins and have never been observed in the plasma membrane. Abolition of FA transport by deletion of *FAA1* and *FAA4*, together with inhibition of *de novo* FA synthesis (by fas1p inhibition with the inhibitor cerulenine) leads to non-growth of the strain in minimal media supplemented with FA. In this genetic background, overexpression of *FAT1* (an homologue to the murin FA transporter) restores a slow growth (Faergeman et al. 1997). Similarly, a Δ*fat1* strain shows also a slow growth in the same media. Taking together, these results demonstrate a genetic interaction between *FAT1*, *FAA1* and *FAA4*, involved in FA transport. However, to the best of our knowledge, fatty acid transport has not been clearly demonstrated for yeasts.

For *alkane*s, transport/export issues have been approached. Mauersberger and colleagues isolated tagged mutants affected in HS utilisation and in dimorphism (Mauersberger et al. 2001). Mutants which were unable to utilise HS were identified. Tagged mutagenesis used in this study allowed the easy identification of the disrupted genes. Several mutants including ones deficient in isocitrate lyase, isocitrate dehydrogenase, pyruvate dehydrogenase kinase, glycerol-3-phosphate dehydrogenase, and peroxines were sequenced. The results suggested the importance of peroxisome biogenesis, glyoxylate bypass and other re-arrangements of carbon central metabolism. More interestingly, we selected several mutants presenting an alkane chain-length dependent growth. For example mutants N032, N155, Z021,

Z083 and Z110 were able to grow on C16, but could not utilise C10-alkane. On the other hand, mutants N046, Z077, Z080 and B095 were able to grow on C10, but could not utilise C16 (Thevenieau et al. 2007).

Two genes were identified in such mutants presenting a chain length preference. The first gene, *ANT1*, which was disrupted in Z110 mutant (C10$^-$, C16$^+$) encoded a peroxisomal membrane localized adenine nucleotide transporter protein which provide ATP for the activation of short-chain fatty acids by acyl-CoA synthetase II in peroxisomes, whereas long-chain fatty acids derived from the terminal oxidation of long-chain alkanes are known to be activated in the cytoplasm. The second gene, *ABC1*, which was disrupted in N046 and B095 mutants (C10 + , C16 $^-$) encoded an ABC transporter. These results revealed a differential processing of alkanes based on their chain length. The short-chain alkanes were transformed through a P450–dependent alkane monooxygenase system (AMOS) to give the corresponding short-chain-FA. The short-chain FA could diffuse into the cells up into peroxisome, where they are activated into FA-CoA. Long-chain alkanes were transformed by the AMOS system to the corresponding long-chain-FA. On the contrary to short-chain-FA which are activated into the peroxisome, long-chain-FA are activated in the cytoplasm (Fig. 5).

**Fig. 5** Growth rates of *ANT1* and *ABC1* mutants on alkanes of different chain length. Mutant disrupted for *ANT1* (Z110, *grey bar*) and mutant disrupted for *ABC1* (N046, *black bar*) were grown in YNBY containing 1% of the corresponding alkane as carbon source

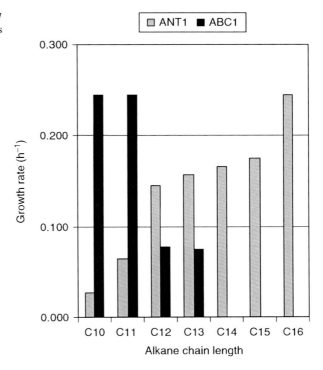

**Fig. 6** Growth of *ABC* deletants on decane (C10). Wild-Type and *ABC* mutants were grown in minimal medium YNBY containing 1% of decane as carbon source. YNBY; yeast nitrogen base media, complemented with 0.15% of yeast extract. Symbols for strains are: Wild Type H222 (*line*), Δabc1 (*square*), Δ*abc2* (*diamond*), Δ*abc3* (*triangle*), and Δ*abc4* (*circle*)

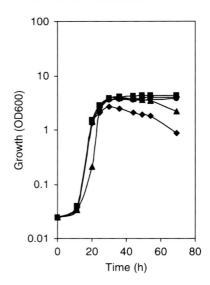

Mechanisms for alkane transport are not yet known. The phenotype of the Δ*abc1* suggested firstly that ABC transporter may be involved in the transport of these substrates. Four genes highly homologous to *ABC1* were identified in *Y. lipolytica* genome: *ABC1* (YALI0E14729g), *ABC2* (YALI0C20265g), *ABC3* (YALI0B02544g) and *ABC4* (YALI0B12980g). Thevenieau (2006), constructed different strains deleted for these ABC transporters. On C16, only Δ*abc1* presented a clear alkane growth defective phenotype (growth was completely abolished). No growth differences were observed for the three other *abc* mutants (data not shown). On C10, no difference could be observed except for Δ*abc2* which presented a decrease in cell density 30 h after transfer into alkane medium (Fig. 6). Preliminary results for C10 and C16 transport in the Δ*abc1* suggested that this mutant is not affected in the alkane entry but rather in the alkane export (Thevenieau 2006). This represents the first evidence for the involvement of an ABC transporter in the export of alkane.

# 4    Modification of HS into the Cells

Transformation of alkanes into their corresponding chain length fatty acids involves the monoterminal oxidation. The transformation of an alkane to the corresponding alcohol is catalysed by the AMOS system which involves a cytochrome P450 coupled to an NADPH-cytochrome P450 reductase (CPR). The alcohol so formed is subsequently converted into the corresponding aldehyde by either a fatty-alcohol dehydrogenase in peroxisome or a fatty alcohol oxidase in the endoplasmic

**Table 1** Substrate specificities of *Y. lipolytica* cytochromes P450 encoded by *ALK* genes

| Genes | Number | Alkane specificity | Fatty acid specificity | Effect on long-chain DCA production | References |
|---|---|---|---|---|---|
| YALI0E25982g | YlALK1 | C10 | | ns | Iida et al. 1998; Thevenieau 2006 |
| YALI0F01320g | YlALK2 | C16 | | + | Iida et al. 1998; Thevenieau 2006 |
| YALI0A23474g | YlALK3 | | C12 | – | Hanley et al. 2003; Thevenieau 2006 |
| YALI0B13838g | YlALK5 | C10 | C12 | + | Hanley et al. 2003; Thevenieau 2006 |
| YALI0A15488g | YlALK7 | | C12 | + | Hanley et al. 2003; Thevenieau 2006 |
| YALI0B06248g | YlALK9 | | | ns | Thevenieau 2006 |
| YALI0B20702g | YlALK10 | | | + | Thevenieau 2006 |
| YALI0C10054g | YlALK11 | C10 | | ns | Thevenieau Ph. D., 2006 |

*ns* non significant, + increase, – decrease

reticulum. The final steps are catalyzed by a fatty-aldehyde dehydrogenase (for more details see Fickers et al. 2005a).

Genome surveys revealed that there is a single gene coding for the NADPH-cytochrome P450 reductase, while there are 12 genes coding for cytochrome P450 isoforms of the CYP52 family. Several studies have demonstrated specific roles of cytochrome P450 isoforms in substrate assimilation depending on the substrate nature and chain length (Table 1). Cytochrome P450s encoded by *ALK1*, *ALK5* and *ALK11* genes are specific for short chain length alkanes, whereas *ALK2* is specific for long chain length alkanes. In addition, the expression of *ALK3*, *ALK5* and *ALK7* genes in *Nicotiana benthaminia* indicated a specific function in short chain fatty acid assimilation. Finally, the overexpression of *ALK2*, *ALK5*, *ALK7* or *ALK10* genes increased the bioconversion rate of long chain fatty acid to corresponding dicarboxylic acid (DCA).

In another study it was demonstrated that hydroxylation by the NADPH-cytochrome P450 reductase was the rate-limiting step in the ω-oxidation pathway.

Indeed, the overexpression of the *CPR* gene resulted in a two-fold increase in DCA production (Thevenieau 2006).

## 5    Degradation Through the β-Oxidation Pathway

Degradation of fatty acids in *Y. lipolytica* takes place only *via* β-oxidation in peroxisomes where the responsible enzymes are present. Six different acyl-CoA oxidases (Aox1–6, encoded by the *POX1– POX6* genes, respectively) catalyze the first and rate limiting step of peroxisomal β-oxidation of activated acids. The second and third step of the β-oxidation is catalysed by the multifunctional enzyme encoded by the *MFE* gene bearing the hydratase and the dehydrogenase activities and the fourth step is performed by the 3-ketoacyl-CoA-thiolase encoded by the *POT1* gene. More recently, a decane-inducible peroxisomal acetoacetyl-CoA thiolase (encoded by *PAT1*) was identified, which together with *POT1* thiolase is thought to be involved in the last step of β−oxidation.

These *POX* genes encoding acyl-CoA oxidases (Aox) exhibit different activities and substrate specificities, which were demonstrated by gene disruption in *Y. lipolytica* (Wang et al. 1999) and by expression and purification in *E. coli* for at least two of these enzymes, Aox2p and Aox3p. Indeed, these two acyl-CoA oxidases possess high activity and specificity either for long-chain fatty acids (Aox2p), or for short-chain fatty acids (Aox3p) as substrates, correspondingly. We have also shown that Aox1p and Aox6p were involved in the degradation of dicarboxylic acids (Thevenieau 2006). Indeed, a bioconversion from alkane C13 with a strain deleted for all *POX* genes (*Δpox1–6*) showed the presence of the corresponding DCA (DC13) as a main product. Shorter chain DCAs (DC12 and 10) were present in lower concentration and were a result of the bioconversion of the fatty acids produced by de novo synthesis (Fig. 7). This confirmed that the β-oxidation pathway was completely blocked in this strain. However, if Aox1p or Aox6p was active, a production of DC11 and DC9 was observed resulting from the β−oxidation of DC13 produced (data not shown).

Apart from activity and specificity, analysis of this gene family by gene disruption revealed striking features. Firstly, the mechanism of their entry into peroxisomes was different and these enzymes did not possess peroxisome targeting sequence (PTS). Indeed, acyl-CoA oxidase (Aox) complex of *Y. lipolytica* was shown to be first preassembled in the cytosol and being then imported into peroxisomes as a hetero-pentameric, cofactor-containing complex with Aox2p and Aox3p playing a pivotal role in the process (Titorenko et al. 2002). Secondly, Aox enzymes were shown to play an important role in peroxisome division. Thus, during maturation of peroxisomes a membrane-bound pool of Aox interacts with a membrane-associated peroxin Pex16p, which otherwise negatively regulates the membrane fission event required for the division of immature peroxisomal vesicles, thereby preventing their excessive proliferation. This was revealed by the presence of giant peroxisomes in *Δpex4* and *Δpex5* mutants (Guo et al. 2003). And thirdly, analysis of the mutant

**Fig. 7** DCA production from C13 alkane by $\Delta$ *pox1–6* mutant. $\Delta$ *pox* mutant was grown in rich media YPD containing 3% of alkane C13. Symbols for DCA product are: DC13 (*close diamond*); DC12 (*open triangle*) and DC10 (*open circle*).

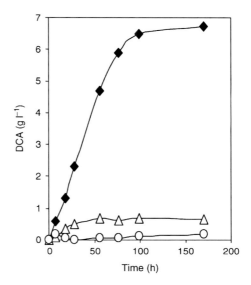

phenotypes with respect to lipid accumulation during growth on oleic acid revealed that *POX* genotype could result in *obese* or *slim* yeast (Mlickova et al. 2004).

# 6    Storage as Triglyceride and Sterol Ester

In microorganisms, excessive carbon source is often accumulated in a *hydrophobic compound-modified form*. These storage molecules represent one of the most efficient energy sources due to their volume/energy balance. After uptake of hydrophobic substrates from the culture medium, the internalized aliphatic chains can be either degraded for growth requirements or accumulated in an unchanged or modified form. In bacteria, they are mainly accumulated as PHA or wax esters. In yeasts, storage molecules are mainly triglycerides (TAG) and steryl esters (SE) which form the hydrophobic core of so-called lipid bodies (LB).

However, storage capability differs between yeast species. The bakers' yeast (*Saccharomyces cerevisiae*), which is a *non oleagenous* yeast, could accumulate less than 20% of lipid (g lipid/g biomass). In contrast, *oleagenous* yeasts, such as *Rhodotorula glutinis* could accumulate up to 70% of lipid (Ratledge 1994). The level of lipid could differ as well as the type of lipid. For example, *S. cerevisiae* accumulates similar proportions of TAG and SE (50, 50%), while in the oleagenous yeast *Y. lipolytica* it represents 75, 25%, respectively.

On the other hand, the biosynthetic pathway is well conserved between yeasts. TAG synthesis follows the Kennedy pathway. Free fatty acids are activated by condensation to coenzyme A (CoA) and used for the acylation of the glycerol backbone to synthesize TAG. In the first step of TAG assembly, glycerol-3-phosphate (G-3-P) is acylated by G-3-P acyltranferase (*SCT1*) to lysophosphatidic acid

(*LPA*), which is then further acylated by lysophosphatidic acid acyltransferase (*SLC1*) to phosphatidic acid (PA). This is followed by dephosphorylation of PA by phosphatidic acid phosphohydrolase (*PAP*) to release diacylglycerol (DAG). In the final step DAG is acylated either by diacylglycerol acyltransferase (*DGA1*- with acyl-CoA as acyl donor) or phospholipid diacylglycerol acyltransferase (*LRO1* – with glycerophospholipids as acyl donor) to produce TAG. Involvement of *DGA1* and *LRO1* in lipid accumulation was confirmed for *Y. lipolytica*. Indeed, by gene disruption, we have shown that a Δ*dga1* and Δ*lro1* mutants present a 40% and 50% decrease in lipid content, respectively. The double Δ*dga1* Δ*lro1* mutant accumulated less than 10% of lipid (Beopoulos et al. to be published). Synthesis of mono-glyceride and di-glyceride may differ between *S. cerevisiae* and *Y. lipolytica*. Indeed, in *S. cerevisiae SLC1* and *SCT1* genes could be deleted, whereas we were not able to disrupt these genes in *Y. lipolytica*.

TAG synthesis in yeast cells is strongly regulated by carbon availability in the culture medium. Mobilization of accumulated lipids occurs as a consequence of three different metabolic states: (1) during exponential phase of growth, where storage lipid compounds are used for membrane lipid synthesis to support cellular growth and division; (2) during stationary phase, upon nutrient depletion, FFA are liberated rather slowly from the TAG and subjected to peroxisomal β-oxidation; (3) when cells exit starvation conditions, e.g., from stationary phase, and enter a vegetative growth cycle upon carbon supplementation, lipid depots are very rapidly degraded to FFA. In addition, nitrogen starvation leads to an increased rate of TAG degradation.

The ability of *Yarrowia lipolytica* to accumulate lipids up to 50% of its dry weight makes this yeast a good candidate for white biotechnology applications. Especially since several molecular genetic tools, such as gene disruption to modify metabolic pathways and vectors for heterologous gene over expression, are available. All these facilitate strain construction for mastering lipid accumulation and in modification of the type of lipid accumulated. Preliminary results towards this direction are already promising. For example, mastering TAG synthesis, lipid accumulation and lipid bioconversion in *Y. lipolytica* could lead to an alternative source of lipids for biofuel production and to PUFA synthesis.

## 7    Research Needs

The complete genome sequence of *Y. lipolytica* has been determined (Dujon et al. 2004) and DNA microarrays are now available. This opens the way to study the regulation of lipid accumulation. Further work will concern the production of complex hydrophobic molecules from *Y. lipolytica* destined for the field of fine chemistry. Additionally, works on the identification of the gene coding for liposan and it biosynthesis pathway need to be done.

Another aspect would be to determine why there are several gene families of the same gene in this yeast that are present far in excess. This fact might be linked with chain length specificity and/or substrate specificity of the corresponding gene products. For example, the specificity of the P450 was partly determined for some members of this family. For the alcohol and aldehyde enzyme, there is biochemical evidence, but the genes have not yet been identified. Also, a new challenge would be to understand the HS traffic inside the cell between the different compartments. Understanding the biogenesis of lipid bodies and of the gene(s) involved in the regulation of lipid accumulation is also a great challenge.

# References

Barth G, Gaillardin C (1996) *Yarrowia lipolytica*. In: Wolf K (ed) Nonconventional yeasts in biotechnology: a handbook. Springer, Heidelberg, pp 313–388

Barth G, Beckerich JM, Dominguez A, Kerscher S, Ogrydziak D, Titorenko V, Gaillardin C (2003) Functional genetics of *Yarrowia lipolytica*. In: de Winde JH (ed) Functional genetics of industrial yeasts. Springer, Berlin, pp 227–271

Cirigliano M, Carman GM (1984) Isolation of a bioemulsifer from Candida lipolytica. Appl Environ Microbiol 48:747–750

Dujon B, Sherman D, Fischer G, Durrens P, Casaregola S, Lafontaine I, De Montigny J, Marck C, Neuveglise C, Talla E, Goffard N, Frangeul L, Aigle M, Anthouard V, Babour A, Barbe V, Barnay S, Blanchin S, Beckerich JM, Beyne E, Bleykasten C, Boisrame A, Boyer J, Cattolico L, Confanioleri F, De Daruvar A, Despons L, Fabre E, Fairhead C, Ferry-Dumazet H, Groppi A, Hantraye F, Hennequin C, Jauniaux N, Joyet P, Kachouri R, Kerrest A, Koszul R, Lemaire M, Lesur I, Ma L, Muller H, Nicaud JM, Nikolski M, Oztas S, Ozier-Kalogeropoulos O, Pellenz S, Potier S, Richard GF, Straub ML, Suleau A, Swennen D, Tekaia F, Wesolowski-Louvel M, Westhof E, Wirth B, Zeniou-Meyer M, Zivanovic I, Bolotin-Fukuhara M, Thierry A, Bouchier C, Caudron B, Scarpelli C, Gaillardin C, Weissenbach J, Wincker P, Souciet JL (2004) Genome evolution in yeasts. Nature 430:35–44

Faergeman NJ, DiRusso CC, Elberger A, Knudsen J, Black PN (1997) Disruption of the *Saccharomyces cerevisiae* homologue to the murine fatty acid transport protein impairs uptake and growth on long-chain fatty acids. J Biol Chem 272:8531–8538

Fickers P, Benetti PH, Wache Y, Marty A, Mauersberger S, Smit MS, Nicaud JM (2005a) Hydrophobic substrate utilisation by the yeast *Yarrowia lipolytica*, and its potential applications. FEMS Yeast Res 5:527–543

Fickers P, Fudalej F, Le Dall MT, Casaregola S, Gaillardin C, Thonart P, Nicaud JM (2005b) Identification and characterisation of *LIP7* and *LIP8* genes encoding two extracellular tri-acylglycerol lipases in the yeast *Yarrowia lipolytica*. Fungal Genet Biol 42:264–274

Guo T, Kit YY, Nicaud JM, Le Dall MT, Sears SK, Vali H, Chan H, Rachubinski RA, Titorenko VI (2003) Peroxisome division in the yeast *Yarrowia lipolytica* is regulated by a signal from inside the peroxisome. J Cell Biol 162:1255–1266

Hadeball W (1991) Production of lipase by *Yarrowia lipolytica*. Acta Biotechnol 11:159–167

Hanley K, Nguyen LV, Khan F, Pogue GP, Vojdani F, Panda S, Pinot F, Oriedo VB, Rasochova L, Subramanian M, Miller B, White EL (2003) Development of a plant viral-vector-based gene expression assay for the screening of yeast cytochrome P450 monooxygenases. Assay Drug Dev Technol 1:147–160

Iida T, Ohta A, Takagi M (1998) Cloning and characterization of an n-alkane-inducible cytochrome P450 gene essential for n-Decane assimilation by *Yarrowia lipolytica*. Yeast 14:1387–1397

Johnson DR, Knoll LJ, Levin DE, Gordon JI (1994) *Saccharomyces cerevisiae* contains four fatty acid activation (FAA) genes: an assessment of their role in regulating protein *N*-myristoylation and cellular lipid metabolism. J Cell Biol 127:751–762

Klug MJ, Markovetz AJ (1967) Degradation of hydrocarbons by members of the genus Candida. II. Oxidation of n-alkanes and l-alkenes by *Candida lipolytica*. J Bacteriol 93:1847–1852

Knoll LJ, Johnson DR, Gordon JI (1995) Complementation of *Saccharomyces cerevisiae* strains containing fatty acid activation gene (FAA) deletions with a mammalian acyl-CoA synthetase. J Biol Chem 270:10861–10867

Kohlwein S, Paltauf F (1984) Uptake of fatty acids by the yeasts, *Saccharomyces uvarum* and *Saccharomycopsis lipolytica*. Biochim Biophys Acta 792:310–317

Mauersberger S, Wang HJ, Gaillardin C, Barth G, Nicaud JM (2001) Insertional mutagenesis in the n-alkane-assimilating yeast *Yarrowia lipolytica*: generation of tagged mutations in genes involved in hydrophobic substrate utilization. J Bacteriol 183:5102–5109

Mlickova K, Roux E, Athenstaedt K, d'Andrea S, Daum G, Chardot T, Nicaud JM (2004) Lipid accumulation, lipid body formation, and acyl coenzyme A oxidases of the yeast *Yarrowia lipolytica*. Appl Environ Microbiol 70:3918–3924

Nuttley WM, Brade A, Gaillardin C, Eitzen G, Glover J, Aitchison J, Rachubinski R (1993) Rapid identification and characterization of peroxisomal assembly mutants in *Yarrowia lipolytica*. Yeast 9:507–517

Osumi M, Fukuzumi F, Yamada N, Nagatani T, Teranishi Y, Tanaka A, Fukui S (1975) Surface structure of some Candida yeast cells grown on n-alkanes. J Ferment Technol 53:244–248

Papanikolaou S, Aggelis G (2003) Modeling lipid accumulation and degradation in *Yarrowia lipolytica* cultivated on industrial fats. Curr Microbiol 46:398–402

Ratledge C (1994) Yeasts, moulds, algae and bacteria as sources of lipids. In: Kamel BS, Kakuda Y (eds) Technological advances in improved and alternative sources of lipids. Blackie Academic and Professional, London, pp 235–291

Tanaka A, Fukui S (1989) Metabolism of n-alkanes. In: Rose AH, Harrison JS (eds) The yeasts, Metabolism and physiology of yeasts, vol 3, 2nd edn. Academic, London, pp 261–287

Thevenieau F (2006) Metabolic engineering of the yeast *Yarrowia lipolytica* for the production of long-chain dicarboxylic acids from renewable oil feedstock. Ph.D. thesis, Institut National Agronomique Paris-Grignon

Thevenieau F, Le Dall MT, Nthangeni B, Mauersberger S, Marchal R, Nicaud JM (2007) Characterization of *Yarrowia lipolytica* mutants affected in hydrophobic substrate utilization. Fungal Genet Biol 44:531–542

Thevenieau F, Nicaud JM, Gaillardin C (2009) Application of the non-conventional yeast *Yarrowia lipolytica*. In: Satyanarayana T, Kunze G (eds) Yeast biotechnology: diversity and applications. Springer part III, chapter 26.

Titorenko VI, Nicaud JM, Wang H, Chan H, Rachubinski RA (2002) Acyl-CoA oxidase is imported as a heteropentameric, cofactor-containing complex into peroxisomes of *Yarrowia lipolytica*. J Cell Biol 156:481–494

Vance-Harrop M, Gusmao N, Campos-Takaki GM (2003) New bioemulsifiers produced by *Candida lipolytica* using D-glucose and babassu oil as carbon sources. Braz J Microbiol 34:120–123

Wang H, Le Dall MT, Wache Y, Laroche C, Belin JM, Nicaud JM (1999) Cloning, sequencing, and characterization of five genes coding for acyl-CoA oxidase isozymes in the yeast *Yarrowia lipolytica*. Cell Biochem Biophys 31:165–174

Zinjarde SS, Pant A (2002) Emulsifier from a tropical marine yeast, *Yarrowia lipolytica* NCIM 3589. J Basic Microbiol 42:67–73

Zinjarde SS, Sativel C, Lachke AH, Pant A (1997) Isolation of an emulsifier from *Yarrowia lipolytica* NCIM 3589 using a modified mini isoelectric focusing unit. Lett Appl Microbiol 24:117–121

# Biodiversity of Biosurfactants and Roles in Enhancing the (Bio)availability of Hydrophobic Substrates

**6**

Amedea Perfumo, Michelle Rudden, Roger Marchant, and Ibrahim M. Banat

## Contents

1   Introduction ................................................................. 76
2   Biodiversity of Biosurfactant Producing Microorganisms ................................. 77
3   Rhamnolipid Biosurfactants in *Pseudomonas* spp. and Related Organisms ............... 79
4   The Lipopeptide Biosurfactants ............................................... 87
5   Biosurfactants Produced by Actinobacteria ................................................. 88
6   Glycolipids Produced by Eukaryotes ....................................... 89
7   Applications of Biosurfactant-Enhanced (Bio)availability of Hydrophobic Substrates .... 91
8   Current State and Research Needs ........................................... 93
References ................................................................. 94

**Abstract**

This chapter focusses on the biodiversity of microbial biosurfactants and the organisms that produce them. Specific attention is given to the low molecular weight glycolipids and lipopeptides produced by bacteria such as *Pseudomonas, Burkholderia, Bacillus, Rhodococcus*, and *Alcanivorax* in addition to other glycolipids synthesized by eukaryotic organisms such as *Starmerella, Pseudozyma*, and *Candida* spp. The applications of microbial surfactants utilizing their

A. Perfumo (✉)
Helmholtz Centre Potsdam, GFZ German Research Centre for Geosciences, Potsdam, Germany
e-mail: amedea.perfumo@gfz-potsdam.de

M. Rudden
Department of Biology, University of York, York, UK
e-mail: michelle.rudden@york.ac.uk

R. Marchant · I. M. Banat (✉)
School of Biomedical Sciences, Ulster University, Coleraine, Northern Ireland, UK
e-mail: r.marchant@ulster.ac.uk; IM.Banat@ulster.ac.uk

© Springer International Publishing AG, part of Springer Nature 2018
T. Krell (ed.), *Cellular Ecophysiology of Microbe: Hydrocarbon and Lipid Interactions*,
Handbook of Hydrocarbon and Lipid Microbiology, https://doi.org/10.1007/978-3-319-50542-8_35

properties for accessing substrates and in microemulsion technology is covered plus reference to potential applications in environmental remediation. Finally a summary of the current state of research and identification of significant areas for further investigation are highlighted.

# 1    Introduction

Interest in microbial biosurfactants has increased significantly in the last few years largely due to their perceived enormous potential as sustainable replacements for chemical surfactants in a wide range of consumer products. The pressure on companies to use sustainable, green resources to produce their products and consumer interest in "natural" products have effectively fuelled the research interest in these compounds (Banat et al. 2010; Campos et al. 2013; Satpute et al. 2016a; Elshikh et al. 2016; De Almeida et al. 2016). Unsurprisingly the main focus of the research has been directed towards the physicochemical properties of biosurfactants and investigations of how they might be employed as at least partial replacements for chemical surfactants in high turnover consumer products such as laundry detergents, surface cleaners, personal care products, cosmetics and even foodstuffs. There has also been interest in their use in more specialised applications such as pharmaceuticals, exploiting their synergistic potential in combination with antibiotics and other bioactive molecules, and even in the bioactivity of the biosurfactants themselves as antimicrobials and anti-cancer agents (Marchant and Banat 2012a, b; Fracchia et al. 2014, 2015; Díaz de Rienzo et al. 2016b).

Thus far one of the main hurdles to exploitation has been the fermentation yields of biosurfactants, which have often been too low for economic commercial use coupled with the difficulty and cost of downstream processing for the production of defined, pure products. As a result of the current emphasis on the commercial exploitation of biosurfactants less effort has been directed towards investigation into why microorganisms produce these surface active molecules and what role they play in the life history of the organisms. Their function must be significant since many organisms direct a large part of their metabolic energy and resources into their production. Many functions have been ascribed to biosurfactants, including maintenance of biofilm structure, motility, cell adhesion and access to hydrophobic substrates (Fracchia et al. 2012, 2014; Díaz de Rienzo et al. 2016a, c). It is the last of these functions that this chapter will address. In the context of hydrophobic substrate degradation e.g., hydrocarbons biosurfactants have attracted interest as augmentation agents to speed the removal of environmental contaminating fuels and crude oil, both in marine and terrestrial situations. Experiments using biosurfactants in microcosms with hydrocarbonoclastic microorganisms have demonstrated some enhancement of the rate and extent of degradation (Rahman et al. 2003), however, this remains an area of investigation with unexplored potential.

## 2     Biodiversity of Biosurfactant Producing Microorganisms

Biosurfactants (BSs) are a diverse class of compounds that are synthesised by a wide variety of microorganisms spanning all domains of life (Menezes Bento et al. 2005; Khemili-Talbi et al. 2015; Roelants et al. 2014; Satpute et al. 2016b). The structural diversity of these metabolites is reflected by the diversity of producers and environments from which they can be isolated. Conventionally BS producing microorganisms have been most frequently isolated from hydrocarbon contaminated sites which are typically dominated by a few main microbial consortia including *Pseudomonas, Burkholdeiria, Bacillus, Streptomyces, Sphingomonas* and *Actinobacteria* in soils and sediments, and *Pseudoalteromonas, Halomonas, Alcanivorax* and *Acinetobacter* in marine ecosystems. However, BS producing microorganisms are ubiquitous in nature and have been isolated from a variety of niches including extreme environments such as high salinity (Donio et al. 2013; Pradhan et al. 2013), high temperatures (Elazzazy et al. 2015; Darvishi et al. 2011), and cold environments (Malavenda et al. 2015; Cai et al. 2014).

*Pseudomonas* and *Bacillus* are the main genera reported for biosurfactant production, however novel biosurfactant producers are continually being reported, recently Hošková et al. 2015 characterised rhamnolipids synthesised by *Acinetobacter calcoaceticus* and *Enterobacter asburiae*. Traditionally efforts to increase the frequency of isolating BS producing microorganisms from hydrocarbon contaminated soils involve enrichment culture techniques, which have been reported to increase recovery of BS producers from 25% to 80% (Thies et al. 2016; Bodour et al. 2003; Steegborn et al. 1999; Rahman et al. 2002; Donio et al. 2013; Pradhan et al. 2013; Walter et al. 2013). Enrichment techniques have been widely applied to members of the genera *Pseudomonas* and *Bacillus* for bioremediation of hydrocarbon contaminates sites (Li et al. 2016). Identification of biosurfactant producing microorganisms still relies on culture dependent techniques and typically qualitative methods for biosurfactant characterisation such as the orcinol assay, drop collapse, oil spreading assay and bath assay (Marchant and Banat 2014). While these methods are applicable for preliminary characterisation of BSs, they are not accurate and often overestimate production yields of biotechnologically important BSs (Marchant and Banat 2014).

In recent years research has focused on metabolic engineering of biosurfactants from a diverse range of microorganisms which has largely been facilitated by the increased number of published genomes of biosurfactant producing organisms (Table 1). Advances in functional metagenomics techniques have enabled the exploitation of these genomes for biosurfactant bio-discovery from diverse environments (Kennedy et al. 2011). Marine environments have proven to be a rich resource for isolating BS producing organisms including from marine sponges (Dhasayan et al. 2015; Kiran et al. 2009, 2010) and hydrocarbon contaminated marine areas (Gutierrez and Banat 2014; Antoniou et al. 2015). Recently Oliveira et al. 2015 have published BioSurfDB, www.biosurfdb.org, a tailored databased specifically for

**Table 1** List of published genomes of common biosurfactant producing microorganisms

| Species | Biosurfactant | References |
|---|---|---|
| *Pseudomonas aeruginosa* PAO1 | Rhamnolipids | (Winsor et al. 2016) |
| *Burkholderia thailandensis* E264 | Rhamnolipids | (Winsor et al. 2008) |
| *Pseudozyma antarctica* T-34 | Mannosylerythritol lipids | (Morita et al. 2013a) |
| *Starmerella bombicola* NBRC10243 | Sophorolipids | (Matsuzawa et al. 2015) |
| *Bacillus amyloliquefaciens* RHNK22 | Surfactin, Iturin, and Fengycin | (Narendra Kumar et al. 2016) |
| *Alcanivorax borkumensis* | Lipopeptide | (Schneiker et al. 2006) |
| *Pseudozyma hubeiensis* SY62 | Mannosylerythritol lipids | (Konishi et al. 2013) |
| *Bacillus safensis* CCMA-560 | Pumilacidin | (Domingos et al. 2015) |
| *Pseudozyma aphidis* DSM 70725 | Mannosylerythritol lipids | (Lorenz et al. 2014) |
| *Rhodococcus sp.* PML026 | Trehalose lipids | (Sambles and White 2015) |
| *Rhodococcus sp.* BS-15 | Trehalose lipids | (Konishi et al. 2014) |
| *Gordonia amicalis* CCMA-559 | Polymeric glycolipids | (Domingos et al. 2013) |
| *Dietzia maris* | Polymeric glycolipids | (Ganguly et al. 2016) |

analysing biosurfactant production in microorganisms. This database unique to biosurfactant production integrates a range of information including metagenomics, biosurfactant producing organisms, biodegradation relevant genes, proteins and their metabolic pathways, results from bioremediation experiments and a biosurfactant-curated list, grouped by producing organism, surfactant name, class and reference. The database contains 3,736 biosurfactant coding genes and offers a unique platform for recombinant engineering of novel/diverse biosurfactant coding genes identified from a range of microorganisms including uncultivable bacteria (Thies et al. 2016).

Functional metagenomics offers enormous opportunities for recombinant production of novel biosurfactant molecules with tailored applications, however it should be noted that several technical challenges still exist. Biosynthetic clusters for natural products tend to be relatively large and highly regulated especially from yeast and fungal sp., for example the biosynthetic gene cluster from *Ustilago* sp. encoding cellobiose lipids spans 58 Kb (Teichmann et al. 2007). Associated challenges include heterologous expression of large biosynthetic clusters from single fragments, toxicity of associated product to host organism, promoter and transcription factor recognition and extracellular transport of gene products (Jackson et al. 2015). Current research is focussed on improving efficiency of functional metagenomics techniques, for example to overcome dynamic regulation by native promoters within natural biosynthetic gene clusters. Kang et al. (2016) used mCRISTAR, a multiplexed CRISPR and TAR technique which allows for precision promoter engineering by specifically inserting inducible synthetic promoters to replace native promoters. Sophisticated techniques like this could be applied to large biosurfactant biosynthetic gene clusters and with collated resources such as BioSurfdb, efficient heterologous production of tailor made biosurfactants looks promising.

## 3 Rhamnolipid Biosurfactants in *Pseudomonas* spp. and Related Organisms

Rhamnolipids (RLs) are low molecular glycolipid secondary metabolites predominantly synthesised by the genera *Pseudomonas* and *Burkholderia* spp. RLs are synthesised as a heterogeneous mixture of congeners comprising a hydrophilic rhamnose (Rha) glycosidically linked to one or more β-hydroxy fatty acids (Fig. 1a). They are produced in two main classes, mono- and di-rhamnolipids, and usually vary in alkyl chain length ranging from $C_8$-$C_{16}$ (Haba et al. 2003a, b; Gunther et al. 2005; Déziel et al. 2000) and degree of saturation, with some polyunsaturated rhamnolipid congeners reported for *P. aeruginosa* (Abalos et al. 2001; Haba et al. 2003a).

The advancement in chromatographic and mass spectrometric techniques in the last few decades has resulted in the identification of up to 60 different RL congeners produced by various microorganisms with *Pseudomonas* spp. and *Burkholderia* spp being the most dominant producers (Abdel-Mawgoud et al. 2010; Mata-Sandoval et al. 1999; Heyd et al. 2008; Dubeau et al. 2009). Among these diverse homologues some more unusual RLs have been detected. Arino et al. (1996) have reported both mono- and di-RL congeners containing mono-lipidic fatty acid only in trace amounts and Andrä et al. (2006) reported a di-RL congener with three $C_{14}$ β-hyroxy fatty acids in *Burkholdieria plantarii*. *P. aeruginosa* preferentially synthesises di-lipidic RLs with alkyl lengths $C_{10}$-$C_{10}$, whereas $C_{14}$ β-hydroxy acids are the most abundant fatty acids for the *Burkholderia* spp. (Funston et al. 2016; Hörmann et al. 2010; Häussler et al. 1998; Howe et al. 2006).

In recent years there has been significant emphasis on RL production from non-pathogenic producers as alternatives to the opportunistic pathogen *Pseudomonas aeruginosa* (Marchant et al. 2014). Alternative producers include phylogenetically related species including *Pseudomonas fluorescens* (Vasileva-Tonkova et al. 2011) and the exclusively mono-RL producing strain *Pseudomonas chlororaphis* (Gunther et al. 2005). RL production has also been reported among the more taxonomically distant species such as Acinetobacter and Enterobacter bacteria (Hošková et al. 2015). However production yields are not comparable to RLs produced from *P. aeruginosa*. Care must always be exercised in evaluating claims for new or unusual producers since the methods used to determine the identity of the producer organism and the characterisation of the product are not always sufficiently rigorous. The most promising non-pathogenic biotechnological RL producer is *Burkholdeira thailandensis* E264. It has been shown that *B. thailandensis* produces RL yields comparable to wild-type *P. aeruginosa* making it a suitable candidate for large scale production of RLs. The composition of RLs from *B. thailandensis* differs only in the length of the di-lipid alkyl chains, in *B. thailandensis* $C_{14}$ is the preferred chain length (Funston et al. 2016), whereas in *P. aeruginosa* $C_{10}$ is selectively incorporated to synthesise RLs (Rudden et al. 2015).

The biosynthetic pathway and complex genetic regulation of RL synthesis has been extensively studied in *P. aeruginosa* (Reis et al. 2011; Schmidberger et al.

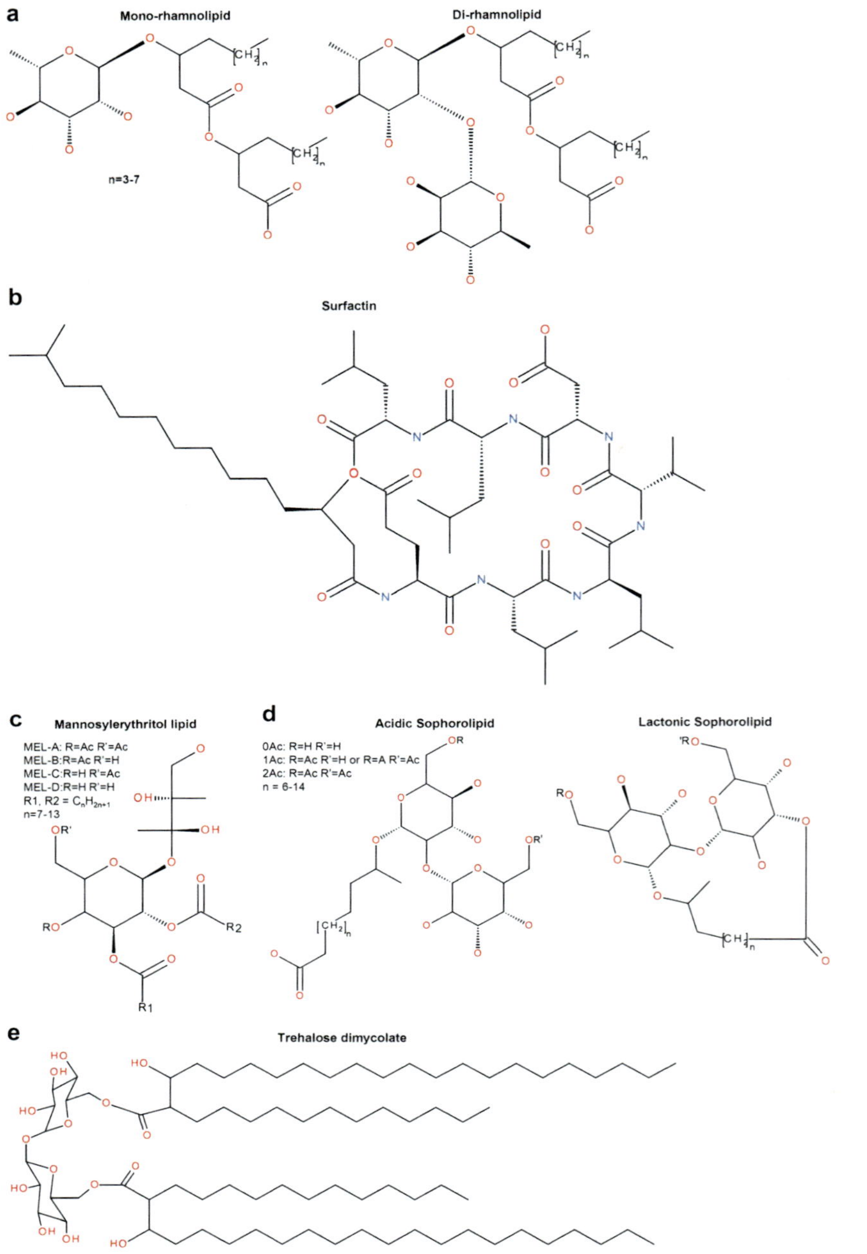

**Fig. 1** Molecular structure of common biosurfactants. (**a**) Rhamnolipids from *Pseudomonas* sp. (**b**) Surfactin from *Bacillus* sp. (**c**) Mannosylerythritol lipids from *Pseudozyma* sp. (**d**) Sophorolipids from *Starmerella bombicola* and (**e**) Trehalose dimycolate from *Rhodococcus* sp.

2013). RL precursor molecules are synthesised de novo from central metabolic pathways, dTDP-L-rhamnose is synthesised by the proteins encoded in the *rmlABCD* operon from the rhamnose pathway (Aguirre-Ramirez et al. 2012) and β-hydroxy fatty acids are derived from both β-oxidation and FAS II pathways (Zhang et al. 2014; Zhu and Rock 2008). RL biosynthesis is catalysed by three enzymes, RhlA, RhlB and RhlC. RhlA (rhamnosyltransferase chain A) is responsible for the supply of the lipid precursors for RLs by the dimerisation of two β-hydroxyacyl-acyl carrier proteins ACPs to form one molecule of hydroxyalkanoic acid (HAA). RhlA is selective for $C_{10}$ carbon intermediates and regulates the fatty acid composition of RLs (Zhu and Rock 2008). Mono-RLs are synthesised by the condensation of the precursors derived from the central metabolic pathway by RhlB (rhamnosyltransferase chain B). RhlB transfers one dTDP-L-rhamnose to the HAA producing a mono-RL. Mono-RL together with another dTDP-L-rhamnose is the substrate for rhamnosyltransferase 2 (RhlC) which synthesises di-RLs. RhlA and RhlB form the heterodimer rhamnosyltransferase 1, with RhlB being the catalytic subunit for RL biosynthesis and RhlA responsible for the lipid precursor synthesis. RhlA and RhlB are co-transcribed from *rhlAB* which is clustered together with the regulatory genes *rhlR/I*.

Expression of *rhlAB* is regulated by Quorum Sensing and other regulatory factors in *P. aeruginosa* (Müller and Hausmann 2011). The RhlAB polypeptide is loosely membrane associated, while RhlA is a soluble protein located in the cytoplasm. RhlB has two putative hydrophobic membrane domains associated with the inner membrane (Ochsner et al. 1994). The second rhamnosyltransferase RhlC, encoded by *rhlC,* is co-transcribed from a different bicistronic operon located with a putative major facilitator superfamily transporter (MFS) protein 2.5 Mb away from the *rhlAB* genes. Ochsner and co-workers (1994) also predicted that RhlC is located in the inner membrane based on a hydrophobic region between amino acid residues 257 and 273 (Rahim et al. 2001). This would suggest that RL biosynthesis occurs at the cytoplasmic membrane and RLs are subsequently extracellularly transported. In contrast to *P. aeruginosa, Burkholderia thailandensis* contains the RL biosynthetic *rhlABC* genes clustered together on a single operon which is duplicated in the chromosome; both operons are functional and contribute to RL biosynthesis (Dubeau et al. 2009). For many decades *P. aeruginosa* has been the main focus for RL production, however, in recent years there has been a significant shift toward alternatives for RL production, namely RL production from non-pathogenic producers and heterologous production of RLs in non-pathogenic hosts (Loeschcke and Thies 2015).

The structural diversity of RL congeners from many different producers can be attributed to significant differences in their RL biosynthetic enzymes. For example *P. chlororaphis* is a mono-RL producing only organism (Gunther et al. 2005), which does not contain an RhlC orthologue and cannot synthesise di-RLs from mono-RLs. It has recently been used for recombinant production of di-RLs with *P. aeruginosa rhlC* (Solaiman et al. 2015). RLs from *P. chlororaphis* are synthesised as a heterogeneous mixture of congers with $C_{12}$-$C_{10}$ or $C_{12:1}$-$C_{10}$ as the predominant β-hydroxy

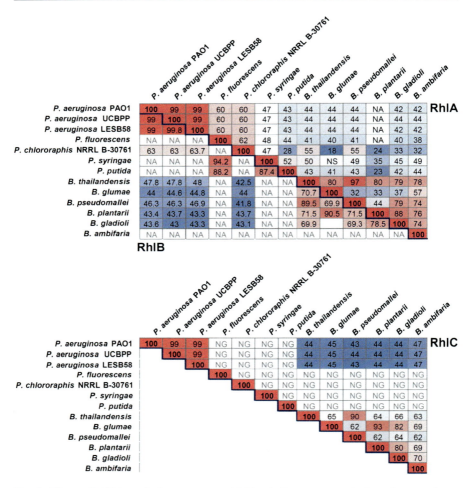

**Fig. 2** Rhamnolipid biosynthetic enzymes are highly similar among producing microorganisms. Amino acid (aa) percent identity matrix for rhamnolipid enzymes RhlA, RhlB and RhlC for reported rhamnolipid producing strains. NA – Not aligned, NG – No homologous gene found

moieties, not surprisingly RhlA from *P. chlororaphis* shares only 60% amino acid identity with *P. aeruginosa* (Fig. 2). *Burkholderia* sp. synthesises predominantly di-RLs with $C_{14}$-$C_{14}$ β-hydroxy fatty acids, both RhlA proteins only share 44% homology with *P. aeruginosa* RhlA (Fig. 2). RhlA functions primarily as an acyltransferase with a conserved α/β-hydrolase domain motif found within the first 100 residues in all RL producing microorganisms. The sequence divergence observed significantly correlates with the specificity of the RhlA enzyme for a certain acyl chain length in the producing microorganism e.g., *P. aeruginosa* RhlA is specific for $C_{10}$ carbon intermediates (Zhu and Rock 2008). Recombinant production of RLs in non-pathogenic organisms could use for example the specificity of RhlA from selected microorganisms to synthesise tailored compositions of mono- and di-RLs (Wittgens et al. 2011; Dobler et al. 2016).

*P. fluorescens* and *P. putida* have previously been reported to produce RLs (Tuleva et al. 2002; Martínez-Toledo et al. 2006), however, their production profiles have not been analytically quantified and it seems unlikely, based on the low sequence homology and lack of a RhlC orthologue, that these strains are capable of producing high yields of RLs comparable to *P. aeruginosa*. However, these organisms have been used for heterologous production of *P. aeruginosa* RLs (Ochsner et al. 1995), with *P. putida* KCTC 1067 (pNE2) reported to produce 7.3 g/L using soybean oil as the carbon source. It should be noted that recombinant production of *P. aeruginosa* RLs has extensively been quantified with the orcinol method which significantly overestimates RLs yields (Marchant et al. 2014). Also RLs produced with hydrophobic substrates must be purified before quantification as excess unutilised hydrophobic substrate is always co-extracted and also overestimates actual yields.

RL production is highly conserved among RL producing microorganisms. Fig 3 shows the multiple aligned sequences coloured by conservation with a minimum threshold for >95% identity for both RhlA and RhlC, and 50% for RhlB. RhlA has >20 of the first 110 residues conserved among all sequences which fall with the specific $\alpha/\beta$-hydrolase domain. In contrast RhlC has >114 residues conserved among all sequences analysed which span the rhamnosyltransferase/glycosytransferase domains. RhlC is a rhamnosyltransferase enzyme that specifically catalyses the transfer of a second dTDP-L-rhamnose to the mono-RL substrate. dTDP-L-rhamnose is a metabolic precursor for other pathways including lipopolysaccharide (LPS) biosynthesis in *P. aeruginosa* (Rahim et al. 2000). Given the distinct location of *rhlC* on a separate operon to *rhlAB* it is possible that this rhamnosyltransferase could have evolved from conserved rhamnosyltransferase/glycosyltransferase enzymes. Interesting to note is the lack of conservation of RhlB between RL producing microorganisms (Fig. 3), suggesting that RhlB has evolved only for RL biosynthesis. The structural differences in RL congeners is mirrored by phylogenetic analysis of the RL enzymes, for all RL biosynthetic enzymes in *P. aeruginosa* distinct clades are separated from the other reported RL producing strains (Fig. 4). Interestingly, all RL biosynthetic genes from *Burkholderia pseuodomallei* share high sequence homology and are closely related to *Burkholderia thailendesis* (Fig. 2, 4). This is consistent with high reports of RLs from *B. pseuodomallei* which are predominantly $C_{14}$-$C_{14}$ di-RL congeners (Dubeau et al. 2009).

For next generation RLs it is essential that all the RL biosynthetic enzymes are structurally characterised with regard to specificity and activity. Combining metabolic engineering of RL biosynthesis precursors with precision protein engineering of RL enzymes for enhanced efficiency/activity could offer novel molecular approaches for increased RL production in recombinant hosts.

In *P. aeruginosa* one of the main physiological roles of RLs is hydrocarbon assimilation by increasing the bioavailability of hydrophobic substrates either by

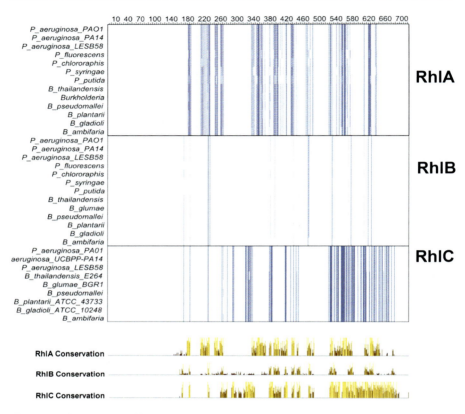

**Fig. 3** Multiple sequence alignment showing high conservation in both RhlA and RhlC among all RL producing microorganisms. Sequences were aligned with Mafft-LiNS, edited and visualised in Jalview. Residues are coloured by percent identity with a minimum threshold of 95% conservation shown for RhlA and RhlC and 50% for RhlB

increasing solubility or by increasing cell surface hydrophobicity by inducing the removal of cell associated lipopolysaccharide (LPS). Upon direct contact RLs accumulate as monomers at the aqueous/hydrophobic interface until the concentration reaches and exceeds a critical level known as the critical micelle concentration (CMC), where the RL monomers aggregate into micelles and larger vesicles. Hydrophobic substrates become incorporated within the hydrophobic core of micelles and this effectively enhances their dispersion into the aqueous phase and hence their bioavailability for uptake by cells. This process has been largely studied with alkanes as model substrates and is referred to as "micelle solubilisation" or "pseudosolubilisation" (Fig. 5a, b) (Zhang and Miller 1992). RLs have been also shown to modify the outer cell membrane of *P. aeruginosa* by inducing the removal of the cell associated lipopolysaccharides (LPS) (Al-Tahhan et al. 2000), subsequently increasing cell surface hydrophobicity (CSH) (Fig. 5a). RLs can also induce changes in the composition of outer membrane proteins (OMP) even at

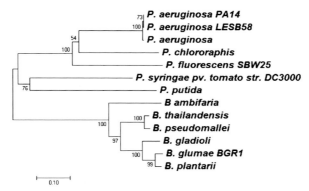

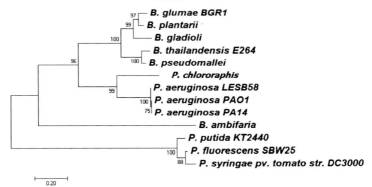

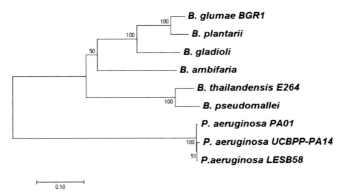

**Fig. 4** Phylogenetic analysis of rhamnolipid biosynthetic enzymes. The phylogenetic tree was inferred using the Neighbor-Joining method and the evolutionary distances were computed using the Poisson correction method. The percentage of replicate trees in which the associated taxa clustered together in the bootstrap test (1000 replicates) are given

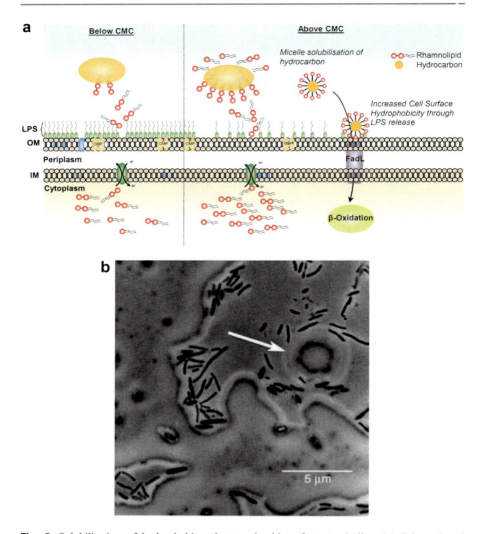

**Fig. 5** Solubilisation of hydrophobic substrates by biosurfactant micelles. (**a**) Schematic of hydrocarbon solubilisation by rhamnolipid (RL) biosurfactants. At concentrations below the critical micelle concentration (CMC) monolayers of RL accumulate at the aqueous-hydrocarbon interface. When RL concentrations is above the CMC the hydrocarbon readily partitions into the hydrophobic core of the micelle thus increasing hydrocarbon bioavailability through solubilisation. RLs also induce cell surface changes that increase cell hydrophobicity via the release of cell associated lipopolysaccharides (LPS). (**b**) Photomicrograph of a mixed population of bacteria growing in the presence of hydrocarbons. Cells tend to occur at the interface between the aqueous and hydrophobic phases and a micelle (arrow) can be also observed

concentrations below the CMC. Sotirova et al. 2009 have shown significant decrease in the major OMPs in the presence of RLs below the CMC. They attributed this change in membrane organisation to the binding of RL monomers which may facilitate uptake of hydrophobic compounds in certain regions of the membrane.

## 4  The Lipopeptide Biosurfactants

Lipopeptide (LPs) are a structurally distinct group of biosurfactants produced by bacteria, mainly *Bacillus* and *Pseudomonas* species but also *Streptomyces* and other Actinobacteria, and fungi such as *Aspergillus* and *Fusarium* (Raaijmakers et al. 2010; Mnif and Ghribi 2015). The diversity of lipopeptide producing organisms and lipopeptide encoding genes has increased remarkably during the last years due to the contribution of molecular approaches such as genome mining and PCR-based detection methods (de Bruijn et al. 2007; Rokni-Zadeh et al. 2011; Domingos et al. 2015). Thus, gene clusters for the biosynthesis of novel lipopeptides were recently discovered in the genome sequences of well-known organisms such as *Bacillus subtilis* (i.e., locillomycins) (Luo et al. 2015a), but also in uncommon organisms such as *Kibdelosporangium* sp., a rare actinobacterium (Ogasawara et al. 2015), and *Janthinobacterium* sp., a soil bacterium (Graupner et al. 2015).

The biosynthetic pathway of lipopeptides, based on non-ribosomal peptide synthetases (NRPSs) – large multi-enzymes that carry out the sequential incorporation of amino acids into the nascent peptide – gives rise to a high diversity of molecular structures, yet built around a conserved template. All lipopeptides consist of a short cyclic or linear oligopeptide – the hydrophilic moiety of the molecule – linked to a fatty acid tail – the hydrophobic moiety, but they can greatly vary in the number, type and order of the amino acid residues as well as the length and branching of the chain (Raaijmakers et al. 2010). Minor changes in the structural organisation can significantly affect the overall physico-chemical properties of the molecule, hence its activity. To add further variability, lipopeptides are often synthesised as mixtures of several congeners and isoforms as, for example, surfactin, bacillomycin L, fengycin and locillomycin that are co-produced in *B. subtilis* and contribute each to a specific trait of the organism phenotype (Luo et al. 2015b).

Lipopeptides have been extensively studied for their broad spectrum of antimicrobial activities, which are due to their exceptional ability to interact with and disrupt cell membranes (Hamley 2015; Cochrane and Vederas 2016). However, lipopeptides are also potent biosurfactants that can stimulate the mobilisation, solubilisation and emulsification of hydrophobic compounds including hydrocarbons, chemicals in general and vegetable or waste oils.

Initially, the use of lipopeptides was limited to microbial enhanced oil recovery (MEOR) and biodegradation. Lipopeptides are able to reduce the interfacial tension (IFT) between crude oil and the water phase to ultra-low values, e.g., below 0.1 mN/m, which is necessary to overcome the capillary forces that trap the oil in porous media, thus enabling its mobilisation and release (Youssef et al. 2007). Both bench-scale and in-situ lipopeptide production by strains of *Bacillus* spp. have proven successful in improving the oil recovery also from wells close to their production limits (Youssef et al. 2013; Al-Wahaibi et al. 2014).

Solubilisation, i.e., the increase in aqueous solubility of hydrophobic compounds, is instead the main mechanism of biodegradation. When present above their critical micelle concentration (CMC), biosurfactants induce a drastic increase in solubilisation of hydrocarbons, heavy metals and chemicals via the micellisation process (Sarubbo et al. 2015), as it was shown to occur for many lipopeptides, including surfactin and fengycin in *B. subtilis* (Singh and Cameotra 2013) and viscosin, massetolide A, putisolvin, and amphisin in *Pseudomonas* spp. (Bak et al. 2015).

Finally, lipopeptides can promote the formation of emulsions, including micro-emulsions, which hold a tremendous potential for cutting-edge applications such as advanced biofuels and drug-delivery systems. The underlying physico-chemical properties have been studied in particular for surfactin, one of the most powerful biosurfactants known so far, capable of reducing the surface tension of water and air from 72 to 27 mN/m at a concentration as low as 10 μM (Seydlová and Svobodová 2008). The values of Hydrophilic-Lipophilic-Balance (HLB) of 10–12 indicate that surfactin can favour the formation of oil-in-water (O/W) microemulsions, and the Critical Packing Parameter (CPP) ($<1/3$) suggests that spherical micelles is the preferred geometry for surfactin self-assembly (Gudiña et al. 2013). Background knowledge of these parameters is essential to guide the further development of novel applications.

## 5    Biosurfactants Produced by Actinobacteria

The class of Actinobacteria, comprising 5 subclasses and 9 orders for a total of 54 families, contains a very large diversity of organisms capable of producing biosurfactants of various types (Khan et al. 2012; Kügler et al. 2015). *Rhodococcus*, *Arthrobacter*, *Mycobacterium*, *Nocardia* and *Corynebacterium* genera produce pre-dominantly trehalose-containing glycolipids, where the hydrophilic moiety consists of trehalose – a two α-glucose unit sugar – and the hydrophobic tail can be either mycolic acids or ester fatty acids and can vary greatly for length (up to $C_{90}$), branching, substitutions and saturation degree. In some strains (e.g., *Arthrobacter parafineous*, *Brevibacterium* sp. and *Nocardia* sp.), the trehalose can be replaced by other sugars such as sucrose and fructose. Most of these trehalose lipids remain cell-bound, making the cell surface highly hydrophobic, while only a minor fraction seems to be released extracellularly (Lang and Philp 1998). Moreover, some Actinobacteria including *Streptomyces* sp., *Actinoplanes* sp. and *Rhodococcus* sp. have also been found capable of synthesising lipopeptide biosurfactants highly variable in both the peptide ring and the hydrophobic tail (Kügler et al. 2015).

Despite such remarkable diversity of producer strains and biosurfactant types, research has only focussed on a few model organisms. *Rhodococcus* sp., for example, has been studied extensively for the ability to degrade hydrocarbons and pollutants in a broad range of situations. Because the cell surface is hydrophobic, the main uptake mode occurs via direct contact with the hydrophobic substrates, while solubilisation mediated by free biosurfactants would have a secondary role (Bouchez-Naïtali and Vandecasteele 2008). It was recently shown that trehalose synthesis through alkane metabolism and gluconeogenesis is an important pathway for biosurfactant synthesis in *Rhodococcus* sp. strain SD-74 that involves three, so far identified, enzymes: the alkane monooxygenase AlkB responsible for the initial step of alkane oxidation, the fructose-biphosphate aldolase Fda for the trehalose backbone synthesis and finally a putative acyl-coenzyme A transferase TlsA that converts trehalose to trehalose lipids (Inaba et al. 2013). Several other genes involved in the biosynthesis of trehalose lipids have been discovered in the genome of *Rhodococcus ruber* IEGM 231, including a fatty acid synthase I, a cyclopropane mycolic acid synthase, three mycolyltransferases, a maltooligosyl trehalose synthase, a maltooligosyl trehalose hydrolase and a trehalose synthase (Ivshina et al. 2014).

Similarly well studied is *Gordonia* sp., a member of the *Corynebacterium/Mycobacterium/Nocardia* (CMN) complex. A dual-step uptake of hydrophobic substrates has been suggested for *Gordonia* strains growing on *n*-hexadecane, in which first, during early exponential phase, cells are highly hydrophobic and adhere directly onto the hydrocarbons, then, at a later phase of growth, they become more hydrophilic and access biosurfactant-solubilised droplets of dispersed hydrocarbons (Franzetti et al. 2008). *Gordonia* is known as a potent biodegrader, capable to grow and use as carbon source a large variety of highly hydrophobic and recalcitrant compounds, including solid alkanes (e.g., hexatriacontane with 36 carbon atoms), plastic additives (e.g., phthalate esters) and natural or synthetic isoprene rubber (e.g., *cis*-1,4-polyisoprene) (Lo Piccolo et al. 2011; Drzyzga 2012).

Widespread amongst Actinobacteria is also their ability to produce biosurfactants from agro-industrial wastes or inexpensive oil derivatives. Several substrates have proven suitable to support biosynthesis of trehalose lipid in various organisms, for example in *Rhodococcus* sp. and *Tsukamurella* sp. growing on sunflower frying oil, rapeseed oil or glyceryltrioleate (White et al. 2013; Ruggeri et al. 2009; Kügler et al. 2014) and in the marine *Brachybacterium paraconglomeratum* on oil seed cake, a pressed mixture of the residues after oil extraction (Kiran et al. 2014). It can be anticipated that the use of renewable material for the synthesis of added value products such as biosurfactants is a growing trend in current and near future research.

# 6   Glycolipids Produced by Eukaryotes

The best known BSs from eukaryotes are the glycolipids from non-pathogenic yeasts, a heterogeneous mixture of structurally diverse compounds that are produced in relatively high yields (claimed to be as high as 400 g/L). Glycolipids from *Starmerella*, *Candida* and *Pseudozyma* sp. are the most extensively studied eukaryotic BSs with a wide range of applications (Roelants et al. 2014) and are currently the most promising for competitive large scale industrial production. Sophorolipids (SLs) are hydrophobic surfactants that comprise a hydrophilic disaccharide sophorose (2-$O$-β-D-glucopyranosyl-D-glucopyranose) β-glycosidically linked to a long chain hydroxyl fatty acid, most commonly $C_{18:1}$. SLs are synthesised in a variety of compounds with >40 structural isomers described in the literature (Van Bogaert et al. 2007). SLs vary by degree of acetylation (mono-, di- or nonacetylated) at $C_6{}'$ and $C_6{}''$, acidic (containing a free carboxylic group) or lactonic (forming a macrocyclic lactone ring at $C_4{}''$) form (Fig. 1d), fatty acid length ranging between $C_{16}$ and $C_{18}$ and saturation. All these structural variations can alter the physicochemical properties of SLs (e.g., di-acetylated acidic SLs are more hydrophilic compared to nonacetylated) which makes them suitable for different applications.

The complete biosynthetic gene cluster has been described in detail for sophorolipids (SLs) in *S. bombicola* (Saerens et al. 2011a, b, c; Van Bogaert et al. 2013). Similar to the glycolipid produced by *Pseudomonas* sp. and *Burkholderia* the fatty acid composition is to a certain degree regulated by the first enzyme in the SL biosynthetic pathway, cytochrome p450 monooxygenase CYP52M1 which has been shown to preferentially hydroxylate oleic acid ($C_{18:1}$) at the terminal or sub-terminal carbon (Huang et al. 2014; Saerens et al. 2015). The hydroxylated fatty acid is the substrate for two UDP-glucosyltransferases (UGTA1 and UGTB1) which covalently link two glucose molecules to produce acidic SLs (Saerens et al. 2011a, b, 2015).

Recently Ciesielska et al. (2016) characterised the lactonesterase responsible for the esterification of the carboxyl group from the hydroxylated fatty acid to the second glucose molecule producing lactonic sophorolipids. Recent advances in our understanding of SL biosynthesis and regulation has enabled recombinant engineering to produce tailor made compositions of SLs for specific applications (Solaiman et al. 2014) and also facilitated the discovery of novel sophorolipids from the non-pathogenic yeasts *Starmerella* and *Candida* sp. (Kurtzman et al. 2010; Price et al. 2012). Van Bogaert et al. 2016 have produced a novel bolaform SL from engineered *S. boimbicola* that is deficient in both the lactonesterase and acetyltransferase. The bolaform SLs are highly water soluble due to the presence of two hydrophilic moieties compared to their nonacetylated hydrophobic counterparts. They have also produced several recombinant engineered strains that synthesise specific compositions of SLs. The next step for SL production will be fine tuning the production process for specifically engineered strains to produce yields comparable or higher than native producers.

Mannosylerythritol lipids (MELs) are surface active glycolipids synthesised by a variety of phylogenetically related basidiomycetous yeasts with *Pseudozyma* sp. and *Ustilago* sp. being the most common producers. MELs are synthesised as a diverse mixture of congeners differing in length of fatty acid chains, degree of saturation and acetylation. MELs are classified by their acetylation at the $C_4$ and $C_6$ positions of the mannose sugar group on the hydrophilic head (Fig. 1c). The number of acetyl groups has a significant impact when considering the physico-chemical and self-assembly properties of MELs, where the more hydrophobic di-acetylated MEL-A has a much lower critical micelle concentration (CMC) compared to the more hydrophilic mono-acetylated counter parts MEL-B and MEL-C (Yu et al. 2015). These significant differences make MELs suitable for a wide range of applications (Morita et al. 2015), for example MEL-A significantly increases the efficiency of gene transfection via membrane fusion while MEL-B/MEL-C have no such effect (Inoh et al. 2010, 2011).

The biosynthetic pathway for mannosylerythritol lipids (MEL) production was originally described in *Ustilago maydis* under nitrogen limiting conditions (Hewald et al. 2006), with five MEL biosynthetic genes clustering together on a single operon. Using comparative genomics Morita et al. (2014) have identified and characterised the MEL biosynthetic gene cluster also in the phylogenetically related yeast *Pseudozyma antarctica*. In contrast to SLs, the first step in MEL biosynthesis proceeds by the stereospecific mannosylation of erythritol catalysed by Emt1 (erythritol/mannose transferase), this is the substrate for the acyltransferases Mac1 and Mac2, which specifically acylate the mannose at positions $C_2$ and $C_3$. Similar to all other biosurfactant biosynthetic pathways discussed here, MELs are regioselectively acylated (at $C_2$ and $C_3$ on mannose) by the affinity of Mac1 for short chain fatty acids ($C_2$–$C_8$) and Mac2 for medium to long chain fatty acids ($C_{10}$–$C_{18}$) (Hewald et al. 2006).

Recently a number of genomes have been published relating to MEL in *Pseudozyma* sp. including *P. antarctica* T-34 and JCM10317 (Morita et al. 2013a; Saika et al. 2014), *P. aphidis* DSM70725 (Lorenz et al. 2014), and *P. hubeiensis* SY62 (Konishi et al. 2013). Similar to SL production, advances in our understanding of the MEL biosynthetic pathway is paving the way for comparative genomics to identify novel taxonomically related producers with unique structural variations (e.g., diastereomer type of MEL-B biosynthesis in *P. tsukubaensis*) (Saika et al. 2016). Morita et al. (2014) have published the transcriptomic profile of the MEL hyper-producer *P. antarctica* providing fundamental insights into regulation of the MEL biosynthetic cluster under specific inducing conditions (i.e., excess of hydrophobic substrates). Combining the transcriptional expression profile of MEL biosynthetic genes with the extracellular composition of MELs (e.g., mass spectrometry) will help elucidate the metabolic state during production, this will be essential for developing a highly efficient and tailored production process. Currently the glycolipids from non-pathogenic yeasts offer the greatest potential for large scale production of biosurfactants at an economic scale that could be competitive with current synthetic production processes. Current advances in the metabolic engineering strategies for these biosurfactants can lead to tailor made production pipelines where biosurfactants with specific compositions and modifications will be produced specifically for the desired application. The future of such tailored pipelines will need to be paralleled by efficient downstream recovery of the desired biosurfactants.

## 7 Applications of Biosurfactant-Enhanced (Bio)availability of Hydrophobic Substrates

Numerous biotechnologies have been developed that are based on the capability of biosurfactants to interact with hydrophobic substrates and cause phenomena such as decrease of surface and interfacial tensions, solubilisation, dispersion, emulsification, desorption and wetting. Applications of biosurfactants can cover a wide variety of fields, with the main areas being environmental bioremediation, microemulsion-based technologies and conversion of renewable resources.

**Environmental Bioremediation** Biosurfactants have been traditionally tested for their activity to support and enhance microbial degradation of hydrocarbons and hydrophobic pollutants. Being less toxic and more biodegradable than synthetic surfactants offers important advantages for in-situ applications, thus biosurfactants can be released with less risk in the environment. Moreover, the capability of natural communities of degrader microorganisms to produce biosurfactants can be exploited directly in the natural environment via biostimulation techniques (Ławniczak et al. 2013). Marine oil spills have been successfully treated in lab-scale or mesocosm experiments where crude oil degradation rates were enhanced through the supply of exogenous rhamnolipids (Chen et al. 2013; Nikolopoulou et al. 2013; Tahseen et al. 2016) or consortia of indigenous microorganisms were stimulated to produce biosurfactants such as rhamnolipids and sophorolipids (Antoniou et al. 2015). Surfactin, because of its long fatty acid chain and hydrophobic amino acid ring, has poorer water solubility, which limits its applications in water systems. However, a surfactin-derivative fatty acyl-glutamate (FA-Glu), consisting of a β-hydroxy fatty acid chain of $C_{12}$–$C_{17}$ linked to a single glutamic acid, was produced in an engineered *B. subtilis* strain (Reznik et al. 2010) and was positively tested for use in marine oil spills, showing much higher water-solubility and reduced toxicity (Marti et al. 2014).

Being highly hydrophobic, when adsorbed onto soil particles, hydrocarbons, chemicals and heavy metals are very resistant to removal. Enhanced removal of such contaminants can be achieved via soil-washing techniques that more and more often use biosurfactants (Lau et al. 2014). Lipopeptides from both *Bacillus* and *Pseudomonas* strains have shown high efficiency of removal of crude oil (from 60% to 90%) and heavy metals (>40% for cadmium and lead) when used as washing agents (Singh and Cameotra 2013; Xia et al. 2014). A rhamnolipid solution similarly worked well to remove petroleum hydrocarbons (>80%) and was found to further support the biodegradation of organic compounds by a consortium of degrader bacteria (Yan et al. 2011).

**Microemulsion-Based Technologies** Microemulsions are thermodynamically stable dispersions of oil, water and surfactants, and are highly desirable in applications such as enhanced oil recovery (Elshafie et al. 2015), biodiesel formulation, drug delivery and food and cosmetic products. Rhamnolipid biosurfactants have shown good performance in generating glycerol-in-diesel microemulsions that are stable for over 6 months, thus opening up the possibility to produce greener biofuels that have reduced combustion emissions and also make use of an industrial inexpensive by-product, i.e., glycerol (Leng et al. 2015). In addition, due to their biocompatibility and low toxicity, biosurfactants are ideal candidates for pharmaceutical and cosmetic applications. Both rhamnolipids and sophorolipids have proven to be effective in producing lecithin-based microemulsions – having high affinity for the phospholipids of cell membranes – with various oils. Microemulsions with isopropyl myristate were stable over a wide range of temperatures (10–40 °C) and electrolyte concentrations (0.9–4.0% w/v), which is very attractive for cosmetic and also drug delivery applications (Nguyen et al. 2010; Rodrigues 2015).

**Conversion of Renewable Resources** Despite the substantial research and huge number of reports about potential applications and benefits from the use of biosurfactants, the commercialisation of these compounds remains at a very early stage. One of the main constraints limiting the diffusion of biosurfactants is their high production costs. One strategy to overcome this problem is the use of inexpensive waste material as substrate for the growth of biosurfactant-producing microorganisms, with the additional important advantage that renewable resources can be converted into high added value products (Makkar et al. 2011; Banat et al. 2014). In this context a lot of research has been done in the last few years, and various waste oils have been demonstrated to support effectively biosurfactant production. The existing surplus of glycerol, for example, derived as co-product from the oleochemical industry and biodiesel production in particular, makes it one of the preferred substrates. However, it is always necessary to bear in mind that a low value waste material rapidly acquires a commercial value once an application for it has been discovered and a demand developed. The other problem with glycerol produced as a by-product from other processes is its variable purity and contamination by other components. Glycerol has, however, been used to stimulate synthesis of various types of biosurfactants at standard yields, including sophorolipids (over 40 g/L) in the yeast *Starmerella bombicola* ATCC2214, glycolipids (32.1 g/L) in *Ustilago maydis*, mannosylerythritol lipids (16.3 g/L) in *Pseudozyma antarctica* JCM 10317 (Nicol et al. 2012), rhamnolipids (in the range 1.0–8.5 g/L) in strains of *Pseudomonas aeruginosa* (Henkel et al. 2012; Perfumo et al. 2013), and surfactin lipopeptide (in the range 230–440 mg/L) in *B. subtilis* (de Faria et al. 2011; Sousa et al. 2012). Olive oil mill waste is another abundant waste from the extraction of olive oil that is of environmental concern due to the difficulty of further processing. Recent work has shown that oil mill waste can be recycled and used as substrate to produce both rhamnolipid and surfactin biosurfactants from *Pseudomonas* and *Bacillus* respectively, and that considerably higher yields (299 mg/L and 25.5 mg/L respectively) can be obtained when using a hydrolysed form of it (Moya-Ramírez et al. 2015, 2016; Radzuan et al. 2017). A number of other inexpensive materials, including frying oils and fats, and vegetable oils (e.g., rapeseed oil, sunflower oil, soybean oil, canola oil, palm oil) have proven to be an attractive alternative paving the way towards a cost-competitive and environmentally friendly production of biosurfactants (Banat et al. 2014).

## 8      Current State and Research Needs

The state of our knowledge concerning the production of microbial biosurfactants has now reached the stage where we can identify a number of key areas for future research and development. One area that has already attracted some interest is the expression of biosurfactant synthetic genes in a non-pathogenic heterologous host organism. Achieving expression of the small number of genes directly involved in the synthesis of, for example, the glycolipids from bacteria or yeasts is not particularly difficult to achieve. What is more difficult is to provide in the host organism a metabolic environment able to provide the levels of precursors sufficient to achieve high yields of the target biosurfactants.

Research on the production of biosurfactants is currently advancing strongly into the area of gene expression studies during the growth cycle of the organism and this is providing important information on the regulation systems operating in these organisms. These data have been valuable in determining how the fermentation production of biosurfactants may be optimised since, although it is often considered that fermentation substrate costs and downstream processing will be cost limiting factors for production, in practice extended fermentation periods require high energy inputs and can be a major cost for the production process. Therefore any reduction in the fermentation duration will be an important step in developing an economic production process. The understanding of how biosurfactant production can be further optimised will depend on the use of metabolomic and systems approaches to the whole problem.

One of the most positive aspects of microbial biosurfactant production is the fact that, in most organisms, the final product is exported from the cells into the growth medium making at least the first steps in separation and purification relatively straightforward. Interestingly, however, we have little information on whether there are specific transporter systems for each biosurfactant or whether general transporter systems are employed. This knowledge will be important as we move to heterologous expression systems so that we can ensure the products are effectively exported from the cells in the same way that they are from the original producers.

The whole field of biosurfactants is an actively expanding one with many new opportunities for both basic research and industrial exploitation.

# References

Abalos A, Pinazo A, Infante MR, Casals M, García F, Manresa A (2001) Physicochemical and antimicrobial properties of new rhamnolipids produced by *Pseudomonas aeruginosa* AT10 from soybean oil refinery wastes. Langmuir 17:1367–1371

Abdel-Mawgoud AM, Lépine F, Déziel E (2010) Rhamnolipids: diversity of structures, microbial origins and roles. Appl Microbiol Biotechnol 86:1323–1336

Aguirre-Ramirez M, Medina G, Gonzalez-Valdez A, Grosso-Becerra V, Soberon-Chavez G (2012) The *Pseudomonas aeruginosa* rmlBDAC operon, encoding dTDP-L-rhamnose biosynthetic enzymes, is regulated by the quorum-sensing transcriptional regulator RhlR and the alternative sigma factor S. Microbiology 158:908–916

Al-Tahhan RA, Sandrin TR, Bodour AA, Maier RM (2000) Rhamnolipid-induced removal of lipopolysaccharide from *Pseudomonas aeruginosa*: effect on cell surface properties and interaction with hydrophobic substrates. Appl Environ Microbiol 66:3262–3268

Al-Wahaibi Y, Joshi S, Al-Bahry S, Elshafie A, Al-Bemani A, Shibulal B (2014) Biosurfactant production by Bacillus subtilis B30 and its application in enhancing oil recovery. Colloids Surf B Biointerfaces 114:324–333

Andrä J, Rademann J, Howe J, Koch MHJ, Heine H, Zähringer U, Brandenburg K (2006) Endotoxin-like properties of a rhamnolipid exotoxin from *Burkholderia (Pseudomonas) plantarii*: immune cell stimulation and biophysical characterization. Biol Chem 387:301–310

Antoniou E, Fodelianakis S, Korkakaki E, Kalogerakis N (2015) Biosurfactant production from marine hydrocarbon-degrading consortia and pure bacterial strains using crude oil as carbon source. Front Microbiol 6:274

Arino S, Marchal R, Vandecasteele J-P (1996) Identification and production of a rhamnolipidic biosurfactant by a *Pseudomonas* species. Appl Microbiol Biotechnol 45:162–168

Bak F, Bonnichsen L, Jørgensen NO, Nicolaisen MH, Nybroe O (2015) The biosurfactant viscosin transiently stimulates n-hexadecane mineralization by a bacterial consortium. Appl Microbiol Biotechnol 99:1475–1483

Banat IM, Franzetti A, Gandolfi I, Bestetti G, Martinotti MG, Fracchia L, Smyth TJ, Marchant R (2010) Microbial biosurfactants production, applications and future potential. Appl Microbiol Biotechnol 87:427–444

Banat IM, Satpute SK, Cameotra SS, Patil R, Nyayanit NV (2014) Cost effective technologies and renewable substrates for biosurfactants' production. Front Microbiol 5:697

Bodour AA, Drees KP, Maier RM (2003) Distribution of biosurfactant-producing bacteria in undisturbed and contaminated arid southwestern soils. Appl Environ Microbiol 69:3280–3287

Bouchez-Naïtali M, Vandecasteele JP (2008) Biosurfactants, an help in the biodegradation of hexadecane? The case of *Rhodococcus* and *Pseudomonas* strains. World J Microbiol Biotechnol 24:1901–1907

Cai Q, Zhang B, Chen B, Zhu Z, Lin W, Cao T (2014) Screening of biosurfactant producers from petroleum hydrocarbon contaminated sources in cold marine environments. Mar Pollut Bull 86:402–410

Campos JM, Montenegro Stamford TL, Sarubbo LA, de Luna JM, Rufino RD, Banat IM (2013) Microbial biosurfactants as additives for food industries. A review. Biotechnol Prog 29:1097–1108

Chen Q, Bao M, Fan X, Liang S, Sun P (2013) Rhamnolipids enhance marine oil spill bioremediation in laboratory system. Mar Pollut Bull 71:269–275

Ciesielska K, Roelants SLKW, Van Bogaert INA, De Waele S, Vandenberghe I, Groeneboer S, Soetaert W and Devreese B (2016) Characterization of a novel enzyme – Starmerella bombicola lactone esterase (SBLE) – responsible for sophorolipid lactonization. Appl Microbiol Biotechnol 100:9529–9541

Cochrane SA, Vederas JC (2016) Lipopeptides from Bacillus and Paenibacillus spp.: a gold mine of antibiotic candidates. Med Res Rev 36:4–31

Darvishi P, Ayatollahi S, Mowla D, Niazi A (2011) Biosurfactant production under extreme environmental conditions by an efficient microbial consortium, ERCPPI-2. Colloids Surf B: Biointerfaces 84:292–300

de Almeida DG, Da Silva RCFS, Luna JM, Rufino RD, Santos VA, Banat IM, Sarubbo LA (2016) Biosurfactants: promising molecules for petroleum biotechnology advances. Front Microbiol. doi:10.3389/fmicb.2016.01718

de Bruijn I, de Kock MJ, Yang M, de Waard P, van Beek TA, Raaijmakers JM (2007) Genome-based discovery, structure prediction and functional analysis of cyclic lipopeptide antibiotics in *Pseudomonas* species. Mol Microbiol 63:417–428

de Faria AF, Teodoro-Martinez DS, de Oliveira Barbosa GN, Vaz BG, Silva IS, Garcia JS, Tótola MR, Eberlin MN, Grossman M, Alves OL, Durrant LR (2011) Production and structural characterization of surfactin (C 14/Leu 7) produced by *Bacillus subtilis* isolate LSFM-05 grown on raw glycerol from the biodiesel industry. Process Biochem 46:1951–1957

Déziel E, Lépine F, Milot S, Villemur R (2000) Mass spectrometry monitoring of rhamnolipids from a growing culture of *Pseudomonas aeruginosa* strain 57RP. Biochim Biophys Acta Mol Cell Biol Lipids 1485:145–152

Dhasayan A, Selvin J, Kiran S (2015) Biosurfactant production from marine bacteria associated with sponge *Callyspongia diffusa*. Biotech 5:443–454

Díaz de Rienzo M, Stevenson P, Marchant R, Banat IM (2016a) Antibacterial properties of biosurfactants against selected Gram-positive and – negative bacteria. FEMS Microbiol Let. doi:10.1093/femsle/fnv224

Díaz de Rienzo M, Stevenson P, Marchant R, Banat IM (2016b) *P. aeruginosa* biofilm disruption using microbial biosurfactants. J Appl Microbiol 120:868–876. doi:10.1111/jam.13049

Díaz de Rienzo M, Stevenson P, Marchant R, Banat IM (2016c) Effect of biosurfactants on *Pseudomonas aeruginosa* and *Staphylococcus aureus* biofilms in a BioFlux channel. Appl Microbiol Biotechnol 100:5773–5779. doi:10.1007/s00253-016-7310-5

Dobler L, Vilela LF, Almeida RV, Neves BC (2016) Rhamnolipids in perspective: gene regulatory pathways, metabolic engineering, production and technological forecasting. New Biotechnol 33:123–135

Domingos DF, Dellagnezze BM, Greenfield P, Reyes LR, Melo IS, Midgley DJ, Oliveira VM (2013) Draft genome sequence of the biosurfactant-producing bacterium *Gordonia amicalis* strain CCMA-559, isolated from petroleum-impacted sediment. Genome Announc 1: e00894–13. e00894–13

Domingos DF, de Faria AF, de Souza Galaverna R, Eberlin MN, Greenfield P, Zucchi TD, Melo IS, Tran-Dinh N, Midgley D, de Oliveira VM (2015) Genomic and chemical insights into biosurfactant production by the mangrove-derived strain *Bacillus safensis* CCMA-560. Appl Microbiol Biotechnol 99:3155–3167

Donio MBS, Ronica FA, Viji VT, Velmurugan S, Jenifer JSCA, Michaelbabu M, Dhar P, Citarasu T, Abalos A, Pinazo A et al (2013) *Halomonas* sp. BS4, a biosurfactant producing halophilic bacterium isolated from solar salt works in India and their biomedical importance. Springerplus 2:149

Drzyzga O (2012) The strengths and weaknesses of *Gordonia*: a review of an emerging genus with increasing biotechnological potential. Crit Rev Microbiol 38:300–316

Dubeau D, Déziel E, Woods DE, Lépine F, Jarvis F, Johnson M, Edwards J, Hayashi J, Kitamoto D, Isoda H et al (2009) *Burkholderia thailandensis* harbors two identical rhl gene clusters responsible for the biosynthesis of rhamnolipids. BMC Microbiol 9:263

Elazzazy AM, Abdelmoneim TS, Almaghrabi OA (2015) Isolation and characterization of biosurfactant production under extreme environmental conditions by alkali-halo-thermophilic bacteria from Saudi Arabia. Saudi J Biol Sci 22:466–475

Elshafie AE, Joshi SJ, Al-Wahaibi YM, Al-Bemani AS, Al-Bahry SN, Al-Maqbali D, Banat IM (2015) Sophorolipids production by *Candida bombicola* ATCC 22214 and its potential application in microbial enhanced oil recovery. Front Microbiol. doi:10.3389/fmicb.2015.01324

Elshikh M, Marchant R, Banat IM (2016) Biosurfactants: promising bioactive molecules for oral-related health applications. FEMS Microbiol Lett 363(18):fnw213. doi:10.1093/femsle/fnw213

Fracchia L, Cavallo M, Martinotti MG, Banat IM (2012) Biosurfactants and bioemulsifiers biomedical and related applications–present status and future potentials, Chapter 14. In: Ghista DN (ed) Biomedical science, engineering and technology. InTech Europe, University Campus STeP Ri, Rijeka, Croatia. pp 326–335. ISBN 978-953-307-471-9

Fracchia L, Ceresa C, Franzetti A, Cavallo M, Gandolfi I, Van Hamme ., Gkorezis P, Marchant R., Banat IM (2014) Industrial applications of biosurfactants. Biosurfactants: production and utilization – processes, technologies, and economics, Chapter 12. In: Kosaric N, Sukan FV (eds) Surfactant science series, vol 159. CRC press, United States. pp 245–260. ISBN 978-14665-9669-6

Fracchia L, Banat JJ, Cavallo M, Ceresa C, Banat IM (2015) Potential therapeutic applications of microbial surface-active compounds. AIMS Bioeng 2(3):144–162. doi:10.3934/bioeng.2015.3.144

Franzetti A, Bestetti G, Caredda P, La Colla P, Tamburini E (2008) Surface-active compounds and their role in the access to hydrocarbons in *Gordonia* strains. FEMS Microbiol Ecol 63:238–248

Funston SJ, Tsaousi K, Rudden M, Smyth TJ, Stevenson PS, Marchant R, Banat IM (2016) Characterising rhamnolipid production in Burkholderia thailandensis E264, a non-pathogenic producer. Appl Microbiol Biotechnol 100:7945–7956

Ganguly S, Jimenez-Galisteo G, Pletzer D, Winterhalter M, Benz R, Viñas M (2016) Draft genome sequence of *Dietzia maris* DSM 43672, a gram-positive bacterium of the Mycolata Group. Genome Announc 4:e00542–16

Graupner K, Lackner G, Hertweck C (2015) Genome sequence of mushroom soft-rot pathogen *Janthinobacterium agaricidamnosum*. Genome Announc 3, pii: e00277–15. doi:10.1128/genomeA.00277-15

Gudiña EJ, Rangarajan V, Sen R, Rodrigues LR (2013) Potential therapeutic applications of biosurfactants. Appl Mar Drugs 14:38

Gunther NW, Nunez A, Fett W, Solaiman DKY (2005) Production of rhamnolipids by *Pseudomonas* chlororaphis, a nonpathogenic bacterium. Appl Environ Microbiol 71:2288–2293

Gutierrez T, Banat IM (2014) Isolation of glycoprotein bioemulsifiers produced by marine bacteria. In: McGenity et al (eds) Hydrocarbon and lipid microbiology protocols, Springer protocols handbooks. Springer, Berlin. doi:10.1007/8623_2014_1

Haba E, Abalos A, Jáuregui O, Espuny MJ, Manresa A (2003a) Use of liquid chromatography-mass spectroscopy for studying the composition and properties of rhamnolipids produced by different strains of *Pseudomonas aeruginosa*. J Surfactant Deterg 6:155–161

Haba E, Pinazo A, Jauregui O, Espuny MJ, Infante MR, Manresa A (2003b) Physicochemical characterization and antimicrobial properties of rhamnolipids produced by *Pseudomonas aeruginosa* 47 T2 NCBIM 40044. Biotechnol Bioeng 81:316–322

Hamley IW (2015) Lipopeptides: from self-assembly to bioactivity. Chem Commun 51:8574–8583

Häussler S, Nimtz M, Domke T, Wray V, Steinmetz I (1998) Purification and characterization of a cytotoxic exolipid of *Burkholderia pseudomallei*. Infect Immun 66:1588–1593

Henkel M, Müller MM, Kügler JH, Lovaglio RB, Contiero J, Syldatk C, Hausmann R (2012) Rhamnolipids as biosurfactants from renewable resources: concepts for next-generation rhamnolipid production. Process Biochem 47:1207–1219

Hewald S, Linne U, Scherer M, Marahiel MA, Kamper J, Bolker M (2006) Identification of a gene cluster for biosynthesis of mannosylerythritol lipids in the basidiomycetous fungus *Ustilago maydis*. Appl Environ Microbiol 72:5469–5477

Heyd M, Kohnert A, Tan T-H, Nusser M, Kirschhöfe F, Brenner-Weiss G, Franzreb M, Berensmeier S (2008) Development and trends of biosurfactant analysis and purification using rhamnolipids as an example. Anal Bioanal Chem 391:1579–1590

Hörmann B, Müller MM, Syldatk C, Hausmann R (2010) Rhamnolipid production by *Burkholderia plantarii* DSM 9509 T. Eur J Lipid Sci Technol 112:674–680

Hošková M, Ježdík R, Schreiberová O, Chudoba J, Šír M, Čejková A, Masák J, Jirků V, Řezanka T (2015) Structural and physiochemical characterization of rhamnolipids produced by *Acinetobacter calcoaceticus*, Enterobacter asburiae and *Pseudomonas aeruginosa* in single strain and mixed cultures. J Biotechnol 193:45–51

Howe J, Bauer J, Andrä J, Schromm AB, Ernst M, Rössle M, Zähringer U, Rademann J, Brandenburg K (2006) Biophysical characterization of synthetic rhamnolipids. FEBS J 273:5101–5112

Huang F-C, Peter A, Schwab W (2014) Expression and characterization of CYP52 genes involved in the biosynthesis of sophorolipid and alkane metabolism from *Starmerella bombicola*. Appl Environ Microbiol 80:766–776

Inaba T, Tokumoto Y, Miyazaki Y, Inoue N, Maseda H, Nakajima-Kambe T, Uchiyama H, Nomura N (2013) Analysis of genes for succinoyl trehalose lipid production and increasing production in *Rhodococcus* sp. strain SD-74. Appl Environ Microbiol 79:7082–7090

Inoh Y, Furuno T, Hirashima N, Kitamoto D, Nakanishi M (2010) The ratio of unsaturated fatty acids in biosurfactants affects the efficiency of gene transfection. Int J Pharm 398:225–230

Inoh Y, Furuno T, Hirashima N, Kitamoto D, Nakanishi M (2011) Rapid delivery of small interfering. RNA by biosurfactant MEL-A-containing liposomes. Biochem Biophys Res Commun 414(3):635–640

Ivshina IB, Kuyukina MS, Krivoruchko AV, Barbe V, Fischer C (2014) Draft genome sequence of propane- and butane-oxidizing Actinobacterium *Rhodococcus ruber* IEGM 231. Genome Announc 2, pii: e01297–14

Jackson SA, Borchert E, O'Gara F, Dobson AD (2015) Metagenomics for the discovery of novel biosurfactants of environmental interest from marine ecosystems. Curr Opin Biotechnol 33:176–182

Kang HS, Charlop-Powers Z, Brady SF (2016) Multiplexed CRISPR/Cas9- and TAR-mediated promoter engineering of natural product biosynthetic gene clusters in yeast. ACS Synth Biol 5:1002

Kennedy J, O'Leary ND, Kiran GS, Morrissey JP, O'Gara F, Selvin J, Dobson ADW (2011) Functional metagenomic strategies for the discovery of novel enzymes and biosurfactants with biotechnological applications from marine ecosystems. J Appl Microbiol 111:787–799

Khan AA, Stocker BL, Timmer MS (2012) Trehaloseglycolipids – synthesis and biological activities. Carbohydr Res 356:25–36

Khemili-Talbi S, Kebbouche-Gana S, Akmoussi-Toumi S, Angar Y, Gana ML (2015) Isolation of an extremely halophilic arhaeon Natrialba sp. C21 able to degrade aromatic compounds and to produce stable biosurfactant at high salinity. Extremophiles 19:1109–1120

Kiran GS, Hema TA, Gandhimathi R, Selvin J, Thomas TA, Rajeetha Ravji T, Natarajaseenivasan K (2009) Optimization and production of a biosurfactant from the sponge-associated marine fungus Aspergillus ustus MSF3. Colloids Surf B: Biointerfaces 73:250–256

Kiran GS, Sabarathnam B, Selvin J (2010) Biofilm disruption potential of a glycolipid biosurfactant from marine Brevibacterium casei. FEMS Immunol Med Microbiol 59:432–438

Kiran GS, Sabarathnam B, Thajuddin N, Selvin J (2014) Production of glycolipid biosurfactant from sponge-associated marine Actinobacterium Brachybacterium paraconglomeratum MSA21. J Surfactant Deterg 17:531–542

Konishi M, Hatada Y, Horiuchi J-I (2013) Draft genome sequence of the basidiomycetous yeast-like fungus Pseudozyma hubeiensis SY62, which produces an abundant amount of the biosurfactant mannosylerythritol lipids. Genome Announc 1:e00409–e00413. e00409–e00413

Konishi M, Nishi S, Fukuoka T, Kitamoto D, Watsuj T, Nagano Y, Yabuki A, Nakagawa S, Hatada Y, Horiuchi J (2014) Deep-sea Rhodococcus sp. BS-15, lacking the phytopathogenic fas genes, produces a novel glucotriose lipid biosurfactant. Mar Biotechnol 16:484–493

Kügler JH, Muhle-Goll C, Kühl B, Kraft A, Heinzler R, Kirschhöfer F, Henkel M, Wray V, Luy B, Brenner-Weiss G, Lang S, Syldatk C, Hausmann R (2014) Trehalose lipid biosurfactants produced by the actinomycetes Tsukamurella spumae and T. pseudospumae. Appl Microbiol Biotechnol 98:8905–8915

Kügler JH, Le Roes-Hill M, Syldatk C, Hausmann R (2015) Surfactants tailored by the class Actinobacteria. Front Microbiol 6:212

Kurtzman CP, Price NPJ, Ray KJ, Kuo T-M (2010) Production of sophorolipid biosurfactants by multiple species of the Starmerella (Candida) bombicola yeast clade. FEMS Microbiol Lett 311:140–146

Lang S, Philp JC (1998) Surface-active lipids in Rhodococci. Antonie Van Leeuwenhoek 74:59–70

Lau EV, Gan S, Ng HK, Poh PE (2014) Extraction agents for the removal of polycyclic aromatic hydrocarbons (PAHs) from soil in soil washing technologies. Environ Pollut 184:640–649

Ławniczak Ł, Marecik R, Chrzanowski Ł (2013) Contributions of biosurfactants to natural or induced bioremediation. Appl Microbiol Biotechnol 97:2327–2339

Leng L, Yuan X, Zeng G, Chen X, Wang H, Li H, Fu L, Xiao Z, Jiang L, Lai C (2015) Rhamnolipid based glycerol-in-diesel microemulsion fuel: formation and characterization. Fuel 147:76–81

Li J, Deng M, Wang Y, Chen W (2016) Production and characteristics of biosurfactant produced by Bacillus pseudomycoides BS6 utilizing soybean oil waste. Int Biodeterior Biodegrad 112:72–79

Lo Piccolo L, De Pasquale C, Fodale R, Puglia AM, Quatrini P (2011) Involvement of an alkane hydroxylase system of Gordonia sp. strain SoCg in degradation of solid n-alkanes. Appl Environ Microbiol 77:1204–1213

Loeschcke A, Thies S (2015) Pseudomonas putida-a versatile host for the production of natural products. Appl Microbiol Biotechnol 99:6197–6214

Lorenz S, Guenther M, Grumaz C, Rupp S, Zibek S, Sohn K (2014) Genome sequence of the basidiomycetous fungus Pseudozyma aphidis DSM70725, an efficient producer of biosurfactant mannosylerythritol lipids. Genome Announc 2:e00053–14. e00053–14

Luo C, Liu X, Zhou X, Guo J, Truong J, Wang X, Zhou H, Li X, Chen Z (2015a) Unusual biosynthesis and structure of locillomycins from Bacillus subtilis 916. Appl Environ Microbiol 81:6601–6609. doi:10.1128/AEM.01639-15

Luo C, Liu X, Zhou H, Wang X, Chen Z (2015b) Nonribosomal peptide synthase gene clusters for lipopeptide biosynthesis in Bacillus subtilis 916 and their phenotypic functions. Appl Environ Microbiol 81:422–431

Makkar RS, Cameotra SS, Banat IM (2011) Advances in utilization of renewable substrates for biosurfactant production. Appl Micorbiol Biotechnol Exp 1:1–5

Malavenda R, Rizzo C, Michaud L, Gerçe B, Bruni V, Syldatk C, Hausmann R, Lo Giudice A (2015) Biosurfactant production by Arctic and Antarctic bacteria growing on hydrocarbons. Polar Biol 38:1565–1574

Marchant R, Banat IM (2012b) Biosurfactants: a sustainable replacement for chemical surfactants? Biotechnol Lett 34:1597–1605

Marchant R, Banat IM (2012a) Microbial biosurfactants: challenges and opportunities for future exploitation. Trends Biotechnol, Trends Biotechnol 30:558–565

Marchant R, Banat IM (2014) Protocols for measuring biosurfactants production in microbial cultures. In: McGenity TJ et al (eds) Hydrocarbon and lipid microbiology protocols, Springer Protocols Handbooks. Humana press, United States. pp 1–10

Marchant R, Funston S, Uzoigwe C, Rahman PKSM, Banat IM (2014) Production of biosurfactants from nonpathogenic bacteria, Chapter 5. In: Kosaric N, Sukan FV (eds) Biosurfactants: production and utilization – processes, technologies, and economics. Surfactant Science series 159. CRC press, United States. pp 73–82. Print ISBN 978-14665-9669-6

Marti ME, Colonna WJ, Patra P, Zhang H, Green C, Reznik G, Pynn M, Jarrell K, Nyman JA, Somasundaran P, Glatz CE, Lamsal BP (2014) Production and characterization of microbial biosurfactants for potential use in oil-spill remediation. Enzym Microb Technol 55:31–39

Martínez-Toledo A, Ríos-Leal E, Vázquez-Duhalt R, González-Chávez M d C, Esparza-García JF, Rodríguez-Vázquez R (2006) Role of phenanthrene in rhamnolipid production by P. putida in different media. Environ Technol 27:137–142

Mata-Sandoval JC, Karns J, Torrents A (1999) High-performance liquid chromatography method for the characterization of rhamnolipid mixtures produced by Pseudomonas aeruginosa UG2 on corn oil. J Chromatogr A 864:211–220

Matsuzawa T, Koike H, Saika A, Fukuoka T, Sato S, Habe H, Kitamoto D, Morita T (2015) Draft genome sequence of the yeast Starmerella bombicola NBRC10243, a producer of sophorolipids, glycolipid biosurfactants. Genome Announc 3:e00176–15

Menezes Bento F, de Oliveira Camargo FA, Okeke BC, Frankenberger WT (2005) Diversity of biosurfactant producing microorganisms isolated from soils contaminated with diesel oil. Microbiol Res 160:249–255

Mnif I, Ghribi D (2015) Review lipopeptides biosurfactants: mean classes and new insights for industrial, biomedical, and environmental applications. Biopolymers 104:129–147

Morita T, Koike H, Koyama Y, Hagiwara H, Ito E, Fukuoka T, Imura T, Machida M, Kitamoto D (2013) Genome sequence of the basidiomycetous yeast Pseudozyma antarctica T-34, a producer of the glycolipid biosurfactants mannosylerythritol lipids. Genome Announc 1: e00064–13

Morita T, Koike H, Hagiwara H, Ito E, Machida M, Sato S, Habe H, Kitamoto D, Boekhout T, Wang Q et al (2014) Genome and transcriptome analysis of the basidiomycetous yeast Pseudozyma antarctica producing extracellular glycolipids, mannosylerythritol lipids. PLoS One 9:359–366

Morita T, Fukuoka T, Imura T, Kitamoto D (2015) Mannosylerythritol lipids: production and applications. J Oleo Sci 64:133–141

Moya-Ramírez I, Tsaousi K, Rudden M, Marchant R, Alameda EJ, García-Román M, Banat IM (2015) Rhamnolipid and surfactin production from olive oil mill waste as sole carbon source. Bioresour Technol 198:231–236

Moya-Ramírez I, Vaz DA, Banat IM, Marchant R, Alameda EJ, García-Román M (2016) Hydrolysis of olive mill waste to enhance rhamnolipids and surfactin production. Bioresour Technol 205:1–6

Müller MM, Hausmann R (2011) Regulatory and metabolic network of rhamnolipid biosynthesis: traditional and advanced engineering towards biotechnological production. Appl Microbiol Biotechnol 91:251–264

Narendra Kumar P, Swapna TH, Sathi Reddy K, Archana K, Nageshwar L, Nalini S, Khan MY, Hameeda B (2016) Draft genome sequence of Bacillus amyloliquefaciens strain RHNK22, isolated from rhizosphere with biosurfactant (Surfactin, Iturin, and Fengycin) and antifungal activity. Genome Announc 4:e01682–15

Nguyen TT, Edelen A, Neighbors B, Sabatini DA (2010) Biocompatible lecithin-based micro-emulsions with rhamnolipid and sophorolipid biosurfactants: formulation and potential applications. J Colloid Interface Sci 348:498–504

Nicol RW, Marchand K, Lubitz WD (2012) Bioconversion of crude glycerol by fungi. Appl Microbiol Biotechnol 93:1865–1875

Nikolopoulou M, Eickenbusch P, Pasadakis N, Venieri D, Kalogerakis N (2013) Microcosm evaluation of autochthonous bioaugmentation to combat marine oil spills. New Biotechnol 30:734–742

Ochsner UA, Fiechter A, Reiser J (1994) Isolation, characterisation, and expression in Escherichia coli of the Pseudomonas aeruginosa rhlAB genes encoding a rhamnosyltransferase involved in rhamnolipid biosurfactant synthesis. J Biol Chem 269:19787–19795

Ochsner UA, Reiser J, Fiechter A, Witholt B (1995) Production of *Pseudomonas aeruginosa* rhamnolipid biosurfactants in heterologous hosts. Appl Environ Microbiol 61:3503–3506

Ogasawara Y, Torrez-Martinez N, Aragon AD, Yackley BJ, Weber JA, Sundararajan A, Ramaraj T, Edwards JS, Melançon CE (2015) High-quality draft genome sequence of *Actinobacterium Kibdelosporangium* sp. MJ126-NF4, producer of type ii polyketide azicemicins, using Illumina and PacBio technologies. Genome Announc 3, pii: e00114–15

Oliveira JS, Araújo W, Lopes Sales AI, de Brito Guerra A, da Silva Araújo SC, de Vasconcelos ATR, Agnez-Lima LF, Freitas AT (2015) BioSurfDB: knowledge and algorithms to support biosurfactants and biodegradation studies. Database J Biol Databases Curation 2015:bav033

Perfumo A, Rudden M, Smyth TJ, Marchant R, Stevenson PS, Parry NJ, Banat IM (2013) Rhamnolipids are conserved biosurfactants molecules: implications for their biotechnological potential. Appl Microbiol Biotechnol 97:7297–7306

Pradhan AK, Pradhan N, Mall G, Panda HT, Sukla LB, Panda PK, Mishra BK (2013) Application of lipopeptide biosurfactant isolated from a Halophile: *Bacillus tequilensis* CH for inhibition of biofilm. Appl Biochem Biotechnol 171:1362–1375

Price NPJ, Ray KJ, Vermillion KE, Dunlap CA, Kurtzman CP (2012) Structural characterization of novel sophorolipid biosurfactants from a newly identified species of *Candida* yeast. Carbohydr Res 348:33–41

Raaijmakers JM, De Bruijn I, Nybroe O, Ongena M (2010) Natural functions of lipopeptides from *Bacillus* and *Pseudomonas*: more than surfactants and antibiotics. FEMS Microbiol Rev 34:1037–1062

Radzuan MZ, Banat IM, Winterburn J (2017) Production and characterization of rhamnolipid using palm oil agricultural refinery waste. Bioresour Technol 225:99–105

Rahim R, Burrows LL, Monteiro MA, Perry MB, Lam JS (2000) Involvement of the rml locus in core oligosaccharide and O polysaccharide assembly in *Pseudomonas aeruginosa*. Microbiology 146:2803–2814

Rahim R, Ochsner UA, Olvera C, Graninger M, Messner P, Lam JS, Soberón-Chávez G (2001) Cloning and functional characterization of the *Pseudomonas aeruginosa* rhlC gene that encodes rhamnosyltransferase 2, an enzyme responsible for di-rhamnolipid biosynthesis. Mol Microbiol 40:708–718

Rahman KSM, Rahman TJ, McClean S, Marchant R, Banat IM (2002) Rhamnolipid biosurfactant production by strains of *Pseudomonas aeruginosa* using low-cost raw materials. Biotechnol Prog 18:1277–1281

Rahman KSM, Rahman TJ, Kourkoutas Y, Petsas I, Marchant R, Banat IM (2003) Enhanced bioremediation of n-alkane in petroleum sludge using bacterial consortium amended with rhamnolipid and micronutrients. Bioresour Technol 90:159–168

Reis RS, Pereira AG, Neves BC, Freire DMG (2011) Gene regulation of rhamnolipid production in *Pseudomonas aeruginosa* – a review. Bioresour Technol 102:6377–6384

Reznik GO, Vishwanath P, Pynn MA, Sitnik JM, Todd JJ, Wu J, Jiang Y, Keenan BG, Castle AB, Haskell RF, Smith TF, Somasundaran P, Jarrell KA (2010) Use of sustainable chemistry to produce an acyl amino acid surfactant. Appl Microbiol Biotechnol 86:1387–1397

Rodrigues LR (2015) Microbial surfactants: fundamentals and applicability in the formulation of nano-sized drug delivery vectors. J Colloid Interface Sci 449:304–316

Roelants SLKW, De Maeseneire SL, Ciesielska K, Van Bogaert INA, Soetaert W (2014) Biosurfactant gene clusters in eukaryotes: regulation and biotechnological potential. Appl Microbiol Biotechnol 98:3449–3461

Rokni-Zadeh H, Mangas-Losada A, De Mot R (2011) PCR detection of novel non-ribosomal peptide synthetase genes in lipopeptide-producing *Pseudomonas*. Microb Ecol 62:941–947

Rudden M, Tsauosi K, Marchant R, Banat IM, Smyth TJ (2015) Development and validation of an ultra-performance liquid chromatography tandem mass spectrometry (UPLC-MS/MS) method for the quantitative determination of rhamnolipid congeners. Appl Microbiol Biotechnol 99:9177–9187

Ruggeri C, Franzetti A, Bestetti G, Caredda P, La Colla P, Pintus M, Sergi S, Tamburini E (2009) Isolation and characterisation of surface active compound-producing bacteria from hydrocarbon-contaminated environments. Int Biodeterior Biodegrad 63:936–942

Saerens KMJ, Roelants SLKW, Van Bogaert INA, Soetaert W, Altschul S, Madden T, Schaffer A, Zhang J, Zhang Z, Miller W et al (2011a) Identification of the UDP-glucosyltransferase gene UGTA1, responsible for the first glucosylation step in the sophorolipid biosynthetic pathway of *Candida bombicola* ATCC 22214. FEMS Yeast Res 11:123–132

Saerens KMJ, Saey L, Soetaert W (2011b) One-step production of unacetylated sophorolipids by an acetyltransferase negative *Candida bombicola*. Biotechnol Bioeng 108:2923–2931

Saerens KMJ, Zhang J, Saey L, Van Bogaer INA, Soetaert W (2011c) Cloning and functional characterization of the UDP-glucosyltransferase UgtB1 involved in sophorolipid production by *Candida bombicola* and creation of a glucolipid-producing yeast strain. Yeast 28:279–292

Saerens KMJ, Van Bogaert INA, Soetaert W, Ashby R, Solaiman D, Foglia T, Asmer HJ, Lang S, Wagner F, Baccile N et al (2015) Characterization of sophorolipid biosynthetic enzymes from *Starmerella bombicola*. FEMS Yeast Res 15:253–260

Saika A, Koike H, Hori T, Fukuoka T, Sato S, Habe H, Kitamoto D, Morita T (2014) Draft genome sequence of the yeast *Pseudozyma antarctica* type strain JCM10317, a producer of the glyco-lipid biosurfactants, mannosylerythritol lipids. Genome Announc 2:e00878–14

Saika A, Koike H, Fukuoka T, Yamamoto S, Kishimoto T, Morita T, Kitamoto D, Isoda H, Nakahara T, Kitamoto D et al (2016) A gene cluster for biosynthesis of mannosylerythritol lipids consisted of 4-*O*-β-D-mannopyranosyl-(2R,3S)-erythritol as the sugar moiety in a basidiomycetous yeast *Pseudozyma tsukubaensis*. PLoS One 11:e0157858

Sambles CM, White DA (2015) Genome sequence of *Rhodococcus* sp. strain PML026, a trehalolipid biosurfactant producer and biodegrader of oil and alkanes. Genome Announc 3: e00433–15

Sarubbo LA, Rocha RB Jr, Luna JM, Rufino RD, Santos VA, Banat IM (2015) Some aspects of heavy metals contamination remediation and role of biosurfactants. Chem Ecol 31:707–723. doi:10.1080/02757540.2015.1095293

Satpute SK, Banpurkar AG, Banat IM, Sangshetti JN, Patil RR, Gade WN (2016a) Multiple roles of biosurfactants in biofilms. Curr Pharm Des 22:1429–1448

Satpute SK, Kulkarni GR, Banpurkar AG, Banat IM, Mone NS, Patil RH, Cameotra SS (2016b) Biosurfactant/s from *Lactobacilli* species: properties, challenges and potential biomedical applications. J Basic Microbiol 56:1–19

Schmidberger A, Henkel M, Hausmann R, Schwartz T (2013) Expression of genes involved in rhamnolipid synthesis in *Pseudomonas aeruginosa* PAO1 in a bioreactor cultivation. Appl Microbiol Biotechnol 97:5779–5791

Schneiker S, dos Santos VAM, Bartels D, Bekel T, Brecht M, Buhrmester J, Chernikova TN, Denaro R, Ferrer M, Gertler C et al (2006) Genome sequence of the ubiquitous hydrocarbon-degrading marine bacterium *Alcanivorax borkumensis*. Nat Biotechnol 24:997–1004

Seydlová G, Svobodová J (2008) Review of surfactin chemical properties and the potential biomedical applications. Cent Eur J Med 3:123–133

Singh AK, Cameotra SS (2013) Efficiency of lipopeptide biosurfactants in removal of petroleum hydrocarbons and heavy metals from contaminated soil. Environ Sci Pollut Res Int 20:7367–7376

Solaiman DKY, Liu Y, Moreau RA, Zerkowski JA (2014) Cloning, characterization, and heterologous expression of a novel glucosyltransferase gene from sophorolipid-producing *Candida bombicola*. Gene 540:46–53

Solaiman DKY, Ashby RD, Gunther NW, Zerkowski JA (2015) Dirhamnose-lipid production by recombinant nonpathogenic bacterium *Pseudomonas chlororaphis*. Appl Microbiol Biotechnol 99:4333–4342

Sotirova A, Spasova D, Vasileva-Tonkova E, Galabova D (2009) Effects of rhamnolipid-biosurfactant on cell surface of *Pseudomonas aeruginosa*. Microbiol Res 164:297–303

Sousa M, Melo VM, Rodrigues S, Sant'ana HB, Gonçalves LR (2012) Screening of biosurfactant-producing Bacillus strains using glycerol from the biodiesel synthesis as main carbon source. Bioprocess Biosyst Eng 35:897–906

Steegborn C, Clausen T, Sondermann P, Jacob U, Worbs M, Marinkovic S, Huber R, Wahl MC (1999) Kinetics and inhibition of recombinant human cystathionine -lyase: toward the rational control of transsulfuration. J Biol Chem 274:12675–12684

Tahseen R, Afzal M, Iqbal S, Shabir S, Khan QM, Khalid ZM, Banat IM (2016) Rhamnolipids and nutrients boost remediation of crude oil contaminated soil by enhancing bacterial colonization and metabolic activities. Int Biodeterior Biodegrad 115:192–198

Teichmann B, Linne U, Hewald S, Marahiel MA, Bölker M (2007) A biosynthetic gene cluster for a secreted cellobiose lipid with antifungal activity from *Ustilago maydis*. Mol Microbiol 66:525–533

Thies S, Rausch SC, Kovacic F, Schmidt-Thaler A, Wilhelm S, Rosenau F, Daniel R, Streit W, Pietruszka J, Jaeger K-E et al (2016) Metagenomic discovery of novel enzymes and biosurfactants in a slaughterhouse biofilm microbial community. Sci Report 6:27035

Tuleva BK, Ivanov GR, Christova NE (2002) Biosurfactant production by a new *Pseudomonas putida* strain. Z Naturforsch C J Biosci 57:356–360

Van Bogaert INA, Saerens K, De Muynck C, Develter D, Soetaert W, Vandamme EJ (2007) Microbial production and application of sophorolipids. Appl Microbiol Biotechnol 76:23–34

Van Bogaert INA, Holvoet K, Roelants SLKW, Li B, Lin Y-C, Van de Peer Y, Soetaert W (2013) The biosynthetic gene cluster for sophorolipids: a biotechnological interesting biosurfactant produced by *Starmerella bombicola*. Mol Microbiol 88:501–509

Van Bogaert INA, Buyst D., Martins JC, Roelants SLKW, Soetaert WK (2016) Synthesis of bolaform biosurfactants by an engineered Starmerella bombicola yeast. Biotechnol Bioeng 113:2644–2651

Vasileva-Tonkova E, Sotirova A, Galabova D (2011) The efect of rhamnolipid biosurfactant produced by *Pseudomonas fluorescens* on model bacterial strains and isolates from industrial wastewater. Curr Microbiol 62:427–433

Walter V, Syldatk C, Hausmann R (2013) Screening concepts for the isolation of biosurfactant producing microorganisms. Adv Exp Med Biol 672:1–13

White DA, Hird LC, Ali ST (2013) Production and characterisation of a trehalolipid biosurfactant produced by the novel marine bacterium *Rhodococcus* sp., strain PML026. J Appl Microbiol 115:744–755

Winsor GL, Khaira B, Van Rossum T, Lo R, Whiteside MD, Brinkman FSL (2008) The *Burkholderia* genome database: facilitating flexible queries and comparative analyses. Bioinformatics 24:2803–2804

Winsor GL, Griffiths EJ, Lo R, Dhillon BK, Shay JA, Brinkman FSL (2016) Enhanced annotations and features for comparing thousands of *Pseudomonas* genomes in the *Pseudomonas* genome database. Nucleic Acids Res 44:D646–D653

Wittgens A, Tiso T, Arndt TT, Wenk P, Hemmerich J, Müller C, Wichmann R, Küpper B, Zwick M, Wilhelm S et al (2011) Growth independent rhamnolipid production from glucose using the non-pathogenic *Pseudomonas putida* KT2440. Microb Cell Factories 10:80

Xia W, Du Z, Cui Q, Dong H, Wang F, He P, Tang Y (2014) Biosurfactant produced by novel *Pseudomonas* sp. WJ6 with biodegradation of n-alkanes and polycyclic aromatic hydrocarbons. J Hazard Mater 276:489–498

Yan P, Lu M, Guan Y, Zhang W, Zhang Z (2011) Remediation of oil-based drill cuttings through a biosurfactant-based washing followed by a biodegradation treatment. Bioresour Technol 102:10252–10259

Youssef NH, Nguyen T, Sabatini DA, McInerney MJ (2007) Basis for formulating biosurfactant mixtures to achieve ultra low interfacial tension values against hydrocarbons. J Ind Microbiol Biotechnol 34:497–507

Youssef N, Randall Simpson D, McInerney MJ, Duncan KE (2013) In-situ lipopeptide biosurfactant production by *Bacillus* strains correlates with improved oil recovery in two oil wells approaching their economic limit of production. Int Biodeterior Biodegrad 81:127–132

Yu M, Liu Z, Zeng G, Zhong H, Liu Y, Jiang Y, Li M, He X, He Y (2015) Characteristics of mannosylerythritol lipids and their environmental potential. Carbohydr Res 407:63–72

Zhang Y, Miller RM (1992) Enhanced octadecane dispersion and biodegradation by a *Pseudomonas* rhamnolipid surfactant (biosurfactant). Appl Environ Microbiol 58:3276–3282

Zhang L, Pemberton JE, Maier RM (2014) Effect of fatty acid substrate chain length on *Pseudomonas aeruginosa* ATCC 9027 monorhamnolipid yield and congener distribution. Process Biochem 49:989–995

Zhu K, Rock CO (2008) RhlA converts beta-hydroxyacyl-acyl carrier protein intermediates in fatty acid synthesis to the beta-hydroxydecanoyl-beta-hydroxydecanoate component of rhamnolipids in *Pseudomonas aeruginosa*. J Bacteriol 190:3147–3154

# Biofilm Stress Responses Associated to Aromatic Hydrocarbons

# 7

Laura Barrientos-Moreno and Manuel Espinosa-Urgel

## Contents

1 Introduction .................................................................................. 106
2 Resistance Mechanisms to Toxic Aromatic Hydrocarbons ............................ 107
3 Toxic Hydrocarbons and Biofilms ....................................................... 108
4 Influence of Toluene on *P. putida* Biofilms .......................................... 109
5 Research Needs ............................................................................. 111
References ...................................................................................... 112

### Abstract

Efficient biotransformations that involve toxic aromatic hydrocarbons and biore-mediation of environmental sites polluted with these compounds rely on the metabolic potential of microorganisms and their survival strategies to cope with their deleterious effects. Biofilm formation is acknowledged as one of the main colonization and persistence mechanisms of bacteria in the environment, provid-ing protection against stress. Many bioremediation systems and bioreactors commonly rely on pure culture or mixed community biofilms. Although reaction parameters and overall population dynamics have been studied in some instances, there is limited information on how toxic aromatic hydrocarbons influence the process of biofilm development, the potentially associated tolerance mechanisms and their interplay with other biofilm stress response mechanisms. In this chapter, we briefly summarize the current information on this topic and present the existing research gaps that could expand the biotechnological exploitation of biofilms in this field.

L. Barrientos-Moreno · M. Espinosa-Urgel (✉)
Department of Environmental Protection, Estación Experimental del Zaidín, CSIC, Granada, Spain
e-mail: laura.barrientos@eez.csic.es; manuel.espinosa@eez.csic.es

© Springer International Publishing AG, part of Springer Nature 2018
T. Krell (ed.), *Cellular Ecophysiology of Microbe: Hydrocarbon and Lipid Interactions*,
Handbook of Hydrocarbon and Lipid Microbiology, https://doi.org/10.1007/978-3-319-50542-8_32

# 1    Introduction

In the environment, bacteria encounter a number of fluctuating factors, including temperature, nutrient and water availability, and the presence of toxic molecules produced in their abiotic and biotic surroundings or resulting from their own metabolism. Survival in such unpredictable conditions requires a wide range of adaptive responses. The acclimatization potential of bacteria relies on many different aspects that include metabolic versatility, the capacity to acquire new genetic information via horizontal transfer mediated by plasmids or other mobile elements, and a variety of stress resistance mechanisms. Microorganisms able to combine all these features for physiological/biochemical adaptation are better suited to colonizing changing niches. At the molecular level, bacterial responses rely on a combination of constitutive, basal elements acting as a first line of defense and the activation of the expression of genes encoding products that deal with a given physicochemical stress in response to environmental or cellular signals.

One of the key strategies for colonization and persistence in many different environments is the ability to form biofilms. These multicellular communities associated to solid surfaces offer protection against predation, stress, and the action of biocides and thus are considered the predominant lifestyle for many bacterial species in the environment. Bacteria growing as biofilms are surrounded by an extracellular matrix with high water content and usually composed of proteins, exopolysaccharides, and DNA (Costerton et al. 1995; Sutherland 2001). The composition of the biofilm matrix varies depending on the bacterial species and the environmental conditions, and it is considered one of the main elements that determine biofilm protection (Balcázar et al. 2015; Gambino and Cappitelli 2016). Different studies have shown that bacteria growing as biofilms show increased tolerance to a wide range of environmental challenges, such as antibiotics, metal toxicity, acid exposure, or dehydration, among others. Whereas antibiotic resistance in biofilms has received significant attention, given the obvious clinical implications, the development of biofilms in the presence of toxic hydrocarbons and the potentially associated tolerance mechanisms are much less characterized. However, this knowledge would be of great interest in terms of biotechnological applications of biofilms, such as remediation of sites contaminated with organic pollutants, the development of biosensors, or biotransformation reactions that could render added value products. The toxicity of compounds (substrates or products) can limit such reactions and thus reduce their biotechnological potential. This is particularly relevant in the case of biotransformations that involve toxic aromatic hydrocarbons (toluene, xylenes, etc.) to produce aromatic organic acids or in the bioproduction of hydroxylated aromatic compounds from carbohydrates or alcohols (Gosset 2009; Vargas-Tah and Gosset 2015; Molina-Santiago et al. 2016). The toxicity of organic solvents is directly related with the logarithm of their partitioning coefficient (log $P_{ow}$) in a defined octanol–water mixture (Sikkema et al. 1995). Compounds with a log $P_{ow}$ value below 4, such as toluene or styrene, are highly toxic because they accumulate in the cytoplasmic membrane, disorganizing its structure and altering its functionality. This results in a loss of ions and metabolites, disruption of the pH and electron gradient, and eventually leads to cell death. Nonetheless, bacterial strains (mainly from the genus *Pseudomonas*) capable of enduring the presence of significant amounts of toxic hydrocarbons have been isolated, and several tolerance mechanisms have been reported (Segura et al. 1999, 2012; Ramos et al. 2015). In this chapter, we will revise the current knowledge and present new information on the role of biofilm formation in the persistence of *Pseudomonas* in the presence of toxic hydrocarbons such as toluene.

## 2    Resistance Mechanisms to Toxic Aromatic Hydrocarbons

Many *P. putida* strains are able to grow in the presence of low concentrations of aromatic hydrocarbons, but their survival is compromised at higher concentrations of these compounds. However, a few *P. putida* strains have been found to tolerate and grow in the presence of high concentrations of such chemicals, which would cause the death of most other microorganisms. Well-studied examples of these strains include *P. putida* DOT-T1E and S12 (Ramos et al. 1995; Weber et al. 1993). DOT-T1E was isolated from a wastewater treatment plant and is an efficient degrader of benzene, ethylbenzene, and toluene, while S12 was isolated in an enrichment experiment for styrene degrading bacteria from soil and water samples. In these organisms, there are several levels of responses to toxic aromatic hydrocarbons that include structural alterations, active extrusion mechanisms, and general stress and metabolic responses (Ramos et al. 2015). Membrane rigidity is increased through *cis-trans* isomerization of unsaturated fatty acids and changes in cardiolipin contents. A set of chaperones are expressed, thus limiting the impact on protein misfolding or denaturation, and scavenging systems directed toward oxidative stress are triggered, likely due to the interference of solvents with the electron transport systems causing increased production of hydrogen peroxide and other reactive oxygen species. There is also an increase in the metabolic rates, and the biodegradation routes for aromatic hydrocarbons are activated, thus contributing to reduce the concentration of the toxic compound while obtaining energy from it. However, the key elements that determine the tolerance of a particular strain correspond to a set of efflux pumps that are in general induced in the presence of the aromatic hydrocarbon and actively remove it from the membrane (Isken and de Bont 1996; Duque et al. 2001). The high tolerance of *P. putida* DOT-T1E is for the most part due to the TtgGHI efflux pump, which is encoded in a self-transmissible plasmid (Rodríguez-Herva et al. 2007). Although two other chromosomally encoded pumps, TtgABC and TtgDEF, also contribute to the resistance phenotype, mutations that inactivate TtgGHI cause the cells to lose the ability to survive in the presence of high concentrations of toluene (Rojas et al. 2001). The regulation of these tolerance mechanisms relies on specific transcriptional repressors associated to each efflux pump (TtgR, TtgT, and TtgV), capable of recognizing a range of molecules as effectors to relieve repression, including in some cases antibiotics and certain plant-derived flavonoids (Guazzaroni et al. 2005; Alguel et al. 2007; Espinosa-Urgel et al. 2015).

## 3 Toxic Hydrocarbons and Biofilms

Bioremediation systems designed for the removal of hydrocarbon pollution commonly rely on biofilms used for bioventing or biofiltration procedures. Several studies have focused on aspects such as modeling and measuring biofilm diffusion-reaction processes that direct the rate of hydrocarbon biodegradation, both in saturated (i.e., submerged) and unsaturated (in direct contact with air) systems (Fan et al. 1990; Alvarez et al. 1991; Holden et al. 1997), or the population dynamics and efficiency of biofilms during biodegradation (Farhadian et al. 2008; Amit et al. 2009). However, the influence of toxic hydrocarbons on the process of biofilm development and the cellular responses in these conditions have received relatively little attention. There has been some research done on the expression of catabolic genes for aromatic hydrocarbons in biofilms, the metabolic relationships, and the potential transfer of catabolic elements in biofilm communities during hydrocarbon degradation, using mixed populations that included a *P. putida* strain harboring derivatives of pWW0 (Christensen et al. 1998; Møller et al. 1998). This 117-kbp self-transmissible plasmid (also called TOL plasmid) carries catabolic and regulatory genes responsible for the complete transformation of toluene into Krebs cycle intermediates (Franklin et al. 1981), organized in two operons corresponding to the upper (toluene to benzoate metabolism) and *meta* (benzoate catabolism to the Krebs cycle) pathways. In those conditions, it was shown that the upper pathway was uniformly induced in the presence of benzyl alcohol in pure culture and mixed biofilms, whereas the *meta* pathway was only active in the mixed community, revealing the importance of community level metabolic interactions (Møller et al. 1998) in biofilm development during hydrocarbon degradation. Rapid spread of TOL-plasmid harboring bacteria was shown to take place in mixed biofilms growing in the presence of benzyl alcohol, but horizontal conjugative transfer of the plasmid seemed to occur with a rather low frequency (Christensen et al. 1998). Interestingly, the presence of this plasmid has been reported to favor cell attachment and biofilm formation of *P. putida* (D'Alvise et al. 2010). This effect was observed in the absence of toluene or other aromatic hydrocarbons that could be substrates for catabolism and was shown to be associated to an increase in the amount of extracellular DNA in the biofilm matrix, a factor that in other microorganisms has been correlated with stress tolerance (Svensson et al. 2014).

General stress responses with a role in tolerance to toxic hydrocarbons have also been described to be relevant in biofilms. Thus, the chaperones DnaK and GroE, which have been shown to be induced in response to toluene or butanol (Ramos et al. 2015), also play a role in surface attachment and biofilm formation in different microorganisms (Lemos et al. 2001; Eaton et al. 2016). Similarly, oxidative stress resistance mechanisms that are triggered by toxic hydrocarbons are increasingly being revealed as key elements in the biofilm lifestyle of bacteria. For example, several genes of the OxyR regulon are induced in *P. putida* in response to toluene (Domínguez-Cuevas et al. 2006), and OxyR has been shown to be relevant for biofilm formation and colonization of abiotic and biotic surfaces by *Vibrio*, *Klebsiella*, and other microorganisms (Hennequin and Forestier 2009; Chung et al. 2015;

Wang et al. 2016). Such oxidative stress responses should be therefore taken into account when analyzing the design and performance of biofilm-based bioremediation strategies.

With respect to the efflux systems involved in toluene tolerance mentioned in the previous section, they belong to the RND ("resistance, nodulation, and cell division") family of transporters, which are responsible for extrusion of a diversity of compounds, including antimicrobials (Poole 2004). This family has been widely studied because of their role in multidrug resistance in clinically relevant strains. An increasing number of reports are providing evidences of the role of this type of efflux pumps in the resistance of biofilm cells against antimicrobial compounds (Soto 2013; Buroni et al. 2014) and even in modulating the process of biofilm development itself (Yoon et al. 2015; Sakhtah et al. 2016). Consistently, some of these pumps have been shown to be preferentially expressed in biofilm populations (Zhang and Mah 2008). In the course of studies analyzing the influence of antimicrobials on bacterial biofilms, it has become apparent that not only are biofilms more resistant to these molecules than planktonic populations but also that sublethal concentrations of certain antibiotics can promote biofilm formation, induce the expression of efflux pumps, and select for mutants with increased efflux rates (Gilbert et al. 2002; Morita et al. 2014). Although the efflux pumps more specifically dedicated to solvent tolerance have not been investigated in detail in this context, it is worth mentioning that exposure of bacterial cultures to subinhibitory concentrations of toluene in the gas phase renders them more tolerant to high levels of the toxic hydrocarbon, as a consequence of the derepression of the efflux pumps and the activation of stress responses. This, and the relative promiscuity of their regulators in terms of effector molecules (including plant-derived antimicrobials), makes it reasonable to think that the Ttg pumps may play a role in biofilms under certain environmental conditions.

## 4      Influence of Toluene on *P. putida* Biofilms

Based on the above evidences, we have done preliminary work to test if the presence of toxic organic compounds such as toluene influences biofilm formation and persistence by three closely related *P. putida* strains: the highly tolerant DOT-T1E; mt-2, which harbors the catabolic plasmid pWW0; and KT2440, a plasmid-free derivative of mt-2 widely used as model organism in studies of biofilm formation. In this strain, two large extracellular proteins, LapA (8682 amino acids) and LapF (6310 amino acids), are the main adhesins required for biofilm formation in different environmental conditions (Hinsa et al. 2003; Yousef-Coronado et al. 2008; Martínez-Gil et al. 2010, 2014).

In the absence of stressor, mt-2 formed more robust biofilms than KT2440, as previously observed (D'Alvise et al. 2010), whereas DOT-T1E showed limited adhesion to solid surfaces compared to the other two strains (Fig. 1a). An *in silico* analysis of extracellular matrix components based on the homology of this strain with *P. putida* KT2440 revealed that DOT-T1E lacks most of the gene encoding

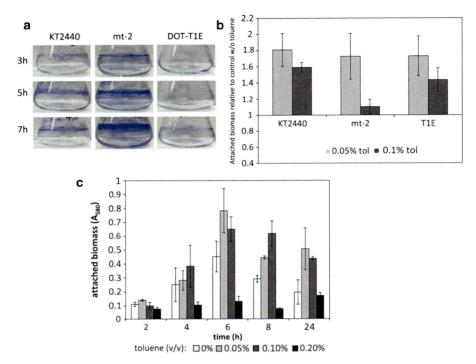

**Fig. 1** (a) Biofilm formation by strains mt-2, KT2440, and DOT-T1E in rich medium. Cultures were incubated in flasks in rich medium at 30 °C under orbital shaking (40 rpm). Biomass attached to the surface was stained with 1% crystal violet for 15′, followed by washing with distilled water. (b) Effect of toluene on biofilm formation: comparison between strains at 6 h of growth. The biomass attached to the surface, relative to the control without toluene (= 1), was measured spectrophotometrically after staining and solubilization of the dye with 10% acetic acid. (c) Details on the effect of increasing concentrations of toluene on biofilm formation by the tolerant strain DOT-T1E. The attached biomass was measured as above (Data in (b) and (c) correspond to averages and standard deviations from two independent experiments with three replicas each)

LapA, rendering a 363 amino acid protein, and also has a shorter homolog of LapF, which explains the poor adhesion characteristics of this strain.

When cultures were not preexposed to low concentrations of toluene in the gas phase, addition of high concentrations of the compound to the medium limited growth and consequently hampered biofilm formation. On the other hand, low concentrations of the hydrocarbon (0.05% v/v) resulted in increased biofilm development in all the strains (Fig. 1b), an effect that was observed in the tolerant strain DOT-T1E at concentration to up to 0.1% v/v toluene (Fig. 1c). In preexposed cultures, the most noticeable effect was that biofilms persisted after 24 h, while in the absence of stressor, they had already dispersed, as it is known to happen in the batch culture conditions used (Martínez-Gil et al. 2010). This could suggest that the hydrocarbon may favor the sessile lifestyle, as a protective state against stress, similarly to the effect of subinhibitory concentrations of antibiotics. Paradoxically,

**Fig. 2** Effect of toluene on biofilms of *P. putida* mt-2 pre-formed in the absence of the hydrocarbon. Cultures were grown in LB in tubes at 30 °C under orbital rotation (40 rpm) for 4 h, and then toluene (0.1% or 0.2%) was added. The evolution of the attached biomass in the absence or presence of the hydrocarbon was followed for two more hours by direct visual inspection and crystal violet staining

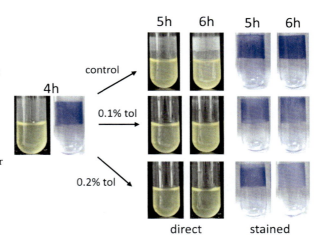

addition of toluene to biofilms of mt-2 and KT2440 pre-formed in the absence of toluene caused accelerated detachment (Fig. 2). A similar but less pronounced effect was observed for DOT-T1E although the limited attachment capacity of this strain in the absence of stressor did not allow a detailed analysis. We interpret these data as indicative that the presence of toluene from the early stages promotes attachment and gives rise to biofilms that are structurally different (in terms of matrix composition and probably hydrophobicity) from those produced under optimal growth conditions, so that when already established, sessile populations are abruptly confronted with the solvent, their matrix is not "prepared," and the biofilm is disorganized. It is worth noting that in *P. putida*, there are four known exopolysaccharides produced that may have different roles and importance depending on the surface colonized and the environmental conditions (Nielsen et al. 2011; Martínez-Gil et al. 2013).

## 5   Research Needs

Several lines of research deserve further exploration that could potentially lead to optimized and expanded biotechnological uses of biofilms in relation with biodegradation or biotransformation of toxic hydrocarbons. In the past two decades, there have been significant advances in understanding the mechanisms of bacterial biofilm formation and unveiling the resistance and tolerance strategies that microbes use to cope with toxic aromatic hydrocarbons. There is also solid information on the different parameters (kinetics, population dynamics, production rates) relevant to biofilm bioreactors applied to hydrocarbon biodegradation. However, these three (biofilm biology, tolerance mechanisms, and bioreactor performance) are still rather isolated bodies of knowledge that need to be integrated for a full exploitation of microbial capacities. Analyzing the expression and activity of tolerance mechanisms such as efflux pumps and other stress responses, or of catabolic genes in biofilms growing in close-to-real situations, would be important to model and predict the behavior of the systems and ultimately to improve biotransformations. There is also little or no information on how the substrates and products of the desired reactions can affect biofilm growth and persistence and hence may limit productivity. From a more basic point of view, investigating the role and activity of tolerance mechanisms in natural environments could offer new information on the natural compounds that act as stressors and trigger these responses.

Finally, the preliminary data obtained with three closely related strains of *P. putida* open intriguing evolutionary questions. Catabolic genes for toluene and the most important element for tolerance to the compound, the TtgGHI pump, are present in mobilizable plasmids, pWW0 and pGRT1, sharing some similar features (Segura et al. 2014), but not combined into one, which one would predict to offer increased advantages. On the other hand, it is noticeable that the most tolerant strain, DOT-T1E, lacks the essential elements for robust biofilm formation present in mt-2 and its derivative KT2440. It is possible that the environmental conditions found by each microorganism during their evolution (wastewater in the case of DOT-T1E and soil of a planted orchard in the case of mt-2) have favored the selection of one or the other as the most efficient survival mechanism, taking into account the energetic burden of maintaining such big plasmids and the intact genes for large adhesins (nearly 27 kb for *lapA* and 19 kb for *lapF*, not counting the genes involved in their transport and regulation). Nonetheless, this information can open the way to construct modified strains that carry all the relevant elements in the most compact possible way and test their performance in biofilm bioreactors. Well-studied, non-hazardous strains such as *Pseudomonas putida* KT2440, generally recognized as safe, and for which many tools for genetic manipulation and gene expression are available, can provide the background for such modifications and their future use in the biotechnology industry.

**Acknowledgments** Work funded by grant P11-CVI-7391 from Junta de Andalucía and EFDR funds.

# References

Alguel Y, Meng C, Terán W, Krell T, Ramos JL, Gallegos MT, Zhang X (2007) Crystal structures of multidrug binding protein TtgR in complex with antibiotics and plant antimicrobials. J Mol Biol 369:829–840

Alvarez PJJ, Anid PJ, Vogel TM (1991) Kinetics of aerobic biodegradation of benzene and toluene in sandy aquifer material. Biodegradation 2:45–51

Amit K, Dewulf J, Wiele TV, Langenhove HV (2009) Bacterial dynamics of biofilm development during toluene degradation by *Burkholderia vietnamiensis* G4 in a gas phase membrane bioreactor. J Microbiol Biotechnol 19:1028–1033

Balcázar JL, Subirats J, Borrego CM (2015) The role of biofilms as environmental reservoirs of antibiotic resistance. Front Microbiol 6:1216

Buroni S, Matthijs N, Spadaro F, Van Acker H, Scoffone VC, Pasca MR, Riccardi G, Coenye T (2014) Differential roles of RND efflux pumps in antimicrobial drug resistance of sessile and planktonic *Burkholderia cenocepacia* cells. Antimicrob Agents Chemother 58:7424–7429

Christensen BB, Sternberg C, Andersen JB, Eberl L, Møller S, Givskov M, Molin S (1998) Establishment of new genetic traits in a microbial biofilm community. Appl Environ Microbiol 64:2247–2255

Chung CH, Fen SY, Yu SC, Wong HC (2015) Influence of *oxyR* on growth, biofilm formation, and mobility of *Vibrio parahaemolyticus*. Appl Environ Microbiol 82:788–796

Costerton JW, Lewandowski Z, Caldwell DE, Korber DR, Lappin-Scott HM (1995) Microbial biofilms. Annu Rev Microbiol 49:711–745

D'Alvise PW, Sjøholm OR, Yankelevich T, Jin Y, Wuertz S, Smets BF (2010) TOL plasmid carriage enhances biofilm formation and increases extracellular DNA content in *Pseudomonas putida* KT2440. FEMS Microbiol Lett 312:84–92

Domínguez-Cuevas P, González-Pastor JE, Marqués S, Ramos JL, de Lorenzo V (2006) Transcriptional tradeoff between metabolic and stress-response programs in *Pseudomonas putida* KT2440 cells exposed to toluene. J Biol Chem 281:11981–11991

Duque E, Segura A, Mosqueda G, Ramos JL (2001) Global and cognate regulators control the expression of the organic solvent efflux pumps TtgABC and TtgDEF of *Pseudomonas putida*. Mol Microbiol 39:1100–1106

Eaton DS, Crosson SD, Fiebig A (2016) Proper control of *Caulobacter crescentus* cell-surface adhesion requires the general protein chaperone, DnaK. J Bacteriol 198:2631–2642

Espinosa-Urgel M, Serrano L, Ramos JL, Fernández-Escamilla AM (2015) Engineering biological approaches for detection of toxic compounds: a new microbial biosensor based on the *Pseudomonas putida* TtgR repressor. Mol Biotechnol 57:558–564

Fan LS, Leyva-Ramos R, Wisecarver KD, Zehner BJ (1990) Diffusion of phenol through a biofilm rrown on activated carbon particles in a draft-tube three-phase fluidized-bed bioreactor. Biotechnol Bioeng 35:279–286

Farhadian M, Duchez D, Vachelard C, Larroche C (2008) Monoaromatics removal from polluted water through bioreactors-a review. Water Res 42:1325–1341

Franklin FC, Bagdasarian M, Bagdasarian MM, Timmis KN (1981) Molecular and functional analysis of the TOL plasmid pWW0 from *Pseudomonas putida* and cloning of genes for the entire regulated aromatic ring meta cleavage pathway. Proc Natl Acad Sci U S A 78:7458–7462

Gambino M, Cappitelli F (2016) Mini-review: biofilm responses to oxidative stress. Biofouling 32:167–178

Gilbert P, Allison DG, McBain AJ (2002) Biofilms in vitro and in vivo: do singular mechanisms imply cross-resistance? J Appl Microbiol 92(Suppl):98S–110S

Gosset G (2009) Production of aromatic compounds in bacteria. Curr Opin Biotechnol 20:651–658

Guazzaroni ME, Krell T, Felipe A, Ruiz R, Meng C, Zhang X, Gallegos MT, Ramos JL (2005) The multidrug efflux regulator TtgV recognizes a wide range of structurally different effectors in solution and complexed with target DNA: evidence from isothermal titration calorimetry. J Biol Chem 280:20887–20893

Hennequin C, Forestier C (2009) OxyR, a LysR-type regulator involved in *Klebsiella pneumoniae* mucosal and abiotic colonization. Infect Immun 77:5449–5457

Hinsa SM, Espinosa-Urgel M, Ramos JL, O'Toole GA (2003) Transition from reversible to irreversible attachment during biofilm formation by *Pseudomonas fluorescens* WCS365 requires an ABC transporter and a large secreted protein. Mol Microbiol 49:905–918

Holden PA, Hunt JR, Firestone MK (1997) Toluene diffusion and reaction in unsaturated *Pseudomonas putida* biofilms. Biotechnol Bioeng 56:656–670

Isken S, de Bont JAM (1996) Active efflux of toluene in a solvent-resistant bacterium. J Bacteriol 178:6056–6058

Lemos JA, Chen YY, Burne RA (2001) Genetic and physiologic analysis of the *groE* operon and role of the HrcA repressor in stress gene regulation and acid tolerance in *Streptococcus mutans*. J Bacteriol 183:6074–6084

Martínez-Gil M, Yousef-Coronado F, Espinosa-Urgel M (2010) LapF, the second largest *Pseudomonas putida* protein, contributes to plant root colonization and determines biofilm architecture. Mol Microbiol 77:549–561

Martínez-Gil M, Quesada JM, Ramos-González MI, Soriano MI, de Cristóbal RE, Espinosa-Urgel M (2013) Interplay between extracellular matrix components of Pseudomonas putida biofilms. Res Microbiol 164:382–389

Martínez-Gil M, Ramos-González MI, Espinosa-Urgel M (2014) Roles of cyclic di-GMP and the Gac system in transcriptional control of the genes coding for the *Pseudomonas putida* adhesins LapA and LapF. J Bacteriol 196:1287–1313

Molina-Santiago C, Cordero BF, Daddaoua A, Udaondo Z, Manzano J, Valdivia M, Segura A, Ramos JL, Duque E (2016) *Pseudomonas putida* as a platform for the synthesis of aromatic compounds. Microbiology. doi:10.1099/mic.0.000333. (in press)

Møller S, Sternberg C, Andersen JB, Christensen BB, Ramos JL, Givskov M, Molin S (1998) In situ gene expression in mixed-culture biofilms: evidence of metabolic interactions between community members. Appl Environ Microbiol 64:721–732

Morita Y, Tomida J, Kawamura Y (2014) Responses of *Pseudomonas aeruginosa* to antimicrobials. Front Microbiol 4:422. doi:10.3389/fmicb.2013.00422

Nielsen L, Li X, Halverson LJ (2011) Cell-cell and cell-surface interactions mediated by cellulose and a novel exopolysaccharide contribute to *Pseudomonas putida* biofilm formation and fitness under water-limiting conditions. Environ Microbiol 13:1342–1356

Poole K (2004) Efflux pumps. In: Ramos JL (ed) *Pseudomonas* vol. I, genomics, life style and molecular architecture. Kluwer, New York, pp 635–674

Ramos JL, Duque E, Huertas MJ, Haïdour A (1995) Isolation and expansion of the catabolic potential of a *Pseudomonas putida* strain able to grow in the presence of high concentrations of aromatic hydrocarbons. J Bacteriol 177:3911–3916

Ramos JL, Sol Cuenca M, Molina-Santiago C, Segura A, Duque E, Gómez-García MR, Udaondo Z, Roca A (2015) Mechanisms of solvent resistance mediated by interplay of cellular factors in *Pseudomonas putida*. FEMS Microbiol Rev 39:555–566

Rodríguez-Herva JJ, García V, Hurtado A, Segura A, Ramos JL (2007) The *ttgGHI* solvent efflux pump operon of *Pseudomonas putida* DOT-T1E is located on a large self-transmissible plasmid. Environ Microbiol 9:1550–1561

Rojas A, Duque E, Mosqueda G, Golden G, Hurtado A, Ramos JL, Segura A (2001) Three efflux pumps are required to provide efficient tolerance to toluene in *Pseudomonas putida* DOT-T1E. J Bacteriol 183:3967–3973

Sakhtah H, Koyama L, Zhang Y, Morales DK, Fields BL, Price-Whelan A, Hogan DA, Shepard K, Dietrich LE (2016) The *Pseudomonas aeruginosa* efflux pump MexGHI-OpmD transports a natural phenazine that controls gene expression and biofilm development. Proc Natl Acad Sci U S A 113:E3538–E3547

Segura A, Duque E, Mosqueda G, Ramos JL, Junker F (1999) Multiple responses of gram-negative bacteria to organic solvents. Environ Microbiol 1:191–198

Segura A, Molina L, Fillet S, Krell T, Bernal P, Muñoz-Rojas J, Ramos JL (2012) Solvent tolerance in gram-negative bacteria. Curr Opin Biotechnol 23:415–421

Segura A, Molina L, Ramos JL (2014) Plasmid-mediated tolerance toward environmental pollutants. Microbiol Spectr 2(6):PLAS-0013–PLAS-2013

Sikkema J, de Bont JA, Poolman B (1995) Mechanisms of membrane toxicity of hydrocarbons. Microbiol Rev 59:201–222

Soto SM (2013) Role of efflux pumps in the antibiotic resistance of bacteria embedded in a biofilm. Virulence 4:223–229

Sutherland IW (2001) The biofilm matrix-an immobilized but dynamic microbial environment. Trends Microbiol 9:222–227

Svensson SL, Pryjma M, Gaynor EC (2014) Flagella-mediated adhesion and extracellular DNA release contribute to biofilm formation and stress tolerance of *Campylobacter jejuni*. PLoS One 9(8):e106063

Vargas-Tah A, Gosset G (2015) Production of cinnamic and *p*-hydroxycinnamic acids in engineered microbes. Front Bioeng Biotechnol 3:116

Wang P, Lee Y, Igo MM, Roper MC (2016) Tolerance to oxidative stress is required for maximal xylem colonization by the xylem-limited bacterial phytopathogen, *Xylella fastidiosa*. Mol Plant Pathol. doi:10.1111/mpp.12456. (in press)

Weber FJ, Ooijkaas LP, Schemen RM, Hartmans S, de Bont JA (1993) Adaptation of *Pseudomonas putida* S12 to high concentrations of styrene and other organic solvents. Appl Environ Microbiol 59:3502–3504

Yoon EJ, Chabane YN, Goussard S, Snesrud E, Courvalin P, Dé E, Grillot-Courvalin C (2015) Contribution of resistance-nodulation-cell division efflux systems to antibiotic resistance and biofilm formation in *Acinetobacter baumannii*. MBio 6(2.) pii:e00309–e00315

Yousef-Coronado F, Travieso ML, Espinosa-Urgel M (2008) Different, overlapping mechanisms for colonization of abiotic and plant surfaces by *Pseudomonas putida*. FEMS Microbiol Lett 288:118–124

Zhang L, Mah TF (2008) Involvement of a novel efflux system in biofilm-specific resistance to antibiotics. J Bacteriol 190:4447–4452

# Part II

# Sensing, Signaling and Uptake

# Sensing, Signaling, and Uptake: An Introduction

**8**

## Tino Krell

## Contents

1 Introduction ................................................................................ 119
2 Sensing and Signaling ..................................................................... 120
3 One-Component Systems ................................................................... 121
4 Two-Component Systems ................................................................... 123
5 Chemoreceptor-Based Signaling ........................................................... 124
6 Hydrocarbon Uptake ....................................................................... 124
7 Research Needs ............................................................................ 124
References ..................................................................................... 125

**Abstract**

The three most frequent sensing and signal transduction mechanisms in bacteria are one- and two-component systems as well as chemosensory pathways, and members of these families were found to be involved in the sensing of hydrocarbons. These systems were shown to modulate the expression of hydrocarbon degradation pathways and efflux pumps as well as to mediate hydrocarbon chemotaxis. Hydrocarbons are thought to cross the outer membrane via specific pores and the inner membrane by diffusion. However, it still remains controversial as to whether there are also active hydrocarbon uptake mechanisms.

T. Krell (✉)
Department of Environmental Protection, Estación Experimental del Zaidín, Consejo Superior de Investigaciones Científicas, Granada, Spain
e-mail: tino.krell@eez.csic.es

T. Krell (ed.), *Cellular Ecophysiology of Microbe: Hydrocarbon and Lipid Interactions*,
Handbook of Hydrocarbon and Lipid Microbiology, https://doi.org/10.1007/978-3-319-50542-8_29

# 1    Introduction

Many hydrocarbons can be considered as "useful toxins": On one hand they are toxic to life, threatening bacterial survival, whereas at the same time many hydrocarbons can be used as growth substrates, and the capacity to degrade them is an evolutionary advantage enhancing the chances of survival. It is therefore essential that bacteria are able to sense hydrocarbons and to use that information to generate specific responses. Hydrocarbon-sensing mechanisms and signaling circuits were primarily shown to alter the expression of hydrocarbon degradation pathways and efflux pumps. In general, chemotaxis permits bacteria to approach compounds of interest, which are in most of the cases C- and N-sources, as well as to escape form toxic compounds. It is therefore not astonishing that mechanisms enabling hydrocarbon-specific chemoattraction as well as chemorepellation have evolved. To understand the physiological consequences of any class of compound, it is essential to establish how these compounds penetrate the cell. However, as discussed in this section, the current state of knowledge on how hydrocarbons enter the cell is still subject to controversy and further studies are necessary to establish a general model.

# 2    Sensing and Signaling

Bacteria have evolved a series of different mechanisms to sense environmental signals and to convert them into a cellular response. The most abundant systems are one- and two-component systems as well as chemosensory signaling pathways (Ulrich et al. 2005; Wuichet and Zhulin 2010). One-component systems represent the most straightforward way of transcriptional regulation. These systems are typically composed of a single protein containing an effector-binding sensor domain and a helix-turn-helix motif containing DNA-binding domain. Effector binding causes conformational alterations that typically alter the affinity of the DNA-binding domain for promoter DNA.

The basic units of a two-component system are the sensor kinase and the response regulator (Zschiedrich et al. 2016). Effector binding to the sensor domain of the kinase modulates its autophosphorylation activity, leading to changes in the trans-phosphorylation rate to its cognate response regulator, which ultimately changes the concentration of phosphorylated regulator. Two-component systems are from the genetic (maintenance of two genes), metabolic (synthesis of two proteins), and energetic (mechanism based on ATP hydrolysis) points of view more costly than one-component systems. Since most sensor kinases possess an extracytoplasmic sensor domain, the concomitant capacity to sense extracytosolic signals is thought to be the advantage that compensates the cost-expensiveness of two-component systems. However, some two-component systems, like those for hydrocarbons (see below), sense their effectors in the cytosol, and the advantage of such systems over one-component systems remains to be established.

Chemosensory signaling cascades are sophisticated versions of two-component systems where effector recognition is achieved through binding at chemoreceptors, which form complexes with the CheA histidine kinase (Bi and Lai 2015; Wuichet and Zhulin 2010). Due to different mechanisms for pathway adaptation to an existing compound concentration, chemosensory cascades permit to sense compound gradients, which is the molecular basis for chemotaxis or the directed movement of bacteria in compound gradients. Whereas the output of one- and two-component systems is primarily at the level of transcriptional regulation, chemosensory pathways, apart from mediating chemotaxis, were also found to exert alternative cellular function or are associated with type IV pili-mediated motility (Wuichet and Zhulin 2010).

## 3    One-Component Systems

The expression of many degradation pathways is controlled by one-component systems. As reviewed in ▶ Chap. 10, "One-Component Systems that Regulate the Expression of Degradation Pathways for Aromatic Compounds" by Durante-Rodríguez et al., regulators either act as activators or repressors. The authors also show that one-component systems that are involved in the regulation of aromatics degradation pathways belong to at least 11 different families of transcriptional regulators.

As discussed in this volume, bacteria have evolved many different strategies to gain resistance against the toxic effects of hydrocarbons. Among these mechanisms, the active compound efflux appears to be the most important strategy (Ramos et al. 2015). The same strategy, based on root-nodulation-cell division (RND) efflux pumps, that is used by bacteria to expulse antibiotics and to gain resistance (Li et al. 2015) is also been used to expulse hydrocarbons (Ramos et al. 2015). In this section data are reviewed indicating that the transcriptional control of pump expression is mediated by the concerted action of global regulators, frequently systems that respond to different stressors, as well as specific one-component regulators with the capacity to sense the corresponding pump substrate. Many of these specific regulators employ an effector mediated de-repression mechanism where the protein is bound at the promoter preventing transcription in the absence of effector. Binding of an effector to the repressor/DNA complex induces protein dissociation enabling for transcription. A well-studied example is the control of the promoter for the TtgGHI hydrocarbon efflux pump by hydrocarbon-sensitive transcriptional regulator TtgV (Fig. 1a). Pumps as well as their cognate specific regulators appear to share a broad ligand spectrum.

Hydrocarbons are compounds composed of hydrogen and carbon atoms. However, several closely related compound classes were found to be signals for intercellular communication and quorum-sensing purposes. These compound classes include different fatty acids and derivatives as well as acyl-homoserine lactones. As cell densities increase these signals accumulate and are recognized by quorum-sensing regulators (one-component systems) that control the expression of genes, whose products show activities that are beneficial when performed by groups of bacteria acting in synchrony (Rutherford and Bassler 2012). In this section the

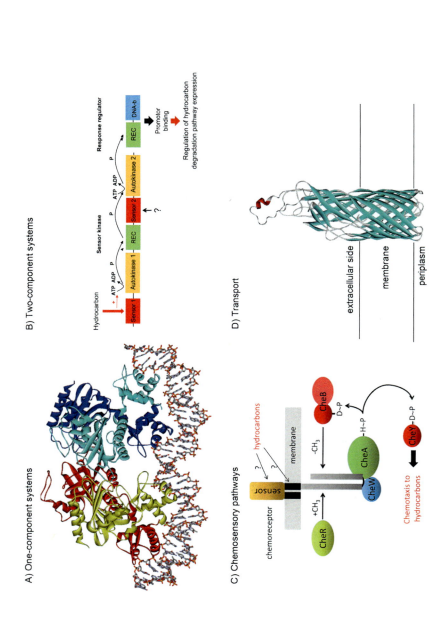

**Fig. 1** Principal mechanisms of hydrocarbon sensing and transport. (**a**) Structure of the TtgV repressor (pdb ID: 2 XRO) bound to a fragment of the promoter $P_{ttgG}$ that controls the expression of the TtgGHI hydrocarbon efflux pump. Hydrocarbon binding to TtgV causes its dissociation from the DNA enabling

different classes of hydrocarbon like quorum-sensing signaling molecules and their cellular functions are reviewed.

# 4      Two-Component Systems

As mentioned above the expression of several degradation pathway operons is mediated by one-component systems. In contrast, other degradation pathways are controlled by two-component systems. A representative example is the transcriptional control of the six different toluene degradation pathways that have so far been identified (Parales et al. 2008). Whereas three of them are controlled by one-component systems, the remaining three are controlled by two-component systems that belong to the TodS/TodT-like family (Silva-Jimenez et al. 2012). The reason for the use of different regulatory machineries for apparently the same purpose is unclear. Hydrocarbon responsive one- and two-component systems have in common that their effectors are sensed in the cytosol. In contrast to the prototypal sensor kinase, which has a periplasmic sensor domain, members of the TodS sensor kinase family lack transmembrane regions and are entirely located in the cytosol. For the TodS sensor kinase it has been shown that aromatic hydrocarbons bind to the N-terminal PAS-domain, leading to a stimulation of the N-terminal autokinase domain (Lacal et al. 2006) (Fig. 1b). The phosphorylgroups are then transferred to an internal receiver domain and subsequently to the C-terminal autokinase module prior to the phosphorylgroup transfer to the response regulator (Busch et al. 2009). The physiological relevance of this phosphorelay as well as potential ligands that may bind to the C-terminal PAS domain remain to be identified. Interestingly, the autokinase activity of TodS is modulated by agonists (for example, toluene) and antagonists (for example, *o*-xylene) (Busch et al. 2007; Koh et al. 2016). Since crude oil is basically an agonist/antagonist mixture, the action of antagonists may account for the reduced pathway expression that has been shown to limit biodegradation efficiency *in situ*.

◄─────────────────────────────────────────────────────────────────

**Fig. 1** (continued) transcription. (**b**) Schematic view of the mechanism of action of TodS/TodT-like two-component systems. Hydrocarbon binding to the sensor kinase triggers a number of phosphorylation events increasing ultimately the phosphorylation state of the response regulator, which in turn activates the expression of hydrocarbon degradation pathways. (**c**) Schematic view of a chemosensory pathway responsible for the chemotaxis towards hydrocarbons. (**d**) Structure of the FadL outer membrane protein (pdb ID: 1T1L) that enables fatty acid uptake. FadL homologues are thought to permit hydrocarbon entry into the periplasm

# 5    Chemoreceptor-Based Signaling

The ambivalent character of hydrocarbons is well reflected in the fact that attractant as well as repellent responses have been reported. Repellent responses to phenol involving several chemoreceptors were observed for *E. coli* (Pham and Parkinson 2011), whereas toluene, nitrobenzoate, and naphthalene degrading *Pseudomonas* strains showed strong chemoattraction to these growth substrates (Grimm and Harwood 1999; Iwaki et al. 2007; Lacal et al. 2011). In addition, chemotactic responses to several other xenobiotics have been reported (Parales et al. 2015).

The mechanism of the prototypal chemotaxis receptor consists in the ligand binding to a periplasmic sensor domain that generates molecular stimuli that ultimately causes chemotaxis (Bi and Lai 2015). The mechanism of hydrocarbon receptors has so far not been elucidated (Fig. 1c), but Pham and Parkinson showed that the response to phenol is not initiated by its binding to the sensor domain, but likely to be caused by the phenol-induced perturbation of the two transmembrane regions of the receptor (Pham and Parkinson 2011). In line with this alternative mechanism for hydrocarbon chemotaxis is the report on the identification of a chemoreceptor that mediates responses to different aromatic compounds. However, these compounds were not recognized by the sensor domain of this receptor, which was found to bind several TCA cycle intermediates (Ni et al. 2013). The authors propose several alternative mechanisms.

# 6    Hydrocarbon Uptake

Although of central importance, the mechanism by which hydrocarbons enter the cell is still subject to controversy and the current knowledge will be reviewed in ▶ Chap. 13, "Chemotaxis to Hydrocarbons" by Parales and Ditty. There is a significant body of evidence showing that hydrocarbons cross the outer membrane of Gram-negative bacteria via pores, similar to that shown in Fig. 1d. Hydrocarbons are then thought to enter the cell by passive diffusion through the inner membrane. As discussed in detail in the chapter, there are a number of reports showing active hydrocarbon transport. However, most of these studies did not assess the possibility that the cellular metabolization of hydrocarbons may create gradients that in turn enhances their passive diffusion. In contrast to the limited knowledge available on the hydrocarbon uptake, the central role of RND efflux pump in hydrocarbon export has been well documented.

# 7    Research Needs

Among the main research needs in the field of hydrocarbon sensing and transport are:

1. Understand the effector promiscuity of hydrocarbon responsive two-component systems. TodS-like sensor kinases show a significant promiscuity in ligand recognition and the majority of compounds that are sensed by TodS and that lead to its activation are not among the substrates of the corresponding degradation pathway. In addition, there are several compounds with antagonistic effects on sensor kinase activity. The physiological relevance of both phenomena remains to be established.
2. Understand how hydrocarbons enter the cell. Although it is generally accepted that hydrocarbons enter the cells primarily by diffusion, it remains to be established whether there are cases of active transport and which are the corresponding transporters.
3. Understand how hydrocarbons activate chemoreceptors. The prototypal mechanism of chemoreceptor activation consists in the either direct or periplasmic binding protein mediated interaction of chemoeffectors with the chemoreceptor sensor domain. However, there is also evidence that the dissolution of some hydrocarbons in the cell membrane causes stimuli that are sensed via the transmembrane regions of the chemoreceptor, leading to its activation. It remains to be established whether this corresponds to a general mechanism for hydrocarbons.

**Acknowledgments** We acknowledge financial support from FEDER funds and Fondo Social Europeo through grants from the Junta de Andalucía (grant CVI-7335) and the Spanish Ministry for Economy and Competitiveness (grants BIO2013-42297 and BIO2016-76779-P).

# References

Bi S, Lai L (2015) Bacterial chemoreceptors and chemoeffectors. Cell Mol Life Sci 72(4):691–708
Busch A et al (2007) Bacterial sensor kinase TodS interacts with agonistic and antagonistic signals. Proc Natl Acad Sci U S A 104(34):13774–13779
Busch A et al (2009) The sensor kinase TodS operates by a multiple step phosphorelay mechanism involving two autokinase domains. J Biol Chem 284(16):10353–10360
Grimm AC, Harwood CS (1999) NahY, a catabolic plasmid-encoded receptor required for chemotaxis of Pseudomonas putida to the aromatic hydrocarbon naphthalene. J Bacteriol 181 (10):3310–3316
Iwaki H et al (2007) Characterization of a pseudomonad 2-nitrobenzoate nitroreductase and its catabolic pathway-associated 2-hydroxylaminobenzoate mutase and a chemoreceptor involved in 2-nitrobenzoate chemotaxis. J Bacteriol 189(9):3502–3514
Koh S et al (2016) Molecular insights into toluene sensing in the TodS/TodT signal transduction system. J Biol Chem 291(16):8575–8590
Lacal J et al (2006) The TodS-TodT two-component regulatory system recognizes a wide range of effectors and works with DNA-bending proteins. Proc Natl Acad Sci U S A 103(21):8191–8196
Lacal J et al (2011) Bacterial chemotaxis towards aromatic hydrocarbons in Pseudomonas. Environ Microbiol 13(7):1733–1744
Li XZ, Plesiat P, Nikaido H (2015) The challenge of efflux-mediated antibiotic resistance in Gram-negative bacteria. Clin Microbiol Rev 28(2):337–418
Ni B et al (2013) Comamonas testosteroni uses a chemoreceptor for tricarboxylic acid cycle intermediates to trigger chemotactic responses towards aromatic compounds. Mol Microbiol 90(4):813–823

Parales RE et al (2008) Diversity of microbial toluene degradation pathways. Adv Appl Microbiol 64:1–73. 2 p following 264

Parales RE et al (2015) Bacterial chemotaxis to xenobiotic chemicals and naturally-occurring analogs. Curr Opin Biotechnol 33:318–326

Pham HT, Parkinson JS (2011) Phenol sensing by Escherichia coli chemoreceptors: a nonclassical mechanism. J Bacteriol 193(23):6597–6604

Ramos JL et al (2015) Mechanisms of solvent resistance mediated by interplay of cellular factors in Pseudomonas putida. FEMS Microbiol Rev 39(4):555–566

Rutherford ST, Bassler BL (2012) Bacterial quorum sensing: its role in virulence and possibilities for its control. Cold Spring Harb Perspect Med 2(11):1–26

Silva-Jimenez H et al (2012) Study of the TmoS/TmoT two-component system: towards the functional characterization of the family of TodS/TodT like systems. Microb Biotechnol 5 (4):489–500

Ulrich LE, Koonin EV, Zhulin IB (2005) One-component systems dominate signal transduction in prokaryotes. Trends Microbiol 13(2):52–56

Wuichet K, Zhulin IB (2010) Origins and diversification of a complex signal transduction system in prokaryotes. Sci Signal 3(128):ra50

Zschiedrich CP, Keidel V, Szurmant H (2016) Molecular mechanisms of two-component signal transduction. J Mol Biol 428(19):3752–3775

# Bioinformatic, Molecular, and Genetic Tools for Exploring Genome-Wide Responses to Hydrocarbons

# 9

O. N. Reva, R. E. Pierneef, and B. Tümmler

## Contents

1  Introduction ................................................................. 127
    1.1  In Silico Analyses of Genes and Oligonucleotide Signatures ...................... 128
    1.2  Omics Analyses of Cellular Constituents in Man-Made Standardized
         Environments .......................................................... 130
2  Genetic and Metagenomic Tools for the Analysis of the Response of Microbial
    Communities to Hydrocarbons in Artificial and Natural Habitats ...................... 132
3  Research Needs ............................................................. 133
References ................................................................... 133

### Abstract

The response profiles of bacteria to hydrocarbons in the wild can be directly assessed by high-throughput cDNA sequencing of metagenomes, tracking the fate or metabolism of labeled cells in the microbial community or can be indirectly inferred from the screening of mutant libraries for key genetic determinants. Transcriptome, proteome, and metabolome data are collected from homogeneous bacterial populations that are exposed to hydrocarbons under strictly controlled culturing conditions.

O. N. Reva (✉) · R. E. Pierneef
Centre for Bioinformatics and Computational Biology, Department of Biochemistry, University of Pretoria, Hillcrest, Pretoria, South Africa
e-mail: oleg.reva@up.ac.za; repierneef@live.com

B. Tümmler
Klinische Forschergruppe, Medizinische Hochschule Hannover, Hannover, Germany
e-mail: tuemmler.burkhard@mh-hannover.de

© Springer International Publishing AG, part of Springer Nature 2018
T. Krell (ed.), *Cellular Ecophysiology of Microbe: Hydrocarbon and Lipid Interactions*,
Handbook of Hydrocarbon and Lipid Microbiology, https://doi.org/10.1007/978-3-319-50542-8_33

# 1      Introduction

The ability to degrade alkanes or hydrocarbons as a source of carbon and energy may be highly beneficial to an organism. The omnipresence of these compounds in conjunction with the high-energy content has been a driving force in the evolution or acquisition of alkane-degrading mechanisms (Yakimov et al. 2007).

The analysis of the genome-wide responses of a microorganism to hydrocarbons can be divided into two tasks. First, one should address the issue whether the microorganism of interest is capable of metabolizing alkanes. This task is accomplished by data mining of genomic sequences (if available), homology-driven cloning and sequencing, and straightforward in vitro tests of substrate degradation. Second, any microorganism can be investigated in its global response to hydrocarbons by applying the current omics technologies. One should note, however, that the global profiling of RNA transcripts, proteins, and metabolites will only yield meaningful data if the bacteria are exposed to hydrocarbons in meticulously controlled environments such as chemostats. If one wants to study the responses of microbial communities to hydrocarbons in the wild, genetic or metagenomic approaches should be pursued that are adapted to the particular habitat on a case-to-case basis.

## 1.1      In Silico Analyses of Genes and Oligonucleotide Signatures

Genes encoding enzymes for alkane degradation can be detected by the criteria of protein sequence similarity, operon organization, and conserved protein domains. An established tool is the BlastP algorithm (http://blast.ncbi.nlm.nih.gov/ Blast.cgi?PAGE=Proteins). Knowledge about the abundance of these genes is catalogued in public databases such as BRENDA (http://www.brenda-enzymes. info/) (Chang et al. 2015).

A more demanding task is the search for the promoters. Promoter sequences of alkane degradation operons show rather weak homology that hinders their prima facie identification. The promoter sequences may however be indirectly detected by analysis of oligonucleotide composition. Promoter regions typically exhibit higher DNA curvature, lower base-stacking energy, and are more rigid (Reva et al. 2008). Consensus motifs of promoter regions may be identified by the MEME Suite (http://meme-suite.org) (Bailey et al. 2015).

Horizontal gene transfer contributed to the spread of alkane metabolic activities among γ-Proteobacteria. Besides the search for mobile genetic elements within or adjacent the alkane degradation operons (van Beilen et al. 2001), evidence for horizontal gene transfer is gained from the comparison of oligonucleotide compositional biases between the gene cluster and the surrounding chromosomal sequences by using tools such as the SeqWord Genomic Island Sniffer (Bezuidt et al. 2009). Pre-identified genomic islands and horizontally transferred genes may also be searched through records of the Predicted Genomic Islands Database (Pierneef et al. 2015). In the *Alcanivorax borkumensis* SK2 genome, for example, the two regions comprising *alkB1* and *alkB2* genes are as similar to each other in their

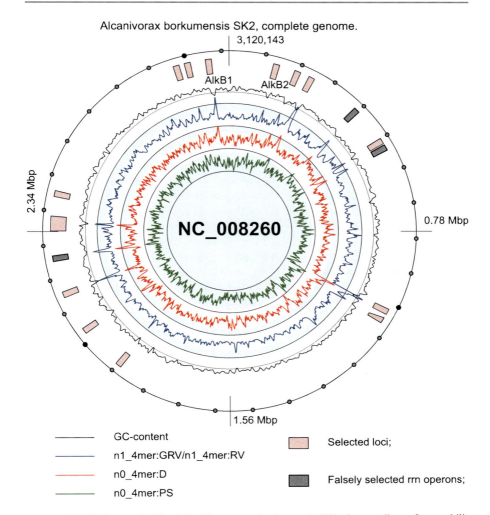

**Fig. 1** Identified genomic islands in *Alcanivorax borkumensis* SK2, the paradigm of mesophilic hydrocarbonoclastic bacteria. The genomic fragment [3,060,396–3,068,240] of *A. borkumensis* SK2 encodes the gene cluster of the *alkB1* regulator, *alkB1*, *alkH1*, *alkH2*, and *alkJ*. The alkane-1-monooxygenase *alkB2* gene is located in the horizontally transferred genomic region [125,500–145,500]. The alkane degradation gene cluster is characterized by below-average GC content and an atypical tetranucleotide usage depicted by peaks of the *n0_4mer:D* and the *n1_4mer: GRV/n1_4mer:RV* lines (The figure was downloaded from the Pre_GI database (http://pregi.bi.up. ac.za/index.php)).

tetranucleotide usage (TU) patterns as each of them is to *Yersinia* species (Fig. 1). The two *alk* genes differ in their TU usage from the bulk sequence of *A. borkumensis* indicating that the *alkB1* and *alkB2* genes were delivered to *A. borkumensis* from an ancestor of the *Yersinia* lineage (Reva et al. 2008).

## 1.2    Omics Analyses of Cellular Constituents in Man-Made Standardized Environments

Transcriptome, proteome, and metabolome analyses can generate comprehensive quantitative profiles of the genome-wide response of a microorganism to hydrocarbons. All these omics technologies determine average values of bulk samples containing millions of bacteria. Consequently the most meaningful data are produced from homogeneous bacterial populations with minimal variation of the expression profiles of individual cells. To minimize spatiotemporal gradients, bacteria should be grown under carefully controlled conditions in batch cultures or – better though – in chemostats. In the latter case, stationary cultures are perturbed by the stimulus of interest, i.e., in this context the exposure to hydrocarbons. Experiments in chemostats will yield reproducible and reliable quantitative data with low statistical spread as it is requested for applications in systems biology or white biotechnology.

With the advent of next-generation sequencing technologies, whole transcriptome analysis by RNA sequencing (RNA-seq) has become the state of the art. The digital RNA-seq datasets provide quantitative information about coding and noncoding transcripts with single nucleotide precision in a strand-specific manner (Creecy and Conway 2015; Hrdlickova et al. 2016). Often the most abundant ribosomal RNAs (rRNAs) are of limited interest, and consequently the unwanted rRNAs are hybridized with biotinylated DNA or locked nucleic acid probes, followed by depletion with streptavidin beads. After rRNA depletion the intact RNAs are reversely transcribed, and the full-length cDNAs are fragmented because of the size limitation of most sequencing platforms. The directionality of the RNA is captured by either the attachment of different adapters to the $5'$ and $3'$ ends or by incorporation of dUTP into the second strand of cDNA followed by adapter ligation and degradation of the second strand with uracil DNA glycosylase. To identify transcription start sites, the RNA pool is depleted from processed $5'$ monophosphate RNAs with $5'$ monophosphate-dependent terminator exonuclease. The remaining mRNAs with $5'$ triphosphate ends represent the primary products of transcription initiation.

The RNA sequence datasets are used to annotate the transcriptional units and to quantify gene expression at the base count level. The relative transcript levels of operons are calculated by averaging the base counts from the promoter to the terminator locations. Attention should be paid to abrupt shifts of base counts indicating alternative transcripts generated by promoter and terminator activities within the operon. To study the influence of hydrocarbons on the bacterial RNA expression profile, time-series experiments will typically require three to four biological replicates per time point.

Alternatively – although less and less frequently used by the scientific community – global mRNA expression profiles may be obtained by the hybridization of bacterial cDNA onto Gene chips that have each open reading frame of the genome represented by either one long (50–70 mers) or numerous pairs of short complementary and single mismatch oligonucleotides (20–35 mers) (Dharmadi and

Gonzalez 2004; Ness 2007). The experimental protocols comprise the RNA preparation from bacterial cells, cDNA synthesis, fragmentation, and labeling and the hybridization of the cDNA preparation onto the microarray. After washing off nonspecific binding, the hybridization signals are scanned and processed. Particular care should be taken for the proper normalization of signals of gene probe sets and the nontrivial identification of up- and downregulated genes. An acceptable rate of false positive signals is calculated by permutation methods like The Significance Analysis of Microarrays (SAM) (freely available at http://statweb.stanford.edu/~tibs/SAM/).

In addition to global RNA expression profiling, the "ChIP-Seq" (Galagan et al. 2013; Myers et al. 2015) and "ChIP-chip" (Liu 2007) approaches may be applied to map the binding and regulatory genomic sites of transcription factors in a global and high-throughput fashion. The transcription factor is crosslinked with genomic DNA, fragmented to approximately 500 bp, and immunoprecipitated with an antibody. The coprecipitated DNA is sequenced or hybridized on a genome-spanning tiling microarray. In addition to identifying binding sites, correlation of ChIP-seq data with expression data can reveal important information about bacterial regulons and regulatory networks.

Considering the short average lifetime of a bacterial mRNA of less than 2 min, the transcriptome will provide a snapshot of the global expression profile at that particular point of time. The longer-living proteins visualize the cellular response to a signal such as hydrocarbons in a broader time frame (Sabirova et al. 2006). With the emergence of mass spectrometry (MS) in protein science and the availability of complete genome sequences, bacterial proteomics has gone through a rapid development. The application of gel-based and gel-free technologies, the analyses of subcellular proteome fractions, and the use of multidimensional capillary HPLC combined with MS/MS have allowed high-qualitative and quantitative coverage of currently more than 60% of the theoretical bacterial proteome (Van Oudenhove and Devreese 2013; Otto et al. 2014; Soufi and Macek 2015; Maaß and Becher 2016).

Metabolomics is the youngest omics discipline that analyzes metabolic profiles in response to environmental compounds and signals such as hydrocarbons (Bargiela et al. 2015). Metabolites are a chemically highly diverse group of compounds. Hence the analysis of microbial metabolomes is a formidable challenge. Protocols need to be more flexible than those for proteomics or transcriptomics. Gas chromatography, liquid chromatography, or capillary electrophoresis are combined with numerous MS methods (Putri et al. 2013). State-of-the-art metabolomics platforms detect more than 1,000 different metabolites but only a minority of which may be identifiable because the mass spectra of metabolites and their volatile derivatives often are not known or not listed in the available spectral libraries. The whole interacting metabolome is typically studied by in vivo perturbation experiments, that is, stimulus response experiments using different setups and quantitative analytical approaches, including dynamic carbon tracing (Vasilakou et al. 2016). In summary, metabolomics is a young and vibrant scientific discipline with workflows that are just being developed leaving still much room for improvement.

## 2    Genetic and Metagenomic Tools for the Analysis of the Response of Microbial Communities to Hydrocarbons in Artificial and Natural Habitats

Omics technologies can differentiate the response of heterogeneous microbial communities to hydrocarbons at the species or genus level. The constituents of the community are resolved by either ribosomal DNA profiling (Yarza et al. 2014) or by whole-metagenome shotgun sequencing (Segata et al. 2013; Garza and Dutilh 2015). Next, deep cDNA sequencing can reveal the metatranscriptome profile of the population (Aylward et al. 2015; Huson et al. 2016). After removal of ambiguous low-complexity reads from the dataset (protocol described by Losada et al. 2016), the remaining reads are aligned to a microbial pangenome so that quantitative transcript profiles of the community are evaluated at the level of the individual gene and species.

As the next step, one may study the metabolic activity of the community. Stable isotopes or radioisotopes can be incorporated into bacterial subpopulations (Neufeld et al. 2007). Subsequent analysis of labeled biomarkers of subpopulations with stable-isotope probing (DNA, RNA, or phospholipid-derived fatty acid) or of individual cells with a combination of fluorescence in situ hybridization and microautoradiography reveals linked phylogenetic and functional information about the organisms that assimilated the compounds of interest such as hydrocarbons.

A complementary approach to the analysis of metabolic activities is the genome-wide search for the key determinants of the bacterial response to hydrocarbons by in vivo expression technology (IVET) (Rediers et al. 2005), signature-tagged mutagenesis (STM) (Mazurkiewicz et al. 2006), differential display using arbitrarily primed PCR (Fislage 1998), subtractive and differential hybridization (Ito and Sakaki 1997), or selective capture of transcribed sequences (SCOTS) (Graham and Clark-Curtiss 1999). We discuss the most widely used IVET and STM strategies.

IVET involves the construction of a conditionally compromised strain that is mutated in a gene encoding an essential growth factor (*egf*). This mutant strain is unable to grow in the environment under study. The second component of IVET is the promoter trap, consisting of a promoterless *egf* gene and a transcriptionally linked reporter gene (*rep*). Bacterial DNA is cloned randomly into the promoter trap and integrated in the chromosome of the *egf* mutant strain. Only in strains that carry a promoter active in the specified niche can the *egf* mutation be complemented. After selection in this environment, bacteria are reisolated and spread on a general growth medium that is suitable for monitoring reporter gene activity in vitro. Accordingly, constitutive promoters are distinguished from promoters that are specifically induced in the wild. Colonies bearing the latter type of transcriptional fusion are subjected to a second IVET screening to eliminate false positives.

STM is a mutation-based screening method that uses a population of isogenic transposon mutants for the identification of essential genes by negative selection. A pool of mutants can simultaneously be examined because they are differentiated by unique DNA marker sequences, the "signature tags." The pools of mutants are exposed to the habitat of interest. One screens for mutants that are unable to survive or to grow because they are inactivated in a key gene for survival in this habitat of interest. The beauty of the technology is an in vivo selection process done by the habitat among a mixed population of mutants.

The investigation of heterogeneous microbial communities requires the spatio-temporal resolution of the signals of individual cells. To visualize the growth and decay of individual species in the population upon exposure to hydrocarbons, the target organisms may be labeled with specific fluorescence markers, for example, ribosomal ribonucleic acid-targeted oligonucleotide probes (Daims and Wagner 2007). Labeled cells are visualized by fluorescence microscopy and are quantified by direct visual cell counting or by digital image analysis. The next step within the foreseeable future will be the multi-omics profiling of single cells (Bock et al. 2016; Kodzius and Gojobori 2016).

## 3  Research Needs

Most genome-wide assays provide averages across large numbers of cells, but recent technological advances promise to overcome this limitation. Pioneering single-cell assays are available but yet have only applied to a few model organisms or mammalian systems (Bock et al. 2016). When having these sophisticated methodologies at hand, the genome-wide response of individual cells to hydrocarbons could be resolved with currently inconceivable spatiotemporal resolution.

There are still substantial research needs within the omics and systems biology fields. Genomics and transcriptomics have reached a mature state, but metagenomics, metatranscriptomics, proteomics, and particularly metabolomics still require substantial improvements in protocols, software, and hardware.

## References

Aylward FO, Eppley JM, Smith JM, Chavez FP, Scholin CA, DeLong EF (2015) Microbial community transcriptional networks are conserved in three domains at ocean basin scales. Proc Natl Acad Sci U S A 112:5443–5448

Bailey TL, Johnson J, Grant CE, Noble WS (2015) The MEME suite. Nucleic Acids Res 43(W1): W39–W49

Bargiela R, Herbst FA, Martínez-Martínez M, Seifert J, Rojo D, Cappello S, Genovese M, Crisafi F, Denaro R, Chernikova TN, Barbas C, von Bergen M, Yakimov MM, Ferrer M, Golyshin PN (2015) Metaproteomics and metabolomics analyses of chronically petroleum-polluted sites reveal the importance of general anaerobic processes uncoupled with degradation. Proteomics 15:3508–3520

Bezuidt O, Lima-Mendez G, Reva ON (2009) SeqWord Gene Island Sniffer: a program to study the lateral genetic exchange among bacteria. World Acad Sci Eng Technol 58:1169–1174

Bock C, Farlik M, Sheffield NC (2016) Multi-omics of single cells: strategies and applications. Trends Biotechnol 34:605–608

Chang A, Schomburg I, Placzek S, Jeske L, Ulbrich M, Xiao M, Sensen CW, Schomburg D (2015) BRENDA in 2015: exciting developments in its 25th year of existence. Nucleic Acids Res 43 (Database issue):D439–D446

Creecy JP, Conway T (2015) Quantitative bacterial transcriptomics with RNA-seq. Curr Opin Microbiol 23:133–140

Daims H, Wagner M (2007) Quantification of uncultured microorganisms by fluorescence microscopy and digital image analysis. Appl Microbiol Biotechnol 75:237–248

Dharmadi Y, Gonzalez R (2004) DNA microarrays: experimental issues, data analysis, and application to bacterial systems. Biotechnol Prog 20:1309–1324

Fislage R (1998) Differential display approach to quantitation of environmental stimuli on bacterial gene expression. Electrophoresis 19:613–616

Galagan J, Lyubetskaya A, Gomes A (2013) ChIP-Seq and the complexity of bacterial transcriptional regulation. Curr Top Microbiol Immunol 363:43–68

Garza DR, Dutilh BE (2015) From cultured to uncultured genome sequences: metagenomics and modeling microbial ecosystems. Cell Mol Life Sci 72:4287–4308

Graham JE, Clark-Curtiss JE (1999) Identification of *Mycobacterium tuberculosis* RNAs synthesized in response to phagocytosis by human macrophages by selective capture of transcribed sequences (SCOTS). Proc Natl Acad Sci U S A 96:11554–11559

Hrdlickova R, Toloue M, Tian B (2016) RNA-Seq methods for transcriptome analysis. Wiley Interdiscip Rev RNA. doi:10.1002/wrna.1364

Huson DH, Beier S, Flade I, Górska A, El-Hadidi M, Mitra S, Ruscheweyh HJ, Tappu R (2016) MEGAN community edition – interactive exploration and analysis of large-scale microbiome sequencing data. PLoS Comput Biol 12:e1004957

Ito T, Sakaki Y (1997) Fluorescent differential display. Methods Mol Biol 85:37–44

Kodzius R, Gojobori T (2016) Single-cell technologies in environmental omics. Gene 576:701–707

Liu XS (2007) Getting started in tiling microarray analysis. PLoS Comput Biol 3:1842–1844

Losada PM, Chouvarine P, Dorda M, Hedtfeld S, Mielke S, Schulz A, Wiehlmann L, Tümmler B (2016) The cystic fibrosis lower airways microbial metagenome. ERJ Open Res 2:00096–02015

Maaß S, Becher D (2016) Methods and applications of absolute protein quantification in microbial systems. J Proteome 136:222–233

Mazurkiewicz P, Tang CM, Boone C, Holden DW (2006) Signature-tagged mutagenesis: barcoding mutants for genome-wide screens. Nat Rev Genet 7:929–939

Myers KS, Park DM, Beauchene NA, Kiley PJ (2015) Defining bacterial regulons using ChIP-seq. Methods 86:80–88

Ness SA (2007) Microarray analysis: basic strategies for successful experiments. Mol Biotechnol 36:205–219

Neufeld JD, Wagner M, Murrell JC (2007) Who eats what, where and when? Isotope-labelling experiments are coming of age. ISME J 1:103–110

Otto A, van Dijl JM, Hecker M, Becher D (2014) The *Staphylococcus aureus* proteome. Int J Med Microbiol 304:110–120

Pierneef R, Cronje L, Bezuidt O, Reva ON (2015) Pre_GI: a global map of ontological links between horizontally transferred genomic islands in bacterial and archaeal genomes. Database 2015:bav058

Putri SP, Nakayama Y, Matsuda F, Uchikata T, Kobayashi S, Matsubara A, Fukusaki E (2013) Current metabolomics: practical applications. J Biosci Bioeng 115:579–589

Rediers H, Rainey PB, Vanderleyden J, De Mot R (2005) Unraveling the secret lives of bacteria: use of in vivo expression technology and differential fluorescence induction promoter traps as tools for exploring niche-specific gene expression. Microbiol Mol Biol Rev 69:217–261

Reva ON, Hallin PF, Willenbrock H, Sicheritz-Ponten T, Tümmler B, Ussery DW (2008) Global features of the *Alcanivorax borkumensis* SK2 genome. Environ Microbiol 10:614–625

Sabirova JS, Ferrer M, Regenhardt D, Timmis KN, Golyshin PN (2006) Proteomic insights into metabolic adaptations in *Alcanivorax borkumensis* induced by alkane utilization. J Bacteriol 188:3763–3773

Segata N, Boernigen D, Tickle TL, Morgan XC, Garrett WS, Huttenhower C (2013) Computational meta'omics for microbial community studies. Mol Syst Biol 9:666

Soufi B, Macek B (2015) Global analysis of bacterial membrane proteins and their modifications. Int J Med Microbiol 305:203–208

van Beilen JB, Panke S, Lucchini S, Franchini AG, Röthlisberger M, Witholt B (2001) Analysis of *Pseudomonas putida* alkane-degradation gene clusters and flanking insertion sequences: evolution and regulation of the *alk* genes. Microbiology 147:1621–1630

Van Oudenhove L, Devreese B (2013) A review on recent developments in mass spectrometry instrumentation and quantitative tools advancing bacterial proteomics. Appl Microbiol Biotechnol 97:4749–4762

Vasilakou E, Machado D, Theorell A, Rocha I, Nöh K, Oldiges M, Wahl SA (2016) Current state and challenges for dynamic metabolic modeling. Curr Opin Microbiol 33:97–104

Yakimov MM, Timmis KN, Golyshin PN (2007) Obligate oil-degrading marine bacteria. Curr Opin Biotechnol 18:257–266

Yarza P, Yilmaz P, Pruesse E, Glöckner FO, Ludwig W, Schleifer KH, Whitman WB, Euzéby J, Amann R, Rosselló-Móra R (2014) Uniting the classification of cultured and uncultured bacteria and archaea using 16S rRNA gene sequences. Nat Rev Microbiol 12:635–645

# One-Component Systems that Regulate the Expression of Degradation Pathways for Aromatic Compounds

# 10

G. Durante-Rodríguez, H. Gómez-Álvarez, J. Nogales, M. Carmona, and E. Díaz

## Contents

| | | |
|---|---|---|
| 1 | Introduction | 138 |
| 2 | Effector-Specific Transcriptional Regulators Involved in the Catabolism of Aromatic Compounds: General Features | 139 |
| 3 | Aromatic Compound Responsive Regulators of the LysR Family | 143 |
| 4 | Aromatic Compound Responsive Regulators of the NtrC Family | 150 |
| 5 | Aromatic Compound Responsive Regulators of the AraC/XylS Family | 152 |
| 6 | Aromatic Compound Responsive Regulators of the CRP/FNR Family | 154 |
| 7 | Aromatic Compound Responsive Regulators of the IclR Family | 156 |
| 8 | Aromatic Compound Responsive Regulators of the TetR Family | 157 |
| 9 | Aromatic Compound Responsive Regulators of the BzdR Subfamily | 159 |
| 10 | Aromatic Compound Responsive Regulators of the GntR Family | 160 |
| 11 | Aromatic Compound Responsive Regulators of the MarR Family | 162 |
| 12 | Aromatic Compound Responsive Regulators that Belong to Other Families | 164 |
| 13 | Research Needs | 164 |
| References | | 166 |

**Abstract**

The expression of pathways for the catabolism of aromatic compounds is energetically expensive, and aromatic compounds are generally toxic even to bacteria that can use them as growth substrates. Hence, complex regulatory circuits that control the expression of the degradation pathways have evolved. Transcriptional

G. Durante-Rodríguez and H. Gómez-Álvarez contributed equally to this work.

G. Durante-Rodríguez · H. Gómez-Álvarez · J. Nogales · M. Carmona · E. Díaz (✉)
Environmental Biology Department, Centro de Investigaciones Biológicas-CSIC, Madrid, Spain
e-mail: gdurante@cib.csic.es; hgomalv@cib.csic.es; jnogales@cib.csic.es; mcarmona@cib.csic.es; ediaz@cib.csic.es

© Springer International Publishing AG, part of Springer Nature 2018
137
T. Krell (ed.), *Cellular Ecophysiology of Microbe: Hydrocarbon and Lipid Interactions*,
Handbook of Hydrocarbon and Lipid Microbiology, https://doi.org/10.1007/978-3-319-50542-8_5

regulation appears to be the most common mechanism for control of gene expression. Effector-specific transcriptional regulation of aromatic catabolic pathways depends on the performance of a specific regulator acting on a specific promoter and responding to a specific effector signal. One-component regulatory systems combine within the same cytosolic protein the effector-binding input domain and a DNA-binding output domain. A great variety of one-component regulatory systems can be classified within different families of prokaryotic transcriptional regulators revealing a wide diversity in their evolutionary origins and showing that a regulatory issue, i.e., having an operon induced in the presence of a given aromatic compound, can be solved through different types of regulators and mechanisms of transcriptional control in different bacteria. The effector-specific regulation can be tightly fine-tuned by the action of certain modulators and is, in turn, under control of overimposed mechanisms that connect the metabolic and energetic status of the cell to the activity of the individual catabolic clusters, leading to complex regulatory networks. Elucidating such regulatory networks will pave the way for a better understanding of the regulatory intricacies that control microbial biodegradation of aromatic compounds, which are key issues that should be taken into account for the rational design of more efficient recombinant biodegraders, bacterial biosensors, and biocatalysts for modern green chemistry.

# 1    Introduction

Aromatic compounds are the second most widely distributed class of organic compounds in nature (Díaz et al. 2013). Although some of these compounds are recalcitrant or toxic for the vast majority of the microorganisms, bacteria usually have evolved biochemical and genetic information that allow them to use the aromatic compounds as a sole carbon and energy sources (Lovley 2003). The production of the multiple enzymes of pathways for the catabolism of aromatic compounds is energetically expensive, and aromatic compounds are generally toxic even to bacteria that can use them as growth substrates. Hence, complex regulatory circuits that control the expression of the degradation pathways have evolved (Lovley 2003; Cases and de Lorenzo 2005; Carmona et al. 2008; Díaz et al. 2013). Although regulation can be carried out at different levels (transcription, translation, posttranslation), transcriptional regulation appears to be the most common mechanism for control of gene expression. Transcriptional regulation of aromatic catabolic pathways is not just dependent on the performance of a specific regulator acting on a specific promoter and responding to a specific environmental signal (effector-specific transcriptional regulation) but is also dependent on overimposed mechanisms that connect the metabolic and energetic status of the cell to the activity of the individual catabolic clusters (Fig. 1) (Díaz and Prieto 2000; Tropel and van der Meer 2004; Carmona et al. 2008; Díaz et al. 2013). In this chapter, we will review the effector-specific transcriptional regulation involved in the catabolism of aromatic compounds.

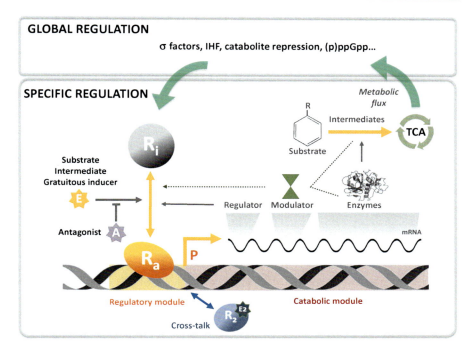

**Fig. 1** Scheme of the regulatory network that controls the expression of genes responsible for the catabolism of aromatic compounds. The effector-specific regulator can be either in its inactive (Ri) or active and bound to DNA (Ra) forms. The interaction with the effector molecule (E) determines the transition between both states, leading to derepression (as in case of a repressor) or activation (as in case of an activator) of gene expression from the target promoter (P). Structural analogues of the effector molecules may act as antagonists (A) preventing the action of the former. Auxiliary regulators (modulator), that interact with the enzymatic machinery or some pathway intermediates, may control the activation of the main regulator modifying its final regulatory output. Additional regulatory loops based on cross-talk regulation of the target promoter by a different regulator (R2) and effector (E2) couple can exist. The specific regulation is generally subjected to a more global level of regulation dependent on the overall physiological state of the cell and that in turn responds to the final metabolic flux derived from the funneling of the cognate aromatic compound to the central metabolism (TCA)

## 2   Effector-Specific Transcriptional Regulators Involved in the Catabolism of Aromatic Compounds: General Features

Signal transduction in prokaryotes is conducted by two major regulatory systems, (i) one-component systems and (ii) two-component systems. Two-component regulatory systems function as a result of phosphotransfer between two key proteins, a sensor histidine kinase (input element) and a cytosolic response regulator (output element). One-component systems combine within the same cytosolic protein the sensor input domain and a functional output domain. Small molecule-binding motifs constitute the majority of the input domains, and helix-turn-helix (HTH) DNA-binding motifs are the most common output domains (Ulrich et al. 2005). In this chapter, however, we will focus only on one-component transcriptional regulatory systems that, in the presence of the inducer molecule (effector), assure the production of the enzymes and transporters involved in the catabolism of aromatic compounds either by activating (activator) or derepressing (repressor) their cognate genes (Fig. 1).

The effector-specific regulator can be either in its inactive (Ri) or active and bound to DNA (Ra) forms. The interaction with the effector molecule (E) determines the transition between both states, leading to derepression (as in case of a repressor) or activation (as in case of an activator) of gene expression from the target promoter (P). Structural analogues of the effector molecules may act as antagonists (A) preventing the action of the former. Auxiliary regulators (modulator), that interact with the enzymatic machinery or some pathway intermediates, may control the activation of the main regulator modifying its final regulatory output. Additional regulatory loops based on cross-talk regulation of the target promoter by a different regulator (R2) and effector (E2) couple can exist. The specific regulation is generally subjected to a more global level of regulation dependent on the overall physiological state of the cell and that in turn responds to the final metabolic flux derived from the funneling of the cognate aromatic compound to the central metabolism (TCA).

Regulators that show similar domain architectures might be responsible for different biological effects depending on the locations of their binding sites (operator regions) in the cognate promoters. Thus, the same regulator can activate some genes when it binds upstream of the RNA polymerase (RNAP)-binding sites while repressing others when it binds downstream of such RNAP-binding regions. The inducer molecule that activates the transcriptional regulator can be the pathway substrate and/or a pathway intermediate or product, or some structural analogues of the natural effector (gratuitous inducer) that may not themselves be a substrate for the corresponding catabolic pathway (Fig. 1). Some specific regulators have more than one effector-binding pockets, and the cognate effector molecules may have peculiar synergistic effects on transcriptional activation (Manso et al. 2009). On the contrary, efficient recognition of molecules (antagonists) that show structural similarity to the inducers (agonists) by certain transcriptional regulators leads to a lack of activation of the target promoter (Fig. 1), which may compromise an efficient degradation response when bacteria are exposed to complex mixtures of aromatic pollutants, some of which behave as agonists and other as antagonists (Silva-Jiménez et al. 2011). To prevent the gratuitous induction by non-metabolizable analogues or nonproductive intermediates, some regulatory proteins (modulators), e.g., ThnY and PaaY (García et al. 2011; Fernández et al. 2013; Ledesma-García et al. 2016), are coupled to the aromatic degradation enzymes in order to induce gene expression when there is an efficient catabolic flux in the cell (Fig. 1).

Although both transcriptional activators and repressors have been shown to regulate aromatic catabolic pathways, those pathways that use CoA-derived aromatic compounds are mainly controlled by transcriptional repressors that recognize CoA-derived effector molecules (Sakamoto et al. 2011; Hirakawa et al. 2012; Valderrama et al. 2012; Juárez et al. 2015). This observation may reflect that repressors are generally preferred to control low-demand genes whose unspecific transcription can decrease the overall fitness of the cell by spending valuable resources, such as CoA and ATP, on futile processes (Sasson et al. 2012). In some cases, the transcription factors control a set of different functionally related metabolic clusters, e.g., the PhhR regulon that assures the homeostasis of aromatic amino acids in *Pseudomonas putida* (Herrera et al. 2010).

The acquisition of specificity for a new inducer in a transcriptional regulator requires a pre-existing regulator with a certain escape (responsiveness to non-legitimate effectors or regulatory noise), upon which new specificity can be built by several rounds of natural or artificial mutagenesis and selection. In fact, regulators of microbial pathways for recent compounds (e.g., aromatic xenobiotics) are not too specific for their substrates, which may reflect an ongoing evolution of as yet not entirely optimized regulation in response to unusual nutrients (Cases and de Lorenzo 2005). In *Burkholderia* sp. DNT, the regulation of the 2,4-dinitrotoluene (DNT) degradation pathway is in an earlier stage of evolution since the NtdR regulator still recognizes salicylate, an effector of its NagR-like ancestor, but does not respond to 2,4-DNT. That a useless but still active transcriptional factor operates along enzymes that have already evolved a new substrate specificity points to the fact that the emergence of novel catalytic activities precedes the setting of a specific regulatory device for their expression, not vice versa. This shades some light into the chicken-and-the-egg dilemma between regulators and enzymes that recognize the same compounds (de las Heras et al. 2011). The evolution of transcriptional regulators has also been assessed by in vitro experimental evolution/selection setups. For instance, the XylR regulator from *P. putida* was evolved first to an effector-promiscuous variant and then to a more specific regulator where the natural response to *m*-xylene was decreased and the nonnative acquired response to the synthetic 2,4-DNT was increased. The new XylR28 version may be used to develop more efficient 2,4-DNT responsive reporter systems to engineer whole-cell biosensors for explosives (de las Heras and de Lorenzo 2011). The promiscuity or specificity of inducer recognition might be also tuned in a regulatory network just by changing the promoter architecture and without requiring the evolution of new transcription factors with altered inducer specificity, e.g., the 3-methylbenzoate-dependent induction of the *ben* operon for benzoate degradation by the BenR regulator (Silva-Rocha and de Lorenzo 2012), or the participation of some global regulators in the activation of certain promoters, e.g., the ppGpp−/DksA-independent co-stimulation of the *dmpR* regulatory gene that controls phenol degradation (del Peso-Santos et al. 2011) in *P. putida*.

Transcriptional regulators may form regulatory cascades that involve the interplay between two or more proteins controlling a pathway for the degradation of aromatic compounds (Fig. 1). Although these circuits are mainly based on an activation strategy, e.g., the XylR/XylS (Silva-Rocha et al. 2011), AadR/HbaR (Egland and Harwood 2000), and PhcT/PhcR (Teramoto et al. 2001) regulatory pairs, there are also examples of cross regulation between two transcriptional repressors that control subsequent steps in a catabolic pathway (Jiménez et al. 2011). These regulatory networks may help to maintain the catabolic pathways as an autonomous metabolic machinery that interacts only minimally with the central carbon consumption routes of the host cells (orthogonality), thus facilitating a quick spread of such degradation routes through the microbial population under suitable environmental pressure (Silva-Rocha et al. 2011).

Cross regulation between different aromatic catabolic pathways may assure a tight control of gene expression (Valderrama et al. 2012), prevent the expensive expression of funneling pathways that produce the cognate compound used as growth substrate (vertical regulation) (del Peso-Santos et al. 2006), or select for a specific pathway when mixtures of substrates that are feeding into different pathways are provided as carbon source (Bleichrodt et al. 2010). Usually there is a hierarchical use of aromatic compounds when bacteria grow in mixtures of these carbon sources in the environment, e.g., benzoate is usually a preferred carbon source over 4-hydroxybenzoate. Whereas the 4-hydroxybenzoate transport gene (*pcaK*) has been proposed as the main target of the repression in *Acinetobacter baylyi* (Brzostowicz et al. 2003) and *P. putida* (Cowles et al. 2000), the 4-hydroxybenzoate hydroxylase gene (*pobA*) is the key controlled element in *C. necator*, being benzoate itself the molecule mediating the repression through a possible interaction with the PobR regulator (Donoso et al. 2011; Pérez-Pantoja et al. 2015). Interestingly, the aromatic preference profile can change even between closely related strains (Jõesaar et al. 2010). Nevertheless, there are, of course, examples of simultaneous degradation of aromatic compound-containing mixtures, including synergistic interactions as those described in *Sagittula stellata* where increased growth rates were observed when cells were provided with benzoate/4-hydroxybenzoate mixtures compared to cells grown singly with an equimolar concentration of either substrate alone (Gulvik and Buchan 2013). Remarkably, the recent observation that there is cross regulation between aerobic and anaerobic degradation pathways could be an adaptive advantage for certain bacteria that thrive in changing oxygen environments (Valderrama et al. 2012).

Most promoters in an environmental context are regulated as part of complex circuits involving several global transcription factors (e.g., σ factors) (Fig. 1). In bacteria, transcription factors are usually present at a few copies per cell, which unavoidably leads to fluctuations in protein abundance and thus to cell heterogeneity in isogenic cell populations. This means that individual cells in the population, which may either exist in different growth phases and thus expressing different σ factors or in the same growth phase but expressing different levels of transcription factors, can activate transcription at the same promoter to different degrees. The existence of subpopulations that express differentially certain catabolic pathways can in turn favor the adaptation of the cell community to unpredictable environmental changes (Guantes et al. 2015).

One-component regulatory systems can be classified within different families of prokaryotic transcriptional regulators based on the sequence and structure of the DNA-binding motif of the output domains. Examples of regulators involved in controlling the expression of aromatic catabolic pathways reveal a wide diversity in their evolutionary origins and show that a regulatory problem, i.e., having an operon induced in the presence of a given aromatic compound, can be solved through different types of regulators and mechanisms of transcriptional control in different bacteria (Table 1). Moreover, this global view confirms that the regulatory networks have an extraordinary degree of plasticity and adaptability, reinforcing the idea that catabolic and regulatory genes in aromatic degradation pathways have evolved independently (Cases and de Lorenzo 2005; Carmona et al. 2008).

Below, we briefly explain the general features of the different families of one-component transcriptional regulators controlling the catabolism of aromatic compounds, paying special attention to those examples that have been characterized in the last 10 years. Other reviews on various aspects of the regulation of aromatic degradation pathways have been previously published (Díaz and Prieto 2000; Tropel and van der Meer 2004; Carmona et al. 2008). Two-component transcriptional regulators are extensively discussed in a different chapter of this series.

## 3   Aromatic Compound Responsive Regulators of the LysR Family

The LysR family of transcriptional regulators (LTTRs) represents the largest family of bacterial transcription factors (Pareja et al. 2006; Maddocks and Oyston 2008), and family members regulate the expression of a wide variety of biological functions (Schell 1993; Tropel and van der Meer 2004; Maddocks and Oyston 2008). LTTRs were firstly described as transcriptional activators; however, several LTTRs have been shown to act as repressors (Jourlin-Castelli et al. 2000; Kim et al. 2003), and some LTTRs, such as GltC, act both as activators and repressors (Picossi et al. 2007).

LTTRs controlling the catabolism of aromatic compounds are co-inducer-responsive transcriptional regulators that have been described exclusively as activators (Table 1) and that were associated initially with the classic β-ketoadipate pathway. Some examples include CatR and ClcR involved in the degradation of catechol and chlorocatechol, respectively, in *P. putida* (McFall et al. 1998); BenM/CatM and SalR involved in benzoate and salicylate degradation, respectively, by the soil bacterium *Acinetobacter baylyi* strain ADP1 (Collier et al. 1998; Jones et al. 2000; Vaneechoutte et al. 2006); and PcaQ from *Agrobacterium tumefaciens* controlling the conversion of protocatechuate into β-ketoadipate (Parke 1996) (Table 1). However, it is now evident that LTTRs control a broad spectrum of aromatic degradation pathways (Table 1). For instance, HcaR regulates the expression of *hca* genes for the initial catabolism of 3-phenylpropionic acid in *E. coli* K12 (Díaz et al. 1998), TsaR regulates the *tsa* operon encoding the first steps in the degradation of *p*-toluenesulfonate in *Comamonas testosteroni* T-2 (Tralau et al. 2003b), AphT controls the *meta*-cleavage pathway of phenol degradation in *Comamonas testosteroni* TA441 (Arai et al. 2000), and DntR activates the expression of the *dnt* genes for 2,4-DNT degradation in *Burkholderia* sp. DNT (de las Heras et al. 2011). Recently, the DbdR protein from *Thauera aromatica* strain AR-1 was shown to control the

**Table 1** Some examples of regulatory proteins of aromatic catabolic pathways in bacteria

| Regulator[a] | Family | Microorganism | Activity[a] | Pathway[b] | Accession no. |
|---|---|---|---|---|---|
| CatM | LysR | A. baylyi ADP1 | Activator | Catechol | P07774 |
| ClcR | LysR | R. opacus 1CP (pAC25) | Activator | 3-chlorobenzoate | AAC38250 |
| CatR | LysR | P. putida | Activator | Catechol | A35118 |
| CatR | LysR | A. lwoffi | Activator | Phenol/Catechol | O33945 |
| CdoR | LysR | Comamonas sp. JS765 | Activator | Catechol | AAC79916 |
| AphT | LysR | C. testosteroni TA441 | Activator | Phenol | BAA88500 |
| NahR | LysR | P. putida (pNAH7) | Activator | Naphthalene/Salicylate | A31382 |
| NahR | LysR | P. stutzeri AN10 | Activator | Naphthalene/Salicylate | AAD02145 |
| SalR | LysR | A. baylyi ADP1 | Activator | Salicylate | AAF04311 |
| PhnS | LysR | Burkholderia sp. RP007 | Activator | Phenanthrene/Naphthalene | AAD09867 |
| TcbR | LysR | Pseudomonas sp. P51 (pP51) | Activator | Trichlorobenzene | A38861 |
| CbnR | LysR | R. eutropha NH9 | Activator | Chlorocatechol | BAA74529 |
| PcaQ | LysR | A. tumefaciens A348 | Activator | Protocatechuate (ortho-cleavage) | AAA91130 |
| LigR | LysR | Sphingobium sp. SYK-6 | Activator | Protocatechuate (meta-cleavage) | BAB88739 |
| BenM | LysR | A. baylyi ADP1 | Activator | Benzoate | AAC46441 |
| TfdT | LysR | R. eutropha JMP134 | Activator | 3-,4-chlorocatechol(2-,3-chlorobenzoate) | AAC44724 |
| NtdR | LysR | Acidovorax sp. JS42 | Activator | 2,4- and 2,6-dinitrotoluene | AAP70492 |
| DntR | LysR | Burkholderia sp. DNT | Activator | 2,4-dinitrotoluene | AAP70493 |
| NbzR | LysR | Comamonas sp. JS765 | Activator | Salicylate, anthranilate | AAL76198 |
| LinR | LysR | Sphingomonas paucimobilis UT26 | Activator | 2,5 and 2,6-dichlorohydroquinone | BAA36280 |
| DbdR | LysR | Thauera aromatica AR-1 | Activator | 3,5-dihydroxybenzoate (α-resorcylate; anaerobic) | AIO06107 |
| ThnR | LysR | Sphingopysis granuli TFA | Activator | Tetralin | AAU12855 |
| Orf3 | LysR | Pseudomonas sp. HR199 | Activator | 4-hydroxybenzoate | Y18527 |
| HcaR | LysR | E. coli K-12 | Activator | 3-phenylpropionic acid | Q47141 |
| TdnR | LysR | P. putida UCC22 (pTDN1) | Activator | Aniline | BAA12810 |
| TsaR | LysR | C. testosteroni T-2 | Activator | p-toluene sulfonate | AAC44806 |
| PucR | LysR | B. subtilis | Activator | Purines | O32146 |
| AtzR | LysR | Pseudomonas sp. ADP | Activator | Cyanuric acid | NP_862536 |

| Regulator | Family | Organism | Function | Substrate | Accession |
|---|---|---|---|---|---|
| XylR | NtrC | *P. putida* mt-2 (pWW0) | Activator | Toluene | AAA26028 |
| TouR | NtrC | *P. stutzeri* OX1 | Activator | Toluene | CAB52211 |
| TbuT | NtrC | *B. pickettii* PKO1 | Activator | Toluene | AAC44567 |
| TbmR | NtrC | *Burkholderia* sp. JS150 | Activator | Toluene | |
| DmpR | NtrC | *Pseudomonas* sp. CF600 (pVI150) | Activator | Phenol | A47078 |
| PhhR | NtrC | *P. putida* P35X | Activator | Phenol | S47095 |
| AphR | NtrC | *C. testosteroni* TA 441 | Activator | Phenol | BAA34177 |
| MopR | NtrC | *A. calcoaceticus* NCIB8250 | Activator | Phenol | CAA93242 |
| MphR | NtrC | *A. pittii* PHEA-2 | Activator | Phenol | ACO88925 |
| PhlR | NtrC | *P. putida* H (pPGH1) | Activator | Phenol | CAA62584 |
| PhlR | NtrC | *R. eutropha* JMP134 | Activator | Phenol | AAC77386 |
| PhcR | NtrC | *C. testosteroni* R5 | Activator | Phenol | BAA87867 |
| PoxR | NtrC | *R. eutropha* E2 | Activator | Phenol | AAC32451 |
| PheR | NtrC | *P. putida* BH | Activator | Phenol | D63814 |
| PdeR | NtrC | *T. aromatica* K172 | Activator | Phenol (anaerobic) | CAC12685 |
| PdeR | NtrC | "*Aromatoleum aromaticum*" EbN1 | Activator | Phenol (anaerobic) | CAI07889 |
| PhnR | NtrC | *Burkholderia* sp. RP007 | Activator | Phenanthrene/Naphthalene | AAD09866 |
| TmbR | NtrC | *P. putida* TMB | Activator | Trimethylbenzene | U41301 |
| HbpR | NtrC | *P. azelaica* HBP1 | Activator | 2-hydroxybiphenyl | AAA84988 |
| BphR | NtrC | *S. aromaticivorans* F199 (pLN1) | Activator | Biphenyl | AAD03979 |
| AreR | NtrC | *A. baylyi* ADP1 | Activator | Aryl esters | AF1509028 |
| EtpR | NtrC | "*Aromatoleum aromaticum*" EbN1 | Activator | *p*-ethylphenol/ *p*-hydroxyacetophenone (anaerobic) | CAI06292 |
| XylS | AraC/XylS | *P. putida* mt-2 (pWW0) | Activator | 3-methylbenzoate/Benzoate | AAA26029 |
| BenR | AraC/XylS | *P. putida* | Activator | Benzoate | AF218267 |
| HpaA | AraC/XylS | *E. coli* W | Activator | 3-, 4-hydroxyphenylacetate | Z37980 |
| PobR | AraC/XylS | *Azotobacter chroococcum* | Activator | 4-hydroxybenzoate | AAF03756 |
| MaoB | AraC/XylS | *E. coli* K-12 | Activator | Tyramine | BAA11058 |
| IpbR | AraC/XylS | *P. putida* RE204 | Activator | Isopropylbenzene | AF006691 |
| CbdS | AraC/XylS | *Burkholderia* sp. TH2 | Activator | 2-halobenzoate | BAB21583 |

*(continued)*

**Table 1** (continued)

| Regulator[a] | Family | Microorganism | Activity[a] | Pathway[b] | Accession no. |
|---|---|---|---|---|---|
| OxoS | AraC/XylS | *P. putida* 86 | Activator | Quinoline | CAA73202 |
| AntR | AraC/XylS | *P. resinovorans* CA10 | Activator | Carbazole, anthranilate | BAC41529 |
| IifR | AraC/XylS | *A. baumannii* | Activator | Indole | ENW75084 |
| PheR | AraC/XylS | *R. erythropolis* | Activator | Phenol | CAJ01323 |
| AraC | AraC/XylS | *Mycobacterium gilvum* PYR-GCK | Activator | Pyrene | ABP47731 |
| FeaR | AraC/XylS | *E. coli* | Activator | Phenylethylamine | CTY06606 |
| AadR | CRP/FNR | *R. palustris* | Activator | *Benzoate/4-hydroxybenzoate/Cyclohexanecarboxylate (anaerobic)* | Q01980 |
| HbaR | CRP/FNR | *R. palustris* | Activator | *4-hydroxybenzoate (anaerobic)* | AAF04013 |
| CprK | CRP/FNR | *D. hafniense* | Activator | *o-chlorophenol (anaerobic)* | AAL87770 |
| GlxR | CRP/FNR | *Corynebacterium glutamicum* | Repressor | 3-hydroxybenzoate/Gentisate | Q79VI7 |
| CRP | CRP/FNR | *E. coli* K-12 | Activator | 3-hydroxyphenylpropionate/4-hydroxyphenylacetate/phenylacetate | J01598 |
| AcpR | CRP/FNR | *Azoarcus* sp. CIB | Activator | *Benzoate/3-methylbenzoate (anaerobic)* | AAY81959 |
| CymR | TetR | *P. putida* F1 | Repressor | *p*-cymene | AAB62296 |
| PaaR | TetR | *C. glutamicum* | Repressor | Phenylacetate | CCF55037 |
| PaaR | TetR | *T. thermophilus* HB8 | Repressor | Phenylacetate | BAD70796 |
| PaaR | TetR | *Streptomyces pristinaespiralis* | Repressor | Phenylacetate | EDY64415 |
| PfmR | TetR | *T. thermophilus* HB8 | Repressor | Phenylacetate | PDB: 3VPR_D |
| HdnoR | TetR | *A. nicotinovorans* | Repressor | Nicotine | ABA41004 |
| NicS | TetR | *P. putida* | Repressor | Nicotinate | Q88FX7 |
| RolR | TetR | *C. glutamicum* | Repressor | Resorcinol | BAB97394 |
| MbdR | TetR | *Azoarcus* sp. CIB | Repressor | *3-methylbenzoate (anaerobic)* | CCH23038 |
| BzdR | BzdR | *Azoarcus* sp. CIB | Repressor | *Benzoate (anaerobic)* | AAQ08805 |
| BoxR | BzdR | *Azoarcus* sp. CIB | Repressor | Benzoate | CCD33120 |
| PaaX | GntR | *E. coli* W | Repressor | Phenylacetate | CAA66101 |
| PaaN | GntR | *P. putida* U | Repressor | Phenylacetate | AF029714 |
| BphS | GntR | *R. eutropha* A5 | Repressor | Biphenyl/4-chlorobiphenyl | CAC05302 |
| BphR1 | GntR | *P. pseudoalcaligenes* KF707 | Activator | Biphenyl | BAA12882 |
| AphS | GntR | *C. testosteroni* TA441 | Repressor | Phenol | BAA8295 |
| PhcS | GntR | *C. testosteroni* R5 | Repressor | Phenol | BAB61103 |

| | | | | | |
|---|---|---|---|---|---|
| VanR | GntR | A. baylyi ADP1 | Repressor | Vanillate | O24839 |
| VanR | GntR | P. putida WCS358 | Repressor | Vanillate | AJ252091 |
| VanR | GntR | Caulobacter crescentus CB15 | Repressor | Vanillate | AAK24363 |
| CarR$_{J3}$ | GntR | Janthinobacterium sp. 13 | Repressor | Carbazole | BAC56739 |
| MeqR2 | GntR | Arthrobacter sp. Rue61a | Repressor | Quinaldine | AFR31129 |
| PbaR | IclR | Sphingobium wenxiniae JZ-1T | Activator | 3-phenoxybenzoate | A0A0K2CTW1 |
| IphR | IclR | Comamonas sp. E6 | Repressor | Isophthalate | C4TNS6 |
| TphR | IclR | Comamonas sp. E6 | Activator | Terephthalate | AB238679 |
| MhpR | IclR | C. testosteroni TA441 | Activator | 3-hydroxyphenylpropionate | Q9S159 |
| MhpR | IclR | E. coli K-12 | Activator | 3-hydroxyphenylpropionate | P77569 |
| TsaQ | IclR | C. testosteroni T-2 | Activator | p-toluenesulfonate | Q6XL52 |
| PobR | IclR | A. baylyi ADP1 | Activator | 4-hydroxybenzoate | Q43992 |
| PcaU | IclR | A. baylyi ADP1 | Activator/Repressor | Protocatechuate | AAC37157 |
| PcaR | IclR | P. putida PRS2000 | Activator | Protocatechuate | Q52154 |
| HmgR | IclR | P. putida KT2440 | Repressor | Homogentisate | Q88E46 |
| TsdR | IclR | R. jostii RHA1 | Repressor | γ-Resorcylate | Q0SFL4 |
| OphR | IclR | Rhodococcus sp DK17 | Repressor | Phthalate | A4ZXZ2 |
| CatR | IclR | R. erythropolis CCM2595 | Repressor | Catechol | T1VW59 |
| NpdR | IclR | Rhodococcus opacus HL PM-1 | Repressor | 2,4,6-trinitrophenol (picric acid) | Q9AH06 |
| GenR | IclR | C. glutamicum | Activator | 3-hydroxybenzoate /Gentisate | Q8NLB8 |
| HpaR | MarR | E. coli W | Repressor | Homoprotocatechuate | AFH14229 |
| CbaR | MarR | C. testosteroni BR60 | Repressor | Chlorobenzoate | NP_869726 |
| NbzR | MarR | P. putida HS12 (pNB1) | Repressor | Aminophenol | AAK26517 |
| HcaR | MarR | A. baylyi ADP1 | Repressor | p-hydroxycinnamates | AAP78949 |
| CouR | MarR | R. jostii RHA1 | Repressor | p-hydroxycinnamates | PDB: 5CYV_B |
| CouR | MarR | R. palustris | Repressor | p-coumarate (anaerobic) | CAE27235 |
| FerR | MarR | P. fluorescens BF13 | Repressor | Ferulate | CAD60265 |
| FerC | MarR | Sphingobium sp. SYK-6 | Repressor | Ferulate | SLG_25040 |
| PcaV | MarR | Streptomyces coelicolor | Repressor | Protocatechuate (ortho-cleavage) | PDB: 4G9Y_A |
| GenR | MarR | C. testosteroni CNB-1 | Repressor | Gentisate | ACY33523 |
| IacR | MarR | Acinetobacter baumannii | Repressor | Indole-3-acetate | |
| BadR | MarR | R. palustris | Repressor | Cyclohexanecarboxylate (anaerobic) | CAE26099 |

(continued)

**Table 1** (continued)

| Regulator[a] | Family | Microorganism | Activity[a] | Pathway[b] | Accession no. |
|---|---|---|---|---|---|
| NicR | MarR | *P. putida* KT2440 | Repressor | Nicotinate | Q88FY0 |
| BadM | Rrf2 | *R. palustris* | Repressor | *Benzoate (anaerobic)* | CAE26107 |
| BgeR | Rrf2 | *Geobacter bemidjiensis* | Repressor | *Benzoate (anaerobic)* | ACH38458 |
| VanR | PadR | *C. glutamicum* | Repressor | Vanillate | NP_601583 |

[a] Activators are indicated in green. Repressors are indicated in red
[b] Anaerobic pathways are indicated in italics

expression of the genes involved in the 3,5-dihydroxybenzoate (alpha-resorcylate) anaerobic degradation pathway, thus expanding the scope of LTTRs to the control of the anaerobic metabolism of aromatic compounds (Molina-Fuentes et al. 2015).

LTTRs involved in the degradation of aromatic compounds typically activate divergently transcribed catabolic promoters in response to inducers, usually intermediates of the corresponding catabolic pathways, and they repress their own synthesis (Schell 1993; Tropel and van der Meer 2004; Maddocks and Oyston 2008). This preference for an intermediate of the catabolic pathway rather than by the initial substrate of the pathway has been suggested as a mechanism to avoid gratuity expression. An unprecedented example of a regulatory system to prevent gratuitous induction has been described for the tetralin biodegradation genes (*thn*) in *Sphingopyxis granuli* strain TFA. ThnR is an LTTR that activates transcription in response to tetralin, but its activity is under the control of ThnY (López-Sánchez et al. 2009; García et al. 2011). ThnY is an iron-sulfur flavoprotein which, in the absence of an efficient substrate acting as an electron sink, is reduced by the ThnA3 ferredoxin of the tetralin dioxygenase complex avoiding expression of the *thn* genes irrespective of the ThnR activation by gratuitous inducers (Ledesma-García et al. 2016). This scenario occurs in the presence of an inducer molecule that is not a substrate for the dioxygenase (deficient electron flux to the dioxygenase) or when the concentration of the inducer is very low (Ledesma-García et al. 2016).

Aromatic compound responsive LTTRs bind within the target promoter, independently of the presence of the inducer molecule, to a long sequence of approximately 50–60 bp which contains two distinct sites, RBS and ABS. RBS (recognition binding site) is a high-affinity binding site centered at position $-66$ and encompassing a characteristic inverted repeat motif including a T-N11-A consensus sequence. ABS (activation binding site) is a low-affinity binding site with half-dyad symmetry and located at positions $-27$ to $-32$. Type I LTTRs bind to ABS only in the presence of the inducer molecule in contrast to type II LTTRs which bind ABS irrespective of inducer. Based on this plasticity, the so-called sliding dimer model has been proposed as the mechanism of activation of the target promoter by LTTRs. This mechanism has been well studied in AtzR, an LTTR responsible for activation of the *atzDEF* cyanuric acid utilization operon in *Pseudomonas* sp. ADP (Porrúa et al. 2007). According to this model, in the absence of inducer, a type I LTTR dimer binds only the RBS site, while the ABS site is also occupied by a second dimer in type II LTTRs, causing in both cases a DNA bending. A conformational change upon inducer binding causes a shift from a more proximal ABS subside to a more distal ABS subside in type II LTTRs or the occupancy of the ABS site by the second dimer in type I LTTRs. In both cases, a relaxation of the DNA bending allows the formation of an active complex with the RNAP leading to transcription activation (Tropel and van der Meer 2004; Porrúa et al. 2007; Maddocks and Oyston 2008). However, an activation mechanism based on bend induction rather than bend relaxation has been found in LigR, an LTTR involved in regulation of the protocatechuate *meta*-cleavage pathway in *Sphingobium* sp. strain SYK-6 (Kamimura et al. 2010). Thus, the mechanistic versatility of LTTRs is still far to be completely understood.

LTTRs act as tetramers in its biologically active form; however, variable oligo-meric states, ranging from monomers to homotetramers, are found when LTTRs are in solution (Schell 1993, Tropel and van der Meer 2004; Maddocks and Oyston 2008). LTTRs display two domains, an N-terminal domain that contains a winged helix-turn-helix (wHTH) motif for DNA binding and a C-terminal domain that provides the effector-binding and multimerization functions. The three-dimensional structures of the C-terminal domains of BenM and CatM have been determined and were found to consist of nine α-helices and nine β-strands with Rossmann-like folds (Ezezika et al. 2007). Other aromatic responsive LTTRs, such as DntR, appear to have similar structure (Smirnova et al. 2004). The first full-length LTTR crystal structure resolved was that of CbnR which controls the degradation of chlorocatechols in *Ralstonia eutropha* NH9 (Ogawa et al. 1999; Muraoka et al. 2003). CbnR was crystallized as a tetramer consisting of two dimers. Each dimer is composed by a short-form and an extended-form subunit (protomers), resulting in a tetrameric molecule with asymmetrical ellipsoidal shape. The increasing number of full-length LTTR structures resolved in the last years has contributed to a better understanding of the structure-function relationships of this group of transcriptional regulators (Monferrer et al. 2010). The protomers are composed of two domains, the N-terminal domain which harbors the DNA-binding elements and a C-terminal domain responsible of binding to inducer and connected by a large linker helix. Two different conformations, extended and compact, were found in the protomers of the asymmetric dimer (Monferrer et al. 2010). The full-length structure of DntR has revealed that while apo-DntR maintains an inactive compact configuration in solu-tion, the inducer-bound holo-DntR adopts an expanded conformation. These obser-vations are consistent with the known shifting of LTTR DNA-binding sites upon activation and the consequent relaxation in the bend of the promoter-operator region DNA, thus strongly supporting the sliding dimer model of activation proposed for LTTRs (Lerche et al. 2016).

# 4    Aromatic Compound Responsive Regulators of the NtrC Family

NtrC-like regulators are activators of promoters that utilize the alternative sigma factor $\sigma^{54}$. The $\sigma^{54}$ promoters display a particular architecture defined by a highly conserved $-12/-24$ sequence recognized by the $\sigma^{54}$-RNAP holoenzyme, upstream activator sequences (UASs) situated more than 100 bp upstream the transcriptional start site and that bind to the NtrC-like regulator, and a DNA intrinsic or protein-induced curvature that promote the DNA looping required for the specific contact between the activator and the $\sigma^{54}$-RNAP subunit (Beck et al. 2007). Typically, NtrC-like activators consist of three different domains: (i) an N-terminal regulatory domain, (ii) a central domain responsible of multimerization into a hexamer and of ATP-hydrolyzing (activating) activity, and (iii) a C-terminal DNA-binding domain. Upon effector binding, NtrC regulators oligomerize and bind to the UAS. After the loop formation and the contact between the activator and $\sigma^{54}$-RNAP holoenzyme are established, the energy produced by ATP hydrolysis is invested in remodeling the transcription complex from its closed configuration into an open transcriptionally active form (Bush and Dixon 2012).

Several NtrC-like activators respond to aromatic compounds (Table 1), but the best characterized are the XylR and DmpR proteins from *Pseudomonas* strains. XylR is the transcriptional regulator of the upper operon controlled by the *Pu* promoter and that encodes the upper route of the TOL catabolic pathway for toluene, *m*-xylene, and *p*-xylene degradation in *P. putida* (Ramos et al. 1997, Galvão and de Lorenzo 2006). The N-terminal domain of XylR generates an intramolecular repression on the central activating domain of the protein. Two additional domains described in XylR are a Q-linker of 20 residues that might be involved in protein oligomerization and a DNA-binding domain that shares similarity to that of the Fis protein, a well-known global regulator (Garmendia and de Lorenzo 2000; O'Neill 2001). The binding to the N-terminal domain of XylR of a surprising variety of alkylbenzene effectors (Galvão and de Lorenzo 2006) releases its repression on the activating domain and enables the binding to the target promoter and ATP hydrolysis (Perez-Martin and de Lorenzo 1995), which constitutes the molecular basis for the activation of the $\sigma^{54}$-dependent *Pu* promoter. The *xylR* gene is transcribed from the *Pr* promoter, and XylR levels in the cell are negatively regulated at the transcriptional level by a XylR self-repression that requires the participation of the IHF host factor forming an unusual feed-forward regulatory loop (Guantes et al. 2015). Moreover, XylR production is also subject of a complex posttranscriptional mechanism in which the so-called catabolic repressor control protein (Crc) acts as a translational co-repressor along with the RNA-binding factor Hfq (Moreno et al. 2015). Recently it was shown that the levels of the global regulators IHF and Crc are subject to growth phase-dependent control which in turn originates a bimodal regime of *Pu* expression in exponential phase, where a fraction of the population remains inactive at any one time after induction by *m*-xylene, and an unimodal response in stationary phase, where the whole population is induced at comparable rates. These results highlight the importance of cell physiology and internal composition and its impact on phenotypic variability that may be advantageous in competitive environmental settings (Guantes et al. 2015).

The ability of XylR to recognize several alkylbenzene compounds has been exploited to develop biosensors for BTEX, the more abundant aromatic mixture in the oil industry (Kim et al. 2005; de las Heras and de Lorenzo 2011). XylR was also engineered using synthetic biology approaches to detect nitrotoluenes for bio-detection of landmines (Garmendia et al. 2008; de las Heras and de Lorenzo 2011). Novel XylR variants that allow the implementation of single Boolean logic operation were also generated, and they can be used for biosensor development (Calles and Lorenzo 2013).

DmpR and PhlR are closely related NtrC-like regulators of aerobic phenol catabolism in *Pseudomonas* (Table 1). DmpR has been extensively studied, and it regulates the catabolism of phenols and methyl-phenols in *Pseudomonas* sp. strain CF600 by controlling the transcription of the $\sigma^{54}$-dependent *Po* promoter that drives the expression of the catabolic *dmp*-operon (Shingler 2004; Gupta et al. 2012). The nonoverlapping $\sigma^{70}$-dependent promoter (*Pr*) controls the production of DmpR.

Transcription-driven supercoiling arising from the $\sigma^{54}$-promoter allows inter-promoter communication that results in stimulation of the activity of the $\sigma^{70}$-promoter without it possessing a cognate binding site for the $\sigma^{54}$-RNAP holoenzyme. This mode of control has the potential to be a prevalent, but hitherto unappreciated, mechanism by which bacteria adjust promoter activity to gain appropriate transcriptional control (del Peso-Santos and Shingler 2016). Recently, it was shown that the 5′-leader region (5′-LR) of the *dmpR* gene functions as a regulatory hub to control DmpR levels by two distinct mechanisms. At the level of transcription, inhibition of full-length transcripts was traced to an A-rich DNA-binding motif located downstream of the *Pr* promoter. At the translational level, Hfq aids Crc to bind to a catabolite activity motif overlapping the ribosome-binding site at the mRNA facilitating the Crc-dependent repression in intact cells. Interestingly, the entire 5′-LR of *dmpR* is highly conserved in closely related phenolic catabolic systems, suggesting a strong evolutionary pressure to maintain these regulatory motifs as well as additional potential regulatory features that remain to be elucidated (Madhushani et al. 2014).

Although the molecular architecture of DmpR is similar to that of XylR, they show different effector specificities (Galvão and de Lorenzo 2006), and the residues involved in effector recognition are confined to a stretch of 75 amino acids defined as the effector-specifying region (Skarfstad et al. 2000). 3D models of XylR and DmpR predicted structural features for shaping an effector-binding pocket and interaction with the central domain (Suresh et al. 2010). Three other phenolic compound-sensing NtrC-like regulators are those controlling the anaerobic degradation of phenol in *Thauera aromatica* K172 and *Aromatoleum aromaticum* EbN1 (PdeR) and the one involved in the anaerobic *p*-ethylphenol degradation in *A. aromaticum* EbN1 (EtpR) (Table 1) (Breinig et al. 2000; Wöhlbrand et al. 2007; Büsing et al. 2015). All these five regulators share eight residues that may be involved in the recognition of the phenolic moiety of the effector molecule (Büsing et al. 2015).

Other examples of regulators of the NtrC family are TbuT and TbmR, which control toluene monooxygenase gene expression in two strains of *Burkholderia* (Byrne and Olsen 1996; Leahy et al. 1997); TouR, which controls the degradation of phenol and toluene in *Pseudomonas stutzeri* OX1 (Solera et al. 2004); and AreR, which is involved in the aryl ester degradation pathway in *Acinetobacter baylyi* ADP1 (Jones and Williams 2001) (Table 1). Cross talk regulation has been demonstrated between XylR/DmpR and TbuT/TbmR for activation of their mutual promoters while maintaining their inducer specificity (Leahy et al. 1997; Arenghi et al. 1999).

# 5 Aromatic Compound Responsive Regulators of the AraC/XylS Family

Members of the AraC/XylS family have two structural domains, i.e., the C-terminal DNA-binding domain and the N-terminal signaling domain, connected by a relatively unstructured linker (Seedorff and Schleif 2011). The more variable N-terminal region is responsible for cofactor binding and/or multimerization. The C-terminal domain includes two tetra-helical HTH DNA-binding motifs. One or both HTH motifs bind DNA upstream, and sometimes downstream, of the target promoters (Seedorff and Schleif 2011).

The AraC/XylS family members responsible for the control of the catabolism of aromatic compounds are widely distributed in prokaryotes (Table 1). The XylS protein is the best characterized AraC member that controls the expression of an aromatic catabolic pathway. The pWW0-encoded XylS regulator mediates transcriptional activation of the *Pm* promoter driving the expression of the *meta*-pathway genes in *P. putida* mt-2, in response to 3-methylbenzoate (*m*-toluate) and benzoate as inducers (Gallegos et al. 1993). In addition to its known influence favoring protein dimerization, the effector is able to modify XylS conformation to trigger N-terminal domain intramolecular derepression (Domínguez-Cuevas et al. 2008). It has been suggested that the presence of the effector *m*-toluate triggers a cell response similar to the heat-shock response (Marqués et al. 1999), which explains that XylS-mediated transcription activation from the *Pm* promoter is driven by the $\sigma^{32}$ heat-shock sigma factor in the early exponential growth phase. Activation of *Pm* transcription is achieved through a switch to the $\sigma^{38}$ sigma factor when cultures reach the stationary phase (Marqués et al. 1999). By using a recombinant XylS-CTD soluble monomeric variant devoid of the N-terminal domain, it was shown that binding to *Pm* occurred sequentially. Firstly, a XylS-CTD monomer binds to the proximal site overlapping the RNAP-binding sequence to form complex I. This first event increased *Pm* bending to 50 degrees and was followed by the binding of the second monomer, which further increased the observed global curvature to 98 degrees (Domínguez-Cuevas et al. 2010). Despite the lack of information about the structure of XylS, mutagenesis studies have successfully generated regulators with altered inducer specificity (Michán et al. 1992).

BenR is a XylS-like activator able to trigger the activity of *Pben* promoter by recognition of benzoate as effector, allowing the expression of the *ben* operon that encodes the benzoate dioxygenase which converts benzoate into catechol in *P. putida* (Table 1) (Cowles et al. 2000). The N-terminal regions of BenR and XylS share about 65% amino acid identity, and both regulators respond to benzoate as an effector molecule (Cowles et al. 2000). The similarity also extends to their C-terminal DNA-binding domains suggesting that cross activation of their target promoters could take place. In fact, the ability of BenR to activate the *Pm* promoter of the *meta*-cleavage pathway operon of the TOL catabolic plasmid in response to benzoate has been described. Therefore, BenR behaves as an activator of benzoate degradation via *ortho*-ring fission, as an activator of benzoate and methylbenzoate degradation via *meta*-ring fission, and it is also involved in the benzoate-dependent repression of 4-hydroxybenzoate degradation by controlling the expression of the *pcaK* gene (encodes the 4-hydroxybenzoate transporter) in *P. putida* (Cowles et al. 2000). Although the cross activation of *Pben* by XylS had been previously shown (Cowles et al. 2000; Domínguez-Cuevas et al. 2006), these studies were performed using multicopy *Pben-lacZ* transcriptional fusions. In fact, any cross activation of *Pben* promoter by XylS will cause a metabolic conflict during the degradation of *m*-xylene because the produced 3-methylbenzoate could be channeled through the *ortho*-pathway and generate toxic dead-end metabolites. Recently, it was shown that the natural expression ranges of XylS are insufficient to cause a significant cross regulation of *Pben* if cells face either endogenous or exogenous 3-methylbenzoate. This lack of cross regulation relies on the fact that the *Pben* promoter has evolved to avoid a strong interaction with XylS, likely by lacking the A box in the proximal operator. This scenario reveals how a simple genetic tinkering facilitates the recruitment of catabolic pathways (the *meta*-pathway) in a host that harbors a non-fully compatible metabolism (the *ortho*-pathway) and suggests strategies for orthogonalization of new pathways implanted in a pre-existing metabolic chassis (Pérez-Pantoja et al. 2015).

Some other examples of AraC/XylS family members involved in the control of aromatic catabolic pathways (Table 1) are: PobR controls the *p*-hydroxybenzoate hydroxylase in many bacteria (Quinn et al. 2001; Donoso et al. 2011); OxoS is required for quinoline-dependent growth of *P. putida* 86 (Carl and Fetzner 2005); AntR controls the expression of the *antABC* operon coding for anthranilate 1,2-dioxygenase as well as of the *car* operon involved in the conversion of carbazole to anthranilate in *P. resinovorans* strain CA10 (Urata et al. 2004); HpaA regulates the *hpaBC* operon of *E. coli* W, which produces the hydroxylase activity for the catabolism of 4-hydroxyphenylacetic acid, in response to this aromatic acid, 3-hydroxyphenylacetic acid, or phenylacetic acid (Prieto and García 1994); IifR activates de *iif* operon involved in the indole degradation in response to indole (Lin et al. 2015); PheR activates the *pheA2* promoter that controls phenol degradation genes in *Rhodococcus* strains (Szőköl et al. 2014).

# 6    Aromatic Compound Responsive Regulators of the CRP/FNR Family

CRP/FNR proteins stand out in responding to a broad spectrum of intracellular and exogenous signals such as cAMP, anoxia, redox state, oxidative and nitrosative stress, nitric oxide (NO), carbon monoxide (CO), 2-oxoglutarate, or temperature (Körner et al. 2003). Within the CRP/FNR superfamily, functionally distinct transcriptional regulators (both activators and repressors) have evolved based on a common modular design. The N-terminal domain comprises a β-barrel, responsible for ligand recognition, while the C-terminal domain possesses a four-stranded wHTH motif for DNA binding. Both parts are connected by an α-helix frequently implicated in protein dimerization (Townsend et al. 2014). To accomplish their roles, CRP/FNR members might also have prosthetic groups such as an iron-sulfur group or heme, designed for the interaction with oxygen, NO, or CO (Körner et al. 2003). Regardless of the common structure, however, allosteric networks leading to the regulator activation after ligand binding are diverse within the CRP-FNR superfamily. There are reported cases of both negative (Townsend et al. 2014) and positive (Levy et al. 2008) cooperativity for ligand binding, and also ligand-independent regulators have been described (Agari et al. 2012). In the last case, regulator abundance would determine the extent of the transcription regulation effect (Agari et al. 2010). The vast majority of FNR-regulated promoters contain a consensus FNR-binding site centered around 41.5 bp upstream of the transcriptional start site, and they are termed class II FNR-dependent promoters (Busby and Ebright 1999).

Among the functions of CRP/FNR proteins is that of regulating the expression of metabolic pathways for the use of aromatic compounds (Table 1). Two of these proteins, HbaR and AadR, have been described in the anaerobic catabolism of aromatic compounds in *Rhodopseudomonas palustris*. HbaR regulates the anaerobic 4-hydroxybenzoate catabolism by activating the expression of the gene encoding the first enzyme of the pathway (4-hydroxybenzoate-CoA ligase) in the presence of the 4-HBA inducer (Egland and Harwood 2000). The expression of the *hbaR* gene is, in turn, under oxygen control since it requires activation by AadR under anaerobic conditions. Accordingly, AadR contains some of the essential conserved Cys residues for iron-sulfur coordination as in the FNR protein (Dispensa et al. 1992). AadR, together with the BadR protein (see below), also modulates the expression of genes involved in the anaerobic metabolism of benzoate and cyclohexanecarboxylate, thus representing an oxygen sensor that regulates anaerobic catabolism of aromatic compounds in *R. palustris* (Dispensa et al. 1992; Egland and Harwood 2000). At the top of this regulatory cascade is another CRP/FNR family member, the FixJ regulator, that controls the expression of the *aadR* gene (Rey and Harwood 2010).

In the β-proteobacterium *Azoarcus* sp. CIB, the AcpR protein is required for the expression of the *bzd* and *mbd* genes that encode the central pathways for the anaerobic catabolism of benzoate and 3-methylbenzoate, respectively (Table 1) (Durante-Rodríguez et al. 2006; Juárez et al. 2012). AcpR favors the activation of the $P_N$ promoter, which drives the expression of the *bzd* genes, in the absence of oxygen through contacts with the $\sigma^{70}$ and the α-subunit of the RNAP (Durante-Rodríguez et al. 2006). Despite the predicted structural similarity between FNR and AcpR, the two proteins do not have the same regulatory functions within the cell. Thus, whereas in *E. coli* the lack of the FNR protein has a pleiotropic effect on the expression of a moderate number of genes, the lack of AcpR in *Azoarcus* sp. CIB does alter the ability to catabolize aromatic compounds through the benzoyl-CoA pathway but does not affect the anaerobic growth on nonaromatic carbon sources. In this sense, the physiological role of AcpR in *Azoarcus* would be equivalent to that of AadR in *R. palustris* (Durante-Rodríguez et al. 2006).

In *E. coli* and *Rhodococcus* sp. TFB, CRP regulators have been reported to mediate carbon catabolite repression of several aromatic acids and tetralin catabolic pathways, respectively (Table 1) (Díaz et al. 2001; Torres et al. 2003; Tomás-Gallardo et al. 2012). In contrast to CRP from *E. coli* that acts as an activator in the absence of the preferred carbon source (glucose), CRP from *Rhodococcus* sp. TFB acts as a repressor in the presence of the preferred carbon source. A third CRP-type protein, GlxR, is also a global regulator that represents a central control point in the *Corynebacterium glutamicum* response to different nutrient sources. GlxR represses, among others, the genes for 3-hydroxybenzoate and gentisate metabolism, and it responds to cAMP levels (Table 1) (Chao and Zhou 2014; Townsend et al. 2014).

A branch of the CRP-FNR family, CprK proteins, includes transcriptional regulators that mediate the response to halogenated aromatic compounds in dehalorespiration of different strains of *Desulfitobacterium*. CprK proteins activate transcription from promoters containing a 14-bp inverted repeat (dehalobox) that closely resembles the FNR-box (Pop et al. 2006). The unusually high occurrence of CprK paralogs is likely to be correlated with the relatively large number of halogenated compounds that these organisms can accept as terminal electron acceptors, enabling a specific response by each regulator to a specific group of halogenated compounds (Gabor et al. 2008). In *Desulfitobacterium hafniense*, for instance, CprK1 induces expression of halorespiratory genes upon binding of *o*-chlorophenol ligands and is reversibly inactivated by oxygen through disulfide bond formation (Pop et al. 2006). Crystal structures of CprK1 in the ligand-free (both oxidation states), ligand-bound (reduced), and DNA-bound states allowed a complete structural description of both redox-dependent and allosteric molecular rearrangements (Levy et al. 2008).

## 7    Aromatic Compound Responsive Regulators of the IclR Family

The IclR family is an extended type of prokaryotic transcription regulators (activators, repressors, and proteins with a dual role) that have been described to be involved in the control of different bacterial processes (Krell et al. 2006; Molina-Henares et al. 2006; Chao and Zhou 2013). The N-terminal domain of these regulators comprises a wHTH DNA-binding motif responsible for its positioning on target promoters as a dimer or as a pair of dimers. However, no clear consensus exists on the architecture of DNA-binding sites within the IclR-targeted promoters (Cheng et al. 2015). The C-terminal domain of IclR-like regulators is the effector-binding domain and regulates subunit multimerization after recognition of the effector molecule. In case of working as transcriptional activators, IclR proteins bind to their target promoters in the absence of the effector molecule (DiMarco and Ornston 1994; Gerischer et al. 1998; Guo and Houghton 1999; Torres et al. 2003), but they need the inducer to recruit RNAP to the promoter (Guo and Houghton 1999). For IclR-negative regulators, transcription is prevented either by occluding the RNAP-binding site or by destabilizing the open complex, and the presence of the effector molecule abolishes that behavior (Yamamoto and Ishihama 2002).

An important number of IclR members regulate catabolic pathways for the degradation of aromatic compounds (Table 1). In *A. baylyi* ADP1 the PobR and PcaU proteins have been described to be indispensable for the induction of 4-hydroxybenzoate and protocatechuate metabolic pathways, respectively, and both of them act as repressors of their own expression (DiMarco and Ornston 1994; Gerischer et al. 1998; Trautwein and Gerischer 2001). Interestingly, it is known that PcaU acts on the promoter of the catabolic genes both as a transcriptional activator, in the presence of the cognate inducer (protocatechuate), and as a repressor, in the absence of the inducer (Popp et al. 2002). A homologue of PcaU, PcaR, exists in *Pseudomonas putida*, working in this case exclusively as activator of the *pca* genes for protocatechuate degradation (Guo and Houghton 1999).

The MhpR protein is necessary for the induction of the genes responsible of 3-hydroxyphenylpropionic acid metabolism in *E. coli*. In contrast to most aromatic compound responsive IclR-type regulators, the expression of *mhpR* from *Pr* promoter is constitutive and independent of self-regulation. Moreover, MhpR seems to be essential for recruiting a second activator, the global cAMP receptor protein (CRP) regulator, to the cognate *Pa* catabolic promoter (Torres et al. 2003), a feature that has not been reported for other IclR-type regulators. A MhpR-like protein involved in 3-hydroxyphenylpropionic acid metabolism has also been described in *Comamonas testosteroni* (Arai et al. 1999). Two other IclR-type activators, i.e., TphR and TsaQ for terephthalate and *p*-toluenesulfonate metabolism, respectively, have been reported in *C. testosteroni* strains (Tralau et al. 2003a; Kasai et al. 2010). The PbaR activator from *Sphingobium wenxiniae* JZ-1 T controls the degradation of 3-phenoxybenzoate and is the only IclR-type regulator so far described that binds downstream to the translation start site of the regulated gene (Cheng et al. 2015).

In Gram-negative bacteria, most IclR-type regulators that control aromatic catabolic pathways behave as transcriptional activators. Some exceptions are the HmgR protein that controls homogentisate degradation in *P. putida* (Arias-Barrau et al. 2004) and the IphR regulator for isophthalate metabolism in *C. testosteroni* (Table 1) (Fukuhara et al. 2009). In contrast, in *Actinobacteria* most IclR-type regulators described behave as transcriptional repressors (Table 1), e.g., TsdR (γ-resorcylate pathway), OphR (phthalate pathway), CatR (catechol pathway), and NpdR (2,4,6-trinitrophenol pathway) in *Rhodococcus* strains (Nga et al. 2004; Veselý et al. 2007; Choi et al. 2015; Kasai et al. 2015); and only one transcriptional activator, the GenR regulator that controls the catabolism of 3-hydroxybenzoate and gentisate, was reported in *Corynebacterium glutamicum* (Chao and Zhou 2013).

## 8   Aromatic Compound Responsive Regulators of the TetR Family

The TetR family is well characterized and widely distributed in bacteria (Ramos et al. 2005). The 3D structure of the prototype TetR reveals the existence of two domains, a N-terminal domain which contains the tetra-helical HTH-DNA-binding motif and a C-terminal domain involved in effector (tetracycline) binding (Orth et al. 2000). Members of the TetR family exhibit a high conservation of sequences for the DNA-binding domain. The TetR family regulators are mostly repressors that bind their operators, composed of 10–30-bp palindromic sequences, to repress the target genes and are released from the DNA when bound to their cognate ligands (Ramos et al. 2005).

Some TetR-type regulators involved in the regulation of aromatic catabolic pathways have been described (Table 1). CymR is a transcriptional repressor involved in the control of the gene expression for *p*-cymene (*cym*) and *p*-cumate (*cmt*) degradation in *P. putida* F1 (Eaton 1997). The CymR protein is a dimer in solution, and it imposes its repressing effect by inhibiting RNAP access to the promoter, being *p*-cumate the effector molecule that avoids binding of CymR to its operator site (Eaton 1997). The PaaR protein is a transcriptional repressor of the *paa* genes for phenylacetate degradation in *Thermus thermophilus* HB8. Phenylacetyl-CoA is the effector molecule for effective transcriptional derepression (Sakamoto et al. 2011). Moreover, it has been described a new regulator, PfmR, that weakly cross regulated PaaR in vitro and that has an additional function in regulating the fatty acid metabolism in strain HB8 (Agari et al. 2012). The X-ray crystal structure of the N-terminal DNA-binding domain of PfmR and the nucleotide sequence of the predicted PfmR-binding site are quite similar to those of the TetR family repressor QacR from *Staphylococcus aureus*. Similar to QacR, two PfmR dimers bound per target DNA. The center of the PfmR molecule contains a tunnellike pocket, which may be the ligand-binding site of this regulator (Agari et al. 2012). PaaR is also the repressor of the *paa* genes involved in phenylacetate catabolism, in *Corynebacterium glutamicum* (Chen et al. 2012). An imperfect palindromic motif (5-″-ACTNACCGNNCGNNCGGTNAGT-3″, 22 bp) was identified in the upstream regions of *paa* genes. In addition, GlxR-binding sites were found, and binding to GlxR was confirmed. Therefore, phenylacetate catabolism in *C. glutamicum* is regulated by the pathway-specific repressor PaaR, which responds to phenylacetyl-CoA, and also likely by the global transcription regulator GlxR. By comparative genomic analysis, orthologous PaaR regulons were identified in 57 species, including species of *Actinobacteria*, *Proteobacteria*, and *Flavobacteria* that carry phenylacetate utilization genes and operate by conserved binding motifs, suggesting that PaaR-like regulation might commonly exist in these bacteria (Chen et al. 2012). In this sense, PaaR-like proteins controlling the expression of *paa* genes for the catabolism of phenylacetate have been also described in *Burkholderia cenocepacia* (Yudistira et al. 2011) and proposed in *Azoarcus* strains (Mohamed et al. 2002). In *Streptomyces pristinaespiralis*, PaaR is also involved in controlling the expression of the *paa* genes and plays a positive role in the regulation of the biosynthesis of pristinamycin I by affecting the levels of phenylacetyl-CoA as a supply of L-phenylglycine, one of the seven amino acid precursors of this antibiotic (Zhao et al. 2015).

HdnoR is a transcriptional repressor of the 6-hydroxy-D-nicotine oxidase, and it is encoded on the catabolic plasmid pAO1 responsible for nicotine degradation in *Arthrobacter nicotinovorans* (Sandu et al. 2003). The inducers 6-hydroxy-D-nicotine and 6-hydroxy-L-nicotine prevent the binding of HdnoR to its operator site allowing the expression of the 6-hydroxy-D-nicotine oxidase (Sandu et al. 2003). RolR may represent the first member of a new subfamily of TetR proteins involved in resorcinol degradation in *Corynebacterium glutamicum* (Li et al. 2011), and it shows generally low sequence similarities to other TetR family members, especially at its C-terminal end (Li et al. 2011). A 29-bp operator *rolO* was located at the intergenic region of *rolR* and the catabolic *rolHMD* genes, and it contains two overlapping inverted repeats that are essential for RolR binding and repression of both operons. The binding of RolR to *rolO* was avoided by resorcinol and hydroxyquinol, which are the starting compounds of the resorcinol catabolic pathway, leading to the induction of *rol* genes (Li et al. 2012). A novel resorcinol-inducible expression system based on the RolR regulator and the cognate promoter fused with the operator (*rolO*) has been developed for *Streptomyces* and other *Actinobacteria* (Horbal et al. 2014). NicS is a repressor that controls the expression of the *nicAB* genes responsible for the conversion of nicotinic acid to 6-hydroxynicotinic acid in *P. putida* (Jiménez et al. 2011). Both aromatic heterocycles behave as NicS inducers. Interestingly, the expression of *nicS* is under control of a second regulator, NicR, that responds to 6-hydroxynicotinic acid, thus generating a peculiar regulatory loop (see below) (Jiménez et al. 2011).

The MbdR protein is a transcriptional repressor of the *mbd* genes for 3-methylbenzoate degradation in *Azoarcus* sp. CIB (Juárez et al. 2015). The 3D structure of MbdR revealed a conformation similar to that of other TetR family transcriptional regulators. 3-Methylbenzoyl-CoA, the first intermediate of the catabolic pathway, but not benzoyl-CoA, was shown to interact with MbdR and avoid binding to the operator region at the target promoters, leading to derepression of *mbd* genes. These results highlight the importance of recruiting the MbdR-based regulatory circuit to evolve a distinct central catabolic pathway that is only induced for the anaerobic degradation of aromatic compounds that generate 3-methylbenzoyl-CoA as central intermediate (Juárez et al. 2015).

## 9    Aromatic Compound Responsive Regulators of the BzdR Subfamily

The BzdR-like proteins (Table 1) constitute a new subfamily of aromatic transcriptional regulators belonging to the widely distributed HTH-XRE family of transcriptional regulators that includes the well-known Cro and cI lambda repressors. BzdR is the transcriptional repressor of the *bzd* cluster responsible for the anaerobic catabolism of benzoate in *Azoarcus* strains (Barragán et al. 2005). The BzdR protein exhibits two domains separated by a linker region, i.e., the N-terminal domain with a tetra-helical HTH-DNA-binding motif similar to that of the lambda repressor and the C-terminal domain similar to shikimate kinases (Barragán et al. 2005). Benzoyl-CoA, the first intermediate of the anaerobic benzoate degradation pathway, is the effector molecule of BzdR. Benzoyl-CoA interacts with the C-terminal domain of BzdR and prevents the binding of this protein to the three operator regions of the target $P_N$ promoter without affecting its oligomeric state. The linker region of BzdR is required to transfer the conformational changes induced by benzoyl-CoA to the DNA-binding domain. The predicted structures of the respective N- and C-terminal domains could be fitted into a 3D reconstruction of the BzdR homodimer obtained by electron microscopy (Durante-Rodríguez et al. 2010). BzdR has been proposed as a model to study the evolution of transcriptional regulators. In this sense, an active BzdR-like regulator was engineered by fusing the DNA-binding domain of BzdR to the shikimate kinase I of *E. coli*, supporting the notion that an ancestral shikimate kinase domain could have been involved in the evolutionary origin of BzdR (Durante-Rodríguez et al. 2013). On the other hand, the C-terminal domain of

BzdR has been fused to the N-terminal domain of CI protein of the lambda phage to design a chimeric regulator, termed Qλ, able to reprogram the lytic/lysogenic lambda phage decision according to the intracellular production of benzoyl-CoA in *E. coli* (Durante-Rodríguez et al. 2016).

BoxR is a transcriptional repressor of the *box* genes involved in the aerobic hybrid pathway to degrade benzoate via coenzyme A derivatives in bacteria (Valderrama et al. 2012). The BoxR protein shows a significant sequence identity to BzdR. In *Azoarcus* sp. CIB, the paralogous BoxR and BzdR regulators act synergistically to assure a tight repression of the *bzd* and *box* genes in the absence of the common intermediate and inducer molecule benzoyl-CoA. Moreover, the observed expression of the *box* genes under anaerobic conditions (Valderrama et al. 2012) may constitute an alternative oxygen-scavenging mechanism when the cells face low-oxygen tensions that could inactivate the highly oxygen-sensitive anaerobic reductase and also a strategy to rapidly shift to the aerobic degradation if oxygen levels become high.

# 10  Aromatic Compound Responsive Regulators of the GntR Family

The proteins of the GntR superfamily are 239–254 amino acids long and share a similar N-terminal wHTH DNA-binding domain. This output domain is coupled to the C-terminal effector-binding and oligomerization domain that responds to a range of stimuli in the form of different small molecules. The C-terminal domain imposes steric constraints on the DNA-binding domain, hence influencing the HTH motif and thus playing an important role in regulation (Hoskisson and Rigali 2009, Suvorova et al. 2015). The structural data show that FadR from *E. coli* and AraR from *B. subtilis* bind as dimers to the target DNA through their N-terminal domains, but only few base pairs are specifically recognized within the complex (van Aalten 2001, Xu et al. 2001, Jain and Nair 2012).

There are some GntR family members related with aromatic catabolic pathways (Table 1). Most of them behave as repressors in the absence of effector with the exception of BphR1 (Orf0) of *P. pseudoalcaligenes* KF707 which acts as a repressor of salicylate catabolic genes but activates its own expression and that of biphenyl catabolic genes (Fujihara et al. 2006). The repressors PhcS and AphS regulate the expression of the phenol degradation genes in *Comamonas testosteroni* strains R5 and TA441, respectively (Arai et al. 1999; Teramoto et al. 2001). The VanR protein represses the vanillate demethylase (*vanAB*) genes in *A. baylyi* ADP1 and in diverse *Pseudomonas* strains (Morawski et al. 2000). Expression of the *vanAB* operon is repressed by VanR and induced by vanillate as well as, to a smaller degree, by its reduced derivatives vanillin and vanillyl alcohol in *Caulobacter crescentus* (Thanbichler et al. 2007). BphS controls the biphenyl degradation in *Ralstonia eutropha* A5 (Mouz et al. 1999) and *Pseudomonas* sp. strain KKS102 (Ohtsubo et al. 2001). CarR$_{J3}$ protein binds to two operator sequences (TtGTAGAACAA) in the absence of its inducer, which was identified as 2-hydroxy-6-oxo-6-

(2''-aminophenyl)hexa-2,4-dienoate, an intermediate of the carbazole degradation pathway, and represses the *car* operon in *Janthinobacterium* sp. 13 (Miyakoshi et al. 2006).

PaaX-like regulators are transcriptional repressors that control phenylacetic acid degradation gene clusters in several *Proteobacteria* (García et al. 2000; del Peso-Santos et al. 2006). Since they are bigger and do not show a significant sequence similarity with other members of the GntR family, they may constitute a new GntR subfamily. The PaaX repressor from *E. coli* recognizes the operator palindromic sequence (TGATTC-$N_{26-28}$-GAATCa) (Ferrández et al. 2000; Galán et al. 2004; Kim et al. 2004). Phenylacetyl-CoA specifically inhibited binding of PaaX to the target sequences, confirming the first intermediate of the pathway as the true inducer (Ferrández et al. 2000). Whereas the mechanism of repression of PaaX on the regulatory *Px* and the catabolic *Pz* promoters involves competition with the RNAP binding, the catabolic *Pa* promoter appears to be controlled by PaaX at a later stage of the transcription initiation process (Fernández et al. 2013). The PaaX repressor links the catabolism of aromatic compounds with the metabolism of penicillins since it is also a repressor of the *pac* gene encoding the penicillin G acylase (Galán et al. 2004). A role for the PaaX regulator in repressing the expression of the *sty* genes for the catabolism of styrene in *Pseudomonas* sp. Y2 has been reported, suggesting that PaaX is a major regulatory protein in the phenylacetyl-CoA catabolon through its response to the levels of this central metabolite (del Peso-Santos et al. 2006). Crystallization and preliminary X-ray diffraction studies on some PaaX-like regulators have been reported (Rojas-Altuve et al. 2011). In *E. coli*, the *paaX* gene is co-transcribed with *paaY*, which encodes a thioesterase, forming a regulatory operon. The PaaY protein is necessary for the efficient degradation of phenylacetate in *E. coli*, and two different roles for this protein can be envisioned. At the metabolic level, PaaY helps to prevent that phenylacetic acid catabolism might collapse cell growth by hydrolyzing some CoA derivatives whose accumulation may lead to the inhibition of the first steps of the *paa* pathway (Teufel et al. 2012). Moreover, PaaY plays a second role by facilitating the induction of the *paa* genes likely by its thioesterase activity that reduces the amount of some CoA-derived intermediate(s) originated during the catabolism of PA and that may behave as antagonists of the effect caused by the phenylacetyl-CoA inducer molecule on the PaaX repressor. This regulatory function mediated by PaaY constitutes an additional regulatory checkpoint that makes the circuit that controls the transcription of the *paa* genes more complex than previously anticipated, and it could represent a general strategy present in most bacterial *paa* gene clusters that also harbor the *paaY* gene (Fernández et al. 2013).

MeqR2 is a PaaX-type transcriptional repressor involved in the regulation of the genes responsible of quinaldine catabolism in *Arthrobacter* sp. strain Rue61a. MeqR2 forms a dimer in solution and binds to a palindromic operator whose core sequence (TGACGNNCGTcA) does not resemble that of PaaX operators. As some other GntR family regulators, such as PaaX and FadR that bind CoA thioesters, MeqR2 shows a high specificity for anthraniloyl-CoA, a downstream metabolite of the Meq pathway for quinaldine degradation, as effector. A binding stoichiometry of one effector molecule per MeqR2 monomer and a high affinity ($K_D$ of 22 nM) were determined (Niewerth et al. 2012).

## 11 Aromatic Compound Responsive Regulators of the MarR Family

MarR-type regulators are relatively small proteins (148–196 amino acids), and their 3D structures reveal a common triangular shape with a wHTH-DNA-binding motif. These transcription factors are typically homodimers and bind to palindromic DNA operators located within the target promoters. In most cases, in the absence of ligand, apo-MarR proteins bind to specific DNA operators, and upon ligand binding they show diminished DNA affinity. Usually, MarR regulators are promiscuous and can accommodate a variety of aromatic ligands with not very high affinity (Grove 2013; Kim et al. 2016).

Some members of this family are involved in specific responses to aromatic compounds (Table 1). HpaR is the transcriptional repressor of the 3,4-dihydroxyphenylacetate (homoprotocatechuate) *hpa* cluster of *E. coli* W. The *hpaR* gene is located upstream and divergently oriented with respect to the catabolic operon. HpaR negatively regulates not only the expression of the *hpa-meta* operon but also its own expression, with homoprotocatechuate, 4-hydroxyphenylacetate, and 3-hydroxyphenylacetate being the inducer molecules (Galán et al. 2003). Two DNA operators, OPR1 and OPR2, have been identified in the intergenic region located between the *hpa-meta* operon and the *hpaR* gene. The binding of HpaR to OPR2 displays a clear cooperativity with OPR1 binding (Galán et al. 2003). The CbaR repressor controls the *cbaABC* operon of plasmid pBRC60 required for chlorobenzoate degradation in *C. testosteroni* BR60. 3-Chlorobenzoate and protocatechuate are effectors for CbaR, with their binding leading to derepression (Providenti and Wyndham 2001). NbzR is a repressor that regulates the *nbz* operon for aminophenol degradation encoded on plasmid pNB1 of *P. putida* HS12, but the chemical inducer for the pathway has not yet been identified (Park and Kim 2001).

HcaR is the repressor of the *hca* genes responsible for hydroxycinnamates degradation in *A. baylyi* ADP1, with hydroxycinnamoyl-CoA thioesters being the effector molecules (Parke and Ornston 2003). The crystal structure of the apo-HcaR protein was recently determined in complexes with hydroxycinnamates and a specific 23-bp palindromic DNA operator. HcaR appears to be a tetramer, a dimer of dimers, in solution, and each dimer binds separate DNA-binding sites (*hca1* and *hca2*) using probably a DNA-loop formation mechanism that interferes with RNAP binding. HcaR recognizes four different ligands, i.e., ferulate, *p*-coumarate, vanillin, and 3,4-dihydroxybenzoate (which are substrate, intermediates, and products of ferulic acid processing by *hca* gene products) using the same binding site and rendering this repressor unproductive in recognizing a specific DNA target. These studies are consistent with a mechanism of HcaR derepression based on stabilization of a compact protein conformation that is unproductive in recognizing and binding a specific DNA operator (Kim et al. 2016).

CouR (FerC, FerR) regulates *p*-hydroxycinnamates (e.g., ferulate, *p*-coumarate) catabolism in different bacteria such as *Sphingobium* sp. SYK-6 (Kasai et al. 2012), *R. palustris* (Hirakawa et al. 2012), *P. fluorescens* (Calisti et al. 2008), and *R. jostii* RHA1 (Otani et al. 2015). In these repressors, DNA binding is abolished by *p*-hydroxycinnamoyl-CoA, the first metabolite of the pathway, allowing expression of the catabolic *cou* genes. Recent structural data with the CouR protein from *R. jostii* RHA1 establish that the CouR dimer binds two *p*-coumaroyl-CoA molecules in nonequivalent configuration, but this ligand binding did not lead to a significant conformational change in the repressor. Interestingly, the anionic bulky CoA moiety of *p*-hydroxycinnamoyl-CoA prevents the binding of DNA by steric occlusion of key DNA-binding residues and charge repulsion of the DNA backbone (Otani et al. 2015). In *R. palustris*, *p*-coumarate is not only a carbon source but the precursor of an unusual acyl-homoserine lactone (HSL) quorum-sensing signal, *p*-coumaroyl-HSL. A quantitative proteome and microarray study suggested that at least 40 genes and their encoded proteins are upregulated during growth on *p*-coumarate compared to succinate. Some of these genes are regulated by *p*-coumaroyl-HSL and the transcription protein RpaR, and others are regulated by CouR. In this bacterium, CouR controls not only the expression of *couAB* genes for *p*-coumarate degradation but also transport systems that are likely involved in the uptake of *p*-coumarate and structurally related compounds into cells (Phattarasukol et al. 2012).

PcaV is a MarR family regulator that represses transcription of genes encoding the central β-ketoadipate pathway in *Streptomyces coelicolor*. Structural data revealed that PcaV binds the β-ketoadipate pathway substrate protocatechuate with a high affinity and in a 1:1 stoichiometry, leading to a change in protein conformation incompatible with DNA binding. PcaV exhibits an unusually high degree of ligand selectivity and is one of the few MarR homologues incapable of binding salicylate. The Arg15 residue is critical for coordinating the protocatechuate ligand and plays a key role in binding DNA, thus functioning as a gatekeeper residue for regulating PcaV transcriptional activity (Davis et al. 2013).

GenR is a MarR-type transcriptional regulator that, in the absence of effectors, represses the *gen* cluster encoding the gentisate pathway in *Comamonas testosteroni* CNB-1. When effectors such as gentisate, 3-hydroxybenzoate, and benzoyl-CoA are present, the GenR protein is released from its DNA-binding site, and the repression of transcription is abolished. The finding that benzoyl-CoA can be recognized as GenR effector explains why the gentisate dioxygenase was induced when CNB-1 grew on benzoate using the *box* aerobic hybrid pathway that activates benzoate to benzoyl-CoA (Chen et al. 2014). IacR regulates negatively the *iac* genes responsible for indole-3-acetate catabolism in *Acinetobacter baumannii*, being this aromatic acid the potential effector that induces *iac* expression (Shu et al. 2015). BadR has been recently reassigned as a repressor that controls the genes involved in cyclohexane-carboxylate degradation in *R. palustris*. Some of these genes are also involved in the anaerobic degradation of benzoate. 2-Ketocyclohexane-1-carboxyl-CoA, an intermediate of cyclohexanecarboxylate degradation, interacts with BadR to abrogate repression (Hirakawa et al. 2015).

In *P. putida* the NicR repressor controls three nicotinic acid inducible catabolic operons, i.e., *nicAB*, encoding the upper pathway that converts nicotinic acid into 6-hydroxynicotinic acid, *niccCDEFTP*, and *nicXR* operons, responsible for channeling the latter to the central metabolism, which are driven by the *Pa*, *Pc*, and *Px* promoters, respectively (Jiménez et al. 2011). The *nicR* regulatory gene encodes a MarR-like protein that represses the activity of the divergent *Pc* and *Px* promoters being 6-hydroxynicotinic acid the inducer molecule. An additional gene, *nicS*, which is associated to the *nicAB* genes in the genomes of different γ- and β-*Proteobacteria*, encodes a TetR-like regulator that represses the activity of *Pa* in the absence of the nicotinic/6-hydroxynicotinic acids as inducers. The nicotinic acid regulatory circuit in *P. putida* has evolved an additional repression loop based on the NicR-dependent cross regulation of the *nicS* gene, thus assuring a tight transcriptional control of the catabolic genes that may prevent depletion of nicotinic acid (vitamin B3) when needed for the synthesis of essential cofactors (Jiménez et al. 2011).

## 12 Aromatic Compound Responsive Regulators that Belong to Other Families

BadM is a member of the Rrf2 family of transcription factors that acts as a repressor of the *bad* genes involved in the anaerobic degradation of benzoate in *R. palustris*. In vivo data suggest that benzoate- or benzoyl-CoA is the effector for BadM (Hirakawa et al. 2015). Another aromatic compound responsive regulator of the Rrf2 family is the BgeR protein that represses expression of the *bamA* gene encoding the hydrolase for the ring-cleavage step during the anaerobic degradation of benzoate in *Geobacter bemidjiensis*. It was suggested that BgeR plays a key role in regulating the genes involved in the anaerobic degradation of aromatic compounds in *Geobacter* species (Ueki 2011).

The PadR-type regulators contain a wHTH domain with about 80–90 residues that is responsible for the binding to target DNA. The variable C-terminal domain in PadR-like proteins is involved in dimerization through a leucine zipper-like structure. In *Corynebacterium glutamicum* the *van* operon involved in vanillate degradation is regulated by VanR, a PadR-type repressor. VanR forms a dimer and binds cooperatively to two overlapping 24-bp inverted repeats of the target promoter, being vanillate the effector molecule that avoids formation of the protein-DNA complex. It is proposed that VanR-DNA complexes contain two VanR dimers at the VanR operator (Morabbi Heravi et al. 2014).

## 13 Research Needs

While catabolism of aromatic compounds is relatively well conserved in different organisms, gene regulation shows a wider diversity, and therefore, the whole understanding of the complex regulatory network that controls the expression of the genes involved in a particular degradation pathway is a challenging task. A large number of aromatic sensing transcriptional regulator sequences have been deposited in databases, but their structure and function remain unknown for most of them. On the other hand, the available metagenomic libraries are a source of still unknown aromatic regulators. Substrate-induced gene expression (SIGEX) is a promoter trap method based on single-cell sorting of clones from a plasmid library using flow cytometry, where transcriptional regulators are identified by the increased expression of a downstream fluorescent reporter gene in the presence, but not in the absence, of an inducing compound. Using SIGEX of a metagenomic library, several transcriptional regulators with different compound specificities and induction rates have been successfully identified (Uchiyama and Miyazaki 2013, Meier et al. 2015). However, SIGEX is limited in several important ways, e.g., distal location of the regulators from the target promoters, library sizes, the substrate of a pathway is not always the cognate inducer, etc., and there is a need for novel methods and strategies for high-throughput screening of aromatic regulators.

A structural understanding of effector binding to a regulatory protein and the molecular mechanisms by which ligands affect derepression/activation at the target promoter is critical. Recent advances in identifying ligand-binding pockets in some regulators may furnish a tool toward identifying the ligands for homologous regulators for which the effector remains unknown. The biological role of antagonists modulating the effect of the agonists (effectors) on the cognate regulators when bacteria are exposed to complex mixtures of aromatic substrates should be also addressed. The characterization of novel modulators that fine-tune the activation of the specific regulators, e.g., by preventing gratuitous induction, requires further studies. An obvious question still unanswered is why the regulatory proteins have evolved so divergently despite regulating very similar pathways for degradation of similar compounds.

The complex regulatory network underlying the hierarchical use of aromatic compounds when bacteria grow in mixtures of these carbon sources in the environment needs to be unraveled. Moreover, the ecophysiological meaning of the diversity found in the regulation of the hierarchical utilization of aromatic compounds among closely related strains sharing ecological niches should be addressed. In this sense, a more complete view of the molecular mechanisms underlying carbon catabolite repression and, in general, other ways of global regulation that sense the physiological status of the cell and overtake the effector-specific regulation of a particular aromatic catabolic pathway should be explored further. Then, computational tools should be used to study the logic structure of the intricate regulatory networks and to formalize it as a digital circuit by converting all known molecular interactions into binary logic operations (logicome) (Silva-Rocha et al. 2011). The integration of future regulatory models with the current genome-scale metabolic models should be a further step for a more accurate in silico reconstruction of bacterial metabolism. On the other hand, it is currently known that microbial populations exploit metabolic diversification of single cells to achieve phenotypic diversity and survive to unpredictable adverse changes in environmental conditions. Studying the regulatory circuits that drive gene expression in individual cells is, therefore, warranted.

From a biotechnology point of view, the in vitro evolution or de novo synthesis of new regulators exhibiting novel specificities and effector-binding affinities is an interesting way to track the evolutionary roadmap of these proteins and to engineer new synthetic regulatory circuits or to develop genetic traps to survey new enzymatic activities in metagenomic libraries.

In summary, a deeper understanding of the complex regulatory network that controls aromatic metabolism will pave the way for the forward engineering of bacteria as efficient biocatalysts for bioremediation of chemical waste and/or biotransformation to biofuels and renewable chemicals, for detection of toxic molecules (biosensors), and for biomedical applications.

**Acknowledgments** Work in E. Díaz laboratory was supported by Ministry of Economy and Competitiveness of Spain Grant BIO2012-39501, BIO2016-79736-R and PCIN2014-113, European Union FP7 Grant 311815, and Fundación Ramón-Areces XVII CN.

# References

Agari Y, Kuramitsu S, Shinkai A (2010) Identification of novel genes regulated by the oxidative stress-responsive transcriptional activator SdrP in *Thermus thermophilus* HB8. FEMS Microbiol Lett 313:127–134

Agari Y, Sakamoto K, Kuramitsu S, Shinkai A (2012) Transcriptional repression mediated by a TetR family protein, PfmR, from *Thermus thermophilus* HB8. J Bacteriol 194:4630–4641

Arai H, Chang MY, Kudo T, Ohishi T (2000) Arrangement and regulation of the genes for *meta*-pathway enzymes required for degradation of phenol in *Comamonas testosteroni* TA441. Microbiology 146:1707–1715

Arai H, Kudo T, Yamamoto T, Ohishi T, Shimizu T, Nakata T (1999) Genetic organization and characteristics of the 3-(3-hydroxyphenyl)propionic acid degradation pathway of *Comamonas testosteroni* TA441. Microbiology 145:2813–2820

Arenghi FL, Pinti M, Galli E, Barbieri P (1999) Identification of the *Pseudomonas stutzeri* OX1 toluene-*o*-xylene monooxygenase regulatory gene (*touR*) and of its cognate promoter. Appl Environ Microbiol 65:4057–4063

Arias-Barrau E, Olivera ERR, Luengo JMM, Fernández C, Galán B, García JL, Díaz E, Miñambres B (2004) The homogentisate pathway: a central catabolic pathway involved in the degradation of L-phenylalanine, L-tyrosine, and 3-hydroxyphenylacetate in *Pseudomonas putida*. J Bacteriol 186:5062–5077

Barragán MJL, Blázquez B, Zamarro MT, Mancheño JM, Jl G, Díaz E, Carmona M (2005) BzdR, a repressor that controls the anaerobic catabolism of benzoate in *Azoarcus* sp. CIB, is the first member of a new subfamily of transcriptional regulators. J Biol Chem 280: 10683–10694

Beck LL, Smith TG, Hoover TR (2007) Look, no hands! Unconventional transcriptional activators in bacteria. Trends Microbiol 15:530–537

Bleichrodt FS, Fischer R, Gerischer UC (2010) The beta-ketoadipate pathway of *Acinetobacter baylyi* undergoes carbon catabolite repression, cross-regulation and vertical regulation, and is affected by Crc. Microbiology 156:1313–1322

Breinig S, Schiltz E, Fuchs G (2000) Genes involved in anaerobic metabolism of phenol in the bacterium *Thauera aromatica*. J Bacteriol 182:5849–5863

Brzostowicz PC, Reams AB, Clark TJ, Neidle EL (2003) Transcriptional cross-regulation of the catechol and protocatechuate branches of the β-ketoadipate pathway contributes to carbon source-dependent expression of the *Acinetobacter* sp. strain ADP1 *pobA* gene. Appl Environ Microbiol 69:1598–1606

Busby S, Ebright RH (1999) Transcription activation by catabolite activator protein (CAP). J Mol Biol 293:199–213

Bush M, Dixon R (2012) The role of bacterial enhancer binding proteins as specialized activators of σ54-dependent transcription. Microbiol Mol Biol Rev 76:497–529

Büsing I, Kant M, Dörries M, Wöhlbrand L, Rabus R (2015) The predicted σ54-dependent regulator EtpR is essential for expression of genes for anaerobic *p*-ethylphenol and *p*-hydroxya-cetophenone degradation in "*Aromatoleum aromaticum*" EbN1. BMC Microbiol 15:251

Byrne AM, Olsen RH (1996) Cascade regulation of the toluene-3-monooxygenase operon (*tbuA1UBVA2C*) of *Burkholderia pickettii* PKO1: role of the *tbuA1* promoter (*PtbuA1*) in the expression of its cognate activator, TbuT. J Bacteriol 178:6327–6337

Calisti C, Ficca AG, Barghini P, Ruzzi M (2008) Regulation of ferulic catabolic genes in *Pseudomonas fluorescens* BF13: involvement of a MarR family regulator. Appl Microbiol Biotechnol 80:475–483

Calles B, de Lorenzo V (2013) Expanding the boolean logic of the prokaryotic transcription factor XylR by functionalization of permissive sites with a protease-target sequence. ACS Synth Biol 2:594–603

Carl B, Fetzner S (2005) Transcriptional activation of quinoline degradation operons of *Pseudomonas putida* 86 by the AraC/XylS-type regulator OxoS and cross-regulation of the PqorM promoter by XylS. Appl Environ Microbiol 71:8618–8626

Carmona M, Prieto MA, Galán B, García JL, Díaz E (2008) Signaling networks and design of pollutants biosensors. In: Díaz E (ed) Microbial Biodegradation: genomics and molecular biology. Caister Academic Press, Norfolk, pp 97–143

Cases I, de Lorenzo V (2005) Promoters in the environment: transcriptional regulation in its natural context. Nat Rev Microbiol 3:105–118

Chao H, Zhou N (2014) Involvement of the global regulator GlxR in 3-hydroxybenzoate and gentisate utilization by *Corynebacterium glutamicum*. Appl Environ Microbiol 80:4215–4225

Chao H, Zhou N (2013) GenR, an IclR-type regulator, activates and represses the transcription of genes involved in 3-hydroxybenzoate and gentisate catabolism in *Corynebacterium glutamicum*. J Bacteriol 195:1598–1609

Chen DW, Zhang Y, Jiang CY, Liu SJ (2014) Benzoate metabolism intermediate benzoyl-coenzyme a affects gentisate pathway regulation in *Comamonas testosteroni*. Appl Environ Microbiol 80:4051–4062

Chen X, Kohl TA, Ruckert C, Rodionov DA, Li LH, Ding JY, Kalinowski J, Liu SJ (2012) Phenylacetic acid catabolism and its transcriptional regulation in *Corynebacterium glutamicum*. Appl Environ Microbiol 78:5796–5804

Cheng M, Chen K, Guo S, Huang X, He J, Li SJJ (2015) PbaR, an IclR family transcriptional activator for the regulation of the 3-phenoxybenzoate 1′,2′-dioxygenase gene cluster in *Sphingobium wenxiniae* JZ-1T. Appl Environ Microbiol 81:8094–8092

Choi KY, Kang BS, Nam MH, Sul WJ, Kim E (2015) Functional identification of OphR, an IclR family transcriptional regulator involved in the regulation of the phthalate catabolic operon in *Rhodococcus* sp. strain DK17. Indian J Microbiol 55:313–318

Collier LS, Gaines GL, Neidle EL (1998) Regulation of benzoate degradation in *Acinetobacter* sp. strain ADP1 by BenM, a LysR-type transcriptional activator. J Bacteriol 180:2493–2501

Cowles CE, Nichols NN, Harwood CS (2000) BenR, a XylS homologue, regulates three different pathways of aromatic acid degradation in *Pseudomonas putida*. J Bacteriol 182:6339–6346

Davis JR, Brown BL, Page R, Sello JK (2013) Study of PcaV from *Streptomyces coelicolor* yields new insights into ligand-responsive MarR family transcription factors. Nucleic Acids Res 41: 3888–3900

de las Heras A, Chavarría M, de Lorenzo V (2011) Association of *dnt* genes of *Burkholderia* sp. DNT with the substrate-blind regulator DntR draws the evolutionary itinerary of 2,4-dinitrotoluene biodegradation. Mol Microbiol 82:287–299 de las Heras A, de Lorenzo V (2011) Cooperative amino acid changes shift the response of the σ54-dependent regulator XylR from natural *m*-xylene towards xenobiotic 2,4-dinitrotoluene. Mol Microbiol 79:1248–1259

del Peso-Santos T, Shingler V (2016) Inter-sigmulon communication through topological promoter coupling. Nucleic Acids Res 44:9638–9649

del Peso-Santos T, Bartolome-Martín D, Fernández C, Alonso S, García JL, Díaz E, Shingler V, Perera J (2006) Coregulation by phenylacetyl-coenzyme a-responsive PaaX integrates control of the upper and lower pathways for catabolism of styrene by *Pseudomonas* sp. strain Y2. J Bacteriol 188:4812–4821

del Peso-Santos T, Bernardo LMD, Skarfstad E, Holmfeldt L, Togneri P, Shingler V (2011) A hyper-mutant of the unusual σ70-*Pr* promoter bypasses synergistic ppGpp/DksA co-stimulation. Nucleic Acids Res 39:5853–5865

Díaz E, Ferrández A, García JL (1998) Characterization of the *hca* cluster encoding the dioxygenolytic pathway for initial catabolism of 3-phenylpropionic acid in *Escherichia coli* K-12. J Bacteriol 180:2915–2923

Díaz E, Jiménez JI, Nogales J (2013) Aerobic degradation of aromatic compounds. Curr Opin Biotechnol 24:431–442

Díaz E, Prieto MA (2000) Bacterial promoters triggering biodegradation of aromatic pollutants. Curr Opin Biotechnol 11:467–475

Díaz E, Ferrández A, Prieto MA, García JL (2001) Biodegradation of aromatic compounds by Escherichia coli. Microbiol Mol Biol Rev 65:523–569

DiMarco AA, Ornston LN (1994) Regulation of *p*-hydroxybenzoate hydroxylase synthesis by PobR bound to an operator in *Acinetobacter calcoaceticus*. J Bacteriol 176:4277–4284

Dispensa M, Thomas CT, Kim MK, Perrotta JA, Gibson J, Harwood CS (1992) Anaerobic growth of *Rhodopseudomonas palustris* on 4-hydroxybenzoate is dependent on AadR, a member of the cyclic AMP receptor protein family of transcriptional regulators. J Bacteriol 174:5803–5813

Domínguez-Cuevas P, González-Pastor JE, Marqués S, Ramos JL, de Lorenzo V (2006) Transcriptional tradeoff between metabolic and stress-response programs in *Pseudomonas putida* KT2440 cells exposed to toluene. J Biol Chem 281:11981–11991

Domínguez-Cuevas P, Marín P, Busby S, Ramos JL, Marqués S (2008) Roles of effectors in XylS-dependent transcription activation: intramolecular domain derepression and DNA binding. J Bacteriol 190:3118–3128

Domínguez-Cuevas P, Ramos JL, Marqués S (2010) Sequential XylS-CTD binding to the *Pm* promoter induces DNA bending prior to activation. J Bacteriol 192:2682–2690

Donoso RA, Pérez-Pantoja D, González B (2011) Strict and direct transcriptional repression of the *pobA* gene by benzoate avoids 4-hydroxybenzoate degradation in the pollutant degrader bacterium *Cupriavidus necator* JMP134. Environ Microbiol 13:1590–1600

Durante-Rodríguez G, Mancheño JM, Díaz E, Carmona M (2016) Refactoring the λ phage lytic/lysogenic decision with a synthetic regulator. Microbiol Open 5:575–581

Durante-Rodríguez G, Mancheño JM, Rivas G, Alfonso C, García JL, Díaz E, Carmona M (2013) Identification of a missing link in the evolution of an enzyme into a transcriptional regulator. PLoS One 8:e57518

Durante-Rodríguez G, Valderrama JA, Mancheño JM, Rivas G, Alfonso C, Arias-Palomo E, Llorca O, García JL, Díaz E, Carmona M (2010) Biochemical characterization of the transcriptional regulator BzdR from *Azoarcus* sp. CIB J Biol Chem 285:35694–35705

Durante-Rodríguez G, Zamarro MT, García JL, Díaz E, Carmona M (2006) Oxygen-dependent regulation of the central pathway for the anaerobic catabolism of aromatic compounds in *Azoarcus* sp. strain CIB. J Bacteriol 188:2343–2354

Eaton RW (1997) *p*-cymene catabolic pathway in *Pseudomonas putida* F1: cloning and characterization of DNA encoding conversion of *p*-cymene to *p*-cumate. J Bacteriol 179:3171–3180

Egland PG, Harwood CS (2000) HbaR, a 4-hydroxybenzoate sensor and FNR-CRP superfamily member, regulates anaerobic 4-hydroxybenzoate degradation by *Rhodopseudomonas palustris*. J Bacteriol 182:100–106

Ezezika OC, Haddad S, Clark TJ, Neidle EL, Momany C (2007) Distinct effector-binding sites enable synergistic transcriptional activation by BenM, a LysR-type regulator. J Mol Biol 367: 616–629

Fernández C, Díaz E, García JL (2013) Insights on the regulation of the phenylacetate degradation pathway from *Escherichia coli*. Environ Microbiol Rep 6:239–250

Ferrández A, García JL, Díaz E (2000) Transcriptional regulation of the divergent *paa* catabolic operons for phenylacetic acid degradation in *Escherichia coli*. J Biol Chem 275:12214–12222

Fujihara H, Yoshida H, Matsunaga T, Goto M, Furukawa K (2006) Cross-regulation of biphenyl- and salicylate-catabolic genes by two regulatory systems in *Pseudomonas pseudoalcaligenes* KF707. J Bacteriol 188:4690–4697

Fukuhara Y, Inakazu K, Kodama N, Kamimura N, Kasai D, Katayama Y, Fukuda M, Masai E (2009) Characterization of the isophthalate degradation genes of *Comamonas* sp. strain E6. Appl Environ Microbiol 76:519–527

Gabor K, Hailesellasse Sene K, Smidt H, de Vos WM, van der Oost J (2008) Divergent roles of CprK paralogues from *Desulfitobacterium hafniense* in activating gene expression. Microbiology 154:3686–3696

Galán B, García JL, Prieto MA (2004) The PaaX repressor, a link between penicillin G acylase and the phenylacetyl-coenzyme A catabolon of *Escherichia coli* W. J Bacteriol 186: 2215–2220

Galán B, Kolb A, Sanz J, García JL, Prieto MA (2003) Molecular determinants of the *hpa* regulatory system of *Escherichia coli*: the HpaR repressor. Nucleic Acids Res 31:6598–6609

Gallegos MT, Michán C, Ramos JL (1993) The XylS/AraC family of regulators. Nucleic Acids Res 21:807–810

Galvão TC, de Lorenzo V (2006) Transcriptional regulators à la carte: engineering new effector specificities in bacterial regulatory proteins. Curr Opin Biotechnol 17:34–42

García B, Olivera ER, Minambres B, Carnicero D, Muniz C, Naharro G, Luengo JM (2000) Phenylacetyl-coenzyme a is the true inducer of the phenylacetic acid catabolism pathway in *Pseudomonas putida* U. Appl Environ Microbiol 66:4575–4578

García LL, Rivas-Marín E, Floriano B, Bernhardt R, Ewen KM, Reyes-Ramírez F, Santero E (2011) ThnY is a ferredoxin reductase-like iron-sulfur flavoprotein that has evolved to function as a regulator of tetralin biodegradation gene expression. J Biol Chem 286:1709–1718

Garmendia J, de las Heras A, Galvão TC, de Lorenzo V (2008) Tracing explosives in soil with transcriptional regulators of *Pseudomonas putida* evolved for responding to nitrotoluenes. Microb Biotechnol 1:236–246

Garmendia J, de Lorenzo V (2000) The role of the interdomain B linker in the activation of the XylR protein of *Pseudomonas putida*. Mol Microbiol 38:401–410

Gerischer U, Segura A, Ornston LN (1998) PcaU, a transcriptional activator of genes for protocatechuate utilization in *Acinetobacter*. J Bacteriol 180:1512–1524

Grove A (2013) MarR family transcription factors. Curr Biol 23:R142–R143

Guantes R, Benedetti I, Silva-Rocha R, de Lorenzo V (2015) Transcription factor levels enable metabolic diversification of single cells of environmental bacteria. ISME J 10:1122–1133

Gulvik CA, Buchan A (2013) Simultaneous catabolism of plant-derived aromatic compounds results in enhanced growth for members of the *Roseobacter* lineage. Appl Environ Microbiol 79:3716–3723

Guo Z, Houghton JE (1999) PcaR-mediated activation and repression of *pca* genes from *Pseudomonas putida* are propagated by its binding to both the −35 and the −10 promoter elements. Mol Microbiol 32:253–263

Gupta S, Saxena M, Saini N, Mahmooduzzafar KR, Kumar A (2012) An effective strategy for a whole-cell biosensor based on putative effector interaction site of the regulatory DmpR protein. PLoS One 7:e43527

Herrera MC, Duque E, Rodríguez-Herva JJ, Fernández-Escamilla AM, Ramos JL (2010) Identification and characterization of the PhhR regulon in *Pseudomonas putida*. Environ Microbiol 12 (6):1427–1438

Hirakawa H, Hirakawa Y, Greenberg EP, Harwood CS (2015) BadR and BadM proteins transcriptionally regulate two operons needed for anaerobic benzoate degradation by *Rhodopseudomonas palustris*. Appl Environ Microbiol 81:4253–4262

Hirakawa H, Schaefer AL, Greenberg EP, Harwood CS (2012) Anaerobic *p*-coumarate degradation by *Rhodopseudomonas palustris* and identification of CouR, a MarR repressor protein that binds *p*-coumaroyl coenzyme A. J Bacteriol 194:1960–1967

Horbal L, Fedorenko V, Luzhetskyy A (2014) Novel and tightly regulated resorcinol and cumate-inducible expression systems for *Streptomyces* and other actinobacteria. Appl Microbiol Biotechnol 98:8641–8655

Hoskisson PA, Rigali S (2009) Chapter 1: Variation in Form and Function: the helix-turn-helix regulators of the GntR superfamily. Adv Appl Microbiol 69:1–22

Jain D, Nair DT (2012) Spacing between core recognition motifs determines relative orientation of AraR monomers on bipartite operators. Nucleic Acids Res 41:639–647

Jiménez JI, Juárez JF, García JL, Díaz E (2011) A finely tuned regulatory circuit of the nicotinic acid degradation pathway in *Pseudomonas putida*. Environ Microbiol 13:1718–1732

Jõesaar M, Heinaru E, Viggor S, Vedler E, Heinaru A (2010) Diversity of the transcriptional regulation of the *pch* gene cluster in two indigenous *p*-cresol-degradative strains of *Pseudomonas fluorescens*. FEMS Microbiol Ecol 72:464–475

Jones RM, Pagmantidis V, Williams PA (2000) *sal* genes determining the catabolism of salicylate esters are part of a supraoperonic cluster of catabolic genes in *Acinetobacter* sp. strain ADP1. J Bacteriol 182:2018–2025

Jones RM, Williams PA (2001) *areCBA* is an operon in *Acinetobacter* sp. strain ADP1 and is controlled by AreR, a σ54-dependent regulator. J Bacteriol 183:405–409

Jourlin-Castelli C, Mani N, Nakano MM, Sonenshein AL (2000) CcpC, a novel regulator of the LysR family required for glucose repression of the *citB* gene in *Bacillus subtilis*. J Mol Biol 295:865–878

Juárez JF, Liu H, Zamarro MT, McMahon S, Liu H, Naismith JH, Eberlein C, Boll M, Carmona M, Díaz E (2015) Unraveling the specific regulation of the central pathway for anaerobic degradation of 3-methylbenzoate. J Biol Chem 290:12165–12183

Juárez JF, Zamarro MT, Eberlein C, Boll M, Carmona M, Díaz E (2012) Characterization of the *mbd* cluster encoding the anaerobic 3-methylbenzoyl-CoA central pathway. Environ Microbiol 15:148–166

Kamimura N, Takamura K, Hara H, Kasai D, Natsume R, Senda T, Katayama Y, Fukuda M, Masai E (2010) Regulatory system of the protocatechuate 4,5-cleavage pathway genes essential for lignin downstream catabolism. J Bacteriol 192:3394–3405

Kasai D, Araki N, Motoi K, Yoshikawa S, Iino T, Imai S, Masai E, Fukuda M (2015) γ-resorcylate catabolic-pathway genes in the soil Actinomycete *Rhodococcus jostii* RHA1. Appl Environ Microbiol 81:7656–7665

Kasai D, Kamimura N, Tani K, Umeda S, Abe T, Fukuda M, Masai E (2012) Characterization of FerC, a MarR-type transcriptional regulator, involved in transcriptional regulation of the ferulate catabolic operon in *Sphingobium* sp. strain SYK-6. FEMS Microbiol Lett 332:68–75

Kasai D, Kitajima M, Fukuda M, Masai E (2010) Transcriptional regulation of the terephthalate catabolism operon in *Comamonas* sp. strain E6. Appl Environ Microbiol 76:6047–6055

Kim HS, Kang TS, Hyun JS, Kang HS (2004) Regulation of penicillin G acylase gene expression in *Escherichia coli* by repressor PaaX and the cAMP- receptor protein complex. J Biol Chem 279:33253–33262

Kim MN, Park HH, Lim WK, Shin HJ (2005) Construction and comparison of *Escherichia coli* whole-cell biosensors capable of detecting aromatic compounds. J Microbiol Methods 60:235–245

Kim SI, Jourlin-Castelli C, Wellington SR, Sonenshein AL (2003) Mechanism of repression by *Bacillus subtilis* CcpC, a LysR family regulator. J Mol Biol 334:609–624

Kim Y, Joachimiak G, Bigelow L, Babnigg G, Joachimiak A (2016) How aromatic compounds block DNA binding of HcaR catabolite regulator. J Biol Chem 291:13243–13256

Körner H, Sofia HJ, Zumft WG (2003) Phylogeny of the bacterial superfamily of Crp-Fnr transcription regulators: exploiting the metabolic spectrum by controlling alternative gene programs. FEMS Microbiol Rev 27:559–592

Krell T, Molina-Henares AJ, Ramos JL (2006) The IclR family of transcriptional activators and repressors can be defined by a single profile. Protein Sci 15:1207–1213

Leahy JG, Johnson GR, Olsen RH (1997) Cross-regulation of toluene monooxygenases by the transcriptional activators TbmR and TbuT. Appl Environ Microbiol 63:3736–3739

Ledesma-García L, Sánchez-Azqueta A, Medina M, Reyes-Ramírez F, Santero E (2016) Redox proteins of hydroxylating bacterial dioxygenases establish a regulatory cascade that prevents gratuitous induction of tetralin biodegradation genes. Sci Rep 6:23848

Lerche M, Dian C, Round A, Lönneborg R, Brzezinski P, Leonard GA (2016) The solution configurations of inactive and activated DntR have implications for the sliding dimer mechanism of LysR transcription factors. Sci Rep 6:19988

Levy C, Pike K, Heyes DJ, Joyce MG, Gabor K, Smidt H, van der Oost J, Leys D (2008) Molecular basis of halorespiration control by CprK, a CRP-FNR type transcriptional regulator. Mol Microbiol 70:151–167

Li DF, Zhang N, Hou YJ, Huang Y, Hu Y, Zhang Y, Liu SJ, Wang DC (2011) Crystal structures of the transcriptional repressor RolR reveals a novel recognition mechanism between inducer and regulator. PLoS One 6:e19529

Li T, Zhao K, Huang Y, Li D, Jiang CY, Zhou N, Fan Z, Liu SJ (2012) The TetR-type transcriptional repressor RolR from *Corynebacterium glutamicum* regulates resorcinol catabolism by binding to a unique operator, *rolO*. Appl Environ Microbiol 78:6009–6016

Lin GH, Chen HP, Shu HY (2015) Detoxification of indole by an indole-induced flavoprotein oxygenase from *Acinetobacter baumannii*. PLoS One 10:e0138798

López-Sánchez A, Rivas-Marín E, Martínez-Pérez O, Floriano B, Santero E (2009) Co-ordinated regulation of two divergent promoters through higher-order complex formation by the LysR-type regulator ThnR. Mol Microbiol 73:1086–1100

Lovley DR (2003) Cleaning up with genomics: applying molecular biology to bioremediation. Nat Rev Microbiol 1:35–44

Maddocks SE, Oyston PCF (2008) Structure and function of the LysR-type transcriptional regulator (LTTR) family proteins. Microbiology 154:3609–3623

Madhushani A, del Peso-Santos T, Moreno R, Rojo F, Shingler V (2014) Transcriptional and translational control through the 5′-leader region of the *dmpR* master regulatory gene of phenol metabolism. Environ Microbiol 17:119–133

Manso I, Torres B, Andreu JM, Menéndez M, Rivas G, Alfonso C, Díaz E, García JL, Galán B (2009) 3-Hydroxyphenylpropionate and phenylpropionate are synergistic activators of the MhpR transcriptional regulator from *Escherichia coli*. J Biol Chem 284:21218–21228

Marqués S, Manzanera M, González-Pérez MM, Gallegos MT, Ramos JL (1999) The XylS-dependent *Pm* promoter is transcribed in vivo by RNA polymerase with σ32 or σ38 depending on the growth phase. Mol Microbiol 31:1105–1113

McFall SM, Chugani SA, Chakrabarty AM (1998) Transcriptional activation of the catechol and chlorocatechol operons: variations on a theme. Gene 223:257–267

Meier MJ, Paterson ES, Lambert IB (2015) Use of substrate-induced gene expression in meta-genomic analysis of an aromatic hydrocarbon-contaminated soil. Appl Environ Microbiol 82:897–909

Michán C, Zhou L, Gallegos MT, Timmis KN, Ramos JL (1992) Identification of critical amino-terminal regions of XylS. The positive regulator encoded by the TOL plasmid. J Biol Chem 267:22897–22901

Miyakoshi M, Urata M, Habe H, Omori T, Yamane H, Nojiri H (2006) Differentiation of carbazole catabolic operons by replacement of the regulated promoter via transposition of an insertion sequence. J Biol Chem 281:8450–8457

Mohamed M, Ismail W, Heider J, Fuchs G (2002) Aerobic metabolism of phenylacetic acids in *Azoarcus evansii*. Arch Microbiol 178:180–192

Molina-Fuentes A, Pacheco D, Marín P, Philipp B, Schink B, Marqués S (2015) Identification of the gene cluster for the anaerobic degradation of 3,5-dihydroxybenzoate (α-resorcylate) in *Thauera aromatica* strain AR-1. Appl Environ Microbiol 81:7201–7214

Molina-Henares AJ, Krell T, Eugenia Guazzaroni M, Segura A, Ramos JL (2006) Members of the IclR family of bacterial transcriptional regulators function as activators and/or repressors. FEMS Microbiol Rev 30:157–186

Monferrer D, Tralau T, Kertesz MA, Dix I, Solà M, Usón I (2010) Structural studies on the full-length LysR-type regulator TsaR from *Comamonas testosteroni* T-2 reveal a novel open conformation of the tetrameric LTTR fold. Mol Microbiol 75:1199–1214

Morabbi Heravi K, Lange J, Watzlawick H, Kalinowski J, Altenbuchner J (2014) Transcriptional regulation of the vanillate utilization genes (*vanABK* operon) of *Corynebacterium glutamicum* by VanR, a PadR-like repressor. J Bacteriol 197:959–972

Morawski B, Segura A, Ornston LN (2000) Repression of *Acinetobacter* vanillate demethylase synthesis by VanR, a member of the GntR family of transcriptional regulators. FEMS Microbiol Lett 187:65–68

Moreno R, Hernández-Arranz S, la Rosa R, Yuste L, Madhushani A, Shingler V, Rojo F (2015) The Crc and Hfq proteins of *Pseudomonas putida* cooperate in catabolite repression and formation of ribonucleic acid complexes with specific target motifs. Environ Microbiol 17:105–118

Mouz S, Merlin C, Springael D (1999) A GntR-like negative regulator of the biphenyl degradation genes of the transposon Tn*4371*. Mol Gen Genet 262:790–799

Muraoka S, Okumura R, Ogawa N, Nonaka T, Miyashita K, Senda T (2003) Crystal structure of a full-length LysR-type transcriptional regulator, CbnR: unusual combination of two subunit forms and molecular bases for causing and changing DNA bend. J Mol Biol 328:555–566

Nga DP, Altenbuchner J, Heiss GS (2004) NpdR, a repressor involved in 2,4,6-trinitrophenol degradation in *Rhodococcus opacus* HL PM-1. J Bacteriol 186:98–103

Niewerth H, Parschat K, Rauschenberg M, Ravoo BJ, Fetzner S (2012) The PaaX-type repressor MeqR2 of *Arthrobacter* sp. strain Rue61a, involved in the regulation of quinaldine catabolism, binds to its own promoter and to catabolic promoters and specifically responds to anthraniloyl coenzyme A. J Bacteriol 195:1068–1080

O'Neill E (2001) An active role for a structured B-linker in effector control of the σ54-dependent regulator DmpR. EMBO J 20:819–827

Ogawa N, McFall SM, Klem TJ, Miyashita K, Chakrabarty AM (1999) Transcriptional activation of the chlorocatechol degradative genes of *Ralstonia eutropha* NH9. J Bacteriol 181:6697–6705

Ohtsubo Y, Delawary M, Kimbara K, Takagi M, Ohta A, Nagata Y (2001) BphS, a key transcriptional regulator of *bph* genes involved in polychlorinated biphenyl/biphenyl degradation in *Pseudomonas* sp. KKS102. J Biol Chem 276:36146–36154

Orth P, Schnappinger D, Hillen W, Saenger W, Hinrichs W (2000) Structural basis of gene regulation by the tetracycline inducible Tet repressor-operator system. Nat Struct Biol 7:215–219

Otani H, Stogios PJ, Xu X, Nocek B, Li SN, Savchenko A, Eltis LD (2015) The activity of CouR, a MarR family transcriptional regulator, is modulated through a novel molecular mechanism. Nucleic Acids Res 44:595–607

Pareja E, Pareja-Tobes P, Manrique M, Pareja-Tobes E, Bonal J, Tobes R (2006) ExtraTrain: a database of extragenic regions and transcriptional information in prokaryotic organisms. BMC Microbiol 6:29

Park HS, Kim HS (2001) Genetic and structural organization of the aminophenol catabolic operon and its implication for evolutionary process. J Bacteriol 183:5074–5081

Parke D (1996) Characterization of PcaQ, a LysR-type transcriptional activator required for catabolism of phenolic compounds, from *Agrobacterium tumefaciens*. J Bacteriol 178:266–272

Parke D, Ornston LN (2003) Hydroxycinnamate (*hca*) catabolic genes from *Acinetobacter* sp. strain ADP1 are repressed by *hcaR* and are induced by hydroxycinnamoyl-coenzyme A thioesters. Appl Environ Microbiol 69:5398–5409

Pérez-Martín J, de Lorenzo V (1995) The amino-terminal domain of the prokaryotic enhancer-binding protein XylR is a specific intramolecular repressor. Proc Natl Acad Sci USA 92: 9392–9396

Pérez-Pantoja D, Leiva-Novoa P, Donoso RA, Little C, Godoy M, Pieper DH, González B (2015) Hierarchy of carbon source utilization in soil bacteria: hegemonic preference for benzoate in complex aromatic compound mixtures degraded by *Cupriavidus pinatubonensis* strain JMP134. Appl Environ Microbiol 81:3914–3924

Phattarasukol S, Radey MC, Lappala CR, Oda Y, Hirakawa H, Brittnacher MJ, Harwood CS (2012) Identification of a *p*-coumarate degradation regulon in *Rhodopseudomonas palustris* by Xpression, an integrated tool for prokaryotic RNA-Seq data processing. Appl Environ Microbiol 78:6812–6818

Picossi S, Belitsky BR, Sonenshein AL (2007) Molecular mechanism of the regulation of *Bacillus subtilis gltAB* expression by GltC. J Mol Biol 365:1298–1313

Popp R, Kohl T, Patz P, Trautwein G, Gerischer U (2002) Differential DNA binding of transcriptional regulator PcaU from *Acinetobacter* sp. strain ADP1. J Bacteriol 184:1988–1997

Pop SM, Gupta N, Raza AS, Ragsdale SW (2006) Transcriptional activation of dehalorespiration. Identification of redox-active cysteines regulating dimerization and DNA binding. J Biol Chem 8:26382–26390

Porrúa O, García-Jaramillo M, Santero E, Govantes F (2007) The LysR-type regulator AtzR binding site: DNA sequences involved in activation, repression and cyanuric acid-dependent repositioning. Mol Microbiol 66:410–427

Prieto MA, García JL (1994) Molecular characterization of 4-hydroxyphenylacetate 3-hydroxylase of *Escherichia coli*. A two-protein component enzyme. J Biol Chem 269:22823–22829

Providenti MA, Wyndham RC (2001) Identification and functional characterization of CbaR, a MarR-Like modulator of the *cbaABC*-encoded chlorobenzoate catabolism pathway. Appl Environ Microbiol 67:3530–3541

Quinn JA, McKay DB, Entsch B (2001) Analysis of the *pobA* and *pobR* genes controlling expression of *p*-hydroxybenzoate hydroxylase in *Azotobacter chroococcum*. Gene 264:77–85

Ramos JL, Marqués S, Timmis KN (1997) Transcriptional control of the *Pseudomonas tol* plasmid catabolic operons is achieved through an interplay of host factors and plasmid-encoded regulators. Annu Rev Microbiol 51:341–373

Ramos JL, Martínez-Bueno M, Molina-Henares AJ, Teran W, Watanabe K, Zhang X, Gallegos MT, Brennan R, Tobes R (2005) The TetR family of transcriptional repressors. Microbiol Mol Biol Rev 69:326–356

Rey FE, Harwood CS (2010) FixK, a global regulator of microaerobic growth, controls photosynthesis in *Rhodopseudomonas palustris*. Mol Microbiol 75:1007–1020

Rojas-Altuve A, Carrasco-López C, Hernández-Rocamora VM, Sanz JM, Hermoso JA (2011) Crystallization and preliminary X-ray diffraction studies of the transcriptional repressor PaaX, the main regulator of the phenylacetic acid degradation pathway in *Escherichia coli* W. Acta Cryst Sect F 67:1278–1280

Sakamoto K, Agari Y, Kuramitsu S, Shinkai A (2011) Phenylacetyl coenzyme A is an effector molecule of the TetR family transcriptional repressor PaaR from *Thermus thermophilus* HB8. J Bacteriol 193:4388–4395

Sandu C, Chiribau CB, Brandsch R (2003) Characterization of HdnoR, the transcriptional repressor of the 6-hydroxy-D-nicotine oxidase gene of *Arthrobacter nicotinovorans* pAO1, and its DNA-binding activity in response to L- and D-nicotine derivatives. J Biol Chem 278:51307–51315

Sasson V, Shachrai I, Bren A, Dekel E, Alon U (2012) Mode of regulation and the insulation of bacterial gene expression. Mol Cell 46:399–407

Schell MA (1993) Molecular biology of the LysR family of transcriptional regulators. Annu Rev Microbiol 47:597–626

Seedorff J, Schleif R (2011) Active role of the interdomain linker of AraC. J Bacteriol 193: 5737–5746

Shingler V (2004) Transcriptional regulation and catabolic strategies of phenol degradative pathways. In: Ramos JL (ed) *Pseudomonas*, vol 2. Kluwer Academic, New York, pp 451–477

Shu HY, Lin LC, Lin TK, Chen HP, Yang HH, Peng KC, Lin GH (2015) Transcriptional regulation of the *iac* locus from *Acinetobacter baumannii* by the phytohormone indole-3-acetic acid. Antonie Van Leeuwenhoek 107:1237–1247

Silva-Jiménez H, García-Fontana C, Cadirci BH, Ramos-González MI, Ramos JL, Krell T (2011) Study of the TmoS/TmoT two-component system: towards the functional characterization of the family of TodS/TodT like systems. Microb Biotechnol 5:489–500

Silva-Rocha R, de Jong H, Tamames J, de Lorenzo V (2011) The logic layout of the TOL network of *Pseudomonas putida* pWW0 plasmid stems from a metabolic amplifier motif (MAM) that optimizes biodegradation of *m*-xylene. BMC Syst Biol 5:191

Silva-Rocha R, de Lorenzo V (2012) Broadening the signal specificity of prokaryotic promoters by modifying *cis*-regulatory elements associated with a single transcription factor. Mol BioSyst 8:1950

Skarfstad E, O'Neill E, Garmendia J, Shingler V (2000) Identification of an effector specificity subregion within the aromatic-responsive regulators DmpR and XylR by DNA shuffling. J Bacteriol 182:3008–3016

Smirnova IA, Dian C, Leonard GA, McSweeney S, Birse D, Brzezinski P (2004) Development of a bacterial biosensor for nitrotoluenes: the crystal structure of the transcriptional regulator DntR. J Mol Biol 340:405–418

Solera D, Arenghi FLG, Woelk T, Galli E, Barbieri P (2004) TouR-mediated effector-independent growth phase-dependent activation of the σ54 *Ptou* promoter of *Pseudomonas stutzeri* OX1. J Bacteriol 186:7353–7363

Suresh PS, Kumar R, Kumar A (2010) Three dimensional model for N-terminal A domain of DmpR (2-dimethylphenol) protein based on secondary structure prediction and fold recognition. In Silico Biol 10:223–233

Suvorova IA, Korostelev YD, Gelfand MS (2015) GntR family of bacterial transcription factors and their DNA binding motifs: structure, positioning and co-evolution. PLoS One 10:e0132618

Szőköl J, Rucká L, Šimčíková M, Halada P, Nešvera J, Pátek M (2014) Induction and carbon catabolite repression of phenol degradation genes in *Rhodococcus erythropolis* and *Rhodococcus jostii*. Appl Microbiol Biotechnol 98:8267–8279

Teramoto M, Harayama S, Watanabe K (2001) PhcS represses gratuitous expression of phenol-metabolizing enzymes in *Comamonas testosteroni* R5. J Bacteriol 183:4227–4234

Teufel R, Friedrich T, Fuchs G (2012) An oxygenase that forms and deoxygenates toxic epoxide. Nature 483:359–362

Thanbichler M, Iniesta AA, Shapiro L (2007) A comprehensive set of plasmids for vanillate- and xylose-inducible gene expression in *Caulobacter crescentus*. Nucleic Acids Res 35:e137–e137

Tomás-Gallardo L, Santero E, Floriano B (2012) Involvement of a putative cyclic AMP receptor protein (CRP)-like binding sequence and a CRP-like protein in glucose-mediated catabolite repression of *thn* genes in *Rhodococcus* sp. strain TFB. Appl Environ Microbiol 78:5460–5462

Torres B, Porras G, García JL, Díaz E (2003) Regulation of the *mhp* cluster responsible for 3-(3-hydroxyphenyl)propionic acid degradation in *Escherichia coli*. J Biol Chem 278:27575–27585

Townsend PD, Jungwirth B, Pojer F, Bußmann M, Money VA, Cole ST, Pühler A, Tauch A, Bott M, Cann MJ, Pohl E (2014) The crystal structures of Apo and cAMP-bound GlxR from *Corynebacterium glutamicum* reveal structural and dynamic changes upon cAMP binding in CRP/FNR family transcription factors. PLoS One 9:e113265

Tralau T, Cook AM, Ruff J (2003a) An additional regulator, TsaQ, is involved with TsaR in regulation of transport during the degradation of *p*-toluenesulfonate in *Comamonas testosteroni* T-2. Arch Microbiol 180:319–326

Tralau T, Mampel J, Cook AM, Ruff J (2003b) Characterization of TsaR, an oxygen-sensitive LysR-type regulator for the degradation of *p*-toluenesulfonate in *Comamonas testosteroni* T-2. Appl Environ Microbiol 69:2298–2305

Trautwein G, Gerischer U (2001) Effects exerted by tanscriptional regulator PcaU from Acinetobacter sp. strain ADP1. J Bacteriol 183:873–881

Tropel D, van der Meer JR (2004) Bacterial transcriptional regulators for degradation pathways of aromatic compounds. Microbiol Mol Biol Rev 68:474–500

Uchiyama T, Miyazaki K (2013) Metagenomic screening for aromatic compound-responsive transcriptional regulators. PLoS One 8:e75795

Ueki T (2011) Identification of a transcriptional repressor involved in benzoate metabolism in *Geobacter bemidjiensis*. Appl Environ Microbiol 77:7058–7062

Ulrich LE, Koonin EV, Zhulin IB (2005) One-component systems dominate signal transduction in prokaryotes. Trends Microbiol 13:52–56

Urata M, Miyakoshi M, Kai S, Maeda K, Habe H, Omori T, Yamane H, Nojiri H (2004) Transcriptional regulation of the *ant* operon, encoding two-component anthranilate 1,2-dioxygenase, on the carbazole-degradative plasmid pCAR1 of *Pseudomonas resinovorans* strain CA10. J Bacteriol 186:6815–6823

Valderrama JA, Durante-Rodríguez G, Blázquez B, García JL, Carmona M, Díaz E (2012) Bacterial degradation of benzoate: cross-regulation between aerobic and anaerobic pathways. J Biol Chem 287:10494–10508

van Aalten DMF (2001) The structural basis of acyl coenzyme A-dependent regulation of the transcription factor FadR. EMBO J 20:2041–2050

Vaneechoutte M, Young DM, Ornston LN, de Baere T, Nemec A, Van Der Reijden T, Carr E, Tjernberg I, Dijkshoorn L (2006) Naturally transformable *Acinetobacter* sp. strain ADP1 belongs to the newly described species *Acinetobacter baylyi*. Appl Environ Microbiol 72:932–936

Veselý M, Knoppová M, Nešvera J, Pátek M (2007) Analysis of *catRABC* operon for catechol degradation from phenol-degrading *Rhodococcus erythropolis*. Appl Microbiol Biotechnol 76:159–168

Wöhlbrand L, Kallerhoff B, Lange D, Hufnagel P, Thiermann J, Reinhardt R, Rabus R (2007) Functional proteomic view of metabolic regulation in "*Aromatoleum aromaticum*" strain EbN1. Proteomics 7:2222–2239

Xu Y, Heath RJ, Li Z, Rock CO, White SW (2001) The FadR.DNA complex: transcriptional control of fatty acid metabolism in *Escherichia coli*. J Biol Chem 276:17373–17379

Yamamoto K, Ishihama A (2002) Two different modes of transcription repression of the *Escherichia coli* acetate operon by IclR. Mol Microbiol 47:183–194

Yudistira H, McClarty L, Bloodworth RAM, Hammond SA, Butcher H, Mark BL, Cardona ST (2011) Phenylalanine induces *Burkholderia cenocepacia* phenylacetic acid catabolism through degradation to phenylacetyl-CoA in synthetic cystic fibrosis sputum medium. Microb Pathog 51:186–193

Zhao Y, Feng R, Zheng G, Tian J, Ruan L, Ge M, Jiang W, Lu Y (2015) Involvement of the TetR-type regulator PaaR in the regulation of pristinamycin I biosynthesis through an effect on precursor supply in *Streptomyces pristinaespiralis*. J Bacteriol 197:2062–2071

# Transcriptional Regulation of Hydrocarbon Efflux Pump Expression in Bacteria

**11**

Cauã Antunes Westmann, Luana de Fátima Alves, Tiago Cabral Borelli, Rafael Silva-Rocha, and María-Eugenia Guazzaroni

## Contents

1   Introduction ................................................................................ 178
2   Efflux Pumps that Increase Hydrocarbon Tolerance ..................................... 179
3   Regulatory Networks Involved in Hydrocarbon Efflux Pump ........................... 180
    3.1   An Overview of HAE-RND Pumps in *E. coli*, *Salmonella* sp., and
          *P. aeruginosa* ................................................................ 180
    3.2   General Genetic Architecture of RND Systems ................................... 182
    3.3   RND Regulatory Networks Are Multi-hierarchical .............................. 182
4   Regulation of Efflux Pumps in *P. putida* DOT-T1E as a Case of Study .................. 188
    4.1   Identification of the Efflux Pumps Involved in Solvent Tolerance ................. 189
    4.2   Study of the Transcriptional Repressors Involved in the Regulation of the
          Pumps ......................................................................... 189
5   Research Needs ........................................................................... 191
References ................................................................................... 191

C. A. Westmann · R. Silva-Rocha
Departamento de Biologia Celular, FMRP – University of São Paulo, Ribeirão Preto, SP, Brazil
e-mail: cauawest@hotmail.com; silvarochar@gmail.com

L. de Fátima Alves
Departamento de Biologia, FFCLRP – University of São Paulo, Ribeirão Preto, SP, Brazil

Departamento de Bioquímica e Imunologia, FMRP – University of São Paulo, Ribeirão Preto, SP, Brazil
e-mail: luanadfalves@gmail.com

T. C. Borelli
Departamento de Biologia, FFCLRP – University of São Paulo, Ribeirão Preto, SP, Brazil
e-mail: tiago.cabral42@hotmail.com

M.-E. Guazzaroni (✉)
Departamento de Biologia, FFCLRP – University of São Paulo, Ribeirão Preto, SP, Brazil

Faculdade de Filosofia, Ciências e Letras de Ribeirão Preto, Universidade de São Paulo, Ribeirão Preto, São Paulo, Brazil
e-mail: meguazzaroni@gmail.com

© Springer International Publishing AG, part of Springer Nature 2018                 177
T. Krell (ed.), *Cellular Ecophysiology of Microbe: Hydrocarbon and Lipid Interactions*,
Handbook of Hydrocarbon and Lipid Microbiology, https://doi.org/10.1007/978-3-319-50542-8_4

**Abstract**
Efflux pumps were found to be the most efficient mechanism of hydrocarbon tolerance in several bacterial strains resistant to solvents and other toxic chemicals. This involves an energy-dependent process that ensures the active removal of toxic compounds from the bacterial cytoplasm to the external medium. In order to achieve the maximal response of the process, regulatory networks of RND efflux pumps (resistance-nodulation-cell division family of bacterial transporters) are complex and modulated by the simultaneous coordination of many transcription factors in response to perturbations and cellular states. Several studies in the literature report the identification and molecular characterization of regulatory genes of RND efflux pumps, acting as global or specific regulators, activating or repressing transcription. In this sense, the main objective of this chapter is to provide a general view of the regulatory networks used by bacteria to modulate the response to toxic hydrocarbons mediated by efflux pumps. We also explore conceptual properties that remain conserved in different regulatory systems and outline common principles of RND regulation in gram-negative bacteria.

# 1    Introduction

Bacteria have been exposed during evolution to aromatic hydrocarbons since they have been present in the environment for millions of years as they are the products of the natural pyrolysis of organic material (Dagley 1971). Additionally, different natural toxic compounds with related structures (such toxins, secondary metabolites, or antibiotics, to cite some) also have been part of the ambient to which microorganisms have been exposed. One-ring aromatic compounds such as benzene, xylenes, ethylbenzene, and toluene have a $logP_{ow}$ (logarithm of the octanol/water partition coefficient, normally used to quantify hydrophobicity) of 2.5–3.5 and are thus toxic for microorganisms and other living cells because they partition preferentially in the cytoplasmic membrane, disorganizing its structure and impairing vital functions (Sikkema et al. 1995). In this context, microorganisms have developed different strategies to detoxify or eliminate these compounds (Ramos et al. 2002). Some strategies are directed to prevent the entrance of toxic compounds into the cells, such as the rigidification of the cell membrane via alteration of the phospholipid composition or alteration of the cell surface (Heipieper et al. 1992; Junker and Ramos 1999). On the other hand, energy-dependent processes ensure the active removal of recalcitrant compounds from the bacterial cytoplasm (or inner part of the cytoplasmic membrane) to the external medium by using efflux pumps (Ramos et al. 2002). The latter mechanism was found to be the most efficient hydrocarbon tolerance system in several strains resistant to solvents and other toxic chemicals (Inoue et al. 1991; Isken and De Bont 1996; Aires et al. 1999). Both members of the resistance-nodulation-cell division (RND) family of bacterial transporters and the ATP-binding cassette (ABC) transporter family have been associated to hydrocarbon tolerance (Kim et al. 1998; García et al. 2010).

Several families of transcriptional regulators have been shown to be involved in the regulation of RND efflux pumps conferring resistance to aromatic and aliphatic hydrocarbons (Ramos et al. 2002; Terán et al. 2003). In addition, numerous examples in the literature report the identification and molecular characterization of these regulatory genes, acting as global or specific regulators, activating or repressing transcription. Therefore, in this chapter we provide a broad view of the complex regulatory networks used by bacteria to modulate the response to toxic hydrocarbons mediated by efflux pumps, with special emphasis in the common principles of the regulatory systems most extensively studied in the literature.

## 2 Efflux Pumps that Increase Hydrocarbon Tolerance

Hydrocarbon compounds cover a vast number of molecules with different aromatic and aliphatic structures, and some of these compounds are toxic to living organisms. Environmentally the most significant are hydrocarbon alkanes, main components of fossil fuels; organic solvents such as benzene, toluene, ethylbenzene, and xylenes (BTEX) are the most harmful to microorganisms (Fillet et al. 2012). The main mechanism used by microorganisms to overcome this toxicity is the efflux systems, generally named multidrug resistance pumps (MDR pumps). Additionally, these pumps are responsible for the extrusion of a broad range of compounds, such as antibiotics, heavy metals, free fatty acids, and organic solvents (Jones et al. 2015).

Multidrug resistance efflux pumps are found in a large number of prokaryotic and eukaryotic organisms. In bacteria, efflux pumps are prevalent in gram-negatives, and they are classified into five families: (i) major facilitator superfamily (MFS), (ii) the small multidrug resistance (SMR) family, (iii) the multidrug and toxic compound extrusion (MATE) family, (iv) the resistance-nodulation-cell division (RND) superfamily, and (v) the ATP-binding cassette (ABC) superfamily. These pumps use different energy sources. For instance, the RND, MSF, and SMR couple substrate extrusion to proton-motive force; in other words, these pumps are dependent on the pH gradient. Pumps belonging to the ABC superfamily use ATP hydrolysis as energy source, whereas MATE pumps are driven by Na+/H+ for extrusion of compounds (Putman et al. 2000).

The RND members are the main efflux pumps involved in gram-negative bacteria to provide hydrocarbon tolerance (Fernandes et al. 2003). These pumps consist of a trimeric system: (i) an inner membrane transporter operating as an energy-dependent pump, (ii) an outer membrane channel, and (iii) a membrane fusion protein. Several examples in the literature showed the relevance of the RND transporters for the tolerance to hydrocarbons (see the next section). Additionally, efflux pumps belonging to the ABC superfamily involved in hydrocarbon tolerance mechanisms have also been described in gram-negative bacteria. For instance, in *Escherichia coli*, the MsbA pump is involved in the extrusion of isoprenoids, and its overexpression increases two- to fourfold the extrusion of these molecules in a strain that produces zeaxanthin, canthaxanthin, and β-carotene (Doshi et al. 2013). In *Pseudomonas putida* DOT-T1E, although the TtgGHI efflux pump (RND family) is the main responsible for the high solvent tolerance of this strain, the TtgABC transporter (ABC superfamily) is also involved in toluene tolerance (García et al. 2010).

Even though most of the studies related to hydrocarbon tolerance mediated by efflux pumps were carried out in gram-negative bacteria, some studies show that gram-positives such as strains belonging to the genera *Staphylococcus*, *Bacillus*, *Rhodococcus*, and *Arthrobacter* have also developed similar mechanisms (Sardessai and Bhosle 2002; Truong-Bolduc et al. 2014; Alnaseri et al. 2015). This is the case of the pathogenic strain *Staphylococcus aureus* USA300. This bacterium colonizes human skin and is the cause of some human infections that conduce to abscess formation. When *S. aureus* USA300 is exposed to antimicrobial fatty acids on the skin, an efflux pump named FarE promotes the extrusion of antimicrobial fatty acids, conferring in this way tolerance to these compounds (Alnaseri et al. 2015). Additionally, the expression of FarE in *S. aureus* USA300 is strongly induced by the presence of arachidonic and linoleic acids (Alnaseri et al. 2015). In general, these findings demonstrated that efflux pumps are not only very relevant for the resistance of microorganism to environmental injuries but also play major roles in the increase in antibiotic resistance in clinically relevant strains.

# 3    Regulatory Networks Involved in Hydrocarbon Efflux Pump

Regulatory networks of RND efflux pumps in bacteria are complex and modulated by the simultaneous coordination of many transcription factors in response to perturbations and cellular states. The combined analysis of some RND systems belonging to *E. coli*, *Salmonella* sp., *Pseudomonas aeruginosa*, and *P. putida* revealed that although the components of the RND regulatory system may differ between phylogenetic groups, many conceptual properties remain conserved. In this section of the chapter, we explore those properties and outline common principles of RND regulation in gram-negative bacteria.

## 3.1    An Overview of HAE-RND Pumps in *E. coli*, *Salmonella* sp., and *P. aeruginosa*

In gram-negative bacteria, most of the multidrug resistance efflux pumps related to hydrophobic and amphiphilic transport (e.g., antibiotics, biocides, and hydrocarbons) are from the hydrophobe-amphiphile efflux-1 (HAE1) subdivison of the RND superfamily (Saier and Paulsen 2001). In *E. coli*, the best characterized HAE-RND efflux pumps are the orthologues of the acriflavine resistance (Acr) group (Ma et al. 1993): *acrB-acrAB* operon (Murakami et al. 2002, 2006; Das et al. 2007; Nikaido 2009; Pos 2009), *acrD*-single gene (Rosenberg et al. 2000; Nishino and Yamaguchi 2001; Aires and Nikaido 2005), and *acrF-acrEF* operon (Ma et al. 1993; Nishino and Yamaguchi 2001; Lin et al. 2005). The *acrAB* operon, which encodes the AcrA membrane fusion protein (MFP) and the AcrB inner membrane transporter (IMT), together with the *tolC* gene, which encodes for the outer membrane protein (OMP), provides the three components of the RND pump (Ma et al. 1995). Those genes are induced by different types of stress (i.e., osmotic, ethanol, or entrance into stationary phase) and are independently transcribed from the local regulator AcrR (Ma et al. 1995, 1996). Furthermore, multiple studies have shown the importance of MarA, SoxS, and Rob global regulators in the expression of *acrAB/tolC* genes (Ma et al. 1995; Sulavik et al. 1995; Okusu et al. 1996; Tanaka et al. 1997; White et al. 1997; Alekshun and Levy 1999), suggesting a complex network of regulatory signals for the induction of this system.

Analogous to the structure and regulation of its *E. coli* homolog, the *Salmonella* sp. AcrAB-TolC system is controlled through several regulatory pathways, such as AcrR, MarA, and SoxS (Koutsolioutsou et al. 2001; Randall and Woodward 2001; Eaves et al. 2004; Li and Nikaido 2009; Zheng et al. 2009). A recent study also showed that the expression level of the *acrB*, *acrD*, and/or *acrF* gene was increased when one or multiple *acr* genes were deleted (Blair et al. 2015), and this observation was similar to the situation found among the m*ex* genes of *P. aeruginosa* (Li et al. 2000). Another gene locus, *ramRA*, that is widespread in *Enterobacteriaceae* except *E. coli*, also significantly influences the expression of not only *acrAB* but also *acrEF* and *mdtABC* in *Salmonella* (Li and Nikaido 2009; Zheng et al. 2009; Bailey et al. 2010; Baucheron et al. 2014). RamR is a TetR repressor and can be inhibited by multiple agents (Yamasaki et al. 2013), allowing the expression of *ramA* that serves as a small activator protein to the *acrAB* operon (Nikaido et al. 2008; Sun et al. 2011; Lawler et al. 2013; Baucheron et al. 2014). A recent review also highlighted the role of RamRA in the regulation of *acrAB* (Piddock 2014).

*P. aeruginosa* PAO1 has 12 RND efflux pumps described in its genome, and the ones known as multidrug efflux (Mex) are homologs to the *E. coli* AcrAB. The pattern of organization of the *mex* genes is very similar to the one observed in *E. coli*, consisting of one local regulator – like the MexT activator – upstream the *mex* pump operon. However, in contrast to *E. coli*, the tripartite efflux pump genes are usually clustered in a single operon. From the 12 RND systems, four have been already well characterized due to their clinical importance: MexAB-OprM (Poole et al. 1993), MexCD-OprJ (Poole et al. 1996a), MexEF-OprN (Köhler et al. 1997), and MexXY (Mine et al. 1999). Among these, both MexAB-OprM and MexXY-OprM seemed to contribute to the intrinsic multidrug and hydrocarbon resistance in wild-type *P. aeruginosa PAO1* (Li et al. 1995; Yoneyama et al. 1997; Aires et al. 1999; Masuda et al. 2000; Morita et al. 2001). Whereas MexAB-OprM is a constitutive pump, the MexXY pump is induced by subinhibitory concentrations of its substrates, such as tetracycline and gentamicin (Masuda et al. 2000). The MexCD-OprJ and MexEF-OprN pumps seem to be functional only in mutants of *nfxB* and *nfxC*, respectively, and contribute to an increased resistance to many antimicrobial agents and organic solvents (Hirai et al. 1987; Fukuda et al. 1995; Poole et al. 1996a; Köhler et al. 1997). Furthermore, a complex network of global and local regulators (Li et al. 2015) controls the expression levels of each Mex system in a dependent and coordinated way, as evidenced by the decreased expression of specific pumps in response to an increased expression of other Mex paralogs in *P. aeruginosa* (Li et al. 2000).

## 3.2 General Genetic Architecture of RND Systems

As mentioned before, the RND genes are usually organized as clusters and found in multiple paralogs in the genomes of gram-negative bacteria (Li and Nikaido 2004, 2009; Martinez et al. 2009; Nikaido 2011; Alvarez-Ortega et al. 2013; Li et al. 2015). Examples for this are the Acr systems of *E. coli* (Anes et al. 2015), Mex systems of *P. aeruginosa* (Schweizer 2003; Lister et al. 2009), and Ttg systems of *P. putida* (Ramos et al. 2015). Each cluster is usually comprised by one local regulator upstream the RND operon, being both divergently transcribed.

Adjacent transcription regulators (Aires et al. 1999) or two-component systems (BrlR-SagS) tightly regulate the expression of the MDR operons in *P. aeruginosa* (Gupta et al. 2013). Regulatory cross talks, in which local regulators can modify the expression of distant operons, can be observed in the promiscuous regulators AcrS (Hirakawa et al. 2008), MexT (Maseda et al. 2004; Uwate et al. 2013), and TtgV/ TtgT paralogs (Terán et al. 2007) from *E. coli*, *P. aeruginosa*, and *P. putida*, respectively.

The genes encoding RND operons usually follow a particular transcriptional order, being the membrane fusion protein (MFP) the first gene, followed by the inner membrane transporter (IMT) and the outer membrane protein (OMP). In some cases, the operon lacks the OMP gene, and the pump uses a protein that is encoded by another operon (for instance, in the MexJK-OprM system of *P. aeruginosa*) (Chuanchuen et al. 2002) or an independent gene (e.g., *tolC* in *E. coli*) (Nishino et al. 2003). However, despite the degree of cluster fragmentation, genes composing the same RND pump share the same transcription regulators (Li et al. 2015).

From an evolutionary perspective, the conservation of multiple RND operons in gram-negative bacteria could be linked to an increase in their resilience against environmental perturbations by functional redundancy (Girvan et al. 2005). Furthermore, as the substrate specificity among RND paralogs is not the same, which was evidenced directly (Elkins and Nikaido 2002; Anes et al. 2015; Dreier and Ruggerone 2015) or indirectly, by the study of interactions between effectors and local regulators (Guazzaroni et al. 2005), the combination of specific stressors, quorum-sensing molecules, and regulatory cross talks may act as context-dependent molecular switches for the differential expression of RND systems (Maseda et al. 2004; Hirakawa et al. 2008; Martinez et al. 2009; Wong et al. 2014).

## 3.3 RND Regulatory Networks Are Multi-hierarchical

Accumulation of toxic organic solvents in the membrane can disrupt its structure, allowing the leakage of macromolecules and the malfunction of membrane-associated proteins (Segura et al. 2008). Thus, stressors (such as antibiotics, organic solvents, and hydrocarbons) are the key components for the first layer in the regulatory network (Li et al. 2015). Furthermore, physiological markers for growth-phase state and cellular density (such as quorum-sensing molecules) can modulate the system, triggering essential processes as secretion of virulence factors and biofilm formation (Rahmati et al. 2002; Martinez et al. 2009; Kim et al. 2014).

On a second layer of regulation, the activity of global regulators is modulated by stressors (see Table 1) (Grkovic et al. 2002; Li and Nikaido 2009). In *E. coli*, those regulators can be directly (e.g., Rob, CpxRA, PhoPQ, EvgAS) or indirectly activated (e.g., antirepressors for MarR and SoxR), allowing the propagation of stress signals inside the cell (Ma et al. 1995; Li and Nikaido 2004, 2009; Lennen et al. 2012; Li et al. 2015). Cross talk in the same layer among global regulators is very common as represented by the interactions between PhoPQ-EvgAS (Eguchi et al. 2011) and SoxR/SoxS-MarRAB-Rob in *E. coli* and *S. enterica* (Duval and Lister 2013).

The third layer of the regulatory network is composed by local regulators of RND operons (Table 1, Fig. 1) that can be modulated by global regulators and that directly respond to ligands and various cellular states (Grkovic et al. 2002; Li et al. 2015). Those ligands usually inhibit the activity of RND local repressors – as observed in AcrR (Gu et al. 2008), RamRA (Baucheron et al. 2012), MexR (Schweizer 2003), and TtgV (Guazzaroni et al. 2007a; Ramos et al. 2015) from *E. coli*, *S. enterica*, *P. aeruginosa*, and *P. putida*, respectively – and promote the action of local activators as seen in MexT and MexS from *P. aeruginosa* (Schweizer 2003). Moreover, in *P. aeruginosa* an additional layer of protein antirepressors for local regulators – ArmR for MexR (Starr et al. 2012) and ArmZ for MexZ (Hay et al. 2013) – coordinates the expression of *mexAB-oprM* and *mexXY-oprA* RND genes.

As a final point, the combination of stressors, global activators, and local regulators modulates a fourth layer in the system: the expression of RND efflux pumps and the repression of the membrane porin genes related to the influx of toxic compounds (e.g., OmpF porin in *E. coli*) (see Fig. 1). As presented before, cross talk between local regulators and RND operons from different RND systems may also occur, which depends on the nature of the substrate or cellular states. The so-called bottom-up interactions (i.e., connections from a lower level of a hierarchy or process to a higher one; see Fig. 1) are also found at local regulators modulating the expression of global regulators, as seen in the repression of the SoxR/SoxS and MarRAB systems by the AcrR repressor protein in *E. coli* (Li et al. 2015).

Additionally, posttranscriptional and posttranslational layers can be included into this process when the effects of small RNAs and protein-protein interactions are considered. In *E. coli*, the small RNA *micF* (regulated by the same transcription factors controlling the *acrAB* operon) represses the translation of the OmpF porin (Andersen and Delihas 1990; Delihas and Forst 2001), whereas the small RNA *dsrA* represses the global repressor H-NS, acting as a molecular switch for the expression of specific RND pumps (Nishino et al. 2011). Posttranslational regulation was observed in the degradation of both MarA and SoxS proteins by the Lon protease (Griffith et al. 2004) and in the sequestering of MarA by its cognate MarB in *E. coli* (Nichols et al. 2011). In *P. aeruginosa*, modulation of MexR activity by ArmR (Starr et al. 2012) is also regulated by protein-protein sequestering.

**Table 1** Global and local transcription factors of RND operons involved in transcriptional regulation of hydrocarbon efflux pump expression

| TF | Global/local | Function | Active oligomeric state | Family | Effector | 3D structure (PDB ID) | References |
|---|---|---|---|---|---|---|---|
| *E. coli* K12 | | | | | | | |
| AcrAB/TolC | | | | | | | |
| AcrR | Local | Repressor (*acrAB*) | Dimer | TetR | Rhodamine 6G; ethidium bromide; proflavin | Yes (2QOP) | Li et al. (2007) |
| AcrS | Local | Repressor (*acrAB* and *acrEF*) | Dimer | TetR | Unknown | No | Hirakawa et al. (2008) |
| H-NS | Global | Repressor (*tolC*) | Dimer | Histone-like protein H-NS | Unknown | Yes (1HNR) | Shindo et al. (1995) and Esposito et al. (2002) |
| SdiA | Global | Activator (*acrAB*) | Dimer | LuxR | Quorum-sensing molecules; indole | Yes (4LFU) | Rahmati et al. (2002) and Kim et al. (2014) |
| MarA | Global | Activator (*mar* genes) | Monomer | AraC/XylS | Unknown | Yes (1BL0) | Rhee et al. (1998), Martin et al. (1999), and Martin and Rosner (2011) |
| SoxS | Global | Activator (mar regulon, *acrAB*, *tolC*) | Monomer | AraC/XylS | Unknown | No | Martin et al. (1999) |
| Rob | Global | Activator (mar regulon, *acrAB*, *tolC*) | Monomer | AraC/XylS | Dipyridil, fatty acids, bile salts | Yes (1D5Y) | Martin et al. (1999) and Kwon et al. (2000) |
| EvgA (TCS)[a] | Global | Activator (*acrAB/tolC*) | Dimer | LuxR | EvgS (pH 5–6, high concentrations of alkali metals) | Yes (3F6C) | Perraud et al. (2000), Eguchi et al. (2003), Bachhawat and Stock (2007), Zhang et al. (2008), and Eguchi and Utsumi (2014) |

| | Global/Local | Function | Oligomeric state | Family | Signal | Structure | References |
|---|---|---|---|---|---|---|---|
| PhoP (TCS) | Global | Activator (*acrAB/tolC*) | Dimer | OmpR/PhoB | PhoQ (low $Mg^{2+}$ concentrations) | Yes (2PKX) | Bachhawat and Stock (2007) and Zhang et al. (2008) |
| CpxR (TCS) | Global | Activator (*mar* genes) | Dimer | OmpR/PhoB | CpxA (alkaline pH, cell-envelope stress, high osmolarity) | Yes (4UHJ) (4UHT) | Sulavik et al. (1995), Brooun et al. (1999), Dorel et al. (2006), Fleischer et al. (2007), and Rosner and Martin (2013) |
| MprA | Local | Repressor (*emrAB*) | Dimer | MarR-like | 2,4-DNP, CCCP, and CCPP[b] | No | Sulavik et al. (1995) and Brooun et al. (1999) |
| *P. aeruginosa* PAO1 | | | | | | | |
| MexAB-OprM | | | | | | | |
| MexR | Local | Operon repressor | Dimer | MarR | Redox state, antibiotics, phenol | Yes (1LNW) | Poole et al. (1996b), Lim et al. (2002), Chen et al. (2008), and Starr et al. (2012) |
| RocA2 (TCS) | Global | Operon repressor | Unknown | Lux-R | RocS2 (c-di-GMP) | No | Sivaneson et al. (2011) |
| NalD | Local | Operon repressor | Dimer | TetR | Novobiocin | Yes (5DAJ) | Morita et al. (2006) and Chen et al. (2016) |
| MexT | Local | Repressor | Oligomer | LysR | Redox state; s-nitrosoglutathione; chloramphenicol | No | Köhler et al. (1999), Fetar et al. (2011), Fargier et al. (2012), and Uwate et al. (2013) |
| BrlR | Global | Operon activator | Dimer | MerR | C-di-GMP | No | Liao et al. (2013) and Chambers et al. (2014) |
| AmpR | Global | Repressor (*mexR*) | Tetramer | LysR | UDP-MurNAc-pentapeptide, 1,6-anhydroMurNAc-pentapeptide | Yes (4WKM) | Balasubramanian et al. (2012) and Vadlamani et al. (2015) |
| MexCD-OprJ | | | | | | | |
| NfxB | Local | Repressor | Multimer/tetramer | TetR | Unknown | No | Shiba et al. (1995) and Pursell and Poole (2013) |
| EsrC | Local | Repressor | Monomer | AraC | Unknown | No | Pursell et al. (2015) |

*(continued)*

**Table 1** (continued)

| TF | Global/local | Function | Active oligomeric state | Family | Effector | 3D structure (PDB ID) | References |
|---|---|---|---|---|---|---|---|
| AlgU | Global | Operon activator | Dimer | Sigma E-like | Cell-envelope stress | No | Fraud et al. (2008) |
| MexEF-OprN | | | | | | | |
| Mvat | Global | Repressor | Oligomer | H-NS | Environmental factors | Yes (2MXE) | Vallet et al. (2004), Castang and Dove (2010), and Ding et al. (2015) |
| MexT | Local | Activator (*mexS* and operon) | Unknown | LysR | Redox state; *s*-nitrosoglutathione; chloramphenicol | No | Köhler et al. (1999), Fetar et al. (2011), Fargier et al. (2012), and Uwate et al. (2013) |
| MexS | Local | Activator (operon) | Unknown | ADH_N | Unknown | No | Sobel et al. (2005) and Uwate et al. (2013) |
| ParR (TCS) | Local | Activator (*mexT* and operon); repressor (*oprD*) | Unknown | OmpR/PhoB | ParS (polymyxins) | No | Fernández et al. (2010) and Wang et al. (2013) |
| AmpR | Global | Repressor | Tetramer | LysR | UDP-MurNAc-pentapeptide, 1,6-anhydroMurNAc-pentapeptide | Yes; Yes (4WKM) | Balasubramanian et al. (2012) and Vadlamani et al. (2015) |
| BrlR | Global | Activator (operon) | Dimer | MerR | C-di-GMP | No | Liao et al. (2013) and Chambers et al. (2014) |
| MexXY-OprA | | | | | | | |
| MexZ | Local | Repressor | Dimer | TetR | ArmZ (aberrant protein synthesis) | Yes (2WUI) | Alguel et al. (2010) |
| AmgR (TCS) | Global | Activator (operon) | Unknown | OmpR/PhoB | AmgS (aminoglycosides, aberrant protein synthesis) | No | Lau et al. (2015) |

| | | | | | | | |
|---|---|---|---|---|---|---|---|
| ParR (TCS) | Local | Activator (*mexT* and operon); repressor (*oprD*) | Unknown | OmpR/PhoB | ParS (polymyxins) | No | Fernández et al. (2010) and Wang et al. (2013) |
| MexJK | | | | | | | |
| MexL | Local | Repressor | Tetramer | TetR | Unknown | No | Chuanchuen et al. (2005) |
| *P. putida* DOT-T1E | | | | | | | |
| TtgABC | | | | | | | |
| TtgR | Local | Repressor | Dimer | TetR | Antibiotics, butanol, flavonoids | Yes (2UXH) | Alguel et al. (2007) and Fernandez-Escamilla et al. (2015) |
| TtgDEF | | | | | | | |
| TtgT | Local | Repressor | Tetramer | IclR | Toluene and diverse aliphatic and aromatic hydrocarbons | No | Terán et al. (2007) |
| TtgGHI | | | | | | | |
| TtgV | Local | Repressor | Tetramer | IclR | Toluene, and diverse aromatic hydrocarbons | Yes (2XRN) | Guazzaroni et al. (2005) |

[a]*TCS* two-component system. Information of TCS is given in two columns: the Global/local column for the response regulator protein and the Effector column for the sensor protein. In this latter, the effectors of the sensor protein are given in parentheses

[b]Abbreviations: *2,4-DNP* 2,4-dinitrophenol, *CCCP* carbonyl cyanide m-chlorophenylhydrazone, *CCCP* carbonyl cyanide m-chlorophenylhydrazone, *CCCP* carbonyl cyanide-p-trifluoromethoxyphenylhydrazone

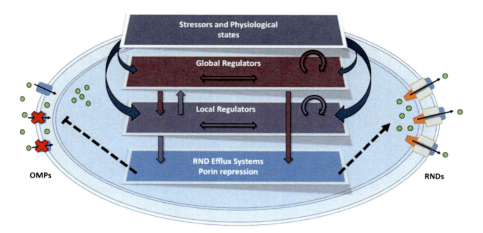

**Fig. 1** Schematic representation of the hierarchical regulatory network of RND efflux systems. Stressors and cellular states compose the first layer in the system and modulate the activity of both local and global regulators. Interactions within the same layer are represented by *semicircles* (autoregulatory loops) and *horizontal bidirectional arrows* (cross talks between components). The *vertical arrows* represent the direction of interactions between the hierarchical layers. The output of the system, the coordinated efflux pump response, is represented by the *dashed hammer* on the left (repression of OMPs expression) and the *dashed arrow* on the right (expression of RND efflux pumps)

## 4    Regulation of Efflux Pumps in *P. putida* DOT-T1E as a Case of Study

*P. putida* DOT-T1E is a peculiar hydrocarbon-tolerant microorganism since it can grow in the presence of unusually high concentrations of toxic and harmful compounds such as aromatic hydrocarbons (Ramos et al. 1995, 2002). One remarkable feature of this strain is its capacity of growing in a medium saturated with toluene, being able to use this toxic hydrocarbon as a sole carbon and energy source through the induction of the enzymes of the toluene dioxygenase (TOD) catabolic pathway for the conversion of toluene into Krebs cycle intermediates (Zylstra and Gibson 1989). To survive the deleterious effects of the presence of aromatic hydrocarbons in the cytoplasm, the coordinated expression of three RND efflux pumps, termed TtgABC, TtgDEF, and TtgGHI, is of imperative importance (Ramos et al. 2015). As previously cited, most of the regulatory genes that encode for proteins involved in the control of the expression of the efflux pumps belonging to the RND family are located adjacent to the structural genes of the pump, and generally the regulatory gene is transcribed divergently from the efflux pump operon. Accordingly, the three transcriptional repressors involved in the regulation of the pumps of the strain DOT-T1E (called TtgR, TtgT, and TtgV, respectively) show a similar organization (Segura et al. 2012).

## 4.1 Identification of the Efflux Pumps Involved in Solvent Tolerance

The first efflux pump identified in *P. putida* DOT-T1E to be involved in solvent tolerance was the TtgABC efflux pump (Ramos et al. 1998). Physiological experiments done with a knockout mutant of the gene encoding for the inner membrane pump component (the *ttgB* gene) suggested that this efflux pump was involved in the noninduced intrinsic tolerance (Ramos et al. 1998). This mutant strain did not withstand a sudden toluene shock (0.3% vol/vol), and only a small fraction of cells (about 1 out of $10^5$) survived the shock if preexposed to low toluene concentrations. On the basis of this observation, the existence of other efflux pump (s) involved in toluene extrusion was postulated (Ramos et al. 1998).

Therefore, new studies were directed toward the identification of additional molecular elements involved in the solvent tolerance phenomena of strain DOT-T1E. To this end, Mosqueda and Ramos (2000) sequenced upstream from the toluene dioxygenase (*tod*) operon of *P. putida* DOT-T1E and identified three open reading frames (*ttgDEF*) that encode for the three components of an efflux pump sharing homology with other efflux pumps of *Pseudomonas*. Sudden toluene shock experiments were done with a knockout mutant of one of the constituents of the pump (the *ttgD* gene) that showed the involvement of this efflux pump in solvent tolerance (Mosqueda and Ramos 2000). However, studies of the expression of the *ttgDEF* operon at the transcriptional level revealed that this pump was not expressed during growth under normal laboratory conditions and demonstrated its inducible character in the presence of aromatic hydrocarbons (Mosqueda and Ramos 2000). In this case, toluene shock experiments with *ttgD* and *ttgB* knockout mutants showed that there was still a significant fraction of cells that survived the shock if preexposed to low toluene concentrations, pointing to the presence of yet another pump.

Finally, a third efflux pump (TtgGHI) was shown to complete the entire mechanism of solvent tolerance in the strain DOT-T1E and was shown to be responsible for the phenotype of high resistance, since this pump plays a major role in solvent removal (Rojas et al. 2001). In contrast to knockout mutants of the other efflux pumps that were still able to survive the toluene shock to different levels, a knockout mutant of the TtgGHI efflux pump was unable to survive the toluene shock regardless of the growth conditions (Rojas et al. 2001), suggesting that this efflux pump is involved in intrinsic as well as inducible resistance to organic solvents. Interestingly, subsequent studies showed that the operon encoding for the structural genes of this efflux pump and its regulator TtgV were located on the large self-transmissible plasmid pGRT1 (Rodríguez-Herva et al. 2007).

## 4.2 Study of the Transcriptional Repressors Involved in the Regulation of the Pumps

Regardless of the high amino acid sequence similarity of each of the efflux pumps identified in strain DOT-T1E, the substrate profiles of the three toluene efflux pumps seem to be different. Several in vivo and in vitro experiments showed that the TtgABC pump recognizes a wide range of hydrocarbons, antibiotics, and secondary plant metabolites (Duque et al. 2001; Rojas et al. 2001; Terán et al. 2006), whereas the TtgDEF and TtgGHI pumps mainly efflux hydrocarbons (Rojas et al. 2001; Guazzaroni et al. 2005). However, the complex response to hydrocarbon exposure is only reached by the coordinated expression at transcriptional level of the chromosomal (*ttgABC* and *ttgDEF*) and the plasmid-encoded *ttgGHI* operons (Ramos et al. 2015).

The transcriptional repressor TtgR (Terán et al. 2003), which belongs to the TetR family, controls the expression of the operon *ttgABC* (Rojas et al. 2001), whereas the IclR-type repressors TtgT and TtgV (both sharing 63% amino acid sequence identity) regulate the expression of operons for the other two pumps (Rojas et al. 2003; Terán et al. 2007; Guazzaroni et al. 2007b).The three regulatory systems were widely studied aiming to unravel the molecular bases involved in DNA-protein recognition and effector modulation. By using several in vivo and in vitro approaches, including biophysical techniques such as atomic force microscopy (AFM), isothermal titration calorimetry (ITC), and analytical ultracentrifugation (AUC), it was possible to establish important features of the molecular mechanisms related to DNA-regulator interactions, effector-regulator interactions, and intramolecular signal transmission (Guazzaroni et al. 2005, 2007a, b; Terán et al. 2006, 2007).The three repressor proteins function by the same ligand-induced derepression mechanism. In the absence of effector molecules, they bind to the $-10/-35$ region of their respective promoter thereby preventing transcriptional initiation by the RNA polymerase. In the presence of hydrocarbons (or other effector molecules) in the cytoplasm, binding of small effector molecules results in the release of the DNA-bound repressor allowing RNA polymerase binding and consequent transcription of the genes encoding the pump components (Rojas et al. 2003; Guazzaroni et al. 2004).

Remarkably, regulation of the two efflux pumps mainly involved in tolerance of both aliphatic and aromatic hydrocarbons (i.e., TtgDEF and TtgGHI) was shown to be cross regulated by TtgT and TtgV, with TtgV being predominant for both operons (Terán et al. 2007). However, TtgT and TtgV have similar but not identical effector profiles (Terán et al. 2007). In addition, it is important to highlight that the regulatory activity of TtgT is only detectable in a TtgV-deficient background (Terán et al. 2007). This latter molecular scenario could be feasibly reached in nature in the case of the loss of the *ttgV* harboring plasmid pGRT1 that occurs under nonselective environmental conditions. Moreover, in the presence of TtgT-specific effectors, the chromosomally encoded TtgT repressor should gain a key role in the modulation of the TtgDEF efflux pump expression. In this sense, authors suggest that the presence of both proteins would allow the recognition of a wider range of effector molecules than when only one of them was present, which certainly represents an evolutionary advantage (Terán et al. 2007). Moreover, authors highlight that this is an interesting example of evolution of a simple regulatory circuit into a more complex network, as represented in Fig. 1, which reflects the ability of bacterial regulatory systems to evolve in order to respond more efficiently to environmental signals (Terán et al. 2007).

# 5   Research Needs

Aromatic hydrocarbons have promoted the development of complex regulatory networks aiming to enhance chances of survival in bacteria. In this sense, combination of different levels of transcriptional and posttranscriptional regulation allows a fine modulation of the response of the system, involving efflux pumps and other mechanisms of detoxification (Ramos et al. 2002).

A comparison between the regulatory networks of RND efflux pumps operating in *E. coli*, *Salmonella* sp., *P. aeruginosa*, and *P. putida* allowed us to elucidate common conceptual features that can be outlined: (i) RND genetic systems are usually clustered and composed by a local regulator upstream the RND operon, both divergently transcribed; (ii) multiple nonidentical copies of RND clusters can be found in the genome of gram-negative bacteria as adaptive switches; (iii) RND regulatory networks consist of multi-hierarchical and interconnected layers of stressors, global regulators, local regulators, and target genes coding for RND efflux pumps; (iv) regulatory cross talks are common, occurring horizontally (inside a single regulation layer) and vertically (between layers); and (v) additional layers can be included as posttranscriptional and posttranslational regulation, making the regulatory network even more complex.

Although there are several works in literature reporting comprehensive studies on transcriptional regulation of RND systems involved in hydrocarbon extrusion, further systemic and comparative studies on RND regulatory networks should be done. The latter should provide both, essential information regarding the intrinsic logic of decision-making in bacteria and principles of design for the generation of novel genetic circuits and functionalities that can be used in several fields, such as synthetic biology or metabolic engineering (Khalil et al. 2012).

**Acknowledgments** This work was supported by the National Counsel of Technological and Scientific Development (CNPq 472893/2013-0 and 441833/2014-4) and by the Sao Paulo State Foundation (FAPESP, grant number 2015/04309-1 and 2012/21922-8). CAW and LFA are beneficiaries of CAPES fellowship. Authors have no conflict of interest to declare.

# References

Aires JR, Nikaido H (2005) Aminoglycosides are captured from both periplasm and cytoplasm by the AcrD multidrug efflux transporter of *Escherichia coli* aminoglycosides are captured from both periplasm and cytoplasm by the AcrD multidrug efflux transporter of *Escherichia coli*. J Bacteriol. https://doi.org/10.1128/JB.187.6.1923

Aires JR, Köhler T, Nikaido H (1999) Involvement of an active efflux system in the natural resistance of *Pseudomonas aeruginosa* to aminoglycosides involvement of an active efflux system in the natural resistance of *Pseudomonas aeruginosa* to aminoglycosides. Antimicrob Agents 43:2624–2628

Alekshun MN, Levy SB (1999) The mar regulon: multiple resistance to antibiotics and other toxic chemicals. Trends Microbiol 7:410–413

Alguel Y, Meng C, Terán W et al (2007) Crystal structures of multidrug binding protein TtgR in complex with antibiotics and plant antimicrobials. J Mol Biol 369:829–840. https://doi.org/10.1016/j.jmb.2007.03.062

Alguel Y, Lu D, Quade N et al (2010) Crystal structure of MexZ, a key repressor responsible for antibiotic resistance in *Pseudomonas aeruginosa*. J Struct Biol 172:305–310. https://doi.org/10.1016/j.jsb.2010.07.012

Alnaseri H, Arsic B, Schneider JET et al (2015) Inducible expression of a resistance-nodulation-division-type efflux pump in *Staphylococcus aureus* provides resistance to linoleic and arachidonic acids. J Bacteriol 197:1893–1905. https://doi.org/10.1128/JB.02607-14

Alvarez-Ortega C, Olivares J, Martínez JL (2013) RND multidrug efflux pumps: what are they good for? Front Microbiol. https://doi.org/10.3389/fmicb.2013.00007

Andersen J, Delihas N (1990) micF RNA binds to the 5′ end of ompF mRNA and to a protein from *Escherichia coli*. Biochemistry 29:9249–9256. https://doi.org/10.1021/bi00491a020

Anes J, McCusker MP, Fanning S, Martins M (2015) The ins and outs of RND efflux pumps in *Escherichia coli*. Front Microbiol. https://doi.org/10.3389/fmicb.2015.00587

Bachhawat P, Stock AM (2007) Crystal structures of the receiver domain of the response regulator PhoP from *Escherichia coli* in the absence and presence of the phosphoryl analog beryllofluoride. J Bacteriol 189:5987–5995. https://doi.org/10.1128/JB.00049-07

Bailey AM, Ivens A, Kingsley R et al (2010) RamA, a member of the AraC/XylS family, influences both virulence and efflux in *Salmonella enterica* Serovar Typhimurium. J Bacteriol 192:1607–1616. https://doi.org/10.1128/JB.01517-09

Balasubramanian D, Schneper L, Merighi M et al (2012) The regulatory repertoire of *Pseudomonas aeruginosa* AmpC β-lactamase regulator AmpR includes virulence genes. PLoS One. https://doi.org/10.1371/journal.pone.0034067

Baucheron S, Coste F, Canepa S et al (2012) Binding of the RamR repressor to wild-type and mutated promoters of the ramA gene involved in efflux-mediated multidrug resistance in *Salmonella enterica* serovar typhimurium. Antimicrob Agents Chemother 56:942–948. https://doi.org/10.1128/AAC.05444-21

Baucheron S, Nishino K, Monchaux I et al (2014) Bile-mediated activation of the acrAB and tolC multidrug efflux genes occurs mainly through transcriptional derepression of ramA in *Salmonella enterica* serovar Typhimurium. J Antimicrob Chemother 69:2400–2406. https://doi.org/10.1093/jac/dku140

Blair JMA, Smith HE, Ricci V et al (2015) Expression of homologous RND efflux pump genes is dependent upon AcrB expression: implications for efflux and virulence inhibitor design. J Antimicrob Chemother 70:424–431. https://doi.org/10.1093/jac/dku380

Brooun A, Tomashek JJ, Lewis K (1999) Purification and ligand binding of EmrR, a regulator of a multidrug transporter. J Bacteriol 181:5131–5133

Castang S, Dove SL (2010) High-order oligomerization is required for the function of the H-NS family member MvaT in *Pseudomonas aeruginosa*. Mol Microbiol 78:916–931. https://doi.org/10.1111/j.1365-2958.2010.07378.x

Chambers JR, Liao J, Schurr MJ, Sauer K (2014) BrlR from pseudomonas aeruginosa is a c-di-GMP-responsive transcription factor. Mol Microbiol 92:471–487. https://doi.org/10.1111/mmi.12562

Chen H, Hu J, Chen PR et al (2008) The *Pseudomonas aeruginosa* multidrug efflux regulator MexR uses an oxidation-sensing mechanism. Proc Natl Acad Sci U S A 105:13586–13591. https://doi.org/10.1073/pnas.0803391105

Chen W, Wang D, Zhou W et al (2016) Novobiocin binding to NalD induces the expression of the MexAB-OprM pump in *Pseudomonas aeruginosa*. Mol Microbiol. https://doi.org/10.1111/mmi.13346

Chuanchuen R, Narasaki CT, Schweizer HP (2002) The MexJK efflux pump of *Pseudomonas aeruginosa* requires OprM for antibiotic efflux but not for efflux of triclosan. J Bacteriol 184:5036–5044. https://doi.org/10.1128/JB.184.18.5036-5044.2002

Chuanchuen R, Gaynor JB, Karkhoff-Schweizer R, Schweizer HP (2005) Molecular characterization of MexL, the transcriptional repressor of the mexJK multidrug efflux operon in *Pseudomonas aeruginosa*. Antimicrob Agents Chemother 49:1844–1851. https://doi.org/10.1128/AAC.49.5.1844-2851.2005

Dagley S (1971) Catabolism of aromatic compounds by micro-organisms. Adv Microb Physiol 6:1–46. https://doi.org/10.1016/S0065-2911(08)60066-1

Das D, Xu QS, Lee JY et al (2007) Crystal structure of the multidrug efflux transporter AcrB at 3.1A resolution reveals the N-terminal region with conserved amino acids. J Struct Biol 158:494–502. https://doi.org/10.1016/j.jsb.2006.12.004

Delihas N, Forst S (2001) MicF: an antisense RNA gene involved in response of *Escherichia coli* to global stress factors. J Mol Biol 313:1–12

Ding P, McFarland KA, Jin S et al (2015) A novel AT-rich DNA recognition mechanism for bacterial xenogeneic silencer MvaT. PLoS Pathog 11:e1004967. https://doi.org/10.1371/journal.ppat.1004967

Dorel C, Lejeune P, Rodrigue A (2006) The Cpx system of *Escherichia coli*, a strategic signaling pathway for confronting adverse conditions and for settling biofilm communities? Res Microbiol 157:306–314. https://doi.org/10.1016/j.resmic.2005.12.003

Doshi R, Nguyen T, Chang G (2013) Transporter-mediated biofuel secretion. Proc Natl Acad Sci U S A 110:7642–7647. https://doi.org/10.1073/pnas.1301358110

Dreier J, Ruggerone P (2015) Interaction of antibacterial compounds with RND efflux pumps in *Pseudomonas aeruginosa*. Front Microbiol. https://doi.org/10.3389/fmicb.2015.00660

Duque E, Segura A, Mosqueda G, Ramos JL (2001) Global and cognate regulators control the expression of the organic solvent efflux pumps TtgABC and TtgDEF of *Pseudomonas putida*. Mol Microbiol 39:1100–1106. https://doi.org/10.1046/j.1365-2958.2001.02310.x

Duval V, Lister IM (2013) MarA, SoxS and Rob of *Escherichia coli* – global regulators of multidrug resistance, virulence and stress response. Int J Biotechnol Wellness Ind 2:101–124. https://doi.org/10.6000/1927-3037.2013.02.03.2

Eaves DJ, Ricci V, Piddock LJV (2004) Expression of acrB, acrF, acrD, marA, and soxS in Salmonella enterica serovar Typhimurium: role in multiple antibiotic resistance. Antimicrob Agents. https://doi.org/10.1128/AAC.48.4.1145

Eguchi Y, Utsumi R (2014) Alkali metals in addition to acidic pH activate the EvgS histidine kinase sensor in *Escherichia coli*. J Bacteriol 196:3140–3149. https://doi.org/10.1128/JB.01742-14

Eguchi Y, Oshima T, Mori H et al (2003) Transcriptional regulation of drug efflux genes by EvgAS, two-component system in *Escherichia coli*. Microbiology 149:2819–2828. https://doi.org/10.1099/mic.0.26460-0

Eguchi Y, Ishii E, Hata K, Utsumi R (2011) Regulation of acid resistance by connectors of two-component signal transduction systems in *Escherichia coli*. J Bacteriol 193:1222–1228. https://doi.org/10.1128/JB.01124-20

Elkins CA, Nikaido H (2002) Substrate specificity of the RND-type multidrug efflux pumps AcrB and AcrD of *Escherichia coli* is determined predominately by two large periplasmic loops. J Bacteriol 184:6490–6498. https://doi.org/10.1128/JB.184.23.6490-6499.2002

Esposito D, Petrovic A, Harris R et al (2002) H-NS oligomerization domain structure reveals the mechanism for high order self-association of the intact protein. J Mol Biol 324:841–850. https://doi.org/10.1016/S0022-2836(02)01141-5

Fargier E, Mac AM, Mooij MJ et al (2012) Mext functions as a redox-responsive regulator modulating disulfide stress resistance in *Pseudomonas aeruginosa*. J Bacteriol 194:3502–3511. https://doi.org/10.1128/JB.06632-11

Fernandes P, Sommer Ferreira B, Sampaio Cabral JM (2003) Solvent tolerance in bacteria: role of efflux pumps and cross-resistance with antibiotics. Int J Antimicrob Agents 22:211–216. https://doi.org/10.1016/S0924-8579(03)00209-7

Fernández L, Gooderham WJ, Bains M et al (2010) Adaptive resistance to the "last hope" antibiotics polymyxin B and colistin in *Pseudomonas aeruginosa* is mediated by the novel two-component regulatory system ParR-ParS. Antimicrob Agents Chemother 54:3372–3382. https://doi.org/10.1128/AAC.00242-10

Fernandez-Escamilla AM, Fernandez-Ballester G, Morel B et al (2015) Molecular binding mechanism of TtgR repressor to antibiotics and antimicrobials. PLoS One 10:e0138469. https://doi.org/10.1371/journal.pone.0138469

Fetar H, Gilmour C, Klinoski R et al (2011) mexEF-oprN multidrug efflux operon of *Pseudomonas aeruginosa*: regulation by the MexT activator in response to nitrosative stress and chloramphenicol. Antimicrob Agents Chemother 55:508–514. https://doi.org/10.1128/AAC.00830-10

Fillet S, Daniels C, Pini C et al (2012) Transcriptional control of the main aromatic hydrocarbon efflux pump in *Pseudomonas*. Environ Microbiol Rep 4:158–167

Fleischer R, Heermann R, Jung K, Hunke S (2007) Purification, reconstitution, and characterization of the CpxRAP envelope stress system of *Escherichia coli*. J Biol Chem 282:8583–8593. https://doi.org/10.1074/jbc.M605785200

Fraud S, Campigotto AJ, Chen Z, Poole K (2008) MexCD-OprJ multidrug efflux system of *Pseudomonas aeruginosa*: involvement in chlorhexidine resistance and induction by membrane-damaging agents dependent upon the AlgU stress response sigma factor. Antimicrob Agents Chemother 52:4478–4482. https://doi.org/10.1128/AAC.01072-08

Fukuda H, Hosaka M, Iyobe S et al (1995) nfxC-type quinolone resistance in a clinical isolate of *Pseudomonas aeruginosa*. Antimicrob Agents Chemother 39:790–792. https://doi.org/10.1128/AAC.39.3.790

García V, Godoy P, Daniels C et al (2010) Functional analysis of new transporters involved in stress tolerance in *Pseudomonas putida* DOT-T1E. Environ Microbiol Rep 2:389–395. https://doi.org/10.1111/j.1758-2229.2009.00093.x

Girvan MS, Campbell CD, Killham K et al (2005) Bacterial diversity promotes community stability and functional resilience after perturbation. Environ Microbiol 7:301–313. https://doi.org/10.1111/j.1462-2920.2004.00695.x

Griffith KL, Shah IM, Wolf RE (2004) Proteolytic degradation of *Escherichia coli* transcription activators SoxS and MarA as the mechanism for reversing the induction of the superoxide (SoxRS) and multiple antibiotic resistance (Mar) regulons. Mol Microbiol 51:1801–1816. https://doi.org/10.1046/j.1365-2958.2003.03952.x

Grkovic S, Brown MH, Skurray RA (2002) Regulation of bacterial drug export systems. Microbiol Mol Biol Rev 66:671–701 . https://doi.org/10.1128/MMBR.66.4.671 table of contents

Gu R, Li M, Su CC et al (2008) Conformational change of the AcrR regulator reveals a possible mechanism of induction. Acta Crystallogr Sect F Struct Biol Cryst Commun 64:584–588. https://doi.org/10.1107/S1744309108016035

Guazzaroni ME, Terán W, Zhang X et al (2004) TtgV bound to a complex operator site represses transcription of the promoter for the multidrug and solvent extrusion TtgGHI pump. J Bacteriol 186:2921–2927. https://doi.org/10.1128/JB.186.10.2921-2927.2004

Guazzaroni ME, Krell T, Felipe A et al (2005) The multidrug efflux regulator TtgV recognizes a wide range of structurally different effectors in solution and complexed with target DNA: evidence from isothermal titration calorimetry. J Biol Chem 280:20887–20893. https://doi.org/10.1074/jbc.M500783200

Guazzaroni M-E, Gallegos M-T, Ramos JL, Krell T (2007a) Different modes of binding of mono- and biaromatic effectors to the transcriptional regulator TTGV: role in differential derepression from its cognate operator. J Biol Chem 282:16308–16316. https://doi.org/10.1074/jbc.M610032200

Guazzaroni ME, Krell T, Gutiérrez del Arroyo P et al (2007b) The transcriptional repressor TtgV recognizes a complex operator as a tetramer and induces convex DNA bending. J Mol Biol 369:927–939. https://doi.org/10.1016/j.jmb.2007.04.022

Gupta K, Marques CNH, Petrova OE, Sauer K (2013) Antimicrobial tolerance of *Pseudomonas aeruginosa* biofilms is activated during an early developmental stage and requires the two-component hybrid sagS. J Bacteriol 195:4975–4987. https://doi.org/10.1128/JB.00732-13

Hay T, Fraud S, Lau CHF et al (2013) Antibiotic inducibility of the mexXY multidrug efflux operon of *Pseudomonas aeruginosa*: involvement of the MexZ anti-repressor ArmZ. PLoS One. https://doi.org/10.1371/journal.pone.0056858

Heipieper HJ, Diefenbach R, Keweloh H (1992) Conversion of *cis* unsaturated fatty acids to *trans*, a possible mechanism for the protection of phenol-degrading *Pseudomonas putida* P8 from substrate toxicity. Appl Environ Microbiol 58:1847–1852

Hirai K, Suzue S, Irikura T et al (1987) Mutations producing resistance to norfloxacin in *Pseudomonas aeruginosa*. Antimicrob Agents Chemother 31:582–586. https://doi.org/10.1128/AAC.31.4.582

Hirakawa H, Takumi-Kobayashi A, Theisen U et al (2008) AcrS/EnvR represses expression of the acrAB multidrug efflux genes in *Escherichia coli*. J Bacteriol 190:6276–6279. https://doi.org/10.1128/JB.00190-08

Inoue A, Yamamoto M, Horikoshi K (1991) *Pseudomonas putida* which can grow in the presence of toluene. Appl Environ Microbiol 57:1560–1562

Isken S, De Bont JAM (1996) Active efflux of toluene in a solvent-resistant bacterium. J Bacteriol 178:6056–6058

Jones CM, Hernández Lozada NJ, Pfleger BF (2015) Efflux systems in bacteria and their metabolic engineering applications. Appl Microbiol Biotechnol 99:9381–9393. https://doi.org/10.1007/s00253-015-6963-9

Junker F, Ramos JL (1999) Involvement of the cis/trans isomerase Cti in solvent resistance of *Pseudomonas putida* DOT-T1E. J Bacteriol 181:5693–5700

Khalil AS, Lu TK, Bashor CJ et al (2012) A synthetic biology framework for programming eukaryotic transcription functions. Cell 150:647–658. https://doi.org/10.1016/j.cell.2012.05.045

Kim K, Lee S, Lee K, Lim D (1998) Isolation and characterization of toluene-sensitive mutants from the toluene-resistant bacterium *Pseudomonas putida* GM73. J Bacteriol 180:3692–3696

Kim T, Duong T, Wu CA et al (2014) Structural insights into the molecular mechanism of *Escherichia coli* SdiA, a quorum-sensing receptor. Acta Crystallogr Sect D Biol Crystallogr 70:694–707. https://doi.org/10.1107/S1399004713032355

Köhler T, Michéa-Hamzehpour M, Henze U et al (1997) Characterization of MexE-MexF-OprN, a positively regulated multidrug efflux system of *Pseudomonas aeruginosa*. Mol Microbiol 23:345–354. https://doi.org/10.1046/j.1365-2958.1997.2281594.x

Köhler T, Epp SF, Curty LK, Pechère JC (1999) Characterization of MexT, the regulator of the MexE-MexF-OprN multidrug efflux system of *Pseudomonas aeruginosa*. J Bacteriol 181:6300–6305

Koutsolioutsou A, Martins EA, White DG et al (2001) A soxRS -constitutive mutation contributing to antibiotic resistance in a clinical isolate of *Salmonella enterica* (Serovar A soxRS -constitutive mutation contributing to antibiotic resistance in a clinical isolate of *Salmonella enterica*) Serovar Typhimu. Antimicrob Agents. https://doi.org/10.1128/AAC.45.1.38

Kwon HJ, Bennik MH, Demple B, Ellenberger T (2000) Crystal structure of the *Escherichia coli* Rob transcription factor in complex with DNA. Nat Struct Biol 7:424–430. https://doi.org/10.1038/75213

Lau CHF, Krahn T, Gilmour C et al (2015) AmgRS-mediated envelope stress-inducible expression of the mexXY multidrug efflux operon of *Pseudomonas aeruginosa*. Microbiol Open 4:121–135. https://doi.org/10.1002/mbo3.226

Lawler AJ, Ricci V, Busby SJW, Piddock LJV (2013) Genetic inactivation of acrAB or inhibition of efflux induces expression of ramA. J Antimicrob Chemother 68:1551–1557. https://doi.org/10.1093/jac/dkt069

Lennen RM, Politz MG, Kruziki MA, Pfleger BF (2012) Identification of transport proteins involved in free fatty acid efflux in *Escherichia coli*. J Bacteriol 195:135–144. https://doi.org/10.1128/JB.01477-12

Li X-Z, Nikaido H (2004) Efflux-mediated drug resistance in bacteria. Drugs 64:159–204. https://doi.org/10.2165/11317030-000000000-00000

Li XZ, Nikaido H (2009) Efflux-mediated drug resistance in bacteria: an update. Drugs 69:1555–1623. https://doi.org/10.2165/11317030-000000000-00000

Li XZ, Nikaido H, Poole K (1995) Role of MexA-MexB-OprM in antibiotic efflux in *Pseudomonas aeruginosa*. Antimicrob Agents Chemother 39:1948–1953. https://doi.org/10.1128/AAC.39.9.1948

Li XZ, Barré N, Poole K (2000) Influence of the MexA-MexB-oprM multidrug efflux system on expression of the MexC-MexD-oprJ and MexE-MexF-oprN multidrug efflux systems in *Pseudomonas aeruginosa*. J Antimicrob Chemother 46:885–893. https://doi.org/10.1128/JB.182.5.1410-1414.2000

Li M, Gu R, Su CC et al (2007) Crystal structure of the tanscriptional regulator AcrR from *Escherichia coli*. J Mol Biol 374:591–603. https://doi.org/10.1016/j.jmb.2007.09.064

Li XZ, Plesiat P, Nikaido H (2015) The challenge of efflux-mediated antibiotic resistance in Gram-negative bacteria. Clin Microbiol Rev 28:337–418. https://doi.org/10.1128/CMR.00117-14

Liao J, Schurr MJ, Sauera K (2013) The merR-like regulator brlR confers biofilm tolerance by activating multidrug efflux pumps in *Pseudomonas aeruginosa* biofilms. J Bacteriol 195:3352–3363. https://doi.org/10.1128/JB.00318-13

Lim D, Poole K, Strynadka NCJ (2002) Crystal structure of the MexR repressor of themexRAB-oprM multidrug efflux operon of *Pseudomonas aeruginosa*. J Biol Chem 277:29253–29259

Lin J, Akiba M, Sahin O, Zhang Q (2005) CmeR functions as a transcriptional repressor for the multidrug efflux pump CmeABC in *Campylobacter jejuni*. Antimicrob Agents Chemother 49:1067–1075. https://doi.org/10.1128/AAC.49.3.1067-1075.2005

Lister PD, Wolter DJ, Hanson ND (2009) Antibacterial-resistant *Pseudomonas aeruginosa*: clinical impact and complex regulation of chromosomally encoded resistance mechanisms. Clin Microbiol Rev 22:582–610. https://doi.org/10.1128/CMR.00040-09

Ma D, Cook DN, Alberti M et al (1993) Molecular cloning and characterization of acrA and acrE genes of *Escherichia coli*. J Bacteriol 175:6299–6313

Ma D, Cook DN, Alberti M et al (1995) Genes acrA and acrB encode a stress-induced efflux system of *Escherichia coli*. Mol Microbiol 16:45–55. https://doi.org/10.1111/j.1365-2958.1995.tb02390.x

Ma D, Alberti M, Lynch C et al (1996) The local repressor AcrR plays a modulating role in the regulation of acrAB genes of *Escherichia coli* by global stress signals. Mol Microbiol 19:101–112. https://doi.org/10.1046/j.1365-2958.1996.357881.x

Martin RG, Rosner JL (2011) Promoter discrimination at class I mara regulon promoters mediated by glutamic acid 89 of the mara transcriptional activator of *escherichia coli*. J Bacteriol 193:506–515. https://doi.org/10.1128/JB.00360-10

Martin RG, Gillette WK, Rhee S, Rosner JL (1999) Structural requirements for marbox function in transcriptional activation of mar/sox/rob regulon promoters in *Escherichia coli*: sequence, orientation and spatial relationship to the core promoter. Mol Microbiol 34:431–441. https://doi.org/10.1046/j.1365-2958.1999.01599.x

Martinez JL, Sánchez MB, Martínez-Solano L et al (2009) Functional role of bacterial multidrug efflux pumps in microbial natural ecosystems. FEMS Microbiol Rev 33:430–449

Maseda H, Sawada I, Saito K et al (2004) Enhancement of the mexAB-oprM efflux pump expression by a quorum-sensing autoinducer and its cancellation by a regulator, MexT, of the mexEF-oprN efflux pump operon in Pseudomonas aeruginosa. Antimicrob Agents Chemother 2004(48):1320–1328. https://doi.org/10.1128/AAC.48.4.1320

Masuda N, Sakagawa E, Ohya S (2000) Substrate specificities of MexAB-OprM, MexCD-OprJ, and MexXY-oprM efflux pumps in Pseudomonas aeruginosa. Antimicrob Agents Chemother 44:3322–3327. https://doi.org/10.1128/AAC.44.12.3322-3327.2000.Updated

Mine T, Morita Y, Kataoka A et al (1999) Expression in *Escherichia coli* of a new multidrug efflux pump, MexXY, from *Pseudomonas aeruginosa*. Antimicrob Agents Chemother 43:415–417

Morita Y, Kimura N, Mima T et al (2001) Roles of MexXY- and MexAB-multidrug efflux pumps in intrinsic multidrug resistance of *Pseudomonas aeruginosa* PAO1. J Genet Appl Microbiol 47:27–32. https://doi.org/10.2323/jgam.47.27

Morita Y, Cao L, Gould VC et al (2006) nalD encodes a second repressor of the mexAB-oprM multidrug efflux operon of *Pseudomonas aeruginosa*. J Bacteriol 188:8649–8654. https://doi.org/10.1128/JB.01342-06

Mosqueda G, Ramos JL (2000) A set of genes encoding a second toluene efflux system in *Pseudomonas putida* DOT-T1E is linked to the tod genes for toluene metabolism. J Bacteriol 182:937–943. https://doi.org/10.1128/JB.182.4.937-943.2000

Murakami S, Nakashima R, Yamashita E, Yamaguchi A (2002) Crystal structure of bacterial multidrug efflux transporter AcrB. Nature 419:587–593. https://doi.org/10.5940/jcrsj.45.256

Murakami S, Nakashima R, Yamashita E et al (2006) Crystal structures of a multidrug transporter reveal a functionally rotating mechanism. Nature 443:173–179. https://doi.org/10.2142/biophys.47.309

Nichols RJ, Sen S, Choo YJ et al (2011) Phenotypic landscape of a bacterial cell. Cell 144:143–156. https://doi.org/10.1016/j.cell.2010.11.052

Nikaido H (2009) Multidrug resistance in bacteria. Annu Rev Biochem 78:119–146. https://doi.org/10.1146/annurev.biochem.78.082907.145923

Nikaido H (2011) Structure and mechanism of RND-type multidrug efflux pumps. Adv Enzymol Relat Areas Mol Biol 77:1–60. https://doi.org/10.1111/j.1365-2958.2011.07544.x.Chlorinated

Nikaido E, Yamaguchi A, Nishino K (2008) AcrAB multidrug efflux pump regulation in *Salmonella enterica* serovar Typhimurium by RamA in response to environmental signals. J Biol Chem 283:24245–24253. https://doi.org/10.1074/jbc.M804544200

Nishino K, Yamaguchi A (2001) Analysis of a complete library of putative drug transporter genes in *Escherichia coli*. J Bacteriol 183:5803–5812. https://doi.org/10.1128/JB.183.20.5803-5812.2001

Nishino K, Yamada J, Hirakawa H et al (2003) Roles of TolC-dependent multidrug transporters of *Escherichia coli* in resistance to β-lactams. Antimicrob Agents Chemother 47:3030–3033. https://doi.org/10.1128/AAC.47.9.3030-3033.2003

Nishino K, Yamasaki S, Hayashi-Nishino M, Yamaguchi A (2011) Effect of overexpression of small non-coding DsrA RNA on multidrug efflux in *Escherichia coli*. J Antimicrob Chemother 66:291–296. https://doi.org/10.1093/jac/dkq420

Okusu H, Ma D, Nikaido H (1996) AcrAB efflux pump plays a major role in the antibiotic resistance phenotype of *Escherichia coli* multiple-antibiotic-resistance (Mar) mutants. J Bacteriol 178:306–308

Perraud AL, Rippe K, Bantscheff M et al (2000) Dimerization of signalling modules of the EvgAS and BvgAS phosphorelay systems. Biochim Biophys Acta Protein Struct Mol Enzymol 1478:341–354. https://doi.org/10.1016/S0167-4838(00)00052-2

Piddock L (2014) Understanding the basis of antibiotic resistance: QConnect results. Microbiol Soc. http://mic.microbiologyresearch.org/content/journal/micro/10.1099/mic.0.082412-0. Accessed 30 Jun 2016

Poole K, Krebes K, McNally C, Shadi N (1993) Multiple antibiotic resistance in *Pseudomonas aeruginosa*: evidence for involvement of an efflux operon. J Bacteriol 175:7363–7372

Poole K, Gotoh N, Tsujimoto H et al (1996a) Overexpression of the mexC-mexD-oprJ efflux operon in nfxB type multidrug-resistant strains of *Pseudomonas aeruginosa*. Mol Microbiol 21:713–725. https://doi.org/10.1046/j.1365-2958.1996.281397.x

Poole K, Tetro K, Zhao Q et al (1996b) Expression of the multidrug resistance operon mexA-mexB-oprM in *Pseudomonas aeruginosa*: mexR encodes a regulator of operon expression. Antimicrob Agents Chemother 40:2021–2028

Pos KM (2009) Drug transport mechanism of the AcrB efflux pump. Biochim Biophys Acta 1794:782–793. https://doi.org/10.1016/j.bbapap.2008.12.015

Purssell A, Poole K (2013) Functional characterization of the NfxB repressor of the mexCD-oprJ multidrug efflux operon of *Pseudomonas aeruginosa*. Microbiology 159:2058–2073. https://doi.org/10.1099/mic.0.069286-0

Purssell A, Fruci M, Mikalauskas A et al (2015) EsrC, an envelope stress-regulated repressor of the mexCD-oprJ multidrug efflux operon in *Pseudomonas aeruginosa*. Environ Microbiol 17:186–198. https://doi.org/10.1111/1462-2920.12602

Putman M, van Veen HW, Konings WN (2000) Molecular properties of bacterial multidrug transporters. Microbiol Mol Biol Rev 64:672–693. https://doi.org/10.1128/MMBR.64.4.672-693.2000

Rahmati S, Yang S, Davidson AL, Zechiedrich EL (2002) Control of the AcrAB multidrug efflux pump by quorum-sensing regulator SdiA. Mol Microbiol 43:677–685. https://doi.org/10.1046/j.1365-2958.2002.02773.x

Ramos JL, Duque E, Huertas MJ, Haidour A (1995) Isolation and expansion of the catabolic potential of a *Pseudomonas putida* strain able to grow in the presence of high concentrations of aromatic hydrocarbons. J Bacteriol 177:3911–3916

Ramos JL, Duque E, Godoy P, Segura A (1998) Efflux pumps involved in toluene tolerance in *Pseudomonas putida* DOT- T1E. J Bacteriol 180:3323–3329 doi: 00219193/98/$04.00+0

Ramos JL, Duque E, Gallegos MT et al (2002) Mechanisms of solvent tolerance in gram-negative bacteria. Annu Rev Microbiol 56:743–768. https://doi.org/10.1146/annurev.micro.56.012302.161038

Ramos JL, Cuenca MS, Molina-Santiago C et al (2015) Mechanisms of solvent resistance mediated by interplay of cellular factors in *Pseudomonas putida*. FEMS Microbiol Rev 39:555–566. https://doi.org/10.1093/femsre/fuv006

Randall LP, Woodward MJ (2001) Multiple antibiotic resistance (mar) locus in *Salmonella enterica* serovar typhimurium DT104. Appl Environ Microbiol 67:1190–1197. https://doi.org/10.1128/AEM.67.3.1190-1197.2001

Rhee S, Martin RG, Rosner JL, Davies DR (1998) A novel DNA-binding motif in MarA: the first structure for an AraC family transcriptional activator. Proc Natl Acad Sci U S A 95:10413–10418. https://doi.org/10.1073/pnas.95.18.10413

Rodríguez-Herva JJ, García V, Hurtado A et al (2007) The ttgGHI solvent efflux pump operon of *Pseudomonas putida* DOT-T1E is located on a large self-transmissible plasmid. Environ Microbiol 9:1550–1561. https://doi.org/10.1111/j.1462-2920.2007.01276.x

Rojas A, Duque E, Mosqueda G et al (2001) Three efflux pumps are required to provide efficient tolerance to toluene in *Pseudomonas putida* DOT-T1E. J Bacteriol 183:3967–3973. https://doi.org/10.1128/JB.183.13.3967-3973.2001

Rojas A, Segura A, Guazzaroni ME et al (2003) In vivo and in vitro evidence that TtgV is the specific regulator of the TtgGHI multidrug and solvent efflux pump of *Pseudomonas putida*. J Bacteriol 185:4755–4763. https://doi.org/10.1128/JB.185.16.4755-4763.2003

Rosenberg EY, Ma D, Nikaido H (2000) AcrD of *Escherichia coli* is an aminoglycoside efflux pump. J Bacteriol 182:1754–1756. https://doi.org/10.1128/JB.182.6.1754-2756.2000

Rosner JL, Martin RG (2013) Reduction of cellular stress by TolC-dependent efflux pumps in *Escherichia coli* indicated by baeSR and CpxARP activation of spy in efflux mutants. J Bacteriol 195:1042–1050. https://doi.org/10.1128/JB.01996-12

Saier MH, Paulsen IT (2001) Phylogeny of multidrug transporters. Semin Cell Dev Biol 12:205–213. https://doi.org/10.1006/scdb.2000.0246

Sardessai Y, Bhosle S (2002) Tolerance of bacteria to organic solvents. Res Microbiol 153:263–268

Schweizer HP (2003) Efflux as a mechanism of resistance to antimicrobials in *Pseudomonas aeruginosa* and related bacteria: unanswered questions. Genet Mol Res 2:48–62 [pii] S01

Segura A, Hurtado A, Rivera B, Lazaroaie MM (2008) Isolation of new toluene-tolerant marine strains of bacteria and characterization of their solvent-tolerance properties. J Appl Microbiol 104:1408–1416. https://doi.org/10.1111/j.1365-2672.2007.03666.x

Segura A, Molina L, Fillet S et al (2012) Solvent tolerance in gram-negative bacteria. Curr Opin Biotechnol 23:415–421. https://doi.org/10.1016/j.copbio.2011.11.015

Shiba T, Ishiguro K, Takemoto N et al (1995) Purification and characterization of the *Pseudomonas aeruginosa* NfxB protein, the negative regulator of the nfxB gene. J Bacteriol 177:5872–5877

Shindo H, Iwaki T, Ieda R et al (1995) Solution structure of the DNA binding domain of a nucleoid-associated protein, H-NS, from *Escherichia coli*. FEBS Lett 360:125–131. https://doi.org/10.1016/0014-5793(95)00079-O

Sikkema J, de Bont JA, Poolman B (1995) Mechanisms of membrane toxicity of hydrocarbons. Microbiol Rev 59:201–222

Sivaneson M, Mikkelsen H, Ventre I et al (2011) Two-component regulatory systems in *Pseudomonas aeruginosa*: an intricate network mediating fimbrial and efflux pump gene expression. Mol Microbiol 79:1353–1366. https://doi.org/10.1111/j.1365-2958.2010.07527.x

Sobel ML, Neshat S, Poole K (2005) Mutations in PA2491 (mexS) promote MexT-dependent mexEF-oprN expression and multidrug resistance in a clinical strain of *Pseudomonas aeruginosa*. J Bacteriol 187:1246–1253. https://doi.org/10.1128/JB.187.4.1246-1253.2005

Starr LM, Fruci M, Poole K (2012) Pentachlorophenol induction of the *Pseudomonas aeruginosa* mexAB-oprM efflux operon: involvement of repressors NalC and MexR and the antirepressor ArmR. PLoS One. https://doi.org/10.1371/journal.pone.0032684

Sulavik MC, Gambino LF, Miller PF (1995) The MarR repressor of the multiple antibiotic resistance (mar) operon in *Escherichia coli*: prototypic member of a family of bacterial regulatory proteins involved in sensing phenolic compounds. Mol Med 1:436–446

Sun Y, Dai M, Hao H et al (2011) The role of RamA on the development of ciprofloxacin resistance in *Salmonella enterica* serovar Typhimurium. PLoS One. https://doi.org/10.1371/journal.pone.0023471

Tanaka T, Horii T, Shibayama K et al (1997) RobA-induced multiple antibiotic resistance largely depends on the activation of the AcrAB efflux. Microbiol Immunol 41:697–702

Terán W, Felipe A, Segura A et al (2003) Antibiotic-dependent induction of *Pseudomonas putida* DOT-T1E TtgABC efflux pump is mediated by the drug binding repressor TtgR. Antimicrob Agents Chemother 47:3067–3072. https://doi.org/10.1128/AAC.47.10.3067-3072.2003

Terán W, Krell T, Ramos JL, Gallegos MT (2006) Effector-repressor interactions, binding of a single effector molecule to the operator-bound TtgR homodimer mediates derepression. J Biol Chem 281:7102–7109. https://doi.org/10.1074/jbc.M511095200

Terán W, Felipe A, Fillet S et al (2007) Complexity in efflux pump control: cross-regulation by the paralogues TtgV and TtgT. Mol Microbiol 66:1416–1428. https://doi.org/10.1111/j.1365-2958.2007.06004.x

Truong-Bolduc QC, Villet RA, Estabrooks ZA, Hooper DC (2014) Native efflux pumps contribute resistance to antimicrobials of skin and the ability of *staphylococcus aureus* to colonize skin. J Infect Dis 209:1485–1493. https://doi.org/10.1093/infdis/jit660

Uwate M, ki IY, Shirai A et al (2013) Two routes of MexS-MexT-mediated regulation of MexEF-OprN and MexAB-OprM efflux pump expression in *Pseudomonas aeruginosa*. Microbiol Immunol 57:263–272. https://doi.org/10.1111/1348-0421.12032

Vadlamani G, Thomas MD, Patel TR et al (2015) The β-lactamase gene regulator AmpR is a tetramer that recognizes and binds the D-Ala-D-Ala motif of its repressor UDP-*N*-acetylmuramic acid (MurNAc)-pentapeptide. J Biol Chem 290:2630–2643. https://doi.org/10.1074/jbc.M114.618199

Vallet I, Diggle SP, Stacey RE et al (2004) Biofilm formation in *Pseudomonas aeruginosa*: fimbrial cup gene clusters are controlled by the transcriptional regulator MvaT. J Bacteriol 186:2880–2890. https://doi.org/10.1128/JB.186.9.2880-2890.2004

Wang D, Seeve C, Pierson LS, Pierson EA (2013) Transcriptome profiling reveals links between ParS/ParR, MexEF-OprN, and quorum sensing in the regulation of adaptation and virulence in *Pseudomonas aeruginosa*. BMC Genomics 14:618. https://doi.org/10.1186/1471-2164-24-618

White DG, Goldman JD, Demple B, Levy SB (1997) Role of the acrAB locus in organic solvent tolerance mediated by expression of marA, soxS, or robA in *Escherichia coli*. J Bacteriol 179:6122–6126

Wong K, Ma J, Rothnie A et al (2014) Towards understanding promiscuity in multidrug efflux pumps. Trends Biochem Sci 39:8–16

Yamasaki S, Nikaido E, Nakashima R et al (2013) The crystal structure of multidrug-resistance regulator RamR with multiple drugs. Nat Commun 4:2078. https://doi.org/10.1038/ncomms3078

Yoneyama H, Ocaktan A, Tsuda M, Nakae T (1997) The role of mex-gene products in antibiotic extrusion in *Pseudomonas aeruginosa*. Biochem Biophys Res Commun 233:611–618. https://doi.org/10.1006/bbrc.1997.6506

Zhang A, Rosner JL, Martin RG (2008) Transcriptional activation by MarA, SoxS and Rob of two tolC promoters using one binding site: a complex promoter configuration for tolC in *Escherichia coli*. Mol Microbiol 69:1450–1455. https://doi.org/10.1111/j.1365-2958.2008.06371.x

Zheng J, Cui S, Meng J (2009) Effect of transcriptional activators RamA and SoxS on expression of multidrug efflux pumps AcrAB and AcrEF in fluoroquinolone-resistant *Salmonella* Typhimurium. J Antimicrob Chemother 63:95–102. https://doi.org/10.1093/jac/dkn448

Zylstra GJ, Gibson DT (1989) Toluene degradation by *Pseudomonas putida* F1. Nucleotide sequence of the todC1C2BADE genes and their expression in *Escherichia coli*. J Biol Chem 264:14940–14946

# The Family of Two-Component Systems That Regulate Hydrocarbon Degradation Pathways

# 12

Andreas Busch, Noel Mesa-Torres, and Tino Krell

## Contents

1 Introduction ................................................................... 202
2 The TodS-TodT System ....................................................... 204
   2.1 TodS: Signal Sensing and Transfer Through Multiple Domains .................... 204
   2.2 Mechanistic Insight into Agonist/Antagonist Recognition ......................... 207
   2.3 TodT Forms an Intricate Nucleoprotein Complex for $P_{todX}$ Promotor Activation ... 209
3 Other TodS-TodT–Like SHKs .................................................. 211
   3.1 The TmoS-TmoT System ................................................... 211
   3.2 The StyS-StyR System ..................................................... 211
   3.3 The TutC/TutB and NodV/NodW Systems .................................... 214
   3.4 Identification of Additional TodS/TodT-Like Systems and Phylogenetic
      Distribution ............................................................. 214
4 Biotechnological Exploitation of TodS/TodT-Like Systems ........................... 216
5 Research Needs ............................................................... 216
References ..................................................................... 217

### Abstract

Many bacteria have been isolated in the past that are able to live on toxic aromatic compounds as their sole carbon source. These strains are of great biotechnological interest for bioremediation purposes, since aromatic compounds are components of crude oil and are released, e.g., after a forest fire during the combustion of cellulose and lignins. The degradation pathways involved in braking down these

A. Busch (✉)
Confo Therapeutics, VIB Campus Technologiepark, Zwijnaarde, Belgium
e-mail: andreas.busch@confotherapeutics.com

N. Mesa-Torres · T. Krell
Department of Environmental Protection, Estación Experimental del Zaidín – CSIC, Granada, Spain
e-mail: noel.mesa@eez.csic.es; t.krell@eez.csic.es; tino.krell@eez.csic.es

© Springer International Publishing AG, part of Springer Nature 2018
T. Krell (ed.), *Cellular Ecophysiology of Microbe: Hydrocarbon and Lipid Interactions*,
Handbook of Hydrocarbon and Lipid Microbiology, https://doi.org/10.1007/978-3-319-50542-8_6

compounds are under tight transcriptional control. However, the mechanisms of regulation differ significantly: some degradation routes are regulated by one-component systems (OCS, transcriptional regulators), whereas other pathways are under the control of two-component systems (TCS). This chapter summarizes knowledge available on the TCS-subfamily that is involved in hydrocarbon degradation. The sensor histidine kinases (SHK) of this subfamily differ significantly from a prototypal SHK in its subcellular localization, size, and domain arrangement. We will focus on data available on the TodS/TodT and TmoS/TmoT systems controlling toluene degradation, and the StyS/StyR system regulating a styrene breakdown pathway. Interestingly, the former two systems are controlled by the concerted action of agonists and antagonists, a fact that is of great interest for the development of efficient bioremediation strategies. A phylogenetic sequence analysis indicates that TCSs with a domain arrangement identical to TodS/TodT are predominantly found in strains of the β- and γ-Proteobacteria that sense and degrade aromatic hydrocarbons or that are involved in processes such as the nodulation, where polyaromatic hydrocarbons (PAHs, flavonoids) are sensed by the TodS homolog NodV.

# 1    Introduction

Bacteria are continuously exposed to changing environmental conditions and hence require efficient cellular responses to enable quick adaptation permitting survival in harsh environments. TCSs play a major role in processing signals into virtually all types of adaptive cellular control mechanisms (Galperin 2005; Mascher 2006; Krell et al. 2010). However, little is known about the signaling molecules triggering those responses and the underlying molecular processes. The elementary components of a prototypal TCS are (see Fig. 1a): (i) a sensor histidine kinase (SHK) with a periplasmic sensing and an intracellular phosphoaccepting/kinase domain and (ii) a cognate response regulator (RR) composed of a signal receiver (REC) and an output domain. The RR output domain predominantly regulates DNA transcription but can equally show enzymatic activities or bind other proteins or RNA (Galperin 2010). The elementary molecule acting as a transmitter of a response to an incoming signal is a phosphoryl group. In a prototypal TCS, signal binding to the SHK's input domain modulates the autophosphorylation activity of the kinase domain, which in turn modulates the phosphoryl-group transfer to the cognate RR's receiver domain by a mechanism proposed by Casino et al. (2009, 2014). The RR's output domain usually possesses a helix-turn-helix (HTH) DNA-binding motif that mediates the final response at the transcriptional level by binding to promotors.

Functional and genomic approaches of bacterial transduction systems carried out by Galperin et al. (Galperin 2005) determined that the amount and complexity of TCSs is directly proportional to the array of challenges microbes face. Thus, soil and other free-living bacteria facing unstable milieus developed highly complex hybrid systems. Here, we will discuss such bacteria involved in the degradation of xenobiotic compounds.

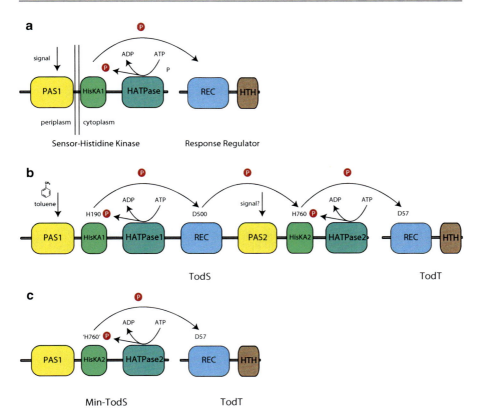

**Fig. 1** Domain arrangement of a prototypal two-component system (**a**), the TodS/TodT system (**b**) and its minimalistic version, Min-TodS/TodT (**c**). PAS, *Per-Arnt-Sim* type sensor domain; HisKA, dimerization/phosphoacceptor domain; HATPase, histidine kinase catalytic domain; REC, response regulator receiver domain; HTH, helix-turn-helix DNA-binding domain; P, phosphoryl group. The phosphoryl group-accepting residues and the phosphor-transfer pathway of the TodS/TodT system are indicated as established in Busch et al. (2009). The Min-TodS/TodT system was reported by Silva-Jímenez et al. (2012b)

Monocyclic aromatic compounds (*b*enzene, *t*oluene, *e*thylbenzene, *x*ylenes, BTEX) are recalcitrant petroleum derivatives and represent an important environmental issue. A significant number of bacterial isolates from contaminated soils or wastewater treatment plants are highly resistant to such aromatic hydrocarbons and are able to metabolize them into Krebs cycle intermediates. These organisms have been extensively studied in the field of bioremediation and the pathways described for degradation of BTEX compounds are under control of either single transcriptional regulators like XylS or XylR (OCS) or complex TCSs (Parales et al. 2008). It is currently not well understood why some pathways are regulated by OCS whereas other by TCSs. We will focus on the most extensively studied TCSs, which regulate the *to*luene *di*oxygenase (TOD) pathway in *Pseudomonas putida* (TodS/TodT TCS)

(Lau et al. 1997), the anaerobic toluene degradation pathway in *Thauera aromatica* (TutC/TutB TCS) (Coschigano and Young 1997) or the toluene-4-monooxygenase pathway in *Pseudomonas mendocina* RK1 (TmoS/TmoT TCS) (Ramos-González et al. 2002).

## 2  The TodS-TodT System

The enzymes responsible for the degradation of benzene, toluene, and ethylbenzene are encoded by the *tod* pathway gene cluster, which was initially described in *P. putida* F1 by the Gibson Lab (Zylstra et al. 1988; Zylstra and Gibson 1989). Lau et al. (1997) later identified the TCS responsible for the *tod* pathway regulation, the SHK TodS and the RR TodT, encoded by the *todST* operon. *P. putida* F1 was unable to grow on toluene when either *todS* or *todT* genes were knocked out. Since TCS activity is modulated by the level of phosphorylation in response to a signal, the authors replaced in the TodT the conserved residue predicted to accept phosphoryl-groups (Asp56) by an asparagine. As a result, the *tod* operon could not be activated in the presence of toluene (Lau et al. 1997). Further experiments by the same authors showed that purified and unphosphorylated TodT is able to bind to the $P_{todX}$ promoter of the *tod* operon. Considering the great potential of *P. putida* in bioremediation strategies, these results motivated additional studies to understand the molecular mechanism of *tod* pathway regulation.

## 2.1  TodS: Signal Sensing and Transfer Through Multiple Domains

In contrast to prototypal TCSs and as illustrated in Fig. 1a, TodS lacks transmembrane regions, is very large in size (almost 1,000 amino acids), and shows a complex domain arrangement (Fig. 1b): the central response regulator receiver domain (REC) is flanked on each side by a segment comprising a PAS sensor domain and an ATPase/kinase module.

In their initial work on the TodS/TodT system, Lau et al. (1997) predicted TodS to be composed of an N-terminal leucine zipper motif, followed by a SHK core, a REC domain, an oxygen sensing domain and a second SHK core. However, more recent homology-based predictions and results have led to a revision of the domain arrangement to that shown in Fig. 1b, namely, PAS1-HK1-REC-PAS2-HK2 (Lacal et al. 2006; Busch et al. 2007).

The Ramos lab initially took on the question of which compounds were able to activate the *tod* operon expression. Using a $P_{todX}$-*lacZ* fusion they demonstrated that the *tod* operon transcription is strongest induced by toluene and to a lower degree by the other pathway substrates benzene and ethylbenzene (Lacal et al. 2006). Surprisingly, a variety of differentially substituted monoaromatic compounds that are not mineralized by the TOD pathway were also able to activate transcription from the $P_{todX}$ promotor.

In contrast to prototypal SHKs, TodS was predicted to be a cytosolic, soluble protein. Recombinantly produced TodS and TodT (Lacal et al. 2006) allowed drawing a picture of the mechanism of action of this unusually complex TCS. Initially, binding of the main TOD pathway substrate toluene to purified TodS was tested by means of Isothermal Titration Calorimetry (ITC) (Krell 2008) and showed that toluene binds to purified TodS with submicromolar affinity and 1:1 stoichiometry (Lacal et al. 2006). In vitro autophosphorylation assays revealed that TodS has basal autophosphorylation activity and that toluene increases phosphorylation levels of TodS about fivefold. This increase in TodS activity is consequently translated into higher transphosphorylation rates towards TodT (Lacal et al. 2006).

Two binding sites for toluene were plausible: PAS1 and/or PAS2. Truncated versions of recombinant TodS (PAS1-HK1 and PAS2-HK2) were titrated with toluene and binding was only observed for the N-terminal PAS1-HK1 construct, suggesting that toluene binds to the PAS1 domain (Lacal et al. 2006). Single mutations performed on full-length TodS of conserved residues likely to be involved in the interaction with the effector were designed based on a homology model of the PAS1 domain. ITC experiments performed with these mutants allowed to pinpoint PAS1 as the toluene-binding domain (Busch et al. 2007). Furthermore, autophosphorylation assays with both truncated TodS versions showed that toluene stimulates autokinase activity of exclusively the HK1 domain (Busch et al. 2009).

Next, the in vivo effector profile determined with help of the $P_{todX}$-$lacZ$ fusion was assessed in vitro: both ITC and autophosphorylation assays were performed with a wide array of aromatic compounds. One main conclusion was that only monoaromatic compounds bound to the PAS1 pocket. Unexpectedly, some of these compounds that bound tightly to purified TodS were unable to stimulate $P_{todX}$ expression in vivo (Busch et al. 2007). For example, the three xylene isomers (*ortho*, *meta*, and *para*) bound to purified TodS with submicromolar affinities in vitro (Busch et al. 2007), whereas only *meta*- and *para*-xylene, but not the *ortho*-isomer, induced transcription from $P_{todX}$. Autophosphorylation assays proved that *ortho*-xylene was unable to increase TodS autophosphorylation (Busch et al. 2007). In summary, there was as direct correlation between the level of autophosphorylation of TodS and the transcription activity from $P_{todX}$ but no correlation between binding and any of the previous. The compounds that bound but did not stimulate TodS autokinase activity were termed antagonists, whereas those that bound TodS and stimulated *tod* operon expression were referred to as agonists. Busch et al. (2007) showed that agonists and antagonists compete in vitro and in vivo (Fig. 2) for binding at the PAS1 domain of TodS. $P_{todX}$ expression was measured in the presence of 1:1 and 1:5 agonist:antagonist mixtures, which revealed that the presence of antagonists reduced the magnitude of agonist-mediated stimulation. This indicates that antagonists are competitive inhibitors of $P_{todX}$ expression. This is a crucial fact to be taken into account in any bioremediation or biosensor prospects, since any with hydrocarbons contaminated site (e.g., crude oil) is likely to contain both, agonists and antagonist.

The so far mentioned experiments only associate a role in response to toluene to the N-terminal PAS1-HK1 domains. Therefore, what is the role of the REC, PAS2,

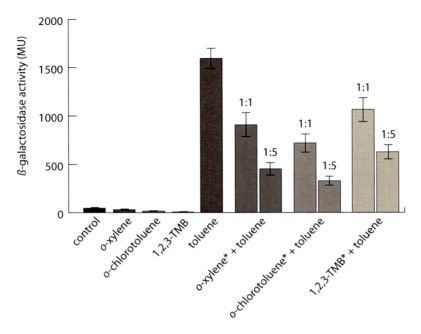

**Fig. 2** Effect of agonists and antagonists of the TodS/TodT system on $P_{todX}$ expression. The initial four columns are β-galactosidase measurements of $P_{todX}$ expression in the presence of the agonist toluene and the antagonists *ortho*-xylene, *ortho*-chlorotoluene and 1,2,3-trimethylbenzene (1,2,3-TMB). All four compounds were found to bind with high affinities to purified TodS. The paired columns represent β-galactosidase measurements in the presence of 1:1 and 1:5 mixtures of toluene (agonist) with each of the three antagonists, respectively. This is a modified version of the figure reproduced with permission from Busch et al. (2007)

and HK2 domains? Busch et al. (2009) first addressed the question whether TodS operates by an intramolecular phosphotransfer mechanism. Single mutations of the predicted phosphoaccepting residues allowed establishing that TodS operates by a phosphorelay mechanism as depicted in Fig. 1b (Busch et al. 2009). Phosphoryl groups are transferred from HK1 via the REC to the C-terminal HK2 domain. As previously mentioned, HK1 was unable to transphosphorylate TodT, with HK2 being the exclusive phosphodonor for TodT (Busch et al. 2009). This was confirmed by the construction of a PAS1-HK2 prototype-like TodS (Min-TodS, Fig. 1c) (Silva-Jímenez et al. 2012b) that was able to phosphorylate TodT. Min-TodS showed similar binding affinities to both agonists and antagonist; however, the stimulation of autophosphorylation by agonists was significantly reduced as compared to full-length TodS. In addition, basal autophosphorylation levels of Min-TodS were much higher, which is probably related to the lack of the REC domain, which was shown to be essential to keep the dephosphorylation rate high and therefore the basal activity of TodS low (Busch et al. 2009). These results point out the relevance of the REC and HK2 in an efficient response of TodS to toluene. As to the role of PAS2, Lau et al. (1997) pointed out that PAS2 shares some sequence similarities with the oxygen sensing PAS domain of the FixL sensor kinase. Up to date, however, the role of PAS2 remains to be established.

TCSs have separate input (SHK) and output (RR) domains and require ATP for signal transmission, while a prototypal OCS is a direct fusion of an input with an output domain that does not require ATP. It seems plausible that the high energy costs involved in TCS-based regulation is largely compensated by the capacity to sense environmental cues at the cell surface through their peri- or extracytoplasmic sensor domain and to translate them into an adaptive response. In this context, the existence of entirely cytosolic TCS, like TodS/TodT, is not well understood. The advantage of cytosolic SHKs, as compared to prototypal transmembrane SHKs, could be related to the fact that they show more complex domain arrangements, including multiple sensor domains that may permit the sensing of additional cues (Mascher 2006; Cheung and Hendrickson 2010; Preu et al. 2012). It has been suggested that the response integration of several cytoplasmic and/or diffusible external signal molecules in one single cytoplasmic protein could be, despite the ATP consumption involved, an additional advantage over OCSs (Krell et al. 2009). How and which additional signals are integrated into one regulatory response has been recently addressed in the TodS/TodT system.

It has previously been shown that some complex kinases recognize different quinones, which are indicators for the cellular redox state (Georgellis et al. 2001; Bock and Gross 2002). The Krell group conducted autokinase assays with TodS in the presence of different oxidative stress agents (Silva-Jímenez et al. 2014) and concluded that menadione reduced the autokinase activity in vitro and gene expression from $P_{todX}$ in vivo. When the sole, strictly conserved cysteine residue (Cys320) in TodS was mutated, the kinase became insensitive to menadione. Since the exposure of *P. putida* to toluene was shown to induce an oxidative stress (Domínguez-Cuevas et al. 2006), menadione was proposed to downregulate TodS activity in an internal feedback mechanism (Silva-Jímenez et al. 2014).

## 2.2 Mechanistic Insight into Agonist/Antagonist Recognition

The fact that agonists and antagonists compete for the same binding site at TodS and its implications in $P_{todX}$ activation (Busch et al. 2007), triggered additional efforts to understand the underlying molecular differences upon either agonist or antagonist binding to the PAS1 domain and its effect on autophosphorylation and transcription activity. Koh et al. (2016) crystallized TodS PAS1 (residues 43–164) in the apo conformation, as well as in complex with the agonist toluene or the antagonist 1,2,4-trimethylbenzene.

All three structures show essentially very similar conformations with a canonical PAS-fold: a five-stranded antiparallel β-sheet and three α-helices spanning residues 45–149 (β2α3β3, Fig. 3) forming a hydrophobic ligand-binding pocket. The main structural difference among the three structures lies within the signal transfer region (STR, residues 150–163) of molecule B in the PAS1 dimer: the STR

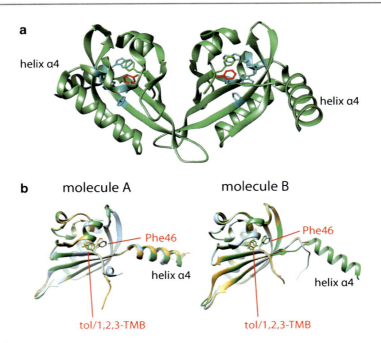

**Fig. 3** Structure of the PAS1 domain of TodS in apo-, agonist-, and antagonist bound form. (**a**) Structure of the PAS1 dimer with bound agonist toluene (*red*). Crucial residues for toluene binding are shown in stick conformation (*cyan*). (**b**) Superposed structures of PAS1 molecules A and B in apo form (*blue*), in complex with the agonist toluene (*green*) and in complex with the antagonist 1,3,5-TMB (*orange*). Phe46, suggested to be the trigger for the conformational changes in the STR is shown in stick mode, as well as the ligands toluene (*green*) and 1,3,5-TMB (*orange*)

is located at the C-terminal end of the PAS fold and appears to be disordered in both the apo- and the antagonist-bound forms. However, in the toluene bound form (Fig. 3a), the STR adopts an α-helical structure (α4), which was postulated to be the trigger for increased autophosphorylation activity of TodS at the HisKA/ATPase core located proximal to the PAS1 domain. Koh et al. (2016) suggest that Phe46 might be involved in signal transmission, since this residue undergoes major changes depending on the nature of the bound ligand (see Fig. 3b). The PAS structures were dimers and interestingly the agonist-induced formation of helix α4 was only observed in one (molecule B) of the two monomers, whereas helix α4 is formed in molecule A in all three structures. This raises the question as to the functionality or role of molecule A within the dimer: is molecule A responsible for the basal autophosphorylation levels of TodS of the apo or antagonist-bound forms, as reported previously (Lacal et al. 2006; Busch et al. 2007)? This "standby" activity of TodS could allow quicker response to agonist binding, triggering a fast increase in autophosphorylation levels only and readily when molecule B forms the α4 helix.

Koh et al. (2016) removed the potential dimerization domain for crystallization purposes. However, the PAS1 domain crystallized as an artificial antiparallel face-to-face dimer, which could simply be the optimal packing during crystal formation. In order to assess the relevance of the dimeric conformation in terms of functionality, the authors modeled the potential dimerization domain and mutated hydrophobic residues potentially interacting between molecule A and B. All mutations abolished TodS activity in the presence of toluene, suggesting these residues to be essential for dimer stability and TodS functionality. Further experiments with SEC-MALS and visualization of immunogold-labeled TodS via TEM reinforced the view of a native dimeric conformation of TodS. These images, on the other hand also suggested major ligand-dependent structural changes of the TodS dimer.

## 2.3 TodT Forms an Intricate Nucleoprotein Complex for P$_{todX}$ Promotor Activation

Investigations with recombinantly produced full-length TodT allowed to shed light on the interaction of TodT with the promotor P$_{todX}$, as depicted in Fig. 4: Monomeric, unphosphorylated TodT binds with high affinity, positive cooperativity, and a sequential manner to two pseudopalindromes (box 1 and box 2) and to one half-palindrome (box 3) (Lacal et al. 2008a, b). Unfortunately, phosphorylated TodT could not be studied due to the very low phosphorylation half-life. Initial data, however, suggested that phosphorylation does not alter affinity in a significant manner and protein phosphorylation was proposed to facilitate the recruitment of the RNA polymerase.

As established for the OmpR promoter of *Escherichia coli* (Yoshida et al. 2006), the model of positive cooperativity in the TodT-P$_{todX}$ interaction was established as follows using different promotor lengths and performing protein-DNA microcalorimetric titrations (Lacal et al. 2008a): first, two TodT monomers bind with positive cooperativity to the initial pseudopalindrome (box 1), which causes additional cooperative effects leading to TodT binding at the vicinal pseudopalindrome (box 2) and finally the half-palindrome (box 3) at elevated TodT concentrations. Higher-order DNA structures (i.e., bending) and DNA-protein complexes are at the basis of many regulatory processes, such as promotor activity control (Pérez-Martín et al. 1994). Atomic force microscopy was used to evaluate the impact of increasing TodT concentrations on the structure of P$_{todX}$ (Lacal et al. 2008a): at low TodT concentration only slight DNA bending was observed, whereas at high TodT concentrations a DNA hairpin bend was introduced between boxes 2 and 3. Most likely, protein binding at box 3, which occurs at high concentrations, induces this bend.

Another regulatory checkpoint in TOD pathway function is the promotor of the *todST* operon, P$_{todST}$. This promotor is under catabolite control, since glucose was shown to significantly reduce P$_{todST}$ activity (Busch et al. 2010). Thus, P$_{todX}$ expression depends firstly on global regulatory mechanisms such as catabolite repression and the impact of oxidative stress (menadione), followed by the presence of agonists/

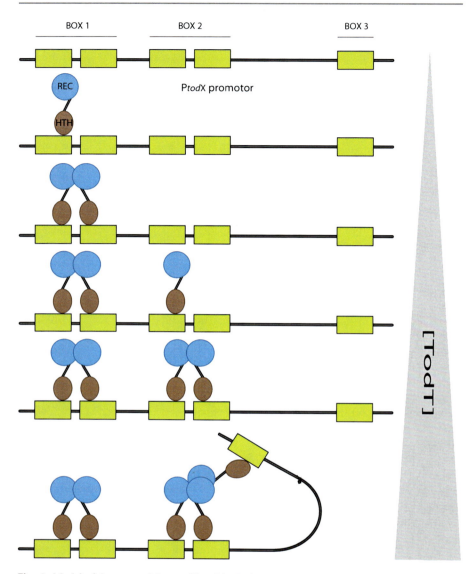

**Fig. 4** Model of the sequential assembly of the TodT-P$_{todX}$ promoter nucleoprotein complex. This sequential and cooperative model has been established based on the data reported in Lacal et al. (2008a, b). The P$_{todX}$ promotor is composed of two pseudopalindromic boxes (BOX 1 and BOX 2) and one half-palindromic BOX 3

antagonists. Strikingly, as described above, only a few of the agonists are TOD pathway substrates. In order to discern whether these unexpected features of TodS apply to the family of TodS-like proteins or if they are TodS specific, further studies on homologous systems were carried out that are described below.

# 3   Other TodS-TodT–Like SHKs

Two other homologous systems to TodS/TodT have been extensively studied: The TmoS/TmoT system, which regulates the TMO (*toluene-monooxygenase*) pathway for toluene degradation and was first described in *P. mendocina* KR1 (Ramos-González et al. 2002; Parales et al. 2008), and the StyS/StyR system for styrene catabolism, first described in *Pseudomonas flourescens* ST and later in several additional *Pseudomonas* sp. strains (Marconi et al. 1996; Velasco et al. 1998; Panke et al. 1998).

## 3.1   The TmoS-TmoT System

TmoS/TmoT TCS is the closest described homologous system to TodS/TodT, with about 85% protein sequence identity. In analogy to studies performed on TodS (Lacal et al. 2006; Busch et al. 2007), purified TmoS was submitted to a screening for possible effector molecules. Isothermal titration calorimetry assays showed that TmoS equally recognized a wide range of aromatic hydrocarbons with affinities in most cases higher than those observed for the corresponding ligands in TodS (Silva-Jímenez et al. 2012a). In TmoS as well as in TodS, toluene was shown to be the effector inducing highest promotor transcription activity. Interestingly, as for TodS, TmoS effectors could also be classified as agonists and antagonists. The profile of TmoS agonist/antagonist profile was almost identical to that of TodS. The physiological relevance of the concerted action of agonists and antagonists in these systems is not understood. An experiment where a *P. putida* DOT-T1E *todST* knockout mutant was complemented by *tmoST*, showed that transcription of $P_{todX}$ can be mediated by TmoS/TmoT in the presence of toluene, revealing the possibility of cross-regulation of these two catabolic pathways (Ramos-González et al. 2002).

## 3.2   The StyS-StyR System

The StyS-StyR TCS regulates the genes of the upper styrene catabolic pathway, which converts styrene into phenylacetate (Velasco et al. 1998; O'Leary et al. 2001; Santos et al. 2002). Phenylacetate is the substrate for the lower styrene catabolic pathway and acts as a repressor of the upper pathway genes even in the presence of styrene by preventing the expression of StyS and StyR (O'Leary et al. 2001). However, the first intermediate of the lower pathway, phenylacetyl-CoA (PA-CoA), inactivates the repressor of the Sty upper and lower pathways, PaaX (del Peso-Santos et al. 2008). The first step of this lower degradation pathway is catalyzed by the *paaF*-encoded phenylacetyl-coenzyme A ligase. Interestingly, coregulation by PaaX provides a mechanism that integrates responses to both, a pathway substrate (styrene, via the StyS/StyR regulatory system) and a pathway intermediate (PA-CoA, through PaaX) for a rapid metabolic coupling of the upper and lower pathways (del Peso-Santos et al. 2008). In addition, *Pseudomonas* sp. strain Y2 is unique in having three *paaF* genes, two located within two complete copies of *paa* gene clusters and one copy (*paaF2*) forming part of the *sty* regulon (upper pathway). PaaF2 is not subject to repression by PaaX (as PaaF and PaaF3) and is regulated through StyR binding to a STY box (STY4) of the $P_{paaF2}$ promoter to provide a system to accelerate and aid coupling of expression of the two enzyme sets required for styrene catabolism.

Studies of the StyS/StyR TCS have been particularly focused on the RR StyR. StyR alone was shown to suffice to initiate transcription of the upper pathway (Panke et al. 1998). Phosphorylation of StyR by small molecular weight phosphoryl group-donors like acetylphosphate induces protein dimerization and increases binding affinity to the $P_{styA}$ promotor (Leoni et al. 2003). Moreover, dimeric StyR binding to the target DNA is cooperative and it was suggested that the initial binding of the RR to the $P_{styA}$ promotor facilitates binding of the RR to additional, lower-affinity sites (Leoni et al. 2003). Three pseudopalindromic StyR binding sequences (STY1–3) and one IHF binding site (URE) were identified in the upper pathway promoter $P_{styA}$. Phosphorylated StyR (StyR-P) was shown to act either as an activator or as a repressor of upper pathway transcription (Leoni et al. 2005; Rampioni et al. 2008). When grown on styrene as sole carbon source, $P_{styA}$ promotor activity is fully induced and StyR-P bound to the highest-affinity site STY2 and IHF to the URE upstream of STY2, forming the so-called open conformation (Rampioni et al. 2008). It is important to note that the URE overlaps with the STY1 binding site, since when grown under catabolite repression conditions (growth on styrene plus glucose), cellular levels of StyR-P increase and STY1 can only be occupied by displacing IHF. This has a negative regulatory effect on $P_{styA}$ promotor expression, the nucleoprotein complex formed by PstyA and StyR-P probably adopting a "closed conformation" not accessible to the RNA-polymerase.

The three-dimensional structure of monomeric and unphosphorylated StyR has been solved at 2.2 Å resolution. It is the only full-length protein structure solved up to date in the family of TCS under review here (Fig. 5) (Milani et al. 2005). Like the vast majority of RRs (Stock et al. 2000), StyR is composed of two domains: a N-terminal regulatory domain and a C-terminal, DNA binding domain (Fig. 5). However, the structure of StyR shows a distinct feature: a long and straight α-helical Q-linker, which keeps the N and C-terminal domains 16 Å apart and with no direct contact. This contrasts with the molecular structures of other full-length RRs that are characterized by interdomain contacts and an essentially unstructured Q-linker region (Djordjevic et al. 1998; Maris et al. 2002). It was suggested that this latter, more tightly packed conformation hinders the access of the C-terminal domain to DNA and that only a hinge bending motion of the Q-linker upon phosphorylation would enable DNA binding (Baikalov et al. 1996; Zhang et al. 2003). The fact that StyR is able to bind to its cognate promotor in an unphosphorylated, however, "active-like" conformation supports the idea of the Q-linker playing an essential role in DNA binding.

In other RRs, the helix analogous to helix α4 in StyR (Fig. 5) is involved in the transition from an inactive (monomeric and unphosphorylated) to an active (dimeric

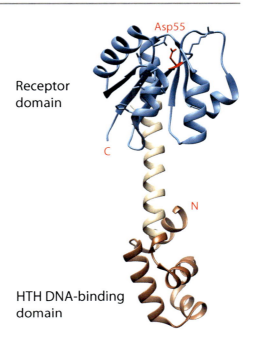

**Fig. 5** The three-dimensional structure of the StyR response regulator. The structure was reported in Milani et al. (2005) (PDB ID: 1YIO). The phosphoryl-group acceptor residue Asp55 on the REC domain and helix α4, in other RRs identified as crucial for dimerization, as well as the N- and C-terminal ends are labeled

and phosphorylated) state. Since this helix is unusually short in StyR, there might be other, unidentified structural features involved in shifting from an "active-like" to an "active" conformation.

In a more recent study, the SHK StyS has been functionally characterized (O'Leary et al. 2014). Though with different approaches, some of the results were in line with the studies performed on TodS: (1) PAS1 and HK2 are essential to trigger a response in the presence of styrene, since there is complete loss of transcriptional activity from $P_{styA}$ in mutants lacking either of these domains; (2) transcription levels were at par with wildtype StyS when either PAS2 or HK1 were deleted, suggesting a nonessential role of PAS2 or HK1 as in the case of TodS. Interestingly, when the mutant lacking only PAS1 was co-expressed with a membrane-embedded ABC transporter StyE, the loss of transcription activity could be reverted and even increased by 2.8 times with respect to wildtype StyS in the presence of styrene. This suggests two different activation mechanisms via either PAS1 presumably in a similar fashion as in TodS or the interaction with StyE, where the PAS1 domain would be dispensable for activity. On the other hand, the authors observed that a construct lacking the PAS2 and HK1 domains at the same time, failed to induce transcription of the *styABC* genes, assuming that the overall structural arrangement of StyS domains may be critical.

StyS and StyR from *Pseudomonas* sp. Y2 share >40% protein sequence identity and the same domain arrangement with TodS and TodT, respectively. Since styrene is also an agonist of TodS, but not a substrate for the TOD pathway, this regulatory system may have been recruited independently and evolved to control different catabolic pathways for the aerobic degradation of pathway-specific monoaromatic compounds (Ramos-González et al. 2002).

## 3.3    The TutC/TutB and NodV/NodW Systems

Apart from the three systems discussed so far, other family members include the TutC/TutB TCS that controls the anaerobic toluene degradation pathway in *T. aromatica* (Coschigano and Young 1997), as well as a number of mostly uncharacterized homologues from other hydrocarbon degrading bacteria like *Dechloromonas aromatica* (Salinero et al. 2009) *or Methylibium petroleiphilum* (Kane et al. 2007) that will be discussed below.

One homologous system though is not associated with anabolic pathways degrading aromatic compounds: NodV/NodW (Göttfert et al. 1990; Loh et al. 1997) and its homolog NwsA/NwsB (Grob et al. 1993) of *Bradyrhizobium japonicum*, a plant endosymbiont. The NodV/NodW TCS positively regulates *nod* gene expression in response to a large variety of plant-produced polyphenolic signal compounds, the isoflavones. These *nod* gene products encode the biosynthesis of substituted lipo-chitin Nod signals that induce many of the early nodulation events in the host. By analogy to the other systems described in this chapter, NodV (SHK) and NodW (RR) most likely activate transcription via a series of phosphorylation steps in response to plant signals. As for StyR, NodW can be phosphorylated *in vitro* by both acetyl phosphate and its cognate SHK, NodV. Genistein, an isoflavone known to induce the *nod* gene expression, is the only effector identified so far known to promote phosphorylation of NodV, presumably via NodW (Loh et al. 1997).

## 3.4    Identification of Additional TodS/TodT-Like Systems and Phylogenetic Distribution

The best-studied examples of TCSs with a TodS-like SHKs are all involved in sensing aromatic hydrocarbons: StyR, TmoS, TutC, and NodV. To identify further TodS homologues, we performed a BLAST analysis with TodS as search query. We only took into account output sequences with exactly the same domain distribution as TodS (see Fig. 1b) to build the phylogenetic tree shown in Fig. 6. The output showed protein sequences found in organisms of the β- and γ-Proteobacteria (as TodS, StyR, TmoS, and TutC), the most closely related classes within the phylum of the Proteobacteria. In agreement with the lifestyle of the β-Proteobacteria – many of them live in areas of anaerobic decomposition of organic matter – the output sequences for these classes are mostly related to bacteria able to use polyaromatic hydrocarbons (PAH) as nutrient source.

Although the class of the γ-Proteobacteria is the largest subgroup, mainly members of the order *Pseudomonadales* are represented in the phylogenetic tree. *Pseudomonas* species are widely known for their capacity of living on monoaromatic hydrocarbons. To this group belong the SHKs TodS, TmoS, and StyR. *Thiotrix*

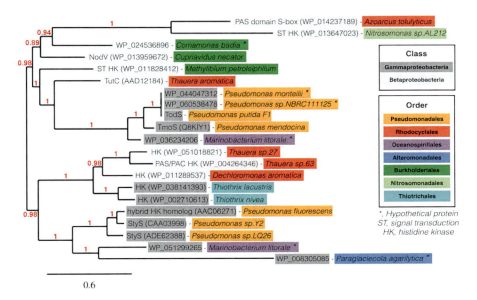

**Fig. 6** Phylogenetic distribution of proteins with the same domain arrangement as the SHK TodS from *P. putida*. Unrooted phylogenetic tree of protein sequences: the lengths of the branches represent the amount of change estimated to have occurred between a pair of nodes; numbers on branches represent the posterior probability, which is considered as the amount of time a single tree is visited during the tree generation. The analysis was performed on the Phylogeny.fr platform (Dereeper et al. 2010). Sequences were aligned with MUSCLE (v3.8.31) (Edgar 2004) configured for highest accuracy. The phylogenetic tree was reconstructed using the Bayesian inference method implemented in MrBayes (v3.2.3) (Huelsenbeck and Ronquist 2001) with number of substitution types fixed to 6. Amino acid substitution was evaluated by a Poisson model and rates variation across sites was fixed to "invgamma." Four Markov Chain Monte Carlo (MCMC) chains were run for 10,000 generations, sampling every ten generations, with the first 250 sampled trees discarded as "burn-in." Finally, a 50% majority rule consensus tree was constructed. Graphical representation and edition were performed with TreeDyn (v198.3) (Chevenet et al. 2006)

species, also represented in this group, live primarily in flowing water and grow on sulfides. However, they are also present in activated sludge wastewater treatment systems (Kim et al. 2002), where aromatic hydrocarbons are abundantly present (Wu et al. 2013; Huang and Li 2014). Many species used for bioremediation purposes related to toxic aromatic compounds, such as from the order of the *Pseudomonadales*, have been isolated from such wastewater treatment plants (Baggi et al. 1987; Ramos et al. 1995). However, TodS-like proteins in *Thiotrix* as well as the other representatives of the γ-Proteobacteria *Marinobacterium litorale* and *Paraglaciecola agarilytica* have yet to be explored.

Among the bacterial strains identified as bearing TodS-homologues within the class of the β-Proteobacteria, most of them are established degraders of aromatic compounds: *Azoarcus tolulyticus* is an anaerobic denitrifying toluene degrader (Zhou et al. 1995), *Comamonas badia* was isolated from activated sludge and is proven to degrade phenol (Felföldi et al. 2010), *Cupriavidus necator* metabolizes dichlorphenols (Kumar et al. 2016), *M. petroleiphilum* degrades BTEX, among other (Kane et al. 2007), and *D. aromatica* is able to anaerobically break down BTEX (Salinero et al. 2009). *Nitrosomonas AL212* is an ammonium-oxidizing bacterium isolated from a wastewater treatment plant (Suwa et al. 2011), however, with no proven capacity to degrade aromatic compounds.

The search for homologous systems showed that these systems are most likely inherent to a very specific niche of organisms sensing mono- and polyaromatic compounds, and there is in most cases evidence that they are not only able to sense but also to degrade those aromatic compounds. Therefore, it seems plausible that TodS/TodT-like proteins are likely to be involved in the regulation of pathways degrading those alternative carbon sources.

## 4     Biotechnological Exploitation of TodS/TodT-Like Systems

Many mono- or polyaromatic hydrocarbons such as BTEX or chlorinated benzene derivatives are classified as priority pollutants. The fact that the SHKs are able to bind a broad array of these compounds has already triggered important efforts to exploit these complex TCSs for biotechnological applications such as for protein-based biosensors to effectively detect and monitor toxins in the environment or wastewater treatment plants. For instance, the construction of a styrene biosensor has been reported (Alonso et al. 2003). The most recent efforts though are based on the TodS/TodT TCS of *P. putida* T-57 and DOT-T1E. Vangnai et al. (2012) engineered a GFP-based double plasmid bioreporter system for anilins and chlorinated anilins (CAs) in *E. coli*: a $P_{todX(T57)}$:*gfp* and a $P_{lac}$:*todST*(T57) fusion. They showed that, in addition to the compounds described to interact with TodS in *P. putida* DOT-T1E, anilin and several CAs are powerful inducers of $P_{todX(T57)}$ transcriptional activity. Hernández-Sánchez et al. (2016) used a similar experimental approach to detect BTEX with a $P_{todX(DOT-T1E)}$:*gfp* and a $P_m$:*todST*(DOT-T1E) fusion on one single plasmid named pKST-1. The biosensor was validated in different environments such as soil, fresh and marine water, being transferred into three different bacterial strains that were isolated from these three different environments.

## 5     Research Needs

Future efforts will aim at miniaturizing biosensors in flow-cell based systems which will allow the monitoring of bioremediation strategies and *in situ* detection of contamination in a fast, quick, and economical way. The fact that the PAS1 domains of the different TodS homologues are characterized by a broad ligand spectrum offers the additional possibility of protein evolution to tailor the binding pocket to recognize specific contaminants.

We learned as well that removing or impairing the REC domain through mutation of the phosphoraccepting residue very significantly increases the phosphorylation half-life (Busch et al. 2009), which could translate into more robust bioreporters with longer signal periods of, e.g., GFP-coupled promotors. The action of antagonists is certainly a handicap for the development of efficient bioremediation strategies. However, the recent structure of the PAS1 domain and the involved conformational changes upon agonist or antagonist binding (Koh et al. 2016) may provide a base for the development of TodS derivatives that respond exclusively to agonists. A number of questions remain on these systems and, clearly, additional structural information on interaction surfaces between domains of the SHKs described and between SHKs and their cognate RRs would aid in further optimizing signal sensing and signal stability through computational protein design approaches. Other questions concern the physiological reason of there being antagonists or agonists that are not pathway substrates and whether the complex phosphorelay mechanism is necessary for the interaction with other regulatory networks in the cell. Finally, although the phylogenetic distribution of TodS-like sensor kinases seems relatively narrow, we believe exploring some of these yet uncharacterized proteins might lead to the discovery of new signaling molecules with potential environmental and consequently biotechnological impact.

**Acknowledgments**  We acknowledge financial support from FEDER funds and Fondo Social Europeo through grants from the Junta de Andalucía (grants P09-RNM-4509 and CVI-7335) and the Spanish Ministry for Economy and Competitiveness (grants BIO2010-16937 and BIO2013-42297).

# References

Alonso S, Navarro-Llorens JM, Tormo A, Perera J (2003) Construction of a bacterial biosensor for styrene. J Biotechnol 102:301–306

Baggi G, Barbieri P, Galli E, Tollari S (1987) Isolation of a *Pseudomonas stutzeri* strain that degrades o-xylene. Appl Environ Microbiol 53(9):2129–2132

Baikalov I, Schröder I, Kaczor-Grzeskowiak M et al (1996) Structure of the *Escherichia coli* response regulator NarL. Biochemistry 35:11053–11061. https://doi.org/10.1021/bi960919o

Bock A, Gross R (2002) The unorthodox histidine kinases BvgS and EvgS are responsive to the oxidation status of a quinone electron carrier. Eur J Biochem 269:3479–3484

Busch A, Lacal J, Martos A et al (2007) Bacterial sensor kinase TodS interacts with agonistic and antagonistic signals. Proc Natl Acad Sci U S A 104:13774–13779. https://doi.org/10.1073/pnas.0701547104

Busch A, Guazzaroni M-E, Lacal J et al (2009) The sensor kinase TodS operates by a multiple step phosphorelay mechanism involving two autokinase domains. J Biol Chem 284:10353–10360. https://doi.org/10.1074/jbc.M900521200

Busch A, Lacal J, Silva-Jímenez H et al (2010) Catabolite repression of the TodS/TodT two-component system and effector-dependent transphosphorylation of TodT as the basis for toluene dioxygenase catabolic pathway control. J Bacteriol 192:4246–4250. https://doi.org/10.1128/JB.00379-10

Casino P, Rubio V, Marina A (2009) Structural insight into partner specificity and phosphoryl transfer in two-component signal transduction. Cell 139:325–336. https://doi.org/10.1016/j.cell.2009.08.032

Casino P, Miguel-Romero L, Marina A (2014) Visualizing autophosphorylation in histidine kinases. Nat Commun 5:3258. https://doi.org/10.1038/ncomms4258

Cheung J, Hendrickson WA (2010) Sensor domains of two-component regulatory systems. Curr Opin Microbiol 13:116–123. https://doi.org/10.1016/j.mib.2010.01.016

Chevenet F, Brun C, Bañuls A-L et al (2006) TreeDyn: towards dynamic graphics and annotations for analyses of trees. BMC Bioinformatics 7:439. https://doi.org/10.1186/1471-2105-7-439

Coschigano PW, Young LY (1997) Identification and sequence analysis of two regulatory genes involved in anaerobic toluene metabolism by strain T1. Appl Environ Microbiol 63:652–660

Dereeper A, Audic S, Claverie J-M, Blanc G (2010) BLAST-EXPLORER helps you building datasets for phylogenetic analysis. BMC Evol Biol 10:8. https://doi.org/10.1186/1471-2148-10-8

Djordjevic S, Goudreau PN, Xu Q et al (1998) Structural basis for methylesterase CheB regulation by a phosphorylation-activated domain. Proc Natl Acad Sci U S A 95:1381–1386

Domínguez-Cuevas P, González-Pastor J-E, Marqués S et al (2006) Transcriptional tradeoff between metabolic and stress-response programs in *Pseudomonas putida* KT2440 cells exposed to toluene. J Biol Chem 281:11981–11991. https://doi.org/10.1074/jbc.M509848200

Edgar RC (2004) MUSCLE: multiple sequence alignment with high accuracy and high throughput. Nucleic Acids Res 32:1792–1797. https://doi.org/10.1093/nar/gkh340

Felföldi T, Székely AJ, Gorál R et al (2010) Polyphasic bacterial community analysis of an aerobic activated sludge removing phenols and thiocyanate from coke plant effluent. Bioresour Technol 101:3406–3414. https://doi.org/10.1016/j.biortech.2009.12.053

Galperin MY (2005) A census of membrane-bound and intracellular signal transduction proteins in bacteria: bacterial IQ, extroverts and introverts. BMC Microbiol 5:35. https://doi.org/10.1186/1471-2180-5-35

Galperin MY (2010) Diversity of structure and function of response regulator output domains. Curr Opin Microbiol 13(2):150–9. https://doi.org/10.1016/j.mib.2010.01.005

Georgellis D, Kwon O, Lin EC (2001) Quinones as the redox signal for the arc two-component system of bacteria. Science 292:2314–2316. https://doi.org/10.1126/science.1059361

Göttfert M, Grob P, Hennecke H (1990) Proposed regulatory pathway encoded by the nodV and nodW genes, determinants of host specificity in *Bradyrhizobium japonicum*. Proc Natl Acad Sci U S A 87:2680–2684

Grob P, Michel P, Hennecke H, Göttfert M (1993) A novel response-regulator is able to suppress the nodulation defect of a *Bradyrhizobium japonicum* nodW mutant. Mol Gen Genet 241:531–541

Hernández-Sánchez V, Molina L, Ramos J-L, Segura A (2016) New family of biosensors for monitoring BTX in aquatic and edaphic environments. Microb Biotechnol 9(6):858–867. https://doi.org/10.1111/1751-7915.12394

Huang Y, Li L (2014) Biodegradation characteristics of naphthalene and benzene, toluene, ethyl benzene, and xylene (BTEX) by bacteria enriched from activated sludge. Water Environ Res 86:277–284

Huelsenbeck JP, Ronquist F (2001) MRBAYES: Bayesian inference of phylogenetic trees. Bioinformatics 17:754–755

Kane SR, Chakicherla AY, Chain PSG et al (2007) Whole-genome analysis of the methyl tert-butyl ether-degrading beta-proteobacterium *Methylibium petroleiphilum* PM1. J Bacteriol 189:1931–1945. https://doi.org/10.1128/JB.01259-06

Kim SB, Goodfellow M, Kelly J et al (2002) Application of oligonucleotide probes for the detection of *Thiothrix* spp. in activated sludge plants treating paper and board mill wastes. Water Sci Technol 46:559–564

Koh S, Hwang J, Guchhait K et al (2016) Molecular insights into toluene sensing in the TodS/TodT signal transduction system. J Biol Chem 291(16):8575–8590. https://doi.org/10.1074/jbc.M116.718841

Krell T (2008) Microcalorimetry: a response to challenges in modern biotechnology. Microb Biotechnol 1:126–136. https://doi.org/10.1111/j.1751-7915.2007.00013.x

Krell T, Busch A, Lacal J et al (2009) The enigma of cytosolic two-component systems: a hypothesis. Environ Microbiol Rep 1:171–176. https://doi.org/10.1111/j.1758-2229.2009.00020.x

Krell T, Lacal J, Busch A et al (2010) Bacterial sensor kinases: diversity in the recognition of environmental signals. Annu Rev Microbiol 64:539–559. https://doi.org/10.1146/annurev.micro.112408.134054

Kumar A, Trefault N, Olaniran AO (2016) Microbial degradation of 2,4-dichlorophenoxyacetic acid: insight into the enzymes and catabolic genes involved, their regulation and biotechnological implications. Crit Rev Microbiol 42:194–208. https://doi.org/10.3109/1 040841X.2014.917068

Lacal J, Busch A, Guazzaroni M-E et al (2006) The TodS–TodT two-component regulatory system recognizes a wide range of effectors and works with DNA-bending proteins. Proc Natl Acad Sci U S A 103:8191–8196. https://doi.org/10.1073/pnas.0602902103

Lacal J, Guazzaroni M-E, Gutiérrez-del-Arroyo P et al (2008a) Two levels of cooperativeness in the binding of TodT to the tod operon promoter. J Mol Biol 384:1037–1047. https://doi.org/ 10.1016/j.jmb.2008.10.011

Lacal J, Guazzaroni ME, Busch A et al (2008b) Hierarchical binding of the TodT response regulator to its multiple recognition sites at the tod pathway operon promoter. J Mol Biol 376:325–337. https://doi.org/10.1016/j.jmb.2007.12.004

Lau PC, Wang Y, Patel A et al (1997) A bacterial basic region leucine zipper histidine kinase regulating toluene degradation. Proc Natl Acad Sci U S A 94:1453–1458

Leoni L, Ascenzi P, Bocedi A et al (2003) Styrene-catabolism regulation in *Pseudomonas fluorescens* ST: phosphorylation of StyR induces dimerization and cooperative DNA-binding. Biochem Biophys Res Commun 303:926–931

Leoni L, Rampioni G, Di Stefano V, Zennaro E (2005) Dual role of response regulator StyR in styrene catabolism regulation. Appl Environ Microbiol 71:5411–5419. https://doi.org/10.1128/ AEM.71.9.5411-5419.2005

Loh J, Garcia M, Stacey G (1997) NodV and NodW, a second flavonoid recognition system regulating nod gene expression in *Bradyrhizobium japonicum*. J Bacteriol 179:3013–3020

Marconi AM, Beltrametti F, Bestetti G et al (1996) Cloning and characterization of styrene catabolism genes from *Pseudomonas fluorescens* ST. Appl Environ Microbiol 62:121–127

Maris AE, Sawaya MR, Kaczor-Grzeskowiak M et al (2002) Dimerization allows DNA target site recognition by the NarL response regulator. Nat Struct Biol 9:771–778. https://doi.org/10.1038/ nsb845

Mascher T (2006) Intramembrane-sensing histidine kinases: a new family of cell envelope stress sensors in *Firmicutes* bacteria. FEMS Microbiol Lett 264:133–144. https://doi.org/10.1111/ j.1574-6968.2006.00444.x

Milani M, Leoni L, Rampioni G et al (2005) An active-like structure in the unphosphorylated StyR response regulator suggests a phosphorylation-dependent allosteric activation mechanism. Structure 13:1289–1297. https://doi.org/10.1016/j.str.2005.05.014

O'Leary ND, O'Connor KE, Duetz W, Dobson AD (2001) Transcriptional regulation of styrene degradation in *Pseudomonas putida* CA-3. Microbiology 147:973–979. https://doi.org/10.1099/ 00221287-147-4-973

O'Leary ND, Mooney A, O'Mahony M, Dobson AD (2014) Functional characterization of a StyS sensor kinase reveals distinct domains associated with intracellular and extracellular sensing of styrene in *P. putida* CA-3. Bioengineered 5:114–122. https://doi.org/10.4161/bioe.28354

Panke S, Witholt B, Schmid A, Wubbolts MG (1998) Towards a biocatalyst for (S)-styrene oxide production: characterization of the styrene degradation pathway of *Pseudomonas* sp. strain VLB120. Appl Environ Microbiol 64:2032–2043

Parales RE, Parales JV, Pelletier DA, Ditty JL (2008) Diversity of microbial toluene degradation pathways. Adv Appl Microbiol 64:1–73. https://doi.org/10.1016/S0065-2164(08)00401-2. 2 p following 264

Pérez-Martín J, Rojo F, de Lorenzo V (1994) Promoters responsive to DNA bending: a common theme in prokaryotic gene expression. Microbiol Rev 58:268–290

del Peso-Santos T, Shingler V, Perera J (2008) The styrene-responsive StyS/StyR regulation system controls expression of an auxiliary phenylacetyl-coenzyme A ligase: implications for rapid metabolic coupling of the styrene upper- and lower-degradative pathways. Mol Microbiol 69:317–330. https://doi.org/10.1111/j.1365-2958.2008.06259.x

Preu J, Panjikar S, Morth P et al (2012) The sensor region of the ubiquitous cytosolic sensor kinase, PdtaS, contains PAS and GAF domain sensing modules. J Struct Biol 177:498–505. https://doi. org/10.1016/j.jsb.2011.11.012

Ramos J-L, Duque E, Huertas MJ, Haïdour A (1995) Isolation and expansion of the catabolic potential of a *Pseudomonas putida* strain able to grow in the presence of high concentrations of aromatic hydrocarbons. J Bacteriol 177:3911–3916

Ramos-González MI, Olson M, Gatenby AA et al (2002) Cross-regulation between a novel two-component signal transduction system for catabolism of toluene in *Pseudomonas mendocina* and the TodST system from *Pseudomonas putida*. J Bacteriol 184:7062–7067

Rampioni G, Leoni L, Pietrangeli B, Zennaro E (2008) The interplay of StyR and IHF regulates substrate-dependent induction and carbon catabolite repression of styrene catabolism genes in *Pseudomonas fluorescens* ST. BMC Microbiol 8:92. https://doi.org/10.1186/1471-2180-8-92

Salinero KK, Keller K, Feil WS et al (2009) Metabolic analysis of the soil microbe *Dechloromonas aromatica* str. RCB: indications of a surprisingly complex life-style and cryptic anaerobic pathways for aromatic degradation. BMC Genomics 10:351. https://doi.org/10.1186/1471-2164-10-351

Santos PM, Leoni L, Di Bartolo I, Zennaro E (2002) Integration host factor is essential for the optimal expression of the styABCD operon in *Pseudomonas fluorescens* ST. Res Microbiol 153:527–536

Silva-Jímenez H, García-Fontana C, Cadirci BH et al (2012a) Study of the TmoS/TmoT two-component system: towards the functional characterization of the family of TodS/TodT like systems. Microb Biotechnol 5:489–500. https://doi.org/10.1111/j.1751-7915.2011.00322.x

Silva-Jímenez H, Ramos J-L, Krell T (2012b) Construction of a prototype two-component system from the phosphorelay system TodS/TodT. Protein Eng Des Sel 25:159–169. https://doi.org/10.1093/protein/gzs001

Silva-Jímenez H, Ortega A, García-Fontana C et al (2014) Multiple signals modulate the activity of the complex sensor kinase TodS. Microb Biotechnol 8(1):103–115. https://doi.org/10.1111/1751-7915.12142

Stock AM, Robinson VL, Goudreau PN (2000) Two-component signal transduction. Annu Rev Biochem 69:183–215. https://doi.org/10.1146/annurev.biochem.69.1.183

Suwa Y, Yuichi S, Norton JM et al (2011) Genome sequence of *Nitrosomonas* sp. strain AL212, an ammonia-oxidizing bacterium sensitive to high levels of ammonia. J Bacteriol 193:5047–5048. https://doi.org/10.1128/JB.05521-11

Vangnai AS, Kataoka N, Soonglerdsongpha S et al (2012) Construction and application of an *Escherichia coli* bioreporter for aniline and chloroaniline detection. J Ind Microbiol Biotechnol 39:1801–1810. https://doi.org/10.1007/s10295-012-1180-3

Velasco A, Alonso S, García JL et al (1998) Genetic and functional analysis of the styrene catabolic cluster of *Pseudomonas* sp. strain Y2. J Bacteriol 180:1063–1071

Wu M, Wang L, Xu H, Ding Y (2013) Occurrence and removal efficiency of six polycyclic aromatic hydrocarbons in different wastewater treatment plants. Water Sci Technol 68:1844–1851. https://doi.org/10.2166/wst.2013.433

Yoshida T, Qin L, Egger LA, Inouye M (2006) Transcription regulation of ompF and ompC by a single transcription factor, OmpR. J Biol Chem 281:17114–17123. https://doi.org/10.1074/jbc.M602112200

Zhang JH, Xiao G, Gunsalus RP, Hubbell WL (2003) Phosphorylation triggers domain separation in the DNA binding response regulator NarL. Biochemistry 42:2552–2559. https://doi.org/10.1021/bi0272205

Zhou J, Fries MR, Chee-Sanford JC, Tiedje JM (1995) Phylogenetic analyses of a new group of denitrifiers capable of anaerobic growth of toluene and description of *Azoarcus tolulyticus* sp. nov. Int J Syst Bacteriol 45:500–506. https://doi.org/10.1099/00207713-45-3-500

Zylstra GJ, Gibson DT (1989) Toluene degradation by *Pseudomonas putida* F1. Nucleotide sequence of the todC1C2BADE genes and their expression in *Escherichia coli*. J Biol Chem 264:14940–14946

Zylstra GJ, McCombie WR, Gibson DT, Finette BA (1988) Toluene degradation by *Pseudomonas putida* F1: genetic organization of the tod operon. Appl Environ Microbiol 54:1498–1503

# Chemotaxis to Hydrocarbons

# 13

## Rebecca E. Parales and Jayna L. Ditty

## Contents

1  Introduction .................................................................................. 221
2  Chemotaxis Assays .......................................................................... 223
3  Chemotaxis to Aromatic Hydrocarbons .................................................... 226
4  Chemotaxis to Linear Alkanes and Alkenes ............................................... 228
5  Chemotaxis to Nitroaromatic Compounds and Explosives ............................... 229
6  Chemotaxis to Chlorinated Hydrocarbons ................................................ 231
7  Chemotaxis to (Methyl)Phenols .......................................................... 232
8  Repellent Responses ...................................................................... 233
9  Conclusions and Research Needs .......................................................... 233
References ..................................................................................... 234

### Abstract

Chemotaxis is the ability of organisms to move towards or away from chemical gradients in the environment. Hydrocarbon compounds, which are sources of carbon and energy for many bacterial species, have been shown to be chemoattractants for specific organisms. While much is known about catabolic pathways for the degradation of hydrocarbons and related compounds, less is known about the molecular basis for chemotactic responses to these volatile and toxic chemicals.

R. E. Parales (✉)
Department of Microbiology and Molecular Genetics, College of Biological Sciences, University of California, Davis, CA, USA
e-mail: reparales@ucdavis.edu

J. L. Ditty
Department of Biology, College of Arts and Sciences, University of St. Thomas, St. Paul, MN, USA
e-mail: jlditty@stthomas.edu

© Springer International Publishing AG, part of Springer Nature 2018
T. Krell (ed.), *Cellular Ecophysiology of Microbe: Hydrocarbon and Lipid Interactions*,
Handbook of Hydrocarbon and Lipid Microbiology, https://doi.org/10.1007/978-3-319-50542-8_43

# 1    Introduction

Chemotaxis is the active movement of cells along chemical gradients in the environment. This behavioral response allows bacteria to move towards or away from specific chemicals to locate an optimal environment for growth and survival. Chemotaxis has been characterized in detail in the enteric bacterium *Escherichia coli* (Parkinson et al. 2015). The chemotaxis machinery in *E. coli* consists of six cytoplasmic chemotaxis proteins that transmit signals from four membrane-bound methyl-accepting chemotaxis proteins (MCPs) to the flagellar motors (Fig. 1). Each

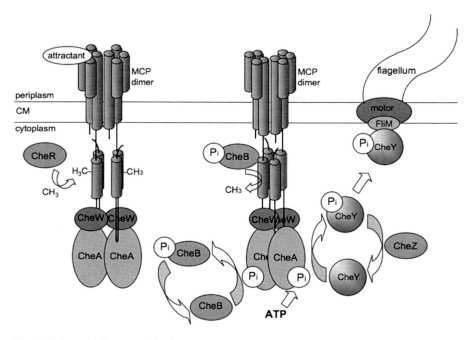

**Fig. 1** Schematic diagram of the chemosensory signaling system of enteric bacteria. MCP dimers with associated CheW and CheA proteins are shown in the presence (*left*) and absence of attractant (*right*). Cells responding to a gradient of attractant will sense the attractant bound to the periplasmic side of the cognate MCP and will continue swimming in the favorable direction due to the inability of CheA to autophosphorylate. In the absence of CheA-P, CheY remains in the inactive unphosphorylated state, and swimming behavior remains unchanged. Cells swimming down a gradient of attractant will sense the decrease in attractant concentration due to decreased occupancy of the MCPs. Under these conditions, the MCPs undergo a conformational change that is transmitted across the cytoplasmic membrane and stimulates CheA kinase activity. CheA-P phosphorylates CheY, which in its phosphorylated state binds to the FliM protein in the flagellar motor and causes a change in the direction of flagellar rotation allowing the cell to randomly reorient and swim off in a new direction. Dephosphorylation of CheY-P is accelerated by the CheZ phosphatase. Under all conditions, the constitutive methyltransferase CheR methylates specific glutamyl residues on the cytoplasmic side of the MCP. Methylated MCPs stimulate CheA autophosphorylation, thus resetting the system such that further increases in attractant concentration can be detected. The methylesterase, CheB, becomes active when it is phosphorylated by CheA-P. CheB-P competes with CheR and removes methyl groups from the MCPs. *CM* cytoplasmic membrane

MCP detects a specific set of chemicals via a periplasmic-sensing domain. Binding of a chemoeffector (or a chemoeffector bound to a periplasmic-binding protein) causes the MCP to undergo a conformational change that is transmitted across the membrane to the cytoplasmic signaling domain. The signal is transmitted to the flagella via CheW and the sensor histidine kinase CheA, which is capable of phosphorylating the response regulator CheY. CheY-P controls swimming behavior by binding to the flagellar switch to reverse the direction of flagellar rotation. This signaling cascade results in directed movement toward or away from the source of the attractant or repellent. Adaptation is mediated by methylation and demethylation of specific glutamyl residues on the MCPs by the methyltransferase CheR and the methylesterase CheB. Additional mechanisms of adaptation have been described for diverse bacteria such as *Bacillus*, *Rhodobacter*, and *Helicobacter* (Roberts et al. 2010). Many chemotactic responses are metabolism-independent, that is, non-metabolizable chemicals can serve as attractants and catabolic mutants remain attracted to the same compounds as wild-type strains. *E. coli* and many other bacteria also exhibit metabolism-dependent energy taxis responses (Alexandre 2010). These responses commonly involve sensing by homologues of the MCP-like protein Aer (Taylor 2007), and signaling via the conserved CheA-CheY phosphorylation cascade (Alexandre 2010; Taylor et al. 2007). Based on studies in a variety of bacteria and the analysis of numerous genome sequences, it appears that the fundamental characteristics of chemotaxis and energy taxis signal transduction systems are conserved among bacteria, although some variations are apparent (Armitage and Schmitt 1997; Roberts et al. 2010; Szurmant and Ordal 2004; Wuichet and Zhulin 2010; Zhulin 2001). However, many nonenteric bacteria, particularly soil bacteria, have many more chemoreceptors than *E. coli* (Krell et al. 2011; Matilla and Krell 2017).

Many hydrocarbon-degrading bacteria have sensory systems that allow cells to detect and respond behaviorally to hydrocarbons and various chemical derivatives. Several reports have suggested that chemotaxis may play a role in biodegradation by bringing cells into contact with the chemicals being degraded (Gkorezis et al. 2016; Hazen 1994; Hazen and Lopez-de-Victoria 1994; Krell et al. 2013; Lacal et al. 2013; Pandey and Jain 2002; Parales and Harwood 2002; Parales et al. 2008; Pieper et al. 1996). However, bacteria capable of degrading hydrocarbons must maintain a balance between acquiring sufficient hydrocarbon for growth and avoiding toxic concentrations of the chemicals (Hanzel et al. 2012). Chemotaxis can result in increased bioavailability of hydrocarbons (Krell et al. 2013; Pandey et al. 2009) and may facilitate the transmission of catabolic plasmids in the environment (Harwood and Ornston 1984). This chapter describes our current understanding of both positive and negative chemotactic responses to hydrocarbons and related compounds, as well as responses to metabolizable compounds via energy taxis.

## 2    Chemotaxis Assays

Bacteria sense and respond to chemical gradients and the result of this behavior is the accumulation of cells near the source of an attractant. Several qualitative and quantitative assays to measure bacterial chemotaxis are based on this fundamental characteristic of the chemotactic response. Some assays are particularly appropriate for the analysis of hydrocarbon taxis while others are more difficult to use with volatile chemicals.

Chemotaxis to metabolizable compounds present in soft agar growth media can be visualized by using the soft agar swim plate assay (Adler 1973; Harwood et al. 1994). Cells are inoculated at the center of a Petri dish containing growth medium solidified with a low concentration of agar (typically 0.3%), and the bacteria generate a chemical concentration gradient as they degrade the attractant, which is a carbon and energy source for the organism. Chemotaxis is visualized as a ring of growth that moves toward the edge of the plate as cells swim through the agar following the self-generated gradient of attractant (Fig. 2a). Only metabolizable compounds can be tested as chemoattractants in this assay, and problems can arise when using volatile attractants because the chemicals can evaporate and redissolve in the medium, interfering with gradient formation.

The gradient plate assay is a variation of the soft agar swim plate assay in which an agar plug serves as the source of the attractant (Pham and Parkinson 2011). The cells grow on an alternative carbon source present throughout the soft agar as they respond to the attractant diffusing from an agar plug (Fig. 2b). In this case, metabolism is not required to generate the gradient or for growth, so the response to nonmetabolizable attractants can be evaluated.

Chemotaxis can also be visualized qualitatively with the agarose plug assay (Yu and Alam 1997). In this assay, a drop of melted agarose containing the chemical of interest is positioned between a microscope slide and a coverslip, and a suspension of motile cells surrounds the plug. The chemotactic response appears as cells accumulate in a ring around the agarose plug, typically within a few minutes (Fig. 2c). This assay is useful for testing responses to volatile compounds such as hydrocarbons because the response is rapid and the chamber is almost completely closed, thus limiting volatilization (Parales et al. 2000). The drop assay (Fahrner et al. 1994; Grimm and Harwood 1997) has some similarities to the swim plate and the agarose plug assays. In this case, cells are suspended in a viscous solution in a small Petri dish and the attractant is dropped into the center of the cell suspension. A gradient forms by diffusion and a ring of cells forms around the attractant within a short period of time (~15 min). Growth is not required for either the agarose plug or drop assay.

Chemotactic behavior can be measured either quantitatively or qualitatively using the capillary assay (Adler 1973; Grimm and Harwood 1997). In this assay, cells respond to a gradient of attractant diffusing out of a microcapillary tube into a suspension of motile bacteria. Chemotactic cells respond by swimming up the gradient and into the tube; the tube is then removed and the number of cells within is enumerated. In the qualitative capillary assay, a solidifying agent such as agarose is included in the capillary, preventing the cells from entering. This results in the accumulation of a cloud of chemotactic cells near the mouth of the capillary, which can be observed under low magnification (Fig. 2d). The qualitative capillary assay works well with both poorly soluble and volatile compounds.

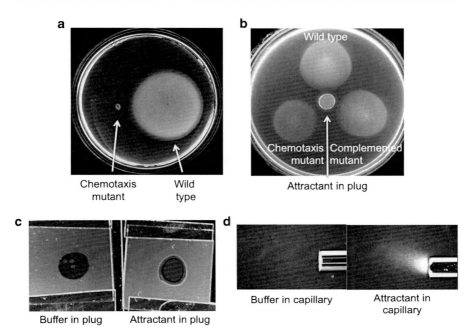

**Fig. 2** Examples of chemotaxis assay results. (**a**) Swim plate assay. Wild type (*right*) and a generally nonchemotactic mutant (*left*) were stabbed into semi-solid agar (0.3%) medium containing an attractant. This assay requires growth on the attractant(s) to generate the gradient. The chemotaxis mutant (*left*) grows at the point of inoculation but does not form a ring of growth that moves out from the point of inoculation. The wild-type strain (*right*) senses the gradient and follows it, forming a large colony. (**b**) Gradient plate assay. Wild type (*top*) and a chemoreceptor mutant (*lower left*) and a complemented mutant (*lower right*) were stabbed into semi-solid agar (0.3%) medium containing glycerol as the carbon source. The attractant diffuses from the central agar plug placed on the surface of the plate and therefore the assay does not require growth on the attractant(s) to generate the gradient. The wild type and complemented mutant bias growth and movement toward the source of the attractant, forming oblong colonies. The mutant that is unable to sense the attractant forms a uniformly round colony. (**c**) Agarose plug assay. A solution of low melting temperature agarose containing the attractant is allowed to solidify between a slide and cover slip and motile bacterial cells suspended in buffer are introduced into the chamber. The attractant diffuses into the suspension and the cells respond to the gradient of attractant forming a ring of cells around the plug. Growth and metabolism are not required and a response is typically seen within 5 min. Left, buffer control; right, plug contains attractant. The response is seen as a white ring of cells accumulating at the optimum concentration of the attractant. (**d**) Modified (qualitative) capillary assay. In this assay, attractant in crystal form or suspended in agarose diffuses from a 1 μl capillary into a motile cell suspension. The cells sense the gradient of attractant and accumulate at the tip of the capillary. The response typically takes place within 5–30 min, and does not require attractant metabolism or cell growth. The response can be monitored under 20–40 X magnification by dark field microscopy. Left, capillary contains buffer in agarose; right, capillary contains an attractant in crystal form. The cells accumulating at the tip of the capillary in response to the attractant appear as a white cloud

Temporal assays are used to monitor the behavior of cells in suspension in response to the addition of attractant. Quantitative assays can be carried out using computer-assisted motion analysis (Harwood et al. 1989, 1990), which requires dedicated software. Manual temporal assays that monitor adaptation of the bacteria to the attractants either directly or following videotaping of cells can also provide quantitative data (Parales 2004; Shioi et al. 1987).

# 3    Chemotaxis to Aromatic Hydrocarbons

The toluene-degrading strains *P. putida* F1, *P. putida* DOT-T1E, *Ralstonia pickettii* PKO1, and *Burkholderia vietnamiensis* (formerly *cepacia*) G4 showed toluene-inducible chemotactic responses to toluene (Lacal et al. 2011; Parales et al. 2000). Benzene and ethylbenzene were also good chemoattractants for *P. putida* F1, which can utilize both substrates as sole sources of carbon and energy. *P. putida* F1 was also attracted to aromatic hydrocarbons that do not serve as growth substrates, including isopropylbenzene and naphthalene (Parales et al. 2000). Mutants of *P. putida* F1 that were unable to degrade toluene remained chemotactic to toluene, indicating that toluene was directly detected as the attractant. Toluene catabolism and chemotaxis are under the control of the same two-component regulatory system encoded by *todST* in *P. putida* F1. Strains with mutations in either *todS* or *todT* were unable to induce the *tod* catabolic operon (Lau et al. 1997) and did not respond to toluene in chemotaxis assays (Parales et al. 2000). These results indicate that toluene chemotaxis and catabolism are genetically linked in *P. putida* F1, although the toluene chemoreceptor in *P. putida* F1 has not yet been identified. However, two nearly identical copies of a gene (*mcpT*) encoding a methyl-accepting chemotaxis protein responsible for a strong chemotactic response to toluene, dubbed "hyper-chemotaxis," were identified on the pGRT1 megaplasmid in the solvent tolerant strain *P. putida* DOT-T1E (Lacal et al. 2011). Inactivation of either copy eliminated the strong chemotactic response to toluene but did not eliminate the response completely. In addition, introduction of *mcpT* cloned on a multicopy vector into *P. putida* KT2440, a strain normally incapable of toluene chemotaxis, resulted in a strong positive taxis response to toluene. McpT was also shown to be responsible for a strong chemotactic response to crude oil (Krell et al. 2012).

Multiple studies have utilized microfluidic devices to study toluene chemotaxis in a model system that better represents bacterial movement through natural aquifer environments. Studies included the use of pore-scale microfluidic chambers to model the dissolution of toluene and microorganisms with pore sizes equivalent to those in sandy soil, the use of a convection-free microfluidic device to investigate chemotactic sensitivity coefficients and chemotactic receptor constants to toluene, and the use of a device to generate a convection-free gradient. In all cases, wild-type *P. putida* F1 was found to accumulate at a greater concentration at the toluene/aqueous interface relative to a nonchemotactic *cheA* mutant (Wang et al. 2012, 2015, 2016).

*P. putida* G7, *Pseudomonas* sp. strain NCIB 9816–4, *P. putida* RKJ1, and *Ralstonia* sp. strain U2 utilize naphthalene as a sole source of carbon and energy, and all four strains have been reported to be chemotactic to naphthalene (Grimm and Harwood 1997; Samanta and Jain 2000; Wood et al. 2006). The response to naphthalene by these *Pseudomonas* strains requires the presence of the resident naphthalene catabolic plasmid (Grimm and Harwood 1997; Samanta and Jain 2000). Naphthalene chemotaxis in strains G7 and NCIB 9816–4 is induced during growth with naphthalene (Grimm and Harwood 1997). NahY, the chemoreceptor for naphthalene in *P. putida* G7, is a MCP that is encoded downstream of the naphthalene catabolic genes on the NAH7 catabolic plasmid. Inactivation of *nahY* resulted in the loss of the chemotactic response to naphthalene (Grimm and Harwood 1999). The chemotactic response of *P. putida* G7 to naphthalene was quantified (Marx and Aitken 1999), and mathematical models describing chemotaxis to naphthalene were developed based on this data (Marx and Aitken 1999; Marx and Aitken 2000b). However, in contrast to aqueous phase naphthalene, the vapor phase form appeared to be a chemorepellant rather than an attractant, which may influence the biodegradation efficiency of chemotactic bacteria responding to volatile organic compounds in environments where the bacteria are at air-water interfaces (Hanzel et al. 2010).

Experiments using wild-type *P. putida* G7, a nonmotile mutant, and a nonchemotactic variant lacking *nahY* demonstrated that chemotaxis enhances biodegradation in a heterogeneous system (Marx and Aitken 2000a). The same set of strains was also used to show that chemotactic bacteria were more efficient at degradation of naphthalene dissolved in a nonaqueous-phase liquid (Law and Aitken 2003). An additional study demonstrated bacterial chemotaxis in water-saturated porous media by evaluating chemotaxis of *P. putida* G7 to naphthalene in an environment of packed glass beads (Pedit et al. 2002). *P. putida* G7 was also used as the model microorganism to investigate residence time in a continuous-flow sand-packed column with a uniform distribution of naphthalene. Wild-type G7 demonstrated a threefold increase in column dispersion, and percent recovery was significantly decreased relative to a chemotaxis mutant when transport was not influenced by the presence of naphthalene (Adadevoh et al. 2016). A recent study also demonstrated that the motile microorganism itself (in this case the eukaryotic ciliate *Tetrahymena pyriformis*) could serve as a vehicle for mass transfer of polycyclic aromatic hydrocarbons, where a 100-fold increase in benzopyrene distribution was seen (Gilbert et al. 2014). These studies provide evidence that motile and chemotactic microorganisms may be more effective for bioremediation applications, especially at sites where contaminants are unevenly distributed or adsorbed to soil particles.

Chemotactic responses to naphthalene and larger polycyclic aromatic hydrocarbons (PAHs) were investigated with three *Pseudomonas* isolates obtained from coal-tar-contaminated sites (Ortega-Calvo et al. 2003). All three strains grew on naphthalene; one (*P. putida* 10D) also grew on phenanthrene and pyrene. All were chemotactic to naphthalene; however, *P. putida* 10D was also chemotactic to phenanthrene but not to pyrene. This strain also showed a repellent response to anthracene, a three-ring PAH that did not serve as a growth substrate.

Chemotaxis to biphenyl has been reported for biphenyl-degrading strains such as *Pseudomonas* sp. strain B4 (Chávez et al. 2006; Gordillo et al. 2007). This strain was chemotactic to biphenyl and monochlorobiphenyls, which serve as growth substrates (Gordillo et al. 2007). The response to biphenyl and chlorobiphenyls did not require induction. *Pseudomonas* sp. strain B4 also responded to benzoate, a growth substrate and intermediate in biphenyl degradation, but not to 2- and 3-chlorobenzoates, which accumulate following growth on 2- and 3-chlorobiphenyls. 3-Chlorobenzoate was toxic to strain B4 and actually appeared to result in a slight repellent response. Benzoate- or biphenyl-grown cells responded weakly to 4-chlorobenzoate, in contrast to glucose-grown cells, which did not respond. Similarly, *Pseudomonas pseudoalcaligenes* KF707, a well-studied biphenyl and polychlorinated biphenyl (PCB) degrader, also showed a constitutive chemotactic response to biphenyl. However, chemotaxis to the biphenyl/PCB degradation intermediates benzoate, and 2- and 3-chlorobenzoate was induced when KF707 cells were grown on either benzoate or biphenyl (Tremaroli et al. 2010). Chemotaxis to biphenyl was also reported for two other biphenyl-degrading strains, *P. putida* P106 and *Rhodococcus erythropolis* NY05 (Wu et al. 2003). Biphenyl is also an attractant for naphthalene-grown *P. putida* G7, although biphenyl is not a growth substrate for this strain. It is likely that biphenyl is sensed by the NAH7 plasmid-encoded naphthalene chemoreceptor NahY in *P. putida* G7 (Grimm and Harwood 1997, 1999).

## 4    Chemotaxis to Linear Alkanes and Alkenes

A metagenomic and metatranscriptomic analysis following the Deepwater Horizon spill in the Gulf of Mexico identified *Oceanospirillales* as dominant members of the microbial community and that genes for alkane metabolism were highly expressed in contaminated sediments (Mason et al. 2012). Synchrotron radiation-based Fourier transform infrared spectromicroscopy suggesting that oil droplets were surrounded by bacterial cells, and the expression of alkane degradation, motility, and chemotaxis genes by *Oceanospirillales* cells were interpreted as evidence for bacterial chemotaxis to hydrocarbons in the plume.

Direct evidence of chemotaxis to hexadecane and hydrocarbons in gas oil (a complex mixture of linear and aromatic hydrocarbons) was reported for a *Flavimonas oryzihabitans* isolate that was obtained from gas oil-contaminated soil (Lanfranconi et al. 2003). The isolate grew on tetradecane, pentadecane, hexadecane, and 2,6,10,14-tetramethyl pentadecane as sole carbon sources. Chemotactic responses to gas oil and hexadecane were demonstrated using qualitative capillary assays and agarose plug assays, but the chemoreceptor has not been identified. A similar response to *n*-hexadecane by *Pseudomonas synxantha* LSH-7′ was reported (Meng et al. 2017). Chemotaxis to hexadecane may be common for hydrocarbon-degrading bacteria, as unpublished data reported by Smits et al. indicated that *P. aeruginosa* PAO1 was also capable of such a response (Smits et al. 2003). In this study a gene designated *tlpS*, which is located downstream of the alkane hydroxylase gene *alkB1* on the *P. aeruginosa* PAO1 genome, was predicted to encode a MCP that may play a role in alkane chemotaxis (Smits et al. 2003). Similarly, the *alkN* gene appears to encode a MCP that could be involved in alkane chemotaxis in *P. putida* GPo1. The *alkN* gene is located in a cluster of genes for alkane degradation on the OCT plasmid (van Beilen et al. 2001). Unfortunately, *P. putida* GPo1 was not motile enough for chemotaxis assays (van Beilen et al. 2001).

In a more recent study, *Pseudomonas* sp. strain H, which was isolated from soil contaminated with chlorophenols, was shown to be chemotactic towards *n*-hexadecane, 1-dodecene, 1-undecene, and kerosene, which are all capable of supporting growth of the bacterium (Nisenbaum et al. 2013). The responses to these compounds were constitutive, but the receptor(s) have not yet been identified.

Five bacterial isolates with antifungal activity (*P. aeruginosa, Paenibacillus jamilae, Brevibacillus brevis, Bacillus sonorensis, Providencia rettgeri*) showed positive chemotactic responses to long chain alkanes (C12–C28) present in banana root extracts (Li et al. 2012). These potentially useful biocontrol strains, which were obtained from a variety of agricultural soil samples, were capable of protecting banana plants from *Fusarium* wilt. Their ability to sense and respond to long chain alkanes present in banana root exudates may provide a colonization advantage in the rhizosphere over nonchemotactic strains. Whether the bacteria are capable of using the alkanes as growth substrates was not tested, and the receptor(s) mediating the responses were not identified.

## 5      Chemotaxis to Nitroaromatic Compounds and Explosives

Nitroaromatic compounds are used as solvents and in the production of pesticides, herbicides, dyes, explosives, and polymers (Spain et al. 2000). *Ralstonia* sp. SJ98 was isolated from pesticide-contaminated agricultural soil for the ability to grow on 4-nitrophenol (Samanta et al. 2000). The strain also grew on 4-nitrocatechol, 3-methyl-4-nitrophenol, and 2- and 4-nitrobenzoate, and was chemotactic to these nitroaromatic growth substrates. It was also chemotactic to other structurally similar nitroaromatic compounds that did not serve as growth substrates but were transformed by strain SJ98 in the presence of an alternative substrate (Bhushan et al. 2000; Pandey et al. 2002; Samanta et al. 2000). In bench-scale experiments, *Ralstonia* sp. SJ98 was capable of generating and following a gradient of 4-nitrophenol in soil and degrading it in the process, suggesting that this strain could be useful for bioremediation of nitroaromatic contaminants in the field (Paul et al. 2006). Strain SJ98 was also found to respond to chloronitrobenzoates and chloronitrophenols that it is capable of completely degrading or cometabolizing, and the response was inducible (Pandey et al. 2012). The observation that structurally similar nonmetabolizable (chloro)nitroaromatic compounds do not serve as attractants for strain SJ98 pointed toward the possibility that the sensing mechanism involved energy taxis, but clear responses to compounds that are transformed but not completely degraded argues against this mechanism. Interestingly, the response to chloronitroaromatic compounds was inhibited in the presence of nitroaromatic attractants, suggesting that the same receptor is used for sensing all (chloro) nitroaromatic attractants (Pandey et al. 2012). Another isolate, *Pseudomonas* sp. strain JHN, which is capable of degradation of 4-chloro-3-nitrophenol and partial metabolism of 4-chloro-2-nitrophenol, was shown to be chemotactic to both compounds (Arora and Bae 2014; Arora et al. 2014).

In contrast to the metabolism-dependent responses of *Ralstonia* sp. strain SJ98 and *Pseudomonas* sp. strain JHN to nitroaromatic compounds, chemotaxis toward *p*-nitrophenol by *Pseudomonas* sp. strain WBC-3 was metabolism-independent and constitutive. The strain also responded to the nitroaromatic compounds nitrobenzene, nitrophenol, 2,6-dinitrophenol, as well as pentachlorophenol and the common intermediates in aerobic aromatic compound catabolism catechol, salicylate, 4-hydroxybenzoate, gentisate, and protocatechuate (Zhang et al. 2008).

The 4-nitrotoluene (4NT) degrading strains *P. putida* TW3 (Rhys-Williams et al. 1993) and *Pseudomonas* sp. strain 4NT (Haigler and Spain 1993) have similar chemotaxis systems for the detection of the 4NT degradation intermediate 4-nitrobenzoate (Parales 2004). The chemotactic response was induced by the catabolic intermediate β-ketoadipate (Parales 2004), and based on this pattern of induction, it is likely that a PcaY ortholog is responsible for the detection of 4-nitrobenzoate in these strains. PcaY was identified as a methyl-accepting chemotaxis protein from *Pseudomonas putida* F1 that mediates the response to a wide range of substituted benzoates, including nitrobenzoates; aminobenzoates; hydroxylated, methylated, and chlorinated benzoates; as well as vanillate, protocatechuate, and the hydroaromatic compounds quinate and shikimate (Luu et al. 2015). In a separate study, the methyl-accepting chemotaxis protein receptor NbaY was shown to be required for chemotaxis to 2-nitrobenzoate in the 2-nitrobenzoate-degrading strain *Pseudomonas fluorescens* KU-7 (Iwaki et al. 2007). There is, however, little sequence similarity between the periplasmic ligand binding domains of PcaY and NbaY.

The response of *Acidovorax* sp. strain JS42 to 2-nitrotoluene was found to result from a combination of metabolism-independent chemotaxis directly to 2-nitrotoluene, metabolism-dependent chemotaxis to the nitrite released during growth on 2-nitrotoluene and energy taxis in response to the catabolism of 2-nitrotoluene (Rabinovitch-Deere and Parales 2012). Chemotaxis to mononitrotoluenes, dinitrotoluenes, and 2,4,6-trinitrotoluene (TNT) by the 2,4-dinitrotoluene degrading strains *B. cepacia* R34 and *Burkholderia* sp. strain DNT was reported (Leungsakul et al. 2005). However, the response to nitroaromatics took up to 3 days to visualize, whereas most chemotactic responses are observable in well under an hour with the assay used. No nitroarene chemoreceptors were identified.

*Clostridium* sp. strain EDB2 is an obligate anaerobe that is chemotactic to the cyclic nitramine explosives hexahydro-1,3,5-trinitro-1,3,5-triazine (RDX), octahydro-1,3,5,7-tetranitro-1,3,5,7-tetrazocine (HMX), and 2,4,6,8,10,12-hexanitro-2,4,6,8,10,12-hexaazaisowurtzitane (CL-20) (Bhushan et al. 2004). The strain transformed RDX, HMX, and CL-20 to nitrite, nitrous oxide, formaldehyde, and formate. Chemotactic responses to RDX, HMX, and CL-20 appeared to be due to the detection of released nitrite rather than the explosives themselves. Previous studies demonstrated nitrate and/or nitrite chemotaxis by bacteria capable of anaerobic respiration (Lee et al. 2002; Taylor et al. 1979). In most cases the response appears to be due to energy taxis (Taylor and Zhulin 1999; Taylor et al. 1999) and not the specific interaction of a chemoreceptor with nitrite. *Clostridium* sp. are generally not capable of anaerobic respiration, and therefore it is not expected that strain EDB2 can use nitrite as an electron acceptor. At this time the receptor for the detection of nitrite released from cyclic nitramine explosives is not known.

## 6    Chemotaxis to Chlorinated Hydrocarbons

As mentioned above, the biphenyl-degrading strain *Pseudomonas* sp. strain B4 is chemotactic to 2-, 3-, and 4-chlorobiphenyl and also to 2,3-dichlorobiphenyl (Gordillo et al. 2007), but chemotaxis to more highly chlorinated polychlorinated biphenyls (PCBs) has not been reported. Chlorobenzoates are intermediates in the degradation of PCBs and other chlorinated aromatic compounds. Although *P. putida* PRS2000 cannot grow on chlorobenzoates, benzoate- or 4-hydroxybenzoate-grown cells were attracted to 3- and 4-chlorobenzoate (Harwood 1989; Harwood et al. 1990). Similarly, *P. putida* F1 responded to 3- and 4-chlorobenzoate and the receptor was identified as the methyl-accepting chemotaxis protein PcaY (Luu et al. 2015).

Interestingly, in *Pseudomonas aeruginosa* PAO1 the inorganic phosphate receptor CtpL was found to be responsible for positive chemotaxis to the toxic pollutant chloroaniline (Vangnai et al. 2013). CtpL also mediated responses to 3-chloroaniline, 3,4-dichloroaniline, benzoate, anthranilate, 4-aminobenzoate, 4-chloronitrobenzene, catechol, and 4-chlorocatechol. Of these attractants, only benzoate, catechol, and anthranilate serve as carbon and energy sources for strain PAO1 (Vangnai et al. 2013).

Chlorinated alkenes such as trichloroethylene (TCE), dichloroethylene (DCE), and perchloroethylene (PCE) are used as solvents, degreasing agents, and cleaning agents in dry cleaning, and they are common groundwater pollutants. Toluene-grown *P. putida* F1 is attracted to TCE, *cis*-1,2-DCE, and PCE (Parales et al. 2000). *P. putida* F1 is unable to grow on these compounds, but toluene dioxygenase, the enzyme that catalyzes the first step in toluene degradation, is capable of oxidizing and detoxifying TCE (Li and Wackett 1992; Wackett and Gibson 1988). Analysis of the chemotactic response of *P. putida* F1 to TCE in a packed column demonstrated bacterial chemotaxis in porous media and provided evidence that chemotaxis is relevant in a variety of environmental conditions and does not just occur in the aqueous phase (Olson et al. 2004). However, use of live-dead staining during chemotaxis plug assays revealed that exposure to high TCE, and to a lesser extent, high toluene concentrations were toxic to *P. putida* F1 cells and resulted in significant cell death over the course of a 20-min assay (Singh and Olson 2010). These findings highlight the importance of considering the toxicity of a pollutant to a bacterial strain capable of its degradation in addition to the ability of the cells to sense and respond to the toxic chemical. *P. putida* F1 was also used to detect benzoate and acetate in a bench-scale aquifer model where sand and aqueous media were used to simulate ground water flow. The center of the wild-type F1 population plume was shown to accumulate approximately 0.74 cm and 0.4 cm closer to the benzoate and acetate attractants, respectively, relative to a non-chemotactic *P. putida* F1 *cheA* mutant. In addition, transverse dispersivity (dispersion perpendicular to the flow) of the wild-type strain was, on average, higher for both the benzoate and acetate attractants relative to the mutant strain. The conclusions from this study indicate that chemotaxis can play a role in the overall improvement of bioremediation in aquifers (Strobel et al. 2011).

Similarly, *o*-xylene-grown *P. stutzeri* OX1 was shown to respond to TCE, DCE, and PCE, as well as *trans*-1,2-DCE, 1,1-DCE, and vinyl chloride. Toluene-grown *P. putida* F1 and *B. vietnamiensis* G4 were also attracted to additional chlorinated ethenes (Vardar et al. 2005). Chlorinated alkenes do not serve as carbon and energy sources for these bacteria, but toluene *o*-monooxygenase from *B. vietnamiensis* G4 is capable of oxidizing TCE and DCE isomers (Shields and Francesconi 1996), and the corresponding enzyme from *P. stutzeri* OX1 oxidizes TCE, PCE, and DCE isomers, as well as chloroform (Chauhan et al. 1998; Ryoo et al. 2000). The genes encoding the *P. putida* F1 toluene dioxygenase and *B. vietnamiensis* G4 toluene mono-oxygenase are induced by TCE (Leahy et al. 1996). Together, these data suggest that if an appropriate carbon and energy source were available, such chemotactic strains could follow and detoxify a moving plume of chlorinated alkene-contaminated groundwater. In contrast to the positive chemotactic responses of these aerobic bacterial strains to TCE, motile anaerobic dechlorinating *Geobacter* in the commercially available mixed culture KB-1™ showed no sign of chemotaxis to TCE (Philips et al. 2014).

*Pseudomonas stutzeri* KC transforms and detoxifies carbon tetrachloride (CT) under anoxic conditions using nitrate as a terminal electron acceptor (Criddle et al. 1990; Dybas et al. 1995; Lewis et al. 2001). Motility and chemotaxis of *P. stutzeri* KC to nitrate were shown to enhance bioremediation of CT by allowing cells to follow the self-generated nitrate gradient in laboratory-scale groundwater-saturated aquifer columns containing CT and nitrate (Witt et al. 1999). The authors concluded that the cells' ability to move toward higher concentrations of nitrate increased CT degradation because the organism requires denitrifying conditions in order to transform CT. The results of this study and a report demonstrating nitrate and acetate chemotaxis by *P. stutzeri* KC in porous media support the proposal that chemotaxis enhances biodegradation in the environment (Roush et al. 2006).

# 7    Chemotaxis to (Methyl)Phenols

*P. putida* harboring the (methyl)phenol (*dmp*) degradation pathway encoded on pVI150 exhibited metabolism-dependent taxis to phenolic compounds (Sarand et al. 2008). The response was mediated by the energy taxis transducer Aer2, a membrane-anchored MCP-like protein that carries a PAS domain. Inactivation of *aer2* eliminated both aerotaxis and metabolism-dependent taxis to metabolizable phenols, succinate, and glucose. This was the first report to demonstrate direct involvement of an energy taxis receptor in a tactic response to aromatic compounds.

In contrast, phenol, catechol, and a series of aromatic acids that serve as carbon and energy sources as well as attractants for *Comamonas testosteroni* CNB-1 are sensed by a different form of metabolism-dependent chemotaxis. In this case, rather than binding these chemicals directly, MCPs bind TCA cycle intermediates that are generated during catabolism of the substrates (Huang et al. 2016; Ni et al. 2013; Ni et al. 2015).

Although it is unable to utilize phenol as a carbon source, *E. coli* has long been known to sense phenol as both an attractant and a repellent via its canonical MCPs Tar (senses phenol as an attractant), Tsr, Trg, and Tap (senses phenol as a repellent) (Imae et al. 1987; Yamamoto et al. 1990). An alternative mechanism of sensing phenol by Tsr and Tar that does not involve binding to the periplasmic ligand-binding domain of an MCP was also reported (Pham and Parkinson 2011). The analysis of a series of hybrid and mutant receptors suggested that phenol entry into the cytoplasmic membrane perturbs the transmembrane helices of the MCPs, and the signal is then transduced to the flagella via the standard chemotaxis signaling pathway.

## 8    Repellent Responses

Hydrophobic hydrocarbons and related chemicals can accumulate in cellular membranes, causing a loss of membrane integrity and dissipation of membrane potential (Sikkema et al. 1995). Negative chemotactic responses (repellent responses) allow motile bacteria to avoid environments with toxic concentrations of these chemicals. In fact, it appears that some bacteria may be capable of exhibiting both positive and negative responses to potentially toxic chemicals depending on the concentration. For example, although *P. putida* F1 exhibits an inducible positive response to toluene, a constitutive repellent response to high concentrations of toluene was observed (Parales et al. 2000).

Negative chemotaxis has been reported for several marine pseudomonads in response to chloroform, toluene, and benzene (Young and Mitchell 1973). In addition, *P. aeruginosa* displayed repellent responses to TCE, PCE, 1,1,1-trichloroethane (TCA), chloroform, and dichloromethane (Shitashiro et al. 2003). The repellent response to TCE and chloroform required the Cluster I chemotaxis genes *cheYZABW* as well as *cheR*, and the three MCP-encoding genes *pctA*, *B*, and *C* (Shitashiro et al. 2005). PctA, PctB, and PctC also mediate positive responses to amino acids (Taguchi et al. 1997).

## 9    Conclusions and Research Needs

Many bacteria have chemotactic responses that allow them to detect and respond behaviorally to hydrocarbons. As a number of studies have demonstrated, chemotaxis may help overcome mass transfer limitations by bringing the biodegradative organisms to sorbed hydrocarbons. The ability of specific bacteria to actively sense and respond to hydrocarbon substrates provides a competitive advantage when concentrations of these compounds are limiting due to low bioavailability. In addition, the ability of an organism to sense and follow a gradient of a particular chemical or class of chemicals that it is *unable* to degrade could bring it into close contact with other organisms carrying relevant transmissible catabolic plasmids for the degradation of these chemicals. In this way, chemotaxis could stimulate the dissemination of catabolic plasmids among environmental bacteria, and hence could enhance the biodegradative capacity of the population.

Although many bacteria that utilize hydrocarbons are also capable of detecting these chemicals, relatively few receptors have been identified to date. Of those that have been identified, some are encoded on catabolic plasmids and others on the chromosome. In addition, receptor genes are frequently located near genes that code for the degradation of the molecule of interest, which is different from *E. coli*, where the receptor genes are located in or near operons devoted to chemotaxis and motility functions. The genetic context of receptor genes on catabolic plasmids or in operons with the catabolic genes suggests that chemotaxis may play an important role in biodegradation. In many cases, chemotactic responses to hydrocarbons are inducible, and in some cases the response is metabolism dependent and involves the participation of an energy taxis receptor. These findings indicate that chemotaxis to hydrocarbons can be linked genetically, physiologically, and/or bioenergetically to metabolism. Regardless of the mechanism, either through sensing the cellular energy state, a catabolic intermediate, or the presence of a hydrocarbon molecule itself in the environment, the end result is useful to the bacteria by bringing them into contact with useful sources of carbon and energy.

Because of the inherent toxicity of hydrocarbons, a relatively narrow range of hydrocarbon concentrations is tolerated by most microorganisms (Sikkema et al. 1995). Because hydrocarbons are known to alter membrane structure and dissipate the pH gradient across the membrane, cells must carefully control the level of hydrocarbons to which they are exposed. Therefore, it seems plausible that some bacteria may have evolved both attractant and repellent responses to hydrocarbons in order to access optimal concentrations for growth and limit exposure to damaging concentrations.

While many hydrocarbon substrates have been identified as attractants for a number of bacterial species, additional work is needed to elucidate the molecular basis of hydrocarbon sensing.

**Acknowledgments** Chemotaxis research in the authors' laboratories has been supported by the National Science Foundation (award MCB-0919930 to REP and JLD) and the University of California Davis Committee on Research New Funding Initiative (to REP).

# References

Adadevoh JS, Triolo S, Ramsburg CA, Ford RM (2016) Chemotaxis increases the residence time of bacteria in granular media containing distributed contaminant sources. Environ Sci Technol 50:181–187

Adler J (1973) A method for measuring chemotaxis and use of the method to determine optimum conditions for chemotaxis by *Escherichia coli*. J Gen Microbiol 74:77–91

Alexandre G (2010) Coupling metabolism and chemotaxis-dependent behaviours by energy taxis receptors. Microbiology 156:2283–2293

Armitage JP, Schmitt R (1997) Bacterial chemotaxis: *Rhodobacter sphaeroides* and *Sinorhizobium meliloti* – variations on a theme? Microbiology 143:3671–3682

Arora PK, Bae H (2014) Biotransformation and chemotaxis of 4-chloro-2-nitrophenol by *Pseudomonas* sp. JHN Microb Cell Fact 13:110

Arora PK, Srivastava A, Singh VP (2014) Degradation of 4-chloro-3-nitrophenol via a novel intermediate, 4-chlororesorcinol by *Pseudomonas* sp. JHN Sci Rep 4:4475

Bhushan B, Halasz A, Thiboutot S, Ampleman G, Hawari J (2004) Chemotaxis-mediated biodegradation of cyclic nitramine explosives RDX, HMX, and CL-20 by *Clostridium* sp. EDB2. Biochem Biophys Res Commun 316:816–821

Bhushan B, Samanta SK, Chauhan A, Chakraborti AK, Jain RK (2000) Chemotaxis and biodegradation of 3-methyl-4-nitrophenol by *Ralstonia* sp. SJ98. Biochem Biophys Res Commun 275:129–133

Chauhan S, Barbieri P, Wood TK (1998) Oxidation of trichloroethylene, 1,1-dichloroethylene, and chloroform by toluene/*o*-xylene monooxygenase from *Pseudomonas stutzeri* OX1. Appl Environ Microbiol 64:3023–3024

Chávez FP, Gordillo F, Jerez CA (2006) Adaptive responses and cellular behaviour of biphenyl-degrading bacteria toward polychlorinated biphenyls. Biotechnol Adv 24:309–320

Criddle CS, DeWitt JT, Grbic-Galic D, McCarty PL (1990) Transformation of carbon tetrachloride by *Pseudomonas* sp. strain KC under denitrification conditions. Appl Environ Microbiol 56:3240–3246

Dybas MJ, Tatara GM, Criddle CS (1995) Localization and characterization of the carbon tetrachloride transformation activity of *Pseudomonas* sp. strain KC. Appl Environ Microbiol 61:758–762

Fahrner KA, Block SM, Krishnaswamy S, Parkinson JS, Berg HC (1994) A mutant hook-associated protein (HAP3) facilitates torsionally induced transformations of the flagellar filament of *Escherichia coli*. J Mol Biol 238:173–186

Gilbert D, Jakobsen HH, Winding A, Mayer P (2014) Co-transport of polycyclic aromatic hydrocarbons by motile microorganisms leads to enhanced mass transfer under diffusive conditions. Environ Sci Technol 48:4368–4375

Gkorezis P, Daghio M, Franzetti A, Van Hamme JD, Sillen W, Vangronsveld J (2016) The interaction between plants and bacteria in the remediation of petroleum hydrocarbons: an environmental perspective. Front Microbiol 7:1836

Gordillo F, Chávez FP, Jerez CA (2007) Motility and chemotaxis of *Pseudomonas* sp. B4 towards polychlorobiphenyls and chlorobenzoates. FEMS Microbiol Ecol 60:322–328

Grimm AC, Harwood CS (1997) Chemotaxis of *Pseudomonas putida* to the polyaromatic hydrocarbon naphthalene. Appl Environ Microbiol 63:4111–4115

Grimm AC, Harwood CS (1999) NahY, a catabolic plasmid-encoded receptor required for chemotaxis of *Pseudomonas putida* to the aromatic hydrocarbon naphthalene. J Bacteriol 181:3310–3316

Haigler BE, Spain JC (1993) Biodegradation of 4-nitrotoluene by *Pseudomonas* sp. strain 4NT. Appl Environ Microbiol 59:2239–2243

Hanzel J, Harms H, Wick LY (2010) Bacterial chemotaxis along vapor-phase gradients of naphthalene. Environ Sci Technol 44:9304–9310

Hanzel J, Thullner M, Harms H, Wick LY (2012) Walking the tightrope of bioavailability: growth dynamics of PAH degraders on vapour-phase PAH. Microb Biotechnol 5:79–86

Harwood CS (1989) A methyl-accepting protein is involved in benzoate taxis in *Pseudomonas putida*. J Bacteriol 171:4603–4608

Harwood CS, Fosnaugh K, Dispensa M (1989) Flagellation of *Pseudomonas putida* and analysis of its motile behavior. J Bacteriol 171:4063–4066

Harwood CS, Nichols NN, Kim M-K, Ditty JL, Parales RE (1994) Identification of the *pcaRKF* gene cluster from *Pseudomonas putida*: involvement in chemotaxis, biodegradation, and transport of 4-hydroxybenzoate. J Bacteriol 176:6479–6488

Harwood CS, Ornston LN (1984) TOL plasmid can prevent induction of chemotactic responses to aromatic acids. J Bacteriol 160:797–800

Harwood CS, Parales RE, Dispensa M (1990) Chemotaxis of *Pseudomonas putida* toward chlorinated benzoates. Appl Environ Microbiol 56:1501–1503

Hazen TC (1994) Chemotactic selection of pollutant degrading soil bacteria. U.S. Patent, 5,324,661

Hazen TC, Lopez-de-Victoria G (1994) Method of degrading pollutants in soil. U.S. Patent, 5,236,703

Huang Z, Ni B, Jiang CY, Wu YF, He YZ, Parales RE, Liu SJ (2016) Direct sensing and signal transduction during bacterial chemotaxis toward aromatic compounds in *Comamonas testosteroni*. Mol Microbiol 101:224–237

Imae Y, Oosawa K, Mizuno T, Kihara M, Macnab RM (1987) Phenol: a complex chemoeffector in bacterial chemotaxis. J Bacteriol 169:371–379

Iwaki H, Muraki T, Ishihara S, Hasegawa Y, Rankin KN, Sulea T, Boyd J, Lau PC (2007) Characterization of a pseudomonad 2-nitrobenzoate nitroreductase and its catabolic pathway-associated 2-hydroxylaminobenzoate mutase and a chemoreceptor involved in 2-nitrobenzoate chemotaxis. J Bacteriol 189:3502–3514

Krell T, Lacal J, Guazzaroni ME, Busch A, Silva-Jiménez H, Fillet S, Reyes-Darías JA, Muñoz-Martínez F, Rico-Jiménez M, García-Fontana C, Duque E, Segura A, Ramos JL (2012) Responses of *Pseudomonas putida* to toxic aromatic carbon sources. J Biotechnol 160:25–32

Krell T, Lacal J, Munoz-Martinez F, Reyes-Darias JA, Cadirci BH, Garcia-Fontana C, Ramos JL (2011) Diversity at its best: bacterial taxis. Environ Microbiol 13:1115–1124

Krell T, Lacal J, Reyes-Darias JA, Jimenez-Sanchez C, Sungthong R, Ortega-Calvo JJ (2013) Bioavailability of pollutants and chemotaxis. Curr Opin Biotechnol 24:451–456

Lacal J, Muñoz-Martínez F, Reyes-Darías JA, Duque E, Matilla M, Segura A, Calvo JJ, Jimenez-Sánchez C, Krell T, Ramos JL (2011) Bacterial chemotaxis towards aromatic hydrocarbons in *Pseudomonas*. Environ Microbiol 13:1733–1744

Lacal J, Reyes-Darias JA, Garcia-Fontana C, Ramos JL, Krell T (2013) Tactic responses to pollutants and their potential to increase biodegradation efficiency. J Appl Microbiol 114:923–933

Lanfranconi MP, Alvarez HM, Studdert CA (2003) A strain isolated from gas oil-contaminated soil displays chemotaxis towards gas oil and hexadecane. Environ Microbiol 5:1002–1008

Lau PCK, Wang Y, Patel A, Labbé D, Bergeron H, Brousseau R, Konishi Y, Rawlings M (1997) A bacterial basic region leucine zipper histidine kinase regulating toluene degradation. Proc Natl Acad Sci U S A 94:1453–1458

Law AM, Aitken MD (2003) Bacterial chemotaxis to naphthalene desorbing from a nonaqueous liquid. Appl Environ Microbiol 69:5968–5973

Leahy JG, Byrne AM, Olsen RH (1996) Comparison of factors influencing trichloroethylene degradation by toluene-oxidizing bacteria. Appl Environ Microbiol 62:825–833

Lee DY, Ramos A, Macomber L, Shapleigh JP (2002) Taxis response of various denitrifying bacteria to nitrate and nitrite. Appl Environ Microbiol 68:2140–2147

Leungsakul T, Keenan BG, Smets BF, Wood TK (2005) TNT and nitroaromatic compounds are chemoattractants for *Burkholderia cepacia* R34 and *Burkholderia* sp. strain DNT. Appl Microbiol Biotechnol 69:321–325

Lewis TA, Paszczynski A, Gordon-Wylie SW, Jeedigunta S, Lee CH, Crawford RL (2001) Carbon tetrachloride dechlorination by the bacterial transition metal chelator pyridine-2,6-bis (thiocarboxylic acid). Environ Sci Technol 35:552–559

Li P, Ma L, Feng YL, Mo MH, Yang FX, Dai HF, Zhao YX (2012) Diversity and chemotaxis of soil bacteria with antifungal activity against *Fusarium* wilt of banana. J Ind Microbiol Biotechnol 39:1495–1505

Li S, Wackett LP (1992) Trichloroethylene oxidation by toluene dioxygenase. Biochem Biophys Res Comm 185:443–451

Luu RA, Kootstra JD, Nesteryuk V, Brunton C, Parales JV, Ditty JL, Parales RE (2015) Integration of chemotaxis, transport and catabolism in *Pseudomonas putida* and identification of the aromatic acid chemoreceptor PcaY. Mol Microbiol 96:134–147

Marx RB, Aitken MD (1999) Quantification of chemotaxis to naphthalene by *Pseudomonas putida* G7. Appl Environ Microbiol 65:2847–2852

Marx RB, Aitken MD (2000a) Bacterial chemotaxis enhances naphthalene degradation in a heterogeneous aqueous system. Environ Sci Technol 34:3379–3383

Marx RB, Aitken MD (2000b) A material-balance approach for modeling bacterial chemotaxis to a consumable substrate in the capillary assay. Biotechnol Bioeng 68:308–315

Mason OU, Hazen TC, Borglin S, Chain PS, Dubinsky EA, Fortney JL, Han J, Holman HY, Hultman J, Lamendella R, Mackelprang R, Malfatti S, Tom LM, Tringe SG, Woyke T, Zhou J, Rubin EM, Jansson JK (2012) Metagenome, metatranscriptome and single-cell sequencing reveal microbial response to Deepwater horizon oil spill. ISME J 6:1715–1727

Matilla MA, Krell T (2017) Chemoreceptor-based signal sensing. Curr Opin Biotechnol 45:8–14

Meng L, Li H, Bao M, Sun P (2017) Metabolic pathway for a new strain *Pseudomonas synxantha* LSH-7′: from chemotaxis to uptake of *n*-hexadecane. Sci Rep 7:39068

Ni B, Huang Z, Fan Z, Jiang CY, Liu SJ (2013) *Comamonas testosteroni* uses a chemoreceptor for tricarboxylic acid cycle intermediates to trigger chemotactic responses towards aromatic compounds. Mol Microbiol 90:813–823

Ni B, Huang Z, Wu YF, Fan Z, Jiang CY, Liu SJ (2015) A novel chemoreceptor MCP2983 from *Comamonas testosteroni* specifically binds to *cis*-aconitate and triggers chemotaxis towards diverse organic compounds. Appl Microbiol Biotechnol 99:2773–2781

Nisenbaum M, Sendra GH, Gilbert GA, Scagliola M, González JF, Murialdo SE (2013) Hydrocarbon biodegradation and dynamic laser speckle for detecting chemotactic responses at low bacterial concentration. J Environ Sci 25:613–625

Olson MS, Ford RM, Smith JA, Fernandez EJ (2004) Quantification of bacterial chemotaxis in porous media using magnetic resonance imaging. Environ Sci Technol 38:3864–3870

Ortega-Calvo JJ, Marchenko AI, Vorobyov AV, Borovick RV (2003) Chemotaxis in polycyclic aromatic hydrocarbon-degrading bacteria isolated from coal-tar- and oil-polluted rhizospheres. FEMS Microbiol Ecol 44:373–381

Pandey G, Chauhan A, Samanta SK, Jain RK (2002) Chemotaxis of a *Ralstonia* sp. SJ98 toward co-metabolizable nitroaromatic compounds. Biochem Biophys Res Commun 299:404–409

Pandey G, Jain RK (2002) Bacterial chemotaxis toward environmental pollutants: role in bioremediation. Appl Environ Microbiol 68:5789–5795

Pandey J, Chauhan A, Jain RK (2009) Integrative approaches for assessing the ecological sustainability of in situ bioremediation. FEMS Microbiol Rev 33:324–375

Pandey J, Sharma NK, Khan F, Ghosh A, Oakeshott JG, Jain RK, Pandey G (2012) Chemotaxis of *Burkholderia* sp. strain SJ98 towards chloronitroaromatic compounds that it can metabolise. BMC Microbiol 12:19

Parales RE (2004) Nitrobenzoates and aminobenzoates are chemoattractants for *Pseudomonas* strains. Appl Environ Microbiol 70:285–292

Parales RE, Ditty JL, Harwood CS (2000) Toluene-degrading bacteria are chemotactic to the environmental pollutants benzene, toluene, and trichloroethylene. Appl Environ Microbiol 66:4098–4104

Parales RE, Harwood CS (2002) Bacterial chemotaxis to pollutants and plant-derived aromatic molecules. Curr Opin Microbiol 5:266–273

Parales RE, Ju K-S, Rollefson J, Ditty JL (2008) Bioavailability, transport and chemotaxis of organic pollutants. In: Diaz E (ed) Microbial Bioremediation. Caister Academic Press, Norfolk, pp 145–187

Parkinson JS, Hazelbauer GL, Falke JJ (2015) Signaling and sensory adaptation in *Escherichia coli* chemoreceptors: 2015 update. Trends Microbiol 3:257–266

Paul D, Singh R, Jain RK (2006) Chemotaxis of *Ralstonia* sp. SJ98 towards *p*-nitrophenol in soil. Environ Microbiol 8:1797–1804

Pedit JA, Marx RB, Miller CT, Aitken MD (2002) Quantitative analysis of experiments on bacterial chemotaxis to naphthalene. Biotechnol Bioeng 78:626–634

Pham HT, Parkinson JS (2011) Phenol sensing by *Escherichia coli* chemoreceptors: a nonclassical mechanism. J Bacteriol 193:6597–6604

Philips J, Miroshnikov A, Haest PJ, Springael D, Smolders E (2014) Motile *Geobacter* dechlorinators migrate into a model source zone of trichloroethene dense non-aqueous phase liquid: experimental evaluation and modeling. J Contam Hydrol 170:28–38

Pieper DH, Timmis KN, Ramos JL (1996) Designing bacteria for the degradation of nitro- and chloroaromatic pollutants. Naturwissenschaften 83:201–213

Rabinovitch-Deere CA, Parales RE (2012) Three types of taxis used in the response of *Acidovorax* sp. strain JS42 to 2-nitrotoluene. Appl Environ Microbiol 78:2308–2315

Rhys-Williams W, Taylor SC, Williams PA (1993) A novel pathway for the catabolism of 4-nitrotoluene by *Pseudomonas*. J Gen Microbiol 139:1967–1972

Roberts MA, Papachristodoulou A, Armitage JP (2010) Adaptation and control circuits in bacterial chemotaxis. Biochem Soc Trans 38:1265–1269

Roush CJ, Lastoskie CM, Worden RM (2006) Denitrification and chemotaxis of *Pseudomonas stutzeri* KC in porous media. J Environ Sci Health A Tox Hazard Subst Environ Eng 41:967–983

Ryoo D, Shim H, Canada K, Barberi P, Wood TK (2000) Aerobic degradation of tetra-chloroethylene by toluene-*o*-monooxygenase of *Pseudomonas stutzeri* OX1. Nat Biotechnol 18:775–778

Samanta SK, Bhushan B, Chauhan A, Jain RK (2000) Chemotaxis of a *Ralstonia* sp. SJ98 toward different nitroaromatic compounds and their degradation. Biochem Biophys Res Commun 269:117–123

Samanta SK, Jain RK (2000) Evidence for plasmid-mediated chemotaxis of *Pseudomonas putida* towards naphthalene and salicylate. Can J Microbiol 46:1–6

Sarand I, Osterberg S, Holmqvist S, Holmfeldt P, Skärfstad E, Parales RE, Shingler V (2008) Metabolism-dependent taxis towards (methyl)phenols is coupled through the most abundant of three polar localized Aer-like proteins of *Pseudomonas putida*. Environ Microbiol 10:1320–1334

Shields MS, Francesconi SC (1996) Microbial degradation of trichloroethylene, dichloroethylenes, and aromatic pollutants. U.S. Patent 5,543,317

Shioi J, Dang CV, Taylor BL (1987) Oxygen as attractant and repellent in bacterial chemotaxis. J Bacteriol 169:3118–3123

Shitashiro M, Kato J, Fukumura T, Kuroda A, Ikeda T, Takiguchi N, Ohtake H (2003) Evaluation of bacterial aerotaxis for its potential use in detecting the toxicity of chemicals to microorganisms. J Biotechnol 101:11–18

Shitashiro M, Tanaka H, Hong CS, Kuroda A, Takiguchi N, Ohtake H, Kato J (2005) Identification of chemosensory proteins for trichloroethylene in *Pseudomonas aeruginosa*. J Biosci Bioeng 99:396–402

Sikkema J, De Bont JAM, Poolman B (1995) Mechanisms of membrane toxicity of hydrocarbons. Microbiol Rev 59:201–222

Singh R, Olson MS (2010) Kinetics of trichloroethylene and toluene toxicity to *Pseudomonas putida* F1. Environ Toxicol Chem 29:56–63

Smits TH, Witholt B, van Beilen JB (2003) Functional characterization of genes involved in alkane oxidation by *Pseudomonas aeruginosa*. Antonie Van Leeuwenhoek 84:193–200

Spain JC, Hughes JB, Knackmuss H-J (2000) Biodegradation of nitroaromatic compounds and explosives. CRC Press, Boca Raton

Strobel KL, McGowan S, Bauer RD, Griebler C, Liu J, Ford RM (2011) Chemotaxis increases vertical migration and apparent transverse dispersion of bacteria in a bench-scale microcosm. Biotechnol Bioeng 108:2070–2077

Szurmant H, Ordal GW (2004) Diversity in chemotaxis mechanisms among the bacteria and archaea. Microbiol Mol Biol Rev 68:301–319

Taguchi K, Fukatomi H, Kuroda A, Kato J, Ohtake H (1997) Genetic identification of chemotactic transducers for amino acids in *Pseudomonas aeruginosa*. Microbiology 143:3223–3229

Taylor BL (2007) Aer on the inside looking out: paradigm for a PAS-HAMP role in sensing oxygen, redox and energy. Mol Microbiol 65:1415–1424

Taylor BL, Miller JB, Warrick HM, Koshland DEJ (1979) Electron acceptor taxis and blue light effect on bacterial chemotaxis. J Bacteriol 140:567–573

Taylor BL, Watts KJ, Johnson MS (2007) Oxygen and redox sensing by two-component systems that regulate behavioral responses: behavioral assays and structural studies of Aer using *in vivo* disulfide cross-linking. Methods Enzymol 422:190–232

Taylor BL, Zhulin IB (1999) PAS domains: internal sensors of oxygen, redox potential, and light. Microbiol Mol Biol Rev 63:479–506

Taylor BL, Zhulin IB, Johnson MS (1999) Aerotaxis and other energy-sensing behavior in bacteria. Annu Rev Microbiol 53:103–128

Tremaroli V, Vacchi Suzzi C, Fedi S, Ceri H, Zannoni D, Turner RJ (2010) Tolerance of *Pseudomonas pseudoalcaligenes* KF707 to metals, polychlorobiphenyls and chlorobenzoates: effects on chemotaxis-, biofilm- and planktonic-grown cells. FEMS Microbiol Ecol 74:291–301

van Beilen JB, Panke S, Lucchini S, Franchini AG, Röthlisberger M, Witholt B (2001) Analysis of *Pseudomonas putida* alkane-degradation gene clusters and flanking insertion sequences: evolution and regulation of the *alk* genes. Microbiology 147:1621–1630

Vangnai AS, Takeuchi K, Oku S, Kataoka N, Nitisakulkan T, Tajima T, Kato J (2013) Identification of CtpL as a chromosomally encoded chemoreceptor for 4-chloroaniline and catechol in *Pseudomonas aeruginosa* PAO1. Appl Environ Microbiol 79:7241–7248

Vardar G, Barbieri P, Wood TK (2005) Chemotaxis of *Pseudomonas stutzeri* OX1 and *Burkholderia cepacia* G4 toward chlorinated ethenes. Appl Microbiol Biotechnol 66:696–701

Wackett LP, Gibson DT (1988) Degradation of trichloroethylene by toluene dioxygenase in whole cell studies with *Pseudomonas putida* F1. Appl Environ Microbiol 54:1703–1708

Wang X, Atencia J, Ford RM (2015) Quantitative analysis of chemotaxis towards toluene by *Pseudomonas putida* in a convection-free microfluidic device. Biotechnol Bioeng 112:896–904

Wang X, Lanning LM, Ford RM (2016) Enhanced retention of chemotactic bacteria in a pore network with residual NAPL contamination. Environ Sci Technol 50:165–172

Wang X, Long T, Ford RM (2012) Bacterial chemotaxis toward a NAPL source within a pore-scale microfluidic chamber. Biotechnol Bioeng 109:1622–1628

Witt ME, Dybas MJ, Worden RM, Criddle CS (1999) Motility-enhanced bioremediation of carbon tetrachloride-contaminated aquifer sediments. Environ Sci Technol 33:2958–2964

Wood PL, Parales JV, Parales RE (2006) Investigation of *Ralstonia* sp. strain U2 chemotaxis to naphthalene. Abstract, 106th general meeting of the American Society for Microbiology

Wu G, Feng Y, Boyd SA (2003) Characterization of bacteria capable of degrading soil-sorbed biphenyl. Bull Environ Contam Toxicol 71:768–775

Wuichet K, Zhulin IB (2010) Origins and diversification of a complex signal transduction system in prokaryotes. Sci Signal 3:50

Yamamoto K, Macnab RM, Imae Y (1990) Repellent response functions of the Trg and Tap chemoreceptors of *Escherichia coli*. J Bacteriol 172:383–388

Young LY, Mitchell R (1973) Negative chemotaxis of marine bacteria to toxic chemicals. Appl Microbiol 25:972–975

Yu HS, Alam M (1997) An agarose-in-plug bridge method to study chemotaxis in the archaeon *Halobacterium salinarum*. FEMS Microbiol Lett 156:265–269

Zhang JJ, Xin YF, Liu H, Wang SJ, Zhou NY (2008) Metabolism-independent chemotaxis of *Pseudomonas* sp. strain WBC-3 toward aromatic compounds. J Environ Sci 20:1238–1242

Zhulin IB (2001) The superfamily of chemotaxis transducers: from physiology to genomics and back. Adv Microbial Phys 45:157–198

# The Potential of Hydrocarbon Chemotaxis to Increase Bioavailability and Biodegradation Efficiency

# 14

Jesús Lacal

## Contents

1  Introduction ................................................................................ 241
2  Most Common Contaminant Hydrocarbons ............................................. 244
3  Chemotactic Bacteria ...................................................................... 246
4  Biodegradation ............................................................................. 249
5  Research Needs ............................................................................ 250
References ..................................................................................... 250

### Abstract

Hydrocarbons are simple organic compounds, containing only carbon and hydrogen, but despite their simplicity, they are common contaminants in our environment. This and the risks they pose to human health require remediation strategies. The decomposition of hydrocarbons by microorganisms into less or nontoxic simpler substances has been under study for many years and important advances have been made in this field. Interestingly, cell adherence and surface hydrophobicity, biosurfactant production, motility, and chemotaxis processes are bacterial abilities that reduce the distance between the microorganisms and solid substrates, enhancing bioavailability. Particularly, chemotaxis may enable hydrocarbon-utilizing bacteria to actively seek new substrates once they are depleted in a given contaminated area increasing their bioavailability and biodegradation. This chapter recapitulates major advances in the potential of hydrocarbon chemotaxis to increase bioavailability and biodegradation efficiency.

J. Lacal (✉)
Department of Microbiology and Genetics, University of Salamanca, Salamanca, Spain
e-mail: jlacal@usal.es

© Springer International Publishing AG, part of Springer Nature 2018
T. Krell (ed.), *Cellular Ecophysiology of Microbe: Hydrocarbon and Lipid Interactions*,
Handbook of Hydrocarbon and Lipid Microbiology, https://doi.org/10.1007/978-3-319-50542-8_3

# 1    Introduction

Soil contamination by hydrocarbons (such as anthropogenic organic chemicals, AOCs) is increasing across the globe due to heavy dependence on petroleum as a major source of energy. For instance, since 2010, over 3300 incidents of crude oil and liquefied natural gas leaks have occurred on US pipelines, releasing over seven million gallons of crude according to the Pipeline and Hazardous Materials Safety Administration (www.phmsa.dot.gov/). Also, motor oils such as diesel or jet fuel contaminate the natural environment with hydrocarbon. Hydrocarbons can spread horizontally on the groundwater surface thereby causing extensive groundwater contamination, whereas in the oceans, oil extraction activities, and the natural seeps associated to them, have increased the production, shipping, and use of oil. As a result of these activities, the exposure of soil and freshwater to oil urge for the need to find ways to decontaminate these areas. These hydrocarbon pollutants may persist for long periods of time causing disruptions of natural equilibrium between the living species and their natural environment, and they represent a real hazard for humans. Indeed, hydrocarbon components that belong to the family of carcinogens and neurotoxic organic pollutants can enter the human food chain (Gibson and Parales 2000; Das and Chandran 2011).

There are physical and chemical remediation techniques to remove hydrocarbons including in situ soil vapor extraction, in situ steam injection vapor extraction, and excavation (Abioye 2011). However, because of the high economic cost of physicochemical strategies, the biological tools for remediation of these persistent pollutants is a promising option (Bisht et al. 2015). Selecting the most appropriate strategy to treat a specific site can be guided by considering the amenability of the pollutant to biological transformation to less toxic products, its accessibility for biotransformation or bioavailability, and bioactivity (Dua et al. 2002). Some other benefits of using bioremediation to fight hydrocarbons include the acceleration of the natural clean-up process and the possibility to treat the contaminants in place. On the other hand, one of the reasons for the lacking efficiency of biodegradation processes is the limited bioavailability of the target compounds. Also, optimal degradation of pollutants trapped in regions of low hydraulic conductivity and permeability remains an issue (Adadevoh et al. 2016). In this context, research over the last decade has shown that chemotaxis increases hydrocarbon bioavailability, which was found to have a beneficial role in bioremediation. Chemotaxis may also increase the bacterial residence time within the pollutant vicinity, potentially long enough for complete contaminant biodegradation (Duffy et al. 1997; Wang et al. 2012).

Chemotaxis is the oriented movement toward (positive chemotaxis) or away (negative chemotaxis) from a particular chemical gradient allowing microorganisms to locate to more advantageous niches for their growth and survival (Pandey and Jain 2002; Parales and Harwood 2002; Alexandre et al. 2004; Ford and Harvey 2007). The ability to modulate movement direction is the outcome of controlled changes in the direction of flagellar rotation, and it is mediated by chemoreceptors that sense chemoeffectors by a specific receptor–ligand interaction that in turn activates intracellular signaling cascades (Lacal et al. 2010). The specificity of a chemotactic response is determined by the chemoreceptor that is typically composed of a periplasmatic ligand-binding region and a cytosolic signaling domain (Bi and Lai 2015). Some studies have shown that the increased number of chemoreceptors in many free-living bacteria enables them to respond to a wider range of compounds, including hydrocarbons, as compared to enterobacteria, suggesting their potential to detect and chemotax toward a wide variety of different hydrocarbons (Sampedro et al. 2015). Therefore, chemotaxis is a selective advantage to the bacteria for guiding them to sense and locate pollutants that are present in the environment. Chemotaxis may play an important role in terrestrial ecosystems (Alexandre et al. 2004; Ford et al. 2007), and it has been demonstrated to enhance bioremediation in order to evaluate its impact in natural systems (Samanta et al. 2000; Scow and Hicks 2005; Wang et al. 2015). Because of the ability of bacteria to locate and degrade hydrocarbons, such bacteria have a selective advantage to survive and grow in various contaminated sites and, as a consequence, reduce the pollutants in these areas (Grimm and Harwood 1997; Samanta et al. 2000; Pandey and Jain 2002; Parales and Harwood 2002; Law and Aitken 2003; Harms and Wick 2006). Therefore, chemotaxis may be a key factor for achieving more efficient bioremediation (Pandey and Jain 2002; Ford and Harvey 2007; Pandey et al. 2009).

Many different bacterial strains exhibit positive chemotaxis in the presence of hydrocarbons, which they also degrade (Grimm and Harwood 1997; Hawkins and Harwood 2002; Law and Aitken 2003; Bhushan et al. 2000; Parales et al. 2000; Samanta et al. 2000). Chemoattraction was observed towards biphenyl, benzoic acid and chlorobenzoic acids (Tremaroli et al. 2010), toluene and its derivatives (Parales et al. 2000; Lacal et al. 2011), naphthalene and its derivatives (Grimm and Harwood 1997; Lacal et al. 2011), nitroaromatics (Iwaki et al. 2007), chloroaromatics (Gordillo et al. 2007), chloronitroaromatics (Pandey et al. 2012), aminoaromatics (Parales 2004), explosives (Leungsakul et al. 2005), aliphatic hydrocarbons (Lanfranconi et al. 2003), and herbicides (Liu and Parales 2009). Similarly, given the toxic potential of most pollutants, some bacteria have also evolved chemorepellent responses. Bacterial repellence has been reported, for example, to hydrogen peroxide, hypochlorite, and N-chlorotaurine (Benov and Fridovich 1996) and the polycyclic aromatic hydrocarbons (PAHs) anthracene and pyrene (Ortega-Calvo et al. 2003). Some chemicals can even be chemoattractants for one bacterial species and be repellent for another (Parales et al. 2000; Shitashiro et al. 2005; Vardar et al. 2005). For instance, *Pseudomonas aeruginosa* is repelled by trichloroethylene (TCE), whereas it serves as a chemoattractant for *Pseudomonas putida* F1 (Singh and Olson 2010). The physical state of the chemical also appears to influence the type of response, because it was shown that the naphthalene degrader *Pseudomonas putida* G7 was repelled by naphthalene in the vapor phase, whereas it was attracted when the compound was dissolved in the aqueous phase (Hanzel et al. 2010). In the light of such results, one has to keep in mind that an observed chemotaxis phenotype can be the result of the action of several, potential antagonistic chemoreceptors that differ in their sensitivity to a given compound. These evidences suggest that understanding how these different chemotactic responses are orchestrated may be important to enhance bacteria-contaminant contacts to potentiate its biodegradation.

## 2    Most Common Contaminant Hydrocarbons

Hydrocarbons are a heterogeneous group (Fig. 1). Lighter hydrocarbons (i.e., less than 16 carbon atoms) tend to be more mobile due to greater solubility, greater volatility, and lower organic partitioning coefficients. Heavier hydrocarbons, referred to as polycyclic aromatic hydrocarbons (PAHs), typically tend to adsorb into the organic fraction of soil having higher toxicity and being more persistent in the environment. Chemotaxis is a relevant mobilizing factor for PAH-degrading rhizosphere bacteria, including moderately hydrophobic PAHs, such as phenan-threne, anthracene, and pyrene (Ortega-Calvo et al. 2003). Chemotaxis may be particularly important for the degradation of the most hydrophobic PAHs, such as benzopyrene. More recently, a study found that the biodegradation of PAHs present in fuel containing nonaqueous-phase liquid (NAPLs) was slow and followed zero-order kinetics, indicating bioavailability restrictions (Tejeda-Agredano et al. 2011). The best example of chemotaxis-enhanced bioavailability (and biodegradation) of a PAH relates to the capacity of *Pseudomonas putida* G7 to degrade naphthalene, and several studies have focused on this strain as a model microorganism (Fig. 1). Using a heterogeneous aqueous system under a slow-diffusion regime, the rate of biodeg-radation of naphthalene by *P. putida* G7 was found to exceed the predictions from a model based on diffusion-limited biodegradation (Marx and Aitken 2000). This indicated that bacterial movement through chemotaxis was faster than substrate mass transfer within the aqueous phase, thus enhancing the rate of substrate acqui-sition. A subsequent study that also used chemotactic and nonchemotactic strains of *P. putida* G7 clearly demonstrated that chemotaxis increased naphthalene degrada-tion when the compound is present in a nonaqueous-phase liquid (Law et al. 2003). On the other hand, no chemotactic bacteria have so far been described with more hydrophobic pollutants, such as high-molecular-mass PAHs.

Another group of frequent hydrocarbon contaminants is the total petroleum hydrocarbons (TPH). TPHs are a mix of several hundred individual hydrocarbon chemicals. This petroleum-based hydrocarbon mixture includes benzene, toluene, xylenes, and naphthalene among many other chemicals (Fig. 1). In certain instances, TPH can be encountered as a phase-separated liquid, which, due to its buoyancy, floats on the surface of the water-table. Commonly, phase separated TPH is referred to as light nonaqueous-phase liquid (LNAPL). A fraction of TPH will also be dissolved into the groundwater or trapped as vapors within the soil "pore-space" in the unsaturated zone. The exact split between phases is linked to the original composition of the source, geological and hydrogeological conditions, and the age since the spillage occurred. Light nonaqueous-phase liquids (NAPLs) are also organic liquids such as fuel oil and gasoline, always associated with human activity, that do not dissolve in water and are frequently found in polluted sites causing severe environmental and health hazards, representing a great challenge for the remediation of contaminated soils and sediments (Lekmine et al. 2017).

Trichloroethene or trichloroethylene (TCE) has become a widespread groundwa-ter pollutant, as a result of its extensive use as a solvent in industry (Fig. 1). Remediation of aquifers contaminated with this chemical is challenging, since

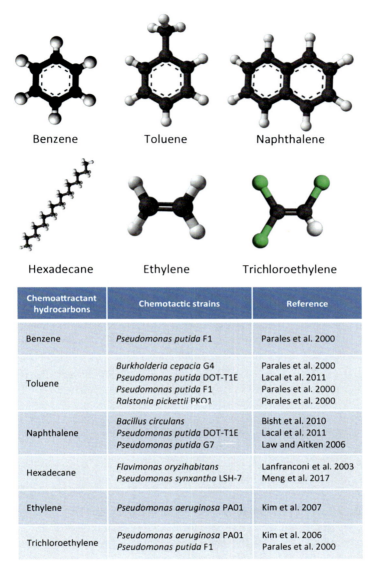

| Chemoattractant hydrocarbons | Chemotactic strains | Reference |
|---|---|---|
| Benzene | *Pseudomonas putida* F1 | Parales et al. 2000 |
| Toluene | *Burkholderia cepacia* G4<br>*Pseudomonas putida* DOT-T1E<br>*Pseudomonas putida* F1<br>*Ralstonia pickettii* PKO1 | Parales et al. 2000<br>Lacal et al. 2011<br>Parales et al. 2000<br>Parales et al. 2000 |
| Naphthalene | *Bacillus circulans*<br>*Pseudomonas putida* DOT-T1E<br>*Pseudomonas putida* G7 | Bisht et al. 2010<br>Lacal et al. 2011<br>Law and Aitken 2006 |
| Hexadecane | *Flavimonas oryzihabitans*<br>*Pseudomonas synxantha* LSH-7 | Lanfranconi et al. 2003<br>Meng et al. 2017 |
| Ethylene | *Pseudomonas aeruginosa* PA01 | Kim et al. 2007 |
| Trichloroethylene | *Pseudomonas aeruginosa* PA01<br>*Pseudomonas putida* F1 | Kim et al. 2006<br>Parales et al. 2000 |

**Fig. 1** Structures of different biodegradable hydrocarbons that were shown to be chemoattractants. This figure illustrates that chemotaxis has evolved to a wide array of different hydrocarbons

TCE forms dense nonaqueous-phase liquids (DNAPL) in the subsoil. Bioremediation can be used to clean up such contaminations, as several anaerobic bacteria grow with the reductive dechlorination of TCE to cisdichloroethene (cis-DCE). It has been shown that the reductive TCE dechlorination enhances the dissolution of the chlorinated ethene DNAPL which, as a consequence, reduces its remediation time (Yang and McCarty 2002). Dechlorination in a contaminated aquifer can be slow or inexistent and, in such cases, bioaugmentation, i.e., the injection of a dechlorinating culture, can initiate bio-enhanced DNAPL dissolution (Adamson et al. 2003; Sleep et al. 2006). Microbial migration towards a DNAPL could increase the migration towards a DNAPL, whereas negative chemotaxis could prevent the exposure to the toxic TCE concentrations adjacent to a TCE DNAPL, eventually accelerating the bioaugmentation of DNAPL source zones (Ford and Harvey, 2007). Chemotaxis towards TCE was also observed for the aerobe *Pseudomonas putida* F1, which metabolically converts TCE (Parales et al. 2000). Among the anaerobic contaminant degraders, motility was observed for several perchloroethene (PCE) and TCE dechlorinating bacteria, including *Geobacterlovleyi* SZ (Sung et al. 2006; Philips et al. 2012).

Hydrophobic organic compounds (HOCs) are common contaminants in soils and sediments including aromatic compounds in petroleum and fuel residue. For hydrophobic organic compounds, sorption is a critical process controlling the fate of these chemicals in groundwater (Oen et al. 2012). HOCs are nearly insoluble in the water phase and their degradation by microorganisms occurs at the interface with water (Mounier et al. 2014). Also, motile microorganisms can therefore function as effective HOC carriers under diffusive conditions which might significantly enhance HOC bioavailability (Gilbert et al. 2014). There is recent experimental evidence that ciliated protozoa can function as fast transport vehicles for hydrophobic organic chemicals. The unique adaptation of microorganisms to move by self-propulsion through aqueous medium makes them effective carriers for HOCs in diffusive boundary layers. Even when only a small fraction of a hydrophobic chemical is bound to a microbial carrier, cotransport with motile microorganisms can significantly enhance the mass transfer of HOCs when transport is diffusion-limited. There is thus increasing body of evidence that some microorganisms beyond their well-known role as contaminant degraders also play an important role in transporting and distributing contaminants (Gilbert et al. 2014).

## 3   Chemotactic Bacteria

In a significant number of cases, the physiological relevance of chemoattraction to pollutants lies in the fact that these compounds serve as carbon and energy sources. This may be exemplified by the chemotaxis towards toluene or naphthalene by *Pseudomonas putida* DOT-T1E and G7, respectively (Tsuda and Iino 1990; Mosqueda et al. 1999). Chemotaxis of *Pseudomonas putida* towards hydrocarbons is perhaps the most studied system (Fig. 2) and a good example to illustrate the link between chemotaxis and biodegradation of hydrocarbons (Parales et al. 2000), with the promise of being a candidate system to tackle groundwater contamination problems. In the two-naphthalene degrading bacteria *Pseudomonas* sp. strain NCIB 9816–4 and *P. putida* G7 (Grimm and Harwood 1997), previous growth of cells in biphenyl-containing media induced chemotaxis toward biphenyl. In contrast, chemotaxis of *Pseudomonas* sp. B4 toward biphenyl is not induced by previous growth in this substrate. Interestingly, it was reported that the nonmetabolizable

| None or very weak chemotaxis | Hyperchemotaxis |
|---|---|
| **Benzene**, styrene, *p*-xylene, *p*-nitrotoluene | Benzene derivatives propylbenzene, butylbenzene |

Toluene and singly substituted derivatives
**toluene**, *o*-xylene, *m*-xylene, *o*-iodotoluene, *m*-iodotoluene, *p*-iodotoluene, *o*-toluidine, *m*-toluidine, *p*-toluidine, *o*-chlorotoluene, *m*-chlorotoluene, *m*-bromotoluene, *m*-fluorotoluene, *p*-fluorotoluene, *o*-fluorotoluene, *m*-nitrotoluene, *o*-nitrotoluene

**Moderate chemotaxis**

Benzene derivatives
chlorobenzene, nitrobenzene, **ethylbenzene**, fluorobenzene, benzonitrile

Multiply substituted benzene derivatives
1,2,4-trimethylbenzene, 1,3,5-trimethylbenzene, 2,3-dimethylphenol

Toluene derivatives
*p*-ethyltoluene, *p*-chlorotoluene, *p*-bromotoluene

Biaromatics
naphthalene, 1,2,3,4-tetrahidronaphthalene

**Fig. 2** Chemotaxis phenotype of the organic-solvent tolerant *Pseudomonas putida* DOT-T1E strain towards aromatic hydrocarbons. Highlighted in *bold* are the compounds that also serve for growth (Data were taken from Lacal et al. 2011)

3- and 4-chlorobenzoate are attractants to *P. putida* PRS2000 previously grown on benzoate or 4-hydroxybenzoate (Harwood et al. 1990). *P. putida* DOT-T1E is able to degrade only benzene, toluene, and ethylbenzene (Fig. 2). However, the ligand spectrum of McpT was found to be large; most McpT ligands are not pathway substrates and the relevance of chemotaxis to these compounds is not clear. It can be hypothesized that the advantages of McpT-mediated taxis to pathway substrates outweighs the consequences of taxis to compounds that do not represent an apparent benefit, which hence caused the maintenance of such broad range hydrocarbon receptors during evolution. The physiological relevance of chemoattraction to compounds that do not serve as growth substrates is not fully understood.

Positive chemotaxis along contaminant gradients overcomes the mass-transfer limitations that impede the bioremediation process and increases the bioavailability of soil contaminants, as it improves the contact between degrader organisms and patchy contaminants (Fig. 3). In saturated systems, chemotaxis may be important for the access of bacteria to heterogeneously distributed contaminations (Velasco-Casal et al. 2008). Bioavailability for biodegradation may be very different from bioavailability for toxic effects, although the bioaccessible pool for both effects may be identical (Semple et al. 2007). Although chemotaxis requires at least some solubility of the target chemical (Ortega-Calvo et al. 2003), it is theoretically possible that also poorly soluble contaminants cause chemotaxis via their gas-phase distribution at very low concentrations when present in unsaturated porous media (Nijland and Burgess 2010). In such habitats, chemotactic swimming along continuous networks such as fungal mycelia (Furuno et al. 2010) may become important. Also, vapor-phase chemical gradients influence the chemotactic dispersal and the spatiotemporal

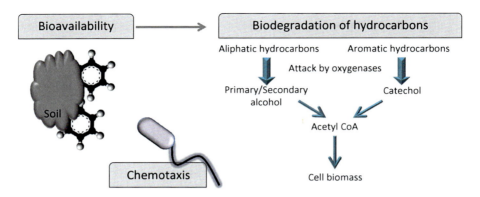

**Fig. 3** Chemotaxis and its capacity to increase the bioavailability and biodegradation efficiency of hydrocarbons. Hydrocarbon properties, such as molecular structure and toxicity, and microbial characteristics including gene regulation, metabolic flexibility, and tolerance determine the bioremediation performance

distribution of contaminant degrading bacteria. Air-exposed bacteria become increasingly important for the biodegradation of vapor-phase contaminants in technical filters as well as in the phyllosphere (Sandhu et al. 2007). Chemotactic dispersal hence can affect the dynamics of microbial ecosystems and the ecology of contaminant biodegradation (Nijland and Burgess 2010).

Chemotaxis enhances the mixing of bacteria with contaminant sources in low-permeability regions, which may not be readily accessible by advection and dispersion alone. The accumulation of chemotactic bacteria around the contaminated sources within the low-permeability region is expected to increase contaminant consumption. That and the increase in the growth rate due to hydrocarbon consumption are expected to accelerate the biodegradation process (Wang et al. 2016). Many studies have demonstrated the importance of chemotaxis towards hydrocarbons under various circumstances and in recent years more quantitative analyses have been undertaken to measure the two essential parameters that characterize the chemotaxis of bioremediation bacteria: the chemotactic sensitivity coefficient $x0$ [m$^2$/s] and the chemotactic receptor constant $Kc$[mol/L]. The chemotactic sensitivity coefficient represents the directed bacterial movement resulting from the chemotactic response, and the chemotactic receptor constant represents the bacterial propensity to bind the chemoattractant to its chemotactic receptors (Ford and Harvey 2007). Capturing reliable values of these two chemotaxis parameters is essential to improve the efficiency of bioremediation. Indeed, a new methodology based on the chemotactic sensitivity coefficient ($x0$) and chemotactic receptor constant ($Kc$) serves to enrich the current database of chemotaxis parameters (Wang et al. 2015).

The major evolutionary driving force for chemotactic movements is considered to be the capacity to access compounds that serve as carbon/energy sources or electron acceptors. This statement is based on data for taxis towards common nontoxic carbon sources like sugars, amino acids, or Krebs cycle intermediates but appears to apply also to taxis towards toxic biodegradable compounds. The observation that many biodegrading bacteria show tactic behaviors towards the cognate biodegradation substrate combined with the observation that specific pollutant chemoreceptors are colocalized on plasmids harboring pollutant degradation plasmids, strongly suggest a link between biodegradation and taxis. There is now a significant body of experimental information available that documents the beneficial effect of taxis on bioavailability. This knowledge may form the basis for a rational engineering to optimize the performance of pollutant-degrading microbial populations (Krell et al. 2011, 2013).

## 4    Biodegradation

In the context of bioremediation, chemotaxis may work to enhance bacterial residence times in zones of contamination, thereby improving treatment (Fig. 3). Chemotaxis could enhance the mass transfer of bacteria to the source of contamination, allowing for better pollutant accessibility and, concomitantly, increased contaminant biodegradation. There is a natural link between chemotaxis and biodegradation by showing the inducible nature of the chemotactic response and the occurrence of a common regulation of chemoreceptor and biodegradation genes. Coordinated regulation of bacterial chemotaxis to many xenobiotic compounds and their respective mineralization and/or transformation indicates that this phenomenon might be an integral feature of degradation. Indeed, with enhanced bacterial migration toward the attractant as a result of chemotaxis, the rate of attractant degradation is increased on close to in situ conditions in contaminated soil and in a reconstituted bench-scale microcosm and under laboratory conditions. It has also been shown that bacterial motility and transport can be controlled through a suitable choice of chemical effectors (Velasco-Casal et al. 2008; Jimenez-Sanchez et al. 2012).

Bhushan et al. (2004) studied the biodegradation of cyclic nitramine explosives that are hydrophobic pollutants with very low water solubility. In sediment and soil environments, they are often attached to solid surfaces and/or trapped in pores and are distributed heterogeneously in aqueous environments. Interestingly, biotransformation reactions by the anaerobic bacterium *Clostridium sp.* EDB2 that led to the production of $NO_2$ caused a chemotactic behavior that in turn was found to increase the biodegradation rate. In this case, it is not chemotaxis to the pollutant perse but to a derived metabolite, $NO_2$, which caused the chemotactic response. Another study revealed the link between chemotaxis and degradation of various nitroaromatic compounds (Samanta et al. 2000). Other studies have attributed a role for chemotaxis in directing bacterial migration towards contaminants in natural porous media under groundwater flow conditions (Wang et al. 2016). Furthermore, deposition of chemotactic bacteria during transport in porous media was investigated as well (Velasco-Casal et al. 2008; Jimenez-Sanchez et al. 2012). These studies concluded that bacteria chemotactic to naphthalene or to salicylate (the latter is an intermediate of naphthalene degradation) are deposited to a lesser extent in porous media than the control strain, which was devoid of chemotactic movement. Taken this information together, it can be concluded that there is a significant link between chemotaxis and the degradation of pollutants. This is consistent with the notion that an optimization of the chemotactic movement has the potential to increase the degradation efficiency (Lacal et al. 2013).

A particular way of bioremediation is rhizoremediation, or the decontamination of polluted soils by the plant rhizosphere. The high levels of microbial biomass associated with the rhizosphere, the carbon turnover caused by root exudates and their microbial utilization as cosubstrates for the cometabolism of PAHs and chemotaxis, as well as the migration of dissolved contaminants within the soil matrix as a result of interactions with DOM are all positive factors that promote the bioavailability of these PAHs. Identification of mutants defective in colonization and survival in the rhizosphere has provided some important insight into the processes involved in rhizosphere colonization, such as the implication of chemotaxis, movement toward and attachment to the roots. In particular, PAH-polluted soil, positive chemotaxis towards PAHs, bacterial components such as LPS, and root exudates may play a stimulatory role in biodegradation due to the existence of a favorable niche for microorganisms around the plant root. There are studies that have determined an enhanced PAH biodegradation rate and an enrichment of PAH-degrading bacteria in the rhizosphere (Miya and Firestone 2000; Binet et al. 2000). Rhizoremediation provides ways to increase the bioavailability of PAHs in a low-risk manner through different mechanisms. Plant rhizospheres could be effective in promoting the bioavailability of PAHs in contaminated soils that have previously undergone extensive bioremediation but that still contain inacceptable PAH levels (Ortega-Calvo et al. 2003).

## 5    Research Needs

Limited mechanistic information exists on bacterial activity and dispersal in the unsaturated zone of subsurface as well as on the precise molecular mechanisms underlying the chemotactic response. Complete understanding of these mechanisms would help to engineer chemotactic hydrocarbon degraders. Further research is necessary to elucidate the potential relevance of chemoattraction towards pollutants that are not metabolized. In addition, there is only limited knowledge available on the dispersal of air-water interface-associated microbes under the influence of vapor-phase contaminant gradients, as they likely exist in the capillary fringe of contaminated aquifers. Understanding how microorganisms move and adjust themselves to environmental cues is integral to the understanding of the complexity of microbial communities. Therefore, in the future, it will be important to work with both, genetically modified laboratory strains as well as newly isolated microbes and to observe them in the context of structured communities to fully appreciate their developmental and degradation potentials.

## References

Abioye OP (2011) Biological remediation of hydrocarbon and heavy metals contaminated soil, soil contamination. MSc Simone Pascucci (Ed.), INTECH, pp 127–142, ISBN: 978-953-307-647-8

Adadevoh JS, Triolo S, Ramsburg CA, Ford RM (2016) Chemotaxis increases the residence time of bacteria in granular media containing distributed contaminant sources. Environ Sci Technol 50:181–187

Adamson DT, McDade JM, Hughes JB (2003) Inoculation of a DNAPL source zone to initiate reductive dechlorination of PCE. Environ Sci Technol 37:2525–2533

Alexandre G, Greer-Phillips S, Zhulin IB (2004) Ecological role of energy taxis in microorganisms. FEMS Microbiol Rev 28:113–126

Benov L, Fridovich I (1996) Escherichia coli Exhibits negative chemotaxis in gradients of hydrogen peroxide, hypochlorite, and N-chlorotaurine: products of the respiratory burst of phagocytic cells. Proc Natl Acad Sci U S A 93:4999–5002

Bhushan B, Chauhan A, Samanta SK, Jain RK (2000) Kinetics of biodegradation of p-nitrophenol by different bacteria. Biochem Biophys Res Commun 274:626–630

Bhushan B, Halasz A, Thiboutot S, Ampleman G, Hawari J (2004) Chemotaxis-mediated biodegradation of cyclic nitramine explosives RDX, HMX, and CL-20 by *Clostridium* sp. EDB2. Biochem Biophys Res Commun 316:816–821

Bi S, Lai L (2015) Bacterial chemoreceptors and chemoeffectors. Cell Mol Life Sci 72:691–708

Binet P, Portal JM, Leyval C (2000) Dissipation of 3–6-ring polycyclic aromatic hydrocarbons in the rhizosphere of ryegrass. Soil Biol Biochem 32:2011–2017

Bisht S, Pandey P, Sood A, Sharma S, Bisht NS (2010) Biodegradation of naphthalene and anthracene by chemo-tactically active rhizobacteria of populusdeltoides. Braz J Microbiol 41:922–930

Bisht S, Pandey P, Bhargava B, Sharma S, Kumar V, Sharma KD (2015) Bioremediation of polyaromatic hydrocarbons (PAHs) using rhizosphere technology. Braz J Microbiol 46:7–21

Das N, Chandran P (2011) Microbial degradation of petroleum hydrocarbon contaminants: an overview. Biotechnol Res Int 2011:941810

Dua M, Singh A, Sethunathan N, Johri AK (2002) Biotechnology and bioremediation: successes and limitations. Appl Microbiol Biotechnol 59:143–152

Duffy K, Ford RM, Cummings PT (1997) Residence time calculation for chemotactic bacteria within porous media. Biophys J 73:2930–2936

Ford RM, Harvey RW (2007) Role of chemotaxis in the transport of bacteria through saturated porous media. Adv Water Resour 30:1608–1617

Furuno S, Pazolt K, Rabe C, Neu TR, Harms H, Wick LY (2010) Fungal mycelia allow chemotactic dispersal of polycyclic aromatic hydrocarbon degrading bacteria in water-unsaturated systems. Environ Microbiol 12:1391–1398

Gibson DT, Parales RE (2000) Aromatic hydrocarbon dioxygenases in environmental biotechnology. Curr Opin Biotechnol 11:236–243

Gilbert D, Jakobsen HH, Winding A, Mayer P (2014) Co-transport of polycyclic aromatic hydrocarbons by motile microorganisms leads to enhanced mass transfer under diffusive conditions. Environ Sci Technol 48:4368–4375

Gordillo F, Chávez FP, Jerez CA (2007) Motility and chemotaxis of Pseudomonas sp. B4 towards polychlorobiphenyls and chlorobenzoates. FEMS Microbiol Ecol 60:322–328

Grimm AC, Harwood CS (1997) Chemotaxis of *Pseudomonas* spp. to the polyaromatic hydrocarbon naphthalene. Appl Environ Microbiol 63:4111–4115

Hanzel J, Harms H, Wick LY (2010) Bacterial chemotaxis along vapor phase gradients of naphthalene. Environ Sci Technol 44:9304–9310

Harms H, Wick LY (2006) Dispersing pollutant-degrading bacteria in contaminated soil without touching it. Eng Life Sci 6:252–260

Harwood CS, Parales RE, Dispensa M (1990) Chemotaxis of *Pseudomonas putida* toward chlorinated benzoates. Appl Environ Microbiol 56:1501–1503

Hawkins AC, Harwood CS (2002) Chemotaxis of *Ralstoniaeutropha* JMP134 (pJP4) to the herbicide 2,4-dichlorophenoxyacetate. Appl Environ Microbiol 68:968–972

Iwaki H, Muraki T, Ishihara S, Hasegawa Y, Rankin KN, Sulea T, Boyd J, Lau PC (2007) Characterization of a pseudomonad 2-nitrobenzoate nitroreductase and its catabolic pathway-associated 2-hydroxylaminobenzoate mutase and a chemoreceptor involved in 2-nitrobenzoate chemotaxis. J Bacteriol 189:3502–3514

Jimenez-Sanchez C, Wick LY, Ortega-Calvo JJ (2012) Chemical effectors cause different motile behavior and deposition of bacteria in porous media. Environ Sci Technol 46:6790–6797

Kim HE, Shitashiro M, Kuroda A, Takiguchi N, Ohtake H, Kato J (2006) Identification and characterization of the chemotactic transducer in Pseudomonas aeruginosa PAO1 for positive chemotaxis to trichloroethylene. J Bacteriol 188:6700–6702

Kim HE, Shitashiro M, Kuroda A, Takiguchi N, Kato J (2007) Ethylene chemotaxis in Pseudomonas aeruginosa and other Pseudomonas species. Microbes Environ 22:186–189

Krell T, Lacal J, Muñoz-Martínez F, Reyes-Darias JA, Cadirci BH, García-Fontana C, Ramos JL (2011) Diversity at its best: bacterial taxis. Environ Microbiol 13:1115–1124

Krell T, Lacal J, Reyes-Darias JA, Jimenez-Sanchez C, Sungthong R, Ortega-Calvo JJ (2013) Bioavailability of pollutants and chemotaxis. Curr Opin Biotechnol 24:451–456

Lacal J, García-Fontana C, Muñoz-Martínez F, Ramos JL, Krell T (2010) Sensing of environmental signals: classification of chemoreceptors according to the size of their ligand binding regions. Environ Microbiol 12:2873–2884

Lacal J, Muñoz-Martínez F, Reyes-Darías JA, Duque E, Matilla M, Segura A, Calvo JJ, Jímenez-Sánchez C, Krell T, Ramos JL (2011) Bacterial chemotaxis towards aromatic hydrocarbons in Pseudomonas. Environ Microbiol 13:1733–1744

Lacal J, Reyes-Darias JA, García-Fontana C, Ramos JL, Krell T (2013) Tactic responses to pollutants and their potential to increase biodegradation efficiency. J Appl Microbiol 114:923–933

Lanfranconi MP, Alvarez HM, Studdert CA (2003) A strain isolated from gas oil-contaminated soil displays chemotaxis towards gas oil and hexadecane. Environ Microbiol 5:1002–1008

Law AMJ, Aitken MD (2003) Bacterial chemotaxis to naphthalene desorbing from a nonaqueous liquid. Appl Environ Microbiol 69:5968–5973

Law AMJ, Aitken MD (2006) The effect of oxygen on chemotaxis to naphthalene by Pseudomonas putida G7. Biotechnol Bioeng 93:457–464

Lekmine G, SookhakLari K, Johnston CD, Bastow TP, Rayner JL, Davis GB (2017) Evaluating the reliability of equilibrium dissolution assumption from residual gasoline in contact with water saturated sands. J Contam Hydrol 196:30–42

Leungsakul T, Keenan BG, Smets BF, Wood TK (2005) TNT and nitroaromatic compounds are chemoattractants for Burkholderiacepacia R34 and Burkholderia sp. strain DNT. Appl Microbiol Biotechnol 69:321–325

Liu X, Parales RE (2009) Bacterial chemotaxis to atrazine and related s-triazines. Appl Environ Microbiol 75:5481–5488

Marx RB, Aitken MD (2000) Bacterial chemotaxis enhances naphthalene degradation in a heterogeneous aqueous system. Environ Sci Technol 34:3379–3383

Meng L, Li H, Bao M, Sun P (2017) Metabolic pathway for a new strain Pseudomonas synxantha LSH-7': from chemotaxis to uptake of n-hexadecane. Sci Rep 7:39068

Miya RK, Firestone MK (2000) Phenanthrene-degrader community dynamics in rhizosphere soil from a common annual grass. J Environ Qual 29:584–592

Mosqueda G, Ramos-González MI, Ramos JL (1999) Toluene metabolism by the solvent-tolerant Pseudomonas putida DOT-T1 strain, and its role in solvent impermeabilization. Gene 232:69–76

Mounier J, Camus A, Mitteau I, Vaysse PJ, Goulas P, Grimaud R, Sivadon P (2014) The marine bacterium Marinobacter hydrocarbonoclasticus SP17 degrades a wide range of lipids and hydrocarbons through the formation of oleolytic biofilms with distinct gene expression profiles. FEMS Microbiol Ecol 90:816–831

Nijland R, Burgess JG (2010) Bacterial olfaction. Biotechnol J 5:974–977

Oen AM, Beckingham B, Ghosh U, Kruså ME, Luthy RG, Hartnik T, Henriksen T, Cornelissen G (2012) Sorption of organic compounds to fresh and field-aged activated carbons in soils and sediments. Environ Sci Technol 46:810–817

Ortega-Calvo JJ, Marchenko AI, Vorobyov AV, Borovick RV (2003) Chemotaxis in polycyclic aromatic hydrocarbon-degrading bacteria isolated from coal-tar- and oil-polluted rhizospheres. FEMS Microbiol Ecol 44:373–381

Pandey G, Jain RK (2002) Bacterial chemotaxis toward environmental pollutants: role in bioremediation. App Environ Microbiol 68:5789–5795

Pandey J, Chauhan A, Jain RK (2009) Integrative approaches for assessing the ecological sustainability of in situ bioremediation. FEMS Microbiol Rev 33:324–375

Pandey J, Sharma NK, Khan F, Ghosh A, Oakeshott JG, Jain RK, Pandey G (2012) Chemotaxis of *Burkholderia* sp. strain SJ98 towards chloronitroaromatic compounds that it can metabolise. BMC Microbiol 12:19

Parales RE (2004) Nitrobenzoates and aminobenzoates are chemoattractants for Pseudomonas strains. Appl Environ Microbiol 70:285–292

Parales RE, Harwood CS (2002) Bacterial chemotaxis to pollutants and plant-derived aromatic molecules. Curr Opin Microbiol 5:266–273

Parales RE, Ditty JL, Harwood CS (2000) Toluene-degrading bacteria are chemotactic towards the environmental pollutants benzene, toluene, and trichloroethylene. Appl Environ Microbiol 66:4098–4104

Philips J, Hamels F, Smolders E, Springael D (2012) Distribution of a dechlorinating community in relation to the distance from a trichloroethene dense nonaqueous phase liquid in a model aquifer. FEMS Microbiol Ecol 81:636–647

Samanta SK, Bhushan B, Chauhan A, Jain RK (2000) Chemotaxis of a *Ralstonia sp.* SJ98 toward different nitroaromatic compounds and their degradation. Biochem Biophys Res Commun 269:117–123

Sampedro I, Parales RE, Krell T, Hill JE (2015) *Pseudomonas* chemotaxis. FEMS Microbiol Rev 39:17–46

Sandhu A, Halverson LJ, Beattie GA (2007) Bacterial degradation of airborne phenol in the phyllosphere. Environ Microbiol 9:383–392

Scow KM, Hicks KA (2005) Natural attenuation and enhanced bioremediation of organic contaminants in groundwater. Curr Opin Biotechnol 16:246–253

Semple KT, Doick KJ, Wick LY, Harms H (2007) Microbial interactions with organic contaminants in soil: definitions, processes and measurement. Environ Pollut 150:166–176

Shitashiro M, Tanaka H, Hong CS, Kuroda A, Takiguchi N, Ohtake H, Kato J (2005) Identification of chemosensory proteins for trichloroethylene in *Pseudomonas aeruginosa*. J Biosci Bioeng 99:396–402

Singh R, Olson M (2010) Kinetics of trichloroethylene and toluene toxicity to *Pseudomonas putida* F1. Environ Toxicol Chem 29:56–63

Sleep BE, Seepersad DJ, Kaiguo MO, Heidorn CM, Hrapovic L, Morrill PL, McMaster ML, Hood ED, Lebron C, Lollar BS, Major DW, Edwards EA (2006) Biological enhancement of tetrachloroethene dissolution and associated microbial community changes. Environ Sci Technol 40:3623–3633

Sung Y, Fletcher KE, Ritalahti KM, Apkarian RP, Ramos-Hernández N, Sanford RA, Mesbah NM, Löffler FE (2006) *Geobacterlovleyi* sp. nov. strain SZ, a novel metal-reducing and tetrachloroethene-dechlorinating bacterium. Appl Environ Microbiol 72:2775–2782

Tejeda-Agredano MC, Gallego S, Niqui-Arroyo JL, Vila J, Grifoll M, Ortega-Calvo JJ (2011) Effect of interface fertilization on biodegradation of polycyclic aromatic hydrocarbons present in nonaqueous-phase liquids. Environ Sci Technol 45:1074–1081

Tremaroli V, VacchiSuzzi C, Fedi S, Ceri H, Zannoni D, Turner RJ (2010) Tolerance of *Pseudomonas pseudoalcaligenes* KF707 to metals, polychlorobiphenyls and chlorobenzoates: effects on chemotaxis-, biofilm- and planktonic-grown cells. FEMS Microbiol Ecol 74:291–301

Tsuda M, Iino T (1990) Naphthalene degrading genes on plasmid NAH7 are on a defective transposon. Mol Gen Genet 223:33–39

Vardar G, Barbieri P, Wood TK (2005) Chemotaxis of *Pseudomonas stutzeri* OX1 and *Burkholderiacepacia* G4 toward chlorinated ethenes. Appl Microbiol Biotechnol 66:696–701

Velasco-Casal P, Wick LY, Ortega-Calvo JJ (2008) Chemoeffectors decrease the deposition of chemotactic bacteria during transport in porous media. Environ Sci Technol 42:1131–1137

Wang X, Long T, Ford RM (2012) Bacterial chemotaxis toward a NAPL source within a pore-scale microfluidic chamber. Biotechnol Bioeng 109:1622–1628

Wang X, Atencia J, Ford RM (2015) Quantitative analysis of chemotaxis towards toluene by *Pseudomonas putida* in a convection-free microfluidic device. Biotechnol Bioeng 112:896–904

Wang X, Lanning LM, Ford RM (2016) Enhanced retention of chemotactic bacteria in a pore network with residual NAPL contamination. Environ Sci Technol 50:165–172

Yang Y, McCarty PL (2002) Comparison between donor substrates for biologically enhanced tetrachloroethene DNAPL dissolution. Environ Sci Technol 36:3400–3404

# Amphiphilic Lipids, Signaling Molecules, and Quorum Sensing

# 15

## M. Dow and L. M. Naughton

## Contents

1  Introduction ................................................................. 256
2  N-AHL-Mediated Signaling ..................................................... 257
   2.1  The Archetypal LuxIR QS System ..................................... 258
   2.2  LuxMN Is a Second QS System ........................................ 259
   2.3  Enzymatic Degradation of N-AHL Signals ............................. 259
3  DSF-Mediated Signaling ....................................................... 260
   3.1  The Rpf Proteins and DSF Signaling in Xanthomonads ................. 261
   3.2  A Second Core Pathway of DSF Signaling ............................. 262
   3.3  RpfF and Signal Synthesis ......................................... 263
   3.4  RpfB and Signal Degradation ....................................... 264
4  Gamma-Butyrolactone-Mediated Signaling ....................................... 264
5  Other Lipid-Based QS Systems ................................................. 266
   5.1  Hydroxylated Fatty Acid Esters in *Ralstonia* ..................... 266
   5.2  *(S)*-3-hydroxytridecan-4-one in *Vibrio cholerae* ................ 267
6  Interspecies and Inter-kingdom Signaling ..................................... 268
7  Research Needs ............................................................... 269
References ...................................................................... 270

### Abstract

Many bacteria communicate with each other through the action of diffusible signal molecules, a process that has been termed quorum sensing (QS). QS acts to regulate diverse processes in different bacteria, to include the formation of biofilms, cellular differentiation, synthesis of antibiotics and other secondary metabolites, and the production of virulence factors in pathogens. Many bacteria use amphiphilic lipids of different chemical classes as signal molecules. N-acyl homoserine lactones, *cis*-2-unsaturated fatty acids, methyl esters of hydroxylated

M. Dow (✉) · L. M. Naughton
School of Microbiology, University College Cork, Cork, Ireland
e-mail: m.dow@ucc.ie; lynn.naughton@ucc.ie

© Springer International Publishing AG, part of Springer Nature 2018
T. Krell (ed.), *Cellular Ecophysiology of Microbe: Hydrocarbon and Lipid Interactions*,
Handbook of Hydrocarbon and Lipid Microbiology, https://doi.org/10.1007/978-3-319-50542-8_31

fatty acids, and tridecanone derivatives have been described in different Gram-negative organisms and gamma-butyrolactones in Gram-positive streptomycetes. A diverse range of mechanisms for perception and transduction of these signals has been described. Here we review these different signals, their mode of biosynthesis, and transduction pathways before going on to discuss interference of QS as a strategy for the control of bacterial disease.

# 1    Introduction

Many bacteria communicate with each other through the action of diffusible signal molecules. Such cell-cell communication allows organisms to monitor aspects of their environment such as population density, a process that has been termed quorum sensing (QS). The elevated levels of signal concentration resulting from a higher local population density lead to activation of specific QS-regulated functions. Other growth conditions such as confinement to particular niches in which diffusion may be limited or exposure to conditions that affect signal production or stability can also affect the local concentration of signal molecules. The term QS is now used to describe cell-cell signaling in this range of different contexts (Platt and Fuqua 2010). QS allows a colony or group of organisms to act in a coordinated fashion to regulate diverse processes such as the formation of biofilms, cellular differentiation, synthesis of antibiotics and other secondary metabolites, and the production of virulence factors in pathogenic bacteria. The signal molecules synthesized by bacteria belong to a wide range of chemical classes to include amphiphilic lipids as well as peptides and carbohydrate derivatives. Multiple systems using different types of signal can often occur within a single organism. Equally, a diverse range of mechanisms for signal perception and transduction has been described.

As noted above, different amphiphilic lipids have been shown to act as bacterial cell-to-cell signals (Fig. 1). Indeed the most common signal molecules found in Gram-negative bacteria are the N-acyl homoserine lactones (N-AHLs) (Waters and Bassler 2005). The plant pathogen *Ralstonia solanacearum* has an additional signaling system mediated by methyl esters of 3-OH palmitic or myristic acids (Kai et al. 2015). Gram-negative bacteria from the order Xanthomonadales, which includes important plant and human pathogens, utilize *cis*-unsaturated fatty acids of the DSF (diffusible signal factor) family as signals (Ryan et al. 2015), whereas *Vibrio cholerae* utilizes (*S*)-3-hydroxytridecan-4-one (Higgins et al. 2007). Although many Gram-positive bacteria use amino acids or modified peptides as signals, actinomycetes such as *Streptomyces* species use gamma-butyrolactones (GBLs) (Takano 2006). In this overview, we will discuss each of these lipid-based signals, the pathways for their synthesis and turnover, and the signal transduction pathways that lead to specific alteration in bacterial behavior. We will go on to briefly discuss interference of QS as a strategy for the control of disease as part of a consideration of outstanding research questions. The reader is also directed to

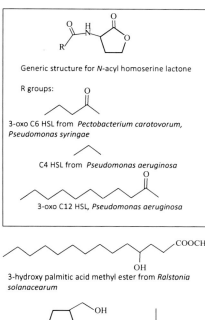

DSF, 11-Methyl-*cis*- dodecenoic acid from *Xanthomonas campestris*

BDSF, *cis* dodecenoic acid, first identified in *Burkholderia cenocepacia*

CDSF, *Cis,cis*-11-methyldodeca-2,5-dienoic acid first identified in *Burkholderia* spp. and in *Xanthomonas oryzae*

12-Me-tetradecanoic acid, tentatively identified as DSF in *Xylella fastidiosa*

CAI-1, (S)-3-hydroxytridecan-4-one first identified in *Vibrio cholerae*

Generic structure for *N*-acyl homoserine lactone

R groups:

3-oxo C6 HSL from *Pectobacterium carotovorum*, *Pseudomonas syringae*

C4 HSL from *Pseudomonas aeruginosa*

3-oxo C12 HSL, *Pseudomonas aeruginosa*

3-hydroxy palmitic acid methyl ester from *Ralstonia solanacearum*

A factor from *Streptomyces griseus*

**Fig. 1** Bacterial signal molecules belong to a range of different chemical classes to include amphiphilic lipids

several other reviews in this area that focus on particular signaling systems, strategies for interference, or their applications in synthetic biology (Deng et al. 2011; LaSarre and Federle 2013; Biarnes-Carrera et al. 2015).

## 2    *N*-AHL-Mediated Signaling

*N*-AHLs comprise an invariant homoserine lactone ring attached to a fatty acid residue through an amide bond (see Fig. 1). The fatty acid moieties differ in chain length from 4 to 18 carbons, some can be unsaturated and often occur with hydroxy or keto groups at position 3 (Ng and Bassler 2009). These signals were first described in the marine bioluminescent bacterium *Vibrio fischeri* (now *Photobacterium fischeri)* in which QS regulates light production. A new class of *N*-AHL first described in *Rhodopseudomonas palustris* has *p*-coumaric acid instead of a fatty acid (Schaefer et al. 2008) but will not be considered further here.

**Fig. 2** The archetypal *N*-AHL signaling system. *N*-AHLs are synthesized by a protein of the LuxI family and sensed by a LuxR family transcriptional regulator. *N*-AHL binding by LuxR enhances binding to the promoters of target genes that can include *luxI*, leading to positive feedback in signal synthesis

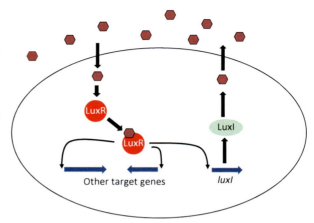

## 2.1    The Archetypal LuxIR QS System

The archetypal *N*-AHL QS system comprises two proteins belonging to the LuxI and LuxR families, respectively (Fuqua et al. 2001; Whitehead et al. 2001; Ng and Bassler 2009; Fig. 2). *N*-AHLs are synthesized by cytoplasmic LuxI family proteins using S-adenosyl methionine and a fatty acyl-acyl carrier protein (fatty acyl-ACP) as substrates. The reaction generates the *N*-AHL and 5′-methylthioadenosine as products. LuxI proteins do not have a strict substrate specificity and will generate a range of *N*-AHLs with fatty acid substituents of similar chain length or additional substitution. The precise mode of action of several LuxI family proteins has been described (Pappas et al. 2004). After synthesis, the signal can move across the bacterial membranes and accumulates both intra- and extracellularly in proportion to cell number. It is not clear whether movement across the cytoplasmic membrane requires facilitation by transporter proteins.

The sensing of the *N*-AHL signal is mediated by a transcriptional regulator of the LuxR family. These proteins comprise two domains: an amino-terminal region involved in *N*-AHL binding and a C-terminal domain implicated in DNA binding (Pappas et al. 2004). Preferential binding of a particular *N*-AHL by its cognate LuxR family protein ensures a good degree of specificity in signal transduction and in most cases results in the formation of homodimers. These complexes can then bind at specific promoter DNA sequences called *lux* boxes affecting the expression of target QS-regulated genes.

The *luxI* and *luxR* genes are in most cases linked within the bacterial genome. Some organisms have several LuxI family proteins that direct the synthesis of *N*-AHL signal molecules sometimes with diverse acyl moieties. Each of these LuxI proteins has an associated LuxR protein, and the different LuxI/R systems usually interact extensively and are hierarchically organized. The best studied of these hierarchical systems is probably that of *Pseudomonas aeruginosa* (Jimenez et al. 2012). This organism has two *N*- AHL-QS systems, the Las and Rhl systems. LasI directs the synthesis of *N*-(3-oxo-dodecanoyl)-L-homoserine lactone (3-oxo-$C_{12}$-HSL) which interacts with

LasR and activates target promoters. RhlI directs the synthesis of *N*-(butanoyl)-L-homoserine lactone ($C_4$-HSL) which interacts with the cognate regulator RhlR, thus activating its gene promoters. The Las and Rhl systems are connected and regulate the production of multiple virulence factors such as rhamnolipid, elastase, and pyocyanin production as well as biofilm formation (Jimenez et al. 2012).

The LasIR and RhlIR system also regulate the synthesis of PQS, (for *Pseudomonas* quinolone signal; 2-heptyl-3-hydroxy-4 (1H)-quinolone) and its precursor 2-heptyl-3-hydroxy-4(1H)-hydroxyquinolone (HHQ). Both PQS and HHQ act as QS signals in *Pseudomonas aeruginosa*, whereas other species of *Pseudomonas* and *Burkholderia* species do not synthesize PQS but use HHQ as a QS signal.

## 2.2    LuxMN Is a Second QS System

A second pathway of *N*-AHL-dependent QS that is distinct from LuxIR has been described in *Vibrio harveyi* (Waters and Bassler 2005). In this organism, synthesis of *N*-AHL (specifically 3-hydroxy butanoyl-homoserine lactone; 3-OH-BHL) is catalyzed by LuxM, which is unrelated to LuxI and sensed by a periplasmic loop of LuxN, a histidine kinase in the cytoplasmic membrane. This pathway acts in concert with two other QS pathways mediated by CAI-I, which in *V. harveyi* is (*Z*)-3-aminoundecan-4-one (see below) and AI-2, a furanosyl borate diester, to activate bioluminescence and inhibit exopolysaccharide production and type III secretion at high cell density. Each pathway involves a different histidine kinase, but they converge at the cytoplasmic phosphotransfer protein LuxU. This protein can exchange phosphoryl groups with the $\sigma^{54}$-dependent activator LuxO. At low cell density, signal concentration is low, and LuxN acts as a kinase resulting in autophosphorylation. The phosphoryl group is then transferred via LuxU to LuxO. This leads to activation of synthesis of small RNA species that together with the protein Hfq inhibit the transcription of *luxR*, which encodes an activator of bioluminescence (hence light is not produced) (Fig. 3).

At high cell density, the 3-hydroxy butanoyl-homoserine lactone, AI-2, and CAI-1 signal molecules are produced at a high level and interact with their cognate sensors. (In the case of AI-2, the signal binds to LuxP, a periplasmic protein that is associated with the sensor kinase LuxQ.) These interactions convert the sensor proteins (LuxN, LuxQ, and CqsS) from kinases to phosphatases, resulting in loss of phosphoryl groups from LuxU and LuxO. Consequently LuxO is inactivated, the small RNAs are not synthesized, and LuxR synthesis can proceed. This allows the LuxR transcriptional activator (which is unrelated to LuxR of *Vibrio fischeri*) to activate expression of bioluminescence genes.

## 2.3    Enzymatic Degradation of *N*-AHL Signals

A number of enzymes capable of degradation of *N*-AHL signals have been described (LaSarre and Federle 2013). The two principal classes are acylases that release the fatty acid from the homoserine moiety by hydrolysis and lactonases that open the

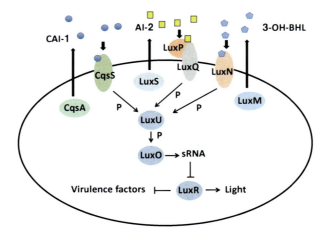

**Fig. 3** QS circuitry in *Vibrio harveyi*. The major QS signals in *Vibrio harveyi* are CAI-1, which is *(S)*-3-hydroxytridecan-4-one, AI-2, a furanosyl borate diester, and 3-OH butanoyl homoserine lactone (3-OH-BHL). Synthesis of CAI-1 requires CqsA, whereas signal perception and transduction require the sensor kinase CqsS. These signaling pathways act in concert to repress virulence factor synthesis and promote bioluminescence when the cognate signals are present, an action mediated by the LuxR regulator. All three systems act via the LuxU phosphotransfer protein and the $\sigma^{54}$ - dependent activator LuxO to modulate *luxR* expression. A "P" next to an arrow indicates the transfer of phosphoryl groups (see text for details)

homoserine lactone ring. A cytochrome P450 oxidoreductase from *Bacillus megaterium* can also act to oxidize *N*-AHLs at the ω-1, ω-2, or ω-3 position, as a first step to their further degradation (Chowdhary et al. 2007). These quorum-quenching enzymes can have diverse roles within the producing organisms. They may be involved in degradation of signals within a species, so that the organism is no longer subject to QS control. In contrast, they may also be involved in competition with other bacteria, where the enzymes degrade the *N*-AHL signals of the competitor to provide an advantage to the producing organism. There is a substantial interest in engineering the use of such enzymes in the control of bacterial disease. The first description of such quorum quenching was the expression of the lactonase AiiA in tobacco and potato to control of symptoms caused by *Erwinia carotovora* (now *Pectobacterium carotovorum*) (Dong et al. 2001). The expression of AiiA reduced the maceration symptoms which are normally caused by extracellular enzymes (pectinases and cellulases) that are under QS control. Further examples are discussed by LaSarre and Federle (2013).

# 3    DSF-Mediated Signaling

Cell-cell signals of the DSF (diffusible signal factor) family are *cis*-2-unsaturated fatty acids of different chain lengths and branching (Fig. 1). The first of these to be described was 11-methyl-*cis*-2-dodecenoic acid (which was named DSF) from the plant pathogen *Xanthomonas campestris* (Barber et al. 1997; Wang et al. 2004; Ryan et al. 2015). The *cis*-unsaturated double bond at the 2 position is a key structural feature for the signaling and regulatory activities; *trans* isomers and saturated derivatives have little or no activity (Wang et al. 2004). Different family members have been described in *Burkholderia cenocepacia* (*cis*-2-dodecenoic acid; BDSF), *Pseudomonas aeruginosa* (*cis*-2-decenoic acid), *Xylella fastidiosa* (*cis*-2-tetra-decenoic acid; XfDSF), and *Xanthomonas oryzae* (*cis,cis*-11 methyldodeca-2,5-dienoic acid; CDSF) (Fig. 1).

## 3.1    The Rpf Proteins and DSF Signaling in Xanthomonads

In *Xanthomonas*, the synthesis and perception of the DSF signal require products of genes within the *rpf* cluster (for regulation of pathogenicity factors) (Barber et al. 1997; Slater et al. 2000). DSF signaling in *Xanthomonas* positively regulates the synthesis of multiple virulence factors including extracellular enzymes and extracellular polysaccharides but negatively regulates biofilm formation/aggregation. The synthesis of DSF is totally dependent on RpfF, which has amino acid sequence relatedness to enoyl CoA hydratase and is a member of the crotonase superfamily of proteins (Barber et al. 1997). The *rpfF* gene is downstream of and transcriptionally linked to *rpfB*, which encodes a long chain fatty acyl CoA ligase, although *rpfF* also has its own promoter. RpfB does not have a role in DSF synthesis but rather in DSF turnover (see below).

The sensing and transduction of the DSF signal in *Xanthomonas* depend upon a two-component regulatory system encoded by the *rpfGHC* operon, which is adjacent to *rpfF* but convergently transcribed (Ryan et al. 2015). RpfC is a complex sensor kinase comprising an N-terminal membrane-associated sensory input domain, histidine kinase and histidine kinase acceptor (HisKA) domains, a CheY-like receiver (REC) domain, and a C-terminal histidine phosphotransfer (HPT) domain. The RpfG regulator comprises a REC domain and an HD-GYP domain, which is a phosphodiesterase involved in degradation of the second messenger cyclic di-GMP (Ryan et al. 2015). RpfH is a novel protein with four transmembrane helices that has amino acid sequence similarity to the sensory input domain of RpfC but no known function (Slater et al. 2000). DSF signal transduction is believed to involve autophosphorylation of RpfC, followed by phosphorelay and finally phosphotransfer to the REC domain of the RpfG regulator. Phosphorylation of RpfG leads to its activation as a cyclic di-GMP phosphodiesterase and consequent alterations in the level of cyclic di-GMP in the cell, which influences the synthesis of virulence factors by diverse mechanisms (Ryan et al. 2015). A simplified scheme is shown in Fig. 4a. RpfC acts to positively regulate synthesis of virulence factors, but to negatively regulate DSF synthesis (Slater et al. 2000; Wang et al. 2004), although this requires neither phosphorelay nor involvement of RpfG (reviewed in He and Zhang 2008).

Genome sequencing indicates the presence of a largely conserved *rpf* gene cluster in all xanthomonads as well as in *Stenotrophomonas* spp. some of which are opportunistic human pathogens. The *rpfH* gene is not fully conserved however.

**Fig. 4** Two "core" pathways of signaling involving DSF family signals are exemplified by *Xanthomonas* and *Burkholderia* species (**a, b**, respectively). In both cases, the DSF signal is synthesized by an RpfF homolog and is linked to the turnover of the second messenger cyclic di-GMP. In xanthomonads (**a**), signal perception involves the sensor kinase RpfC and two-component regulator RpfG, which is an HD-GYP domain cyclic di-GMP phosphodiesterase. A "P" next to the arrow indicates the transfer of phosphoryl groups. In *Burkholderia* (**b**) signal sensing involves RpfR, a cytoplasmic GGDEF-EAL domain protein implicated in cyclic di-GMP degradation (see text for details)

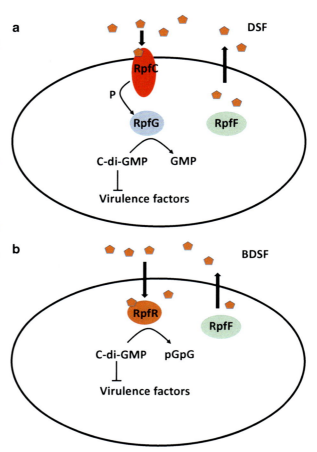

Furthermore a role for DSF signaling in the virulence of a number of these other bacteria has now been described.

Some of these organisms have an alternate sensor for DSF called RpfS, a histidine kinase that is predicted to be cytoplasmic and recognizes DSF through an N-terminal PAS sensor domain (An et al. 2014). RpfS is not widely conserved, however, and has been considered accessory to the core pathway in DSF transduction involving RpfGC.

## 3.2   A Second Core Pathway of DSF Signaling

A second core pathway in DSF family signal transduction was first identified in *Burkholderia*. The synthesis of the *Burkholderia* signal (BDSF) depends on a homolog of RpfF, but in this case signal perception depends upon RpfR, a protein with PAS, GGDEF, and EAL domains (Deng et al. 2012; Fig. 4b). GGDEF and EAL domains are implicated in the synthesis and degradation, respectively, of the second messenger cyclic di-GMP (Römling et al. 2013). In vitro, RpfR exhibits cyclic di-GMP phosphodiesterase activity that is modulated by binding of BDSF to the N-terminal PAS domain (Deng et al. 2012). The *rpfF* and *rpfR* genes are adjacent and convergently transcribed. The RpfR-RpfF system is widely conserved not only in *Burkholderia* species but also in bacteria from related genera such as *Achromobacter* and unrelated Enterobacteriaceae including *Enterobacter*, *Cronobacter*, *Yersinia*, and *Serratia* (Deng et al. 2012). Recently it has been reported that the RpfFR system of *Cronobacter* regulates a diverse range of functions (Suppiger et al. 2016). A second sensing system for BDSF in *B. cenocepacia* involves BCAM0227, a complex sensor kinase that is not a homolog of RpfC of *Xanthomonas* (McCarthy et al. 2010). However unlike RpfR, BCAM0227 is restricted to *B. cenocepacia* suggesting that it is an accessory sensor. Notably the two "core" pathways exemplified by RpfFR in *B. cenocepacia* and RpfFGC in *X. campestris* both link sensing of a DSF family signal to cyclic di-GMP turnover, but the mechanisms are completely different.

*P. aeruginosa* produces *cis*-2-decenoic acid (Fig. 1), a factor that can induce dispersion of biofilms produced by *P. aeruginosa* as well as other bacteria (Davies and Marques 2009). The enzyme responsible for the synthesis of *cis*-2-decenoic acid is an RpfF homolog called DspI (Amari et al. 2013). The *dspI* gene is located in a cluster of genes encoding enzymes implicated in fatty acid metabolism. *P. aeruginosa* does not have an *rpfF-rpfC-rpfG* or *rpfF-rpfR* gene cluster, and the identity of the sensor for this signal is not known. Homologs of DspI occur in over ten *Pseudomonas* species.

### 3.3 RpfF and Signal Synthesis

In vitro studies of the RpfF homolog from *B. cenocepacia* have shown that it is a bifunctional crotonase having both desaturase and thioesterase activity (Bi et al. 2012). This *B. cenocepacia* enzyme acts upon 3-hydroxylated fatty acyl-ACP, an intermediate of fatty acid biosynthesis, to produce BDSF (Bi et al. 2012). The *Xanthomonas* RpfF enzyme also exhibits activity as a broad specificity thioesterase and desaturase in vitro. RpfF is the only member of the crotonase superfamily with both desaturase and thioesterase activity. A model for the action of RpfF is that the enzyme first works as a dehydratase to convert 3-hydroxydodecanoyl-ACP to *cis*-2-dodecenoyl-ACP and then as a thioesterase to release free BDSF (*cis*-2-dodecenoic acid). RpfF can generate free saturated fatty acids from any fatty acyl-ACP substrate through its thioesterase activity. The synthesis of BDSF in the in vitro assay requires the addition of an exogenous acyl-ACP synthetase to reverse the thioesterase reaction. It is unclear how the two actions of dehydratase and thiosterase are coordinated in vivo to produce BDSF (Bi et al. 2012). The dual nature of RpfF is consistent with observations that mutation of *rpfF* affects the appearance in culture supernatants of saturated fatty acids as well as unsaturated fatty acids of the DSF family.

In vivo, individual bacteria can produce multiple DSF family signals that are all dependent on RpfF for their synthesis (see, e.g., He et al. 2010; Zhou et al. 2015a). This suggests that the enzyme does not have a strict specificity for a particular substrate. The available evidence suggests that the pattern of signals produced is not regulated by differences in specificity of different RpfF synthases but rather by the supply of different substrates. Consistent with this contention, in *X. oryzae* pv. *oryzae*, the rice pathogen produces three signals (DSF, BDSF, CDSF) with different time courses during growth and are present in different ratios depending on the culture medium (He et al. 2010). The substrate for DSF synthesis in vivo must be 11-methyl-3-hydroxydodecanoyl-ACP, with the 11-methyl substitution derived from leucine via the branched chain fatty acid synthetic pathway (Bi et al. 2012; Zhou et al. 2015a). Recent work has identified minor signals of the DSF family in *Xanthomonas* including *cis*-9-methyl-2-dodecenoic acid and *cis*-10-methyl-2-dodecenoic acid, where the 10-methyl substitution is probably derived from isoleucine (Deng et al. 2015; Zhou et al. 2015a).

Whether production of multiple DSF family signals by one organism has any biological relevance is unclear. For example, there are no reports that different signals induce different responses in the producing organisms. However the systems for sensing DSF family signals within a particular organism appear to be attuned to the major signal produced by that organism. For example, *Xanthomonas* and *Xylella fastidiosa* generate *cis*-11-methyl-dodecenoic acid and *cis*-2-tetradecenoic acid, respectively, as major signals, each is more responsive to its own signal than the heterologous one (Beaulieu et al. 2013).

## 3.4    RpfB and Signal Degradation

Although RpfB was originally thought to be involved in DSF synthesis, more recent work has established that it has a different role, acting in the mobilization of (saturated) free fatty acids generated by the thioesterase action of RpfF and in the degradation of DSF (Bi et al. 2014). Work in *Xanthomonas* has shown that RpfB, which is a predicted fatty acid CoA ligase, activates free saturated fatty acids allowing their use in phospholipid biosynthesis. In this way RpfB counteracts the thioesterase activity of RpfF. Although RpfB has little activity against BDSF or DSF in vitro (Bi et al. 2014; Zhou et al. 2015b), *Xanthomonas* cells can degrade exogenous DSF, and RpfB has a role in this process (Zhou et al. 2015b). It is suggested that in vivo, the substrate specificity of RpfB can be modulated by additional factors (cofactors, salts, or metals) or by an alteration in conformation which could conceivably involve interactions with other protein (Zhou et al. 2015b).

## 4    Gamma-Butyrolactone-Mediated Signaling

Gamma-butyrolactones (GBLs) have been identified as signaling molecules in Actinobacteria and principally in *Streptomyces* species (Takano 2006; Polkade et al. 2016). GBL signaling within different streptomycetes acts to regulate morphological differentiation and antibiotic production (Takano 2006). The first such molecule described was the autoregulatory factor or A-factor (2-isocapryloyl-3Rhydroxymethyl-gamma-butyrolactone) from *Streptomyces griseus* (Fig. 1). A-factor regulates streptomycin production and sporulation. The determination of the structure of a number of GBLs involved in signaling shows that these share the 3*R*-hydroxymethyl-gamma-butyrolactone moiety but differ in the length, branching, and stereochemistry of the fatty acid side chain which in general is specific for each species (Polkade et al. 2016). Nevertheless most of these molecules regulate the production of different antibiotics.

Different GBL signaling circuits occur in different *Streptomyces* species (Biarnes-Carrera et al. 2015). In *S. griseus*, A-factor synthesis is catalyzed by AfsA, and signal recognition involves ArpA, a repressor of the TetR family. Binding of the A-factor to the ArpA homodimer causes its release from the promoter of *adpA*, which encodes a master regulator of streptomycin synthesis and morphological differentiation (Fig. 5). The genes encoding the synthase and receptor for A-factor (*afsA* and *arpA*, respectively) are separated by 100 kb in the genome. This contrasts with what is seen in other *Streptomyces* species such as *S. coelicolor*, where the genes encoding the synthase and receptor are divergently transcribed and have overlapping promoters (Biarnes-Carrera et al. 2015).

The proposed synthetic mechanism of A-factor synthesis by AfsA involves 8-methyl-3-oxononanoyl-acyl carrier protein and dihydroxyacetone phosphate (DHAP) as substrates (Kato et al. 2007). Beta-ketoacyl transfer to the hydroxyl group of DHAP catalyzed by AfsA produces 8-methyl-3-oxononanoyl-DHAP ester as a product. This ester is nonenzymatically converted to a butenolide phosphate by intramolecular aldol condensation. The butenolide phosphate is then reduced by BprA that was encoded just downstream of *afsA*. The phosphate group on the resultant butanolide is finally removed by a phosphatase, resulting in formation of A-factor. The 8-methyl-3-oxononanoyl-DHAP ester is also converted to A-factor by a second pathway. In this scheme, the phosphate group on the ester

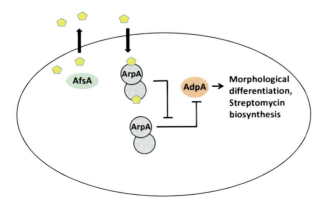

**Fig. 5** A-factor (Gamma-butyrolactone) signaling in *Streptomyces griseus*. A-factor synthesis is catalyzed by AfsA, and signal recognition involves ArpA, a repressor of the TetR family. Binding of the A-factor to the ArpA homodimer causes its release from the promoter of *adpA*, which encodes a master regulator of streptomycin synthesis and morphological differentiation (see text for details)

is first removed by a phosphatase, and the dephosphorylated product is converted nonenzymatically to a butenolide. Reduction of this butenolide by a reductase (different from BprA) generates A-factor. The ability of cloned *afsA* to direct production of an A-factor activity in *Escherichia coli* (Kato et al. 2007) suggests that AfsA is the key enzyme for the biosynthesis of GBLs and that the reductase (s) and phosphatase(s) are commonly present in bacteria and hence are not specific for A-factor biosynthesis.

As outlined above, the organization of genes involved in GBL signaling is different within different *Streptomyces* genomes. For example, in *S. coelicolor*, the gene encoding the receptor (*scbR*) is divergently transcribed from *scbA*, which encodes the synthase (Takano et al. 2001). Furthermore, the promoters of the two genes overlap, indicating the possible occurrence of transcriptional interference. Such an organization may be required to allow tight regulation of synthesis of prodigiosin and actinorhodin at relatively low GBL concentration.

## 5    Other Lipid-Based QS Systems

### 5.1    Hydroxylated Fatty Acid Esters in *Ralstonia*

The plant pathogen *Ralstonia solanacearum* produces the fatty acid esters 3-OH palmitic acid methyl ester (3-OH PAME) and (*R*)-methyl 3-hydroxymyristate [(*R*)-3-OH MAME] as QS signals (Clough et al. 1997; Kai et al. 2015). 3-OH PAME was originally detected in strain AW1 (Clough et al. 1997) but is not found in other strains that produce (*R*)-3-OH MAME instead (Kai et al. 2015). QS mediated by these molecules acts to regulate the synthesis of extracellular enzymes, extracellular polysaccharides, and secondary metabolites called ralfuranones, all which are virulence factors. The signals are synthesized by the methyltransferase PhcB, using S-adenosyl methionine and the 3-hydroxylated fatty acid-ACP as substrates. Signal sensing and transduction involve the membrane-associated histidine kinase PhcS and the two-component regulator PhcR. Interaction of PhcR and the transcriptional regulator PhcA acts to regulate virulence factor synthesis. The available evidence suggests marked differences in mechanistic detail between the AW1 strain, which responds to 3-OH PAME, and strains that respond to (*R*)-3-OH MAME. In the former, it is proposed that in the absence of signal, PhcS acts to phosphorylate PhcR thus negatively regulating PhcA. The presence of a threshold level of 3-OH PAME reduces the kinase activity of PhcS, so that PhcR becomes dephosphorylated thus relieving repression of PhcA. By contrast, in strains that respond to (*R*)-3-OH MAME, it is proposed that sensing of the signal leads to activation of autophosphorylation and phosphotransfer to PhcR. Phosphorylated PhcR is suggested to activate PhcA although the mechanism remains obscure. Intriguingly, the phylogenetic trees of the Phc proteins from *R. solanacearum* strains were divided into two groups, according to their QS signal types: (*R*)-3-OH MAME or (*R*)-3-OH PAME. An added complexity is that in the AW1 strain, PhcA also regulates an *N*-AHL-based QS system called SolIR (Whitehead et al. 2001).

## 5.2 *(S)*-3-hydroxytridecan-4-one in *Vibrio cholerae*

The major QS signal in *Vibrio cholerae* designated CAI-1 has been identified as *(S)*-3-hydroxytridecan-4-one (Higgins et al. 2007). This CAI-1-mediated system works in concert with a QS system mediated by AI-2 (a furanosyl borate diester) to regulate biofilm formation and virulence factor production. Synthesis of CAI-1 requires CqsA, whereas signal perception and transduction require the sensor kinase CqsS. This signaling system influences the phosphorylation level of LuxU and LuxO, which are also involved in AI-2 signaling. These signaling pathways act in concert to repress virulence factor synthesis and promote biofilm formation when the cognate signals are present (Fig. 6). These effects are exerted through an influence on transcription of the HapR regulator, which occurs in an analogous fashion to that described above for LuxR regulation in *Vibrio harveyi*.

CqsA uses S-adenosyl methionine and decanoyl-CoA to produce 3-aminotridec-2-en-4-one, a reaction that depends upon pyridoxal phosphate as a cofactor (Wei et al. 2011). 3-aminotridec-2-en-4-one is converted to CAI-1 in two steps: spontaneous conversion to tridecane-3,4-dione followed by an enzyme-catalyzed conversion to CAI-1. Intriguingly, the CAI-1 signal produced by *V. harveyi* has been isolated as the CqsS ligand and identified as *(Z)*-3-aminoundecan-4-one (Ng et al. 2011). *V. harveyi* CqsA and CqsS are extremely selective for the production and detection, respectively, of this molecule, whereas the *V. cholerae* CqsA/CqsS can produce and sense both *(Z)*-3-aminoundecan-4-one and *(S)*-3-hydroxytridecan-4-one (Ng et al. 2011).

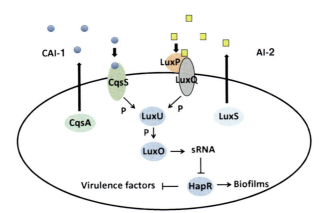

**Fig. 6** QS circuitry in *Vibrio cholerae*. The major QS signals in *Vibrio cholerae* are CAI-1, which is *(S)*-3-hydroxytridecan-4-one and AI-2, a furanosyl borate diester. Synthesis of CAI-1 requires CqsA, whereas signal perception and transduction require the sensor kinase CqsS. This signaling system influences the phosphorylation level of LuxU and LuxO, which are also involved in AI-2 signaling. A "P" next to an arrow indicates the transfer of phosphoryl groups These signaling pathways act in concert to repress virulence factor synthesis and promote biofilm formation when the cognate signals are present, an action mediated by the HapR regulator (see text for details)

**Table 1** A summary of lipid signaling molecules in bacteria, together with signal synthases and their substrates and components involved in signal perception and transduction

| Signal molecule | Synthase | Substrates | Sensor |
|---|---|---|---|
| $N$-acyl homoserine lactone | LuxI family protein | S-adenosyl methionine and fatty acyl-ACP | LuxR, transcriptional regulator |
| $N$-acyl homoserine lactone | LuxM | ? | LuxN histidine kinase |
| Cis-2-unsaturated fatty acid (DSF family) | RpfF | 3-OH fatty acyl-ACP | RpfC histidine kinase (*Xanthomonas*) or RpfR GGDEF-EAL domain containing protein (*Burkholderia*) |
| A-factor (a gamma-butyrolactone) | AfsA | 8-methyl-3-oxononanoyl-ACP and dihydroxyacetone phosphate | ArpA transcriptional repressor |
| CAI-1 (*Vibrio cholerae*) | CqsA | S-adenosyl methionine and decanoyl-CoA | CqsS histidine kinase |
| CAI-1 (*Vibrio harveyi*) | CqsA | S-adenosyl methionine and octanoyl-CoA | CqsS histidine kinase |
| 3-OH-PAME | PhcB | S-adenosyl methionine and 3-OH palmitoyl-ACP | PhcS histidine kinase |

The examples discussed above (summarized in Table 1) deal with QS within individual organisms. In the next section we will consider the role of these molecules in interspecies and inter-kingdom signaling.

# 6    Interspecies and Inter-kingdom Signaling

It is now appreciated that many bacteria occur in polymicrobial communities; many human diseases are polymicrobial in nature (Short et al. 2014). Interactions between organisms in such communities can be mediated by a number of factors, to include the same signals that individual bacteria use for QS. The possibility of interplay between species that utilize the same class of QS signal is evident. However some bacteria can sense signals that they do not themselves synthesize, a process that has been termed eavesdropping. Such interspecies signaling can act to influence bacterial behavior, including biofilm formation and antibiotic tolerance, reflecting the impact that interspecies signaling may have on the efficacy of antibiotic therapies. It is anticipated that research in the next few years will reveal more of these mechanisms.

The next few years should also see an expansion of understanding of the role of QS signals in inter-kingdom signaling. This can occur between bacteria and microbes such as yeast, plants, or mammalian cells. For example, QS signals can act to inhibit morphological transitions in the dimorphic fungus *Candida albicans*

(Hogan et al. 2004; Boon et al. 2008), can trigger or modulate host defense responses in both mammalian and plant cells (Teplitski et al. 2011; Schenk et al. 2014), and in some cases can act as virulence factors in their own right (Cooley et al. 2008).

## 7    Research Needs

Since QS acts in regulation of biofilm formation, antibiotic tolerance, and virulence factor synthesis in many pathogenic bacteria, interference with QS has received a great deal of attention as a route toward reducing the virulence of pathogens or making them more susceptible to existing antibiotics. Inhibition of QS could be effected at several steps: inhibition of signal synthesis, signal sequestration, signal degradation by enzymes (quorum quenching), or the inhibition of signal perception and transduction by small molecules. It has been proposed that since small molecule QS inhibitors would not kill bacterial cells, the target organisms would not develop resistance, although this view has been recently challenged (García-Contreras et al. 2016).

The determination of the structure of different signal synthases may further the rational design of inhibitors of their action. Currently a number of structures of LuxI family N-AHL synthases have been described; there is some information for DSF synthases of the RpfF family and a report on the structure CqsA.

In the field of quorum-quenching enzymes, interest has been largely focused on those degrading N-AHLs. Some work on degradation of DSF family signals has been reported, where a role for RpfB in Xanthomonas and CarAB in Pseudomonas has been indicated, although mechanistic details are sketchy. Strains that can degrade DSF can reduce severity of disease symptoms caused by Xanthomonas in brassica when applied to the leaves (Newman et al. 2008). Overexpression of QS signals causing pathogen confusion has also been suggested as a route to disease control (Fray et al. 1999; Mae et al. 2001). Accordingly, expression of RpfF in grape and citrus can reduce virulence and symptom production by Xylella fastidiosa and Xanthomonas citri pv. citri, respectively (Caserta et al. 2014; Lindow et al. 2014). Deployment of such methods (transgenic plants, strains with enhanced capacity for production or degradation of QS signals) in agriculture must take into account the broader issues of environmental impact, including beneficial plant-microbe interactions.

In addition to inhibition of signal synthesis, the identification of small molecule inhibitors of key steps in signal sensing and transduction may define lead compounds for new drugs (Curtis et al. 2014). Such compounds have been identified by screening libraries of structural analogues of inter- and intracellular signal molecules for action against particular signaling components or by screening of much larger libraries of chemical compounds. As indicated above, these molecules may not act as antibiotics per se but rather as inhibitors of virulence factor synthesis or potentiators of antibiotic action.

Finally the definition of elements of QS circuitry may have applications in synthetic biology (see, e.g., Wu et al. 2014; Biarnes-Carrera et al. 2015). In many nonpathogens, QS acts to regulate the synthesis of important secondary metabolites.

*N*-AHL- and GBL-based circuitry may find applications in the field of synthetic biology in approaches for tight control and improved production of important and valuable microbial products.

# References

Amari DT, Marques CNH, Davies DG (2013) The putative enoyl-coenzyme A hydratase DspI is required for production of the *Pseudomonas aeruginosa* biofilm dispersion autoinducer cis-2-decenoic acid. J Bacteriol 195:4600–4610

An SQ, Allan JH, McCarthy Y, Febrer M, Dow JM, Ryan RP (2014) The PAS domain-containing histidine kinase RpfS is a second sensor for the diffusible signal factor of *Xanthomonas campestris*. Mol Microbiol 92:586–597

Barber CE, Tang JL, Feng JX, Pan MQ, Wilson TJ, Slater H, Dow JM, Williams P, Daniels MJ (1997) A novel regulatory system required for pathogenicity of *Xanthomonas campestris* is mediated by a small diffusible signal molecule. Mol Microbiol 24:555–566

Beaulieu ED, Ionescu M, Chatterjee S, Yokota K, Trauner D, Lindow S (2013) Characterization of a diffusible signaling factor from *Xylella fastidiosa*. MBio 4:e00539–e00512

Bi H, Christensen QH, Feng Y, Wang H, Cronan JE (2012) The *Burkholderia cenocepacia* BDSF quorum sensing fatty acid is synthesized by a bifunctional crotonase homologue having both dehydratase and thioesterase activities. Mol Microbiol 83:840–855

Bi H, Yu Y, Dong H, Wang H, Cronan JE (2014) *Xanthomonas campestris* RpfB is a fatty acyl-CoA ligase required to counteract the thioesterase activity of the RpfF diffusible signal factor (DSF) synthase. Mol Microbiol 93:262–275

Biarnes-Carrera M, Breitling R, Takano E (2015) Butyrolactone signalling circuits for synthetic biology. Curr Opin Chem Biol 28:91–98

Boon C, Deng Y, Wang LH, He Y, Xu JL, Fan Y, Pan SQ, Zhang LH (2008) A novel DSF-like signal from *Burkholderia cenocepacia* interferes with *Candida albicans* morphological transition. ISME J 2:27–36

Caserta R, Picchi SC, Takita MA, Tomaz JP, Pereira WE, Machado MA, Ionescu M, Lindow S, De Souza AA (2014) Expression of *Xylella fastidiosa* RpfF in citrus disrupts signaling in *Xanthomonas citri* subsp. *citri* and thereby its virulence. Mol Plant-Microbe Interact 27:1241–1252

Chowdhary PK, Keshavan N, Nguyen HQ, Peterson JA, González JE, Haines DC (2007) *Bacillus megaterium* CYP102A1 oxidation of acyl homoserine lactones and acyl homoserines. Biochemistry 46:14429–14437

Clough SJ, Lee KE, Schell MA, Denny TP (1997) A two-component system in *Ralstonia* (*Pseudomonas*) *solanacearum* modulates production of PhcA-regulated virulence factors in response to 3-hydroxypalmitic acid methyl ester. J Bacteriol 179:3639–3648

Cooley M, Chhabra SR, Williams P (2008) *N*-acylhomoserine lactone-mediated quorum sensing: a twist in the tail and a blow for host immunity. Chem Biol 15:1141–1147

Curtis MM, Russell R, Moreira CG, Adebesin AM, Wang C, Williams NS, Taussig R, Stewart D, Zimmern P, Lu B, Prasad RN, Zhu C, Rasko DA, Huntley JF, Falck JR, Sperandio V (2014) QseC inhibitors as an antivirulence approach for gram-negative pathogens. MBio 5:e02165

Davies DG, Marques CNH (2009) A fatty acid messenger is responsible for inducing dispersion in microbial biofilms. J Bacteriol 191:1393–1403

Deng Y, Wu J, Tao F, Zhang LH (2011) Listening to a new language: DSF-based quorum sensing in gram-negative bacteria. Chem Rev 111:160–173

Deng Y, Schmid N, Wang C, Wang J, Pessi G, Wu D, Lee J, Aguilar C, Ahrens CH, Chang C, Song H, Eberl L, Zhang LH (2012) Cis-2-dodecenoic acid receptor RpfR links quorum-sensing signal perception with regulation of virulence through cyclic dimeric guanosine monophosphate turnover. Proc Natl Acad Sci U S A 109:15479–15484

Deng Y, Liu X, Wu J, Lee J, Chen S, Cheng Y, Zhang C, Zhang LH (2015) The host plant metabolite glucose is the precursor of diffusible signal factor (DSF) family signals in *Xanthomonas campestris*. Appl Environ Microbiol 81:2861–2868

Dong YH, Wang LH, Xu JL, Zhang HB, Zhang XF, Zhang LH (2001) Quenching quorum-sensing-dependent bacterial infection by an *N*-acyl homoserine lactonase. Nature 411:813–817

Fray RG, Throup JP, Daykin M, Wallace A, Williams P, Stewart GS, Grierson D (1999) Plants genetically modified to produce *N*-acylhomoserine lactones communicate with bacteria. Nat Biotechnol 17:1017–1020

Fuqua C, Parsek MR, Greenberg EP (2001) Regulation of gene expression by cell-to-cell communication: acyl-homoserine lactone quorum sensing. Annu Rev Genet 35:439–468

García-Contreras R, Maeda T, Wood TK (2016) Can resistance against quorum-sensing interference be selected? ISME J 10:4–10

He YW, Zhang LH (2008) Quorum sensing and virulence regulation in *Xanthomonas campestris*. FEMS Microbiol Rev 32:842–857

He Y-W, Je W, Cha J-S, Zhang L-H (2010) Rice bacterial blight pathogen *Xanthomonas oryzae* pv. *oryzae* produces multiple DSF-family signals in regulation of virulence factor production. BMC Microbiol 10:187

Higgins DA, Pomianek ME, Kraml CM, Taylor RK, Semmelhack MF, Bassler BL (2007) The major *Vibrio cholerae* autoinducer and its role in virulence factor production. Nature 450:883–886

Hogan DA, Vik A, Kolter R (2004) A *Pseudomonas aeruginosa* quorum-sensing molecule influences *Candida albicans* morphology. Mol Microbiol 54:1212–1223

Jimenez PN, Koch G, Thompson JA, Xavier KB, Cool RH, Quax WJ (2012) The multiple signaling systems regulating virulence in *Pseudomonas aeruginosa*. Microbiol Mol Biol Rev 76:46–65

Kai K, Ohnishi H, Shimatani M, Ishikawa S, Mori Y, Kiba A, Ohnishi K, Tabuchi M, Hikichi Y (2015) Methyl 3-hydroxymyristate, a diffusible signal mediating *phc* quorum sensing in *Ralstonia solanacearum*. Chembiochem 16:2309–2318

Kato JY, Funa N, Watanabe H, Ohnishi Y, Horinouchi S (2007) Biosynthesis of gamma-butyrolactone autoregulators that switch on secondary metabolism and morphological development in *Streptomyces*. Proc Natl Acad Sci U S A 104:2378–2383

LaSarre B, Federle MJ (2013) Exploiting quorum sensing to confuse bacterial pathogens. Microbiol Mol Biol Rev 77:73–111

Lindow S, Newman K, Chatterjee S, Baccari C, Lavarone AT, Ionescu M (2014) Production of *Xylella fastidiosa* diffusible signal factor in transgenic grape causes pathogen confusion and reduction in severity of Pierce's disease. Mol Plant-Microbe Interact 27:244–254

Mae A, Montesano M, Koiv V, Palva ET (2001) Transgenic plants producing the bacterial pheromone *N*-acyl-homoserine lactone exhibit enhanced resistance to the bacterial phytopathogen *Erwinia carotovora*. Mol Plant-Microbe Interact 14:1035–1042

McCarthy Y, Yang L, Twomey KB, Sass A, Tolker-Nielsen T, Mahenthiralingam E, Dow JM, Ryan RP (2010) A sensor kinase recognizing the cell-cell signal BDSF (cis-2-dodecenoic acid) regulates virulence in *Burkholderia cenocepacia*. Mol Microbiol 77:1220–1236

Newman KL, Chatterjee S, Ho KA, Lindow SE (2008) Virulence of plant pathogenic bacteria attenuated by degradation of fatty acid cell-to-cell signaling factors. Mol Plant-Microbe Interact 21:326–334

Ng WL, Bassler BL (2009) Bacterial quorum-sensing network architectures. Annu Rev Genet 43:197–222

Ng WL, Perez LJ, Wei Y, Kraml C, Semmelhack MF, Bassler BL (2011) Signal production and detection specificity in *Vibrio* CqsA/CqsS quorum-sensing systems. Mol Microbiol 79:1407–1417

Pappas KM, Weingart CL, Winans SC (2004) Chemical communication in proteobacteria: biochemical and structural studies of signal synthases and receptors required for intercellular signalling. Mol Microbiol 53:755–769

Platt TG, Fuqua C (2010) What's in a name? The semantics of quorum sensing. Trends Microbiol 18:383–387

Polkade AV, Mantri SS, Patwekar UJ, Jangid K (2016) Quorum sensing: an under-explored phenomenon in the phylum actinobacteria. Front Microbiol 7:131

Römling U, Galperin MY, Gomelsky M (2013) Cyclic di-GMP: the first 25 years of a universal bacterial second messenger. Microbiol Mol Biol Rev 77:1–52

Ryan RP, An SQ, Allan JH, McCarthy Y, Dow JM (2015) The DSF family of cell-cell signals: an expanding class of bacterial virulence regulators. PLoS Pathog 11:e1004986

Schaefer AL, Greenberg EP, Oliver CM, Oda Y, Huang JJ, Bittan-Banin G, Peres CM, Schmidt S, Juhaszova K, Sufrin JR, Harwood CS (2008) A new class of homoserine lactone quorum-sensing signals. Nature 454:595–599

Schenk ST, Hernández-Reyes C, Samans B, Stein E, Neumann C, Schikora M, Reichelt M, Mithöfer A, Becker A, Kogel KH, Schikora A (2014) N-acyl-homoserine lactone primes plants for cell wall reinforcement and induces resistance to bacterial pathogens via the salicylic acid/oxylipin pathway. Plant Cell 26:2708–2723

Short FL, Murdoch SL, Ryan RP (2014) Polybacterial human disease: the ills of social networking. Trends Microbiol 22:508–516

Slater H, Alvarez-Morales A, Barber CE, Daniels MJ, Dow JM (2000) A two-component system involving an HD-GYP domain protein links cell-cell signalling to pathogenicity gene expression in *Xanthomonas campestris*. Mol Microbiol 38:986–1003

Suppiger A, Eshwar AK, Stephan R, Kaever V, Eberl L, Lehner A (2016) The DSF type quorum sensing signalling system RpfF/R regulates diverse phenotypes in the opportunistic pathogen cronobacter. Sci Rep 6:18753

Takano E (2006) Gamma-butyrolactones: streptomyces signalling molecules regulating antibiotic production and differentiation. Curr Opin Microbiol 9:287–294

Takano E, Chakraburtty R, Nihira T, Yamada Y, Bibb MJ (2001) A complex role for the gamma-butyrolactone SCB1 in regulating antibiotic production in *Streptomyces coelicolor* A3(2). Mol Microbiol 41:1015–1028

Teplitski M, Mathesius U, Rumbaugh KP (2011) Perception and degradation of *N*-acyl homoserine lactone quorum sensing signals by mammalian and plant cells. Chem Rev 111:100–116

Wang LH, He Y, Gao Y, Wu JE, Dong YH, He C, Wang SX, Weng LX, Xu JL, Tay L, Fang RX, Zhang LH (2004) A bacterial cell-cell communication signal with cross-kingdom structural analogues. Mol Microbiol 51:903–912

Waters CM, Bassler BL (2005) Quorum sensing: cell-to-cell communication in bacteria. Annu Rev Cell Dev Biol 21:319–346

Wei Y, Perez LJ, Ng WL, Semmelhack MF, Bassler BL (2011) Mechanism of *Vibrio cholerae* autoinducer-1 biosynthesis. ACS Chem Biol 6:356–365

Whitehead NA, Barnard AM, Slater H, Simpson NJ, Salmond GP (2001) Quorum-sensing in gram-negative bacteria. FEMS Microbiol Rev 25:365–404

Wu F, Menn DJ, Wang X (2014) Quorum-sensing crosstalk-driven synthetic circuits: from unimodality to trimodality. Chem Biol 21:1629–1638

Zhou L, Yu Y, Chen X, Diab AA, Ruan L, He J, Wang H, He YW (2015a) The multiple DSF-family QS signals are synthesized from carbohydrate and branched-chain amino acids via the FAS elongation cycle. Sci Rep 5:13294

Zhou L, Wang XY, Sun S, Yang LC, Jiang BL, He YW (2015b) Identification and characterization of naturally occurring DSF-family quorum sensing signal turnover system in the phytopathogen *Xanthomonas*. Environ Microbiol 17:4646–4658

# Fatty Acids as Mediators of Intercellular Signaling

# 16

Manuel Espinosa-Urgel

## Contents

1   Introduction ................................................................................ 274
2   Fatty Acids in AHL Signaling Systems .................................................. 275
3   Fatty Acids and the *Pseudomonas* Quinolone Signal (PQS) ............................. 278
4   Rhamnolipids .............................................................................. 278
5   Fatty Acids and Fatty Acid Derivatives as Signals ..................................... 279
6   Research Needs ............................................................................ 281
References .................................................................................... 282

**Abstract**

Mechanisms of intercellular communication as a function of population density exist in many bacteria. These signaling circuits are based on the release of diffusible molecules to the extracellular medium and their detection and subsequent alteration of global gene expression above certain concentration thresholds. Fatty acids are structural parts of different signal molecules, such as acyl homoserine lactones, where the length and modifications of the acyl side chains play a role as determinants of signal specificity. Yet, fatty acids and fatty acid derivatives are increasingly being reported as intra- and interspecies cell-cell communication signals and also mediate interactions of bacteria with other organisms. These signals appear to be particularly relevant in plant-associated bacteria, but are also present in other microorganisms, and could offer a chance to develop new strategies to combat pathogens.

M. Espinosa-Urgel (✉)
Department of Environmental Protection, Estación Experimental del Zaidín, CSIC, Granada, Spain
e-mail: manuel.espinosa@eez.csic.es

© Springer International Publishing AG, part of Springer Nature 2018                          273
T. Krell (ed.), *Cellular Ecophysiology of Microbe: Hydrocarbon and Lipid Interactions*,
Handbook of Hydrocarbon and Lipid Microbiology, https://doi.org/10.1007/978-3-319-50542-8_7

# 1    Introduction

It is increasingly evident that bacteria are highly interactive and social organisms capable of displaying cooperative traits and coordinated behavior. Collectively and coordinately, bacteria act far more efficiently than they could as single cells. Thus, responses at the population level endow a species with additional mechanisms of adaptation. Fuqua et al. (1994) introduced the term "quorum sensing" (QS) to describe a process of sophisticated cell-to-cell communication that bacteria employ to monitor population density and to change their gene expression and behavior accordingly. Essentially, bacteria actively "talk" to one another, using chemical signals as information units. QS is based on the production of small diffusible molecules that bacteria release into their environment, the extracellular accumulation of which is related to the population density of the producing organisms. These signaling molecules can be detected by bacteria, and in this way, individual cells can sense the local density of bacteria, thus allowing the population as a whole to initiate a concerted action once a minimal threshold stimulatory concentration ("quorum") has been reached (Whitehead et al. 2001; Miller and Bassler 2001; Fuqua and Greenberg 2002; Bassler and Losick 2006).

Extensive research on QS has led to the identification of an increasing – and surely still incomplete – number of structurally diverse chemical signals. Most QS signals are either small ($<1000$ Da) organic molecules or peptides with 5–20 amino acids (Lazazzera 2001; Williams 2007), some of which interact with receptors at the cell surface, while others act following internalization. Representative microbial QS signals include $N$-acyl homoserine lactones (AHLs), where the acyl chain ranges from C4 to C18; structural analogs of AHLs, such as $p$-coumaric acid homoserine lactones and γ-butyrolactones, 2-alkyl-4-quinolones (AQs), certain fatty acids and fatty acid methyl esters, furanosyl diesters, and linear, modified, or cyclic peptides. Besides being responsible for QS responses in a single-species population, at least several of these signals have also been shown to mediate interspecies and interkingdom communication (Williams 2007; Pacheco and Sperandio 2009; Venturi and Fuqua 2013). Also, molecular mimics produced by plants and interfering with bacterial signaling have been described (Gao et al. 2003; Corral-Lugo et al. 2016). In this chapter, the role of fatty acids as chemical signals per se and as constituents of other QS signal molecules will be briefly revised.

Holding back gene expression before the quorum is achieved is thought to be primordial to QS's biological functions in many organisms. One hypothesis is that QS enables bacteria to coordinate activities so they can operate in groups. In this respect, QS is attracting significant interest from the perspective of social evolution, fitness, and the benefits at the population level associated with costly cooperative behaviors (Diggle et al. 2007a). Inhibiting gene expression when population density is low could serve this purpose, for example, by delaying virulence factor production until enough cells amass to produce effective levels (Rumbaugh et al. 2000, 2009). Restrained gene expression may also benefit groups by enabling coordinated stealth attack during infection (Winzer and Williams 2001; Fuqua and Greenberg 2002). This may be an advantage because QS-controlled factors would be hidden until a large force assembles.

Other situations where bacteria employ diffusible chemical signals have been described. In complex environments, for example, the size of the quorum is not fixed but will vary according to the relative rates of production and loss of signal molecule, which depend on many naturally fluctuating environmental parameters. Factors influencing the chemical gradients of QS molecules, such as local flow and diffusion rates, can determine their accumulation and concentration and thereby the expression of QS-regulated phenotypes. It is also possible for a single bacterial cell to switch from the "non-quorate" to the "quorate" state as has been observed for *Staphylococcus aureus* trapped within an endosome in endothelial cells (Qazi et al. 2001). Consequently, quorum sensing cannot be defined simply in terms of cell population density, and some researchers have proposed new theories, each emphasizing different adaptive functions, for the QS process. In this way, QS can also be considered in the context of "diffusion sensing" (DS) or "compartment sensing" (CS) or even "efficiency sensing" (ES), where the signal molecule supplies information with respect to the local environment and spatial distribution of the cells rather than, or as well as, cell population density (Redfield 2002; Hense et al. 2007). Therefore, benefits of QS gene regulation do not only require that bacteria engage in group or social activities. According to these ideas, QS signals are also used to gauge the rate at which secreted products would be lost by diffusion and flow. The expression threshold of QS could also serve this function by enabling bacteria to conserve energy for exoproduct synthesis until conditions permit signal (and hence exoproduct) accumulation. Thus, restrained gene expression in pre-quorum conditions is critical to the postulated benefits of QS for bacterial groups and individual cells.

## 2  Fatty Acids in AHL Signaling Systems

The chemical structure of the first AHL signal was unveiled more than 2 decades ago (Eberhard et al. 1981) from the luminous marine bacterium *Vibrio fischeri*, whose QS system had been the first to be identified (Nealson and Hastings 1979). These bacteria live in the light organs of the Hawaiian squid *Euprymna scolopes* and produce luminescence, which helps the animal to mask its shadow by counter illumination and to escape from predators. In return, the bacteria benefit from nutrients and shelter from their host. Interestingly, bioluminescence is exhibited by the bacteria when they proliferate and achieve high number, thus, only when they are in the symbiotic mode of life and not in the free-living state.

Regulation of bioluminescence in *V. fischeri* is mediated by a QS system that involves two key components: *luxI*, the gene encoding the AHL synthase enzyme which produces the *N*-(3-oxohexanoyl)-L-homoserine lactone (3-oxo-C6-HSL) signaling molecule (Eberhard et al. 1981), and the cytoplasmic autoinducer receptor/DNA-binding transcriptional activator (Engebrecht et al. 1983) encoded by *luxR*. At low cell densities, the bacterial signal synthase (LuxI) constitutively produces basal amounts of 3-oxo-C6-HSL, which immediately diffuses in and out of the cell into the surrounding environment, and increases in concentration with increasing cell density. Once a critical threshold concentration of this signal molecule has been reached, it can bind the regulatory protein LuxR inside the cell, and the LuxR/3-oxo-C6-HSL complex then binds to its target promoter, activating transcription of the*luxICDABEG* operon (Stevens et al. 1994). An interesting feature of this system is that *luxI* is a target for the LuxR-AHL complex. Thus, activation of the QS cascade results in increased expression of the LuxI synthase, leading to the production of more AHLs. This acts as a positive feedback loop and significantly amplifies the QS effect. Therefore, the process was originally called autoinduction, and consequently the QS signal molecules are also referred to as "autoinducers" (Miller and Bassler 2001).

Similar QS networks involving AHLs as autoinducer signals have been described in a large number of Gram-negative proteobacteria including *Agrobacterium* spp., *Rhizobium* spp., *Pseudomonas* spp., *Brucella* spp., and *Vibrio* spp. AHLs are components of QS systems implicated in biofilm formation (Rice et al. 2005), pathogenicity (mutants that are defective in QS are usually avirulent or significantly reduced in virulence) (Camara et al. 2002), plant-microbial interactions (Cha et al. 1998), swarming motility (Daniels et al. 2004), conjugation (Whitehead et al. 2001), and growth inhibition (Gonzalez and Marketon 2003).

Thus, proteins of the LuxI and LuxR families constitute the canonical QS circuits in many bacteria (Fig. 1). The conserved homoserine lactone ring moiety of AHLs is linked to a fatty acid side chain that varies in length (Fig. 1). AHL signals with side chains ranging from 4 to 18 carbon atoms have been described (Whitehead et al. 2001). Variability also exists in the third carbon position of the acyl chain, where a hydrogen, hydroxyl, or oxo-substitution can be found. Unsaturated chains have been also identified (Von Bodman et al. 2003), as well as non-fatty acid substituents (Schaefer et al. 2008). The variations in the fatty acid chain of AHL indicate that each LuxI-type protein

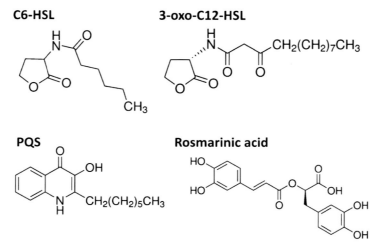

**Fig. 1** Sample structures of fatty acid-containing signal molecules of the AHL (C6-HSL; 3-oxo-C12-HLS) and quinolone (PQS) families. The structure of the plant-derived AHL mimic rosmarinic acid is also shown

possesses an acyl-binding pocket that precisely fits a particular side-chain moiety. The structures and amino-acid sequences of different LuxR homologs also vary considerably, with acyl-binding pockets that until recently were thought to allow each LuxR to bind and be activated specifically by its cognate signal. Hence, these structural features would determine the specificity in signal production, perception, and response. In fact, with the exception of *Pseudomonas aeruginosa* and several other bacteria that produce more than one AHL molecule, a particular bacterial species characteristically produces a specific AHL variant and responds only to AHLs within a certain size range and/or modification in their side chains. However, some LuxR homologs have been found to recognize and bind more than one AHL molecule. This flexibility may be the basis for cross-talk phenomena between different bacterial species and interkingdom signaling. Recently, the *P. aeruginosa* LuxR homolog RhlR, which responds to C4-HSL produced by the bacterium, has been shown to recognize also the plant secondary metabolite rosmarinic acid ((*R*)-*O*-(3,4-dihydroxycinnamoyl)-3-(3,4-dihydroxyphenyl)lactic acid), even with higher affinity than its cognate AHL (Corral-Lugo et al. 2016). Despite their structural differences (Fig. 1), rosmarinic acid acts as a mimic of C4-HSL, both in vitro and in vivo, causing similar cellular responses as the bacterial QS system.

Although in most cases AHLs are synthesized from S-adenosyl methionine (SAM) and particular acyl carrier proteins by LuxI-type enzymes (Parsek et al. 1999), a different class of AHL synthases has also been found in the genus *Vibrio*: LuxM and AinS. Despite their lack of homology with the LuxI family, these enzymes catalyze AHL formation from the same substrates as the LuxI proteins (Milton et al. 2001). A third potential AHL synthase, HdtS, was described in *Pseudomonas fluorescens* F113. HdtS is related to the lysophosphatidic acid acyltransferase protein family, and at present it is not clear whether HdtS is really involved in AHL production (Laue et al. 2000; Cullinane et al. 2005), although evidence suggests that AHLs may be produced as by-products of the reaction carried out by HdtS.

One question that still is requires detailed study is how AHLs are released by producing cells and detected and incorporated into receiving cells of the same or different species. The general notion is that short-chain AHLs are able to diffuse freely across the bacterial cell envelope, whereas long-chain AHLs would need to be transported, both in and out of the cell. In *P. aeruginosa*, the multidrug efflux pump MexA-MexB-OprM is involved in active export of 3-oxo-C12-HSL (Pearson et al. 1999), but the uptake mechanism of long-chain AHLs is still unclear in many cases, since the outer membrane is an effective barrier against diffusion of long-chain fatty acids and likely structurally similar molecules. Recently, the fatty acid transporter FadL of *Sinorhizobium meliloti* was found to participate in detection and transport of long-chain AHLs (3-oxo-C16:1- and 3-oxo-C14-HSLs), but not of AHLs with shorter acyl chains (Krol and Becker 2014). This, and the previously mentioned flexibility shown for some LuxR homologs in terms of signal molecule binding, could indicate that fatty acids determine the specificity of the response not only through the interaction with the transcriptional regulator but also, and perhaps even more importantly, at the level of recognition by internalization systems.

## 3        Fatty Acids and the *Pseudomonas* Quinolone Signal (PQS)

Besides AHLs, the 2-alkyl-4-quinolone signals (AQ) also contain fatty acids in their structure (Fig. 1). This family of molecules was first chemically identified for the antimicrobial activity of many of its members (Hays et al. 1945). Although more than 50 different AQs are known to be produced by *Pseudomonas aeruginosa*, only some of them function as QS signals, being the two major 2-heptyl-3-hydroxy-4-quinolone ("*Pseudomonas* quinolone signal," PQS) and its immediate precursor 2-heptyl-4-quinolone (HHQ). Other important AQs produced by this organism are the 2-nonyl-4-quinolone (NHQ) and the 2-heptyl-4-quinolone N-oxide (HHQNO). AQs such as PQS are synthesized from anthranilate and fatty acid biosynthesis derivatives as precursors. AQs are more restricted in bacteria than AHLs and have so far only been detected in *P. aeruginosa* and certain *Burkholderia* and *Alteromonas* species (Diggle et al. 2006). Besides its role in cell-to-cell communication, PQS has also been studied for its relevance in other biologically important functions, including antibiotic activity, oxidative stress response, and iron homeostasis. PQS exhibits iron-chelating properties at physiological pH, being able to act as an iron trap rather than a siderophore (Diggle et al. 2007b).

The genes controlling AQ biosynthesis and signal transduction are arranged in the *pqsABCDE* operon, which is under the positive control of the transcriptional regulator PqsR (MvfR) and negatively regulated by RhlR (Gallagher et al. 2002). In turn, PqsR shows positive control by LasR and negative regulation by RhlR, thus indicating that quinolone signaling is finely tuned by the AHL QS cascade of *P. aeruginosa*. The *pqsABCD* gene products are responsible for the conversion of anthranilate into HHQ, which is in turn oxidized to PQS by the action of PqsH, a predicted monooxygenase coded by *pqsH*, a gene located downstream of the *pqs* cluster. PQS and HHQ both function as autoinducers since they can both trigger the expression of the *pqsA* promoter in a *PqsR*-dependent manner. The first four genes of the *pqsABCDE* operon are responsible for HHQ biosynthesis, while *pqsE* is involved in AQs response (Déziel et al. 2004; Farrow et al. 2008; Yu et al. 2009). PQS has been shown to interfere with biofilm formation, swarming motility, and iron uptake by *Pseudomonas putida*, which does not produce this signal molecule but shows altered gene expression in response to it (Fernández-Piñar et al. 2011).

## 4        Rhamnolipids

Rhamnolipids are a complex group of extracellular tensoactive glycolipids produced by different bacterial species. They were first identified in *Pseudomonas aeruginosa* as biosurfactants that consist in a mix of rhamnose-bound alkyl chains, the di-rhamnolipidsα-L-rhamnopyranosyl-α-L-rhamnopyranosyl-β-hydroxydecanoyl-β-hydroxydecanoate (Rha-Rha-$C_{10}$-$C_{10}$) and α-L-rhamnopyranosyl-α-L-rhamnopyranosyl-β-hydroxydecanoate (Rha-Rha-$C_{10}$), and their mono-rhamnolipid congeners Rha-$C_{10}$-$C_{10}$ and Rha-$C_{10}$ (Soberón-Chávez et al. 2005). A variety of rhamnolipid molecules with varying length in their alkyl chains have later been reported in *Pseudomonas* and *Burkholderia* strains. The structural diversity, origins, and roles of different rhamnolipids have been thoroughly reviewed by Abdel-Mawgoud et al. (2010). Rhamnolipid production depends on environmental conditions and is a sequential process where the acyltransferase RhlA participates in the synthesis of 3-(3-hydroxyalkanoyloxy) alkanoic acids (HAA), to which TDP-rhamnose is transferred by RhlB. A second rhamnosyltransferase, RhlC, then transfers the rhamnosyl group to mono-rhamnolipids to form di-rhamnolipids. Transcription of both the *rhlAB* operon and of*rhlC* is positively regulated by the Rhl QS system in *P. aeruginosa.*

Although strictly speaking they probably cannot be considered signaling molecules, rhamnolipids are part of a complex regulatory network that involves QS, and they influence collective behaviors such as swarming motility and biofilm development (Dobler et al. 2016). Hence, rhamnolipids have received significant interest as potential modulators of multicellular lifestyles of bacterial pathogens. These molecules are involved in maintaining the architecture of *P. aeruginosa* biofilms by preventing attachment of incoming cells to the water channels of mature biofilms (Davey et al. 2003), and it has been reported that alterations in rhamnolipid production can also alter the synthesis of exopolysaccharides and thus modulate biofilm development (Wang et al. 2014).

## 5    Fatty Acids and Fatty Acid Derivatives as Signals

In different bacteria, fatty acids and modified fatty acids can function as cell-cell communication signals (Fig. 2). The first example was discovered in the plant pathogen *Xanthomonas campestris* pv. campestris and initially called "diffusible signal factor," DSF. The chemical structure of DSF was later identified as the unsaturated fatty acid *cis*-11-methyl-2-dodecenoic acid (Wang et al. 2004). In this QS system, DSF biosynthesis is modulated by a posttranslational autoinduction mechanism that involves *rpfF* and *rpfB*, encoding an enoyl-CoA hydratase and a long-chain fatty acyl-CoA ligase, respectively. Detection of DSF involves the sensor RpfC, which transduces the QS signal to its cognate response regulator, RpfG, through a conserved phosphorelay mechanism. Activated RpfG functions as a phosphodiesterase, degrading the second messenger cyclic diguanylate (c-di-GMP) to GMP. It has been shown that the DSF system is coupled to intracellular regulatory networks through c-di-GMP and the global regulator Clp (Tao et al. 2010). Genomic and genetic analyses show that the DSF QS signaling pathway regulates different biological functions including virulence, biofilm dispersal and formation, and ecological competence, thus playing key roles in the ecology of the organisms that produce them (Ham 2013).

Fatty acid signal molecules of the DSF family have been reported in different plant pathogens such as in several other *Xanthomonas* species, including *X. oryzae* pv. oryzae, *X. axonopodis* pv. citri (He and Zhang 2008; He et al. 2010), and *Xylella fastidiosa* (Chatterjee et al. 2008). This bacterium colonizes the xylem of plants, attaching and forming biofilms on the vessel walls that in susceptible plants such as

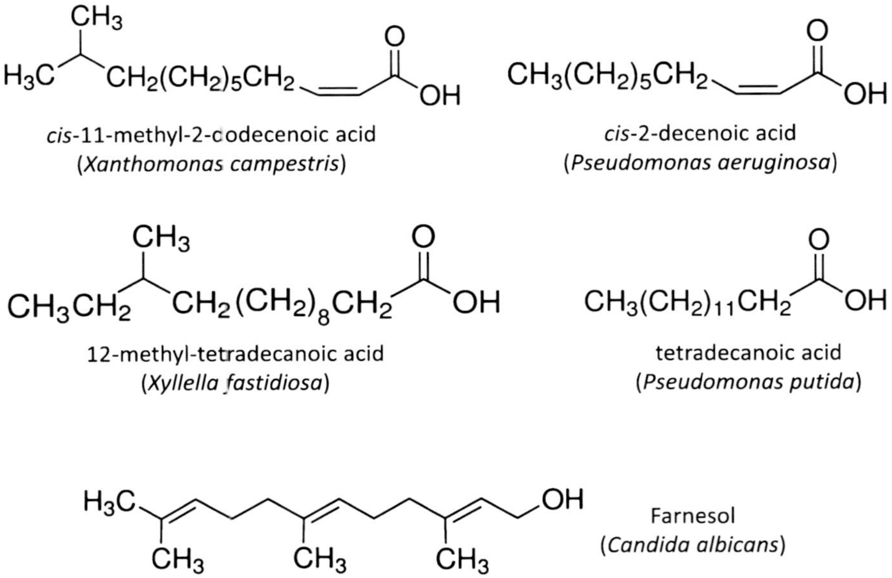

cis-11-methyl-2-dodecenoic acid
(*Xanthomonas campestris*)

cis-2-decenoic acid
(*Pseudomonas aeruginosa*)

12-methyl-tetradecanoic acid
(*Xyllella fastidiosa*)

tetradecanoic acid
(*Pseudomonas putida*)

Farnesol
(*Candida albicans*)

**Fig. 2** DSF (*cis*-11-methyl-2-dodecenoic acid) and DSF-like fatty acid signals identified in different bacteria. The structure of the fungal molecule farnesol is also included

grape can expand and grow as big as to occlude the xylem and block water transport. *X. fastidiosa* produces 12-methyl-tetradecanoic acid as cell-cell communication signal. Interestingly, a mutant in *rpfF*, the gene responsible for the synthesis of the signal, showed increased colonization and spread through the plant and formed larger colonies (Chatterjee et al. 2008). This has led to the idea that accumulation of 12-methyl-tetradecanoic acid in this bacterium is a means to restrain virulence activities within host plants and serves rather as an adaptation mechanism for the endophytic lifestyle of this bacterium. This contrasts with the role of DSF in *Xanthomonas*, where it exerts a positive role on virulence and pathogenicity. It is worth mentioning that transgenic plants expressing *rpfF* have been produced and shown to be able to "confuse" invading pathogens, thus showing the potential of QS control and manipulation as a method to fight plant diseases (Lindow et al. 2014).

Although the DSF system seemed at first to be a hallmark of plant pathogenic bacteria, evidence is accumulating that DSF-type fatty acid signals constitute a broad family of cell-cell communication signals, which are widespread among Gram-negative bacteria and can function in intra- and interspecific signaling. DSF and other structural derivatives have been identified in *Stenotrophomonas maltophilia* and in *Burkholderia* species (Fouhy et al. 2007; Deng et al. 2010). In *P. aeruginosa*, the synthesis and release of *cis*-2-decenoic acid to the extracellular medium was found to inhibit biofilm formation by different bacteria (Davies and Marques 2009). Furthermore, it has been reported that *cis*-2-decenoic acid promotes the reversion of persister to metabolically active cells in biofilms (Marques et al. 2014). All this makes it a potentially useful molecule in clinical treatment and removal of pathogenic biofilms (Marques et al. 2015).

Fatty acids and fatty acid-related molecules have been shown to participate in other interspecies and interkingdom signaling processes (Ryan et al. 2008, 2015). The plant beneficial bacterium *P. putida* KT2440 produces and responds to fatty acids between 12 and 14 carbons in length, which have also been identified in plant root exudates. These fatty acids induce expression of a gene that is involved in seed and root colonization and thus could be a means for plants to "attract" beneficial bacteria. Tetradecanoic acid was also found in fractions of cell-free supernatants of *P. aeruginosa* cultures that induced expression of the same gene (Fernández-Piñar et al. 2012).

Fatty acid-related signaling might show a close connection to other QS circuits. It is worth noting that enzymes involved in the turnover of AHL molecules in one species, or produced by competing species as "quorum quenching" mechanisms, include lactonases that disrupt the lactone ring and acylases, which release the fatty acid side chain. Thus, it may be that some instances of fatty acid-mediated communication function as a "post-quorum" signaling system. In the example mentioned above, the response of *P. putida* to *P. aeruginosa* extracts was diminished in a mutant in *pvdQ*, a gene encoding an acylase (Fernández-Piñar et al. 2012), suggesting that this activity can contribute to fatty acid signaling as a by-product of AHL systems. Interplay between QS systems has been described in the interaction between *Candida albicans* and *Pseudomonas aeruginosa*. *C. albicans* and other fungi produce the fatty acid-related molecules farnesol (3,7,11-trimethyl-2,6,10-dodecatrienol), dodecanol, and 3(R)-hydroxy-tetradecanoic acid as QS signals (Singh and Del Poeta 2011; Albuquerque and Casadevall 2012). Farnesol produced by *C. albicans* induces the production of quinolones in *P. aeruginosa* in a LasR-independent manner (Cugini et al. 2010). On the other hand, DSF-like molecules of bacterial origin have an influence on *C. albicans* morphogenesis (Boon et al. 2008; Vílchez et al. 2010).

## 6   Research Needs

Traditionally considered only from the point of view of structural elements relevant for the buildup of cellular membranes, fatty acids are revealing themselves as a complex and intriguing class of biological signals that participate in bacterial communication and in cross-kingdom interactions. While the number and diversity of molecules identified keeps increasing, their precise roles in these interactions, the population responses they control, and the mechanisms regulating their production and involved in their detection are still far from being fully explored. This knowledge, combined with further studies that tackle the potential manipulation of fatty acid signaling, would contribute to understand the basis of virulence and host-bacteria interactions; it may provide new approaches to prevent and combat biofilms and persistent infections and could even open the way to develop strategies to promote beneficial interactions between organisms, such as selective combinations of plants and plant growth-promoting bacteria.

**Acknowledgments** Work in the author's group on cellular responses and regulatory mechanisms in bacterial populations and biofilms is funded by grant BFU2013-43469-P from Plan Estatal de I +D+I and EFDR funds.

# References

Abdel-Mawgoud AM, Lépine F, Déziel E (2010) Rhamnolipids: diversity of structures, microbial origins and roles. Appl Microbiol Biotechnol 86:1323–1336

Albuquerque P, Casadevall A (2012) Quorum sensing in fungi-a review. Med Mycol 50:337–345

Bassler BL, Losick R(2006) Bacterially speaking. Cell 125:237–246.

Boon C, Deng Y, Wang LH, He Y, JL X, Fan Y, Pan SQ, Zhang LH (2008) A novel DSF-like signal from *Burkholderia cenocepacia* interferes with *Candida albicans* morphological transition. ISME J 2:27–36

Camara M, Williams P, Hardman A (2002) Controlling infection by tuning in and turning down the volume of bacterial small-talk. Lancet Infect Dis 2:667–676

Cha C, Gao P, Chen YC, Shaw PD, Farrand SK (1998) Production of acyl-homoserine lactone quorum-sensing signals by gram-negative plant-associated bacteria. Mol Plant-Microbe Interact 11:1119–1129

Chatterjee S, Wistrom C, Lindow SE (2008) A cell-cell signaling sensor is required for virulence and insect transmission of *Xylella fastidiosa*. Proc Natl Acad Sci USA 105:2670–2675

Corral-Lugo A, Daddaoua A, Ortega A, Espinosa-Urgel M, Krell T (2016) Rosmarinic acid is a homoserine lactone mimic produced by plants that activates a bacterial quorum-sensing regulator. Sci Signal 9(409):ra1

Cugini C, Morales DK, Hogan DA (2010) *Candida albicans*-produced farnesol stimulates *Pseudomonas* quinolone signal production in LasR-defective *Pseudomonas aeruginosa* strains. Microbiology 156:3096–3107

Cullinane M, Baysse C, Morrissey JP, O'Gara F (2005) Identification of two lysophosphatidic acid acyltransferase genes with overlapping function in *Pseudomonas fluorescens*. Microbiology 151:3071–3080

Daniels R, Vanderleyden J, Michiels J (2004) Quorum sensing and swarming migration in bacteria. FEMS Microbiol Rev 28:261–289

Davey ME, Caiazza NC, O'Toole GA (2003) Rhamnolipid surfactant production affects biofilm architecture in *Pseudomonas aeruginosa* PAO1. J Bacteriol 185:1027–1036

Davies DG, Marques CN (2009) A fatty acid messenger is responsible for inducing dispersion in microbial biofilms. J Bacteriol 191:1393–1403

Deng Y, Wu J, Eberl L, Zhang LH (2010) Structural and functional characterization of diffusible signal factor family quorum-sensing signals produced by members of the *Burkholderia cepacia* complex. Appl Environ Microbiol 76:4675–4683

Déziel E, Lépine F, Milot S, He J, Mindrinos MN, Tompkins RG, Rahme LG (2004) Analysis of Pseudomonas aeruginosa 4-hydroxy-2-alkylquinolines (HAQs) reveals a role for 4-hydroxy-2-heptylquinoline in cell-to-cell communication. Proc Natl Acad Sci USA 101:1339–1344

Diggle SP, Lumjiaktase P, Dipilato F, Winzer K, Kunakorn M, Barrett DA, Chhabra SR, Cámara M, Williams P (2006) Functional genetic analysis reveals a 2-Alkyl-4-quinolone signaling system in the human pathogen *Burkholderia pseudomallei* and related bacteria. Chem Biol 13:701–710

Diggle SP, Griffin AS, Campbell GS, West SA (2007a) Cooperation and conflict in quorum-sensing bacterial populations. Nature 450:411–414

Diggle SP, Matthijs S, Wright VJ, Fletcher MP, Chhabra SR, Lamont IL, Kong X, Hider RC, Cornelis P, Cámara M, Williams P (2007b) The *Pseudomonas aeruginosa* 4-quinolone signal molecules HHQ and PQS play multifunctional roles in quorum sensing and iron entrapment. Chem Biol 14:87–96

Dobler L, Vilela LF, Almeida RV, Neves BC (2016) Rhamnolipids in perspective: gene regulatory pathways, metabolic engineering, production and technological forecasting. New Biotechnol 33:123–135

Eberhard A, Burlingame AL, Eberhard C, Kenyon GL, Nealson KH, Oppenheimer NJ (1981) Structural identification of autoinducer of *Photobacterium fischeri* luciferase. Biochemistry 20:2444–2449

Engebrecht J, Nealson K, Silverman M (1983) Bacterial bioluminescence: isolation and genetic analysis of functions from *Vibrio fischeri*. Cell 32:773–781

Farrow JM 3rd, Sund ZM, Ellison ML, Wade DS, Coleman JP, Pesci EC (2008) PqsE functions independently of PqsR-*Pseudomonas* quinolone signal and enhances the rhl quorum-sensing system. J Bacteriol 190:7043–7051

Fernández-Piñar R, Cámara M, Dubern JF, Ramos JL, Espinosa-Urgel M (2011) The *Pseudomonas aeruginosa* quinolone quorum sensing signal alters the multicellular behaviour of *Pseudomonas putida* KT 2440. Res Microbiol 162:773–781

Fernández-Piñar R, Espinosa-Urgel M, Dubern JF, Heeb S, Ramos JL, Cámara M (2012) Fatty acid-mediated signalling between two *Pseudomonas* species. Environ Microbiol Rep 4:417–423

Fouhy Y, Scanlon K, Schouest K, Spillane C, Crossman L, Avison MB, Ryan RP, Dow JM (2007) Diffusible signal factor-dependent cell-cell signaling and virulence in the nosocomial pathogen *Stenotrophomonas maltophilia*. J Bacteriol 189:4964–4968

Fuqua C, Greenberg EP (2002) Listening in on bacteria: acyl-homoserine lactone signaling. Nat Rev Mol Cell Biol 3:685–695

Fuqua WC, Winans SC, Greenberg EP (1994) Quorum sensing in bacteria: the LuxR-LuxI family of cell density-responsive transcriptional regulators. J Bacteriol 176:269–275

Gallagher LA, McKnight SL, Kuznetsova MS, Pesci EC, Manoil C (2002) Functions required for extracellular quinolone signaling by *Pseudomonas aeruginosa*. J Bacteriol 184:6472–6480

Gao M, Teplitski M, Robinson JB, Bauer WD (2003) Production of substances by Medicago truncatula that affect bacterial quorum sensing. Mol Plant Microbe Interact 16:827–334.

Gonzalez JE, Marketon MM (2003) Quorum sensing in nitrogen-fixing rhizobia. Microbiol Mol Biol Rev 67:574–592

Ham JH (2013) Intercellular and intracellular signalling systems that globally control the expression of virulence genes in plant pathogenic bacteria. Mol Plant Pathol 14:308–322

Hays E, Wells E, Katzman I, Cain C, Jacobs CK, Thayer FA, Doisy SA, Gaby EA, Roberts WL, Muir EC, Carroll RD, Jones CJ, Wade NJ (1945) Antibiotic substances produced by *Pseudomonas aeruginosa*. J Biol Chem 159:725–750

He YW, Zhang LH (2008) Quorum sensing and virulence regulation in *Xanthomonas campestris*. FEMS Microbiol Rev 32:842–857

He YW, Wu J, Cha JS, Zhang LH (2010) Rice bacterial blight pathogen *Xanthomonas oryzae* pv. *oryzae* produces multiple DSF-family signals in regulation of virulence factor production. BMC Microbiol 10:187

Hense BA, Kuttler C, Müller J, Rothballer M, Hartmann A, Kreft JU (2007) Does efficiency sensing unify diffusion and quorum sensing? Nat Rev Microbiol 5:230–239

Krol E, Becker A (2014) Rhizobial homologs of the fatty acid transporter FadL facilitate perception of long-chain acyl-homoserine lactone signals. Proc Natl Acad Sci USA 111:10702–10707

Laue BE, Jiang Y, Chhabra SR, Jacob S, Stewart GS, Hardman A, Downie JA, O'Gara F, Williams P (2000) The biocontrol strain *Pseudomonas fluorescens* F113 produces the *Rhizobium* small bacteriocin, *N*-(3-hydroxy-7-cis-tetradecenoyl)homoserine lactone, via HdtS, a putative novel *N*-acylhomoserine lactone synthase. Microbiology 146:2469–2480

Lazazzera BA (2001) The intracellular function of extracellular signaling peptides. Peptides 22:1519–1527

Lindow S, Newman K, Chatterjee S, Baccari C, Lavarone AT, Ionescu M (2014) Production of *Xylella fastidiosa* diffusible signal factor in transgenic grape causes pathogen confusion and reduction in severity of Pierce's disease. Mol Plant-Microbe Interact 27:244–254

Marques CN, Morozov A, Planzos P, Zelaya HM (2014) The fatty acid signaling molecule cis-2-decenoic acid increases metabolic activity and reverts persister cells to an antimicrobial-susceptible state. Appl Environ Microbiol 80:6976–6991

Marques CN, Davies DG, Sauer K (2015) Control of biofilms with the fatty acid signaling molecule cis-2-decenoic acid. Pharmaceuticals 8:816–835

Miller MB, Bassler BL (2001) Quorum sensing in bacteria. Annu Rev Microbiol 55:165–199

Milton DL, Chalker VJ, Kirke D, Hardman A, Cámara M, Williams P (2001) The LuxM homologue VanM from *Vibrio anguillarum* directs the synthesis of *N*-(3-hydroxyhexanoyl)homoserine lactone and *N*-hexanoylhomoserine lactone. J Bacteriol 183:3537–3547

Nealson KH, Hastings JW (1979) Bacterial bioluminescence: its control and ecological significance. Microbiol Rev 43:496–518

Pacheco AR, Sperandio V (2009) Inter-kingdom signaling: chemical language between bacteria and host. Curr Opin Microbiol 12:192–198

Parsek MR, Val DL, Hanzelka BL, Cronan JE Jr, Greenberg EP (1999) Acyl homoserine-lactone quorum-sensing signal generation. Proc Natl Acad Sci USA 96:4360–4365

Pearson JP, Van Delden C, Iglewski BH (1999) Active efflux and diffusion are involved in transport of *Pseudomonas aeruginosa* cell-to-cell signals. J Bacteriol 181:1203–1210

Qazi SN, Counil E, Morrissey J, Rees CE, Cockayne A, Winzer K, Chan WC, Williams P, Hill PJ (2001) agr expression precedes escape of internalized *Staphylococcus aureus* from the host endosome. Infect Immun 69:7074–7082

Redfield RJ (2002) Is quorum sensing a side effect of diffusion sensing? Trends Microbiol 10:365–370

Rice SA, Koh KS, Queck SY, Labbate M, Lam KW, Kjelleberg S (2005) Biofilm formation and sloughing in *Serratia marcescens* are controlled by quorum sensing and nutrient cues. J Bacteriol 187:3477–3485

Rumbaugh KP, Griswold JA, Hamood AN (2000) The role of quorum sensing in the in vivo virulence of *Pseudomonas aeruginosa*. Microbes Infect 2:1721–1731

Rumbaugh KP, Diggle SP, Watters CM, Ross-Gillespie A, Griffin AS, West SA (2009) Quorum sensing and the social evolution of bacterial virulence. Curr Biol 19:341–345

Ryan RP, Fouhy Y, Garcia BF, Watt SA, Niehaus K, Yang L, Tolker-Nielsen T, Dow JM (2008) Interspecies signalling via the *Stenotrophomonas maltophilia* diffusible signal factor influences biofilm formation and polymyxin tolerance in *Pseudomonas aeruginosa*. Mol Microbiol 68:75–86

Ryan RP, An SQ, Allan JH, McCarthy Y, Dow JM (2015) The DSF family of cell-cell signals: an expanding class of bacterial virulence regulators. PLoS Pathog 11(7):e1004986

Schaefer AL, Greenberg EP, Oliver CM, Oda Y, Huang JJ, Bittan-Banin G, Peres CM, Schmidt S, Juhaszova K, Sufrin JR, Harwood CS (2008) A new class of homoserine lactone quorum-sensing signals. Nature 454:595–599

Singh A, Del Poeta M (2011) Lipid signalling in pathogenic fungi. Cell Microbiol 13:177–185

Soberón-Chávez G, Lépine F, Déziel E (2005) Production of rhamnolipids by *Pseudomonas aeruginosa*. Appl Microbiol Biotechnol 68:718–725

Stevens AM, Dolan KM, Greenberg EP (1994) Synergistic binding of the *Vibrio fischeri* LuxR transcriptional activator domain and RNA polymerase to the lux promoter region. Proc Natl Acad Sci USA 91:12619–12623

Tao F, He YW, DH W, Swarup S, Zhang LH (2010) The cyclic nucleotide monophosphate domain of *Xanthomonas campestris* global regulator Clp defines a new class of cyclic di-GMP effectors. J Bacteriol 192:1020–1029

Venturi V, Fuqua C (2013) Chemical signaling between plants and plant-pathogenic bacteria. Annu Rev Phytopathol 51:17–37

Vílchez R, Lemme A, Ballhausen B, Thiel V, Schulz S, Jansen R, Wagner-Döbler I, Sztajer H (2010) *Streptococcus mutans* inhibits *Candida albicans* hyphal formation by the fatty acid signaling molecule trans-2-decenoic acid (SDSF). Chembiochem 11:1552–1162

Von Bodman SB, Bauer WD, Coplin DL (2003) Quorum sensing in plant-pathogenic bacteria. Annu Rev Phytopathol 41:455–482

Wang LH, He Y, Gao Y, JE W, Dong YH, He C, Wang SX, Weng LX, JL X, Tay L, Fang RX, Zhang LH (2004) A bacterial cell–cell communication signal with cross-kingdom structural analogues. Mol Microbiol 51:903–912

Wang S, Yu S, Zhang Z, Wei Q, Yan L, Ai G, Liu H, Ma LZ (2014) Coordination of swarming motility, biosurfactant synthesis, and biofilm matrix exopolysaccharide production in *Pseudomonas aeruginosa*. Appl Environ Microbiol 80:6724–6732

Whitehead NA, Barnard AM, Slater H, Simpson NJ, Salmond GP (2001) Quorum-sensing in Gram-negative bacteria. FEMS Microbiol Rev 25:365–404

Williams P (2007) Quorum sensing, communication and cross-kingdom signaling in the bacterial world. Microbiology 153:3923–3938

Winzer K, Williams P (2001) Quorum sensing and the regulation of virulence gene expression in pathogenic bacteria. Int J Med Microbiol 291:131–143

Yu S, Jensen V, Seeliger J, Feldmann I, Weber S, Schleicher E, Häussler S, Blankenfeldt W (2009) Structure elucidation and preliminary assessment of hydrolase activity of PqsE, the *Pseudomonas* quinolone signal (PQS) response protein. Biochemistry 48:10298–10307

# Substrate Transport

# 17

Rebecca E. Parales and Jayna L. Ditty

## Contents

1    Introduction ................................................................................ 288
2    Hydrocarbon Transport Across the Outer Membrane of Gram-Negative Bacteria ........ 289
3    Hydrocarbon Transport Across the Cytoplasmic Membrane ............................. 291
4    Transport of Aromatic Acids Across the Cytoplasmic Membrane ....................... 292
5    Other Mechanisms to Enhance Hydrocarbon Acquisition ............................... 295
6    Hydrocarbon Efflux Pumps ................................................................ 296
7    Research Needs ............................................................................. 297
References ........................................................................................ 298

### Abstract

Hydrocarbon compounds are known to passively diffuse across bacterial cytoplasmic membranes, and this may be the primary mechanism of hydrocarbon entry into most bacteria. The participation of active transport systems has been suggested in some bacterial strains, but solid evidence for active transport of hydrocarbons is currently lacking. In contrast, many active transport systems have been identified for the energy-dependent uptake of aromatic acids in both Gram-negative and Gram-positive bacteria. In addition, Gram-negative bacteria often harbor specific inducible outer membrane channels that allow entry of various aromatic hydrocarbon substrates.

R. E. Parales (✉)
Department of Microbiology and Molecular Genetics, College of Biological Sciences, University of California, Davis, CA, USA
e-mail: reparales@ucdavis.edu

J. L. Ditty
Department of Biology, College of Arts and Sciences, University of St. Thomas, St. Paul, MN, USA
e-mail: jlditty@stthomas.edu

© Springer International Publishing AG, part of Springer Nature 2018                287
T. Krell (ed.), *Cellular Ecophysiology of Microbe: Hydrocarbon and Lipid Interactions*,
Handbook of Hydrocarbon and Lipid Microbiology, https://doi.org/10.1007/978-3-319-50542-8_44

# 1 Introduction

A particularly intriguing question about the utilization of hydrocarbons by microorganisms as sources of carbon and energy is how these cells manage the problem of growth with a toxic compound. Because hydrocarbons are highly hydrophobic and lipophilic molecules, they are capable of unregulated entry into cells through the cytoplasmic membrane and can cause intracellular toxicity (reviewed in Sikkema et al. (1995)). Therefore, cells that utilize these toxic molecules must acquire sufficient amounts of hydrocarbons to allow growth but must somehow limit their intracellular concentrations to sub-toxic levels.

In Gram-negative cells, the outer membrane provides a significant barrier to hydrocarbon entry (reviewed in Nikaido (2003)). Structurally, the outer membrane is very different from the cytoplasmic membrane, primarily due to the presence of lipopolysaccharide (LPS) in the outer leaflet. The outer membrane prevents the diffusion of molecules due to low fluidity of the tightly packed saturated fatty acids and strong lateral molecular interactions between LPS molecules. The polyanionic lipid of the LPS is further stabilized by divalent cations. These features make the outer membrane an extremely efficient permeability barrier for Gram-negative cells. Passage of substrates across the outer membrane is accomplished via non-specific porins and specific outer membrane protein channels, which are necessary for the entry of nutrients (Fig. 1). Porins typically have substrate preferences based on the size and charge of the molecules (Nikaido 2003).

The cytoplasmic membrane provides a critical permeability barrier to polar and charged molecules, but hydrophobic compounds such as hydrocarbons can readily penetrate the lipid bilayer (Sikkema et al. 1995). At this time, there is no definitive answer to the question of whether specific transport systems for active uptake of hydrocarbons across bacterial cytoplasmic membranes exist. General arguments for the absence of hydrocarbon transporters are as follows: (1) the hydrophobic nature of hydrocarbons is expected to allow these molecules to permeate the cytoplasmic membrane, (2) hydrocarbon transport mutants have not been isolated, (3) genes encoding cytoplasmic transport proteins have not been identified in known gene clusters for hydrocarbon catabolism, and (4) expression and functional activity of genes for (substituted) hydrocarbon degradation in *E. coli* and other heterologous hosts do not require the introduction of transport genes, indicating that hydrocarbons gain entry into cells without expression of specific heterologous transport proteins. However, most of these generalizations are based on the absence of data rather than on any solid empirical evidence.

Due to the differences in cytoplasmic and outer membrane structures, there are potentially different requirements for the presence of transport mechanisms for hydrocarbons. Unfortunately, relatively few studies have been undertaken to characterize hydrocarbon uptake by bacteria across either membrane, and the available studies seem to present conflicting evidence. In contrast, several types of active transport systems for uptake of aromatic acids, which share some of the properties of aromatic hydrocarbons, have been identified. The purpose of this chapter is to review our current understanding of hydrocarbon uptake by bacterial cells, and we also provide a summary of the known transporters for aromatic acids.

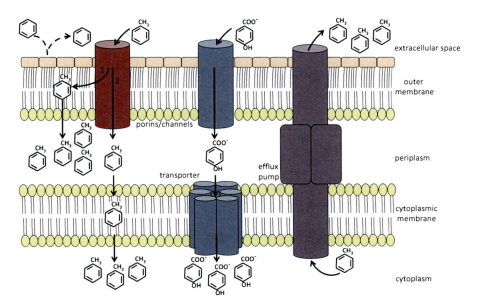

**Fig. 1** Cartoon of the Gram-negative bacterial cell envelope. The outer membrane serves as an extremely efficient permeability barrier to the diffusion of aromatic molecules. Based on our current understanding of hydrocarbon transport, cells have either specific (e.g., TodX or OmpW) or nonspecific outer membrane channels to allow passage of hydrocarbons across the outer membrane into the periplasm, either by (1) lateral diffusion through the barrel wall into the lipid bilayer or (2) through both the LPS and phospholipid leaflets via the channel. Once in the periplasm, hydrocarbons do not need transport proteins for passage across the cytoplasmic membrane. Aromatic acids are also most likely facilitated across the outer membrane via specific or nonspecific porins. Once in the periplasm, MFS, ABC, and tripartite transporters have been shown to enable accumulation of aromatic acids in the cytoplasm. Some bacteria also have specific or nonspecific efflux pumps to reduce hydrocarbon concentrations within the cell and prevent toxic effects. See the text for additional details on these various transport mechanisms

## 2   Hydrocarbon Transport Across the Outer Membrane of Gram-Negative Bacteria

The outer membrane of Gram-negative bacteria provides an effective permeability barrier, and all substrates must enter the Gram-negative cell through outer membrane channels (Nikaido 2003). Substrate-specific channels of the FadL family are widespread in Gram-negative bacteria and appear to be involved in the uptake of a variety of hydrophobic substrates across the outer membrane (van den Berg 2010a). For example, several bacterial hydrocarbon degradation gene clusters have been found to contain a gene that encodes a putative outer membrane protein with homology to FadL, an outer membrane protein that is required for long chain fatty acid transport across the outer membrane in *E. coli* (Black 1991; DiRusso and Black 2004; van den Berg 2005). Crystal structures of FadL demonstrated that it is a monomeric β-barrel comprised of 14 antiparallel β-strands (van den Berg et al. 2004). It has low- and high-affinity binding sites for fatty acids, and a channel that allows these ligands to enter the periplasmic space. Available evidence suggests that the FadL homologs TodX, TbuX, TmoX, XylN, and StyE function as channels that allow aromatic hydrocarbons such as toluene, *m*-xylene, and styrene to cross the outer membrane (Fig. 1). In the toluene-degrading strain *P. putida* F1, *todX* is the first gene in the *tod* operon (Wang et al. 1995). This toluene-inducible operon encodes all of the enzymes required for toluene degradation in this organism (Lau et al. 1994; Menn et al. 1991; Wang et al. 1995; Zylstra and Gibson 1989). When expressed in *E. coli*, TodX localized to the outer membrane, and a *todX* mutant of *P. putida* F1 had a decreased growth rate on low levels of toluene (Wang et al. 1995). Together, these data suggested that TodX facilitates entry of toluene into the periplasm of the cell (Wang et al. 1995). At higher toluene concentrations, toluene can efficiently cross the outer membrane in the absence of TodX, presumably through nonspecific channels or porins.

Several other *todX* homologs have been identified in various hydrocarbon degradation gene clusters. For example, in *Ralstonia pickettii* PKO1, which grows on toluene using a toluene 3-monooxygenase-mediated pathway (Olsen et al. 1994), the TodX homolog TbuX (34% amino acid identity) was shown to be required for toluene utilization and induction of the toluene degradation genes. Based on sequence analysis, TbuX was predicted to be an outer membrane protein and was postulated to facilitate passage of toluene across the outer membrane (Kahng et al. 2000). The *xylN* gene is another *todX* homolog. It is the last gene in the TOL pathway upper operon (for toluene and xylene degradation), and its product shares 38% amino acid identity with TodX (Kasai et al. 2001). XylN was localized to the outer membrane of wild-type *P. putida* carrying the TOL plasmid, and inactivation of *xylN* resulted in a reduced growth rate on *m*-xylene provided in the vapor phase. In contrast, when *m*-xylene was added directly to the medium, the *xylN* mutant entered the exponential growth phase almost immediately, while the wild type had a ~40 h lag phase. Together these data indicate that XylN allows *m*-xylene to cross the outer membrane and makes the cell sensitive to *m*-xylene at high concentrations. In contrast, in *P. putida* CA-3, the StyE protein was required for growth of the strain on the aromatic hydrocarbon styrene (Mooney et al. 2006). StyE was localized to the outer membrane of *P. putida* CA-3 and was proposed to facilitate the transport of styrene across the outer membrane. Other putative outer membrane hydrocarbon transporters (e.g., CymD, CumH, TmoX) were identified based on sequence analysis and gene location within characterized aromatic hydrocarbon catabolic operons (Eaton 1997; Eaton and Timmis 1986; Habe et al. 1996; Ramos-Gonzalez et al. 2002). The structures of TodX and TbuX indicate that like FadL, they are 14-stranded β-barrel proteins with an internal hydrophobic channel that allows binding and passage of specific hydrophobic substrates (Hearn et al. 2008). Recent evidence suggests a "lateral transport" mechanism (Fig. 1) in which the hydrophobic substrate enters the channel in order to cross the outer LPS layer and exits through a pore in the channel wall into the outer membrane rather than being directly released into the periplasm (van den Berg 2010b).

The alkane-degrading *Pseudomonas putida* strains GPo1 and P1 carry *alkL* genes in alkane degradation gene clusters (van Beilen et al. 1994, 2001). The *alkL* gene encodes an outer membrane protein with homology to the *Vibrio cholerae* OmpW outer membrane protein of unknown function (van Beilen et al. 1992). AlkL has been proposed to be involved in uptake of alkane substrates; however, the deletion of *alkL* did not affect the metabolism of alkanes (van Beilen et al. 1994). In contrast, deletion of *ompW* from the naphthalene-degrading strain *Pseudomonas fluorescens* Uper-1 resulted in a mutant that was unable to grow on naphthalene (Neher and Lueking 2009). The *ompW* gene, which was identified in a subtractive hybridization analysis and localized to a large plasmid required for growth on naphthalene, was proposed to encode an outer membrane porin for uptake of hydrophobic substrates. The mutation, however, was generated by a transposon insertion, and complementation was not reported, so it is unclear if the phenotype was due solely to inactivation of *ompW* or if the insertion also affected expression of naphthalene catabolic genes. OmpW is ~90% identical at the amino acid level to its counterpart in *E. coli*. The eight-stranded beta-barrel structure of OmpW from *E. coli* has a hydrophobic channel, suggesting that members of the OmpW/AlkL family function in the transport of hydrophobic molecules (Hearn et al. 2009; Hong et al. 2006). In the case of both FadL and OmpW/AlkL family proteins, the mechanism of hydrophobic substrate transport appears to be by lateral diffusion through the barrel wall into the lipid bilayer (Hearn et al. 2009; van den Berg 2010a, b).

## 3    Hydrocarbon Transport Across the Cytoplasmic Membrane

A study of naphthalene uptake by *P. putida* PpG1 concluded that naphthalene entered cells by passive diffusion across the cytoplasmic membrane based on several lines of evidence. Induction of *P. putida* PpG1 cells was not required for naphthalene accumulation, indicating that the production of a specific protein was not required for this process. In addition, various inhibitors of ATP synthesis, or PMF generation, had no effect on naphthalene uptake, and naphthalene uptake was not saturable (Bateman et al. 1986). In contrast to these results, it was reported that naphthalene uptake by *Pseudomonas fluorescens* Uper-1 may involve an active transport system (Whitman et al. 1998). This study demonstrated that naphthalene transport only occurred in the wild-type naphthalene-degrading strain; a mutant defective in the first enzyme in the naphthalene degradation pathway was incapable of naphthalene uptake. However, the transport assays were carried out over a period of hours rather than minutes, and significant growth and metabolism of $^{14}$C-naphthalene were occurring during this period. Therefore, the results of this study remain inconclusive because it is not possible to separate uptake from metabolism under these conditions. In another study, although biphenyl uptake was inducible and required an energized cytoplasmic membrane, the process was shown to be driven by biphenyl metabolism (Master et al. 2005). Biphenyl uptake by the biphenyl degraders *Burkholderia xenovorans* LB400 (formerly *Burkholderia* sp. LB400) and *Pseudomonas* sp. strain Cam-1 was completely inhibited when cells were treated with ionophores that dissipate the proton gradient across the cytoplasmic membrane. However, a mutant that was blocked in the first step in biphenyl catabolism was unable to transport biphenyl. The authors therefore concluded that biphenyl uptake was a passive process that was driven by biphenyl catabolism in the wild-type strain (Master et al. 2005).

Bugg et al. demonstrated that phenanthrene entry into cells by passive diffusion was counteracted by an active efflux system in *P. fluorescens* LP6a (Bugg et al. 2000). In contrast, two reports have suggested that phenanthrene is actively transported by the Gram-positive isolates *Mycobacterium* sp. strain RJGII-135 (Miyata et al. 2004) and *Arthrobacter* sp. strain Sphe3 (Kallimanis et al. 2007). In both strains, phenanthrene appeared to enter uninduced cells by passive diffusion; however, induced cells showed saturable, energy-dependent phenanthrene uptake. Kinetic analyses suggested that phenanthrene is specifically bound to induce *Mycobacterium* sp. strain RJGII-135 cells (Miyata et al. 2004). In both studies, the bacteria were actively metabolizing phenanthrene; in fact essentially all of the $^{14}$C-phenanthrene taken up by *Mycobacterium* sp. strain RJGII-135 was converted to metabolites within a minute. Therefore, the contribution of metabolism to the measured uptake cannot be ruled out, and further studies with catabolic mutants and/or the identification of specific transport proteins will be necessary to confirm whether active transport of phenanthrene is occurring. Energy-dependent uptake of $^{14}$C- fluoranthene was also demonstrated in *Rhodococcus* sp. strain BAP-1, but again the strain was actively metabolizing fluoranthene and releasing $^{14}$CO$_2$ (Li et al. 2014).

A Gram-positive strain, *Rhodococcus erythropolis* S+14He, which grows on linear alkanes, was reported to actively and selectively transport *n*-hexadecane (Kim et al. 2002); similarly hexadecane uptake by *Pseudomonas aeruginosa* PG201 was shown to be energy dependent (Beal and Betts 2000). However, as in the other studies, cells were both accumulating and metabolizing the compounds during the assays. More importantly, no transport proteins for hydrocarbon uptake have been identified to date in any of these hydrocarbon-degrading bacteria. One study did identify a putative ABC transport system that was induced by *n*-tetradecane (Noda et al. 2003). When the *hcuABC* genes in *P. aeruginosa* were inactivated by random insertion of a transposon carrying desulfurization genes, the recombinant strain was able to desulfurize dibenzothiophene in aqueous culture but not in a two-phase system with the substrate in *n*-tetradecane. However, it seems equally likely that the *hcuABC* genes encode a tetradecane-inducible efflux pump that may prevent accumulation of toxic levels of hydrocarbons from the oil phase (see Sect. 6 below).

# 4    Transport of Aromatic Acids Across the Cytoplasmic Membrane

While aromatic acids are not considered to be hydrocarbons, which by definition contain only carbon and hydrogen atoms, aromatic acids have many of the same properties as aromatic hydrocarbons. Due to the low pKa values of the carboxylic acid groups on aromatic acids (generally in the 4–5 range), at neutral pH the carboxylic acid group tends to be protonated, and as a result these molecules are hydrophobic, have relatively low aqueous solubility, and tend to partition into biological membranes, causing membrane damage and cellular toxicity. Interestingly, although specific cytoplasmic transporters for aromatic hydrocarbons appear to be rare, transporters for aromatic acids are quite common in bacteria, and genes encoding such proteins are often colocalized with genes for aromatic acid catabolism.

Early evidence for the transport of aromatic acids (Allende et al. 1993; Groenewegen et al. 1990; Higgins and Mandelstam 1972) was further supported during an investigation of chemotaxis to 4-hydroxybenzoate by *P. putida* PRS2000 in which a transporter for an aromatic acid (PcaK) was identified (Harwood et al. 1994). PcaK, which is responsible for uptake of 4-hydroxybenzoate and protocatechuate, is an integral membrane protein that is a member of the major facilitatory superfamily (MFS) (Pao et al. 1998; Saier et al. 1999). PcaK was not required for growth on 4-hydroxybenzoate, but growth rates of a *pcaK* mutant were slower at high pH (pH 8.1) when less of the 4-hydroxybenzoate is protonated (Harwood et al. 1994). In the absence of PcaK, 4-hydroxybenzoate enters *P. putida* PRS2000 cells by simple diffusion, and a concentration gradient is maintained by metabolism of 4-hydroxybenzoate via the β-ketoadipate pathway (Harwood and Parales 1996). When PcaK is produced in cells, 4-hydroxybenzoate accumulates to saturating levels even in catabolic mutants or heterologous hosts that are unable to metabolize 4-hydroxybenzoate (Nichols and Harwood 1997). In the environment, transport of 4-hydroxybenzoate is therefore likely to be important when concentrations of 4-hydroxybenzoate are low or when the pH is above neutral.

PcaK was also shown to be required for chemotaxis to 4-hydroxybenzoate (Harwood et al. 1994), and mutations in PcaK that decreased 4-hydroxybenzoate transport also decreased chemotaxis to 4-hydroxybenzoate (Ditty and Harwood 1999, 2002). The function of PcaK in the chemotactic response to 4-hydroxybenzoate was unclear until recently. Several hypotheses had been suggested, including PcaK playing a direct role as a membrane-bound chemoreceptor, its interaction with a membrane-bound methyl-accepting chemotaxis protein (MCP) during 4-hydroxybenzoate transport, or detection of 4-hydroxybenzoate by a cytoplasmic chemoreceptor once 4-hydroxybenzoate was transported into the cell (Ditty and Harwood 1999; Harwood et al. 1994). However, PcaK was shown to play an indirect role in chemotaxis following the identification of a MCP (PcaY) that senses 4-hydroxybenzoate and related aromatic acids in *P. putida* F1 (Luu et al. 2015). It was found that *pcaY* expression was coordinately controlled with genes of the *pca* regulon, which includes *pcaK* and genes for 4-hydroxybenzoate catabolism. PcaK was shown to facilitate 4-hydroxybenzoate chemotaxis by allowing sufficient 4-hydroxybenzoate to enter cells such that the inducer β-ketoadipate (an intermediate in 4-hydroxybenzoate catabolism) built up to a concentration that was detected by the transcriptional activator PcaR, which then allowed induction of the *pcaY* gene (Luu et al. 2015).

MFS transporters transport small molecules via uniport, symport, or antiport mechanisms and use the membrane potential as an energy source. PcaK is a member of the aromatic acid:$H^+$ symporter (AAHS) subfamily of MFS transporters (Harwood et al. 1994; Nichols and Harwood 1997; Pao et al. 1998). Since the identification of PcaK, several aromatic acid transporters that utilize symport with the proton motive force for substrate accumulation have been identified (Saier 2006). Some have been functionally characterized, but many have been inferred as transporters for specific substrates based on sequence comparisons and the location of the encoding gene in a catabolic operon. In *Acinetobacter baylyi* ADP1 (formerly *Acinetobacter calcoaceticus* ADP1), VanK and PcaK have overlapping specificity for 4-hydroxybenzoate and protocatechuate uptake. The inactivation of both *vanK* and *pcaK* severely decreased the ability of this strain to grow on protocatechuate; however, single mutants had no growth defect (D'Argenio et al. 1999). BenK mediates benzoate uptake in *Acinetobacter baylyi* ADP1 (Collier et al. 1997; Vaneechoutte et al. 2006), and TfdK was shown to transport 2,4-dichlorophenoxy-acetate (2,4-D) in *R. eutropha* JMP134 (Leveau et al. 1998), but neither transporter was essential for growth of the host organism under standard laboratory conditions. OphD from *B. cepacia* ATCC 17616 mediated uptake of phthalate (benzene 1,2-dicarboxylate) when expressed in *E. coli* (Chang and Zylstra 1999), MhbT from *Klebsiella pneumoniae* M5a1 and *P. putida* PAW340 was shown to specifically transport 3-hydroxybenzoate (Xu et al. 2012a), and MhpT and HpbX from *E. coli* K12 and *E. coli* W were demonstrated to transport 3-(3-hydroxyphenyl)propionic acid and 4-hydroxyphenylacetate, respectively (Prieto and García 1997; Xu et al. 2013). Additional putative aromatic acid MFS transporters have been identified in various Proteobacteria based on the location of their genes in catabolic operons. Some examples include HcaT, a putative transporter for 3-phenylpropionic acid in *E. coli* (Díaz et al. 1998), MhbT homologs that were found in *Burkholderia xenovorans* LB400 (Romero-Silva et al. 2013), and CadK, a homolog of TfdK that presumably transports 2,4-D in a *Bradyrhizobium* isolate (Kitagawa et al. 2002).

MFS transporters for aromatic acids have also been identified in Gram-positive organisms. GenK was shown to be an MFS transporter for gentisate in *Corynebacterium glutamicum* (Xu et al. 2012b), and the transport functions of VanK (vanillate), PcaK (4-hydroxybenzoate and protocatechuate), and BenK (benzoate) homologs were demonstrated in the same organism (Chaudhry et al. 2007). Other putative MFS transporters in Gram-positive aromatic acid degraders have been identified based on their sequence and gene context, including a BenK homolog in *Rhodococcus* sp. strain RHA1 (Barnes et al. 1997) and HppK, which was predicted to transport 3-(3-hydroxyphenyl)propionic acid in *Rhodococcus globerulus* PWD1 (Kitagawa et al. 2001).

Although most of the identified aromatic acid transporters are members of the MFS, additional aromatic acid transporters outside this superfamily have been documented. For example, in *Rhodococcus* sp. strain RHA1, the *patDABC* genes encode an ABC transporter that mediates transport of phthalate and mono-alkylphthalate esters (Hara et al. 2010). Subsequently, solute-binding protein components of ABC transporters that presumably transport phenylpropanoids and structurally related compounds in *Rhodopseudomonas palustris* were identified by bioinformatic searches followed by binding and structural analyses (Tan et al. 2013). Finally, identification of a putative terephthalate binding protein gene (*tphC*) in the catabolic operon for terephthalate degradation in *Comamonas* sp. strain E6 led to the demonstration of terephthalate uptake by a tripartite aromatic acid transporter composed of TpiA, TpiB, and TphC, which was essential for growth of the strain on terephthalate (Hosaka et al. 2013).

## 5    Other Mechanisms to Enhance Hydrocarbon Acquisition

Bacteria have developed various strategies to enhance the uptake of hydrophobic substrates such as hydrocarbons. These adaptations seem to increase the solubility of poorly accessible hydrocarbons or bring bacteria into direct physical contact with the substrates, thereby contributing to more efficient uptake and metabolism of these hydrophobic substrates.

For example, many hydrocarbon-degrading strains, particularly pseudomonads, produce rhamnolipids, which are biosurfactants that improve the bioavailability of poorly soluble hydrocarbon substrates (Kiran et al. 2016). *P. aeruginosa* PG201 was shown to be more efficient at hexadecane uptake than a mutant incapable of producing rhamnolipids (Beal and Betts 2000). Evidence for a species-specific energy-dependent rhamnolipid-facilitated uptake system for hexadecane was reported for *P. aeruginosa* UG2 (Noordman and Janssen 2002). Strain UG2 was also capable of controlling attachment to hydrocarbon substrates and had an active efflux system (see below) to prevent accumulation of toxic concentrations of hydrocarbons. Addition of rhamnolipids to bacterial cultures has also been shown to decrease the toxicity of certain aromatic compounds (in this case phenol and chlorophenols), presumably by sequestering the toxic molecules in rhamnolipid micelles (Chrzanowski et al. 2009).

Hydrocarbons elicit many physiological changes in bacteria, including alterations to the cell envelope that affect membrane fluidity and permeability by varying the ratios of saturated/unsaturated and *cis/trans* fatty acids (Ramos et al. 2015). In addition, specific changes to the outer cell surface have been reported. For example, *Mycobacterium* sp. LB501T has been shown to attach to anthracene crystals, which are poorly soluble in aqueous solution, and form biofilms. The organism changes its cell surface properties when grown with anthracene in order to bind more efficiently to hydrophobic surfaces (Wick et al. 2002). Similarly, *P. putida* ATCC17514 was shown to adhere to solid polycyclic aromatic hydrocarbons (fluorene and phenanthrene), and it also produced more exopolysaccharide when grown with phenanthrene (Rodrigues et al. 2005). Cells of the alkane-degrading strain *P. aeruginosa* PG201 were more hydrophobic when grown with *n*-hexadecane than with glucose (Beal and Betts 2000). Another alkane degrader, *Rhodococcus erythropolis* S+14He, was highly hydrophobic and adhered to *n*-hexadecane when grown in liquid medium (Kim et al. 2002). This strain, as well as other alkane degraders, has been shown to accumulate alkane substrates in internal membrane-bound inclusions (Kim et al. 2002; Scott and Finnerty 1976). Extracellular structures termed nanopods, which are strings of outer membrane vesicles enclosed in a protein sheath, have been implicated in facilitating phenanthrene degradation by *Delftia acidovorans* CS1-4 (Shetty et al. 2011; Shetty and Hickey 2014). Nanopod formation was induced by phenanthrene, and mutants unable to form nanopods had extended lag times on phenanthrene compared to the wild type. The nanopod-negative mutants also accumulated colored metabolites during growth in the presence of phenanthrene (Shetty and Hickey 2014). However, phenanthrene uptake by the wild-type and mutant strains was not significantly different, indicating that these structures are not directly involved in substrate transport. Current evidence suggests that the nanopods may function to retain phenanthrene intermediates that are secreted into outer membrane vesicles (Shetty and Hickey 2014).

## 6    Hydrocarbon Efflux Pumps

Hydrocarbon-degrading bacteria must balance substrate acquisition with inherent hydrocarbon toxicity. To deal with this problem, many hydrocarbon-degrading bacteria have both specific and nonspecific efflux pumps (Fig. 1) to expel excess solvent from within cells (Ramos et al. 2002; Ramos et al. 2015). The solvent efflux pumps that have been described to date are members of the resistance-nodulation-cell division (RND) family of transporters (Saier and Paulsen 2001). These pumps function in direct opposition to entry of hydrocarbons into cells and are presumed to limit the accumulation of solvents to tolerable levels, thus reducing membrane toxicity (Sikkema et al. 1995). Efflux pumps for toluene, phenanthrene, and anthracene have been described, suggesting that efflux pumps for toxic aromatic hydrocarbons may be widely distributed (Bugg et al. 2000; Kieboom et al. 1998; Phoenix et al. 2003; Ramos et al. 1998; Rojas et al. 2001). Genes encoding three-component pumps have been identified in several strains of *P. putida*, including the highly solvent-resistant strains *P. putida* S12 and DOT-T1E. Mutations in the genes encoding efflux pumps resulted in reduced solvent resistance (reviewed in Ramos et al. (2002)).

A different series of studies demonstrated that phenanthrene entered cells by passive diffusion but was actively extruded by the chromosomally encoded EmhABC efflux system in *P. fluorescens* LP6a (Adebusuyi et al. 2012; Bugg et al. 2000). The efflux system was selective for phenanthrene, anthracene, and fluoranthene and selected antibiotics, but it did not pump out naphthalene (Bugg et al. 2000; Hearn et al. 2003). Interestingly, mutants lacking the efflux system grew more efficiently on phenanthrene than the wild-type strain (Adebusuyi et al. 2012). It therefore appears that the EmhABC efflux system functions to maintain sub-toxic levels of various toxic chemicals within the cell, but its presence is counterproductive when phenanthrene is the sole carbon source, at least under laboratory conditions.

Further details about the mechanisms mediating solvent tolerance are presented in ► Chap. 23, "Extrusion Pumps for Hydrocarbons: An Efficient Evolutionary Strategy to Confer Resistance to Hydrocarbons."

# 7      Research Needs

Although genes and proteins have been identified for passage of hydrocarbons across the outer membrane, there is still relatively little compelling evidence for active transport of hydrocarbons across the cytoplasmic membrane. These studies are particularly difficult due to the volatile nature of hydrocarbons and the proficiency of these molecules to adsorb to plastic and glassware. The available studies provide conflicting data, although strain-to-strain or substrate-to-substrate differences are certainly possible. More importantly, most experiments reported to date that claim active transport of hydrocarbons have failed to clearly separate transport from metabolism. Thus careful studies with specific catabolic mutants must be undertaken in order to study transport in the absence of metabolism.

In aqueous solution, hydrocarbons preferentially partition into cell membranes. Therefore, it seems energetically and physiologically counterproductive to maintain active transport mechanisms that expend energy to accumulate chemicals that can freely pass through membranes and accumulate to potentially toxic levels. In fact, many cells expend energy to export excess hydrocarbons using efflux pumps in order to minimize the toxic effects of these chemicals. It is interesting that no specific genes or proteins have been identified and demonstrated to play a role in hydrocarbon transport across the cytoplasmic membrane. It seems likely that such genes would be located near the catabolic gene clusters (much like aromatic hydrocarbon outer membrane channel encoding genes and aromatic acid transport genes) and would therefore have been identified through bioinformatic searches. With the gigabases of genome and catabolic plasmid sequences currently available, it is noteworthy that such genes have not been found.

In no other metabolic situation is there such a fine line between substrate availability and toxicity. The essential functions of the cytoplasmic membrane in energy generation and maintenance, regulation of internal cellular pH, solute transport, and signaling can be disrupted by the accumulation of hydrocarbons in the bilayer. Various hydrocarbons have been shown to alter the membrane structure, causing leakage of ions and macromolecules, impairment of ATP synthesis, dissipation of the pH gradient across the membrane, and inability to control internal pH (reviewed in Sikkema et al. (1995)). Hydrocarbons are generally poorly soluble in aqueous liquids, and studies have shown that microbes are capable of utilizing hydrocarbons only when they are dissolved in the aqueous phase (Wodzinski and Bertolini 1972; Wodzinski and Coyle 1974). In some cases, the low bioavailability of hydrocarbons results in the rate of metabolism exceeding the rate of hydrocarbon entry into cells, thus limiting the growth rate but also preventing toxicity (Volkering et al. 1992). Maintaining mechanisms for both hydrocarbon entry and efflux may modulate the delicate balance between uptake and toxicity.

The absence of conclusive evidence for hydrocarbon transport across the cytoplasmic membrane does not mean that such active transport mechanisms do not exist, but perhaps they are rare and only required under certain conditions. However, the definitive studies to determine if cytoplasmic transport proteins are required for hydrocarbon accumulation will necessitate the isolation of transport mutants and the identification of specific genes and proteins that are required for the process.

**Acknowledgments** Research in the Parales and Ditty laboratories has been supported by the National Science Foundation (awards MCB 0627248 (REP), MCB 0919930 (REP and JLD), MCB 1022362 (REP)).

# References

Adebusuyi AA, Smith AY, Gray MR, Foght JM (2012) The EmhABC efflux pump decreases the efficiency of phenanthrene biodegradation by *Pseudomonas fluorescens* strain LP6a. Appl Microbiol Biotechnol 95:757–766

Allende JL, Suarez M, Gallego M, Garrido-Pertierra A (1993) 4-Hydroxybenzoate uptake in *Klebsiella pneumoniae* is driven by electric potential. Arch Biochem Biophys 300:142–147

Barnes MR, Duetz W, Williams PA (1997) A 3-(3-hydroxyphenyl)propionic acid catabolic pathway in *Rhodococcus globerulus* PWD1: cloning and characterization of the *hpp* operon. J Bacteriol 179:6145–6153

Bateman JN, Speer B, Feduik L, Hartline RA (1986) Naphthalene association and uptake in *Pseudomonas putida*. J Bacteriol 166:155–161

Beal R, Betts WB (2000) Role of rhamnolipid biosurfactants in the uptake and mineralization of hexadecane in *Pseudomonas aeruginosa*. J Appl Microbiol 89:158–168

Black PN (1991) Primary sequence of *Escherichia coli fadL* gene encoding an outer membrane protein required for long-chain fatty acid transport. J Bacteriol 173:435–442

Bugg T, Foght JM, Pickard MA, Gray MR (2000) Uptake and active efflux of polycyclic aromatic hydrocarbons by *Pseudomonas fluorescens* LP6a. Appl Environ Microbiol 66:5387–5392

Chang H-K, Zylstra GJ (1999) Characterization of the phthalate permease OphD from *Burkholderia cepacia* ATCC 17616. J Bacteriol 181:6197–6199

Chaudhry MT, Huang Y, Shen X-H, Poetsch A, Jiang C-Y, Liu S-J (2007) Genome-wide investigation of aromatic acid transporters in *Corynebacterium glutamicum*. Microbiology 153:857–865

Chrzanowski L, Wick LY, Meulenkamp R, Kaestner M, Heipieper HJ (2009) Rhamnolipid biosurfactants decrease the toxicity of chlorinated phenols to *Pseudomonas putida* DOT-T1E. Lett Appl Microbiol 48:756–762

Collier LS, Nichols NN, Neidle EL (1997) *benK* encodes a hydrophobic permease-like protein involved in benzoate degradation by *Acinetobacter* sp. strain ADP1. J Bacteriol 179:5943–5946

D'Argenio DA, Segura A, Coco WM, Bunz PV, Ornston LN (1999) The physiological contribution of *Acinetobacter* PcaK, a transport system that acts upon protocatechuate, can be masked by overlapping specificity of VanK. J Bacteriol 181:3505–3515

Díaz E, Ferrández A, García JL (1998) Characterization of the *hca* cluster encoding the dioxygenolytic pathway for initial catabolism of 3-phenylpropionic acid in *Escherichia coli* K-12. J Bacteriol 180:2915–2923

DiRusso CC, Black PN (2004) Bacterial long chain fatty acid transport: gateway to a fatty acid-responsive signaling system. J Biol Chem 279:49563–49566

Ditty JL, Harwood CS (1999) Conserved cytoplasmic loops are important for both the transport and chemotaxis functions of PcaK, a protein from *Pseudomonas putida* with 12-membrane-spanning regions. J Bacteriol 181:5068–5074

Ditty JL, Harwood CS (2002) Charged amino acids conserved in the aromatic acid/H$^+$ symporter family of permeases are required for 4-hydroxybenzoate transport by PcaK from *Pseudomonas putida*. J Bacteriol 184:1444–1448

Eaton RW (1997) *p*-Cymene catabolic pathway in *Pseudomonas putida* F1: cloning and characterization of DNA encoding conversion of *p*-cymene to *p*-cumate. J Bacteriol 179:3171–3180

Eaton RW, Timmis KN (1986) Characterization of a plasmid-specified pathway for catabolism of isopropylbenzene in *Pseudomonas putida* RE204. J Bacteriol 168:123–131

Groenewegen PEJ, Driessen AJM, Konigs WN, de Bont JAM (1990) Energy-dependent uptake of 4-chlorobenzoate in the corneyform bacterium NTB-1. J Bacteriol 172:419–423

Habe H, Kasuga K, Nojiri H, Yamane H, Omori T (1996) Analysis of cumene (isopropylbenzene) degradation genes from *Pseudomonas fluorescens* IP01. Appl Environ Microbiol 62:4471–4477

Hara H, Stewart GR, Mohn WW (2010) Involvement of a novel ABC transporter and monoalkyl phthalate ester hydrolase in phthalate ester catabolism by *Rhodococcus jostii* RHA1. Appl Environ Microbiol 76:1516–1523

Harwood CS, Parales RE (1996) The β-ketoadipate pathway and the biology of self-identity. Annu Rev Microbiol 50:533–590

Harwood CS, Nichols NN, Kim M-K, Ditty JL, Parales RE (1994) Identification of the *pcaRKF* gene cluster from *Pseudomonas putida*: involvement in chemotaxis, biodegradation, and transport of 4-hydroxybenzoate. J Bacteriol 176:6479–6488

Hearn EM, Dennis JJ, Gray MR, Foght JM (2003) Identification and characterization of the *emhABC* efflux system for polycyclic aromatic hydrocarbons in *Pseudomonas fluorescens* cLP6a. J Bacteriol 185:6233–6240

Hearn EM, Patel DR, van den Berg B (2008) Outer-membrane transport of aromatic hydrocarbons as a first step in biodegradation. Proc Natl Acad Sci U S A 105:8601–8606

Hearn EM, Patel DR, Lepore BW, Indic M, van den Berg B (2009) Transmembrane passage of hydrophobic compounds through a protein channel wall. Nature 458:367–370

Higgins SJ, Mandelstam J (1972) Evidence for induced synthesis of an active transport factor for mandelate in *Pseudomonas putida*. Biochem J 126:917–922

Hong H, Patel DR, Tamm LK, van den Berg B (2006) The outer membrane protein OmpW forms an eight-stranded β-barrel with a hydrophobic channel. J Biol Chem 28:7568–7577

Hosaka M, Kamimura N, Toribami S, Mori K, Kasai D, Fukuda M, Masai E (2013) Novel tripartite aromatic acid transporter essential for terephthalate uptake in *Comamonas* sp. strain E6. Appl Environ Microbiol 79:6148–6155

Kahng H-Y, Byrne AM, Olsen RH, Kukor JJ (2000) Characterization and role of *tbuX* in utilization of toluene by *Ralstonia pickettii* PKO1. J Bacteriol 182:1232–1242

Kallimanis A, Frillingos S, Drainas C, Koukkou AI (2007) Taxonomic identification, phenanthrene uptake activity, and membrane lipid alterations of the PAH degrading *Arthrobacter* sp. strain Sphe3. Appl Microbiol Biotechnol 76:709–717

Kasai Y, Inoue J, Harayama S (2001) The TOL plasmid pWWO *xylN* gene product from *Pseudomonas putida* is involved in *m*-xylene uptake. J Bacteriol 183:6662–6666

Kieboom J, Dennis JJ, de Bont JA, Zylstra GJ (1998) Identification and molecular characterization of an efflux pump involved in *Pseudomonas putida* S12 solvent tolerance. J Biol Chem 273:85–91

Kim IS, Foght JM, Gray MR (2002) Selective transport and accumulation of alkanes by *Rhodococcus erythropolis* S+14He. Biotechnol Bioeng 80:650–659

Kiran GS, Ninawe AS, Lipton AN, Pandian V, Selvin J (2016) Rhamnolipid biosurfactants: evolutionary implications, applications and future prospects from untapped marine resource. Crit Rev Biotechnol 36:399–415

Kitagawa W, Miyauchi K, Masai E, Fukuda M (2001) Cloning and characterization of benzoate catabolic genes in the Gram-positive polychlorinated biphenyl degrader *Rhodococcus* sp. strain RHA1. J Bacteriol 183:6598–6606

Kitagawa W, Takami S, Miyauchi K, Masai E, Kamagata Y, Tiedje JM, Fukuda M (2002) Novel 2,4-dichlorophenoxyacetic acid degradation genes from oligotrophic *Bradyrhizobium* sp. strain HW13 isolated from a pristine environment. J Bacteriol 184:509–518

Lau PCK, Bergeron H, Labbe D, Wang Y, Brousseau R, Gibson DT (1994) Sequence and expression of the *todGIH* genes involved in the last three steps of toluene degradation by *Pseudomonas putida* F1. Gene 146:7–13

Leveau JH, Zehnder AJ, van der Meer JR (1998) The *tfdK* gene product facilitates uptake of 2,4-dichlorophenoxyacetate by *Ralstonia eutropha* JMP134(pJP4). J Bacteriol 180: 2237–2243

Li Y, Wang H, Hua F, Su M, Zhao Y (2014) Trans-membrane transport of fluoranthene by *Rhodococcus* sp. BAP-1 and optimization of uptake process. Bioresour Technol 155: 213–219

Luu RA, Kootstra JD, Nesteryuk V, Brunton C, Parales JV, Ditty JL, Parales RE (2015) Integration of chemotaxis, transport and catabolism in *Pseudomonas putida* and identification of the aromatic acid chemoreceptor PcaY. Mol Microbiol 96:134–147

Master ER, McKinlay JJ, Stewart GR, Mohn WW (2005) Biphenyl uptake by psychrotolerant *Pseudomonas* sp. strain Cam-1 and mesophilic *Burkholderia* sp. strain LB400. Can J Microbiol 51:399–404

Menn F-M, Zylstra GJ, Gibson DT (1991) Location and sequence of the *todF* gene encoding 2-hydroxy-6-oxohepta-2,4-dienoate hydrolase in *Pseudomonas putida* F1. Gene 104:91–94

Miyata N, Iwahori K, Foght JM, Gray MR (2004) Saturable, energy-dependent uptake of phenanthrene in aqueous phase by *Mycobacterium* sp. strain RJGII-135. Appl Environ Microbiol 70:363–369

Mooney A, O'Leary ND, Dobson AD (2006) Cloning and functional characterization of the *styE* gene, involved in styrene transport in *Pseudomonas putida* CA-3. Appl Environ Microbiol 72:1302–1309

Neher TM, Lueking DR (2009) *Pseudomonas fluorescens ompW*: plasmid localization and requirement for naphthalene uptake. Can J Microbiol 55:553–563

Nichols NN, Harwood CS (1997) PcaK, a high-affinity permease for the aromatic compounds 4-hydroxybenzoate and protocatechuate from *Pseudomonas putida*. J Bacteriol 179:5056–5061

Nikaido H (2003) Molecular basis of bacterial outer membrane permeability. Microbiol Mol Biol Rev 67:593–656

Noda K, Watanabe K, Maruhashi K (2003) Isolation of the *Pseudomonas aeruginosa* gene affecting uptake of dibenzothiophene in *n*-tetradecane. J Biosci Bioeng 95:504–511

Noordman WH, Janssen DB (2002) Rhamnolipid stimulates uptake of hydrophobic compounds by *Pseudomonas aeruginosa*. Appl Environ Microbiol 68:4502–4508

Olsen RH, Kukor JJ, Kaphammer B (1994) A novel toluene-3-monooxygenase pathway cloned from *Pseudomonas pickettii* PKO1. J Bacteriol 176:3749–3756

Pao SS, Paulsen IT, Saier MH Jr (1998) Major facilitator superfamily. Microbiol Mol Rev 62:1–34

Phoenix P, Keane A, Patel A, Bergeron H, Ghoshal S, Lau PCK (2003) Characterization of a new solvent-responsive gene locus in *Pseudomonas putida* F1 and its functionalization as a versatile biosensor. Environ Microbiol 12:1309–1327

Prieto MA, García JL (1997) Identification of the 4-hydroxyphenylacetate transport gene of *Escherichia coli* W: construction of a highly sensitive cellular biosensor. FEBS Lett 414:293–297

Ramos JL, Duque E, Godoy P, Segura A (1998) Efflux pumps involved in toluene tolerance in *Pseudomonas putida* DOT-T1E. J Bacteriol 180:3323–3329

Ramos JL, Duque E, Gallegos MT, Godoy P, Ramos-Gonzalez MI, Rojas A, Teran W, Segura A (2002) Mechanisms of solvent tolerance in Gram-negative bacteria. Annu Rev Microbiol 56:743–768

Ramos JL, Sol Cuenca M, Molina-Santiago C, Segura A, Duque E, Gómez-García MR, Udaondo Z, Roca A (2015) Mechanisms of solvent resistance mediated by interplay of cellular factors in *Pseudomonas putida*. FEMS Microbiol Rev 39:555–566

Ramos-Gonzalez MI, Olson M, Gatenby AA, Mosqueda G, Manzanera M, Campos MJ, Vichez S, Ramos JL (2002) Cross-regulation between a novel two-component signal transduction system for catabolism of toluene in *Pseudomonas mendocina* and the TodST system from *Pseudomonas putida*. J Bacteriol 184:7062–7067

Rodrigues AC, Wuertz S, Brito AG, Melo LF (2005) Fluorene and phenanthrene uptake by *Pseudomonas putida* ATCC 17514: kinetics and physiological aspects. Biotechnol Bioeng 90:281–289

Rojas A, Duque E, Mosqueda G, Golden G, Hurtado A, Ramos JL, Segura A (2001) Three efflux pumps are required to provide efficient tolerance to toluene in *Pseudomonas putida* DOT-T1E. J Bacteriol 183:3967–3973

Romero-Silva MJ, Méndez V, Agulló L, Seeger M (2013) Genomic and functional analyses of the gentisate and protocatechuate ring-cleavage pathways and related 3-hydroxybenzoate and 4-hydroxybenzoate peripheral pathways in *Burkholderia xenovorans* LB400. PLoS One 8: e56038

Saier MH Jr (2006) Transport classification database. http://www.tcdb.org/

Saier MH Jr, Paulsen IT (2001) Phylogeny of multidrug transporters. Semin Cell Dev Biol 12:205–213

Saier MH Jr, Beatty JT, Goffeau A, Harley KT, Heijne WH, Huang SC, Jack DL, Jahn PS, Lew K, Liu J, Pao SS, Paulsen IT, Tseng TT, Virk PS (1999) The major facilitator superfamily. J Mol Microbiol Biotechnol 1:257–279

Scott CC, Finnerty WR (1976) Characterization of intracytoplasmic hydrocarbon inclusions from the hydrocarbon-oxidizing *Acinetobacter* species HO1-N. J Bacteriol 127:481–489

Shetty A, Hickey WJ (2014) Effects of outer membrane vesicle formation, surface-layer production and nanopod development on the metabolism of phenanthrene by *Delftia acidovorans* Cs1-4. PLoS One 9:e92143

Shetty A, Chen S, Tocheva EI, Jensen GJ, Hickey WJ (2011) Nanopods: a new bacterial structure and mechanism for deployment of outer membrane vesicles. PLoS One 6:e20725

Sikkema J, De Bont JAM, Poolman B (1995) Mechanisms of membrane toxicity of hydrocarbons. Microbiol Rev 59:201–222

Tan K, Chang C, Cuff M, Osipiuk J, Landorf E, Mack JC, Zerbs S, Joachimiak A, Collart FR (2013) Structural and functional characterization of solute binding proteins for aromatic compounds derived from lignin: *p*-coumaric acid and related aromatic acids. Proteins 81:1709–1726

van Beilen JB, Eggink G, Enequist H, Bos R, Witholt B (1992) DNA sequence determination and functional characterization of the OCT-plasmid-encoded *alkJKL* genes of *Pseudomonas oleovorans*. Mol Microbiol 6:3121–3136

van Beilen JB, Wubbolts MG, Witholt B (1994) Genetics of alkane oxidation by *Pseudomonas oleovorans*. Biodegradation 5:161–174

van Beilen JB, Panke S, Lucchini S, Franchini AG, Röthlisberger M, Witholt B (2001) Analysis of *Pseudomonas putida* alkane-degradation gene clusters and flanking insertion sequences: evolution and regulation of the *alk* genes. Microbiology 147:1621–1630

van den Berg B (2005) The FadL family: unusual transporters for unusual substrates. Curr Opin Struct Biol 15:401–407

van den Berg B (2010a) Bacterial cleanup: lateral diffusion of hydrophobic molecules through protein channel walls. Biomol Concepts 1:263–270

van den Berg B (2010b) Going forward laterally: transmembrane passage of hydrophobic molecules through protein channel walls. Chembiochem 11:1339–1343

van den Berg B, Black PN, Clemons WMJ, Rapoport TA (2004) Crystal structure of the long-chain fatty acid transporter FadL. Science 304:1506–1509

Vaneechoutte M, Young DM, Ornston LN, De Baere T, Nemec A, Van Der Reijden T, Carr E, Tjernberg I, Dijkshoorn L (2006) Naturally transformable *Acinetobacter* sp. strain ADP1 belongs to the newly described species *Acinetobacter baylyi*. Appl Environ Microbiol 72:932–936

Volkering F, Breure AM, Sterkenberg A, van Andel JG (1992) Microbial degradation of polycyclic aromatic hydrocarbons: effect of substrate availability on bacterial growth kinetics. Appl Microbiol Biotechnol 36:548–552

Wang Y, Rawlings M, Gibson DT, Labbé D, Bergeron H, Brousseau R, Lau PCK (1995) Identification of a membrane protein and a truncated LysR-type regulator associated with the toluene degradation pathway in *Pseudomonas putida* F1. Mol Gen Genet 246:570–579

Whitman BE, Lueking DR, Mihelcic JR (1998) Naphthalene uptake by a *Pseudomonas fluorescens* isolate. Can J Microbiol 44:1086–1093

Wick LY, de Munain AR, Springael D, Harms H (2002) Responses of *Mycobacterium* sp. LB501T to the low bioavailability of solid anthracene. Appl Microbiol Biotechnol 58:378–385

Wodzinski RS, Bertolini D (1972) Physical state in which naphthalene and bibenzyl are utilized by bacteria. Appl Microbiol 23:1077–1081

Wodzinski RS, Coyle JE (1974) Physical state of phenanthrene for utilization by bacteria. Appl Microbiol 27:1081–1084

Xu Y, Gao X, Wang S-H, Liu H, Williams PA, Zhou N-Y (2012a) MhbT is a specific transporter for 3-hydroxybenzoate uptake by Gram-negative bacteria. Appl Environ Microbiol 78:6113–6120

Xu Y, Wang S-H, Chao H-J, Liu S-J, Zhou N-Y (2012b) Biochemical and molecular characterization of the gentisate transporter GenK in *Corynebacterium glutamicum*. PLoS One 7:e38701

Xu Y, Chen B, Chao H-J, Zhou N-Y (2013) *mhpT* encodes an active transporter involved in 3-(3-hydroxyphenyl)propionate catabolism by *Escherichia coli* K-12. Appl Environ Microbiol 79:6362–6368

Zylstra GJ, Gibson DT (1989) Toluene degradation by *Pseudomonas putida* F1: nucleotide sequence of the *todC1C2BADE* genes and their expression in *E. coli*. J Biol Chem 264:14940–14946

# Strategies to Increase Bioavailability and Uptake of Hydrocarbons

# 18

J. J. Ortega-Calvo

## Contents

1   Introduction ................................................................. 303
2   The State of the Art in Bioavailability Science ............................. 304
    2.1   The Bioavailability Concept ......................................... 304
    2.2   Strategies to Enhance Bioavailability for Biodegradation ............. 306
3   Enhanced Pollutant Phase Exchange ......................................... 307
4   Microbial Mobilization .................................................... 309
5   Microbial Attachment ...................................................... 310
6   Research Needs ............................................................ 311
References .................................................................... 312

### Abstract

The biodegradation of hydrocarbons in the environment is often slow due to restricted bioavailability. Research performed during the last 20 years has shown possible pathways to increase the bioavailability of hydrocarbons without necessarily increasing the risk to the environment. Pollutant solubilization through (bio)surfactants, microbial transport, and attachment to pollutant interfaces can increase bioavailability, which translates into an enhancement of biodegradation rates. These strategies can not only be integrated into optimized bioremediation protocols that lead to lower decontamination endpoints in soils and sediments but also help to improve biodegradation in other environmental contexts, such as wastewater treatment and natural attenuation.

J. J. Ortega-Calvo (✉)
Instituto de Recursos Naturales y Agrobiologia de Sevilla, CSIC, Sevilla, Spain
e-mail: jjortega@irnase.csic.es

© Springer International Publishing AG, part of Springer Nature 2018                303
T. Krell (ed.), *Cellular Ecophysiology of Microbe: Hydrocarbon and Lipid Interactions*,
Handbook of Hydrocarbon and Lipid Microbiology, https://doi.org/10.1007/978-3-319-50542-8_10

# 1    Introduction

With our deep knowledge of how hydrocarbons undergo biodegradation in the environment, it is not surprising that the application of biological technologies to the treatment of hydrocarbon pollution is well established. For example, of the 16 polycyclic aromatic hydrocarbons (PAHs) listed by the US Environmental Protection Agency (US-EPA) as priority pollutants, those with a molecular weight of up to 202 g/mol, including the high-molecular-weight PAHs pyrene and fluoranthene can be degraded microbially through growth-linked aerobic reactions, while the rest of the PAHs in that list, such as benzo[a]pyrene, are susceptible to cometabolic removal (Niqui-Arroyo et al. 2011). These reactions are the basis of a variety of approaches applied to biologically treated soils contaminated with hydrocarbons, which include landfarming, composting, bioreactor treatments, and phytoremediation (Ortega-Calvo et al. 2013). However, the benefits derived from the biodegradation of these chemicals are also accompanied by the uncertainties that surround the use of biological treatment when the bioavailability of these chemicals remains unpredictable. The current poor predictability of endpoints associated with the bioremediation of hydrocarbons is a large limitation when evaluating its viability for treating contaminated soils and sediments.

Three questions must be addressed when incorporating bioavailability science into bioremediation: (1) What is meant by "bioavailability?" (2) How should it be measured? and (3) Is it possible to increase bioavailability without enhancing environmental risks of the pollutants? In this chapter, we examine from an environmental perspective the potential of bioavailability-promoting strategies to enhance the microbial biodegradation of hydrocarbons. Therefore, the focus will be on mechanisms that eventually allow the integration of a bioavailability-efficient technology into current bioremediation practices.

# 2    The State of the Art in Bioavailability Science

## 2.1    The Bioavailability Concept

Over the last 30 years, numerous publications have discussed the concepts and definitions of bioavailability of organic chemicals. These have been summarized recently in the context of risk assessment and regulation (Ortega-Calvo et al. 2015) and are illustrated in Fig. 1. The main schools of thought consider bioavailability (focusing on the aqueous or dissolved contaminant), bioaccessibility (incorporating the rapidly desorbing contaminant in the exposure), and chemical activity (determining the potential of the dissolved contaminant for biological effects). Using the same framework, the figure places these different schools (Ehlers and Luthy 2003; Semple et al. 2004; Reichenberg and Mayer 2006) that have dissected bioavailability into the different processes that are involved (A to E), the dissimilar endpoints (bioaccessibility and chemical activity), and the different methodologies (desorption extraction, passive sampling, and biological tests). Each of these processes,

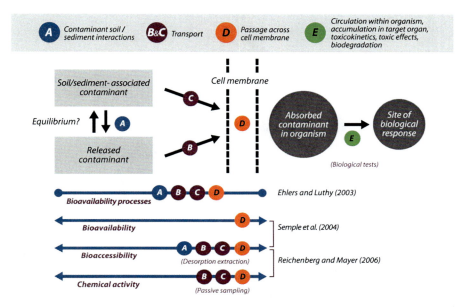

**Fig. 1** Overview of scientific concepts of the bioavailability of organic chemicals (Reproduced with permission from Ortega-Calvo et al. 2015)

endpoints, and methods has been considered differently in a wide variety of bioavailability scenarios. Depending on the processes investigated, bioavailability can be examined through chemical activity (by including the processes B, C, and D) or bioaccessibility measurements which incorporate to B–D the time-dependent release of the contaminant from the soil/sediment. Depending on the biological complexity, the passage of the contaminant molecule across the cell membrane (process D) may represent multiple stages within a given organism before the site of biological response is reached (process E).

Chemical and biological approaches can be used to measure the bioavailability of organic chemicals. The results of infinite sink methods using Tenax and cyclodextrin extraction are currently used to predict toxicity and biodegradation and are in the process of being standardized. The results of these methods represent and define what is referred to as the rapidly desorbing fraction. The second complementary approach is the use of passive sampling to determine the freely dissolved concentration as a measure of the chemical activity of organic chemicals in soils and sediments. This approach proposes that chemical activity drives bioavailability (Fig. 1). Finally, several (mostly standardized by the International Organization for Standardization and the Organization for Economic Cooperation and Development) ecotoxicological test methods are available to determine bioavailability in the soil and sediment compartments to invertebrates, plants, and microorganisms (Ortega-Calvo et al. 2015).

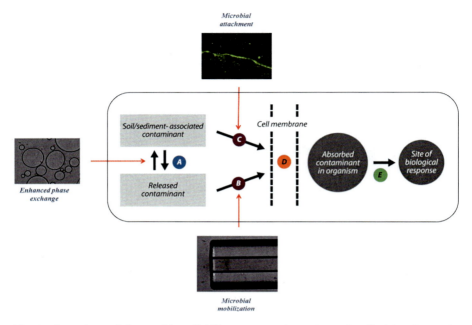

**Fig. 2** Strategies to influence bioavailability processes (*A* to *E*, as described in Fig. 1), in connection with the microbial degradation of hydrocarbons

## 2.2    Strategies to Enhance Bioavailability for Biodegradation

Several strategies can act on bioavailability processes to enhance the microbial degradation of hydrocarbons (Fig. 2). Enhanced phase exchange (represented in the figure by the surfactant action – process A), microbial mobilization (represented by a band of chemotactic bacteria attracted by a chemical gradient inside a capillary – process B), and attachment to interfaces, that allows the direct acquisition of the soil/sediment-associated contaminant (process C), can increase bioavailability to hydrocarbon-degrading microorganisms and, therefore, biodegradation. In Fig. 2, process D represents the pollutant uptake by the microbial cells, necessary for biodegradation processing (process E), what may involve the mineralization of the chemical and the incorporation of substrate carbon into biochemical components and metabolites.

In a bioremediation context, these strategies should be integrated in such a way that the risks to the environment are minimized, such as the modulation of pollutant release relative to the actual biodegradation potential; the use of environmentally acceptable agents, such as (bio)surfactants, microorganisms, and plants; and the application of treatment methods requiring minimal handling of the polluted soil or sediment. A conservative approach is needed, because in some circumstances, bioremediation may even increase the risks of the pollutants (Ortega-Calvo et al. 2013). Polluted soils and sediments treated by bioremediation may increase their toxic potential, due to the release of sorbed hydrocarbons and the formation of toxic metabolites, although prolonged treatments may help to minimize these risks. This knowledge has also implications in other environmental contexts where the lack of biodegradation plays a relevant role in the persistence of hydrocarbons, such as wastewater treatment and natural attenuation (Niqui-Arroyo et al. 2011).

## 3    Enhanced Pollutant Phase Exchange

The positive effects of promoting the phase exchange of hydrocarbons on their biodegradation in soil have been known for some time (Ortega-Calvo et al. 1995). For example, surfactants can increase the biodegradation rates of hydrocarbons that desorb slowly from soils (process A in Fig. 2). In general, nonionic surfactants are the most frequently used types of surfactants in biodegradation studies, mainly because they are uncharged, which minimizes any eventual toxic effects (Martín et al. 2014). In a study focused on the bioavailability of slow desorption hydrocarbons, the nonionic surfactant Brij 35 was applied to a soil originating directly from a site polluted by creosote and a soil from a manufactured gas plant (MGP) that had been treated by bioremediation (Bueno-Montes et al. 2011). In the creosote-polluted soil, biodegradation was inhibited by the surfactant, but biodegradation in the bioremediated MGP soil, which was enriched in slow desorption hydrocarbons, was enhanced. The different outcomes were likely a consequence of the balance of two effects, namely an increase in the bioaccessibility of the chemicals and an enhancement of the consumption rate of other PAHs present in the soil at the same time, yielding subsequent competition effects. Therefore, the selective use of surfactants in soils enriched in hydrocarbons that desorb slowly would not only avoid the inhibition of biodegradation through competition mechanisms. This strategy would also allow for the reduction of risks associated with an increased chemical activity and toxicity of the hydrocarbons and metabolites as a result of solubilization, at concentrations in excess of the metabolic potential of the microorganisms. Other advantages of this different way of surfactant action may relate to the biodegradable nature of the surfactant, which is obviously necessary to minimize environmental impacts. The use of a surfactant involves eventual increases in nutrient and oxygen demands, which would be less important when pollutant loads are attenuated and less bioaccessible, as well as the decrease in concentration of surfactant after extended biodegradation periods, when it is required to solubilize the slow desorption hydrocarbons. However, the use of surfactants in environmental remediation may be potentially risky because they may be toxic and diminish the soil quality, leading to the need for other economic considerations. As a result, it may potentially be better to use biosurfactants or other naturally occurring stimulants.

Electrokinetics can also be successfully employed to enhance the bioavailability of hydrocarbons, particularly for the surfactant-assisted bioremediation of soils that are rich in clay fractions and/or aged contaminants (Niqui-Arroyo et al. 2006; Niqui-Arroyo and Ortega-Calvo 2007, 2010). These studies made use of the surfactant Brij

35 to optimize the process performance. The residual concentrations of the total biodegradable PAHs remaining after bioremediation in soil slurries were twofold lower in electrokinetically pretreated soils than in untreated soils (Niqui-Arroyo and Ortega-Calvo 2010). It is conceivable that physicochemical changes produced in polluted soil particles that were exposed to electric fields may promote the bioavailability of these PAHs, thus improving the bioremediation performance. The application of an electric field to the soil in the presence of the surfactant, through the electroosmotic flow that was consequently generated, enhanced the bioavailability of the PAHs by changing their desorption kinetics. The development of an electroosmotic flow through soil aggregates could probably have caused an increase in the rate of slow desorption (Shi et al. 2008). The induced physical and chemical changes could have resulted in the mobility of the pollutant fraction entrapped within the soil nanopores and/or strongly sorbed to black carbon.

Microbial and plant biosurfactants, if properly managed, are also able to improve bioremediation performance through an enhanced phase exchange. Many different surface-active compounds synthesized by a wide variety of microorganisms, such as *Pseudomonas*, *Bacillus*, *Acinetobacter*, and *Mycobacterium*, have been identified (Banat et al. 2010). The exact physiological roles of microbial surfactants are apparently not restricted to the solubilization of hydrophobic carbon sources, as surfactants can also be produced when the microorganisms are grown with water-soluble substrates, such as glucose. These biosurfactants are important in a number of ecological processes and have been linked to microbial adhesion, antagonistic effects toward other microorganisms, heavy metal sequestration, and cell-cell communication. Although the production of biosurfactants is not universal among all microbes, their effect on the bioavailability of hydrocarbons in the natural environment causes these agents to be an important factor that should be considered when optimizing bioremediation. Indeed, rhamnolipid biosurfactants can dissolve pure, solid PAHs, such as phenanthrene, thus increasing their rate of biodegradation (Garcia-Junco et al. 2001, 2003; Resina-Pelfort et al. 2003), and enhance the desorption and biodegradation of slowly desorbing and aged pyrene present in soil (Congiu and Ortega-Calvo 2014; Congiu et al. 2015).

In addition to biosurfactants, there is a wide variety of other natural organic compounds derived from either microbes or plants that can potentially increase the bioavailability of hydrocarbons. For example, cyclodextrins (Garon et al. 2004) and unsaturated fatty acids (Yi and Crowley 2007) have been proposed to stimulate the biodegradation of PAHs in soil through this mechanism. The bioavailability of hydrocarbons can be enhanced by the general capacity of dissolved organic matter (DOM) to mobilize hydrophobic chemicals (Tejeda-Agredano et al. 2013). This enhancement can occur through a variety of process: (1) enhanced desorption (Haderlein et al. 2001), (2) direct access to DOM-sorbed PAHs due to the physical association of bacteria and DOM (Ortega-Calvo and Saiz-Jimenez 1998), and (3) increased diffusional flux through unstirred boundary layers around bacterial cells (Haftka et al. 2008), similarly to the enhanced uptake of cadmium by spinach plantlets originated from seed germination in the presence of labile metal complexes (Degryse et al. 2006). However, DOM can also inhibit biodegradation in exposure scenarios governed by a low chemical activity of the pollutants, by preventing the bacterial attachment to the hydrocarbon source (Tejeda-Agredano et al. 2014).

## 4 Microbial Mobilization

The bioavailability of hydrocarbons can be increased not only by mobilizing the pollutants but also by promoting the dispersal of microorganisms throughout the polluted matrix, to have a better access to the released contaminant (process B in Fig. 2). However, movement of pollutant-degrading microorganisms in porous media is often restricted by their high deposition rates and adhesion to soil surfaces (Ortega-Calvo et al. 1999; Lahlou et al. 2000). To this end, chemotactic mobilization of flagellated bacteria has gained attention in bioremediation (Marx and Aitken 2000; Krell et al. 2013). A technological innovation based on this concept makes use of the mobilization potential of pollutant-degrading microorganisms from rhizospheres, with chemotactic responses demonstrated toward moderately hydrophobic PAHs, such as phenanthrene, anthracene, and pyrene, and are able to move chemotactically at speeds of approximately 1 mm/min (Ortega-Calvo et al. 2003). It has also been shown in later research that motility and transport of hydrocarbon-degrading bacteria can be controlled through a suitable choice of chemical effectors, including carbon sources and nanomaterials (Velasco-Casal et al. 2008; Ortega-Calvo et al. 2011; Jimenez-Sanchez et al. 2012, 2015). In well-controlled column systems, we assessed the influence of different effectors on the deposition of a chemotactic, naphthalene-degrading bacterial species, *Pseudomonas putida* G7, in selected porous environments (sand, forest soil, and clay aggregates). Cellular deposition, however, was concomitantly dependent on the cellular motility (hypermotility, attraction, or repulsion), the sorption of the effector to the column packing material, and the resulting pore-water concentration. For example, an exposure of the cells to salicylate and naphthalene induced a smooth movement with few acceleration events and positive taxis, while cells exposed to silver nanoparticles (AgNPs) exhibited tortuous movement and repulsion. Although glucose was metabolized by strain G7, it did not cause any attraction, but it induced the cells to go into a hypermotile mode, characterized by a high frequency of acceleration events, a high swimming speed ($>60$ μm s$^{-1}$), and a high tortuosity in the trajectories. The chemically induced motility behaviors demonstrated a distinct affinity of cells for sand particles in batch assays, resulting in the development of breakthrough curves in percolation column experiments. Salicylate and naphthalene significantly reduced the deposition of G7 cells in the column experiments, while glucose and AgNPs enhanced the attachment and caused a blocking of the filter, which resulted in a progressive decrease in deposition. This work was later extended to show distinctive tactic behaviors and mobilization as a result from exposure to zero-valent iron nanoparticles (Ortega-Calvo et al. 2016) and chemoeffector-containing DOM (Jimenez-Sanchez et al. 2015).

Another approach for facilitated microbial dispersal, in connection with hydrocarbon-polluted scenarios, is based on bacterial/oomycete interactions

(Sungthong et al. 2015, 2016). Our results show that zoospores can act as ecological amplifiers of fungal and oomycete actions extending the concept of "mycelial pathways" for dispersal of pollutant-degrading bacteria (Kohlmeier et al. 2005; Furuno et al. 2010). This strategy may be of relevance for nonflagellated bacterial PAH degraders, such as *Mycobacterium* species, which may constitute a significant fraction of the functional microbiome in PAH-polluted environments (Uyttebroek et al. 2006a). Although these bacteria seem to be less well transported through mycelial pathways than self-propelled bacteria (Kohlmeier et al. 2005), the absence of motility is, in relative terms, a positive factor for the biomobilization caused by zoospores. This can be explained by the link of this mobilization mode with flow dynamics. The thrust force created by zoospore swimming mobilizes more efficiently immotile bacteria than actively swimming bacteria. Besides, flagellated (and therefore chemotactically active) bacterial groups, such as *Pseudomonas* and *Achromobacter*, can be dispersed through their own chemotactic navigation along mycelial pathways (Furuno et al. 2010), but they could also be biomobilized by zoospores at the cell growth phases when flagellar motility is limited or not existing.

## 5    Microbial Attachment

Microorganisms can also increase the bioavailability of hydrocarbons when in direct contact with the pollutant source through attachment (Fig. 2, process C), thereby enabling biodegradation to occur, or even to proceed more rapidly (Ortega-Calvo and Alexander 1994; Garcia-Junco et al. 2003; Uyttebroek et al. 2006b; Tejeda-Agredano et al. 2011; Zhang et al. 2016). A feasible strategy to enhance bioavailability is, therefore, to promote the growth and activity of attached bacteria. The main goal of the study focused on PAHs present in nonaqueous-phase liquids or NAPLs (Tejeda-Agredano et al. 2011) was to target the potential nutritional limitations of microorganisms to enhance the biodegradation of PAHs at the interface between the NAPL and the water phase. The results indicated that the biodegradation of PAHs by bacterial cells attached to NAPLs can be limited by nutrient availability as a result of the simultaneous consumption of PAHs within the NAPLs, but this limitation can be overcome by the addition of an oleophilic biostimulant that act as an interface fertilizer, by providing nutrients to attached bacteria, thus promoting bioavailability.

Bacteria attached to hydrocarbon/water interfaces can access to, theoretically unaccessible, submicrometer pores through proliferation and penetration (Akbari and Ghoshal 2015; Akbari et al. 2016). In a series of studies with the hexadecane-degrading, immotile bacterium *Dietzia maris*, Ghoshal et al. determined that the hydrocarbon remained non-bioavailable when it was separated from bacteria by membranes with pores that were 3 μm or smaller. Given the size of individual cells (0.9 μm), their strong tendency to form aggregates (6.0 μm), and the need for attachment to the hydrocarbon phase as a prerequisite for biodegradation of this sparingly soluble compound, the cause for this poor bioavailability was the lack of physical contact between the hydrocarbon phase and the bacterial aggregates (Akbari and Ghoshal 2015). However, during growth in hexadecane-wetted membranes, morphological changes of attached cells allowed the penetration in the membranes through pores that were smaller than the cell size (Akbari et al. 2016). Authors postulated that this penetration was caused by physical forces imposed by growth, colony extension, and biosurfactant production. In this way, bacterial translocation and passage through the pores allowed the mineralization of $^{14}C$–labeled hexadecane that was separated from the filter by an aqueous phase. Therefore, bacterial plasticity may not only play a role in the attachment and colonization of distant, separate phase hydrocarbons located in submicron porosities (process C in Fig. 2) but also in the biodegradation of the dissolved fractions of the chemicals which are only available through diffusion and transport through these small pores (process B).

## 6   Research Needs

Bioavailability was recognized as a research priority in bioremediation as early as 25 years ago (Alexander 1991). Over these years, significant advancements have been made to understand how pollutant phase exchange, microbial mobilization, and cell attachment to interfaces affects bioavailability during biodegradation of organic chemicals and, particularly, hydrocarbons. In spite of these advancements, significant gaps of knowledge exist between bioavailability science and bioremediation. Still today, it is difficult to predict bioavailability of hydrocarbons, for example, solely on the basis of basic parameters such as organic matter, black carbon or clay contents of a given soil or sediment, and the physicochemical constants of the chemicals (such as solubility in water and octanol-water or organic-carbon based distribution coefficients). This limitation even remains with improved assessments through determinations of chemical activity and bioaccessibility. This uncertainty not only applies to biodegradability in natural environments but also to engineered systems where bioavailability-promoting strategies are implemented. Therefore, research is needed to provide a solid link among chemical activity, bioaccessibility, and biodegradation of hydrocarbons. A main research need is also to understand how biological networks react in situ to sustainable, bioavailability-promoting approaches involving biostimulants, specialized microorganisms, and plants for an enhanced pollutant carbon turnover. The findings achieved with simple laboratory or microcosm systems need to be translated to field operations, and this scale-up faces serious difficulties, given the complexity of intra- and interspecific biological interactions occurring in polluted sites. The assessment of possible risks, such as increases in compound toxicity due to bioremediation, should be also part of future research.

**Acknowledgements**   This study was supported by the Spanish Ministry of Science and Innovation (CGL2013-44554-R and CGL2016-77497-R), the Andalusian Government (RNM 2337), and the European Commission (LIFE15 ENV/IT/000396).

# References

Akbari A, Ghoshal S (2015) Bioaccessible porosity in soil aggregates and implications for biodegradation of high molecular weight petroleum compounds. Environ Sci Technol 49:14368–14375

Akbari A, Rahim AA, Ehrlicher AJ, Ghoshal S (2016) Growth and attchment-facilitated entry of bacteria into submicrometer pores can enhance bioremediation and oil recovery in low-permeability and microporous media. Environ Sci Technol Lett 3:399–403

Alexander M (1991) Research needs in bioremediation. Environ Sci Technol 25:1972–1973

Banat IM, Franzetti A, Gandolfi I, Bestetti G, Martinotti MG, Fracchia L, Smyth TJ, Marchant R (2010) Microbial biosurfactants production, applications and future potential. Appl Microbiol Biotechnol 87:427–444

Bueno-Montes M, Springael D, Ortega-Calvo JJ (2011) Effect of a non-ionic surfactant on biodegradation of slowly desorbing PAHs in contaminated soils. Environ Sci Technol 45:3019–3026

Congiu E, Ortega-Calvo J-J (2014) Role of desorption kinetics in the rhamnolipid-enhanced biodegradation of polycyclic aromatic hydrocarbons. Environ Sci Technol 48:10869–10877

Congiu E, Parsons JR, Ortega-Calvo J-J (2015) Dual partitioning and attachment effects of rhamnolipid on pyrene biodegradation under bioavailability restrictions. Environ Pollut 205:378–384

Degryse F, Smolders E, Merckx R (2006) Labile Cd complexes increase Cd availability to plants. Environ Sci Technol 40:830–836

Ehlers LJ, Luthy RG (2003) Contaminant bioavailability in soil and sediment. Environ Sci Technol 37:295A–302A

Furuno S, Pazolt K, Rabe C, Neu TR, Harms H, Wick LY (2010) Fungal mycelia allow chemotactic dispersal of polycyclic aromatic hydrocarbon-degrading bacteria in water-unsaturated systems. Environ Microbiol 12:1391–1398

Garcia-Junco M, De Olmedo E, Ortega-Calvo JJ (2001) Bioavailability of solid and non-aqueous phase liquid (NAPL)-dissolved phenanthrene to the biosurfactant-producing bacterium *Pseudomonas aeruginosa* 19SJ. Environ Microbiol 3:561–569

Garcia-Junco M, Gomez-Lahoz C, Niqui-Arroyo JL, Ortega-Calvo JJ (2003) Biodegradation- and biosurfactant-enhanced partitioning of polycyclic aromatic hydrocarbons from nonaqueous-phase liquids. Environ Sci Technol 37:2988–2996

Garon D, Sage L, Wouessidjewe D, Seigle-Murandi F (2004) Enhanced degradation of fluorene in soil slurry by *Absidia cylindrospora* and maltosyl-cyclodextrin. Chemosphere 56:159–166

Haderlein A, Legros R, Ramsay B (2001) Enhancing pyrene mineralization in contaminated soil by the addition of humic acids or composted contaminated soil. Appl Microbiol Biotechnol 56:555–559

Haftka JJH, Parsons JR, Govers HAJ, Ortega-Calvo JJ (2008) Enhanced kinetics of solid-phase microextraction and biodegradation of polycyclic aromatic hydrocarbons in the presence of dissolved organic matter. Environ Toxicol Chem 27:1526–1532

Jimenez-Sanchez C, Wick LY, Ortega-Calvo JJ (2012) Chemical effectors cause different motile behavior and deposition of bacteria in porous media. Environ Sci Technol 46:6790–6797

Jimenez-Sanchez C, Wick LY, Cantos M, Ortega-Calvo JJ (2015) Impact of dissolved organic matter on bacterial tactic motility, attachment, and transport. Environ Sci Technol 49:4498–4505

Kohlmeier S, Smits THM, Ford RM, Keel C, Harms H, Wick LY (2005) Taking the fungal highway: mobilization of pollutant-degrading bacteria by fungi. Environ Sci Technol 39:4640–4646

Krell T, Lacal J, Reyes-Darías JA, Jimenez-Sanchez C, Sungthong R, Ortega-Calvo JJ (2013) Bioavailability of pollutants and chemotaxis. Curr Opin Biotechnol 24:451–456

Lahlou M, Harms H, Springael D, Ortega-Calvo JJ (2000) Influence of soil components on the transport of polycyclic aromatic hydrocarbon-degrading bacteria through saturated porous media. Environ Sci Technol 34:3649–3656

Martín VI, de la Haba RR, Ventosa A, Congiu E, Ortega-Calvo JJ, Moyá ML (2014) Colloidal and biological properties of cationic single-chain and dimeric surfactants. Colloids Surf B: Biointerfaces 114:247–254

Marx RB, Aitken MD (2000) Bacterial chemotaxis enhances naphthalene degradation in a heterogeneous aqueous system. Environ Sci Technol 34:3379–3383

Niqui-Arroyo JL, Ortega-Calvo JJ (2007) Integrating biodegradation and electroosmosis for the enhanced removal of polycyclic aromatic hydrocarbons from creosote-polluted soils. J Environ Qual 36:1444–1451

Niqui-Arroyo JL, Ortega-Calvo JJ (2010) Effect of electrokinetics on the bioaccessibility of polycyclic aromatic hydrocarbons in polluted soils. J Environ Qual 39:1993–1998

Niqui-Arroyo JL, Bueno-Montes M, Posada-Baquero R, Ortega-Calvo JJ (2006) Electrokinetic enhancement of phenanthrene biodegradation in creosote-polluted clay soil. Environ Pollut 142:326–332

Niqui-Arroyo JL, Bueno-Montes M, Ortega-Calvo JJ (2011) Biodegradation of anthropogenic organic compounds in natural environments. In: Xing B, Senesi N, Huang PM (eds) Biophysico-chemical processes of anthropogenic organic compounds in environmental systems, IUPAC series on Biophysico-chemical processes in environmental systems, vol 3. Wiley, Chichester, pp 483–501

Ortega-Calvo JJ, Alexander M (1994) Roles of bacterial attachment and spontaneous partitioning in the biodegradation of naphthalene initially present in nonaqueous-phase liquids. Appl Environ Microbiol 60:2643–2646

Ortega-Calvo JJ, Saiz-Jimenez C (1998) Effect of humic fractions and clay on biodegradation of phenanthrene by a *Pseudomonas fluorescens* strain isolated from soil. Appl Environ Microbiol 64:3123–3126

Ortega-Calvo JJ, Birman I, Alexander M (1995) Effect of varying the rate of partitioning of phenanthrene in nonaqueous-phase liquids on biodegradation in soil slurries. Environ Sci Technol 29:2222–2225

Ortega-Calvo JJ, Fesch C, Harms H (1999) Biodegradation of sorbed 2,4-dinitrotoluene in a clay-rich, aggregated porous medium. Environ Sci Technol 33:3737–3742

Ortega-Calvo JJ, Marchenko AI, Vorobyov AV, Borovick RV (2003) Chemotaxis in polycyclic aromatic hydrocarbon-degrading bacteria isolated from coal-tar- and oil-polluted rhizospheres. FEMS Microbiol Ecol 44:373–381

Ortega-Calvo JJ, Molina R, Jimenez-Sanchez C, Dobson PJ, Thompson IP (2011) Bacterial tactic response to silver nanoparticles. Environ Microbiol Rep 3:526–534

Ortega-Calvo JJ, Tejeda-Agredano MC, Jimenez-Sanchez C, Congiu E, Sungthong R, Niqui-Arroyo JL, Cantos M (2013) Is it possible to increase bioavailability but not environmental risk of PAHs in bioremediation? J Hazard Mater 261:733–745

Ortega-Calvo J-J, Harmsen J, Parsons JR, Semple KT, Aitken MD, Ajao C, Eadsforth C, Galay-Burgos M, Naidu R, Oliver R, Peijnenburg WJGM, Roembke J, Streck G, Versonnen B (2015) From bioavailability science to regulation of organic chemicals. Environ Sci Technol 49:10255–10264

Ortega-Calvo JJ, Jimenez-Sanchez C, Pratarolo P, Pullin H, Scott TB, Thompson IP (2016) Tactic response of bacteria to zero-valent iron nanoparticles. Environ Pollut 213:438–445

Reichenberg F, Mayer P (2006) Two complementary sides of bioavailability: accessibility and chemical activity of organic contaminants in sediments and soils. Environ Toxicol Chem 25:1239–1245

Resina-Pelfort O, García-Junco M, Ortega-Calvo JJ, Comas-Riu J, Vives-Rego J (2003) Flow cytometry discrimination between bacteria and clay humic acid particles during growth-linked biodegradation of phenanthrene by *Pseudomonas aeruginosa* 19SJ. FEMS Microbiol Ecol 43:55–61

Semple KT, Doick KJ, Jones KC, Burauel P, Craven A, Harms H (2004) Defining bioavailability and bioaccessibility of contaminated soil and sediment is complicated. Environ Sci Technol 38:228A–231A

Shi L, Harms H, Wick LY (2008) Electroosmotic flow stimulates the release of alginate-bound phenanthrene. Environ Sci Technol 42:2105–2110

Sungthong R, van West P, Cantos M, Ortega-Calvo JJ (2015) Development of eukaryotic zoospores within polycyclic aromatic hydrocarbon (PAH)-polluted environments: a set of behaviors that are relevant for bioremediation. Sci Total Environ 511:767–776

Sungthong R, Van West P, Heyman F, Jensen DF, Ortega-Calvo JJ (2016) Mobilization of pollutant-degrading bacteria by eukaryotic zoospores. Environ Sci Technol 50:7633–7640

Tejeda-Agredano MC, Gallego S, Niqui-Arroyo JL, Vila J, Grifoll M, Ortega-Calvo JJ (2011) Effect of interface fertilization on biodegradation of polycyclic aromatic hydrocarbons present in nonaqueous-phase liquids. Environ Sci Technol 45:1074–1081

Tejeda-Agredano MC, Gallego S, Vila J, Grifoll M, Ortega-Calvo JJ, Cantos M (2013) Influence of sunflower rhizosphere on the biodegradation of PAHs in soil. Soil Biol Biochem 57:830–840

Tejeda-Agredano MC, Mayer P, Ortega-Calvo JJ (2014) The effect of humic acids on biodegradation of polycyclic aromatic hydrocarbons depends on the exposure regime. Environ Pollut 184:435–442

Uyttebroek M, Breugelmans P, Janssen M, Wattiau P, Joffe B, Karlson U, Ortega-Calvo JJ, Bastiaens L, Ryngaert A, Hausner M, Springael D (2006a) Distribution of the *Mycobacterium* community and polycyclic aromatic hydrocarbons (PAHs) among different size fractions of a long-term PAH-contaminated soil. Environ Microbiol 8:836–847

Uyttebroek M, Ortega-Calvo JJ, Breugelmans P, Springael D (2006b) Comparison of mineralization of solid-sorbed phenanthrene by polycyclic aromatic hydrocarbon (PAH)-degrading *Mycobacterium* spp. and *Sphingomonas* spp. Appl Microbiol Biotechnol 72:829–836

Velasco-Casal P, Wick LY, Ortega-Calvo JJ (2008) Chemoeffectors decrease the deposition of chemotactic bacteria during transport in porous media. Environ Sci Technol 42:1131–1137

Yi H, Crowley DE (2007) Biostimulation of PAH degradation with plants containing high concentrations of linoleic acid. Environ Sci Technol 41:4382–4388

Zhang M, Shen XF, Zhang HY, Cai F, Chen WX, Gao Q, Ortega-Calvo JJ, Tao S, Wang XL (2016) Bioavailability of phenanthrene and nitrobenzene sorbed on carbonaceous materials. Carbon 110:404–413

# The Mycosphere as a Hotspot for the Biotransformation of Contaminants in Soil

# 19

Lukas Y. Wick and Hauke Harms

## Contents

1    Introduction ................................................................... 315
2    Fungi Cope with Heterogeneous and Contaminated Environments ...................... 316
3    Fungi Transform Complex Contaminants Even at Low Concentrations ................. 318
4    Fungi Promote Contaminant Accessibility and Bioavailability ........................... 319
5    Fungi Create Functional Habitats for Bacterial Contaminant Degradation ............... 320
6    Research Needs ................................................................ 321
References ...................................................................... 321

### Abstract

In order to cope with heterogeneous environments mycelial fungi have developed a unique network-based growth form. Unlike bacteria, hyphae efficiently spread in heterogeneous habitats such as soil, penetrate air-water interfaces and cross over air-filled pores. Here we discuss the prevalent role of the mycosphere (i.e., the microhabitat that surrounds fungal hyphae and mycelia) as a hotspot for the degradation of organic contaminants. We highlight the impact of hyphal networks on the transport of chemicals and bacteria and discuss its effects on contaminant availability and degradation. Given the ubiquity and length of hyphae, we propose that the mycosphere is a hotspot for contaminant transformation and attenuation in soil.

L. Y. Wick (✉) · H. Harms
Department of Environmental Microbiology, Helmholtz Centre for Environmental Research – UFZ, Leipzig, Germany
e-mail: lukas.wick@ufz.de; hauke.harms@ufz.de

© Springer International Publishing AG, part of Springer Nature 2018
T. Krell (ed.), *Cellular Ecophysiology of Microbe: Hydrocarbon and Lipid Interactions*,
Handbook of Hydrocarbon and Lipid Microbiology, https://doi.org/10.1007/978-3-319-50542-8_36

# 1    Introduction

Fungi exist on Earth since 600 million years, occur in nearly every habitat and typically develop a branched and spatially extensive network called mycelium. By forming a major fraction of the soil microbial biomass they are cornerstone of the terrestrial carbon and nutrient cycling by driving the degradation of wood or plant leaf polymeric constituents (Kendrick 2000). High fungal abundance and diversity of up to 300 taxa per 0.25 g of soil (Taylor and Sinsabaugh 2015) are for instance found in moist, aerobic terrestrial habitats containing complex organic carbon. Fungi are also central to the degradation of anthropogenic contaminants in aquatic and terrestrial ecosystems (Harms et al. 2011). Due to a high surface-to-volume-ratio and extensive fractal structures, fungal mycelia take up nutrients and energy sources efficiently and significantly influence the structure and microbial diversity of the surrounding soil via physical, biological and biochemical processes. They also share their microhabitats with bacteria. Warmink and van Elsas (2008), for instance, showed that the bacterial abundance in the mycosphere can be significantly elevated and exhibit distinct diversity as compared to neighbouring bulk soil. Mycelia also provide networks for efficient extra-hyphal dispersal of otherwise immobilised bacteria ('fungal highways' (Kohlmeier et al. 2005)) or allow hyphal transport of nutrients, contaminants and water ('fungal pipelines', (Furuno et al. 2012)) in structurally and chemically heterogeneous soil habitats. Similar to the management of goods and people ('logistics') in human societies, contaminant-degrading eco-systems need to allow sufficient dispersal of degrading microorganisms or to provide appropriate fluxes of matter and energy to and between degrading microbes ('micro-bial logistics' (Fester et al. 2014)). Unlike most bacteria, many fungi tend to oxidize hydrophobic organics co-metabolically by the use of extracellular enzymes, which remain active even at conditions not allowing for bacterial growth. Complementary to previous reviews on the fungal bioremediation (Harms et al. 2011), ecology (e.g., (Taylor and Sinsabaugh 2015; van der Heijden et al. 2008)) or biochemical versatility (Baldrian 2008; Ullrich and Hofrichter 2007; Hofrichter et al. 2010), we outline here briefly major characteristics that make fungi and their mycosphere particularly suitable for handling (e.g., hydrocarbon) contamination (Fig. 1).

# 2    Fungi Cope with Heterogeneous and Contaminated Environments

An often overlooked aspect in the biodegradation of contaminants is the spatially extensive physical structure of fungal mycelia. Fungi embody up to 75% of the subsurface microbial biomass and their hyphae create fractal mycelial networks of $10^2$ to $10^4$ m length per g of topsoil (Ritz and Young 2004). Fungi further penetrate micro-aggregates by wedge-shaped hyphae, disperse the soil and thereby allow maximal pervasion and contact with contaminants and other resources. Many hyphae expose small hydrophobic proteins, so-called hydrophobins (Wessels 1997) that allow a rupture of water-air interfaces permitting growth in the pores between soil

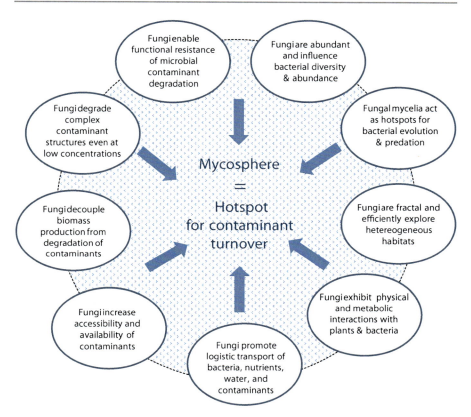

**Fig. 1** Main drivers of the microbial transformation of contaminants in the mycosphere, i.e., the microhabitat that surrounds fungal hyphae and mycelia

aggregates (Wosten et al. 1999). The combination of an adaptive mycelial morphology in response to heterogeneous environmental conditions and a bi-directional cytoplasmic streaming enables an effective mycelial maintenance (Boddy et al. 2010) and foraging strategy (Nakagaki et al. 2004). It further enables to link growth of feeder hyphae in optimal environments with explorative hyphal expansion to areas of possibly poor nutrient conditions. The challenges that these competing demands place on the network organisation, are similar to those of man-made infrastructure or logistic networks (Tero et al. 2010). Mycelial lifestyle may also explain why many fungi tolerate extreme environmental conditions (e.g., temperatures of $-5$ to $+60\ °C$; pH of 1–9) or grow at low water activity such as 0.65, or in the presence of 0.2% of oxygen (for a review cf. Cantrell 2017). Less information, however, is available on the ability of fungi to cope with contaminated environments. Although fungal communities and their biodegradation potential have often been analysed in contaminated habitats (e.g., (Harms et al. 2011)), there are only few reported comparisons between contaminated and uncontaminated reference sites. Existing data on the influence of organic contaminants on fungal diversity and biomass in terrestrial environments are still fragmentary and may not allow for generalisations (Harms et al. 2016), all the more since only a small part of all fungal diversity is known at present. Recent studies indicate that the type and concentration of organic contaminants such oil hydrocarbons may influence the fungal diversity as well as their functional redundancy or abundance (e.g., (Bourdel et al. 2016; Stefani et al. 2015; Ferrari et al. 2011)).

## 3    Fungi Transform Complex Contaminants Even at Low Concentrations

Fungi facilitate the decomposition, sequestration, and production of organic matter. Thereby they often uncouple biomass production from contaminant transformation by attacking complex compounds with a range of extracellular, relatively unspecific, radical generating oxidoreductases under aerobic conditions. Anaerobic fungal pollutant breakdown, however, has also been reported (Russell et al. 2011). With their enzymes, many fungi are able to transform a large range of organic pollutants, including complex and unusual chemical structures such, for example, those found in pharmaceuticals or personal care products. Most fungal contaminant degraders belong to the phyla *Ascomycota* and *Basidiomycota* and the subphylum *Mucoromycotina*. Bacteria on the other hand, transform chemicals by specific biochemical pathways and are able to use varying terminal electron acceptors. Unlike the unspecific fungal enzymes, specific degradation pathways of bacteria may not be expressed for chemicals of either complex chemical structures or chemicals present at low concentrations. Bacterial degradative pathways may solely evolve if they lead to a selective benefit for the encoding organisms. In contrast to fungi, bacteria rely on a positive feedback loop between contaminant uptake and formation of their biomass for efficient decontamination. Torneman et al. (2008) for instance, showed that the phospholipid fatty acid (PLFA)-inferred abundances of bacteria in soil positively correlated with polycyclic aromatic hydrocarbons (PAH) concentration gradients, whereas fungal and actinomycete PLFA markers were lowered in PAH hotspots (Torneman et al. 2008). Given that the contaminant is readily available, it hence appears that the nutritional strategies differing between many bacteria and fungi make bacteria benefit more from a readily bioavailable input of the contaminant's carbon. At oligotrophic conditions as typically found in soils, plant root exudates, however, can be a major driver of co-metabolic fungal degradation. Mycorrhiza, a symbiotic association between a fungus and the roots of vascular host plants, rely on photosynthates (that form up to 30% of a host plant's net carbon fixation) in exchange of transfer of mineral nutrients to its symbiont. Although commonly termed as 'phytoremediation' the degradation of contaminants in presence of plants hence is to be regarded as a result of the rhizosphere ecosystem and rather depends on the functional stability and the metabolic and physical interactions of a large range of organisms including bacteria, fungi, and plants (El Amrani et al. 2015; Fester et al. 2014).

## 4    Fungi Promote Contaminant Accessibility and Bioavailability

The accessibility and availability of a contaminant to degrader microorganisms ('bioavailability') are central factors in the biodegradation (Johnsen et al. 2005) of any chemical. In contrast to mycelial fungi that employ unspecific exo-enzymes and uncouple biomass formation from contaminant degradation, low matter and energy fluxes to bacterial degrader cells exist, when (i) contaminant concentrations are very low (e.g., micropollutants), (ii) contaminants are bound to matrices (e.g., poorly water soluble hydrocarbons), (iii) contaminants contain very little energy (e.g., highly oxidized chemicals) or (iv) contaminants cannot be transformed by the specific bacterial degradative pathways. Fungal exo-enzymatic transformation may also render hydrophobic compounds more water soluble and allow the transfer of more hydrophilic and more bioavailable metabolites to catabolically interacting bacteria in the hyphosphere (Johnsen et al. 2005); i.e., may allow for microbial kleptoparasitism (Johnston et al. 2016). Fungal hyphae have also been found to grow into soil pores with a diameter as little as 2 μm or to mobilise a wide range of PAH by vesicle-bound cytoplasmic transport (Furuno et al. 2012). By doing so, they increase the accessibility and availability of soil-bound chemicals to degrader bacteria (Schamfuss et al. 2013). Moreover, mycelia act as 'fungal highways' and allow for effective dispersal of a wide range of bacteria (Kohlmeier et al. 2005; Pion et al. 2013; Warmink and van Elsas 2009). The fungal foraging of a contaminated soil for resources combined with bacterial migration along hyphae (Furuno et al. 2010; Nazir et al. 2010) may promote the (random or tactic) access of migrator bacteria to new soil habitats (Furuno et al. 2010; Nazir et al. 2010). In order to maintain their extending networks, mycelia exhibit a highly polarised internal cellular organization to allow cytoplasmic transport of nutrients and carbon over distances of centimetres (Darrah et al. 2006) between spatially separated source and sink regions (Bebber et al. 2007; Heaton et al. 2012). The velocity of such cytoplasmic streaming typically ranges from 0.1 to 600 μm min$^{-1}$ (Ross 1976; Suelmann et al. 1997) depending on the type of organism, its physiological state and the location within the mycelium. Not much is known about the translocation in non-mycorrhizal fungi of substances other than nutrients. It appears likely that the mycelial lifestyle of fungi has also consequences for their exposure to chemicals contaminating their habitats. Uptake of contaminants by fungi has been reported for various chemicals including organic pollutants. Verdin et al. (2005), for instance, showed that the uptake of PAH into a fungal mycelium is passive and that PAH are accumulated in lipid vesicles. Time lapse video micrographs of the oomycete *Pythium ultimum* demonstrated a translocation of phenanthrene (PHE) by cytoplasmic streaming at an average velocity of $13 \pm 9$ μm min$^{-1}$ resulting in estimated PHE transport rates of $\approx$ 0.02–1.1 pmol ($\approx$ 4–200 pg) of PHE per hypha per hour over a distance of 1 cm (Furuno et al. 2012). The significance for the ecology of contaminant biodegradation might be illustrated by the fact that these masses correspond to 12 - 600 times the dry weight of a typical bacterium ($\approx$ 0.3 pg). Quantification of solute movement and predictive simulation models further showed that the intravacuolar transport forms a functionally important transport pathway over distances of millimetres up to centimetres and may influence the bioavailability and biodegradation of contaminants (Banitz et al. 2013; Schamfuss et al. 2013). Simulation modelling demonstrates that fungus-mediated bacterial dispersal can considerably improve the bioavailability of organic pollutants under in heterogeneous habitats typical for unsaturated soils (Banitz et al. 2011a, b).

## 5  Fungi Create Functional Habitats for Bacterial Contaminant Degradation

Bacteria in soil often face conditions that are suboptimal for contaminant degradation. Low contaminant bioavailability, the lack of nutrients or predation pressure may put stress on bacteria that they must meet in order to survive and remain active. Bacteria and fungi co-inhabit and interact in a wide variety of soil environments and microhabitats (Nazir et al. 2010) and many bacterial-fungal interactions (BFI) can take place in the mycosphere. Both, bacteria and fungi may compete therein for the same substrates and depending on the organisms and the specific conditions, BFI may be either ecologically neutral, competitive or cooperative (Ul Haq et al. 2014) and range in quality from apparently random physical interactions to specific commensal or symbiotic associations (Frey-Klett et al. 2011). Bacteria may thereby influence fungal development and spore production by mycophagy; i.e., bacterial lysis of fungal hyphae or the extracellular biotrophy of fungal products, such as low molecular weight peptides, organic acids (e.g., oxalic, citric, acetic and amino acids), sugars or sugar alcohols that are secreted as waste products, metal-mobilising or antimicrobial compounds (Leveau and Preston 2008). Extracellular biotrophy (i.e., feeding on a living host organism without killing it) may also be an effective bacterial strategy to obtain nutrients and carbon while migrating along hyphae in the mycosphere (Banitz et al. 2013). Mycophagous activity, however, may also be beneficial for fungi; e.g., in situations, where fungiphilic bacteria provide specific nutrients or degrade antifungal toxins in exchange of fungal products. From such examples it becomes clear that BFI may also have profound consequences on the physiology, life cycles, and survival of contaminant degrading mycosphere microorganisms. Fungi respond to environmental resource heterogeneity by translocating nutrient resources between different parts of their mycelium. Mycelia thereby may not only promote efficient contaminant degradation by (i) translocation of bacteria and nutrients, but simultaneously also by (ii) redistributing water from wet to dry habitats and shaping suitable local matric potentials for improved fungal carbon mineralisation (Guhr et al. 2015; Worrich et al. 2017), or by (iii) sustaining the degradation capacity of microbial ecosystems at low osmotic and matric potentials (Worrich et al. 2016). The physical structure of mycelia further enables efficient foraging and shaping of prey populations and nutrient and carbon turnover in heterogeneous environments (Otto et al. 2016). Recent experimental and theoretical studies have shown that hyphae promote bacterial horizontal gene transfer (HGT) by enabling preferential contact of spatially distinct bacteria; i.e., hyphae may act as focal point for the evolution of bacterial diversity and their potential to degrade contaminants (Berthold et al. 2016; Zhang et al. 2014).

## 6   Research Needs

Numerous studies have shown that bacterial communities associated with fungal hyphae display fungi-specific differences in composition. This suggests that bacteria are under selection and develop traits that give them a competitive advantage in the mycosphere (Leveau and Preston 2008). Hence, detailed knowledge of functional co-occurrence of bacteria and fungi in contaminated habitats is needed. Despite the rising interest in BFI over the last ten years, there is only limited knowledge on the bacterial fungal 'meta-organism' (Olsson et al. 2016) and its metabolic interaction in the degradation of complex (emerging pollutants) or pollutants at low concentrations (micropollutants). Even less is known on their interactions in complex constructed and natural ecosystems at larger scales. Given the fact that at present more than 135 million unique chemical substances http://support.cas.org/content/chemical-sub stances and complex chemical mixtures are available that potentially end up in natural habitats, the prediction of their fate as a prerequisite for the management of environmental pollutants remains demanding. This task will be facilitated by 'omics' (metagenomics, metatranscriptomics, metaproteomics and metabolomics), stable isotope and chemical microscopy (e.g., ToF- or NanoSIMS) approaches that will portrait the genomic potential and functional interactions of fungi and co-occurring bacteria at given environmental habitats and conditions. Such knowledge will be needed to predict and model the spatio-temporal functional stability of bacterial-fungal associations under multiple stresses at all scales and to develop novel attenuation approaches based on ecological theory and principles. The utilisation of plants for the alimentation of mycorrhizal or other fungi in the mycorrhizosphere of terrestrial and water plants thereby seems to be a promising tool for enhanced natural attenuation of hydrophobic and emerging contaminants even at low concentrations.

**Acknowledgements**  This work contributes to the research topic Chemicals in the Environment (CITE) within the research program Terrestrial Environment of the Helmholtz Association.

## References

Baldrian P (2008) Wood-inhabiting ligninolytic basidiomycetes in soils: ecology and constraints for applicability in bioremediation. Fun Ecol 1:4–12

Banitz T, Fetzer I, Johst K, Wick LY, Harms H, Frank K (2011a) Assessing biodegradation benefits from dispersal networks. Ecol Model 222:2552–2560

Banitz T, Wick LY, Fetzer I, Frank K, Harms H, Johst K (2011b) Dispersal networks for enhancing bacterial degradation in heterogeneous environments. Environ Pollut 159:2781–2788

Banitz T, Johst K, Wick LY, Schamfuss S, Harms H, Frank K (2013) Highways versus pipelines: contributions of two fungal transport mechanisms to efficient bioremediation. Environ Microbiol Rep 5:211–218

Bebber DP, Hynes J, Darrah PR, Boddy L, Fricker MD (2007) Biological solutions to transport network design. Proc R Soc B Biol Sci 274:2307–2315

Berthold T, Centler F, Hubschmann T, Remer R, Thullner M, Harms H et al (2016) Mycelia as a focal point for horizontal gene transfer among soil bacteria. Sci Rep 6:8

Boddy L, Wood J, Redman E, Hynes J, Fricker MD (2010) Fungal network responses to grazing. Fungal Genet Biol 47:522–530

Bourdel G, Roy-Bolduc A, St-Arnaud M, Hijri M (2016) Concentration of petroleum-hydrocarbon contamination shapes fungal endophytic community structure in plant roots. Front Microbiol 7:11

Cantrell SA (2017) Fungi in extreme and stressful environments. In: Dighton J, White JF (eds) The fungal community : its organization and role in the ecosystem, 4th revised edn. CRC Press, Boca Raton, pp 459–469

Darrah PR, Tlalka M, Ashford A, Watkinson SC, Fricker MD (2006) The vacuole system is a significant intracellular pathway for longitudinal solute transport in basidiomycete fungi. Eukaryot Cell 5:1111–1125

El Amrani A, Dumas AS, Wick LY, Yergeau E, Berthome R (2015) "Omics" insights into PAH degradation toward improved green remediation biotechnologies. Environ Sci Technol 49:11281–11291

Ferrari BC, Zhang CD, Dorst J (2011) Recovering greater fungal diversity from pristine and diesel fuel contaminated sub-Antarctic soil through cultivation using both a high and a low nutrient media approach. Front Microbiol 2:14

Fester T, Giebler J, Wick LY, Schlosser D, Kastner M (2014) Plant-microbe interactions as drivers of ecosystem functions relevant for the biodegradation of organic contaminants. Curr Opin Biotechnol 27:168–175

Frey-Klett P, Burlinson P, Deveau A, Barret M, Tarkka M, Sarniguet A (2011) Bacterial-fungal interactions: hyphens between agricultural, clinical, environmental, and food microbiologists. Microbiol Mol Biol Rev 75:583–609

Furuno S, Pazolt K, Rabe C, Neu TR, Harms H, Wick LY (2010) Fungal mycelia allow chemotactic dispersal of polycyclic aromatic hydrocarbon-degrading bacteria in water-unsaturated systems. Environ Microbiol 12:1391–1398

Furuno S, Foss S, Wild E, Jones KC, Semple KT, Harms H et al (2012) Mycelia promote active transport and spatial dispersion of polycyclic aromatic hydrocarbons. Environ Sci Technol 46:5463–5470

Guhr A, Borken W, Spohn M, Matzner E (2015) Redistribution of soil water by a saprotrophic fungus enhances carbon mineralization. Proc Natl Acad Sci U S A 112:14647–14651

Harms H, Schlosser D, Wick LY (2011) Untapped potential: exploiting fungi in bioremediation of hazardous chemicals. Nat Rev Microbiol 9:177–192

Harms H, Wick LY, Schlosser D (2016) The fungal community in organically polluted systems. In: John Dighton JFW (ed) The fungal community: its organization and role in the ecosystem, 4th edn. RC Press, Boca Raton

Heaton L, Obara B, Grau V, Jones N, Nakagaki T, Boddy L et al (2012) Analysis of fungal networks. Fungal Biol Rev 26:12–29

van der Heijden MGA, Bardgett RD, van Straalen NM (2008) The unseen majority: soil microbes as drivers of plant diversity and productivity in terrestrial ecosystems. Ecol Lett 11:296–310

Hofrichter M, Ullrich R, Pecyna MJ, Liers C, Lundell T (2010) New and classic families of secreted fungal heme peroxidases. Appl Microbiol Biotechnol 87:871–897

Johnsen AR, Wick LY, Harms H (2005) Principles of microbial PAH-degradation in soil. Environ Pollut 133:71–84

Johnston SR, Boddy L, Weightman AJ (2016) Bacteria in decomposing wood and their interactions with wood-decay fungi. FEMS Microbiol Ecol 92:12

Kendrick B. 2000. *The Fifth Kingdom*. 3rd edn. Newburyport, MA: Focus Publishing

Kohlmeier S, Smits THM, Ford RM, Keel C, Harms H, Wick LY (2005) Taking the fungal highway: mobilization of pollutant-degrading bacteria by fungi. Environ Sci Technol 39:4640–4646

Leveau JHJ, Preston GM (2008) Bacterial mycophagy: definition and diagnosis of a unique bacterial-fungal interaction. New Phytol 177:859–876

Nakagaki T, Kobayashi R, Nishiura Y, Ueda T (2004) Obtaining multiple separate food sources: behavioural intelligence in the *Physarum plasmodium*. Proc R Soc B Biol Sci 271:2305–2310

Nazir R, Warmink JA, Boersma H, van Elsas JD (2010) Mechanisms that promote bacterial fitness in fungal-affected soil microhabitats. FEMS Microbiol Ecol 71:169–185

Olsson S, Bonfante P, Pawlowska TE (2016) Ecologey and evolution of fungal-bacterial interactions. In: Dighton J, White JF, Oudemans P (eds) The fungal community: its organization and tole in the ecosystem. 4th revised ed. CRC Press, Boca Raton, FL. 2017

Otto S, Bruni EP, Harms H, Wick LY (2016) Catch me if you can: dispersal and foraging of *Bdellovibrio* bacteriovorus 109J along mycelia. ISME J 11:386

Pion M, Bshary R, Bindschedler S, Filippidou S, Wick LY, Job D et al (2013) Gains of bacterial flagellar motility in a fungal world. Appl Environ Microbiol 79:6862–6867

Ritz K, Young IM (2004) Interactions between soil structure and fungi. Mycologist 18:52–59

Ross IK (1976) Nuclear migration rates in coprinus-congregatus – new record. Mycologia 68:418–422

Russell JR, Huang J, Anand P, Kucera K, Sandoval AG, Dantzler KW et al (2011) Biodegradation of polyester polyurethane by endophytic fungi. Appl Environ Microbiol 77:6076–6084

Schamfuss S, Neu TR, van der Meer JR, Tecon R, Harms H, Wick LY (2013) Impact of mycelia on the accessibility of fluorene to PAH-degrading bacteria. Environ Sci Technol 47:6908–6915

Stefani FOP, Bell TH, Marchand C, de la Providencia IE, El Yassimi A, St-Arnaud M et al (2015) Culture-dependent and -independent methods capture different microbial community fractions in hydrocarbon-contaminated soils. PLoS One 10:16

Suelmann R, Sievers N, Fischer R (1997) Nuclear traffic in fungal hyphae: in vivo study of nuclear migration and positioning in Aspergillus nidulans. Mol Microbiol 25:757–769

Taylor DL, Sinsabaugh RL (2015) The soil fungi: occurrence, phylogeny and ecology. In: Paul EA (ed) Soil microbiology, ecology and biochemistry. Elsevier, Amsterdam, pp 77–109

Tero A, Takagi S, Saigusa T, Ito K, Bebber DP, Fricker MD et al (2010) Rules for biologically inspired adaptive network design. Science 327:439–442

Torneman N, Yang XH, Baath E, Bengtsson G (2008) Spatial covariation of microbial community composition and polycyclic aromatic hydrocarbon concentration in a creosote-polluted soil. Environ Toxicol Chem 27:1039–1046

Ul Haq I, Zhang MZ, Yang P, van Elsas JD (2014) The interactions of bacteria with fungi in soil: emerging concepts. In: Sariaslani S, Gadd GM (eds) Advances in applied microbiology, vol 89. Elsevier/Academic, San Diego, pp 185–215

Ullrich R, Hofrichter M (2007) Enzymatic hydroxylation of aromatic compounds. Cell Mol Life Sci 64:271–293

Verdin A, Sahraoui ALH, Newsam R, Robinson G, Durand R (2005) Polycyclic aromatic hydrocarbons storage by *Fusarium solani* in intracellular lipid vesicles. Environ Pollut 133:283–291

Warmink JA, van Elsas JD (2008) Selection of bacterial populations in the mycosphere of *Laccaria proxima*: is type III secretion involved? ISME J 2:887–900

Warmink JA, van Elsas JD (2009) Migratory response of soil bacteria to *Lyophyllum* sp strain Karsten in soil microcosms. Appl Environ Microbiol 75:2820–2830

Wessels JGH (1997) Hydrophobins: proteins that change the nature of the fungal surface. In: Poole RK (ed) Advances in microbial physiology, vol 38. Academic/Elsevier, London, pp 1–45

Worrich A, Konig S, Miltner A, Banitz T, Centler F, Frank K et al (2016) Mycelium-like networks increase bacterial dispersal, growth, and biodegradation in a model ecosystem at various water potentials. Appl Environ Microbiol 82:2902–2908

Worrich A, Stryhanyuk H, Musat N, König S, Banitz T, Centler F, Frank K, Thullner M, Harms H, Richnow HH, Miltner A, Kästner M, Wick LY (2017) Mycelium-mediated transfer of water and nutrients stimulates bacterial activity in dry and oligotrophic environments. Nat. Commun. 8:15472

Wosten HAB, van Wetter MA, Lugones LG, van der Mei HC, Busscher HJ, Wessels JGH (1999) How a fungus escapes the water to grow into the air. Curr Biol 9:85–88

Zhang MZ, Silva M, Maryam CD, van Elsas JD (2014) The mycosphere constitutes an arena for horizontal gene transfer with strong evolutionary implications for bacterial-fungal interactions. FEMS Microbiol Ecol 89:516–526

# Part III

# Problems of Solventogenicity, Solvent Tolerance

# Problems of Solventogenicity, Solvent Tolerance: An Introduction

**20**

Miguel A. Matilla

## Contents

1   Introduction ............................................................................. 327
2   Mechanisms of Solvent Tolerance in Bacteria: An Overview ........................... 329
3   Catabolism of Toxic Organic Solvents ................................................ 331
4   Research Needs ......................................................................... 332
References ................................................................................. 333

**Abstract**

Many organic solvents are toxic to prokaryotic and eukaryotic organisms. This general toxicity mainly derives from their ability to preferentially partition into cell membranes, a process that finally can impair their normal functioning. However, multiple microorganisms have evolved different strategies to overcome the effects of toxicity. Thus, the mechanisms of tolerance in Gram-negative bacteria are the result of a multifactorial process that involves a set of changes at both physiological and gene expression levels. These changes include the alteration in the composition of cell membranes to reduce their permeability, the activation of general stress responses, or the induction of efflux pumps and catabolic pathways. The evaluation of these solvent tolerance strategies may lay the basis for the development of effective *in situ* and *ex situ* bacteria-based biodegradation strategies.

M. A. Matilla (✉)
Department of Environmental Protection, Estación Experimental del Zaidín, Consejo Superior de Investigaciones Científicas, Granada, Spain
e-mail: miguel.matilla@eez.csic.es

© Springer International Publishing AG, part of Springer Nature 2018
T. Krell (ed.), *Cellular Ecophysiology of Microbe: Hydrocarbon and Lipid Interactions*,
Handbook of Hydrocarbon and Lipid Microbiology, https://doi.org/10.1007/978-3-319-50542-8_14

# 1    Introduction

Organic solvents are a compound class characterized by an extraordinary diversity of chemical structures, including alcohols, esters, ketones, or hydrocarbons (Nicolaou et al. 2010). Many of these solvents are extremely toxic to microorganisms, plants, animals, and humans. Among these chemicals, organic and inorganic hydrocarbons are within the most common solvents, and the current "Cellular Ecophysiology: Problems of Solventogenicity, Solvent Tolerance" section will mainly cover different aspects of the resistance mechanisms of microorganisms to this structurally diverse group of chemical products.

Hydrocarbons can be classified as saturated, unsaturated, and aromatic compounds (reviewed in this Encyclopedia by Wilkes and Schwarzbauer). In nature, hydrocarbons mainly occur as a result of multiple biosynthetic activities and geochemical processes (Abbasian et al. 2015). Thus, they are the main components of petroleum crude oils and, consequently, very abundant in geological systems. Importantly, our petroleum-based society demands increasing amounts of crude oil for energy supply and the production of raw materials. Furthermore, the increase in the social necessities for certain agrochemicals, pharmaceuticals, and polymers has promoted the production of multiple solvents, and hydrocarbons such as benzene, toluene, or xylenes are some of the most globally produced chemicals (Segura et al. 2012; Abbasian et al. 2015). Unfortunately, and as a consequence of these anthropogenic activities, solvent contaminants are reaching the biosphere due to leaks during crude oil extraction, storage, transportation, industrial accidents, or evaporation. As a result, these pollutants are frequently found in soils and river water, groundwater, and seawater, mainly in the proximity of production factories and petroleum refineries. Therefore, the biological treatment of contaminated areas and residues is a promising strategy for the removal of solvent contaminants. Within the candidate biological treatments, bioremediation uses microorganisms for the *in situ* or *ex situ* conversion of toxic chemicals into neutral or less toxic compounds (de Lorenzo 2008; Zhao and Poh 2008; Nicolaou et al. 2010; Ramos et al. 2011).

The effectiveness of the natural biodegradation of pollutants (a process known as natural attenuation) is a reflection of the persistence of these contaminants in the environment. The persistence of organic solvents in nature highly depends on their physicochemical properties, including their hydrophobicity and the presence of functional groups (e.g., epoxides and aldehydes, among others), properties that may confer toxicity to the chemical. However, in general, hydrocarbons are considered highly recalcitrant to degradation due to their high hydrophobicity. Thus, problems derived from the toxicity of hydrophobic solvents mainly result from their ability to partition into biological membranes, which subsequently causes an unspecific increase in their fluidity and permeability (reviewed in this volume by Heipieper and Martínez, ▶ Chap. 21, "Toxicity of Hydrocarbons to Microorganisms"). In addition to this unspecific toxicity, different transcriptomic, proteomic, and metabolomic approaches allowed determining the involvement of some toxic hydrocarbons in the damage of biological molecules such as proteins, DNA, and RNA (Nicolaou et al. 2010; Ramos et al. 2015). As a consequence, the antimicrobial activity of these solvents is the main cause for their low biodegradation rates. However, multiple microorganisms have acquired different strategies that allow them to protect themselves against toxic hydrocarbons (Fig. 1). This is reflected in the different microbial diversity and community composition found when comparing polluted and non-polluted field samples. Thus, the concentration and diversity of hydrocarbons in contaminated areas shape the composition of bacterial communities by favoring the presence of highly resistant and biodegradative microbial strains (Bargiela et al. 2015; Acosta-González and Marqués 2016). Importantly, "omics" approaches revealed an increased expression of genes involved in the degradation of toxic hydrocarbons *in situ* in contaminated areas (Lamendella et al. 2014). However, and contrary to what was initially anticipated, no correlation was observed between the tolerance to toxic solvents and the catabolic abilities of microorganisms to degrade these solvents (Mosqueda et al. 1999; Ramos et al. 2015). What is therefore the main benefit derived from the microbial catabolism of toxic hydrocarbons? Multiple microorganisms can use toxic solvents as sources of carbon and energy, a process that may allow them to proliferate in environments where energy and carbon sources are scarce. This is somehow reflected in the chemotactic behavior that some solvent-tolerant bacterial strains show toward toxic hydrocarbons (Grimm and Harwood 1999; Gordillo et al. 2007; Lacal et al. 2011). Chemotaxis represents an important energy cost for microbial cells, and such energetic investment can be explained for the increased capacity of chemotactic microorganisms in order to gain access to additional nutritional sources (see ▶ Chap. 22, "Genetics of Sensing, Accessing, and Exploiting Hydrocarbons" reviewed in this volume by Matilla et al.).

## 2   Mechanisms of Solvent Tolerance in Bacteria: An Overview

The accumulation of organic solvents in cell membranes leads to an increase in their fluidity and to multiple structural changes. These changes may result in the loss of ions and solutes, as well as modifications in the electric potential (which will prevent the correct functioning of the membrane) and intracellular pH, alterations that can finally cause growth inhibition and ultimately cell death (Sikkema et al. 1995). To compensate these potentially lethal alterations, bacteria have developed different mechanisms that allow them to adapt to environmental challenges (Fig. 1). Within these membrane-protective mechanisms, bacteria can respond by decreasing the fluidity and permeability of their membranes by altering: (i) the saturated/unsaturated fatty acid ratio, (ii) the length of acyl chains, and (iii) the head groups of phospholipids (see ▶ Chap. 24, "Membrane Composition and Modifications in Response to Aromatic Hydrocarbons in Gram-Negative Bacteria" reviewed in this volume by Ortega et al.). Additionally, in the early 1990s, an additional adaptive mechanism was identified in two strains of *Vibrio* and *Pseudomonas* genera (Okuyama et al. 1991; Heipieper et al. 1992). This mechanism was based on the ability of the enzyme *cis-trans* isomerase to convert *cis*- to *trans*-unsaturated fatty acids which results in an increased rigidity of cell membranes (reviewed in this volume by Heipieper et al. ▶ Chap. 25, "Cis–Trans Isomerase of Unsaturated Fatty

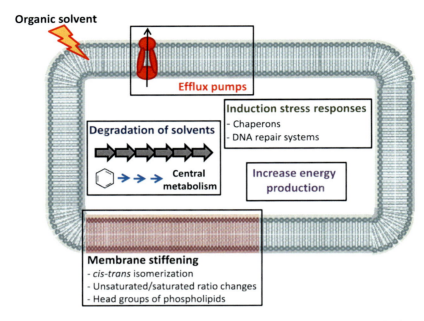

**Fig. 1** Overview of the main solvent tolerance mechanisms described in Gram-negative bacteria

Acids: An Immediate Bacterial Adaptive Mechanism to Cope with Emerging Membrane Perturbation Caused by Toxic Hydrocarbons").

The reduction in the permeability of cell membranes does not entirely prevent the cellular entrance of solvents, and different alternative tolerance mechanisms have been identified. Thus, the induction of DNA repair systems and the expression of diverse chaperons that contribute to the stabilization and correct folding of proteins are well-known bacterial responses associated to the presence of organic solvents (Segura et al. 2005; Dominguez-Cuevas et al. 2006; Wijte et al. 2011). However, the most effective and widespread mechanism for solvent tolerance in Gram-negative bacteria lies in the extrusion of solvents by efflux systems. Thus, the implication of ABC (ATP-binding cassette) transporters and RND (resistance-nodulation-division) efflux pumps in solvent tolerance has been clearly demonstrated (Nikaido and Takatsuka 2009; Segura et al. 2012; Ramos et al. 2015). The physiological role of RND pumps has been associated with the extrusion of intracellularly produced compounds and the protection against toxic compounds present in the environment (e.g., antibiotics, chemotherapeutic agents, heavy metals, and solvents). Therefore, these efflux systems have been shown to improve the fitness and survival of bacteria in the solvent-polluted environments. In this volume (▶ Chap. 23, "Extrusion Pumps for Hydrocarbons: An Efficient Evolutionary Strategy to Confer Resistance to Hydrocarbons" by Fernandez et al.), Fernández and coworkers reviewed the current knowledge in the regulation of the expression of genes encoding RND efflux pumps, besides updating the molecular insights into how toxic hydrocarbons are extruded by RND resistance transporters.

It is worth mentioning that some of the above resistance mechanisms (especially the extrusion of solvents by RND efflux pumps) are energetically expensive, and several transcriptomic and proteomic studies revealed an increased expression of proteins involved in the energy generation in the presence of the toxic hydrocarbon. This response will overcome the high energy demands as a result of the exposure to solvents (Segura et al. 2005; Domínguez-Cuevas et al. 2006; Volkers et al. 2006; Wijte et al. 2011).

## 3    Catabolism of Toxic Organic Solvents

Although many environmental microorganisms have been shown to efficiently catabolize organic solvents, the development of new sequencing technologies and genomics tools has revealed the extraordinary genetic potential to degrade toxic hydrocarbons of multiple uncharacterized microorganisms. The mineralization of such compounds finally results in the decrease of the environmental concentration of the solvent and favors the survival of the microorganism in polluted environments. However, this catabolic-mediated mechanism of resistance has only been shown to be effective when the concentration of the toxic compound is low enough to allow the normal metabolism of the microorganism (Segura et al. 2014).

A major challenge that solvent-degrading microorganisms have to face is the reduced solubility of many hydrocarbons. Thus, in addition to their extraordinary catabolic diversity, microorganisms have also evolved effective strategies to gain access to poorly bioavailable solvents. Some of these adaptive mechanisms include the secretion of biosurfactants and the alteration of the physicochemical properties of their cell surfaces. These strategies will facilitate the interaction between the microorganisms and the hydrophobic substrates (reviewed in this volume by Heipieper et al., ► Chap. 26, "Surface Properties and Cellular Energetics of Bacteria in Response to the Presence of Hydrocarbons").

The catabolic genes responsible for the degradation of multiple toxic solvents such as toluene, xylene, naphthalene, phenol, or nitrobenzene have been described in detail (Yen and Serdar 1988; Powlowski and Shingler 1994; Greated et al. 2002; Ma et al. 2007; Parales et al. 2008). Generally, the first step(s) for the intracellular degradation of these organic solvents is mediated by oxygenases and peroxidases, while different peripheral degradation pathways will enable their conversion into intermediates of the central metabolism (Abbasian et al. 2015; Varjani 2017). The enzymes involved in the degradation of such solvents are frequently organized in transcriptional units, and in general, the expression of these operons is only activated by the presence of an inducer (e.g., the substrate of the respective catabolic pathway). Furthermore, many of these catabolic genes are either associated to mobile elements or encoded in conjugative plasmids, therefore allowing bacteria to rapidly gain new catabolic capabilities through horizontal gene transfer (Segura et al. 2014). Two of the most characterized catabolic systems at both the catabolic and regulatory levels are the toluene (TOL) and toluene dioxygenase (TOD) degradation pathways, involved in the degradation of toluene and xylenes to Krebs cycle intermediates (Mosqueda et al. 1999; Parales et al. 2008; Segura et al. 2014). The expression of the catabolic genes is often tightly regulated as an effective strategy to adjust their metabolism to the elevated energetic cost of solvent degradation. As an example, the expression of TOL and TOD catabolic pathways is activated in the presence of their respective substrates in a regulatory process mediated by one- (i.e., XylS and XylR) and two-component (i.e., TodS/TodT) systems, respectively (see ▶ Chap. 22, "Genetics of Sensing, Accessing, and Exploiting Hydrocarbons" reviewed in this volume by Matilla et al.).

## 4    Research Needs

Since the beginning of the Industrial Revolution, increasing amounts of contaminants have been released into the environment, resulting in the contamination of terrestrial and aquatic ecosystems. Different chemical and physical approaches (i.e., coal agglomeration, chemical oxidation, thermal and ultrasonic desorption, among others) have been employed to remove these pollutants (Agarwal and Liu 2015). However, *in situ* and *ex situ* bioremediation offers an environmentally friendly, versatile, and cost-effective alternative (Varjani 2017).

Although plants and fungi possess good remediation capabilities, bacteria have been shown to be the most active biodegraders (de Lorenzo 2008). Bioremediation exploits naturally derived microorganisms as effective agents for the biological degradation/detoxification of environmental contaminants. Thus, bacteria belonging to the genera *Acinetobacter, Arthrobacter, Azoarcus, Brevibacterium, Flavobacterium, Marinobacter, Micrococcus, Pseudomonas, Stenotrophomonas,* and *Vibrio* were reported to be efficient hydrocarbon degraders. Importantly, the proportion of hydrocarbon-degrading microorganisms in polluted ecosystems was found to be up to 100 times higher as compared with non-polluted environments (Varjani 2017). Thus, the isolation of microorganisms from contaminated areas will, in principle, result in a higher rate of success in the isolation of microorganisms with the desired degradation capabilities. Nevertheless, many of the efforts made in the isolation of microbial biodegraders have been focused on the investigation of genes involved in the catabolism of the contaminants, and more attention needs to be paid to the biodegradation effectiveness of these newly isolated microorganisms (Cases and de Lorenzo 2005; Abbasian et al. 2016).

The rates of biodegradation of solvents are affected by multiple parameters, including the chemical properties of the pollutants and environmental conditions (i.e., temperature, pH, salinity, oxygen). Therefore, it is not abnormal that multiple studies have reported the increased efficiency in the biodegradation capabilities of bacterial consortia as compared with the individual strains (Ghazali et al. 2004; Deppe et al. 2005; Varjani and Upasani 2013; Varjani 2017). With the development of the high-throughput isolation and screening methods, together with new-generation "omics" tools, future research will allow the identification of novel microorganisms with broader range of biodegradation abilities. This investigation will lay the groundwork for the development of efficient microbial consortia with improved biodegradation capabilities. Additionally, synthetic biology strategies can be used to engineer genetically modified microorganisms with increase biodegradation abilities (de Lorenzo 2008). For this purpose, researchers have access to multiple databases with information on bacterial genomes (https://www.ncbi.nlm.nih.gov/genome/) and microbial biocatalytic reactions/biodegradation pathways (University of Minnesota Biocatalysis/Biodegradation Database; http://umbbd.msi.umn.edu).

**Acknowledgments** Miguel A. Matilla was supported by the Spanish Ministry of Economy and Competitiveness Postdoctoral Research Program, Juan de la Cierva (JCI-2012-11815).

# References

Abbasian F, Lockington R, Mallavarapu M, Naidu R (2015) A comprehensive review of aliphatic hydrocarbon biodegradation by bacteria. Appl Biochem Biotechnol 176:670–699

Abbasian F, Palanisami T, Megharaj M, Naidu R, Lockington R, Ramadass K (2016) Microbial diversity and hydrocarbon degrading gene capacity of a crude oil field soil as determined by metagenomics analysis. Biotechnol Prog 32:638–648

Acosta-González A, Marqués S (2016) Bacterial diversity in oil-polluted marine coastal sediments. Curr Opin Biotechnol 38:24–32

Agarwal A, Liu Y (2015) Remediation technologies for oil-contaminated sediments. Mar Pollut Bull 101:483–490

Bargiela R, Mapelli F, Rojo D, Chouaia B, Tornes J, Borin S et al (2015) Bacterial population and biodegradation potential in chronically crude oil-contaminated marine sediments are strongly linked to temperature. Sci Rep 5:11651

Cases I, de Lorenzo V (2005) Genetically modified organisms for the environment: stories of success and failure and what we have learned from them. Int Microbiol 8:213–222

Deppe U, Richnow HH, Michaelis W, Antranikian G (2005) Degradation of crude oil by an arctic microbial consortium. Extremophiles 9:461–470

Domínguez-Cuevas P, González-Pastor JE, Marqués S, Ramos JL, de Lorenzo V (2006) Transcriptional tradeoff between metabolic and stress-response programs in *Pseudomonas putida* KT2440 cells exposed to toluene. J Biol Chem 281:11981–11991

Ghazali FM, Rahman RNZA, Salleh AB, Basri M (2004) Biodegradation of hydrocarbons in soil by microbial consortium. Int Biodeterior Biodegrad 54:61–67

Gordillo F, Chavez FP, Jerez CA (2007) Motility and chemotaxis of *Pseudomonas* sp. B4 towards polychlorobiphenyls and chlorobenzoates. FEMS Microbiol Ecol 60:322–328

Greated A, Lambertsen L, Williams PA, Thomas CM (2002) Complete sequence of the IncP-9 TOL plasmid pWW0 from *Pseudomonas putida*. Environ Microbiol 4:856–871

Grimm AC, Harwood CS (1999) NahY, a catabolic plasmid-encoded receptor required for chemotaxis of *Pseudomonas putida* to the aromatic hydrocarbon naphthalene. J Bacteriol 181:3310–3316

Heipieper HJ, Diefenbach R, Keweloh H (1992) Conversion of cis unsaturated fatty acids to trans, a possible mechanism for the protection of phenol-degrading *Pseudomonas putida* P8 from substrate toxicity. Appl Environ Microbiol 58:1847–1852

Lacal J, Muñoz-Martínez F, Reyes-Darias JA, Duque E, Matilla MA, Segura A et al (2011) Bacterial chemotaxis towards aromatic hydrocarbons in *Pseudomonas*. Environ Microbiol 13:1733–1744

Lamendella R, Strutt S, Borglin S, Chakraborty R, Tas N, Mason OU et al (2014) Assessment of the deepwater horizon oil spill impact on gulf coast microbial communities. Front Microbiol 5:130

de Lorenzo V (2008) Systems biology approaches to bioremediation. Curr Opin Biotechnol 19:579–589

Ma YF, Wu JF, Wang SY, Jiang CY, Zhang Y, Qi SW et al (2007) Nucleotide sequence of plasmid pCNB1 from *Comamonas* strain CNB-1 reveals novel genetic organization and evolution for 4-chloronitrobenzene degradation. Appl Environ Microbiol 73:4477–4483

Mosqueda G, Ramos-González MI, Ramos JL (1999) Toluene metabolism by the solvent-tolerant *Pseudomonas putida* DOT-T1 strain, and its role in solvent impermeabilization. Gene 232:69–76

Nicolaou SA, Gaida SM, Papoutsakis ET (2010) A comparative view of metabolite and substrate stress and tolerance in microbial bioprocessing: from biofuels and chemicals, to biocatalysis and bioremediation. Metab Eng 12:307–331

Nikaido H, Takatsuka Y (2009) Mechanisms of RND multidrug efflux pumps. Biochim Biophys Acta 1794:769–781

Okuyama H, Okajima N, Sasaki S, Higashi S, Murata N (1991) The *cis/trans* isomerization of the double bond of a fatty acid as a strategy for adaptation to changes in ambient temperature in the psychrophilic bacterium, *Vibrio* sp. strain ABE-1. Biochim Biophys Acta 1084:13–20

Parales RE, Parales JV, Pelletier DA, Ditty JL (2008) Diversity of microbial toluene degradation pathways. Adv Appl Microbiol 64:1–73

Powlowski J, Shingler V (1994) Genetics and biochemistry of phenol degradation by *Pseudomonas* sp. CF600. Biodegradation 5:219–236

Ramos JL, Marqués S, van Dillewijn P, Espinosa-Urgel M, Segura A, Duque E et al (2011) Laboratory research aimed at closing the gaps in microbial bioremediation. Trends Biotechnol 29:641–647

Ramos JL, Sol Cuenca M, Molina-Santiago C, Segura A, Duque E, Gómez-García MR et al (2015) Mechanisms of solvent resistance mediated by interplay of cellular factors in *Pseudomonas putida*. FEMS Microbiol Rev 39:555–566

Segura A, Godoy P, van Dillewijn P, Hurtado A, Arroyo N, Santacruz S, Ramos JL (2005) Proteomic analysis reveals the participation of energy- and stress-related proteins in the response of *Pseudomonas putida* DOT-T1E to toluene. J Bacteriol 187:5937–5945

Segura A, Molina L, Fillet S, Krell T, Bernal P, Munoz-Rojas J, Ramos JL (2012) Solvent tolerance in Gram-negative bacteria. Curr Opin Biotechnol 23:415–421

Segura A, Molina L, Ramos JL (2014) Plasmid-mediated tolerance toward environmental pollutants. Microbiol Spectr 2:PLAS-0013–PLAS-2013

Sikkema J, de Bont JA, Poolman B (1995) Mechanisms of membrane toxicity of hydrocarbons. Microbiol Rev 59:201–222

Varjani SJ (2017) Microbial degradation of petroleum hydrocarbons. Bioresour Technol 223:277–286

Varjani SJ, Upasani VN (2013) Comparative studies on bacterial consortia for hydrocarbon degradation. Int J Innovative Res Sci Eng Technol 2:5377–5383

Volkers RJ, de Jong AL, Hulst AG, van Baar BL, de Bont JA, Wery J (2006) Chemostat-based proteomic analysis of toluene-affected *Pseudomonas putida* S12. Environ Microbiol 8:1674–1679

Wijte D, van Baar BL, Heck AJ, Altelaar AF (2011) Probing the proteome response to toluene exposure in the solvent tolerant *Pseudomonas putida* S12. J Proteome Res 10:394–403

Yen KM, Serdar CM (1988) Genetics of naphthalene catabolism in pseudomonads. Crit Rev Microbiol 15:247–268

Zhao B, Poh CL (2008) Insights into environmental bioremediation by microorganisms through functional genomics and proteomics. Proteomics 8:874–881

# Toxicity of Hydrocarbons to Microorganisms

# 21

Hermann J. Heipieper and P. M. Martínez

## Contents

1  Introduction ................................................................. 336
2  General Toxicity of Hydrocarbons on Microorganisms .................................. 336
3  Hydrocarbons in Membranes ............................................................. 338
4  Maximal Membrane Concentration Allows Prediction of Toxicity ....................... 340
References ................................................................. 343

### Abstract

Several classes of organic compounds are toxic for living organisms as they accumulate in and disrupt cell membranes. In these cases, the dose-dependent toxicity of a compound correlates with the logarithm of its partition coefficient between octanol and water (logP). Substances with a logP value between 1 and 5 are, in general, toxic for whole cells. Therefore, toxic effects of hydrocarbons on microorganisms can cause problems in bioremediation of highly contaminated sites. The toxic effect of most hydrocarbons is caused by general, non-specific effects on membrane fluidity due to their accumulation in the lipid bilayer. Only exceptions are hydrocarbons with specific chemically active functional groups such as aldehydes and epoxides that show an additional chemical toxicity.

Most compounds with a higher hydrophobicity than logP of 4 such as e.g., alkanes, PAHs, and biphenyl(s) have very low water solubility, thus their

H. J. Heipieper (✉)
Department Environmental Biotechnology, Helmholtz Centre for Environmental Research – UFZ, Leipzig, Germany
e-mail: hermann.heipieper@ufz.de

P. M. Martínez
Department of Bioremediation, Helmholtz Centre for Environmental Research—UFZ, Leipzig, Germany

© Springer International Publishing AG, part of Springer Nature 2018
T. Krell (ed.), *Cellular Ecophysiology of Microbe: Hydrocarbon and Lipid Interactions*,
Handbook of Hydrocarbon and Lipid Microbiology, https://doi.org/10.1007/978-3-319-50542-8_45

bioavailability is too low to show a toxic effect. By combining the logP value with the water solubility of a compound the maximum membrane concentration (MMC) of a compound can be calculated. By using this parameter it is possible to predict the potential toxicity even of unknown hydrocarbons.

# 1    Introduction

The high persistence of organic pollutants in many contaminated sites such as soils or aquifers can be caused by two major physico–chemical properties of hydrocarbons, namely the low bioavailability (not accessible for microbial degradation) as well as the very high toxicity of several classes of organic hydrocarbons. Especially monoaromatic compounds (e.g., BTEX and phenols), *n*-alkanols and terpeniols are known to be very toxic not only to humans, animals and plants but also to microorganisms that are capable to degrade them. Therefore, environments contaminated with high, bacteriostatic concentrations of such compounds cannot be effectively bioremediated due to the inhibitory effects of the pollutants on the microbiota.

In addition, the toxicity of organic solvents plays also a role in the biotechnological production of fine chemicals in whole cell biotransformations (Schmid et al. 2001). During the production of such fine chemicals using microorganisms as biocatalysts, organic solvents applied in form of a second phase work as a source for toxic water immiscible substrates as well as a sink for synthesized products (Leon et al. 1998; Salter and Kell 1995). Such reaction system allows a much higher volume specific productivity enabling the economically sound synthesis using whole cell biocatalysis. However, problems resulting from the toxicity of hydrophobic organic solvents for whole cells are still an important drawback for the application of these compounds in biocatalysis (Salter and Kell 1995). Thus, the toxicity of organic solvents plays an important role in environmental microbiology as well as in the biotechnological application of microorganisms. Therefore, a systematic knowledge of the toxicity of existing but also of emerging chemicals that are planned to be released in the environment is of fundamental importance.

# 2    General Toxicity of Hydrocarbons on Microorganisms

Several classes of organic compounds are toxic for living organisms as they accumulate in and disrupt cell membranes. In these cases, the toxicity of a compound correlates with the logarithm of its partition coefficient between octanol and water (logP). Substances with a logP value between 1 and 5 are, in general, toxic for whole cells (Liu et al. 1982). Such toxic hydrocarbons can only be degraded at low rate and consequently stay often as persistent pollutants in the environment. Therefore, toxic effects of hydrocarbons on microorganisms can cause problems in bioremediation, waste gas and wastewater treatment (Sikkema et al. 1995).

Most microorganisms tolerate water-miscible solvents such as lower alcohols and acids. On the other hand, it has been established that very lipophilic natural solvents including some hydrocarbons are not toxic for whole cells. However, many of the organic solvents used in petrochemical operations occupy a position between the water-soluble alcohols and acids and the lipophilic compounds. For example, solvents of mediate hydrophobicity like aromatic solvents are very toxic to cells. Compounds like alcohols, aromatics, phenols or terpeniols, have been widely used as antimicrobial agents. They have been applied as food preservatives, disinfectants, tools for the permeabilization of cells and also narcotic agents (Heipieper et al. 1991; Sikkema et al. 1992).

The antimicrobial action of several classes of organic hydrocarbons on different microorganisms was tested in a number of systematic investigations (Heipieper et al. 1995; Ingram 1977; Kabelitz et al. 2003; Liu et al. 1982; Sikkema et al. 1994). All showed a correlation between hydrophobicity (logP) and toxicity. It should be emphasized that the toxic effects of solvents, and their dose-response relationships, were found to be similar for a variety of microorganisms. Thus, the toxicity of most hydrophobic organic hydrocarbons is caused by general, nonspecific effects on membrane fluidity due to their accumulation in the lipid bilayer, and no specific chemical reactions are associated with these toxic effects (Cabral 1991; Ferrante et al. 1995; Liu et al. 1982; Saito et al. 1994). However, not all bacteria show the same sensitivity towards a certain organic compound. A first systematic investigation of the toxicity of hydrocarbons belonging to the classes of (chlorinated) phenols, BTEX and $n$-alkanols was carried out for anaerobic bacteria. For all tested compounds the anaerobic bacteria were about three-times more sensitive than often tested aerobic strains (Duldhardt et al. 2007). Figure 1 shows as an example the relations found for *Thauera aromatica* and *Pseudomonas putida*.

The correlations shown in Fig. 1 are valid for compounds that are known of being only membrane active and thus have only the so-called narcotic effect. Compounds with additional functional groups such as e.g., aldehydes, epoxides, nitro-groups are usually much more toxic than predicted by the above described calculations because next to their membrane toxicity they also show an additional and specific chemical toxicity.

On the other hand, only uncharged molecules are able to enter the lipid bilayer of the membrane. The carboxylic groups of most organic acids are present, at physiological pH values, in their dissociated-charged form. This is the reason why compounds such as benzoic acid and its derivatives but also herbicides such as 2,4-D can be added in very high amounts to microbial cells without causing any toxic effect (Cabral et al. 2003).

Another problem especially of complex aromatic compounds but also of certain alkanes is the very high toxicity of intermediates that accumulate during the biodegradation of these compounds (Blasco et al. 1995; Camara et al. 2004). Also here, 2,4-D is an important example as during its degradation the first degradation intermediate 2,4-dichlorophenol can accumulate which is known to be an extremely toxic compound (Cabral et al. 2003; Heipieper et al. 1995). The same effect was observed for the accumulation of 1-octanol as toxic intermediate during the degradation of

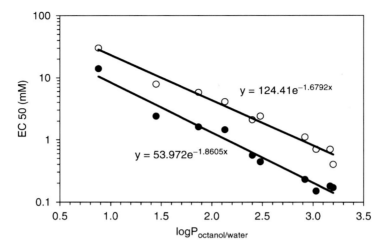

**Fig. 1** Correlation between the hydrophobicity, given as the logP value of different hydrocarbons and growth inhibition of anaerobically grown *Thauera aromatica* K172 (•) and *Pseudomonas putida* as a representative of aerobic bacteria (O). Growth inhibition is presented as the EC 50 concentrations

*n*-octane (Chen et al. 1995a, b). At least for laboratory experiments this should always be taken into consideration.

## 3      Hydrocarbons in Membranes

The major reason why hydrocarbons are toxic to microbial cells is due to their preferential partitioning into membranes causing an increase in the fluidity of the membrane that leads to its non-specific permeabilization. The accumulation of the solvent toluene into bacterial membranes could be made visible by electron microscopy (Aono et al. 1994). These results demonstrate that the membrane is the main target of the toxic effect. This of course does not rule out additional sides of toxic action as they may be caused by the specific properties of a molecule. A systematic relationship exists between the logP value of compounds and their partitioning in a membrane-buffer system (Sikkema et al. 1994). The absolute values of partition coefficients are approximately eight-times lower in membranes than in octanol. Therefore, the logP value of a compound can be taken to calculate its partitioning in the membrane by using the following equilibration from Sikkema et al. (Sikkema et al. 1994):

$$\log P_{M/B} = 0.97 \cdot \log P_{O}/W - 0.64$$

Hence, the logP value is a suitable parameter which describes the accumulation of these solvents in membranes. However, in addition to hydrophobicity, the molecular structure of a compound also influences its membrane solubility. Amphipathic molecules will dissolve relatively well in membranes, as their structure is similar to the one of the phospholipids in the membranes. Consequently, amphiphatic compounds such as chlorinated phenols or alkanols are known to be the most toxic substances. This explains the high toxicity of aromatic structures with hydrophilic substituents, particularly phenols. Additionally, the composition of the membrane also influences the partitioning of a compound into it (Weber and de Bont 1996). The composition of membranes regarding the phospholipid fatty acids plays a major role in toxicity as well as in adaptation. So far, this has only been demonstrated with artificial membranes. As an example the partition coefficient of lindane is for instance 50 times higher in liposomes of dimyristoylphosphate (C14:0) than in liposomes of distearoylphosphatidylcholine (C18:0) (Antunes-Madeira and Madeira 1989).

Although no individual analytical technique is able to determine fully the effects of solvents on a membrane, several mechanisms of the membrane toxicity have been reported. Most important, the accumulation of organic solvents leads to a non-specific permeabilization of the cell membranes. In *Escherichia coli*, it was observed that potassium ions and ATP are released after treatment with phenol (Heipieper et al. 1991). In the case of toluene the leakage from the cell of macromolecules like phospholipids, proteins or even RNA, could be demonstrated. This permeabilization is due to considerable damages of the cytoplasmic membrane, whereas the outer membrane is still intact. Another kind of effect has been observed with the solvent tetraline, which increased the proton permeability in artificial membranes (Heipieper et al. 1994). Also solvents could produced a passive flux of protons and other ions across the membrane (Sikkema et al. 1994). This flux of ions dissipates the proton motive force ($\Delta p$), and affects both the proton gradient ($\Delta pH$) and the electrical potential ($\Delta \psi$) (Sikkema et al. 1994).

Therefore, the second mechanism of the membrane toxicity of organic solvents is to diminish the energy status of the cell. The effect of solvents on the energy transduction of membranes was tested in liposomes reconstituted with cytochrome *c* oxidase, as proton motive force generating mechanism. The presence of the solvent tetraline caused a decrease on both, the proton gradient $\Delta pH$ and the electrical potential $\Delta \Psi$ with 80% and 50%, respectively (Sikkema et al. 1994). In the same study it was established that the dose-response curves for all solvents tested were very similar if the dissolved membrane concentration of the solvent is considered to be the dose. A similar effect on $\Delta pH$ and $\Delta \Psi$ could be observed on intact cells (Sikkema et al. 1992).

The decrease of the proton motive force is not the only reason for lower energy levels of cells in the presence of organic solvents. Additionally, the presence of solvents leads to an impaired ATP synthesis, due to a partial inhibition of the ATPase activity and other proteins engaged in the energy transducing process (Uribe et al. 1990). The accumulation of solvents into a membrane also affects the function of other proteins embedded in the membrane. For example, in *Saccharomyces cerevisiae* toluene leads to an inhibition of the proton-potassium translocation (Uribe et al. 1985).

A further important aspect of the membrane structure, the fluidity which is defined as the reciprocal of viscosity, is also affected by organic solvents (Sikkema et al. 1994). The effect of organic solvents on the fluidity was measured by use of fluorescent probes. The probe 1,6-diphenyl-1,3,5 hexatriene (DPH), partitions into the inner parts of cell membranes while a trimethylamine derivative of this compound (TMA-DPH) anchors its hydrophilic group in the headgroup region of the bilayers. Solvents partitioning into the membrane will affect the polarization of the probe located at the site entered by the solvent. Several solvents with a logP between 1 and 5 affected the polarization of both probes (Sikkema et al. 1994). Thus, the whole membrane was fluidized. An increased fluidity of membranes results in changes in stability, structure and interactions within the membrane.

Membrane active compounds can affect the hydration characteristics of the membrane surface and the thickness of the membrane. One technique to investigate the swelling of the membrane is the use of liposomes labeled with the fluorescent fatty acid, octadecyl rhodamine β-chloride. An expansion of the membrane will lead to a dilution of the fluorescent probe in the membrane, which can be recorded as a reduction in fluorescent self-quenching. Swelling of the membrane is monitored by plotting relative fluorescence against the amount of solvent added to the liposomes (Sikkema et al. 1994). For all solvents tested the surface area of the membranes increases constantly up to a concentration of 0.5 $\mu$mol mg$^{-1}$ of solvent in the membrane. At that concentration, the maximum increase in the membrane swelling was reached. This maximum level corresponds with approximately one solvent molecule per two phospholipid molecules.

It can be concluded that once a solvent has dissolved in a membrane, it will disturb the integrity of the membrane and hence its function as a barrier, as matrix for enzymes and as energy transducer. Figure 2 shows a scheme of these toxic actions of hydrocarbons on cell membranes that, in bacteriostatic concentrations, cause growth inhibition or, at very high bacteriotoxic concentrations, to cell lysis. For membranes of living cells these effects are difficult to quantify due to the complexity and heterogeneous nature of the protein-containing bilayer.

# 4    Maximal Membrane Concentration Allows Prediction of Toxicity

It is important to note that the relation between toxicity and hydrophobicity is only valid until a certain logP value, generally around logP of 4. At higher hydrophobicity a second important parameter of organic compounds plays a role, namely the water solubility or bioavailability. Compounds that are very hydrophobic like e.g., alkanes, PAHs, and biphenyl(s) do not have a toxic effect because of their extremely low water solubility which of course correlates as well with their hydrophobicity. Due to the low water solubility these compounds are not bioavailable enough to reach a certain concentration in the membrane. This relation can be used by predicting the toxicity of hydrocarbons simply by combining the logP value with the water solubility in order to calculate the so-called maximum membrane concentration

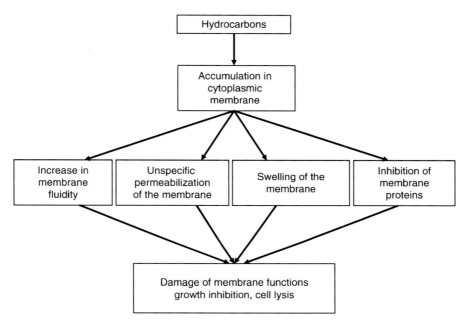

**Fig. 2** Scheme of the toxic actions of hydrocarbons to cell membranes

(MMC) of a compound (de Bont 1998). The solvent's membrane concentration depends on the concentration of solvent in the water phase, the partitioning of the solvent from the water phase into the membrane, and the ratio of the volumes of the two liquid phases. Thus, by using the equation describing the correlation between the logP value of a solvent and its partitioning between membrane and buffer (Sikkema et al. 1994) it is possible to calculate the actual as well as the maximum concentration of a solvent in a membrane if its concentration in the water phase is known (de Bont 1998; Neumann et al. 2005; Sikkema et al. 1994). These calculations were performed with several alkanols and toluene as reference compound and are summarized in Table 1.

Using the calculated MMC also the potential toxicity of a compound to microorganisms can be predicted. A direct relation between the concentrations of organic compounds in the membrane and their toxic concentrations (e.g., their EC50, Effective Concentration inhibiting 50% of cell growth) was also described (Heipieper et al. 1995). In general, the calculated actual concentration in the membrane necessary for an EC50 effect was nearly the same for all compounds tested and ranged around 200 mM (Heipieper et al. 1995). These results support the idea that the toxic effects of organic solvents are independent from the structural features of the molecules, but are strongly related to their ability to accumulate in the membrane (Heipieper et al. 1994). Thus, the MMC needs to be higher than 200 mM to cause toxic effects of the cells.

**Table 1** Relation between partitioning of several hydrocarbons which are important chemical pollutants in octanol/water (logP), membrane/buffer, water solubility and the maximum concentration in the membrane

| Organic compound | $logP_{o/w}$[a] (mM) | $logP_{m/b}$[b] (mM) | $P_{m/b}$[b] | Solubility[c] | MMC[d] | Toxicity |
|---|---|---|---|---|---|---|
| 1-Butanol | 0.89 | 0.21 | 1.60 | 9,701.0 | 1,586 | + |
| 1-Hexanol | 1.87 | 1.17 | 14.80 | 56.9 | 841 | + |
| 1-Octanol | 2.92 | 2.19 | 154.90 | 3.8 | 588 | + |
| 1-Decanol | 3.97 | 3.21 | 1,621.80 | 0.23 | 379 | + |
| 1-Dodecanol | 5.02 | 4.23 | 16,982.40 | 0.015 | 254 | + |
| 1-Tetradecanol | 6.07 | 5.25 | 177,827.90 | 0.0008 | 142 | − |
| Phenol | 1.45 | 0.77 | 5.84 | 880.0 | 5,140 | + |
| 4-Chlorophenol | 2.40 | 1.69 | 48.75 | 186.7 | 9,102 | + |
| 2,4-Dichlorophenol | 3.20 | 2.46 | 291.07 | 27.6 | 8,034 | + |
| 2,4,5-Trichlorophenol | 4.05 | 3.29 | 1,943.12 | 6.1 | 11,795 | + |
| 2,3,4,5-Tetrachlorophenol | 4.59 | 3.81 | 6,490.83 | 0.12 | 779 | + |
| Pentachlorophenol | 5.12 | 4.33 | 21,203.13 | 0.05 | 1,060 | + |
| Benzene | 2.13 | 1.43 | 26.67 | 23.0 | 614 | + |
| Toluene | 2.48 | 1.77 | 58.29 | 6.3 | 367 | + |
| Styrene | 2.94 | 2.21 | 162.85 | 2.95 | 480 | + |
| Ethylbenzene | 3.03 | 2.30 | 199.11 | 1.6 | 319 | + |
| p-Xylene | 3.17 | 2.43 | 272.21 | 1.2 | 327 | + |
| Naphthalene | 3.30 | 2.56 | 363.92 | 0.240 | 87 | − |
| Acenaphtylene | 3.94 | 3.18 | 1519.85 | 0.100 | 152 | − |
| Fluorene | 4.18 | 3.41 | 2,597.77 | 0.010 | 26 | − |
| Phenanthrene | 4.46 | 3.69 | 4,855.12 | 0.006 | 29 | − |
| Biphenyl | 3.98 | 3.22 | 1,661.88 | 0.045 | 75 | − |
| PCE | 2.96 | 2.23 | 170.29 | 1.200 | 204 | − |
| TCE | 2.47 | 1.76 | 57.00 | 7.070 | 403 | + |
| cis-DCE | 1.86 | 1.16 | 14.59 | 50.824 | 742 | + |
| n-Hexane | 3.29 | 2.55 | 355.88 | 0.150 | 53 | − |
| cyclo-Hexane | 3.50 | 2.76 | 568.85 | 0.500 | 284 | + |
| n-Octane | 4.55 | 3.77 | 5,936.08 | 0.0058 | 34 | − |
| n-Decane | 5.61 | 4.80 | 63,343.20 | 0.00035 | 22 | − |

[a]logP octanol/water provided by http://chem.sis.nlm.nih.gov/chemidplus/
[b]logP membrane/buffer and P membrane/buffer calculated according to Sikkema et al. (1994)
[c]Solubility in $H_2O$ (mg/l) at 25°C were provided by http://chem.sis.nlm.nih.gov/chemidplus/
[d]MMC = maximum membrane concentration calculated according to de Bont (de Bont 1998) and Neumann et al. (2005)

For the group of the well-investigated n-alkanols the observed relations may be easier to explain. The higher the chain length of an alkanol, the higher is its hydrophobicity and thus its tendency to accumulate preferentially in membranes. However, as water solubility decreases with increasing chain length, a compound such as dodecanol will not reach a high membrane concentration and is therefore not toxic to an organism. That is the difference with e.g., alkanols as well as other organic compounds in a logP range of 1–4, which are extremely toxic: because they are relatively water soluble and still partition well into the membrane. As a result, the actual membrane concentration of these solvents will be too high (Neumann et al. 2005, 2006). Thus, this calculation method is a useful tool to predict whether a certain hydrocarbon is potentially toxic to a bacterium or whether the bacterium is able to tolerate high-saturating concentrations of it, like in the case of most alkanes and PAHs. This method could be possibly used to predict potential toxicity of emerging compounds, when specific chemical reactive groups are not present in the molecule.

# References

Antunes-Madeira MC, Madeira VMC (1989) Membrane fluidity as affected by the insecticide lindane. Biochim Biophys Acta 982:161–166

Aono R, Kobayashi H, Joblin KN, Horikoshi K (1994) Effects of organic solvents on growth of *Escherichia coli* K–12. Biosci Biotechnol Biochem 58:2009–2014

Blasco R, Wittich RM, Mallavarapu M, Timmis KN, Pieper DH (1995) From xenobiotic to antibiotic, formation of protoanemonin from 4-chlorocatechol by enzymes of the 3-oxoadipate pathway. J Biol Chem 270:29229–29235

Cabral JP (1991) Damage to the cytoplasmic membrane and cell death caused by dodine (dodecylguanidine monoacetate) in *Pseudomonas syringae* ATCC 12271. Antimicrob Agents Chemother 35:341–344

Cabral MG, Viegas CA, Teixeira MC, Sa-Correia I (2003) Toxicity of chlorinated phenoxyacetic acid herbicides in the experimental eukaryotic model *Saccharomyces cerevisiae*: role of pH and of growth phase and size of the yeast cell population. Chemosphere 51:47–54

Camara B, Herrera C, Gonzalez M, Couve E, Hofer B, Seeger M (2004) From PCBs to highly toxic metabolites by the biphenyl pathway. Environ Microbiol 6:842–850

Chen Q, Janssen DB, Witholt B (1995a) Growth on octane alters the membrane lipid fatty acids of *Pseudomonas oleovorans* due to the induction of alkB and synthesis of octanol. J Bacteriol 177:6894–6901

Chen Q, Nijenhuis A, Preusting H, Dolfing J, Janssen DB, Witholt B (1995b) Effects of octane on the fatty acid composition and transition temperature of *Pseudomonas oleovorans* membrane lipids during growth in 2-liquid-phase continuous cultures. Enzym Microb Technol 17:647–652

de Bont JAM (1998) Solvent-tolerant bacteria in biocatalysis. Trends Biotechnol 16:493–499

Duldhardt I, Nijenhuis I, Schauer F, Heipieper HJ (2007) Anaerobically grown *Thauera aromatica*, *Desulfococcus multivorans*, *Geobacter sulfurreducens* are more sensitive towards organic solvents than aerobic bacteria. Appl Microbiol Biotechnol 77:705–711

Ferrante AA, Augliera J, Lewis K, Klibanov AM (1995) Cloning of an organic solvent-resistance gene in *Escherichia coli*: the unexpected role of alkylhydroperoxide reductase. Proc Natl Acad Sci U S A 92:7617–7621

Heipieper HJ, Keweloh H, Rehm HJ (1991) Influence of phenols on growth and membrane permeability of free and immobilized *Escherichia coli*. Appl Environ Microbiol 57:1213–1217

Heipieper HJ, Weber FJ, Sikkema J, Keweloh H, de Bont JAM (1994) Mechanisms behind resistance of whole cells to toxic organic solvents. Trends Biotechnol 12:409–415

Heipieper HJ, Loffeld B, Keweloh H, de Bont JAM (1995) The *cis/trans* isomerization of unsaturated fatty acids in *Pseudomonas putida* S12: an indicator for environmental stress due to organic compounds. Chemosphere 30:1041–1051

Ingram LO (1977) Changes in lipid composition of *Escherichia coli* resulting from growth with organic solvents and with food additives. Appl Environ Microbiol 33:1233–1236

Kabelitz N, Santos PM, Heipieper HJ (2003) Effect of aliphatic alcohols on growth and degree of saturation of membrane lipids in *Acinetobacter calcoaceticus* FEMS Microbiol Lett 220:223–227

Leon R, Fernandes P, Pinheiro HM, Cabral JMS (1998) Whole-cell biocatalysis in organic media. Enzym Microb Technol 23:483–500

Liu D, Thomson K, Kaiser KL (1982) Quantitative structure-toxicity relationship of halogenated phenols on bacteria. Bull Environ Contam Toxicol 29:130–136

Neumann G et al (2005) Prediction of the adaptability of *Pseudomonas putida* DOT-T1E to a second phase of a solvent for economically sound two-phase biotransformations. Appl Environ Microbiol 71:6606–6612

Neumann G et al (2006) Energetics and surface properties of *Pseudomonas putida* DOT-T1E in a two-phase fermentation system with 1-decanol as second phase. Appl Environ Microbiol 72:4232–4238

Saito H, Koyasu J, Shigeoka T, Tomita I (1994) Cytotoxicity of chlorophenols to goldfish GFS cells with the MTT and LDH assays. Toxicol in Vitro 8:1107–1112

Salter GJ, Kell DB (1995) Solvent selection for whole cell biotransformations in organic media. Crit Rev Biotechnol 15:139–177

Schmid A, Dordick JS, Hauer B, Kiener A, Wubbolts M, Witholt B (2001) Industrial biocatalysis today and tomorrow. Nature 409:258–268

Sikkema J, Poolman B, Konings WN, de Bont JA (1992) Effects of the membrane action of tetralin on the functional and structural properties of artificial and bacterial membranes. J Bacteriol 174:2986–2992

Sikkema J, de Bont JA, Poolman B (1994) Interactions of cyclic hydrocarbons with biological membranes. J Biol Chem 269:8022–8028

Sikkema J, de Bont JA, Poolman B (1995) Mechanisms of membrane toxicity of hydrocarbons. Microbiol Rev 59:201–222

Uribe S, Ramirez J, Pena A (1985) Effects of beta pinene on yeast membrane functions. J Bacteriol 161:1195–1200

Uribe S, Rangel P, Espinola G, Aguirre G (1990) Effects of cyclohexane, an industrial solvent, on the yeast *Saccharomyces cerevisiae* and on isolated yeast mitochondria. Appl Environ Microbiol 56:2114–2119

Weber FJ, de Bont JAM (1996) Adaptation mechanisms of microorganisms to the toxic effects of organic solvents on membranes. Biochim Biophys Acta 1286:225–245

# Genetics of Sensing, Accessing, and Exploiting Hydrocarbons

# 22

Miguel A. Matilla, Craig Daniels, Teresa del Castillo, Andreas Busch, Jesús Lacal, Ana Segura, Juan Luis Ramos, and Tino Krell

## Contents

1 Introduction .................................................................. 346
2 Chemotaxis Towards Aromatic Hydrocarbons ............................................. 348
3 The TOD Pathway for Catabolism of Aromatic Hydrocarbons .......................... 350
4 The TOL Pathway for Catabolism of Aromatic Hydrocarbons .......................... 352
5 Catabolite Repression and Expression of the TOL Plasmid Genes ...................... 354
6 Conclusions and Research Needs ......................................................... 355
References ...................................................................... 356

M. A. Matilla · A. Segura · J. L. Ramos · T. Krell (✉)
Department of Environmental Protection, Estación Experimental del Zaidín, Consejo Superior de Investigaciones Científicas, Granada, Spain
e-mail: miguel.matilla@eez.csic.es; ana.segura@eez.csic.es; juanluis.ramos@eez.csic.es; tino.krell@eez.csic.es

C. Daniels
Developmental and Stem Cell Biology Program, Brain Tumour Research Centre, The Hospital for Sick Children, Toronto, ON, Canada
e-mail: craig.daniels@sickkids.ca

T. del Castillo
Group of Physics of Fluids, Interfaces and Colloidal Systems, Department of Applied Physics, Faculty of Science, University of Granada, Granada, Spain
e-mail: tdelcastillo@ugr.es

A. Busch
Confo Therapeutics, VIB Campus Technologiepark, Zwijnaarde, Belgium
e-mail: a.busch@mail.cryst.bbk.ac.uk

J. Lacal
Department of Microbiology and Genetics, University of Salamanca, Salamanca, Spain
e-mail: djlacal@hotmail.com

© Springer International Publishing AG, part of Springer Nature 2018
T. Krell (ed.), *Cellular Ecophysiology of Microbe: Hydrocarbon and Lipid Interactions*,
Handbook of Hydrocarbon and Lipid Microbiology, https://doi.org/10.1007/978-3-319-50542-8_46

**Abstract**

Hydrocarbons abound in the environment and microorganisms are often capable of detecting, assimilating, and degrading these normally recalcitrant molecules. In order to achieve this, bacteria have developed specific sensor proteins and adaptive mechanisms. In the presence of hydrocarbons, the bacterial adaptive response is modulated at the transcriptional and post-transcriptional levels by one- and two-component regulatory systems, global regulators, and DNA-binding proteins. The expressed gene products are then able to degrade the molecules and often take advantage of the stored energy imparted by the physicochemical properties of the hydrocarbon structure. The response of regulators to the presence of hydrocarbons such as toluene in the environment allows initiation or inhibition of transcription, so that the rate of synthesis of metabolically important gene products is adaptively modulated. Microorganisms which mount the most appropriate physiological adaptation are then able to proliferate in the changing environment. Here, we give an overview of the bacterial chemotactic responses towards hydrocarbons and the adaptive regulation of catabolic pathways responsible for the degradation of aromatic hydrocarbons. The use of microorganisms with biodegradative capabilities offers an environmentally friendly alternative for the treatment of hydrocarbon-contaminated environments.

# 1    Introduction

Hydrocarbons are organic compounds that are used as an extraordinary source of energy and employed in the manufacture of drugs, explosives, and raw materials. Based on the bonding between carbons and the nature of the backbone of the carbon chain, they can be classified as aliphatic (i.e., alkanes, alkenes, alkynes), alicyclics, and aromatic compounds. Aromatic hydrocarbons (AHC) possess a large resonance energy that is reflected in a high thermodynamic stability which manifests itself in chemical properties very different from those observed for aliphatic hydrocarbons. AHC are ubiquitous and abundant in nature. In fact, the benzene ring is one of the most widely distributed chemical units in natural environments (Dagley 1981). Although the origin of AHC is controversial, it is commonly accepted that they mainly derive from the natural pyrolysis of organic compounds (Gibson and Subramanian 1984). However, a substantial quantity of AHC found in the environment is of human origin and mainly derived from crude oil. Thus, compounds such as benzene, ethylbenzene, toluene, styrene, and xylenes have been widely used in industry and are among the most produced industrial chemicals worldwide (Segura et al. 2012). Additionally, polycyclic aromatic compounds are often the starting material in the production of agrochemicals (pesticides, herbicides, and fertilizers), polymers, pharmaceuticals, explosives, and a multitude of other everyday products. These man-made AHC are serious contaminants that have been frequently released into the environment causing a serious impact on ecosystems and human health since many of them are toxins, carcinogens, and mutagens

(Pohanish 2011). Unfortunately, plants and animals have none or limited capabilities to degrade these aromatic compounds. However, multiple microorganisms, mainly aerobic and anaerobic bacteria and fungi, have evolved the capacity to tolerate and degrade these compounds (Fuchs et al. 2011; Ramos et al. 2015). Due to the current social and political interest in pollution, global environment, and climate change, increasing attention is being devoted to the isolation and characterization of microorganisms that are able to thrive on these hydrocarbons, which may permit the development of bioremediation strategies (Ramos et al. 2010, 2015; Fuentes et al. 2014).

Different studies showed that there is not an apparent correlation between the tolerance of toxic hydrocarbons and the biodegradative capacity of a particular bacterial strain (Ramos et al. 2015). Thus, bacteria have developed multiple specific adaptive mechanisms to overcome the toxicity of a wide range of hydrocarbons of both natural and human origin that abound in the environment (Segura et al. 2012; Ramos et al. 2015). Some of these defense mechanisms are postsynthetic, such as the release of membrane vesicles or the alteration of membrane fluidity to prevent the entry of hydrocarbons (Baumgarten et al. 2012; Ramos et al. 2015). However, most of the adaptive responses result in the modulation of the expression of target genes, the products of which are able to take advantage of the physicochemical properties of the hydrocarbons. This transcriptional regulation is mediated by the sensing of specific environmental signals (i.e., aromatic hydrocarbons) by a variety of different sensor proteins that ultimately alter the rate of synthesis of metabolically important gene products.

In the late 1980s, a *Pseudomonas putida* strain extremely tolerant to high concentrations of AHC was isolated (Inoue and Horikoshi 1989). Since then, many other similarly adapted bacteria have been identified. One of these highly solvent-tolerant strains, *Pseudomonas putida* DOT-T1E, was isolated in the mid-1990s from a sewage treatment plant in Granada, Spain, and was found to be an efficient degrader of multiple hydrocarbons such as toluene, benzene, ethylbenzene, and xylene that furthermore support bacterial growth as sole carbon source (Ramos et al. 1995). The ability of DOT-T1E to survive in the presence of hydrocarbons is an inducible phenomenon (Ramos et al. 1995). The molecular mechanisms underlying this adaptive response and the subsequent solvent tolerance have been extensively investigated in *P. putida* (reviewed in Segura et al. 2012; Segura and Ramos 2014; Ramos et al. 2015). In the particular case of *P. putida* DOT-T1E, the adaptation to the presence of AHC is a multifaceted process that has been reviewed by Udaondo et al. (2012, 2013) and Ramos et al. (2015) and involves: (i) a reduction in the permeability of the cell membrane through the modification of fatty acids and phospholipid head groups; (ii) synthesis of chaperones that facilitate the correct protein folding in the presence of the hydrocarbons; (iii) activation of an oxidative defense mechanism to reduce the oxidative-mediated damage caused by AHC; (iv) the extrusion of the compounds by three efflux pumps, namely, TtgABC, TtgDEF, and TtgGHI; and (v) the induction of metabolic pathways responsible for the degradation and subsequent use of the hydrocarbons as an energy source. Furthermore, further studies in our laboratory demonstrated that DOT-T1E exhibits chemotaxis towards different AHC, which is a behavior that may facilitate the *in situ* degradation of toxic hydrocarbons (Lacal et al. 2011, 2013).

In this chapter, we present an overview of the ecophysiological adaptation of *P. putida* to AHC, together with its ability to respond chemotactically to these compounds.

## 2    Chemotaxis Towards Aromatic Hydrocarbons

The recognition of environmental signals may result in changes in gene expression at transcriptional and post-transcriptional levels or at the behavioral level. Signal sensing by chemoreceptors that feed the corresponding molecular stimuli into chemosensory pathway permit flagella-based movements in compound gradients, a behavior known as chemotaxis. In the case of *Pseudomonas* strains, this system is rather complex and involves an elevated number of chemoreceptors (frequently 25–45) and in some cases multiple chemosensory pathways (Sampedro et al. 2015).

Historically, *Escherichia coli* is the primary model organism in the study of chemotaxis in bacteria. The genome of this model bacterium encodes five chemoreceptors and its chemoeffector profile includes sugars, amino acids, dipeptides, pyrimidines, electron acceptors, fatty acids, alcohols, and cations (Parkinson et al. 2015). However, the sequencing of thousands of bacterial genomes in the last decade allowed the identification of strains with larger numbers of chemoreceptor-encoding genes. A particular feature of chemoreceptors is the structural diversity of the domain that recognizes the environmental stimuli, which is referred to as ligand-binding domain (Collins et al. 2014; Sampedro et al. 2015).

There are several examples in the literature describing *Pseudomonas* strains that show chemotaxis towards AHC, some of which contain nitro, amino, chloro, or bromo substitutions (reviewed by Sampedro et al. 2015). Pioneering work in this field was carried out by Harwood and coworkers through the characterization of several *Pseudomonas* strains capable to exhibit chemotaxis towards a wide range of aromatic compounds (Harwood et al. 1984; Harwood et al. 1990; Grimm and Harwood 1997, 1999; Parales et al. 2000). Within the identified compounds, aromatic hydrocarbons such as benzene, ethylbenzene, toluene, and naphthalene were shown to act as chemoattractants (Grimm and Harwood 1997, 1999; Parales et al. 2000). More recently, *P. putida* DOT-T1E was also shown to exhibit strong chemotaxis to multiple AHC – including mono- and biaromatic compounds such as toluene, propylbenzene, xylene, or naphthalene (Fig. 1; Lacal et al. 2011). *In silico* analyses of the genome sequence of DOT-T1E revealed that this strain encodes 29 chemoreceptors (Udaondo et al. 2012, 2013). Two of these chemoreceptors, McpT1 and McpT2, were shown to be encoded in the self-transmissible 133-Kbp pGRT1 plasmid (Molina et al. 2011), and the sequences of these two chemoreceptors were found to differ in a single amino acid. The DOT-T1E strain (harboring pGRT1) shows a strong form of chemotaxis (referred to as hyperchemotaxis) towards AHC, either in capillary or agarose plug chemotaxis assays (Fig. 1a). This strong

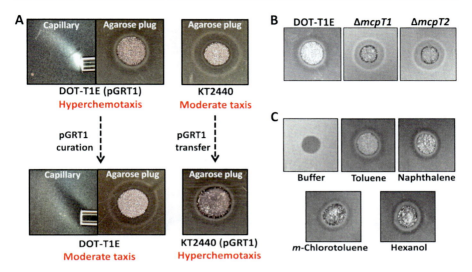

**Fig. 1** Chemotaxis towards aromatic hydrocarbons in *Pseudomonas putida*. (**a**) Capillary and agarose plug chemotaxis assays of *P. putida* strains DOT-T1E and KT2440 towards toluene. (**b**) Chemotaxis towards toluene of the wild type DOT-T1E and mutants in *mcpT1* and *mcpT2*. (**c**) Hyperchemotaxis of DOT-T1E toward different aromatic hydrocarbons. In **b** and **c**, agarose plug chemotaxis assays are used. In **a**–**c**, AHC-filled capillaries or agarose plugs containing chemotaxis buffer (negative control) or different AHC were used. When hyperchemotaxis occurs, a cloud of cells accumulate at the capillary mouth (capillary chemotaxis assay) or form a ring on the surface of the chemoattractant-containing agarose drop (agarose plug chemotaxis assay). Moderate chemotaxis was observed as an accumulation of cells at a defined distance from the capillary mouth or agarose plug

chemotactic behavior was shown to be totally independent of the catabolism of the AHC (Lacal et al. 2011). Importantly, the removal of this plasmid or the mutation of *mcpT1* or *mcpT2* resulted in the disappearance of the hyperchemotaxis phenotype, and the resulting strains show moderate chemotaxis (Fig. 1b) suggesting that hyperchemotaxis is a gene-dose dependent phenomenon. Additionally, the transfer of pGRT1 to *P. putida* KT2440 confers the hyperchemotaxis behavior to this strain indicating that this phenotype may be acquired by the horizontal gene transfer of the pGRT1 plasmid (Fig. 1a; Lacal et al. 2011). A point mutation in the LBD of McPT also abolishes hyperchemotaxis towards AHC – confirming that the activation of the chemoreceptor is initiated by AHC binding at the receptor LBD (Lacal et al. 2011). The use of chemotactically active bacteria has an extraordinary applicability within the bioremediation field since chemoattraction towards toxic pollutants has been demonstrated to increase the efficiency of biodegradation processes (Marx and Aitken 2000a, b; Law and Aitken 2003; Parales et al. 2015). Importantly, DOT-T1E was shown to exhibit hyperchemotaxis to hydrocarbon-rich residues recovered from an oil tanker accident in the Spanish coast, phenotype that was not observed in the *mcpT* mutant (Lacal et al. 2013).

## 3    The TOD Pathway for Catabolism of Aromatic Hydrocarbons

Multiple *Pseudomonas* strains, including *P. putida* DOT-T1E, can utilize benzene, ethylbenzene, and toluene as carbon and energy sources (Mosqueda et al. 1999; Parales et al. 2008). DOT-T1E degrades these AHC to Krebs cycle intermediates through the toluene dioxygenase (TOD) pathway (Zylstra et al. 1988; Mosqueda et al. 1999). The catabolic genes of the TOD pathway are organized in a 9.7-kb operon (*todXFC1C2BADEGIH*) that is transcribed from a single promoter, designed as P$_{todX}$, located immediately upstream from the *todX* gene (Fig. 2a; Mosqueda et al. 1999). The seven-step TOD pathway is initiated by the multicomponent toluene dioxygenase (TDO; encoded by *todC1C2BA*) and terminated by the reactions catalyzed by the TodGHI enzymes, finally resulting in the production of the central metabolism intermediates, pyruvate and acetyl-CoA. Interestingly, TodX is an outer membrane protein suggested to be involved in the transport of AHC from the extracellular environment to the periplasm, as derived from the crystal structure of TodX from *P. putida* F1 (Hearn et al. 2008).

Most aerobic toluene-degrading routes are regulated by one of three existing families of regulators controlling the catabolism of aromatic compounds: (1) LysR transcriptional regulators (Maddocks and Oyston 2008), (2) $\sigma^{54}$-dependent transcriptional activators (Bush and Dixon 2012), and (3) AraC/XylS activators (Gallegos et al. 1997). However, the expression of the *tod* catabolic genes is positively regulated by the TodS/TodT two-component regulatory system (TCS). The *todST* genes are organized in a separate transcriptional unit, constitutively expressed (Mosqueda et al. 1999; Busch et al. 2010), and located immediately downstream of the *tod* catabolic genes (Fig. 2a; Lau et al. 1997; Lacal et al. 2006). TCS allow bacteria to sense multiple environmental signals and mediate changes in gene expression, cellular behavior (e.g., chemotaxis), and biological processes (e.g., catabolism). These systems typically consist of two proteins, an autophosphorylating sensory histidine kinase (HK) and a response regulator (RR). In the basic model of TCS function, signal perception (physical or chemical) by the input domain of the HK modulates its autophosphorylation. When phosphorylated, the phosphoryl group is then transferred to the cognate RR, resulting in conformational changes that activate its output domain, in most cases by altering RR affinities for promoter sequences (Krell et al. 2010; Capra and Laub 2012). TCS-encoding genes are widely distributed within the bacterial kingdom, and the number of TCS per genome has been shown to be elevated in bacteria with complex lifestyles, therefore allowing the recognition of multiple signals and the regulation of multitude of different biological processes (Capra and Laub 2012). In the particular case of DOT-T1E, its genome encodes more than 30 TCS (Udaondo et al. 2012, 2013).

The HK TodS is a 108-kDa protein that belongs to the subfamily of sensor kinases containing two functional modules. These modules are separated by an internal receiver domain and consist of a PAS sensor and histidine kinase domain (Fig. 2b). Its 25-kDa cognate response regulator, TodT, contains an N-terminal input domain for accepting the phosphoryl group from TodS and a C-terminal

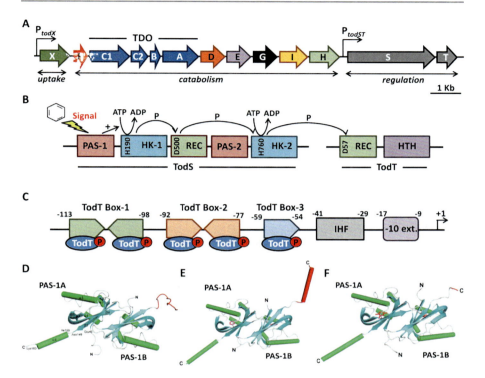

**Fig. 2** Mechanism of regulation of the *tod* degradation pathway mediated by the TodS/TodT two-component system. (**a**) Organization of the *tod* genes in *Pseudomonas putida* DOT-T1E. TDO, toluene deoxygenase. (**b**) Domain organization of TodS/TodT and sequence of phosphoryl group transfer. PAS-1/2, PAS-type sensor domain; HK1/2, histidine kinase domain; REC, response regulator receiver domain; HTH, helix-turn-helix DNA-binding domain. (**c**) Schematic representation of the P*todX* promoter. The three *boxes* (two pseudopalindromes and a half-palindrome) recognized by TodT, the IHF binding site, and the −10 extended region are shown. The binding stoichiometry of TodT is one monomer per half-palindromic element. (**d**–**f**) Crystal structures (crystallizes as dimers) of the PAS-1 domain of TodS in the apo-form (**d**) and in complex with an agonist (**e**; toluene) or an antagonist (**f**; 1,2,4-trimethylbenzene). The binding of the agonist induces a conformational change in the C-terminal region of the PAS domain, shown in *red* (**e**). This conformational change was not observed in the presence of the antagonist (**f**). Toluene and 1,2,4-trimethylbenzene are shown in line mode in *pink* or *red*, respectively (Adapted from Koh et al. (2016) with permission)

helix-turn-helix (HTH) DNA-binding domain (Fig. 2b). Importantly, other TodS/TodT homologues have been identified in several bacterial strains, including StyS/StyR (Leoni et al. 2007) and TmoS/TmoT (Silva-Jiménez et al. 2012) that are involved in the regulation of the styrene or the toluene-4-monooxygenase pathway, respectively.

The TodS/TodT-mediated regulation of the *tod* operon has been extensively studied in *P. putida* F1 and DOT-T1E (Lau et al. 1997; Lacal et al. 2006, 2008a, b; Busch et al. 2007, 2009, 2010; Koh et al. 2016). The expression of the *tod* catabolic genes was shown to be induced by a broad range of AHC including the three Tod pathway substrates, toluene, ethylbenzene, and benzene, as well as other non-substrates such as xylenes or halogen-substituted AHC (Lacal et al. 2006). Ligand recognition occurs by binding to the N-terminal PAS-1 domain, whereas the role of the PAS-2 domain of TodS remains unclear (Lacal et al. 2006; Busch et al. 2007). AHC recognized by TodS were classified as agonists and antagonists based on their ability or inability, respectively, to increase the autophosphorylation activity of TodS (Busch et al. 2007). In the presence of agonists (mono- and biaromatic chemicals such as toluene, benzene, and styrene), the basal level of TodS autophosphorylation increases and the phosphorylation cascade is initiated. This cascade finally results in the transphosphorylation of TodT and an increase in expression from $P_{todX}$ (Fig. 2b). On the contrary, the recognition of antagonists (such as $o$-xylene and other *ortho*-substituted toluenes) by TodS does not stimulate its autophosphorylation and the subsequent induction of the $P_{todX}$ promoter. Recently, the crystal structure of the PAS-1 domain of TodS from *P. putida* F1 revealed that agonistic and antagonistic compounds bind to the same hydrophobic pocket. However, only the binding of an agonist causes a structural change in the signal transfer region involved in signal transmission (Fig. 2d–f; Koh et al. 2016).

The activation of the expression of the *tod* catabolic operon mediated by TodT occurs through the interaction with specific regions of the $P_{todX}$ promoter (Fig. 2c; Lacal et al. 2006, 2008a, b). The $P_{todX}$ promoter is referred to as an "extended promoter" because it has a well-defined $-10$ consensus sequence but no $-35$ consensus (Domínguez-Cuevas and Marqués 2004). TodT was shown to bind to three "TodT boxes" which are centered at base pairs $-106$, $-85$, and $-56$ of the $P_{todX}$ promoter, quite distant from the RNA polymerase binding site (Fig. 2c; Lacal et al. 2008a, b). Maximal expression from $P_{todX}$ also required the binding of the integration host factor (IHF) to a region centered at $-38$ in $P_{todX}$ (Lacal et al. 2006, 2008b). The current regulatory model indicates that IHF induces a bending in the DNA that favors the interaction between TodT and the RNA polymerase. Consequently, the formation of the phosphorylated TodT/IHF/RNA polymerase activation complex results in the induction of the *tod* operon transcription.

## 4 The TOL Pathway for Catabolism of Aromatic Hydrocarbons

The pWW0 TOL plasmid of *Pseudomonas putida* is, in terms of its regulatory and metabolic features, one of the best studied catabolic plasmids. pWW0 contains the genes responsible for the complete biodegradation of toluene and xylenes. These catabolic genes are organized in two adjacent transcriptional units known as the *upper* and *lower* (*meta*) operons. The *upper* operon (*xylUWCMABN*) encodes the enzymes involved in the oxidation of the lateral alkyl chain of toluene/xylenes to their corresponding carboxylic acids, whereas the *lower* operon (*xylXYZLTEGFJQKIH*) encodes the enzymes responsible for the metabolization of benzoates/toluates to

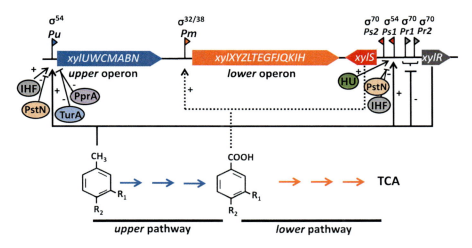

**Fig. 3** Regulatory circuit of the TOL pathway. The genes encoding the *upper* and *lower* pathways are schematically represented, along with those of the regulatory genes *xylS* and *xylR*. The actions of XylR and XylS on the promoters, together with the role of DNA-binding proteins IHF and HU, PprA and TurA in the transcriptional modulation are indicated by *arrows*; *minus* and *plus* signs indicate repression and induction, respectively. Sigma factors required for expression are indicated above each promoter. Recent analyses of the *xylUWCMABN* and *xylXYZLTEGFJQKIH* transcripts by RNA sequencing suggest the absence of additional internal promoters in the *upper* and *lower* operons (Kim et al. 2016)

Krebs cycle intermediates (Fig. 3; Ramos et al. 1997; Silva-Rocha et al. 2011; Kim et al. 2016).

The production of the enzymes encoded by the *upper* and *lower* operons is energetically costly and therefore subjected to complex regulatory circuits that include specific and global regulation in response to hydrocarbons. Expression of the catabolic operons involves the pWW0-encoded XylR and XylS transcriptional regulators, a set of sigma factors ($\sigma^{32}$, $\sigma^{38}$, $\sigma^{54}$ and $\sigma^{70}$), the DNA-bending proteins IHF and HU, and the regulatory proteins TurA and PprA (Fig. 3; Ramos et al. 1997; Silva-Rocha et al. 2011). When bacteria containing the pWW0 plasmid grow in the absence of the substrates of the *upper* or *lower* catabolic pathways, the expression of each operon is turned off until the substrates are present. Thus, toluates (alkylbenzoates) only activate the *lower* pathway; however, xylenes activate both the *upper*- and *lower*-catabolic pathways due to the action of the cascade loop (Fig. 3; Ramos et al. 1997; Silva-Rocha et al. 2011). The expression of the *upper* and *lower* operons is activated by XylR and XylS, respectively – transcriptional regulators that only become activated in the presence of the pathway inducers (Ramos et al. 1997; Silva-Rocha et al. 2011). Briefly, in the presence of toluene/xylenes, XylR directly binds these AHC and drives transcription of the *xylUWCMABN* genes from the *Pu* promoter to convert toluene and *p*- and *m*-xylenes into the corresponding benzoates. Additionally, XylR also increases the expression of *xylS* by inducing transcription from the promoter *Ps1*. Subsequently, this induction results in the transcriptional activation of the *Pm* promoter and the induction of the *lower*-cleavage pathway operon for the oxidation of benzoates to Krebs cycle intermediates (Fig. 3; Ramos et al. 1997; Silva-Rocha et al. 2011). However, the adaptive response of pWW0-containing bacteria to the presence of AHC is also controlled by additional host factors. Thus, the expression of *xylR* is driven from two $\sigma^{70}$-dependent promoters, *Pr1* and *Pr2*. When cells are growing in the presence of toluene/xylenes, the binding of AHC to XylR causes its activation and is consequently able to drive transcription from the *Pu* promoter of the *upper* pathway (Fig. 3; Ramos et al. 1997; Silva-Rocha et al. 2011). The initiation of this transcription from the *Pu* promoter also requires IHF and $\sigma^{54}$ RNA polymerase. On the other hand, the *xylS* gene is constitutively expressed at low levels from its $\sigma^{70}$-dependent promoter, *Ps2*. However, the transcription of *xylS* is further induced by activated XylR and the DNA-bending protein HU that act together to upregulate transcription of the *xylS* gene from a second $\sigma^{54}$-dependent promoter, *Ps1*. Contrary to the transcription from the *Pu* promoter, transcription from *Ps1* was shown to be repressed by IHF (Fig. 3; Ramos et al. 1997; Marqués et al. 1998). The *Pr1/2* and *Ps1/2* promoters are located in the intergenic region between the divergently transcribed *xylR* and *xylS* genes. This proximity complicates the regulatory processes because when the activated XylR binds to the upstream activator sequences of the *Ps1* promoter of *xylS* it blocks access of RNA polymerase to its own *Pr1* and *Pr2* $\sigma^{70}$-dependent promoters, hence reducing its own expression. In the presence of benzoates the *lower*-pathway is rendered fully operational because *xylS* is only transcribed from the constitutive XylR-independent promoter (*Ps2*). Benzoate interacts with XylS leading to the transcriptional activation of the *Pm* promoter in the *lower*-cleavage pathway (Fig. 3).

## 5  Catabolite Repression and Expression of the TOL Plasmid Genes

Catabolite repression (CR) defines the ability of an organism to preferentially metabolize one carbon source over another when both carbon sources are present in the environment (Kremling et al. 2015). CR plays an important role in the control of the expression of the TOL plasmid pathways. However, the CR observed in *P. putida* is normally referred to as "crossed catabolite repression" because the bacteria are able to use two carbon sources (e.g., glucose and toluene) simultaneously (del Castillo and Ramos 2007). This type of control is similar to that observed in *Klebsiella oxytoca,* which can use both sucrose and glycerol simultaneously (Piñar et al. 1998) but is in stark contrast to the strict catabolite repression of *E. coli* since it preferentially consumes glucose prior to the use of lactose (Kremling et al. 2015).

Interestingly, transcription of the TOL catabolic operons is repressed when bacteria are grown in the presence of alternative carbon sources. Thus, when cells are grown in the presence of acetate and *m*-xylene, bacteria contain significantly lower levels of the *upper* pathway proteins than the same cells grown in the presence of *m*-xylene as the sole carbon source (Worsey and Williams 1975). A similar repression was observed when cells are grown with succinate and *o*-xylene since under these conditions of nutrient excess, bacterial cells were unable to induce expression of the TOL catabolic systems (Duetz et al. 1996).

Cyclic AMP (cAMP) is the signal molecule for catabolite repression in *Enterobacteriaceae*, where cAMP levels vary depending on the concentration of glucose and other metabolites in the growth environment (Kremling et al. 2015). However, the cAMP levels in *P. putida* remain stable despite the growth conditions indicating that this second messenger is not involved in catabolite repression in this Pseudomonad (Phillips and Mulfinger 1981; Rojo and Dinamarca 2004). Instead, CR of the TOL plasmid genes in *P. putida* involves the regulators PstN and Crc. PstN has been shown to repress the expression from *Pu* and *Ps1* promoters in rich medium or in the presence of glucose, indicating that PstN prevents the activated XylR from binding to the upstream activator sequences (Aranda-Olmedo et al. 2005, 2006). On the other hand, the global regulator of carbon metabolism, Crc, directly modulates the expression of the TOL genes at the post-transcriptional level. Thus, Crc inhibits the translation not only of the XylR and XylS regulators but also of specific structural genes of the *upper* and *lower* operons. As a result, the levels of the TOL-catabolic proteins are optimized depending on the environmental conditions (Moreno et al. 2010).

Genetic evidence in *P. putida* indicates that the repression caused by glucose on the *Pu* promoter is exerted by catabolites of the Entner-Doudoroff pathway, responsible for the catabolism of glucose to pyruvate. *P. putida* synthesizes 6-phosphogluconate (6PG) by three converging pathways: (i) the glucose dehydrogenase (Gcd) route, (ii) the glucokinase (Glk) route, and (iii) via direct phosphorylation of gluconate mediated by gluconokinase (del Castillo and Ramos 2007). Interestingly, toluene and glucose have mutual regulatory effects in wild-type *P. putida* and also in mutant strains which lack either the Gcd or Glk pathways. Glucose causes repression of toluene metabolism by signaling through 2-dehydro-3-deoxygluconate-6-phosphate and the PtsN global regulator, while toluene causes repression of glucose metabolism via the Glk pathway with the help of the Crc regulator protein. The Crc protein appears to act as a switch for the Glk pathway because in *crc* mutant backgrounds basal glucokinase levels are not repressed in the presence of toluene (del Castillo and Ramos 2007).

# 6    Conclusions and Research Needs

Microorganisms have evolved multiple mechanisms for the sensing, tolerance, and degradation of hydrocarbons. The efficient control of these processes is generally achieved by chemoreceptor-based signaling pathways and one- and two-component systems (Galperin 2005; Krell et al. 2010). Chemotaxis towards aromatic compounds, including AHC, has been demonstrated in multiple bacterial strains (Lacal et al. 2011; Parales et al. 2015; Huang et al. 2016). However, not all the degrading strains were able to show chemotactic responses to hydrocarbons (Parales et al. 2000). Chemotaxis to AHC was shown to increase xenobiotics bioavailability and to increase the rate of their biodegradation. This in turn has important implications in applied biodegradation research since it may lay the groundwork for synthetic biology approaches that allow the engineering of bacterial strains with increased biodegradation properties by conferring or enhancing the capacity to move chemotactically to xenobiotics (Krell et al. 2012; Lacal et al. 2013; Parales et al. 2015).

Although the expression of the *mcpT* genes in DOT-T1E is constitutive (Lacal et al. 2011), studies in multiple bacterial strains provided evidence that genes involved in chemotaxis and biodegradation are coregulated (Parales et al. 2015). In the particular case of the TOL and TOD catabolic pathways in *Pseudomonas putida,* they are regulated by one- (XylS and XylR) and two-component (TodS/TodT) systems, respectively. Phylogenetic analyses indicate that TCS evolved from one-component systems, and it is generally thought that they sense extracytosolic and cytosolic signals, respectively (Krell et al. 2010; Wuichet et al. 2010). Indeed, 88% of TCS sensor kinases were predicted to be transmembrane proteins consistent with the notion that they may sense signals in the extracytoplasmic space. The reason why some hydrocarbon degradation pathways are regulated by one-component systems and others by TCS is unknown, but clearly both systems are efficient in the control of their corresponding gene circuits. Many questions remain to be answered on the mechanism of action of the AHC responsive regulators such as: (i) the determination of the role of PAS-2 domain of TodS in the signaling transduction cascade; (ii) to delve into the observed catabolite repression of the TOD pathway (Busch et al. 2010); (iii) to decipher the conformational changes in TodT following transphosphorylation by TodS; and (iv) to clarify whether Crc directly regulates the glucokinase pathway at the posttranscriptional level or indirectly through additional regulators. Answers to these questions will allow to fully understand the adaptive responses of bacteria to the presence of AHC in their environment.

## References

Aranda-Olmedo I, Ramos JL, Marqués S (2005) Integration of signals through Crc and PtsN in catabolite repression of *Pseudomonas putida* TOL plasmid pWW0. Appl Environ Microbiol 71:4191–4198

Aranda-Olmedo I, Marín P, Ramos JL, Marqués S (2006) Role of the *ptsN* gene product in catabolite repression of the *Pseudomonas putida* TOL toluene degradation pathway in chemostat cultures. Appl Environ Microbiol 72:7418–7721

Baumgarten T, Vazquez J, Bastisch C, Veron W, Feuilloley MG, Nietzsche S, Wick LY, Heipieper HJ (2012) Alkanols and chlorophenols cause different physiological adaptive responses on the level of cell surface properties and membrane vesicle formation in *Pseudomonas putida* DOT-T1E. Appl Microbiol Biotechnol 93:837–845

Busch A, Lacal J, Marcos A, Ramos JL, Krell T (2007) Bacterial sensor kinase TodS interacts with agonistic and antagonistic signals. Proc Natl Acad Sci U S A 104:13774–13779

Busch A, Guazzaroni ME, Lacal J, Ramos JL, Krell T (2009) The sensor kinase TodS operates by a multiple step phosphorelay mechanism involving two autokinase domains. J Biol Chem 284:10353–10360

Busch A, Lacal J, Silva-Jímenez H, Krell T, Ramos JL (2010) Catabolite repression of the TodS/ TodT two-component system and effector-dependent transphosphorylation of TodT as the basis for toluene dioxygenase catabolic pathway control. J Bacteriol 192:4246–4250

Bush M, Dixon R (2012) The role of bacterial enhancer binding proteins as specialized activators of $\sigma^{54}$-dependent transcription. Microbiol Mol Biol Rev 76:497–529

Capra EJ, Laub MT (2012) Evolution of two-component signal transduction systems. Annu Rev Microbiol 66:325–347

Collins KD, Lacal J, Ottemann KM (2014) Internal sense of direction: sensing and signaling from cytoplasmic chemoreceptors. Microbiol Mol Biol Rev 78:672–684

Dagley S (1981) New perspectives in aromatic catabolism. In: Leisinger T, Cook AM, Hütter R, Nüesch J (eds) Degradation of xenobiotics and recalcitrant compounds. Academic Press, New York, pp 181–186

del Castillo T, Ramos JL (2007) Simultaneous catabolite repression between glucose and toluene metabolism in Pseudomonas putida is channelled through different signalling pathways. J Bacteriol 189:6602–6610

Domínguez-Cuevas P, Marqués S (2004) Compiling sigma-70 dependent promoters. In: Ramos JL (ed) Pseudomonas: virulence and gene regulation. Springer, New York, pp 319–345

Duetz WA, Marqués S, Wind B, Ramos JL, van Andel JG (1996) Catabolite repression of the toluene degradation pathway in Pseudomonas putida harboring pWW0 under various conditions of nutrient limitation in chemostat culture. Appl Environ Microbiol 62:601–606

Fuchs G, Boll M, Heider J (2011) Microbial degradation of aromatic compounds – from one strategy to four. Nat Rev Microbiol 9:803–816

Fuentes S, Méndez V, Aguila P, Seeger M (2014) Bioremediation of petroleum hydrocarbons: catabolic genes, microbial communities, and applications. Appl Microbiol Biotechnol 98:4781–4794

Gallegos MT, Schleif R, Bairoch A, Hofmann K, Ramos JL (1997) Arac/XylS family of transcriptional regulators. Microbiol Mol Biol Rev 61:393–410

Galperin MY (2005) A census of membrane-bound and intracellular signal transduction proteins in bacteria: bacterial IQ, extroverts and introverts. BMC Microbiol 5:35

Gibson DT, Subramanian V (1984) Microbial degradation of aromatic hydrocarbons. In: Gibson DT (ed) Microbial degradation of organic compounds. Marcel Dekker, New York, pp 361–369

Grimm AC, Harwood CS (1997) Chemotaxis of Pseudomonas spp. to the polyaromatic hydrocarbon naphthalene. Appl Environ Microbiol 63:4111–4115

Grimm AC, Harwood CS (1999) NahY, a catabolic plasmid-encoded receptor required for chemotaxis of Pseudomonas putida to the aromatic hydrocarbon naphthalene. J Bacteriol 181:3310–3316

Harwood CS, Rivelli M, Ornston LN (1984) Aromatic acids are chemoattractants for Pseudomonas putida. J Bacteriol 160:622–628

Harwood CS, Parales RE, Dispensa M (1990) Chemotaxis of Pseudomonas putida toward chlorinated benzoates. Appl Environ Microbiol 56:1501–1503

Hearn EM, Patel DR, van den Berg B (2008) Outer-membrane transport of aromatic hydrocarbons as a first step in biodegradation. Proc Natl Acad Sci U S A 105:8601–8606

Huang Z, Ni B, Jiang CY, Wu YF, He YZ, Parales RE, Liu SJ (2016) Direct sensing and signal transduction during bacterial chemotaxis towards aromatic compounds in Comamonas testosteroni. Molecular Microbiology 101(2):224–237. https://doi.org/10.1111/mmi.13385

Inoue A, Horikoshi K (1989) A Pseudomonas that thrives in high concentration of toluene. Nature 338:264–266

Kim J, Pérez-Pantoja D, Silva-Rocha R, Oliveros JC, de Lorenzo V (2016) High-resolution analysis of the m-xylene/toluene biodegradation subtranscriptome of Pseudomonas putida mt-2. Environ Microbiol. Environ Microbiol 18(10):3327–3341. https://doi.org/10.1111/1462-2920.13054

Koh S, Hwang J, Guchhait K, Lee EG, Kim SY, Kim S, Lee S, Chung JM, Jung HS, Lee SJ, Ryu CM, Lee SG, TK O, Kwon O, Kim MH (2016) Molecular insights into toluene sensing in the TodS/TodT signal transduction system. J Biol Chem 291:8575–8590

Krell T, Lacal J, Busch A, Silva-Jiménez H, Guazzaroni ME, Ramos JL (2010) Bacterial sensor kinases: diversity in the recognition of environmental signals. Annu Rev Microbiol 64: 539–559

Krell T, Lacal J, Guazzaroni ME, Busch A, Silva-Jiménez H, Fillet S, Reyes-Darías JA, Muñoz-Martínez F, Rico-Jiménez M, García-Fontana C, Duque E, Segura A, Ramos JL (2012) Responses of *Pseudomonas putida* to toxic aromatic carbón sources J Biotechnol 160:25–32

Kremling A, Geiselmann J, Ropers D, de Jong H (2015) Understanding carbon catabolite repression in *Escherichia coli* using quantitative models. Trends Microbiol 23:99–109

Lacal J, Busch A, Guazzaroni ME, Krell T, Ramos JL (2006) The TodS–TodT two-component regulatory system recognizes a wide range of effectors and works with DNA-bending proteins. Proc Natl Acad Sci U S A 103:8191–8196

Lacal J, Guazzaroni ME, Busch A, Krell T, Ramos JL (2008a) Hierarchical binding of the TodT response regulator to its multiple recognition sites at the tod pathway operon promoter. J Mol Biol 376:325–337

Lacal J, Guazzaroni ME, Gutiérrez-del-Arroyo P, Busch A, Vélez M, Krell T, Ramos JL (2008b) Two levels of cooperativeness in the binding of TodT to the *tod* operon promoter. J Mol Biol 384:1037–1047

Lacal J, Muñoz-Martínez F, Reyes-Darías JA, Duque E, Matilla M, Segura A, Calvo JJ, Jímenez-Sánchez C, Krell T, Ramos JL (2011) Bacterial chemotaxis towards aromatic hydrocarbons in *Pseudomonas*. Environ Microbiol 13:1733–1744

Lacal J, Reyes-Darias JA, García-Fontana C, Ramos JL, Krell T (2013) Tactic responses to pollutants and their potential to increase biodegradation efficiency. J Appl Microbiol 114:923–933

Lau PC, Wang Y, Patel A, Labbé D, Bergeron H, Brousseau R, Konishi Y, Rawlings M (1997) A bacterial basic region leucine zipper histidine kinase regulating toluene degradation. Proc Natl Acad Sci U S A 94:1453–1458

Law AM, Aitken MD (2003) Bacterial chemotaxis to naphthalene desorbing from a nonaqueous liquid. Appl Environ Microbiol 69:5968–5973

Leoni L, Rampioni G, Zennaro E (2007) Styrene, an unpalatable substrate with complex regulatory networks. In: Ramos JL, Filloux A (eds) *Pseudomonas*: a model system in biology. Springer, Dorchester, pp 59–88

Maddocks SE, Oyston PC (2008) Structure and function of the LysR-type transcriptional regulator (LTTR) family proteins. Microbiology 154:3609–3623

Marqués S, Gallegos MT, Manzanera M, Holtel A, Timmis KN, Ramos JL (1998) Activation and repression of transcription at the double tandem divergent promoters for the *xylR* and *xylS* genes of the TOL plasmid of *Pseudomonas putida*. J Bacteriol 180:2889–2894

Marx RB, Aitken MD (2000a) Bacterial chemotaxis enhances naphthalene degradation in a heterogeneous system. Environ Sci Technol 34:3379–3383

Marx RB, Aitken MD (2000b) A material-balance approach for modeling bacterial chemotaxis to a consumable substrate in the capillary assay. Biotechnol Bioeng 68:308–315

Molina L, Duque E, Gómez MJ, Krell T, Lacal J, García-Puente A, García V, Matilla MA, Ramos JL, Segura A (2011) The pGRT1 plasmid of *Pseudomonas putida* DOT-T1E encodes functions relevant for survival under harsh conditions in the environment. Environ Microbiol 13:2315–2327

Moreno R, Fonseca P, Rojo F (2010) The Crc global regulator inhibits the *Pseudomonas putida* pWW0 toluene/xylene assimilation pathway by repressing the translation of regulatory and structural genes. J Biol Chem 285:24412–24419

Mosqueda G, Ramos-González MI, Ramos JL (1999) Toluene metabolism by the solvent-tolerant *Pseudomonas putida* DOT-T1 strain, and its role in solvent impermeabilization. Gene 232:69–76

Parales RE, Ditty JL, Harwood CS (2000) Toluene-degrading bacteria are chemotactic towards the environmental pollutants benzene, toluene, and trichloroethylene. Appl Environ Microbiol 66:4098–4104

Parales RE, Parales JV, Pelletier DA, Ditty JL (2008) Diversity of microbial toluene degradation pathways. Adv Appl Microbiol 64:1–73

Parales RE, Luu RA, Hughes JG, Ditty JL (2015) Bacterial chemotaxis to xenobiotic chemicals and naturally-occurring analogs. Curr Opin Biotechnol 33:318–326

Parkinson JS, Hazelbauer GL, Falke JJ (2015) Signaling and sensory adaptation in *Escherichia coli* chemoreceptors: 2015 update. Trends Microbiol 23:257–266

Phillips AT, Mulfinger LM (1981) Cyclic adenosine $3',5'$-monophosphate levels in *Pseudomonas putida* and *Pseudomonas aeruginosa* during induction and carbon catabolite repression of histidase synthesis. J Bacteriol 145:1286–1292

Piñar G, Kovárová K, Egli T, Ramos JL (1998) Influence of carbon source of nitrate removal by nitratetolerant *Klebsiella oxytoca* CECT 4460 in batch and chemostat cultures. Appl Environ Microbiol 64:2970–2976

Pohanish RP (2011) Sittig's handbook of toxic and hazardous chemicals and carcinogens, 6th edn. William Andrew-Elsevier, Norwich

Ramos JL, Duque E, Huertas MJ, Haïdour A (1995) Isolation and expansion of the catabolic potential of a *Pseudomonas putida* strain able to grow in the presence of high concentrations of aromatic hydrocarbons. J Bacteriol 177:3911–3916

Ramos JL, Marqués S, Timmis KN (1997) Transcriptional control of the Pseudomonas TOL plasmid catabolic operons is achieved through an interplay of host factors and plasmid-encoded regulators. Annu Rev Microbiol 51:341–373

Ramos JL, Duque E, van Dillewijn P, Daniels C, Krell T, Espinosa-Urgel M, Ramos-González MI, Rodríguez S, Matilla MA, Wittich R, Segura A (2010) Removal of hydrocarbons and other related chemicals via the rhizosphere of plants. In: Timmis KN (ed) Handbook of hydrocarbon and lipid microbiology. Springer, Berlin, pp 2575–2581

Ramos JL, Sol Cuenca M, Molina-Santiago C, Segura A, Duque E, Gómez-García MR, Udaondo Z, Roca A (2015) Mechanisms of solvent resistance mediated by interplay of cellular factors in *Pseudomonas putida*. FEMS Microbiol Rev 39:555–566

Rojo F, Dinamarca A (2004) Catabolite repression and physiological control. In: Ramos JL (ed) *Pseudomonas*: virulence and gene regulation. Springer, New York, pp 365–387

Sampedro I, Parales RE, Krell T, Hill JE (2015) *Pseudomonas* chemotaxis. FEMS Microbiol Rev 39:17–46

Segura A, Ramos JL (2014) Toluene tolerance systems in *Pseudomonas*. In: Nojiri H, Tsuda M, Fukuda M, Kamagata Y (eds) Biodegradative bacteria: how bacteria degrade, survive, adapt, and evolve. Springer, Tokyo, pp 227–248

Segura A, Molina L, Fillet S, Krell T, Bernal P, Muñoz-Rojas J, Ramos JL (2012) Solvent tolerance in Gram-negative bacteria. Curr Opin Biotechnol 23:415–421

Silva-Jiménez H, García-Fontana C, Cadirci BH, Ramos-González MI, Ramos JL, Krell T (2012) Study of the TmoS/TmoT two-component system: towards the functional characterization of the family of TodS/TodT like systems. Microb Biotechnol 5:489–500

Silva-Rocha R, Tamames J, dos Santos VM, de Lorenzo V (2011) The logicome of environmental bacteria: merging catabolic and regulatory events with Boolean formalisms. Environ Microbiol 13:2389–2402

Udaondo Z, Duque E, Fernández M, Molina L, de la Torre J, Bernal P, Niqui JL, Pini C, Roca A, Matilla MA, Molina-Henares MA, Silva-Jiménez H, Navarro-Avilés G, Busch A, Lacal J, Krell T, Segura A, Ramos JL (2012) Analysis of solvent tolerance in *Pseudomonas putida* DOT-T1E based on its genome sequence and a collection of mutants. FEBS Lett 586:2932–2938

Udaondo Z, Molina L, Daniels C, Gómez MJ, Molina-Henares MA, Matilla MA, Roca A, Fernández M, Duque E, Segura A, Ramos JL (2013) Metabolic potential of the organic-solvent tolerant *Pseudomonas putida* DOT-T1E deduced from its annotated genome. Microb Biotechnol 6:598–611

Worsey MJ, Williams PA (1975) Metabolism of toluene and xylenes by *Pseudomonas* (*putida*) (*arvilla*) mt-2: evidence for a new function of the TOL plasmid. J Bacteriol 124:7–13

Wuichet K, Cantwell BJ, Zhulin IB (2010) Evolution and phyletic distribution of two-component signal transduction systems. Curr Opin Microbiol 13:219–225

Zylstra GJ, McCombie WR, Gibson DT, Finette BA (1988) Toluene degradation by *Pseudomonas putida* F1: genetic organization of the tod operon. Appl Environ Microbiol 54:1498–1503

# Extrusion Pumps for Hydrocarbons: An Efficient Evolutionary Strategy to Confer Resistance to Hydrocarbons

# 23

Matilde Fernández, Craig Daniels, Vanina García, Bilge Hilal Cadirci, Ana Segura, Juan Luis Ramos, and Tino Krell

## Contents

| | | |
|---|---|---|
| 1 | Introduction | 362 |
| 2 | Hydrocarbon Efflux Pumps Belong to the RND Family | 363 |
| 3 | The Presence of Hydrocarbons in the Cytosol Induces Transcription of Pumps Which Possibly Extrude Compounds from the Periplasm | 365 |
| 4 | Substrate Promiscuity of RND Pumps | 366 |
| 5 | A Rotating Mechanism: A Common Theme for Membrane Proteins? | 367 |
| 6 | RND Pumps Are Proton Antiporters | 368 |
| 7 | Open Questions and Future Work | 369 |
| 8 | Research Needs | 369 |
| References | | 369 |

M. Fernández · A. Segura · J. L. Ramos · T. Krell (✉)
Department of Environmental Protection, Estación Experimental del Zaidín, Consejo Superior de Investigaciones Científicas, Granada, Spain
e-mail: matilde.fernandez@csic.es; ana.segura@eez.csic.es; juanluis.ramos@eez.csic.es; tino.krell@eez.csic.es

C. Daniels
Developmental and Stem Cell Biology Program, Brain Tumour Research Centre, The Hospital for Sick Children, Toronto, ON, Canada
e-mail: craig.daniels@sickkids.ca

V. García
The University of Nottingham, Nottingham, Nottinghamshire, UK
e-mail: Vanina.Garcia@nottingham.ac.uk

B. H. Cadirci
Department of Bioengineering, Gaziosmanpasa University, Tokat, Turkey
e-mail: bilgehilal.cadirci@gop.edu.tr

© Springer International Publishing AG, part of Springer Nature 2018
T. Krell (ed.), *Cellular Ecophysiology of Microbe: Hydrocarbon and Lipid Interactions*,
Handbook of Hydrocarbon and Lipid Microbiology, https://doi.org/10.1007/978-3-319-50542-8_47

**Abstract**

Efflux pumps of the RND family are primarily involved in the extrusion of hydrocarbons. These pumps, specific to gram-negative bacteria, are composed of three components. Two components are transmembrane proteins located in the inner and outer membrane whereas the third one spans the periplasm connecting the other two subunits. The large part of information available on RND pumps is related to their capacity to extrude antibiotics. Structural data indicate that substrate binding may occur preferentially in the periplasm at the inner membrane protein.

# 1    Introduction

The toxicity of hydrocarbons to living organisms is related to their capacity to preferentially dissolve in a hydrophobic rather than a hydrophilic environment. Upon contact of bacteria with hydrocarbons, these compounds preferentially dissolve into the cell membrane leading to its disorganization, impairing vital membrane associated processes such as ATP synthesis which eventually leads to cell death (Ramos et al. 2002; Sikkema et al. 1995).

There are three basic potential defense strategies against the action of hydrocarbons: (1) to degrade internalized hydrocarbons, (2) to reduce their entry, or (3) to promote their expulsion. Evolution has made use of all three possibilities. However, their contribution to achieving solvent resistance differs significantly. Various bacteria which are able to degrade hydrocarbons have been described. This is exemplified by the degradation of aromatic and aliphatic compounds by different strains of *Pseudomonas, Flavobacterium, Alcaligenes,* or *Rhodococcus sp.* However, the analyses of mutant strains deficient in the degradation machinery show that the contribution of degradation to solvent tolerance is modest and that the major evolutionary advantage obtained from the degradation of hydrocarbons is the possibility to use these compounds as carbon and energy sources (Daniels et al. 2009). Several mechanisms have been reported which correspond to the second evolutionary strategy, the reduction of the entry of hydrocarbons. The physical properties of membranes can be altered in a way that they are less permissive to the dissolution of hydrocarbons. This can be achieved by changes in the degree of saturation of fatty acids, *cis/trans* isomerization of unsaturated fatty acids and cyclopropanation of bacterial membrane lipids. However, there is no doubt that the third strategy, the expulsion of hydrocarbons, is the mechanism which quantitatively makes the largest contribution to hydrocarbon tolerance. This is exemplified by the *Pseudomonas putida* strains DOT-T1E and S12 that have the extraordinary capacity to live in a toluene-saturated growth medium (Isken and de Bont 1996; Ramos et al. 1995). These bacteria are able to survive a sudden exposure to 0.3% (v/v) toluene when precultured in low concentrations of toluene (note: preculture in low toluene concentration induces transcription of efflux pumps). The analysis of mutants deficient in one or several efflux pumps showed a dramatic effect on the solvent tolerance of these bacteria. The hydrocarbon efflux pumps TtgGHI of DOT-T1E and SrpABC of S12 were found to be of central importance in maintaining solvent tolerance, since mutants deficient in these transporters exhibited a reduction of several orders in magnitude in the bacterial survival rate following sudden toluene exposure (Kieboom et al. 1998; Rojas et al. 2001). Another efflux pump, formed by MexEF/OprN, was found to be involved in trinitrotoluene tolerance in *P. putida* KT2440 (Fernández et al. 2009).

## 2    Hydrocarbon Efflux Pumps Belong to the RND Family

The efflux pumps thus far characterized that expulse hydrocarbons were found to belong to the RND (*root- nodulation cell division*) family of pumps. Other members of this family were found to transport antibiotics and are associated with a multidrug resistance (MDR) phenotype. Due to the enormous clinical importance of MDR, the majority of studies on RND pumps focus on their capacity to extrude antibiotics. It is generally assumed that RND mediated expulsion of antibiotics and hydrocarbons occur in an analogous fashion.

An RND pump consists of three major components; an integral inner membrane component that forms direct contacts with an integral outer membrane protein. In addition, a periplasmic adaptor protein appears to stabilize this interaction (Fig. 1a). The adaptor protein is palmityolated at its N-terminal extension and in this way anchored to the inner membrane. The most studied RND pump at the structural and functional level is the AcrAB-TolC system of *E. coli*. This pump was initially described as a transporter for the topical antiseptic acriflavin (hence the name Acr) but was shown to transport a large variety of other substrates (Nikaido 1998).

The three dimensional structures of the three individual components of the AcrAB-TolC RND pump have been solved. An interesting feature is that the two integral membrane proteins AcrB (inner membrane) and TolC (outer membrane) protrude well into the periplasmic space where both proteins are thought to contact; a view supported by in vivo cross-linking experiments of both proteins (Symmons et al. 2009; Touzé et al. 2004) and by surface plasmon resonance experiments (Tikhonova et al. 2011), suggesting a tip to tip interaction. This contrasts with electron microscopy analyses that suggested that there is no direct interaction between AcrB and TolC (Du et al. 2014; Xu et al. 2011). Since the adaptor protein AcrA is able to bind to individual AcrB and TolC, structural models for subunit assembly, such as the one depicted in Fig. 1a, have been established. However, there are currently several models available for the assembly of an RND pump. Since the inner and outer membrane proteins crystallized as trimers, all models coincide that an RND pump contains a trimer of each protein. In contrast, the adaptor protein was found to be monomeric in solution at neutral pH, dimeric in crystals grown at acidic pH, and associated as oligomers in crystals grown at neutral pH. The diversity of experimentally observed oligomeric states leaves uncertain the true physiological oligomeric state of the adaptor protein. Hence the models on the arrangement of the RND pump differ in the number of adaptor proteins which are proposed to be 3 or

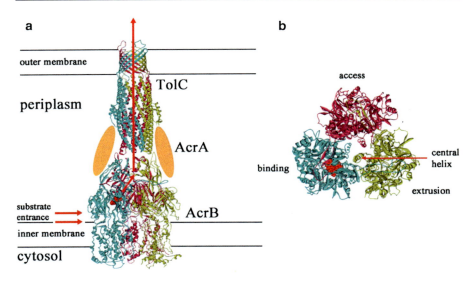

**Fig. 1** The high resolution structures of RND pump subunits provide clues on its molecular mechanism. *Left*: Model of the assembly of the RND pump AcrAB-TolC. Each subunit of the inner membrane protein AcrB and the outer membrane protein TolC is colored differently. The number of adaptor proteins per RND pump has not been determined and therefore AcrA is shown schematically. According to Murakami et al. (2002), only one monomer of AcrB contains bound substrate as illustrated by the bound antibiotic minocycline, which is shown in *red*. The proposed path for substrate expulsion is indicated. Substrate entrance was proposed to occur either from the outer layer of the inner membrane or from the periplasmic space. *Right*: Top view of the AcrB trimer. Bound minocycline is only found in the subunit which is in the "binding conformation." The arrow marks the central helix which opens and closes the release path of substrate into the central pore

6 per functional RND pump, giving rise to AcrB-AcrA-TolC stoichiometries of 3:3:3 or 3:6:3 (Hayashi et al. 2015). In some cases the function of AcrA can be replaced by a paralogue that forms part of another RND pump. For example in *Salmonella typhimurium* AcrA can be replaced by its paralogue AcrE that belongs to the AcrEF pump and shares 68% of protein sequence identity with AcrA (Smith and Blair 2014).

X-ray crystallographic studies of AcrB provided evidence for a fourth component of an RND pump. One copy of helical protein YajC was found to be bound to each monomer of AcrB. This protein is immersed in the membrane making contacts with the transmembrane regions of AcrB. Bacterial mutants lacking YajC showed a modest increase in the susceptibility towards different antibiotics. The protein YajC as well as its binding site at AcrB are highly conserved (Törnroth-Horsefield et al. 2007). However, the contribution of YajC to the function of an RND pump still remains to be established.

The complete physiological role of RND pumps is not well established, but it could be related to both, the extrusion of intracellularly generated compounds and to the defense against toxic compounds present in the environment. The extrusion of a wide range of toxic compounds could be a lateral or a specifically evolved function related to the ecological fitness of bacteria. Whatever it may be, RND pumps have evolved to be multidrug recognition proteins, which implies that a range of structurally different but related molecules serve as substrates for a given pump. However, multidrug recognition is a trade-off between affinity and specificity. This implies that a given pump can only recognize a given number of substrates with physiological affinity. To circumvent these intrinsic chemical limitations of multidrug recognition, bacteria possess various copies of RND pumps with different and in many cases overlapping substrate specificities. *P. putida* DOT-T1E was shown to possess three major RND pumps for hydrocarbons. The TtgABC pump acts primarily on antibiotics and plant secondary metabolites, whereas the TtgDEF and TtgGHI pumps transport mainly organic solvents.

## 3    The Presence of Hydrocarbons in the Cytosol Induces Transcription of Pumps Which Possibly Extrude Compounds from the Periplasm

There is little doubt that the inner membrane protein is the site of substrate recognition for an RND pump. Initial information on the substrate binding region was obtained by the analysis of chimeric forms of the inner membrane protein in which the periplasmic and the transmembrane segments of inner membrane proteins with different substrate profiles were exchanged. It became obvious that the substrate specificity was always associated with the periplasmic part of the inner membrane protein. Based on this observation, it was suggested that RND pumps expulse substrate from the periplasm (Eda et al. 2003; Yu et al. 2005).

A large series of X-ray crystallographic structures of AcrB in complex with different substrates have been solved. However, the ensemble of structures does not identify unambiguously the substrate binding site. Substrates were found to be present at different sites in the periplasmic part of the protein either some 25 Å away from the membrane (Murakami et al. 2002; Yu et al. 2005) or in the immediate vicinity of the membrane (Törnroth-Horsefield et al. 2007) or at the upper part of the central channel formed by the transmembrane helices of the three AcrB monomers (Yu et al. 2003). Since this channel opens to the cytoplasm, the latter report is thus consistent with a substrate expulsion from this cellular compartment, whereas the remaining authors conclude from their studies that substrates bind to the pump in the periplasm. A more recent model proposes an entrance via the inner membrane for hydrophobic compounds as well as an entrance from the periplasm for high molecular weight hydrophilic compounds (Yamaguchi et al. 2015).

The idea that hydrocarbons are expulsed from the periplasm rather than from the cytoplasm has been interpreted as an elegant evolutionary strategy to protect Gramnegative bacteria against toxic compounds. However, the transcription of RND pump genes was found to be under a tight specific control. This has been particularly well studied for the solvent efflux pumps TtgDEF and TtgGHI of *Pseudomonas putida* DOT-T1E (Guazzaroni et al. 2005) which are under the control of the repressor proteins TtgV and TtgT. In the absence of solvents, the transcription of pump genes is kept at a low level; its control being mediated by promoter bound repressor proteins. The binding of hydrocarbons, which are generally pump substrates, to the repressor proteins induce their dissociation from the DNA, promoting transcriptional activation. However, the affinity of these hydrocarbons for the repressor protein is relatively modest. This is exemplified by a $K_D$ of around 120 μM for the binding of toluene to the repressor which regulates the expression of the above mentioned pumps (Guazzaroni et al. 2005). The analysis of bacterial mutants lacking these pumps has clearly indicated that they are responsible for the tolerance of the bacterium towards toluene. Although much of the available data point towards the extrusion of hydrocarbons from the periplasm, toxic compounds need to cross the inner membrane to guarantee the transcriptional activation of pump expression. The compound-induced upregulation of pump production will then promote compound expulsion leading to a steady-state equilibrium of compound entry and expulsion.

In this context the question to be raised concerns the fact that RND pump expression is primarily controlled by cytoplasmic one-component regulatory systems. An alternative would have been to put pump expression under the control of two-component regulator systems, which frequently have periplasmic sensor domains permitting transcriptional control in response to the presence of hydrocarbons in the periplasm. This option, however, has been pursued during evolution only in a limited number of cases.

## 4     Substrate Promiscuity of RND Pumps

Apart from antibiotics and organic solvents, members of the RND efflux pumps were also found to expulse dyes, detergents, bile salts, macrolides, fatty acids, heavy metals, and flavonoids. Furthermore, they play a role in quorum sensing since they were found to extrude quorum sensing signals from other microorganisms (Minagawa et al. 2012). Typically, RND pumps operate on multiple compounds and are characterized by a defined substrate range. The existence of different substrate entrance sites was proposed to be a mechanism for the expansion of pump specificity (Yamaguchi et al. 2015).

The substrate specificity of a given inner membrane protein of an RND transporter is not reflected in an overall sequence similarity. A sequence alignment of 47 inner membrane proteins with known substrate profile revealed that sequences clustered according to their organism rather than to their substrate specificities (Hernández-Mendoza et al. 2007). However, when this clustering was repeated using only amino acids present in the putative substrate binding site in the upper part of the central pore (Yu et al. 2003), clustering occurred according to the substrate specificity. The pumps which were shown to export primarily aromatic hydrocarbons, namely, TtgH and TtgE of *P. putida* DOT-T1E, SrpB of *P. putida* S12, SepB of *P. putida* F1, and TbtB of *P. stutzeri*, were present in the same cluster. The authors concluded that valine 382 and alanine 385 are key residues in the recognition of aromatic hydrocarbons. The authors, however, did not evaluate whether clustering using amino acids present in the other substrate binding sites occurred equally according to the substrate profile.

Godoy et al. (2010) combined in silico and in vivo analyses to classify RND pumps in *P. putida* KT2440. This study resulted in the definition of four functional families, namely, (I) pumps involved in antibiotic resistance, (II) pumps responding to oxidative stress inducing agents, (III) pumps involved in metal resistance, and another less characterized family (IV) composed of only two members, that extrude rubidium, chromate, or tetracycline.

# 5   A Rotating Mechanism: A Common Theme for Membrane Proteins?

The first structure of AcrB in complex with pump substrates was reported by Yu et al. (2003). The crystallographic space group determined for these crystals was of trigonal symmetry, which implies that the trimer consists of three identical monomers. As a consequence the final model contained a trimer consisting of three identical monomers. Three substrate molecules were bound to each AcrB trimer. The authors proposed an elevator mechanism for drug expulsion which results in the concerted up-lifting of the substrates through the periplasmic part of AcrB into the pore formed by TolC and subsequently into the extracellular environment.

Three years later, two independent research laboratories succeeded in obtaining crystals of AcrB in a space group which enabled the consideration of each monomer in the trimer as an individual entity. These crystals permitted the structure to be resolved at a higher resolution (Murakami et al. 2006; Seeger el al. 2006). When inspecting the experimental electron density, both groups noticed that each monomer of the trimer was present in a different conformation. These conformational states were termed access (A), binding (B), and extrusion (E) (Fig. 1b, Murakami et al. 2002; Seeger et al. 2006). Interestingly; only one monomer, the one in the B conformation, contained bound substrate whereas the remaining two trimers were devoid of substrate (Fig. 1). Nakashima et al. (2011) reported the AcrB structure bound to rifampicin and erythromycin. In this structure, substrates were bound to the monomer in A conformation and also in the B conformation but in a different binding pocket. Authors referred to the binding pocket in the A conformation as proximal pocket and the pocket observed in the B conformation as distal pocket. It was proposed that both pockets are present in the A and B conformations but that the proximal pocket is expanded in the A state and reduced in the B state.

A rotating or peristaltic pump mechanism is currently favored where the three conformations are proposed to correspond to consecutive states in a transport cycle. A possible drug-transport mechanism has been proposed in which cycling of each monomer through the conformations A, B, and E is assumed. High molecular weight substrates are recognized by the proximal pocket, which is expanded when the protein is in the A conformation. At this stage, low molecular weight substrates are only weakly bound or unbound. Subsequently a conformational change induces the substrate transfer to the distal pocket, which is expanded during the transition from the A to B conformation. Low molecular weight compounds are now bound in the distal pocket, and high molecular weight compounds are unbound but blocked in the distal pocket since they are unable to return to the proximal pocket. The existence and function of both pockets is furthermore supported by site-directed mutagenesis experiments (Nakashima et al. 2011). Another conformational change then occurs leading to the adaptation of the E state, which is characterized by the closing of the substrate entry channel and the opening of an exit towards the center of the trimer. This opening is mediated by a positional change of a helix located in the central pore through which the substrate is expulsed into the central pore. The conformational changes of the monomers are interdependent, and never two monomers are in the same state at the same time. Biochemical evidence (Seeger et al. 2008) supports this rotating model derived from the analysis of X-ray structures. Asymmetric structures have also been reported for MexB of *P. aeruginosa* (Sennhauser et al. 2009), which suggests that this peristaltic mechanism for drug extrusion in RND pumps is not exclusive to AcrB.

This mode of action proposed for the RND pumps has striking similarities with the mode of action of another membrane protein, the F1-ATPase. This protein complex is also characterized by the asymmetric arrangement of three structural units (Abrahams et al. 1994). In analogy to the RND pump, a catalytic cycle leading to the synthesis of one molecule of ATP includes the consecutive adaptation of three conformational states. Nevertheless, further work is necessary to establish whether this peristaltic rotating mechanism represents a general mode of operation found in other membrane proteins.

## 6   RND Pumps Are Proton Antiporters

The model proposed for the action of RND pumps is based on the adoption of three different states of the trimer. In contrast to the F1-ATPase, little is known about the molecular stimuli which induce the conformational changes necessary to proceed from one conformation to the next, but it is assumed that drug binding as well as proton translocation correspond to these stimulatory events. The pH in the periplasm was estimated to be 1.7 units below that of the cytosol which is generally between 7.5 and 7.7. The substrate export of an RND pump is energized by the influx of protons from the periplasm to the cytosol. It is assumed that the protons traverse the inner membrane protein. However, the exact path of proton transport has not yet been fully established. The transmembrane region of AcrB contains a buried triplet of charged amino acids, namely, two aspartates and a lysine residue, which were proposed to form a segment of the proton transport pathway (Su et al. 2006). This was confirmed by mutagenesis studies which indicate that these residues are essential for AcrB function. The inspection of 3D structures provides no clue on how protons reach this central cluster from the periplasm and how the proton transport continues beyond this cluster to the cytosol. All currently available data support the notion that the paths for substrate export and proton import are fully separated.

# 7    Open Questions and Future Work

The study of membrane proteins is a challenge, and numerous are the questions which remain to be answered regarding RND pumps. RND pumps are critical for solvent tolerance of bacteria and are also largely responsible for MDR, which is of ever growing clinical importance. Designing inhibitors of RND pumps is thus a strategy to fight MDR. In this context, the uncertainty surrounding the physiologically relevant substrate binding site is a major obstacle to overcome. Based on structural studies of AcrB, several sites for substrate recognition were reported of which their physiological relevance is unknown. Biochemical approaches are thus needed to identify the key binding site which might also serve as a target for inhibitor development.

Currently there is no structural information available on an inner membrane pump protein in complex with an aromatic hydrocarbon. The knowledge of the detailed structural and thermodynamic bases of the hydrocarbon interaction with the pump would open the way for protein-engineering approaches aimed at increasing the affinity of a particular hydrocarbon for the pump. An increase in affinity is likely to augment the pump's efficiency in the extrusion of a given compound. Introducing engineered pumps into bacteria might increase their resistance to hydrocarbons which in turn would offer a range of biotechnological applications.

# 8    Research Needs

The physiological relevance of RND pumps is well documented. Studies converge in the conclusion that these pumps are critical for solvent tolerance of bacteria and also largely responsible for the MDR phenotype. RND pumps are thus targets for the development of inhibitors to fight MDR. Alternatively, protein engineering might be employed to create mutant pumps which are more efficient in the extrusion of hydrocarbons. However, these approaches are seriously hampered by the sparseness of biochemical information of which we dispose. Among the questions which wait to be resolved are fundamental issues such as: What is the affinity of substrates for the pump? Where is exactly the substrate recognition site, and which amino acids are directly involved in binding? Does an increase in substrate binding affinity translate into a more efficient expulsion of hydrocarbons? Answering these and other questions will constitute a scientific basis for the biotechnological exploitation of RND pumps.

# References

Abrahams JP, Leslie AG, Lutter R, Walker JE (1994) Structure at 2.8 Å resolution of F1-ATPase from bovine heart mitochondria. Nature 370:621–628

Daniels C, del Castillo T, Krell T, Segura A, Busch A, Lacal J, Ramos JL (2009) Cellular ecophysiology: genetics and genomics of accessing and exploiting hydrocarbons. In: Handbook of hydrocarbons and lipid microbiology, vol 2. Chapter 44, Springer-Verlag Berlin Heidelberg (Germany)

Du D, Wang Z, James NR, Voss JE, Klimont E, Ohene-Agyei T, Venter H, Chiu W, Luisi BF (2014) Structure of the AcrAB–TolC multidrug efflux pump. Nature 509:512–515

Eda S, Maseda H, Nakae T (2003) An elegant means of self-protection in Gram-negative bacteria by recognizing and extruding xenobiotics from the periplasmic space. J Biol Chem 278:2085–2088

Fernández M, Duque E, Pizarro-Tobías P, van Dillewijin P, Wittick RM, Ramos JL (2009) Microbial responses to xenobiotic compounds. Identification of genes that allow *Pseudomonas putida* KT2440 to cope with 2,4,6-trinitrotoluene. Microb Biotechnol 2:287–294

Godoy P, Molina-Henares AJ, de la Torre J, Duque E, Ramos JL (2010) Characterization of the RND family of multidrug efflux pumps: in silico to in vivo confirmation of four functionally distinct subgroups. Microb Biotechnol 3:691–700

Guazzaroni ME, Krell T, Felipe A, Ruiz R, Meng C, Zhang X, Gallegos MT, Ramos JL (2005) The multidrug efflux regulator TtgV recognizes a wide range of structurally different effectors in solution and complexed with target DNA: evidence from isothermal titration calorimetry. J Biol Chem 280:20887–20893

Hayashi K, Nakashima R, Sakurai K, Kitagawa K, Yamasaki S, Nishino K, Yamaguchi A (2015) AcrB-AcrA fusion proteins that act as multidrug efflux transporters. J Bacteriol 198:332–342

Hernández-Mendoza A, Quinto C, Segovia L, Pérez-Rueda E (2007) Ligand-binding prediction in the resistance-nodulation-cell division (RND) proteins. Comput Biol Chem 31:115–123

Isken S, de Bont JA (1996) Active efflux of toluene in a solvent-resistant bacterium. J Bacteriol 178:6056–6058

Kieboom J, Dennis JJ, de Bont JAM, Zylstra G (1998) Identification and molecular characterization of an efflux pump involved in *Pseudomonas putida* S12 solvent tolerance. J Biol Chem 273:85–91

Minagawa S, Inami H, Kato T, Sawada S, Yasuki T, Miyairi S, Horikawa M, Okuda J, Gotoh N (2012) RND-type efflux pump system MexAB-OprM of *Pseudomonas aeruginosa* selects bacterial languages 3-oxo-acyl-homoserine lactones for cell to cell communication. BMC Microbiol 12:70

Murakami S, Nakashima R, Yamashita E, Yamaguchi A (2002) Crystal structure of bacterial multidrug efflux transporter AcrB. Nature 419:587–593

Murakami S, Nakashima R, Yamashita E, Matsumoto T, Yamaguchi A (2006) Crystal structure of a multidrug transporter reveal a functionally rotating mechanism. Nature 443:173–179

Nakashima R, Sakurai K, Yamasaki S, Nishino K, Yamaguchi A (2011) Structures of the multidrug exporter AcrB reveal a proximal multisite drug-binding pocket. Nature 480:565–569

Nikaido H (1998) Multiple antibiotic resistance and efflux. Curr Opin Microbiol 1:516–523

Ramos JL, Duque E, Huertas MJ, Haïdour A (1995) Isolation and expansion of the catabolic potential of a *Pseudomonas putida* strain able to grow in the presence of high concentrations of aromatic hydrocarbons. J Bacteriol 177:3911–3916

Ramos JL, Duque E, Gallegos MT, Godoy P, Ramos-González MI, Rojas A, Teran W, Segura A (2002) Mechanisms of solvent tolerance in gram-negative bacteria. Annu Rev Microbiol 56:743–768

Rojas A, Duque E, Mosqueda G, Golden G, Hurtado A, Ramos JL, Segura A (2001) Three efflux pumps are required to provide efficient tolerance to toluene in *Pseudomonas putida* DOT-T1E. J Bacteriol 183:3967–3973

Seeger MA, Schiefner A, Eicher T, Verrey F, Diederichs K, Pos KM (2006) Structural asymmetry of AcrB trimer suggests a peristaltic pump mechanism. Science 313:1295–1298

Seeger MA, von Ballmoos C, Eicher T, Brandstätter L, Verrey F, Diederichs K, Pos KM (2008) Engineered disulfide bonds support the functional rotation mechanism of multidrug efflux pump AcrB. Nat Struct Mol Biol 15:199–205

Sennhauser G, Bukowska MA, Briand C, Grutter MG (2009) Crystal structure of the multidrug exporter MexB from *Pseudomonas aeruginosa*. J Mol Biol 389:134–145

Sikkema J, de Bont JAM, Poolman B et al (1995) Mechanisms of membrane toxicity of hydrocarbons. Microbiol Rev 59:201–222

Smith HE, Blair JM (2014) Redundancy in the periplasmic adaptor proteins AcrA and AcrE provides resilience and an ability to export substrates of multidrug efflux. J Antimicrob Chemother 69:982–987

Su CC, Li M, Gu R, Takatsuka Y, McDermott G, Nikaido H, Yu EW (2006) Conformation of the AcrB multidrug efflux pump in mutants of the putative proton relay pathway. J Bacteriol 188:7290–7296

Symmons MF, Bokma E, Koronakis E, Hughes C, Koronakis V (2009) The assembled structure of a complete tripartite bacterial multidrug efflux pump. Proc Natl Acad Sci U S A 106(17): 7173–7178

Tikhonova EB, Yamada Y, Zgurskaya HI (2011) Sequential mechanism of assembly of multidrug efflux pump AcrAB-TolC. Chem Biol 18(4):454–463

Törnroth-Horsefield S, Gourdon P, Horsefield R, Brive L, Yamamoto N, Mori H, Snijder A, Neutze R (2007) Crystal structure of AcrB in complex with a single transmembrane subunit reveals another twist. Structure 15:1663–1673

Touzé T, Eswaran J, Bokma E, Koronakis E, Hughes C, Koronakis V (2004) Interactions underlying assembly of the *Escherichia coli* AcrAB-TolC multidrug efflux system. Mol Microbiol 53:697–706

Xu Y, Lee M, Moeller A, Song S, Yoon BY, Kim HM, Jun SY, Lee K, Ha NC (2011) Funnel-like hexameric assembly of the periplasmic adapter protein in the tripartite multidrug efflux pumpin gram-negative bacteria. J Biol Chem 286:17910–17920

Yamaguchi A, Nakashima R, Sakurai K (2015) Structural basis of RND-type multidrug exporters. Front Microbiol 6:327

Yu EW, McDermott G, Zgurskaya HI, Nikaido H, Koshland DE Jr (2003) Structural basis of multiple drug-binding capacity of the AcrB multidrug efflux pump. Science 300:976–980

Yu EW, Aires JR, McDermott G, Nikaido H (2005) A periplasmic drug-binding site of the AcrB multidrug efflux pump: a crystallographic and site-directed mutagenesis study. J Bacteriol 187:6904–6815

# Membrane Composition and Modifications in Response to Aromatic Hydrocarbons in Gram-Negative Bacteria

# 24

Álvaro Ortega, Ana Segura, Patricia Bernal, Cecilia Pini,
Craig Daniels, Juan Luis Ramos, Tino Krell, and Miguel A. Matilla

## Contents

1   Introduction .................................................................... 374
2   Cytoplasmic Membrane Structure and Composition ..................................... 375
3   Biosynthesis of Membrane Phospholipids ............................................. 376
4   Lipid Domains in Bacterial Membranes ............................................... 378
5   Inner Membrane Functions .......................................................... 378
6   Membrane Modifications in Response to Environmental Challenges: Responses to
    Aromatic Hydrocarbons ............................................................. 379
7   Research Needs .................................................................... 382
8   Concluding Remarks ................................................................ 382
References ............................................................................ 382

Á. Ortega (✉) · A. Segura · J. L. Ramos · T. Krell · M. A. Matilla
Department of Environmental Protection, Estación Experimental del Zaidín, Consejo Superior de
Investigaciones Científicas, Granada, Spain
e-mail: alvaro.ortega@csic.es; ana.segura@eez.csic.es; juan.ramos@abengoa.com;
tino.krell@eez.csic.es; miguel.matilla@eez.csic.es

P. Bernal
Imperial College London, London, UK
e-mail: p.bernal@imperial.ac.uk

C. Pini
Shionogi Limited, London, UK
e-mail: piniceci@hotmail.com

C. Daniels
Developmental and Stem Cell Biology Program, Brain Tumour Research Centre, The Hospital for
Sick Children, Toronto, ON, Canada
e-mail: craig.daniels@sickkids.ca

© Springer International Publishing AG, part of Springer Nature 2018                          373
T. Krell (ed.), Cellular Ecophysiology of Microbe: Hydrocarbon and Lipid Interactions,
Handbook of Hydrocarbon and Lipid Microbiology, https://doi.org/10.1007/978-3-319-50542-8_48

**Abstract**
Bacterial cells are surrounded by a cellular envelope composed of the cytoplasmic membrane and the cell wall. The cytoplasmic membrane is a phospholipid bilayer that provides an appropriate matrix for membrane proteins involved in many different cellular processes. Membrane lipid composition can change in response to different environmental challenges such as the presence of toxic compounds (e.g., aromatic hydrocarbons). The changes in membrane fluidity induced by stressors are counteracted by the bacteria through variations in the length of fatty acids, in the degree of saturation, and in the *cis/trans* configuration of the unsaturated fatty acids. The presence of cyclopropane fatty acids and changes in phospholipid head groups has also been shown to be involved in this stress response. The adaptive alterations of the main membrane phospholipids and fatty acids present in the cytoplasmic membrane are the subject of this chapter.

# 1 Introduction

Bacterial cells are surrounded by a cellular envelope composed of two elements: the cytoplasmic membrane and the cell wall. Gram-positive and Gram-negative bacteria differ significantly in the structure and composition of the cell wall (Fig. 1). In Gram-negative bacteria, the cell wall is composed of an outer membrane that surrounds a peptidoglycan layer and the periplasmic space, while in Gram-positive bacteria, there is no outer membrane but instead a wide layer of peptidoglycan and teichoic acids (Fig. 1). Teichoic acids are formed by glycerol and/or ribitol residues connected by phosphodiester bonds. Usually the hydroxyl groups of these polyols are substituted with sugars, amino sugars, or D-alanine. Teichoic acids are anchored to the cytoplasmic membrane (lipoteichoic acids) or covalently bound to the peptidoglycan layer (wall teichoic acids) and constitute the main surface antigens of Gram-positive bacteria. Peptidoglycan is a polymer composed of two sugar derivatives, N-acetylglucosamine and N-acetylmuramic acid, and four amino acids (L-alanine, D-alanine, D-glutamic acid, and either L-lysine or diaminopimelic acid) that form a peptide chain of five amino acids attached to the N-acetylmuramic acid. The glycan chains formed by

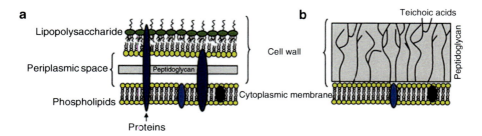

**Fig. 1** Schematic representation of the cell envelope of Gram-negative (**a**) and Gram-positive bacteria (**b**)

alternating β-(1,4) linked sugar residues are reinforced with peptide cross-links formed by the amino acids. Different peptide chains can be cross-linked to each other forming a 3D mesh-like layer. Peptidoglycan is responsible for the maintenance of bacterial rigidity and morphology. The outer membrane of Gram-negative bacteria is an asymmetric bilayer with an inner leaflet of phospholipids and an outer leaflet of lipopolysaccharides (LPS). LPS have a complex structure; they are amphipathic molecules with the hydrophobic region located toward the inner part on the leaflet and a polar region located in the surface of the outer membrane (Whitfield and Trent 2014).

In this chapter, we will describe in detail the structure, composition, and function of the cytoplasmic membrane and the lipid modifications that occur in response to environmental stresses in Gram-negative bacteria. For more information about the cell wall components of Gram-positive and Gram-negative bacteria, we refer the reader to Delcour et al. (1999) and Hancock et al. (1994).

## 2    Cytoplasmic Membrane Structure and Composition

The bacterial cytoplasmic membrane comprises a phospholipid bilayer containing different types of membrane proteins. In this structure, the fatty acid chains (hydrophobic) of the phospholipids are located toward the inner part of the bilayer, while the polar head groups are located toward its outer part. This lipid structure constitutes a hydrophobic barrier that prevents the uncontrolled movement of polar molecules and allows the accumulation and retention of metabolites and proteins within the cytoplasm. The lipid bilayer also provides an appropriate matrix for the function of membrane proteins involved in many different cellular processes. The lipid composition and the interactions between the lipid molecules determine the membrane permeability and influence the topology and function of membrane proteins.

The proteins constitute around 50% of the total membrane surface. Based on their location, we can classify them into three categories: integral membrane proteins (with at least two transmembrane segments), peripheral membrane proteins (anchored to the membrane through transmembrane domains or covalently bound to membrane lipids), and proteins that are only temporarily attached to the membrane.

The phospholipids are the other main component of biological membranes. Phospholipids are molecules in which two fatty acids are attached to a glycerol via ester bonds (in $C_1$ and $C_2$), while the other carbon of the glycerol molecule is linked through a phosphodiester bond to a polar head group.

The most extensively studied bacteria regarding lipid composition among the Gram-negatives are *Escherichia coli*, *Pseudomonas putida*, and *Salmonella typhimurium*. In these bacteria, the main membrane phospholipids are phosphatidylethanolamine (PE), phosphatidylglycerol (PG), and cardiolipin (CL). PE represents about 75% of total phospholipids, while the relative amounts of PG and CL depend on the bacterial growth phase, with PG being more abundant in the exponential phase while CL accumulating during stationary phase. CL is required for osmo-adaptation and oxidative phosphorylation in bacteria (Arias-Cartin et al. 2012).

Some bacteria degrade their phospholipids under conditions of phosphate limitation and replace them with ornithine lipids and glycolipids (Geske et al. 2013; Vences-Guzman et al. 2012). In addition to CL, PE, and PG, other lipids have also been found in bacterial membranes. Examples are sphingolipids, hopanoids, or methylated derivatives like phosphatidylcholine (PC) (Arendt et al. 2012, 2013; Geiger et al. 2013; Hannich et al. 2011).

Fatty acids attached to phospholipids can be saturated or unsaturated. Among the main saturated fatty acids in the membranes of *P. putida* are myristic (tetradecanoic, C14:0), palmitic (hexadecanoic, C16:0), and stearic (octadecanoic, C18:0) acids. The most abundant fatty acids among the unsaturated group are palmitoleic (*cis*-9-hexadecenoic, *cis*-$\Delta^{9,10}$-16:1) and *cis*-vaccenic (*cis*-11-octadecenoic, *cis*-$\Delta^{11,12}$-18:1) acids. Although the outer membranes of Gram-negative bacteria are not the subject of this chapter, it is interesting to note that they contain mainly PE and less unsaturated fatty acids than the inner membrane.

Other fatty acids that are present in some bacterial membranes are the cyclopropane fatty acids. In *E. coli* and *P. putida*, C17:cyclopropane (9,10-methyl-hexadecanoic acid) and the C19:cyclopropane (11,12-methyl-octadecanoic acid) are more abundant. The abundance of the cyclopropane fatty acids is growth-phase dependent and can represent almost 30% of total fatty acids in the late stationary growth phase.

## 3    Biosynthesis of Membrane Phospholipids

Phospholipid synthesis can be divided into two steps: fatty acid biosynthesis and attachment of the fatty acids to *sn*-glycerol-3-phosphate (G3P) followed by the addition of the polar head groups. Fatty acid biosynthesis has been extensively studied in *E. coli* and *S. typhimurium* (DiRusso et al. 1999; Schweizer 2004). The fatty acid synthase systems of these two bacteria belong to the type II family, which implies that the same mechanism as in eukaryotes is used (Fig. 2a).

Phospholipid biosynthesis occurs on the inner membrane of the cell envelope (Cronan and Rock, 1996). The precursor of this synthesis is phosphatidic acid that is converted to CDP-diacylglycerol (CDP-DAP), the intermediate in the biosynthesis of all membrane phospholipids (Fig. 2b). Two different enzyme families generate phosphatidic acid by a double acylation on glycerol-3-phosphate derived from glycolysis (Yao and Rock 2013).

Cardiolipin has various synthesis pathways, in which several CL synthases participate. It is formed from two molecules of PG (ClsA/B as enzymes), but a new phospholipase D enzyme has been described in *E. coli* (ClsC) that uses a PG and a PE molecule (Tan et al. 2012). In bacteria such as *Xanthomonas*, other alternatives exist (Moser et al. 2014). Phosphatidylcholine is synthesized by two well-characterized pathways (Dowhan 2013; Geiger et al. 2013), and the most

**Fig. 2** (**a**) Biosynthesis of saturated fatty acids: there are three potential pathways for the formation of acetoacetyl-ACP in *E. coli*. Only one of them is depicted in the box. (**b**) Phospholipid biosynthesis: *DHAP* dihydroxyacetone phosphate, *G3 P sn*-glycerol 3-phosphate, *PG* phosphatidylglycerol, *ACP* acyl carrier protein

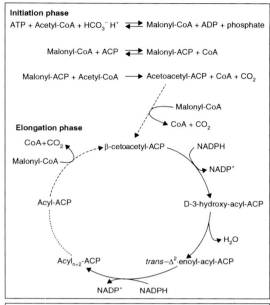

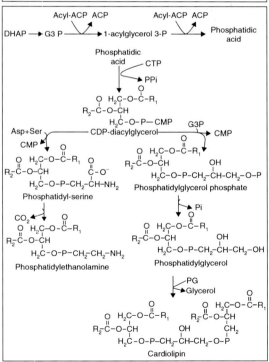

abundant lipid, PE, is obtained through a phosphatidylserine synthase (PssA) followed by a decarboxylase (Psd) acting over the central precursor, CDP-DAG.

## 4      Lipid Domains in Bacterial Membranes

Until recently it was generally accepted that lipids in bacterial membranes were homogeneously distributed; however, it has been shown that membranes produce specific lipid environments for certain membrane proteins. Cardiolipin domains in the membranes of several bacteria (i.e., *E. coli*, *P. putida*, and *Bacillus subtilis*) have been visualized using the fluorescent dye 10-*N*-nonyl acridine orange (NAO) which binds specifically to this phospholipid. It was reported that CL domains were located near the septal membrane and on the polar membrane regions. No such domains were observed in cardiolipin synthase knockout mutants. PE-rich domains were also localized in the septal membranes of *B. subtilis* cells in the exponential growth phase, in the membranes of the polar septal and in engulfment and forespore membranes at various stages in sporulating cells. Both CL and PE have the propensity to form non-bilayer structures, and this property is probably important in the fusion and fission of bilayer membranes which normally occurs where these phospholipids have been visualized. In addition, the negative charge of CL recruits specific peripheral membrane proteins to the membrane surface (i.e., DnaA, FtsY, MinD, and others). Most of the proteins co-localized thus far with the CL domains appear to be involved in the cell division process. In short, it appears that membranes have a complex lipid structure and that lipids are not homogeneously distributed throughout the membrane but form patches of specific lipid molecules (Matsumoto et al. 2006).

## 5      Inner Membrane Functions

The cytoplasmic membrane is a selectively permeable barrier that determines (together with the cell wall) the entry and exit of solutes. Water, dissolved gases, and liposoluble molecules diffuse across the membrane, but other molecules require transport systems in order to cross. Cytoplasmic membranes contain hundreds of different proteins such as flagellar proteins, transporters, or enzymes that participate in different functions. Due to its importance in the cellular cycle, we will highlight five membrane functions (Kadner 1996):

1. Energy functions: the membrane has an important role in energy generation and conservation. Most of the biosynthetic and transport processes in Gram-negative bacteria are driven by the hydrolysis of the high-energy phosphate bonds contained in molecules such as ATP, GTP, and phosphoenolpyruvate or by the proton-motive force formed across the membranes.
2. Transport functions: among them we can include the symporter and antiporter systems, permeases, ATPases, serial transport systems that translocate the substrate

through the inner and outer membrane, and transport systems in which the substrate is chemically altered during transport.

3. Protein translocation: translocation of proteins to the periplasmic space and outer membrane is generally carried out by the Sec system; this system is dependent on ATP hydrolysis, and its activity is enhanced by an electrical potential.

4. Signaling functions: many sensor histidine kinases and chemoreceptors are located in the cytoplasmic membrane. Some of these systems modulate gene expression or cause chemotaxis in response to several environmental stimuli.

5. Cellular division: numerous proteins involved in septum formation as well as in the synthesis of the membrane invaginations and cell constrictions are located in the membrane. Also, some proteins implicated in the replication and segregation of the chromosome are associated at least temporarily with the inner membrane.

It is well known that the membrane lipid composition can change in response to different environmental challenges, and in addition to the five functions depicted above, membranes have an important role in the bacterial adaptation to different environmental conditions.

# 6 Membrane Modifications in Response to Environmental Challenges: Responses to Aromatic Hydrocarbons

Biological membranes constitute the first contact point between microbes and the environment. For this reason, many bacteria have developed different mechanisms to counteract the effects produced by a changing environment. In general, changes in pH, temperature, or the presence of toxic compounds, such as aromatic hydrocarbons, produce an alteration in membrane fluidity. UV-C radiation can also produce these changes (Ghorbal et al. 2013). If bacteria do not maintain the appropriate membrane structure, the membrane-associated functions collapse, and cell death may occur. Toxic chemical compounds intercalate between membrane phospholipids and are able to modify the chemical interactions between fatty acids, leading to an uncontrolled efflux of positive ions that causes a lowering in the proton-motive force and an impairment in energy conservation (Sikkema et al. 1994). Several bacterial strains can survive in the presence of high concentrations of organic compounds such as alkanes, aromatic, or halogenated compounds. Some of them can even use these compounds as a carbon source, attenuating their toxicity in the process (Ramos et al. 1995). The ability of bacteria to compensate for the changes in membrane fluidity produced by environmental challenges is called "homeoviscous adaptation" and is mainly achieved by changes in the fatty acid composition. Variations in the length of fatty acids, the degree of saturation, and the *cis/trans* configuration of the unsaturated fatty acids are the main factors that affect membrane fluidity. The phase transition temperature of a membrane is defined as the temperature required to induce a change in the lipid physical state from the ordered gel phase, where the acyl chains are fully extended and closely packed, to the disordered liquid crystalline phase, where the acyl chains are randomly oriented and fluid.

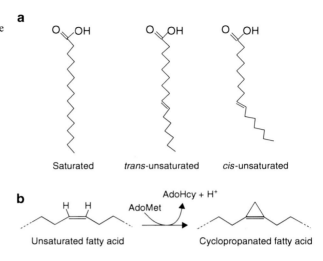

**Fig. 3** (a) Schematic representation of the structure of saturated and unsaturated fatty acids; (b) Schematic representation of cyclopropane fatty acid biosynthesis. *AdoMet* adenosine methionine, *AdoHcy* adenosine homocysteine

Together with fatty acid changes, the charge and head group species of the phospholipids are important in determining the phase transition temperature.

In *Pseudomonas* and *Vibrio* strains, an increase in *trans*-unsaturated fatty acids is detected when the bacteria grow in the presence of organic solvents or at high temperatures (Heipieper et al. 2003). The double bond of a *cis*-unsaturated fatty acid provokes a bend with an angle of approximately 30° in the acyl chain (Fig. 3a). This structure decreases the ordered arrangement of acyl chains in the membranes, which in turn results in lower phase transition temperatures of membranes with a high content of such fatty acids. The *trans* configuration is more extended and lacks the kink so it is able to insert into the membrane similarly to saturated fatty acids. This *cis*-to-*trans* isomerization is catalyzed by the enzyme *cis-trans* isomerase (Cti) which uses the *cis*-fatty acid as a substrate; it is therefore a post-synthetic reaction. In general, when an environmental stress induces the fluidification of the membrane, the cells respond by increasing their relative amounts of saturated fatty acids and in those bacteria that have Cti, by also increasing the amount of *trans*-unsaturated fatty acids. The major disadvantage of adjusting membrane fluidity by increasing the saturated fatty acid content derives from its strict dependency on cell growth and fatty acid biosynthesis. However, *cis-trans* isomerization does not depend on these factors; it has been observed less than 10 min after the addition of the organic solvent, indicating that this is a very rapid adaptation mechanism (Ramos et al. 1997). In fact, the Cti protein is constitutively expressed in cells and only slightly induced in the presence of hydrocarbons such as toluene (Bernal et al. 2007b; Junker and Ramos 1999). One question that remains to be answered is how the enzyme becomes active in the presence of an external stimulus. It has been speculated that under normal conditions, the enzyme cannot reach its target, the fatty acid double bond which is buried within the lipid bilayer. Only when the membrane fluidity is higher than normal the enzyme can access and change the double bond configuration. The Cti sequence has a putative heme-binding motif. Although the reaction mechanism is unknown, it has been proposed that the heme ($Fe^{3+}$) iron attacks the double bond causing the removal of an electron from the double bond to form a transient radical complex covalently bound to iron, enabling the rotation to the *trans* configuration. It seems unlikely that the reaction goes through hydrogenation of the double bond to form an $sp^3$ bond prior to rotation (Heipieper et al. 2004).

The length and the global charge of the fatty acids are other important aspects in the control of membrane fluidity. Longer fatty acid chains decrease membrane fluidity (a decrease in growth temperature generally enhances the level of unsaturation and decreases the average chain length of fatty acids). An example is the effect of the polychlorinated biphenyls (PCBs) on *P. stutzeri* membrane fatty acids (Zoradova et al. 2011). The effects of different toxic compounds having similar toxicity and hydrophobicity on the surface properties and changes in lipopolysaccharide content of the bacterial membrane are not of general nature, which was demonstrated in *P. putida* by comparing the response toward n-alkanols and chlorophenols (Baumgarten et al. 2012).

Cyclopropane fatty acids (CFAs) are also post-synthetically produced by the addition of a methylene group across the double bond of *cis*-unsaturated fatty acids, which occurs during the transition to stationary growth phase (Fig. 3b). The influence of the steric configuration of CFAs on the fluidity of the membrane has not been established unambiguously; some studies indicate that CFAs pack less tightly than the unsaturated fatty acid into the membranes, while others suggest that CFAs may reduce the effects of temperature on membrane fluidity (Loffhagen et al. 2007; Poger and Mark 2015; Zhang and Rock 2008). Direct measurements of membrane fluidity using fluorescent probes suggest that the presence or absence of CFAs did not modify membrane fluidity. As such, a clear physiological role for the presence of CFAs in membranes has not yet been established. Cyclopropanation has been reported to be a major factor in acid resistance and in repeat freeze-thaw treatments of *E. coli*, and in desiccation resistance in *P. putida* (Chang and Cronan 1999; Munoz-Rojas et al. 2006). Recent results from our laboratory showed that a knockout mutant in the cyclopropane synthase (the enzyme responsible for the synthesis of CFAs) is more sensitive to toluene shock than the wild-type strain in stationary phase, indicating that this type of fatty acid is involved in the tolerance toward aromatic hydrocarbons.

Branched chain fatty acids (BCFAs) are another type of molecule that is involved in the maintenance of membrane fluidity. Although they are mainly present in Grampositive bacteria, Gram-negative bacteria also contain these fatty acids (Zhou et al. 2013). BCFAs are synthesized from α-keto acids that are derived from valine and leucine as the precursors of iso-branched chain fatty acids and from isoleucine for anteiso-BCFAs (Murinova and Dercova 2014). Bacteria exposed to sodium benzoate and degrading monochlorophenols, for instance, showed the appearance of BCFAs in their membranes, and this was the most significant change next to fatty acid saturation, hydroxylation, and cyclopropane ring formation (Nowak and Mrozik 2016).

Finally, changes in phospholipid head groups have also been reported to be associated with stress responses. One of these changes is the increase in cardiolipin content observed when *P. putida* DOT-T1E cultures are subjected to a toluene shock. It was demonstrated that the decrease in cardiolipin content observed in a *cls* knockout mutant correlates with a decrease in the solvent tolerance of the strain (Bernal et al. 2007a). In this mutant, the function of the efflux pumps responsible for toluene extrusion was slightly impaired, and the conclusion of the study was that the decrease in cardiolipin content somehow interfered with efflux pump function. Given that CL domains are also present in the *P. putida* DOT-T1E wild-type strain but are absent in the *cls* mutant, it is tempting to speculate that the decrease in solvent tolerance was due to the inappropriate embedding of the efflux pump components in the membrane.

# 7   Research Needs

Changes in membrane fluidity achieved by fatty acid modifications have an important role in the survival of bacteria in a harsh environment. The mechanisms through which the bacteria sense the environment, the regulatory networks controlling membrane lipids, and the interactions between lipids and membrane proteins are interesting questions that remain to be further explored.

# 8   Concluding Remarks

Although most membrane functions have traditionally been related to the proteins that are immersed in the lipid bilayer, recent findings have demonstrated that lipids are not a mere matrix to accommodate proteins. Changes in membrane fluidity achieved by fatty acid modifications have an important role in the survival of bacteria in harsh environments. Phospholipid patches in the membrane have also been demonstrated to exert important physiological functions.

Interestingly, findings have demonstrated the implication of phospholipids in regulatory networks (Inoue et al. 1997). An imbalance in phospholipid composition induced regulatory genes that affected expression of genes encoding porins and those involved in flagellar formation. The discovery of this regulatory network further emphasizes the important biological role of membrane phospholipids.

# References

Arendt W, Hebecker S, Jager S, Nimtz M, Moser J (2012) Resistance phenotypes mediated by aminoacyl-phosphatidylglycerol synthases. J Bacteriol 194:1401–1416

Arendt W, Groenewold MK, Hebecker S, Dickschat JS, Moser J (2013) Identification and characterization of a periplasmic aminoacyl-phosphatidylglycerol hydrolase responsible for *Pseudomonas aeruginosa* lipid homeostasis. J Biol Chem 288:24717–24730

Arias-Cartin R, Grimaldi S, Arnoux P, Guigliarelli B, Magalon A (2012) Cardiolipin binding in bacterial respiratory complexes: structural and functional implications. Biochim Biophys Acta 1817:1937–1949

Baumgarten T, Vazquez J, Bastisch C, Veron W, Feuilloley MG, Nietzsche S, Wick LY, Heipieper HJ (2012) Alkanols and chlorophenols cause different physiological adaptive responses on the level of cell surface properties and membrane vesicle formation in *Pseudomonas putida* DOT-T1E. Appl Microbiol Biotechnol 93:837–845

Bernal P, Munoz-Rojas J, Hurtado A, Ramos JL, Segura A (2007a) A *Pseudomonas putida* cardiolipin synthesis mutant exhibits increased sensitivity to drugs related to transport functionality. Environ Microbiol 9:1135–1145

Bernal P, Segura A, Ramos JL (2007b) Compensatory role of the cis-trans-isomerase and cardiolipin synthase in the membrane fluidity of *Pseudomonas putida* DOT-T1E. Environ Microbiol 9:1658–1664

Chang YY, Cronan JE Jr (1999) Membrane cyclopropane fatty acid content is a major factor in acid resistance of *Escherichia coli*. Mol Microbiol 33:249–259

Cronan JE, Rock CO (1996) Biosynthesis of membrane lipids. In: Neidhart FC, Curtis R III, Ingraham JL, Link ECC, Low KB, Magasanik WS, Reznikoff WS, Riley M, Schaechter M, Umbarger HE (eds) *Escherichia coli* and *Salmonella*: cellular and molecular biology. ASM Press, Washington, DC

Delcour J, Ferain T, Deghorain M, Palumbo E, Hols P (1999) The biosynthesis and functionality of the cell-wall of lactic acid bacteria. Antonie Van Leeuwenhoek 76:159–184

DiRusso CC, Black PN, Weimar JD (1999) Molecular inroads into the regulation and metabolism of fatty acids, lessons from bacteria. Prog Lipid Res 38:129–197

Dowhan W (2013) A retrospective: use of *Escherichia coli* as a vehicle to study phospholipid synthesis and function. Biochim Biophys Acta 1831:471–494

Geiger O, Lopez-Lara IM, Sohlenkamp C (2013) Phosphatidylcholine biosynthesis and function in bacteria. Biochim Biophys Acta 1831:503–513

Geske T, Vom Dorp K, Dormann P, Holzl G (2013) Accumulation of glycolipids and other non-phosphorous lipids in *Agrobacterium tumefaciens* grown under phosphate deprivation. Glycobiology 23:69–80

Ghorbal SK, Chatti A, Sethom MM, Maalej L, Mihoub M, Kefacha S, Feki M, Landoulsi A, Hassen A (2013) Changes in membrane fatty acid composition of *Pseudomonas aeruginosa* in response to UV-C radiations. Curr Microbiol 67:112–117

Hancock R, Karunaratne D, Bernegger-Egli C (1994) Molecular organization and structural role of outer membrane macromolecules. In: Ghuysen JM, Hakenbeck R (eds) Bacterial cell wall. Elsevier, Amsterdam, pp 263–279

Hannich JT, Umebayashi K, Riezman H (2011) Distribution and functions of sterols and sphingolipids. Cold Spring Harb Perspect Biol 3:a004697

Heipieper HJ, Meinhardt F, Segura A (2003) The cis-trans isomerase of unsaturated fatty acids in *Pseudomonas* and *Vibrio*: biochemistry, molecular biology and physiological function of a unique stress adaptive mechanism. FEMS Microbiol Lett 229:1–7

Heipieper HJ, Neumann G, Kabelitz N, Kastner M, Richnow HH (2004) Carbon isotope fractionation during cis-trans isomerization of unsaturated fatty acids in *Pseudomonas putida*. Appl Microbiol Biotechnol 66:285–290

Inoue K, Matsuzaki H, Matsumoto K, Shibuya I (1997) Unbalanced membrane phospholipid compositions affect transcriptional expression of certain regulatory genes in *Escherichia coli*. J Bacteriol 179:2872–2878

Junker F, Ramos JL (1999) Involvement of the cis/trans isomerase Cti in solvent resistance of *Pseudomonas putida* DOT-T1E. J Bacteriol 181:5693–5700

Kadner R (1996) Cytoplasmic membrane. In: Neidhardt FC, Curtis R III, Ingrahanm JL, Link ECC, Low KB, Magasanik WS, Reznikoff WS, Riley M, Schaechter M, Umbarger HE (eds) *Escherichia coli* and *Salmonella*: cellular and Molecular Biology. ASM Press, Washington, DC

Loffhagen N, Härtig C, Geyer W, Voyevoda M, Harms H (2007) Competition between *cis*, *trans* and cyclopropane fatty acid formation and its impact on membrane fluidity. Eng Life Sci 7:67–74

Matsumoto K, Kusaka J, Nishibori A, Hara H (2006) Lipid domains in bacterial membranes. Mol Microbiol 61:1110–1117

Moser R, Aktas M, Fritz C, Narberhaus F (2014) Discovery of a bifunctional cardiolipin/phosphatidylethanolamine synthase in bacteria. Mol Microbiol 92:959–972

Munoz-Rojas J, Bernal P, Duque E, Godoy P, Segura A, Ramos JL (2006) Involvement of cyclopropane fatty acids in the response of *Pseudomonas putida* KT2440 to freeze-drying. Appl Environ Microbiol 72:472–477

Murinova S, Dercova K (2014) Response mechanisms of bacterial degraders to environmental contaminants on the level of cell walls and cytoplasmic membrane. Int J Microbiol 2014:873081

Nowak A, Mrozik A (2016) Facilitation of co-metabolic transformation and degradation of monochlorophenols by *Pseudomonas sp.* CF600 and changes in its fatty acid composition. Water Air Soil Pollut 227:83

Poger D, Mark AE (2015) A ring to rule them all: the effect of cyclopropane fatty acids on the fluidity of lipid bilayers. J Phys Chem B 119:5487–5495

Ramos JL, Duque E, Huertas MJ, Haidour A (1995) Isolation and expansion of the catabolic potential of a *Pseudomonas putida* strain able to grow in the presence of high concentrations of aromatic hydrocarbons. J Bacteriol 177:3911–3916

Ramos JL, Duque E, Rodriguez-Herva JJ, Godoy P, Haidour A, Reyes F, Fernandez-Barrero A (1997) Mechanisms for solvent tolerance in bacteria. J Biol Chem 272:3887–3890

Schweizer HP (2004) Fatty acid biosynthesis and biologically significant acyl transfer reactions in *Pseudomonads*. In: Ramos J-L (ed) *Pseudomonas*. Biosynthesis of macromolecules and molecular metabolism. Kluwer/Plenum Publishers, New York, pp 83–109

Sikkema J, de Bont JA, Poolman B (1994) Interactions of cyclic hydrocarbons with biological membranes. J Biol Chem 269:8022–8028

Tan BK, Bogdanov M, Zhao J, Dowhan W, Raetz CR, Guan Z (2012) Discovery of a cardiolipin synthase utilizing phosphatidylethanolamine and phosphatidylglycerol as substrates. Proc Natl Acad Sci U S A 109:16504–16509

Vences-Guzman MA, Geiger O, Sohlenkamp C (2012) Ornithine lipids and their structural modifications: from A to E and beyond. FEMS Microbiol Lett 335:1–10

Whitfield C, Trent MS (2014) Biosynthesis and export of bacterial lipopolysaccharides. Annu Rev Biochem 83:99–128

Yao J, Rock CO (2013) Phosphatidic acid synthesis in bacteria. Biochim Biophys Acta 1831:495–502

Zhang YM, Rock CO (2008) Membrane lipid homeostasis in bacteria. Nat Rev Microbiol 6:222–233

Zhou A, Baidoo E, He Z, Mukhopadhyay A, Baumohl JK, Benke P, Joachimiak MP, Xie M, Song R, Arkin AP et al (2013) Characterization of NaCl tolerance in *Desulfovibrio vulgaris* Hildenborough through experimental evolution. ISME J 7:1790–1802

Zoradova S, Dudasova H, Lukacova L, Dercova K, Certik M (2011) The effect of polychlorinated biphenyls (PCBs) on the membrane lipids of *Pseudomonas stutzeri*. Int Biodeterior Biodegradation 65:1019–1023

# Cis–Trans Isomerase of Unsaturated Fatty Acids: An Immediate Bacterial Adaptive Mechanism to Cope with Emerging Membrane Perturbation Caused by Toxic Hydrocarbons

**25**

Hermann J. Heipieper, J. Fischer, and F. Meinhardt

## Contents

1   Introduction ............................................................ 386
2   Physiological Function of the *Cis–Trans*-Isomerase ..................................... 386
3   Molecular Biology and Biochemistry ...................................................... 388
4   Occurrence of the *Cis–Trans*-Isomerase Gene in Bacterial Genomes .................... 389
5   Regulation of the *Cis–Trans*-Isomerase Activity ......................................... 389
6   Research Needs ......................................................................... 392
References .................................................................................. 393

**Abstract**

A rather efficient solvent adaptation mechanism enabling several gram-negative bacteria to tolerate and grow in the presence of membrane-disturbing compounds is the isomerization of *cis*- to *trans*-unsaturated membrane fatty acids. The degree of isomerization obviously depends on the toxicity and the concentration of membrane-affecting agents. Synthesis of *trans*-fatty acids comes about by direct isomerization of the respective *cis*-configuration of the double bond without shifting the position. The purpose of the conversion of the *cis*-configuration to *trans* is apparently the rapid adaptation of the membrane fluidity to rising temperature or the presence of toxic organic hydrocarbons.

H. J. Heipieper (✉)
Department Environmental Biotechnology, Helmholtz Centre for Environmental Research – UFZ, Leipzig, Germany
e-mail: hermann.heipieper@ufz.de

J. Fischer
Department of Environmental Biotechnology, Helmholtz Centre for Environmental Research, Leipzig, Germany

F. Meinhardt
Institut für Molekulare Mikrobiologie und Biotechnologie, Westfälische Wilhelms-Universität Münster, Münster, Germany

© Springer International Publishing AG, part of Springer Nature 2018
T. Krell (ed.), *Cellular Ecophysiology of Microbe: Hydrocarbon and Lipid Interactions*,
Handbook of Hydrocarbon and Lipid Microbiology, https://doi.org/10.1007/978-3-319-50542-8_49

The *cis–trans*-isomerase (Cti) is a constitutively expressed periplasmic enzyme that – to exert its action – necessitates neither ATP nor other cofactors, and consistently, is independent of de novo synthesis of lipids. A heme-binding site typical of cytochrome *c*-type proteins is present in the predicted Cti polypeptide indicating a reaction mechanism that renounces temporary saturation of the double bond. Due to its direct correlation with toxicity, *cis–trans*-isomerization is a potential biomarker for recording solvent stress or changes of other environmental conditions.

## 1    Introduction

A number of hydrocarbons are potential environmental pollutants as they are toxic to all living cells by affecting membrane integrity through a nonspecific increase of fluidity eventually leading to the loss of membrane function as a barrier, matrix for enzymes, and energy transducer (Weber and de Bont 1996). In heavily contaminated soil or water, toxic effects may even cover the natural attenuation potential by inhibiting the activity of potential biodegraders. Hence, bacteria not reacting to toxic effects by so-called adaptive responses are drastically growth inhibited and can even be killed. The most important adaptive response concerns maintenance of membrane fluidity at a constant level irrespective of actual environmental conditions. Such homoviscous adaptation is brought about by changes in the fatty acid composition of membrane lipids (Ingram 1977; Suutari and Laakso 1994). However, for a rather long period of time, the *cis*-configuration of the double bond was considered to be the only naturally occurring in bacterial fatty acids. *Trans*-isomers of unsaturated fatty acids were first reported for *Vibrio* and *Pseudomonas* (Guckert et al. 1986, 1987) only about 20 years ago. Soon after, the conversion of *cis*- to *trans*-unsaturated fatty acids as a new adaptive mechanism enabling bacteria to change their membrane fluidity was documented for two species: for the psychrophilic bacterium *Vibrio* sp. strain ABE-1 as a response to a rise in temperature (Okuyama et al. 1991) and for *Pseudomonas putida* P8 as an adaptation to toxic organic hydrocarbons, such as phenol (Heipieper et al. 1992).

## 2    Physiological Function of the *Cis–Trans*-Isomerase

A pronounced increase of the usually minor amounts of *trans*-unsaturated fatty acids was seen when cells were treated with rising temperatures or toxic membrane active hydrocarbons for *Vibrio* sp. strain ABE-1 and *Pseudomonas putida* P8 respectively. *P. putida* cells respond to phenol in a concentration-dependent manner, i.e., by an increase in *trans*- and a parallel decrease in the respective *cis*-unsaturated fatty acids depending on the amount of phenol present in the membrane (Heipieper et al. 1992). The *cis–trans* conversion does not depend on growth, since it happens in non-growing cells, too, in which due to the lack of lipid biosynthesis the content of all the remaining fatty acids cannot be changed (Morita et al. 1993). Consistently, the reaction is not affected in cells in which fatty acid biosynthesis is cut off by cerulenin (Diefenbach et al. 1992; Heipieper and de Bont 1994). *Cis– trans* conversion follows an enzymatic kinetic and reaches its final *trans*- to- *cis*-ratio already 30 min after an addition of membrane-toxic agents. As *cis–trans* conversion is not affected by chloramphenicol, the system operates constitutively and does not depend on de novo protein biosynthesis (Heipieper et al. 1992; Kiran et al. 2005).

*P. putida* incorporates externally provided free fatty acids, such as oleic acid (C18:1Δ9 *cis*), into membrane phospholipids. After an addition of a toxic 4-chlorophenol concentration, this oleic acid was also converted into the *trans*-isomer, namely elaidic acid (C18:1Δ9 *trans*), evidencing that *trans*-fatty acids are formed by a direct isomerization without shifting the position of the double bond (Diefenbach and Keweloh 1994). The increasing levels of *trans*-unsaturated fatty acids were accompanied by the decreasing levels of the respective *cis*-unsaturated fatty acid, and consistently, the total amount of both isomers was kept constant at any toxin concentration (Heipieper et al. 1992). Peculiarly, the system does not require ATP nor any other cofactor (Von Wallbrunn et al. 2003).

Taken together, *cis–trans*-isomerization turned out to be a new adaptive response of bacteria allowing them to withstand rises in temperature or toxic concentrations of membrane-active compounds, which otherwise would perturb membrane fluidity (Weber and de Bont 1996; Heipieper et al. 2003; Kiran et al. 2004).

The advantage of the conversion is due to steric differences between *cis*- and *trans*- unsaturated fatty acids. In general, saturated fatty acids have a higher transition temperature compared with *cis*-unsaturated fatty acids. Phospholipids containing 16:0 saturated fatty acids display a transition temperature that is approximately 63 °C higher than those with 16:1 *cis*-unsaturated fatty acids (Roach et al. 2004; Zhang and Rock 2008). The double bond of a *cis*-unsaturated fatty acid causes an unmoveable 30° bend in the acyl chain (MacDonald et al. 1985; Roach et al. 2004). Hence, unsaturated fatty acids in *cis*-configuration with bended steric structures (caused by the kink in the acyl-chain) provoke a membrane with a high fluidity. On the contrary, the long-extended steric structure of the *trans*-configuration lacks the kink and is able to insert into the membrane as for the saturated fatty acids (MacDonald et al. 1985). This could even be proved on the level of the size of the molecules where phospholipids containing *trans*-unsaturated fatty acids showed a reduced area/molecule when compared with those containing corresponding *cis*-unsaturated fatty acids (Roach et al. 2004).

Routinely, gram-negative bacteria respond to an increase in membrane fluidity by increasing the degree of saturated phospholipid fatty acids (Ingram 1977; Kabelitz et al. 2003; Zhang and Rock 2008). One major drawback of such changes is due to its strict dependency on growth and active fatty acid biosynthesis. Indeed, it has been reported that solvents cause a shift in the ratio of saturated to unsaturated fatty acids only up to concentrations that completely inhibit growth. Thus, higher, i.e., toxic concentrations render the cells unable to adapt or they are even killed (Heipieper and de Bont 1994). The isomerization of *cis*- to *trans*-unsaturated fatty acids – seen to be present exclusively in strains of the genera *Pseudomonas*, including *P. putida* and

*P. aeruginosa* (Keweloh and Heipieper 1996; Heipieper et al. 2003; Von Wallbrunn et al. 2003) and *Vibrio* (Von Wallbrunn et al. 2003) – offers a solution to overcome the drawback of growth dependency as it is also instrumental in nongrowing cells. The conversion from the *cis*- to the *trans*-unsaturated double bond does in fact not have the same quantitative effect on membrane fluidity as the conversion to saturated fatty acids but still substantially influences the rigidity of the membrane (MacDonald et al. 1985; Roach et al. 2004).

Incited by results mainly obtained with phenolic compounds, a number of organic hydrocarbons were checked for their ability to activate the *cis–trans*-isomerase. It was seen that the degree of the isomerization apparently correlates with the toxicity and the concentration of membrane-active compounds (Heipieper et al. 1995). The concentration of such compounds in the membrane depends on the corresponding hydrophobicity given as log $P_{ow}$ value (Sikkema et al. 1995; Isken and de Bont 1998). Organic solvents with log $P_{ow}$ values between 1 and 4 are highly toxic for microorganisms, since they partition preferentially in membranes (Heipieper et al. 1994; Sikkema et al. 1995; Weber and de Bont 1996). There is a direct relation between the toxicity of organic solvents and their *cis–trans*-isomerase activation effects, which is independent from chemical structures of the compounds (Heipieper et al. 2003; Neumann et al. 2005).

Besides organic solvents or the increase in temperature, also membrane-affecting factors such as osmotic stresses (caused by NaCl and sucrose), heavy metals, and membrane-active antibiotics activate the system (Heipieper et al. 1996; Isken et al. 1997), indicating that the *cis–trans*-isomerization is part of a general stress–response mechanism of microorganisms (Segura et al. 1999; Ramos et al. 2001; Ramos et al. 2002; Zhang and Rock 2008).

## 3    Molecular Biology and Biochemistry

The cloning of the gene encoding the *cis–trans*-isomerase allowed isolation of the enzyme as a His-tagged *P. putida* P8 protein heterologously expressed in *E. coli* (Holtwick et al. 1997). The enzyme was also purified from periplasmic fractions of *Pseudomonas oleovorans* (Pedrotta and Witholt 1999) and *Pseudomonas* sp. strain E-3 (Okuyama et al. 1998). Cti is a neutral protein of 87 kDa encoded by a monocistronically transcribed gene that is constitutively expressed (Kiran et al. 2005). Nucleotide sequences of the *cti* genes from *P. putida* P8 (Holtwick et al. 1997), *P. putida* DOT-T1E (Junker and Ramos 1999), and *P. oleovorans* Gpo12 (Pedrotta and Witholt 1999) finally yielded conclusive evidence that the enzyme possesses an N-terminal hydrophobic signal sequence, which is cleaved off after targeting the enzyme to the periplasmic space.

Cti is obviously a cytochrome *c*-type protein as there is a heme-binding site in the predicted Cti polypeptide (Holtwick et al. 1999). For an enzyme preparation from *Pseudomonas* sp. strain E-3, which is presumably homologous to the *cti*-gene product of *P. putida* P8, it was suggested that iron (probably $Fe^{3+}$) plays a crucial role in the catalytic reaction (Okuyama et al. 1998). *Cis–trans*-isomerization is independent of cardiolipin synthase, an enzyme facilitating long-term adaptation of the membrane by enhanced cardiolipin-synthesis (Von Wallbrunn et al. 2002; Bernal et al. 2007).

A molecular mechanism of the isomerization reaction was proposed in which an enzyme–substrate complex is formed in which the electrophilic iron (probably $Fe^{3+}$), provided by the heme domain present in the enzyme, removes an electron from the *cis*-double bond transferring the $sp^2$ linking into a $sp^3$. The double bond is then reconstituted after rotation to the *trans*-configuration (Von Wallbrunn et al. 2003). Such a mechanism is not only in accordance with the suggested role of $Fe^{3+}$(Okuyama et al. 1998) but also agrees with site-directed mutagenesis experiments carried out to destroy the heme-binding motif in Cti of *P. putida* P8 (Holtwick et al. 1999).

## 4    Occurrence of the *Cis–Trans*-Isomerase Gene in Bacterial Genomes

Table 1 shows the results of a BLAST alignment study done with the predicted polypeptide of the *cti* gene of *P. aeruginosa* PAO1 as the reference gene. Next to the intensely investigated representatives of *Pseudomonas* and *Vibrio* the gene is apparently present in several other genera. However, direct physiological or biochemical evidence for the presence of Cti in these bacteria is still lacking. Promising candidates to be included in further work are bacteria belonging to the genera *Methylococcus* and *Nitrosomonas*, because these organisms also contain *trans*-unsaturated fatty acids (Guckert et al. 1991; Keweloh and Heipieper 1996).

Alignments including the known Cti sequences of representatives of *Pseudomonas* so far generally revealed a N-terminal signal sequence to be present, being indicative of the periplasmic localization of the *cis–trans*-isomerase in either case (Junker and Ramos 1999; Pedrotta and Witholt 1999). However, there is no signal peptide in the Cti protein of *V. cholerae*. Multiple sequence alignments of known Cti proteins revealed that proteins from *Pseudomonas* and *Vibrio* strains form a phylogenetic tree composed of three main branches, suggesting a common ancestor of the enzyme. Interestingly, the predicted polypeptide from *V. cholerae* does not constitute a separate group but rather emanates from the diverse group of proteins from *P. aeruginosa* and *Pseudomonas* sp. E-3 (Von Wallbrunn et al. 2003).

## 5    Regulation of the *Cis–Trans*-Isomerase Activity

Cti is encoded by a monocistronically transcribed gene that is constitutively expressed and for which classical transcriptional regulation is not present (Kiran et al. 2005). Therefore, one of the major and most interesting open questions regarding the *cis–trans*-isomerase concerns the regulation of the activity of this constitutively expressed periplasmic enzyme. Since Cti activity is seen even in resting cells and in the utter absence of energy sources (Heipieper et al. 1992), the

**Table 1** Protein BLAST-analysis (Basic Local Alignment Search Tool) of CTI with the protein sequence of *Pseudomonas aeruginosa* PAO1 as reference (National Center for Biotechnology Information, NCBI)

| Strain | Accession number | Identity (%) | Similarity |
|---|---|---|---|
| *Pseudomonas aeruginosa* PAO1 | NP_250537 | 100 | 100 |
| *Pseudomonas aeruginosa* UCBPP-PA14 | YP_791395 | 99 | 99 |
| *Pseudomonas aeruginosa* PA7 | YP_001348806 | 97 | 98 |
| *Pseudomonas fluorescens* Pf-5 | YP_260763 | 70 | 81 |
| *Pseudomonas mendocina* ymp | YP_001187652 | 70 | 80 |
| *Pseudomonas fluorescens* PfO-1 | YP_348835 | 69 | 80 |
| *Azotobacter vinelandii* AvOP | ZP_00417349 | 68 | 80 |
| *Pseudomonas psychrophila* | BAB41104 | 67 | 78 |
| *Pseudomonas syringae* | CAD59690 | 67 | 78 |
| *Pseudomonas stutzeri* A1501 | YP_001172723 | 66 | 77 |
| *Pseudomonas putida* F1 | YP_001268629 | 65 | 78 |
| *Pseudomonas entomophila* L48 | YP_608923 | 65 | 78 |
| *Pseudomonas putida* KT2440 | NP_744525 | 64 | 78 |
| *Pseudomonas syringae* pv. *phaseolicola* 1448A | YP_274814 | 64 | 78 |
| *Pseudomonas syringae* pv. *tomato* str. DC3000 | NP_792539 | 62 | 76 |
| *Alcanivorax borkumensis* SK2 | YP_693420 | 52 | 68 |
| *Nitrosomonas europaea* ATCC 19718 | NP_841379 | 50 | 66 |
| *Janthinobacterium* sp. Marseille | YP_001352887 | 41 | 54 |
| *Methylococcus capsulatus* str. Bath | YP_114244 | 40 | 55 |
| *Saccharophagus degradans* 2-40 | YP_525852 | 39 | 55 |
| *Geobacter lovleyi* SZ | ZP_01594476 | 39 | 54 |
| *Bdellovibrio bacteriovorus* HD100 | NP_968533 | 38 | 56 |
| *Vibrio cholerae* O395 | YP_001215326 | 38 | 55 |
| *Methylococcus capsulatus* str. Bath | YP_114035 | 38 | 54 |
| *Vibrio cholerae* O1 biovar eltor str. N16961 | NP_232942 | 38 | 54 |
| *Vibrio vulnificus* YJ016 | NP_936692 | 37 | 54 |
| *Vibrio fischeri* ES114 | YP_206570 | 37 | 54 |
| *Vibrio harveyi* ATCC BAA-1116 | YP_001448237 | 37 | 54 |
| *Pelobacter propionicus* DSM 2379 | NP_232942 | 37 | 54 |
| *Shewanella baltica* OS195 | YP_001556410 | 37 | 54 |
| *Shewanella baltica* OS155 | YP_001048852 | 37 | 54 |
| *Pseudoalteromonas atlantica* T6c | YP_662860 | 37 | 53 |
| *Shewanella* sp. MR-7 | YP_739567 | 36 | 55 |
| *Vibrio vulnificus* CMCP6 | NP_762110 | 36 | 55 |
| *Shewanella* sp. ANA-3 | YP_868147 | 36 | 55 |
| *Vibrio parahaemolyticus* RIMD 2210633 | NP_800187 | 36 | 54 |
| *Shewanella* sp. MR-4 | YP_732637 | 36 | 54 |
| *Colwellia psychrerythraea* 34H | YP_266855 | 36 | 54 |
| *Vibrio splendidus* 12B01 | ZP_00991480 | 36 | 52 |

(*continued*)

**Table 1**  (continued)

| Strain | Accession number | Identity (%) | Similarity |
|---|---|---|---|
| *Marinomonas* sp. MWYL1 | YP_001340552 | 35 | 52 |
| *Pseudoalteromonas haloplanktis* TAC125 | YP_341269 | 35 | 50 |
| *Arcobacter butzleri* RM4018 | YP_001489204 | 33 | 53 |

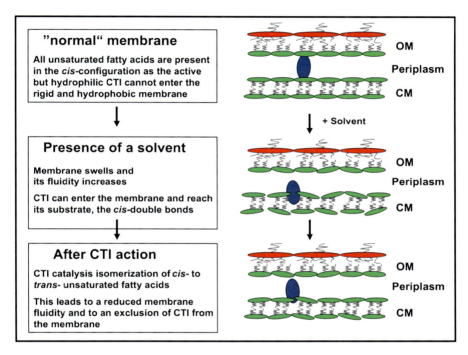

**Fig. 1**  Scheme of a possible regulation of the enzymatic mechanism of the *cis–trans*-isomerase by the fluidity of the membrane. *OM* outer membrane, *CM* cytoplasmic membrane, *Green* phospholipids, *Red* outer membrane lipids, *Blue* Cti

involvement of complex models that include sophisticated enzymatic pathways must be denied.

The currently accepted regulation model is schematically presented in Fig. 1. Enzyme activity is apparently controlled by simply allowing or prohibiting the active center of the enzyme to reach the double bond, which depends, however, on the fluidity of the membrane. The hydrophilic Cti located in the periplasmic space can only reach its double bond target – hidden at a certain depth in the membrane – when the membrane's rigidity is lowered and thus is "opened" by environmental conditions that cause a disintegration of the membrane, e.g., in the presence of toxic hydrocarbons (Heipieper et al. 2001; Hartig et al. 2005). As acyl chain packing is tightened by *cis*- to *trans*-isomerization of the unsaturated fatty acids (Seelig and Waespe-Šarcevic 1978; Chen et al. 1995; Roach et al. 2004; Loffhagen et al. 2007), intrusion of the protein is counteracted, and concomitantly, *cis*- to *trans*-isomerization is impeded, eventually resulting in a proper regulation of acyl chain packing without the involvement of indirect signaling mechanisms or pathways. Upon removal of membrane active compounds, recovery of regular low *trans–cis*-proportion most likely occurs by de novo synthesis of *cis*-fatty acids, since the reverse (*trans* to *cis*) process would require energy input.

The above model for the regulation of Cti activity also sufficiently illuminates the repeatedly reported relation between the degree of *cis–trans*-isomerization and the toxicity caused by different concentrations of specific environmental stress factors (Heipieper et al. 1995, 1996). As a concomitant result of the enzymatic reaction, a reduction of membrane fluidity takes place and the enzyme is forced out of the bilayer. Since Cti cannot reach its target when membrane fluidity has reached its initial level, an immediate and adjustable response to changing environmental conditions is assured (Heipieper et al. 2001).

# 6    Research Needs

The *cis–trans*-isomerization of unsaturated fatty acids is assumedly part of the general stress response in *Pseudomonas* and *Vibrio* cells. As a matter of fact, it constitutes a rapid adaptive mechanism facilitating urgent modifications of membranes to cope with emerging environmental stress. Such an immediate response, acting in terms of minutes, is suited to provide sufficient time for the cell growth-dependent mechanisms to execute their part in the adaptive response; it is the immediate reaction that assures survival under several stress conditions (Heipieper et al. 2007). In other words, the *cis–trans*-isomerase represents a major and urgent system helping the cells to withstand the deathly impact of a toxic hydrocarbon, concomitantly facilitating the induction of further adaptive mechanisms that eventually enable the entire adaptation (Cronan 2002; Hartig et al. 2005; Zhang and Rock 2008).

Because of its simpleness and effectiveness and since it works without complex transcriptional regulation, it is surprising that *cis*- to *trans*-isomerization of fatty acids is not ubiquitously found in gram-negative bacteria. An explanation may be derived from the widespread distribution of members of *Pseudomonas* and *Vibrio*, genera known to be highly adaptable and to have conquered all ecosystems such as soil, human skin, and seawater. Cti activity renders cells extremely flexible and adaptable to rapidly occurring changes in their respective environments. Thus, one future challenge – besides the physiological and biochemical proof of Cti activity – is to provide evidence for the physiological presence of Cti in the bacteria for which the BLAST analysis indicated a possible presence of the *cis–trans*-isomerase gene within the genome (Table 1) in compliance with the conditions. Potential candidates are members of the genera *Methylococcus* and *Nitrosomonas* as the presence of *trans*-unsaturated fatty acids is long known in these bacteria. In addition, these bacteria are also known to occur in a number different ecological habitats. Another potential candidate is *Alcanivorax borkumensis* especially because of its importance for the biodegradation of oil spills in marine environments (Hara et al. 2003). Its similarity (68%, Table 1) to the Cti of *P. aeruginosa*, which is very close to that of *P. putida* strains, makes such an attempt quite promising.

For those bacteria in which Cti is present, it offers the possibilities to use the *trans–cis*-ratio of unsaturated fatty acids as an elegant, reliable, and rapid bioindicator for membrane stress either in experimental setups or in the original environment. Indeed, such an application was already suggested; a *trans– cis*-ratio greater than 0.1 was proposed to serve as an index for starvation or stress in environmental samples (Guckert et al. 1986). Although statistical validation and reliable correlation are to be ensured, the determination of the *trans– cis*-index might, thus, be a valuable option in studying the toxicity status of natural samples.

# References

Bernal P, Segura A, Ramos JL (2007) Compensatory role of the *cis–trans*-isomerase and cardiolipin synthase in the membrane fluidity of *Pseudomonas putida* DOT-T1E. Environ Microbiol 9:1658–1664

Chen Q, Janssen DB, Witholt B (1995) Growth on octane alters the membrane lipid fatty acids of *Pseudomonas oleovorans* due to the induction of alkB and synthesis of octanol. J Bacteriol 177:6894–6901

Cronan JE (2002) Phospholipid modifications in bacteria. Curr Opin Microbiol 5:202–205

Diefenbach R, Keweloh H (1994) Synthesis of *trans* unsaturated fatty acids in *Pseudomonas putida* P8 by direct isomerization of the double bond of lipids. Arch Microbiol 162:120–125

Diefenbach R, Heipieper HJ, Keweloh H (1992) The conversion of *cis*- into *trans*- unsaturated fatty acids in *Pseudomonas putida* P8: evidence for a role in the regulation of membrane fluidity. Appl Microbiol Biotechnol 38:382–387

Guckert JB, Hood MA, White DC (1986) Phospholipid ester-linked fatty acid profile changes during nutrient deprivation of *Vibrio cholerae*: increases in the trans/cis ratio and proportions of cyclopropyl fatty acids. Appl Environ Microbiol 52:794–801

Guckert JB, Ringelberg DB, White DC (1987) Biosynthesis of trans fatty acids from acetate in the bacterium *Pseudomonas atlantica*. Can J Microbiol 33:748–754

Guckert JB, Ringelberg DB, White DC, Hanson RS, Bratina BJ (1991) Membrane fatty acids as phenotypic markers in the polyphasic taxonomy of methylotrophs within the proteobacteria. J Gen Microbiol 137:2631–2641

Hara A, Syutsubo K, Harayama S (2003) *Alcanivorax* which prevails in oil-contaminated seawater exhibits broad substrate specificity for alkane degradation. Environ Microbiol 5:746–753

Hartig C, Loffhagen N, Harms H (2005) Formation of *trans* fatty acids is not involved in growth-linked membrane adaptation of *Pseudomonas putida*. Appl Environ Microbiol 71:1915–1922

Heipieper HJ, de Bont JAM (1994) Adaptation of *Pseudomonas putida* S12 to ethanol and toluene at the level of fatty acid composition of membranes. Appl Environ Microbiol 60:4440–4444

Heipieper HJ, Diefenbach R, Keweloh H (1992) Conversion of *cis* unsaturated fatty acids to *trans*, a possible mechanism for the protection of phenol-degrading *Pseudomonas putida* P8 from substrate toxicity. Appl Environ Microbiol 58:1847–1852

Heipieper HJ, Weber FJ, Sikkema J, Keweloh H, de Bont JAM (1994) Mechanisms behind resistance of whole cells to toxic organic solvents. Trends Biotechnol 12:409–415

Heipieper HJ, Loffeld B, Keweloh H, de Bont JAM (1995) The *cis/trans* isomerization of unsaturated fatty acids in *Pseudomonas putida* S12: an indicator for environmental stress due to organic compounds. Chemosphere 30:1041–1051

Heipieper HJ, Meulenbeld G, VanOirschot Q, De Bont JAM (1996) Effect of environmental factors on the *trans/cis* ratio of unsaturated fatty acids in *Pseudomonas putida* S12. Appl Environ Microbiol 62:2773–2777

Heipieper HJ, de Waard P, van der Meer P, Killian JA, Isken S, de Bont JAM, Eggink G, de Wolf FA (2001) Regiospecific effect of 1-octanol on *cis-trans* isomerization of unsaturated fatty acids in the solvent-tolerant strain *Pseudomonas putida* S12. Appl Microbiol Biotechnol 57:541–547

Heipieper HJ, Meinhardt F, Segura A (2003) The *cis–trans* isomerase of unsaturated fatty acids in *Pseudomonas* and *Vibrio*: biochemistry, molecular biology and physiological function of a unique stress adaptive mechanism. FEMS Microbiol Lett 229:1–7

Heipieper HJ, Neumann G, Cornelissen S, Meinhardt F (2007) Solvent-tolerant bacteria for biotransformations in two-phase fermentation systems. Appl Microbiol Biotechnol 74:961–973

Holtwick R, Meinhardt F, Keweloh H (1997) *Cis–trans* isomerization of unsaturated fatty acids: cloning and sequencing of the *cti* gene from *Pseudomonas putida* P8. Appl Environ Microbiol 63:4292–4297

Holtwick R, Keweloh H, Meinhardt F (1999) *cis/trans* isomerase of unsaturated fatty acids of *Pseudomonas putida* P8: evidence for a heme protein of the cytochrome c type. Appl Environ Microbiol 65:2644–2649

Ingram LO (1977) Changes in lipid composition of *Escherichia coli* resulting from growth with organic solvents and with food additives. Appl Environ Microbiol 33:1233–1236

Isken S, de Bont JAM (1998) Bacteria tolerant to organic solvents. Extremophiles 2:229–238

Isken S, Santos P, de Bont JAM (1997) Effect of solvent adaptation on the antibiotic resistance in *Pseudomonas putida* S12. Appl Microbiol Biotechnol 48:642–647

Junker F, Ramos JL (1999) Involvement of the *cis/trans* isomerase Cti in solvent resistance of *Pseudomonas putida* DOT-T1E. J Bacteriol 181:5693–5700

Kabelitz N, Santos PM, Heipieper HJ (2003) Effect of aliphatic alcohols on growth and degree of saturation of membrane lipids in *Acinetobacter calcoaceticus*. FEMS Microbiol Lett 220:223–227

Keweloh H, Heipieper HJ (1996) *Trans* unsaturated fatty acids in bacteria. Lipids 31:129–137

Kiran MD, Prakash JSS, Annapoorni S, Dube S, Kusano T, Okuyama H, Murata N, Shivaji S (2004) Psychrophilic *Pseudomonas syringae* requires *trans*-monounsaturated fatty acid for growth at higher temperature. Extremophiles 8:401–410

Kiran MD, Annapoorni S, Suzuki I, Murata N, Shivaji S (2005) *Cis–trans* isomerase gene in psychrophilic *Pseudomonas syringae* is constitutively expressed during growth and under conditions of temperature and solvent stress. Extremophiles 9:117

Loffhagen N, Hartig C, Geyer W, Voyevoda M, Harms H (2007) Competition between *cis, trans* and cyclopropane fatty acid formation and its impact on membrane fluidity. Eng Life Sci 7:67–74

MacDonald PM, Sykes BD, McElhaney RN (1985) Flourine-19 nuclear magnetic resonance studies of lipid fatty acyl chain order and dynamics in *Acholeplasma laidlawii* b membranes: a direct comparsion of the effects of *cis* and *trans* cyclopropane ring and double-bond substituents on orientational order. Biochemistry 24:4651–4659

Morita N, Shibahara A, Yamamoto K, Shinkai K, Kajimoto G, Okuyama H (1993) Evidence for cis–trans isomerization of a double bond in the fatty acids of the psychrophilic bacterium *Vibrio* sp. strain ABE-1. J Bacteriol 175:916–918

Neumann G, Kabelitz N, Zehnsdorf A, Miltner A, Lippold H, Meyer D, Schmid A, Heipieper HJ (2005) Prediction of the adaptability of *Pseudomonas putida* DOT-T1E to a second phase of a solvent for economically sound two-phase biotransformations. Appl Environ Microbiol 71:6606–6612

Okuyama H, Okajima N, Sasaki S, Higashi S, Murata N (1991) The cis/trans isomerization of the double bond of a fatty acid as a strategy for adaptation to changes in ambient temperature in the psychrophilic bacterium, *Vibrio* sp. strain ABE-1. Biochim Biophys Acta 1084:13–20

Okuyama H, Ueno A, Enari D, Morita N, Kusano T (1998) Purification and characterization of 9-hexadecenoic acid cis–trans isomerase from *Pseudomonas* sp strain E-3. Arch Microbiol 169:29–35

Pedrotta V, Witholt B (1999) Isolation and characterization of the *cis–trans*-unsaturated fatty acid isomerase of *Pseudomonas oleovorans* GPo12. J Bacteriol 181:3256–3261

Ramos JL, Gallegos MT, Marques S, Ramos-Gonzalez MI, Espinosa-Urgel M, Segura A (2001) Responses of gram-negative bacteria to certain environmental stressors. Curr Opin Microbiol 4:166–171

Ramos JL, Duque E, Gallegos MT, Godoy P, Ramos-Gonzalez MI, Rojas A, Teran W, Segura A (2002) Mechanisms of solvent tolerance in gram-negative bacteria. Annu Rev Microbiol 56:743–768

Roach C, Feller SE, Ward JA, Shaikh SR, Zerouga M, Stillwell W (2004) Comparison of *cis* and *trans* fatty acid containing phosphatidylcholines on membrane properties. Biochemistry 43:6344

Seelig J, Waespe-Šarcevic N (1978) Molecular order in *cis* and *trans* unsaturated phospholipid bilayers. Biochemistry 17:3310–3315

Segura A, Duque E, Mosqueda G, Ramos JL, Junker F (1999) Multiple responses of gram-negative bacteria to organic solvents. Environ Microbiol 1:191–198

Sikkema J, de Bont JA, Poolman B (1995) Mechanisms of membrane toxicity of hydrocarbons. Microbiol Rev 59:201–222

Suutari M, Laakso S (1994) Microbial fatty acids and thermal adaptation. Crit Rev Microbiol 20:285–328

Von Wallbrunn A, Heipieper HJ, Meinhardt F (2002) *Cis/trans* isomerisation of unsaturated fatty acids in a cardiolipin synthase knock-out mutant of *Pseudomonas putida* P8. Appl Microbiol Biotechnol 60:179–185

Von Wallbrunn A, Richnow HH, Neumann G, Meinhardt F, Heipieper HJ (2003) Mechanism of *cis–trans* isomerization of unsaturated fatty acids in *Pseudomonas putida*. J Bacteriol 185:1730–1733

Weber FJ, de Bont JAM (1996) Adaptation mechanisms of microorganisms to the toxic effects of organic solvents on membranes. Biochim Biophys Acta 1286:225–245

Zhang YM, Rock CO (2008) Membrane lipid homeostasis in bacteria. Nat Rev Microbiol 6:222–233

# Surface Properties and Cellular Energetics of Bacteria in Response to the Presence of Hydrocarbons

# 26

Hermann J. Heipieper,
Milva Pepi, Thomas Baumgarten, and Christian Eberlein

## Contents

1   Introduction ................................................................. 398
   1.1   Physicochemical Surface Properties of Bacteria to Adapt to the Presence
      of Hydrocarbons ....................................................... 398
   1.2   *Corynebacterium-Mycobacterium-Nocardia-Rhodococcus* Group ................. 399
   1.3   *Acinetobacter* ........................................................ 399
   1.4   *Pseudomonas* ......................................................... 402
   1.5   Effect of Toxic Hydrocarbons on Bacterial Energetics ...................... 404
2   Research Needs .............................................................. 405
References ..................................................................... 406

**Abstract**

Many toxic hydrocarbons that are present as environmental pollutants are potential substrates for bacteria; other, very hydrophobic hydrocarbons exhibit extremely low water solubility and are poorly bioavailable. The development of specific adaptive mechanisms to the toxicity as well as the low bioavailability of these substrates allows many bacteria to cope with such challenges. Strategies of bacteria to increase the accessibility of these compounds are modifications of

H. J. Heipieper (✉) · C. Eberlein
Department Environmental Biotechnology, Helmholtz Centre for Environmental Research – UFZ, Leipzig, Germany
e-mail: hermann.heipieper@ufz.de; christian.eberlein@ufz.de

M. Pepi
Stazione Zoologica Anton Dohrn, Villa Comunale, Naples, Italy
e-mail: milva.pepi@szn.it

T. Baumgarten
Center for Biomembrane Research, Department of Biochemistry and Biophysics, Stockholm University, Stockholm, Sweden
e-mail: thomas.baumgarten@dbb.su.se

© Springer International Publishing AG, part of Springer Nature 2018
T. Krell (ed.), *Cellular Ecophysiology of Microbe: Hydrocarbon and Lipid Interactions*,
Handbook of Hydrocarbon and Lipid Microbiology, https://doi.org/10.1007/978-3-319-50542-8_50

their cell surfaces or the release of biosurfactants. Both "strategies" aim at an increased accessibility of the compounds, either by the reduction of surface tension or by allowing a direct hydrophobic-hydrophobic interaction between cell surface and the substrates. The toxicity of hydrocarbons is mainly caused by their permeabilizing effect on the cytoplasmic membranes leading also to a loss of ATP and a decrease in the proton gradient. Bacteria are able to modify their cellular energetics in order to adapt to the presence of toxic hydrocarbons by activating their electron transport phosphorylation systems allowing homeostasis of ATP level and energy charge in the presence of the toxic conditions, however, at the price of a reduced growth yield.

# 1     Introduction

Many hydrocarbons such as monoaromatics and alkanols that are present as environmental pollutants are toxic to bacteria. On the other hand, these compounds are also potential substrates for bacteria. Very hydrophobic hydrocarbons exhibit extremely low water solubility and are poorly bioavailable. In order to cope with such challenges, many bacteria capable of degrading potential organic pollutants have developed specific responses to the toxicity as well as to the low bioavailability of the hydrocarbons. One strategy of bacteria to increase the accessibility of these compounds is modifications of their cell envelopes regarding the hydrophobicity in order to allow a direct hydrophobic-hydrophobic interaction with the substrates. Other important responses of bacteria that allow degradation or survival in the presence of hydrocarbons are related to cellular energetics. The toxicity of organic solvents to cells is mainly caused by their permeabilizing effect on the cytoplasmic membranes leading also to a loss of ATP and a decrease in the proton gradient. Therefore, both the surface properties and energetics are important physiological parameters of bacteria to cope with the presence of hydrocarbons. This will be presented in the following chapter.

## 1.1     Physicochemical Surface Properties of Bacteria to Adapt to the Presence of Hydrocarbons

Most potential environmental pollutants can be degraded by at least one group of bacteria. However, even though the degradative potential is present in a contaminated site, the bacteria have to solve the problem that such hydrophobic compounds are poorly bioavailable due to their extremely low water solubility. Bacteria increase the bioavailability of substrates such as polycyclic aromatic hydrocarbons (PAH) by facilitating a direct adhesion to them in order to allow a direct uptake from the solid phase.

Physicochemical cell surface properties of bacteria are measured using the so-called microbial adhesion to hydrocarbon (MATH) test (Rosenberg 2006). This test simply measures the partitioning of cells in a hexadecane/water system and is commonly used for more than 25 years. Next to the MATH test, the water contact angles ($\theta_w$) (Van Loosdrecht et al. 1987b) and zeta potentials ($\zeta$) (Van Loosdrecht et al. 1987a) of cells are measured as values for surface hydrophobicity and surface charge, respectively, and are known to correlate directly to adhesion properties. Cell surface hydrophobicities are derived from the water contact angles ($\theta_w$) on bacterial lawns using a drop shape analysis system. The zeta potential ($\zeta$) as an indirect measure of cell surface charge is approximated from the electrophoretic mobility according to the method of Helmholtz-von Smoluchowski (Van Loosdrecht et al. 1987b).

## 1.2   *Corynebacterium-Mycobacterium-Nocardia-Rhodococcus Group*

Several studies regarding adaptive responses of environmentally relevant bacteria were carried out with Gram-positive bacteria belonging to the *Corynebacterium-Mycobacterium-Nocardia* group, including the genus *Rhodococcus*, bacteria capable of degrading PAH as well as alkanes and terpenes. In order to metabolize such compounds that regularly show a very low bioavailability, these bacteria have developed specific responses. These adaptive mechanisms include the presence of high affinity uptake systems, changes in the surface properties allowing direct adhesion to the substrate, and the excretion of biosurfactant in order to increase the bioavailability of the hydrocarbons (Lang and Philp 1998; Wick et al. 2002a). Next to the production of biosurfactants, this group of bacteria is known to contain large amounts of long-chained $C_{60}$–$C_{90}$ fatty acids (mycolic acids) that are specific for this group of bacteria. These fatty acids are located as a layer on the cell wall and are responsible for the pH-tolerance as well as the high surface hydrophobicity of these bacteria. Changes in the composition of these mycolic acids of *Mycobacterium* strains and other mycolate-containing genera have been described to depend on the availability and structure of the carbon source (Wick et al. 2002b, 2003). Anthracene-grown *Mycobacterium frederiksbergense* LB501T cells respond to the low bioavailability of their substrate by modifying their physicochemical cell wall properties leading to up to 70-fold better adhesion to hydrophobic surfaces in comparison to glucose-grown cells (Wick et al. 2002a). Also, cells of *Rhodococcus erythropolis* show adaptive changes in their surface properties when grown on *n*-alkanes (de Carvalho et al. 2009). In addition, *Rhodococcus opacus* modified its mycolic acid in the presence of both osmotic stress caused by NaCl and toxic concentrations of 4-chlorophenol (de Carvalho et al. 2016).

## 1.3   *Acinetobacter*

In addition, Gram-negative bacteria are known to change their surface properties as an adaptive response to the low bioavailability of potential substrates such as diesel fuel components. Best investigated strains are *Acinetobacter calcoaceticus* RAG-1

(Rosenberg and Rosenberg 1981) and *A. venetianus* VE-C3 (Baldi et al. 1999), the former successively assigned to the same genomic species of VE-C3 strain (Vaneechoutte et al. 1999). The cell surfaces of *A. venetianus* strain RAG-1 are always hydrophobic, even if growth is accomplished in a complex medium without hydrophobic substrate addition. On the contrary, the hydrophilic *A. venetianus* VE-C3 strain became hydrophobic only after exposure to *n*-alkanes. Both strains showed an extremely high hydrophobicity when cultivated with long-chained *n*-alkanes such as *n*-hexadecane or a mixture of *n*-alkanes ($C_{12}$–$C_{28}$) as carbon and energy source. By changing the composition of their outer membrane, especially the lipopolysaccharides (LPS), *Acinetobacter* strains are able to directly adhere to droplets of alkanes in order to allow a direct uptake of these hydrocarbons as substrates (Walzer et al. 2006). Additionally, these bacteria release biosurfactants when they grow in the presence of hydrocarbons that reduce the surface tension and also increase the bioavailability of alkanes.

Electrokinetic properties of the cells pointed out that *Acinetobacter* sp. strain RAG-1 was hydrophobic, with an electrophoretic mobility ($\mu$) of $-0.38 \times 10^{-8}$ m$^2$ V$^{-1}$ s$^{-1}$ and zeta potential ($\zeta$) of $-4.9$ mV in seawater, and MATH test confirmed its hydrophobicity (Baldi et al. 1999). RAG-1 produces emulsan, an extracellular bioemulsifying lipopolysaccharide, which reduces the surface tension of diesel fuel. RAG-1 cells observed in the light microscopy transmission mode break the surface of diesel fuel drops upon contact with the hydrophobic layer. The activity of emulsan is due to a presence of fatty acids that are attached to the polysaccharide backbone via O-ester and N-acyl linkages (Ron and Rosenberg 2002). Because of these properties, *Acinetobacter calcoaceticus* RAG-1 (successively to be emended to *A. venetianus* RAG-1) became well known for the commercial production of the biosurfactant emulsan (Pines and Gutnick 1986; Rosenberg and Ron 1997, 1999). Thus, the efficiency of hydrocarbon degradation in strain RAG-1 is due to the biosurfactant production and surface cell properties.

In the presence of diesel fuel droplets, cells of *A. venetianus* strain VE-C3 formed cell-to-cell aggregates and then adhered to the surface of the *n*-alkane droplets by a polysaccharide-based polymer anchored to the proteins or lipids of the outer membranes (CPS) (Baldi et al. 1999). Microscopic observations showed diesel fuel drops completely colonized by VE-C3 cells, having the effect to decreasing the surface tension. Strain VE-C3 in seawater was hydrophilic, with $\mu$ of $-0.81 \times 10^{-8}$ m$^2$ V$^{-1}$ s$^{-1}$ and $\zeta$ of $-10.5$ mV, and MATH test showed that the hydrophilic VE-C3 strain became hydrophobic only after exposure to *n*-alkanes (Baldi et al. 1999). Thus, strain VE-C3 showed a longer lag phase before starting to grow on diesel fuel as carbon and energy source, due to the fact that this strain formed cell-to-cell aggregates before adhesion to the surface of diesel fuel drops.

Glycosylation activity in strain VE-C3 grown in the presence of diesel fuel produced a polysaccharide capsule, and emulsified diesel fuel nanodroplets were observed at the cell envelope perimeter. A glycoprotein with an apparent molecular mass of 22 kDa, as revealed by ConA-lectin blotting analysis, in the outer membrane was probably involved in the bioemulsifying activity at the cell envelope (Baldi et al. 2003).

Electrochemical techniques allow the calculation of the film formation time ($\tau$), showing that in RAG-1 strain it stabilizes at $\tau_{lim} = 500$ ms, and for VE-C3 $\tau_{lim}$ dropped to zero, pointing out that in the latter strain the film is formed by the adsorption of dissolved biopolymers, whereas with $\tau_{lim}$ greater than zero rate-limiting surface process is involved in film formation. Biofilm formation in VE-C3 thus depends on the adsorption of CPS, which is faster than the coalescence of cell-spreading zones, the rate-limiting surface process in film formation by RAG-1 cells. Differences in $\tau_{lim}$ between RAG-1 and VE-C3 reflect a difference in the flexibility of their outer membrane and cell-to-cell interaction (Baldi et al. 1999).

Different behaviors are observed concerning glycoproteins embedded in outer membranes showing emulsifying activity, as previously observed in *Acinetobacter radioresistens* KA53 with the production of the bioemulsifier alasan, a high-mass complex of proteins and polysaccharides. The major emulsification activity of this complex is associated with a 45 kDa protein (AlnA), which is homologous to the outer membrane protein OmpA (Navon-Venezia et al. 1995; Toren et al. 2002). Different independently isolated *Acinetobacter* strains secrete a structural component of the outer membrane of Gram-negative bacteria, the OmpA protein, suggesting that these proteins have additional functions and are not merely structural outer membrane components (Walzer et al. 2006).

These properties can also be transferred from one cell to another, as experimental evidence indicates that a high-molecular-weight bioemulsifier, which coats the bacterial surface, can be horizontally transferred to other bacteria, thereby changing their surface properties and interactions with the environment (Osterreicher-Ravid et al. 2000).

Another mechanism of adhesion involving the reversible monolayer absorption of a "bald" strain, a mutant of the highly adhesive and hydrophobic bacterium *Acinetobacter* sp. strain Tol 5, to a hydrocarbon surface was described (Hori et al. 2008). This mutant lacks filamentous appendages and thereby agglutinating properties and adsorbs as a monolayer to a hydrocarbon surface. This mechanism of adsorption is important because it allows effective reaction and transport of hydrophobic substrates at oil-water interfaces.

In conclusion, the genus *Acinetobacter* shows different mechanisms of adhesion to hydrocarbons in order to exploit them as carbon and energy source; and this aspect is also of high importance regarding possible environmental biotechnological applications.

Concerning surface properties of *Acinetobacter* strains and Gram-positive bacteria, two bacterial strains, *A. venetianus* RAG-1 and *Rhodococcus erythropolis* 20S-E1-c, showing different macroscopic surface hydrophobicity, were compared using the dynamic pendant drop technique. The results of this study highlight differences between the two bacterial strains in terms of cell-cell interactions at the interface (Kang et al. 2008). Moreover, studies conducted with the atomic force microscopy gave insights into adhesion force distribution, showing the highest forces grouped at one pole of the cell for *R. erythropolis* 20S-E1-c and a random distribution of adhesion forces in the case of *A. venetianus* RAG-1 (Dorobantu et al. 2008).

## 1.4    *Pseudomonas*

Most research on modifications of the surface properties of Gram-negatives as adaptive response to changes in environmental conditions was done with *Pseudomonas aeruginosa* PAO1. This bacterium is the reference type strain for the investigation of biofilm formation as well as adhesion to surfaces and tissues. From this intense investigation, we now have a quite complex picture of the functioning of the very outer layer of Gram-negative bacteria as well as the mechanisms with which the bacteria are able to modify their cell surface properties as a response to changes in the environment, which are also of pathogenic relevance.

The major component in Gram-negative cells that affects surface properties such as charge and hydrophobicity is the composition of the LPS layer of the outer membrane. Here, especially the so-called O-specific region on the very outer cell surface has an effect on surface properties. In the LPS of *P. aeruginosa*, the O-specific region contains two major components. The A-band, a low-molecular-mass LPS, consists of a homopolymer of D-rhamnose with only minor amounts of 2-keto-3-deoxyoctonic acid (KDO). The B-band, a high-molecular-mass LPS, consists of a heteropolymer of mainly uronic acid derivatives and N-acetylfucosamine. In case of stress conditions caused by heat shock, the cells react with a complete loss of B-band LPS compared to the amounts of A-band LPS present on the surface. This leads to cells with a higher hydrophobicity. This loss of the B-band LPS is a rapidly occurring physiological reaction of the cells in order to affect surface charge, surface hydrophobicity, adhesion to hydrophobic surfaces, biofilm formation, as well as susceptibility to antimicrobial agents and host defense (Kelly et al. 1990). This alteration in the LPS composition occurs within 15 min after the addition of stressors such as heat shock (45 °C) (Makin and Beveridge 1996b), the membrane-active antibiotic gentamicin (Kadurugamuwa et al. 1993), and low oxygen stress (Sabra et al. 2003). The very fast physiological response was shown to be related to the formation of membrane vesicles mainly consisting of B-Band LPS that lead to fast and drastic increase in the hydrophobicity of the cells (Kadurugamuwa and Beveridge 1995; Sabra et al. 2003).

The changes in cell surface charge and hydrophobicity of the solvent-tolerant bacterium *Pseudomonas putida* DOT-T1E as an adaptive response to the presence of toxic concentrations of *n*-alkanols were shown. Cells exposed to 1-decanol immediately increased their water contact angles $\theta_w$ from 30 to 85°. This drastic increase in cell surface hydrophobicity was related to an increase in the negative surface charge as the zeta-potentials $\zeta$ decreased from $-15$ to $-30$ mV (Neumann et al. 2006). These corresponding changes of the water contact angle and zeta-potential reflect the negative correlation between cell hydrophobicity and surface charge already observed earlier (Van Loosdrecht et al. 1987a, b; Makin and Beveridge 1996a). The steep increase of cell hydrophobicity and the decrease in $\zeta$-potential of cells grown in the presence of 1-decanol took place within about 15 min after addition of the solvent. These changes were not observed when the 1-decanol was added to cells that had been killed by a 30 min incubation with $HgCl_2$ (Neumann et al. 2006). Thus, it was clearly demonstrated that the observed increase in the hydrophobicity of cells grown in the presence of 1-decanol was caused by physiological changes that could only be carried out by living cells (Neumann et al. 2006). Hence, it was proven that changes in the surface properties also occur as an adaptive response to the presence of toxic hydrocarbons. In addition, one investigated solvent-tolerant strain, *P. putida* Idaho, changed its LPS composition when grown in the presence of *o*-xylene. A higher-molecular-weight LPS band disappeared, whereas it was replaced by a lower-molecular-weight band in the presence of the aromatic compound (Pinkart et al. 1996). However, an explanation for the physiological advantage of a more hydrophobic cell surface as an adaptive response to the presence of very high concentrations of toxic hydrocarbons seems very difficult. This had already been discussed in 1998 by de Bont (1998) who had assumed a decrease in cell hydrophobicity in order to repel the solvent. The outer membrane is known to be a very good barrier for hydrophobic compounds. However, this very low permeability for hydrophobic compounds is usually more affected by the outer membrane porins than by variations in the LPS content. Taking into consideration that a whole cascade of adaptive mechanisms is necessary to allow bacteria to tolerate toxic hydrocarbons, the observed modification of the surface properties makes sense as this hydrophobic layer is able to accumulate more of the solvents (Heipieper et al. 2007). Especially the presence of specific efflux pumps that permanently remove the toxic hydrocarbons from the cytoplasmic membrane and transport them to the outer layer of the outer membrane supports this positive effect of an hydrophobic layer at the surface (Segura et al. 2004). Additionally, it seems that with the release of membrane vesicles leading to a hydrophobic surface, also the solvents accumulated in the membrane are removed from the cells as this was found for toluene (Kobayashi et al. 2000; Mashburn-Warren and Whiteley 2006). In *P. putida*, it was shown that the release of membrane vesicles is caused by different stressors directly leading to an increase in cell surface hydrophobicity and biofilm formation (Baumgarten et al. 2012a, b; Baumgarten and Heipieper 2016). However, the gene regulation as well as the enzymatic mechanism of the membrane vesicle formation is still not completely understood (Mashburn-Warren and Whiteley 2006; Kulp and Kuehn 2010; Tashiro et al. 2010a, b, 2012; Schwechheimer and Kuehn 2015).

Bacteria belonging to the genus *Pseudomonas* are not only famous because of their high solvent tolerance but mainly because of their capability to degrade a wide range of pollutants even at very low concentrations. Also, in the presence of nontoxic crude oil components such as hexadecane, an increase in cell surface hydrophobicity has been observed (Norman et al. 2002). A major reason for this change in the surface properties seems to be an increased adhesion to the surface of the very poorly soluble compounds that leads to an increase in the bioavailability of the compounds (Wick et al. 2002a, 2003). As in the adaptation to poorly water-soluble substrates, also uptake systems seem to be involved, a hydrophobic surface also works as a kind of source for the compounds that accumulates them at the cell surface and allows a better uptake (Arias-Barrau et al. 2005). Thus, a more hydrophobic surface can work as a kind of sink for toxic concentrations of solvents that are excluded by efflux pumps but also as a kind of source for less bioavailable substrates that are transported into the cells by several uptake systems.

## 1.5    Effect of Toxic Hydrocarbons on Bacterial Energetics

Many hydrocarbons are known to be toxic to cells, mainly due to their permeabilizing effect on the membranes (Heipieper et al. 1994; Sikkema et al. 1995). This leads to a loss of important cellular components and ions (Lambert and Hammond 1973; Heipieper et al. 1992), including a decrease in the proton gradient (Sikkema et al. 1994). Additionally, also the efflux of cellular ATP after addition of phenols was described (Heipieper et al. 1991). Next to this loss of energetic potential caused by chemical effects of the solvents, the adaptive mechanisms, especially the activity of energy-dependent efflux pumps (Segura et al. 2004), are consuming ATP that cannot be used for growth. This is reflected in an about 50% reduced growth yield that was observed in cells of *Pseudomonas putida* when grown in the presence of toxic organic solvents such as toluene and 1-decanol (Isken et al. 1999; Neumann et al. 2006). However, cells grown in the presence of solvents showed nearly the same growth rates μ than non-stressed bacteria ($0.63 \ h^{-1}$ compared to $0.70 \ h^{-1}$). The data of both investigations are summarized in Table 1. Taking into consideration all possible negative effects of toxic hydrocarbons on bacteria, one would assume a quite disturbed energetic level of the cells when grown in the presence of solvents (Neumann et al. 2006).

Astonishingly, the ATP content of cells was even slightly higher when grown in the presence of 1-decanol than in the absence of the solvent. The cellular ATP content was 8–9 nmol/mg dry weight, which is similar to results found previously for aerobic bacteria (Tran and Unden 1998).

The lower ATP contents of cells grown in the absence of toxic solvents can be explained by a decreasing specific ATP production at higher growth rates, indicating a higher energetic efficiency of carbon substrate utilization during fermentations in the absence of 1-decanol (Kayser et al. 2005). As the ATP content by itself does not always reflect the actual energy status, also the concentrations of the other adenine nucleotides and the adenylate energy charge that allows an exact expression of the energetic levels of the cells under different growth conditions should be taken into consideration. The adenylate energy charge (E.C.) is given by the following equation:

$$E.C. = \frac{(ATP) + 1/2(ADP)}{(ATP) + (ADP) + (AMP)}$$

Regularly, aerobic bacteria show an E.C. of 0.8–0.95 when grown exponentially (Atkinson and Walton 1967). The E.C. declines to values of 0.3–0.45 when the cells enter late exponential phase and further decreases in the stationary phase; also on the energetic level, viability of the bacterial cells still is maintained before the cells die (Chapman et al. 1971; Loffhagen and Babel 1985; Lundin et al. 1986). However, also the energy charge of the cells showed no difference upon incubation with or without the solvent (Neumann et al. 2006). Apparently, a complete adaptation of the

**Table 1** Growth rates, doubling times, and relative yield for fermentations with two solvent-tolerant strains of *Pseudomonas putida* grown in the presence and absence of a second phase of the solvents toluene and 1-decanol (Isken et al. 1999; Neumann et al. 2006)

| Strain | *P. putida* DOT-T1E | | *P. putida* S12 | |
|---|---|---|---|---|
| Solvent | Control | + 1-decanol | Control | + Toluene |
| C-source | Na$_2$-succinate | | Glucose | |
| Growth rate μ (h$^{-1}$) | 0.70 | 0.63 | 0.71[a] | 0.54[a] |
| Doubling time t$_d$ (min) | 59 | 66 | 58[b] | 76[b] |
| Yield (g protein/g C-source) | 0.21 | 0.11 | 0.34 | 0.20 |
| Relative cell yield (%) | 100 | 52 | 100 | 59 |

[a]Maximum dilution rate (= growth rate) possible in chemostat cultures not leading to washout of the cells
[b]Maximum doubling time possible in chemostat cultures not leading to washout of the cells

bacterial cells can be achieved which is reflected in similar bioenergetics regardless of the presence or absence of toxic organic solvents. Proteomic studies of two *P. putida* strains grown in the presence of toxic solvents revealed an up-regulation of enzymes responsible for the electron chain transport and ATP synthesis (Segura et al. 2005; Volkers et al. 2006).

It can be concluded that although the bacteria need additional energy for their adaptation to the presence of toxic hydrocarbons, they are able to maintain or activate their electron transport phosphorylation allowing homeostasis of ATP level and energy charge in the presence of the solvent, at the price of a reduced growth yield.

## 2   Research Needs

One of the most important open research questions regarding the ability of bacteria to modify their surface properties is the elucidation of the exact physiological and biochemical mechanism of this important process. Here, especially the biology of the fast release of membrane vesicles will be a subject of intense future research, with probable relevance in medical microbiology. A possible field for environmental microbiological research on this topic could be represented by studies on bacterial aggregation and on the hydrocarbon incorporation mechanism by polysaccharides to be carried out in natural samples constituted by aggregates that appear in marine areas. This phenomenon was observed in the Adriatic Sea and, although less evident, in the Tyrrhenian Sea (Misic et al. 2005). Experiments should be conducted with samples collected from marine aggregates or from foam floating at the sea surface in hydrocarbon polluted sites and harbor areas. Physiological studies and microscopic analyses should reveal how the contact between hydrocarbons and bacterial cells is achieved, maybe by still unknown chemicals that are interesting for industrial applications.

# References

Arias-Barrau E, Sandoval A, Olivera ER, Luengo JM, Naharro G (2005) A two-component hydroxylase involved in the assimilation of 3-hydroxyphenyl acetate in *Pseudomonas putida*. J Biol Chem 280:26435

Atkinson DE, Walton GM (1967) Adenosine triphosphate conservation in metabolic regulation. Rat liver citrate cleavage enzyme. J Biol Chem 342:3239–3241

Baldi F et al (1999) Adhesion of *Acinetobacter venetianus* to diesel fuel droplets studied with *in situ* electrochemical and molecular probes. Appl Environ Microbiol 65:2041–2048

Baldi F et al (2003) Envelope glycosylation determined by lectins in microscopy sections of *Acinetobacter venetianus* induced by diesel fuel. Res Microbiol 154:417–424

Baumgarten T, Heipieper HJ (2016) Outer membrane vesicle secretion: from envelope stress to biofilm formation. In: de Bruijn FJ (ed) Stress and environmental control of gene expression in bacteria. Wiley-Blackwell, New York, pp 1322–1327

Baumgarten T et al (2012a) Membrane vesicle formation as a multiple-stress response mechanism enhances *Pseudomonas putida* DOT-T1E cell surface hydrophobicity and biofilm formation. Appl Environ Microbiol 78:6217–6224. https://doi.org/10.1128/aem.01525-12

Baumgarten T et al (2012b) Alkanols and chlorophenols cause different physiological adaptive responses on the level of cell surface properties and membrane vesicle formation in *Pseudomonas putida* DOT-T1E. Appl Microbiol Biotechnol 93:837–845. https://doi.org/10.1007/s00253-011-3442-9

Chapman AG, Fall L, Atkinson DE (1971) Adenylate energy charge in *Escherichia coli* during growth and starvation. J Bacteriol 108:1072–1086

de Bont JAM (1998) Solvent-tolerant bacteria in biocatalysis. Trends Biotechnol 16:493–499

de Carvalho CCCR, Wick LY, Heipieper HJ (2009) Cell wall adaptations of planktonic and biofilm *Rhodococcus erythropolis* cells to growth on C5 to C16 n-alkane hydrocarbons. Appl Microbiol Biotechnol 82:311–320

de Carvalho CCCR, Fischer MA, Kirsten S, Wurz B, Wick LY, Heipieper HJ (2016) Adaptive response of *Rhodococcus opacus* PWD4 to salt and phenolic stress on the level of mycolic acids. AMB Express 6:8. https://doi.org/10.1186/s13568-016-0241-9

Dorobantu LS, Bhattacharjee S, Foght JM, Gray MR (2008) Atomic force microscopy measurement of heterogeneity in bacterial surface hydrophobicity. Langmuir 24:4944–4951

Heipieper HJ, Keweloh H, Rehm HJ (1991) Influence of phenols on growth and membrane permeability of free and immobilized *Escherichia coli*. Appl Environ Microbiol 57:1213–1217

Heipieper HJ, Diefenbach R, Keweloh H (1992) Conversion of *cis*-unsaturated fatty acids to *trans*, a possible mechanism for the protection of phenol-degrading *Pseudomonas putida* P8 from substrate toxicity. Appl Environ Microbiol 58:1847–1852

Heipieper HJ, Weber FJ, Sikkema J, Keweloh H, de Bont JAM (1994) Mechanisms behind resistance of whole cells to toxic organic solvents. Trends Biotechnol 12:409–415

Heipieper HJ, Neumann G, Cornelissen S, Meinhardt F (2007) Solvent-tolerant bacteria for biotransformations in two-phase fermentation systems. Appl Microbiol Biotechnol 74:961–973

Hori K, Watanabe H, Ishii S, Tanji Y, Unno H (2008) Monolayer adsorption of a "bald" mutant of the highly adhesive and hydrophobic bacterium *Acinetobacter* sp strain tol 5 to a hydrocarbon surface. Appl Environ Microbiol 74:2511–2517

Isken S, Derks A, Wolffs PFG, de Bont JAM (1999) Effect of organic solvents on the yield of solvent-tolerant *Pseudomonas putida* S12. Appl Environ Microbiol 65:2631–2635

Kadurugamuwa JL, Beveridge TJ (1995) Virulence factors are released from *Pseudomonas aeruginosa* in association with membrane vesicles during normal growth and exposure to gentamicin: a novel mechanism of enzyme secretion. J Bacteriol 177:3998–4008

Kadurugamuwa JL, Lam JS, Beveridge TJ (1993) Interaction of gentamicin with the A band and B band lipopolysaccharides of *Pseudomonas aeruginosa* and its possible lethal effect. Antimicrob Agents Chemother 37:715–721

Kang ZW, Yeung A, Foght JM, Gray MR (2008) Mechanical properties of hexadecane-water interfaces with adsorbed hydrophobic bacteria. Colloids Surf B: Biointerfaces 62:273–279

Kayser A, Weber J, Hecht V, Rinas U (2005) Metabolic flux analysis of *Escherichia coli* in glucose-limited continuous culture. I. Growth-rate-dependent metabolic efficiency at steady state. Microbiology 151:693–706

Kelly NM, MacDonald MH, Martin N, Nicas T, Hancock REW (1990) Comparison of the outer membrane protein and lipopolysaccharide profiles of mucoid and nonmucoid *Pseudomonas aeruginosa*. J Clin Microbiol 28:2017–2021

Kobayashi H, Uematsu K, Hirayama H, Horikoshi K (2000) Novel toluene elimination system in a toluene-tolerant microorganism. J Bacteriol 182:6451–6455

Kulp A, Kuehn MJ (2010) Biological functions and biogenesis of secreted bacterial outer membrane vesicles. In: Gottesman S, Harwood CS (eds) Annual review of microbiology. Annual Reviews, Palo Alto, pp 163–184

Lambert PA, Hammond SM (1973) Potassium fluxes, first indication of membrane damage in micro-organisms. Biochem Biophys Res Commun 54:796–799

Lang S, Philp JC (1998) Surface-active lipids in rhodococci. Anton Leeuw Int J Gen Mol Microbiol 74:59–70

Loffhagen N, Babel W (1985) pH-linked control of energy charge in *Acetobacter methanolicus* sp. MB 70. J Basic Microbiol 25:575–580

Lundin A, Hasenson M, Persson J, Pousette A (1986) Estimation of biomass in growing cell lines by adenosine triphosphate assay. Methods Enzymol 133:27–44

Makin SA, Beveridge TJ (1996a) The influence of A-band and B-band lipopolysaccharide on the surface characteristics and adhesion of *Pseudomonas aeruginosa* to surfaces. Microbiology 142:299–307

Makin SA, Beveridge TJ (1996b) *Pseudomonas aeruginosa* PAO1 ceases to express serotype-specific lipopolysaccharide at 45°C. J Bacteriol 178:3350–3352

Mashburn-Warren LM, Whiteley M (2006) Special delivery: vesicle trafficking in prokaryotes. Mol Microbiol 61:839–846

Misic C, Giani M, Povero P, Polimene L, Fabiano M (2005) Relationships between organic carbon and microbial components in a Tyrrhenian area (Isola del Giglio) affected by mucilages. Sci Total Environ 353:350–359

Navon-Venezia S et al (1995) Alasan, a new bioemulsifier from *Acinetobacter radioresistens*. Appl Environ Microbiol 61:3240–3244

Neumann G et al (2006) Energetics and surface properties of *Pseudomonas putida* DOT-T1E in a two-phase fermentation system with 1-decanol as second phase. Appl Environ Microbiol 72:4232–4238

Norman RS, Frontera-Suau R, Morris PJ (2002) Variability in *Pseudomonas aeruginosa* lipopolysaccharide expression during crude oil degradation. Appl Environ Microbiol 68:5096–5103

Osterreicher-Ravid D, Ron EZ, Rosenberg E (2000) Horizontal transfer of an exopolymer complex from one bacterial species to another. Environ Microbiol 2:366–372

Pines O, Gutnick D (1986) Role for emulsan in growth of *Acinetobacter calcoaceticus* RAG-1 on crude oil. Appl Environ Microbiol 51:661–663

Pinkart HC, Wolfram JW, Rogers R, White DC (1996) Cell envelope changes in solvent-tolerant and solvent-sensitive *Pseudomonas putida* strains following exposure to o-xylene. Appl Environ Microbiol 62:1129–1132

Ron EZ, Rosenberg E (2002) Biosurfactants and oil bioremediation. Curr Opin Biotechnol 13:249–252

Rosenberg M (2006) Microbial adhesion to hydrocarbons: twenty-five years of doing MATH. FEMS Microbiol Lett 262:129–134

Rosenberg E, Ron EZ (1997) Bioemulsans: microbial polymeric emulsifiers. Curr Opin Biotechnol 8:313–316

Rosenberg E, Ron EZ (1999) High- and low-molecular-mass microbial surfactants. Appl Microbiol Biotechnol 52:154–162

Rosenberg M, Rosenberg E (1981) Role of adherence in growth of *Acinetobacter calcoaceticus* RAG-1 on hexadecane. J Bacteriol 148:51–57

Sabra W, Lunsdorf H, Zeng AP (2003) Alterations in the formation of lipopolysaccharide and membrane vesicles on the surface of *Pseudomonas aeruginosa* PAO1 under oxygen stress conditions. Microbiology 149:2789–2795

Schwechheimer C, Kuehn MJ (2015) Outer-membrane vesicles from Gram-negative bacteria: biogenesis and functions. Nat Rev Microbiol 13:605–619. https://doi.org/10.1038/nrmicro3525

Segura A et al (2004) Enzymatic activation of the *cis-trans* isomerase and transcriptional regulation of efflux pumps in solvent tolerance in *Pseudomonas putida*. In: Ramos JL (ed) The pseudomonads. Kluwer Press, Dordrecht, pp 479–508

Segura A et al (2005) Proteomic analysis reveals the participation of energy- and stress-related proteins in the response of *Pseudomonas putida* DOT-T1E to toluene. J Bacteriol 187:5937–5945

Sikkema J, de Bont JA, Poolman B (1994) Interactions of cyclic hydrocarbons with biological membranes. J Biol Chem 269:8022–8028

Sikkema J, de Bont JA, Poolman B (1995) Mechanisms of membrane toxicity of hydrocarbons. Microbiol Rev 59:201–222

Tashiro Y, Ichikawa S, Nakajima-Kambe T, Uchiyama H, Nomura N (2010a) *Pseudomonas* quinolone signal affects membrane vesicle production in not only Gram-negative but also Gram-positive bacteria. Microbes Environ 25:120–125. https://doi.org/10.1264/jsme2.ME09182

Tashiro Y et al (2010b) Variation of physiochemical properties and cell association activity of membrane vesicles with growth phase in *Pseudomonas aeruginosa*. Appl Environ Microbiol 76:3732–3739. https://doi.org/10.1128/aem.02794-09

Tashiro Y, Uchiyama H, Nomura N (2012) Multifunctional membrane vesicles in *Pseudomonas aeruginosa*. Environ Microbiol 14:1349–1362. https://doi.org/10.1111/j.1462-2920.2011.02632.x

Toren A, Orr E, Paitan Y, Ron EZ, Rosenberg E (2002) The active component of the bioemulsifier alasan from *Acinetobacter radioresistens* KA53 is an OmpA-like protein. J Bacteriol 184:165–170

Tran QH, Unden G (1998) Changes in the proton potential and the cellular energetics of *Escherichia coli* during growth by aerobic and anaerobic respiration or by fermentation. Eur J Biochem 251:538

Van Loosdrecht MCM, Lyklema J, Norde W, Schraa G, Zehnder AJB (1987a) Electrophoretic mobility and hydrophobicity as a measure to predict the initial steps of bacterial adhesion. Appl Environ Microbiol 53:1898–1901

Van Loosdrecht MCM, Lyklema J, Norde W, Schraa G, Zehnder AJB (1987b) The role of bacterial cell wall hydrophobicity in adhesion. Appl Environ Microbiol 53:1893–1897

Vaneechoutte M et al (1999) Oil-degrading *Acinetobacter* strain RAG-1 and strains described as '*Acinetobacter venetianus* sp nov.' belong to the same genomic species. Res Microbiol 150:69–73

Volkers RJM, de Jong AL, Hulst AG, van Baar BLM, de Bont JAM, Wery J (2006) Chemostat-based proteomic analysis of toluene-affected *Pseudomonas putida* S12. Environ Microbiol 8:1674–1679

Walzer G, Rosenberg E, Ron EZ (2006) The *Acinetobacter* outer membrane protein A (OmpA) is a secreted emulsifier. Environ Microbiol 8:1026–1032

Wick LY, de Munain AR, Springael D, Harms H (2002a) Responses of *Mycobacterium* sp LB501T to the low bioavailability of solid anthracene. Appl Microbiol Biotechnol 58:378–385

Wick LY, Wattiau P, Harms H (2002b) Influence of the growth substrate on the mycolic acid profiles of mycobacteria. Environ Microbiol 4:612–616

Wick LY, Pasche N, Bernasconi SM, Pelz O, Harms H (2003) Characterization of multiple-substrate utilization by anthracene-degrading *Mycobacterium frederiksbergense* LB501T. Appl Environ Microbiol 69:6133–6142

# Ultrastructural Insights into Microbial Life at the Hydrocarbon: Aqueous Environment Interface

# 27

Nassim Ataii, Tyne McHugh, Junha Song, Armaity Nasarabadi, and Manfred Auer

## Contents

1   Introduction ............................................................................. 410
2   Fossil Fuel Production and Spills ........................................................ 410
3   Oil Well Souring ......................................................................... 411
4   Sulfate Reduction Using *Desulfovibrio vulgaris* ......................................... 411
5   Biofuel Production ....................................................................... 412
6   The Role of Imaging in Microbe-Hydrocarbon Research ...................................... 413
7   Research Needs ........................................................................... 415
8   Future Studies ........................................................................... 416
References ................................................................................. 416

**Abstract**

Despite the harmful effects observed when bacteria grow in a hydrocarbon-rich environment, some have been able to overcome the potential toxicity; however, specific interactions that operate at the hydrocarbon/aqueous interface remain unknown due to the difficulty of studying these interactions. Fortunately, there have been vast improvements in sample preparation such as the introduction of high-pressure freezing/freeze substitution (HPF/FS) which are able to preserve the ultrastructure while imaging. This process has been a gateway to a greater understanding of the ultrastructure of these interactions which could present deeper insight into the many processes that involve hydrocarbons. These processes include events such as catastrophic oil spills that give the opportunity to study the hydrocarbon/aqueous interface for the potential of utilizing new mechanisms in future disasters. This follows the possibility of reducing industrial oil souring by studying

N. Ataii · T. McHugh · J. Song · A. Nasarabadi · M. Auer (✉)
Molecular Biophysics and Integrated Bioimaging Division, Lawrence Berkeley National Laboratory, Berkeley, CA, USA
e-mail: mauer@lbl.gov

© Springer International Publishing AG, part of Springer Nature 2018
T. Krell (ed.), *Cellular Ecophysiology of Microbe: Hydrocarbon and Lipid Interactions*,
Handbook of Hydrocarbon and Lipid Microbiology, https://doi.org/10.1007/978-3-319-50542-8_11

sulfate-producing bacterium, as well as furthering our understanding in biofuel production, where engineered microbes are used to produce hydrocarbon fuels.

# 1    Introduction

Microbial life at the hydrocarbon/aqueous environment interface is one of the most fascinating yet ill-understood phenomena in microbiology, in part, due to the challenges faced by microbes when interacting with a hydrocarbon surface. Furthermore understanding of the hydrocarbon/aqueous environment interface is of high economic significance, be it in the context of oil spills, oil well souring, or the production of biofuels.

# 2    Fossil Fuel Production and Spills

As a result of fossil fuel drilling, hydrocarbons can be released incrementally from natural leaks in the ocean floor or by catastrophes and can cause havoc on microbial communities in freshwater and oceanic bodies of water which are known to influence the biogeochemical cycles and food webs of the entire planet (Zehr 2010).

The Deepwater Horizon oil spill was the second largest accidental marine oil spill in the history of the oil industry and is an example of how hydrocarbons can devastate bacterial communities. This spill resulted in a huge influx of hydrocarbons (around 780,000 cubic meters or 210 million US gallons) from the Macondo well (MC252) over 3 months of being released into the Gulf of Mexico (Baelum et al. 2012). This tragic accident provided an opportunity to enrich and isolate indigenous hydrocarbon-degrading bacteria and to identify the hydrocarbon concentrations and microbial community composition (Baelum et al. 2012; Hazen et al. 2010). There was also the opportunity to study the effect of the oil dispersant COREXIT and the iron source $FeCl_2$ on floc formation and the microbial degradation of oil (Macondo MC252), *since these reagents are often used in oil spills, yet their effects on indigenous microbes in the environment and their potential influence oil degradation rates had been previously unknown* (Baelum et al. 2012).

When isolated and enriched with Macondo MC252, the indigenous microbial communities demonstrated approximately 60% to 25% degradation of the oil with or without COREXIT, with no negative effects of high amounts of COREXIT on the growth of indigenous microorganisms from the isolation site. Additionally, $FeCl_2$ was shown to increase respiration rates, but not the total amount of hydrocarbons degraded (Baelum et al. 2012). Enrichment with MC252 and/or COREXIT in the absence of $FeCl_2$ also leads to floc formation, with sequences representative of Colwellia becoming dominant. This species specifies was confirmed to rapidly degrade high amounts of MC252, demonstrating the evolution of flocs during exposure to high concentrations of MC252 (Baelum et al. 2012).

The ecological significance of high hydrocarbon amounts on microbial composition was further shown by studies done with the oil on the shore of Grand Isle,

Louisiana, that caused a shift in the community structure toward hydrocarbonoclastic consortia. Gene sequencing demonstrated a diverse array of known petroleum hydrocarbon-degrading microorganisms, including Marinobacter strains (Lamendella et al. 2014).

Petroleum hydrocarbon degradation/catabolism is an energetically favorable process that occurs under both aerobic and anaerobic conditions utilizing pathways that involve terminal oxidation (subterminal oxidation, ω-oxidation, and β-oxidation), reduction, hydroxylation, and dehydrogenation reactions (Varjani 2017; Abbasian et al. 2015). In aerobic conditions, hydrophobic hydrocarbon methyl groups undergo oxidation to form an alcohol, which is then dehydrogenated via an aldehyde into a corresponding carboxylic acid, which can then be metabolized by the β-oxidation pathway of fatty acids (Varjani 2017; Das and Chandran 2011). Corresponding oxygenases and peroxidases to this degradation mechanism have also been identified for the purpose of microbial degradation of hydrocarbon pollutants; examples include alkB gene for alkane monooxygenase, xylE gene for catechol dioxygenase, and nahAc gene for naphthalene dioxygenase (Varjani 2017). We have made progress in elucidating a number of variables that affect biodegradation of hydrocarbons, including oxygen availability, optimal temperature range, and adequate pH (Ratcliffe 2017). However for microbe-, host-, environment-, and location-specific mechanisms of catabolism, the details of the hydrocarbon-microbe interaction remain largely unknown.

## 3    Oil Well Souring

Apart from oil spills, there is also economic value in preventing the souring of oil wells. Sulfate-reducing bacteria can metabolize hydrocarbons present in the aqueous environment of an oil well, using sulfate as a terminal electron acceptor, thus reducing sulfate to hydrogen sulfide gas, with a foul odor and corrosive properties. Oil is considered "sour" if it contains more than 0.5% sulfur and accelerates corrosion of the metal infrastructure used for oil production and processing (Leffler 2008). Similarly, sulfur would damage engine components if it were not removed when oil is refined in fuels. High-sulfur oil is more costly to refine and as a result is less valuable than low-sulfur oil (Leffler 2008). One mechanism for the souring of oil is sulfate-reducing bacteria present in the seawater, which oxidizes hydrocarbons in the aqueous interface and produces sulfides in the process. The *Desulfobulbaceae* family and its notable member *Desulfovibrio vulgaris* (*D. vulgaris*) are known to contribute to oil well souring (Muyzer and Stams 2008).

## 4    Sulfate Reduction Using *Desulfovibrio vulgaris*

Sulfate reducers like *D. vulgaris* are able to survive in a variety of toxic conditions, including hydrocarbon environments by adjusting their metabolism to the point that they can use sulfate as the terminal electron acceptor to generate sulfide, as long as they have hydrogen, formate, lactate, ethanol, or other organic compounds, such as crude oil, as a carbon source (Liamleam and Annachhatre 2007). The US Department of Energy has sponsored detailed research of *D. vulgaris* as a sulfate reducer; as a result, it is a well-studied model system for sulfate reducers. Genome sequence analysis revealed a complex, periplasmic cytochrome network, along with details about its transmembrane electron transport and cytoplasmic sulfate reduction capabilities (Heidelberg et al. 2004).

This research identified cytoplasmic isozymes as a critical component in hydrogen cycling and an unexpectedly high number of formate dehydrogenases. The latter suggests a system of chemiosmotic energy conservation by the diffusion of an uncharged metabolic intermediate, i.e., formate, from the cytosol with subsequent periplasmic oxidation (Heidelberg et al. 2004).

# 5    Biofuel Production

Microbes have a significant role to play in the production of biogenic transportation fuels (Lee et al. 2008). Microbial metabolic pathways are reengineered to produce hydrophobic precursors to biofuels or final products, which when produced in high quantity are either stored inside the cells (due to the hydrophobic effect) or continuously secreted. Hydrocarbons are relatively simple organic substances that, while being comprised of only carbon and hydrogen, constitute a large variety of compounds, each with their own unique chemical properties. For example, hydrocarbons with aliphatic structure will behave differently to aromatic compounds. Thus biodegradability of hydrocarbons can be ranked from highest to lowest as follows: linear alkanes > branched alkanes > low-molecular-weight alkyl aromatics > monoaromatics > cyclic alkanes > polyaromatics ≫ asphaltenes (Atlas 1981; Leahy and Colwell 1990). However, in most cases, hydrocarbons negatively interfere with membrane function: The Douglas fir terpene α-pinene and toluene led to massive leakage of the cytoplasm of *S. cerevisiae* (Andrews et al. 1980) and *E. coli* (de Smet et al. 1978). *Candida* grown in increasing concentration of n-alkanes reduces glucose utilization (Gill and Ratledge 1972). Phospholipid bilayer integrity is compromised, resulting in the breakdown of membrane ion gradients (Sikkema et al. 1994) and thus cell death.

In our own biofuel-related research, we studied *E.coli* and yeast producing various titers of hydrocarbons, and at higher titer, we find microbes to be ultrastructurally highly disturbed (see Fig. 1), with microbes trying to compensate for the toxicity by compartmentalization. While it seems clear that hydrocarbons have a devastating effect on microbes, the exact reason for its toxicity is poorly understood. At the core of the mystery lies how microbes deal with the challenge put forward by hydrophobic, lipophilic surfaces on the microbial cellular membranes and macromolecular bacterial organization. It would seem from a chemical point of view that hydrocarbons could act like an organic solvent, thus posing a real threat to the integrity of lipid bilayer-based membranes and with it to the ion gradients across such membranes needed to allow proper physiological function. It seems almost ironic that microbial cells may literally sit on an infinite carbon source, yet struggling to keep alive and metabolically productive. And while many bacteria are not up for

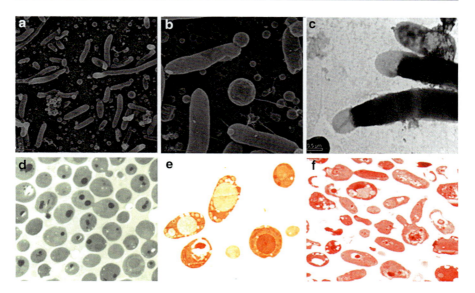

**Fig. 1** Ultrastructural analysis of *E. coli* (top row) and *S. cerevisiae* (bottom row) producing biofuels. Low magnification (**a**) and higher magnification (**b**) scanning electron micrographs of *E. coli* producing biofuels, resulting in the formation of globular structures pinching off from the bacterial membrane surface. Whole-mount TEM imaging of *E. coli* (**c**) confirms that compartmentalization of biofuel into droplets, marked by asterics; (**d–e**) *S. cerevisiae* producing increased titers of biofuels. Note that under low titer conditions (**d**), cells show a healthy uniform shape (note normal-looking uniform dark, electron dense vacuoles), whereas medium (**e**) and high (**f**) titers resulting in cells with increasingly distorted ultrastructural morphology, including distressed overall shape as well as irregular internal features, including partially solvent-extracted vacuoles, a likely storage place in plant cells

the task, there are some that not only will be able to cope with such hostile conditions but also thrive in such hydrophobic hydrocarbon-rich environments. Interestingly, gram-negative bacteria exhibited higher tolerance against lipophilic compound when compared to gram-positive bacteria (Harrop et al. 1989). The higher tolerance of gram-negative bacteria presumably is due to the presence of lipopolysaccharides in its outer membrane, which may constitute a protective layer. Another strategy to increase tolerance against hydrocarbons is the expression of bacterial membrane protein efflux pumps, which can result in rapid removal of toxic hydrocarbon from the inside of microbes. Engineering *E. coli* to express an efflux pump from *A. borkumensis* allowed increased limonene production yield (Dunlop et al. 2011).

# 6    The Role of Imaging in Microbe-Hydrocarbon Research

The mysteries of how exactly microbes have adapted to the challenges of hydrophobicity are complicated by the fact that bacteria rarely exist as individual, independent cells but mostly are found in biofilms. When forming biofilms, a large number of microbes display completely different expression profiles, compared to their planktonic state (Sauer et al. 2002). Therefore it would be expected that hydrocarbon-challenged microbes will form complex microbial communities that are difficult to study.

Imaging may provide unique insight in the hydrocarbon/aqueous environment interface. For instance, synchrotron radiation-based Fourier transform infrared (SR-FTIR) spectromicroscopy studied bacterial floc groups, including their low-resolution structure and chemical composition. Bacterial floc groups are formed by flocculation or the aggregation of oil, biomass, and carbohydrates (Baelum et al. 2012). Strongest peaks indicate the intensity of radiation measured by spectromicroscopy and are presented as absorbance units (a.u.). With higher concentrations of a particular element of a particular chemical bond, the intensity of the emitted IR radiation is higher. This is a measurement of spatial distribution MC252 oil and proteins as compared to measurements taken elsewhere in the sample. Thus the relative location of the non-degraded oil, the oil degradation products, and the bacteria can be determined from such FTIR spectra, revealing that the pollutant oil initially accumulates in the floc groups and then is degraded over time.

When it comes to trying to understand the hydrocarbon/aqueous environment interface, we need to consider a high-resolution imaging mode like electron microscopy. Whole-mount negative stain and cryo-imaging of microbes (2D or 3D via electron tomography) are options but face the challenge that the entire sample thickness (microbe and environment) needs to be thinner than ~0.5–1 μm, which can be difficult to accomplish for individual bacteria and is clearly not achievable for biofilms without sectioning (McDonald and Auer 2006). While cryo-sectioning of frozen-hydrated samples followed by cryo-EM 2D projection or 3D electron tomography is a possibility, this approach suffers from a low success rate and low throughput, with less than ten labs in the world having the necessary background to do it. However, such a sample preparation approach (cryo-sectioning) may be viewed as the gold standard for obtaining unstained, vitrified samples that upon TEM imaging do not suffer from sample preparation artifacts. Fortunately, there is an alternative approach that is technically much less challenging and yields excellent results in terms of contrast and preservation: high-pressure freezing/freeze substitution (HPF/FS), which cryogenically freezes and thus physically immobilizes the biological system within milliseconds at liquid nitrogen temperature, with the high pressure discouraging ice crystal formation (McDonald and Auer 2006; Matias et al. 2003). Not only does this physical fixation allow us to capture the cellular scenery without disturbing the ultrastructure, but it also allows for the cells be viable once thawed (Hunter and Beveridge 2005). The process of high-pressure freezing/freeze substitution is initiated by the immersion of the specimens on Isopore membrane filters in 10% glycerol or 20% bovine serum albumin in CYE medium (Palsdottir et al. 2009). After immersion, the filters are sandwiched between two aluminum planchettes (Palsdottir et al. 2009). Next specimens are cryofixed in a high-pressure freezer such as the Lecia EM PACT2. The Lecia EM PACT2 is the most advanced high-pressure freezer which includes a rapid transfer system (RTS) that allows one to do correlative light and electron microscopy with high time resolution (McDonald et al. 2007). Once cyrofixed specimens are in anhydrous acetone containing 1% osmium tetroxide and 0.1% uranyl acetate and infiltrated with Epon-Araldite, they can be used in a freeze substitution system (McDonald et al. 2007). During the freeze substitution process, samples are kept at low (dry ice) temperature ($-90$ °C), where the cellular water is gradually replaced by an organic solvent by letting the sample sit for 48 h at this temperature (Jhamb and Auer 2015). After the allotted time, the samples should slowly be warmed up until they reach $-30$ °C, and then they should maintain this temperature for 3 h before they are warmed to 0 °C (Jhamb and Auer 2015). During the warm-up period, the samples can undergo heavy metal staining. Such samples can then be resin embedded and ultrathin sectioned, before being examined by room temperature TEM. HPF/FS in combination with electron tomography has been successfully applied to a variety of biofilms, including the well-studied soil bacterium, *Myxococcus xanthus* (Palsdottir et al. 2009), and *Desulfovibrio vulgaris* (manuscript in preparation) with unprecedented insight into microbial community organization and microbial interaction with its extracellular environment. It is expected that with a concerted funding effort, the secrets of microbial interaction with its hydrocarbon environment could be revealed, including the presence of lipopolysaccharides (LPS) on the outer membrane surface or even layers of exopolysaccharides (EPS) that may protect the microbes from external hydrocarbon or the presence of specialized compartments or other strategies that protect from hydrocarbon toxicity from the inside, e.g., in the context of biofuel production.

For transmission electron microscopy, sample preparation is critical to obtain reproducible, reliable results. Despite new techniques, the greatest source of error in thin sections for TEM still lies in sample preparation (Cheville and Stasko 2014). Sample preparation treats specimens of interest in a series of reactions, beginning with primary fixation, postfixation, dehydration, resin infiltration, and embedding.

## 7    Research Needs

Fixation is done with 2% glutaraldehyde at pH 7.2 (McDonald and Zalpuri 1997), postfixation with 1% osmium tetroxide in 0.1 M sodium cacodylate buffer (Cheville and Stasko 2014). Postfixation is followed by rinses with buffer (McDonald and Zalpuri 1997). Dehydration with ethanol series (35% increased to 100%) is critical for removing remaining water in the sample in order for embedding media to effectively infiltrate the specimen during the subsequent treatment (Cheville and Stasko 2014). Resin infiltration (with accelerator BDMA) for samples is done stepwise by increasing the resin-to-ethanol ratio. Samples are then put in three pure resin exchanges (McDonald and Zalpuri 1997). Samples are embedded in epoxy resin. Epoxide resin is intolerant of water and will not polymerize if the sample is not thoroughly dehydrated, resulting in rubbery blocks that cannot be cut into ultrathin sections (Cheville and Stasko 2014). Samples are embedded into molds and put into the oven at 60 °C for 2 days (McDonald and Zalpuri 1997). Sample preparation is followed by ultrathin sectioning of the polymerized blocks in order to allow electrons from the microscope beam to pass through thin cross sections of the sample. Ultramicrotomes are used to cut section into 100 nm or less, optimally 70 nm thick. These sections are then collected on coated grids, followed by positive staining to increase contrast of the images (Cheville and Stasko 2014).

# 8 Future Studies

Research on studying the microbe-hydrocarbon environment would benefit from the subcellular localization of candidate proteins that are suspected to be involved. Such proteins are ideally tagged using green fluorescent protein (GFP and related)-tagged fusion proteins, which work well under aerobic conditions, or SNAP-tagged fusion proteins that also work under anaerobic conditions (Chhabra et al. 2011). In any case, such signals are used as the starting point for photoconversion, which allows subcellular protein localization at higher (TEM) resolution. While this approach works well in some circumstances, in others – e.g., where internal metal deposits can interfere with a clean interpretation of the TEM imaging data (manuscript in preparation) – more work is needed to solve this challenging problem.

# References

Abbasian F, Lockington R, Mallavarapu M, Naidu R (2015) A comprehensive review of aliphatic hydrocarbon biodegradation by bacteria. Appl Biochem Biotechnol 176(3):670–699. https://doi.org/10.1007/s12010-015-1603-5

Andrews RE, Parks LW, Spence KD (1980) Some effects of douglas fir terpenes on certain microorganisms. Appl Environ Microbiol 40(2):301–304. http://www.ncbi.nlm.nih.gov/pubmed/16345609. Accessed 22 Aug 2017

Atlas RM (1981) Microbial degradation of petroleum hydrocarbons: an environmental perspective. Microbiol Rev 45(1):180–209. http://www.ncbi.nlm.nih.gov/pubmed/7012571. Accessed 22 Aug 2017

Baelum J, Borglin S, Chakraborty R et al (2012) Deep-sea bacteria enriched by oil and dispersant from the Deepwater Horizon spill. Environ Microbiol 14(9):2405–2416. https://doi.org/10.1111/j.1462-2920.2012.02780.x

Cheville NF, Stasko J (2014) Techniques in electron microscopy of animal tissue. Veterinary Pathology, 51(1):28–41. https://doi.org/10.1177/0300985813505114

Chhabra et al (2011) Generalized schemes for high-throughput manipulation of *Desulfovibrio vulgaris* genome. Appl Environ Microbiol 77(221):7595-7604. https://doi.org/10.1128/AEM.05495-11

Das N, Chandran P (2011) Microbial degradation of petroleum hydrocarbon contaminants: an overview. Biotechnol Res Int 2011:941810. https://doi.org/10.4061/2011/941810

De Smet MJ, Kingma J, Witholt B (1978) The effect of toluene on the structure and permeability of the outer and cytoplasmic membranes of *Escherichia coli*. Biochim Biophys Acta Biomembr 506(1):64–80. https://doi.org/10.1016/0005-2736(78)90435-2

Dunlop MJ, Dossani ZY, Szmidt HL et al (2011) Engineering microbial biofuel tolerance and export using efflux pumps. Mol Syst Biol 7:487. https://doi.org/10.1038/msb.2011.21

Gill CO, Ratledge C (1972) Effect of n-alkanes on the transport of glucose in *Candida* sp. strain 107. Biochem J 127(3):59P–60P. http://www.ncbi.nlm.nih.gov/pubmed/5076204. Accessed 22 Aug 2017

Harrop AJ, Hocknult MD, Lilly MD (1989) Biotransformations in organic solvents: a difference between gram-positive and gram-negative bacteria. Biotechnol Lett 11:807–810. https://link. springer.com/content/pdf/10.1007/BF01026102.pdf. Accessed 22 Aug 2017

Hazen TC, Dubinsky EA, DeSantis TZ et al (2010) Deep-Sea oil plume enriches indigenous oil-degrading bacteria. Science 330(6001):204. http://science.sciencemag.org/content/330/ 6001/204. Accessed 22 Aug 2017

Heidelberg JF, Seshadri R, Haveman SA et al (2004) The genome sequence of the anaerobic, sulfate-reducing bacterium *Desulfovibrio vulgaris* Hildenborough. Nat Biotechnol 22(5): 554–559. https://doi.org/10.1038/nbt959

Hunter RC, Beveridge TJ (2005) High-resolution visualization of *Pseudomonas aeruginosa* PAO1 biofilms by freeze-substitution transmission electron microscopy. J Bacteriol. https://doi.org/ 10.1128/JB.187.22.7619-7630.2005

Jhamb K, Auer M (2015) Electron microscopy protocols for the study of hydrocarbon-producing and hydrocarbon-decomposing microbes: classical and advanced methods. In: Springer, Berlin/ Heidelberg, Hydrocarbon and Lipid Microbiology Protocols, pp 5–28. https://doi.org/10.1007/ 8623_2015_96

Lamendella R, Strutt S, Borglin S et al (2014) Assessment of the Deepwater Horizon oil spill impact on Gulf coast microbial communities. Front Microbiol 5:130. https://doi.org/10.3389/ fmicb.2014.00130

Leahy JG, Colwell RR (1990) Microbial degradation of hydrocarbons in the environment. Microbiol Rev 54(3):305–315. http://www.ncbi.nlm.nih.gov/pubmed/2215423. Accessed 22 Aug 2017

Lee SK, Chou H, Ham TS, Lee TS, Keasling JD (2008) Metabolic engineering of microorganisms for biofuels production: from bugs to synthetic biology to fuels. Curr Opin Biotechnol 19(6): 556–563. https://doi.org/10.1016/j.copbio.2008.10.014

Leffler WL (2008) Petroleum refining in nontechnical language. PennWell, Tulsa

Liamleam W, Annachhatre AP (2007) Electron donors for biological sulfate reduction. Biotechnol Adv 25(5):452–463. https://doi.org/10.1016/j.biotechadv.2007.05.002

Marchant, Banat (2015) Protocols for measuring biosurfactant production in microbial cultures. In: Hydrocarbon and lipid microbiology protocols – Springer protocols handbooks. https://doi.org/ 10.1007/8623

Matias VRF, Al-Amoudi A, Dubochet J, Beveridge TJ (2003) Cryo-transmission electron microscopy of frozen-hydrated sections of *Escherichia coli* and *Pseudomonas aeruginosa* cryo-transmission electron microscopy of frozen-hydrated sections of *Escherichia coli* and *Pseudomonas aeruginosa*. J Bacteriol. https://doi.org/10.1128/JB.185.20.6112

McDonald KL, Zalpuri R (1997) Electron Microscope Lab. Methods generic processing protocol. University of California, Berkeley, unpublished. http://em-lab.berkeley.edu/EML/protocols/ pgeneric.php

McDonald Z. Electron Microscope Lab, 26 Giannini Hall, University of California, Berkeley, unpublished

McDonald KL, Auer M (2006) High-pressure freezing, cellular tomography, and structural cell biology. Biotechniques. https://doi.org/10.2144/000112226

McDonald KL, Morphew M, Verkade P, Müller-Reichert T (2007) Recent advances in high-pressure freezing: equipment- and specimen-loading methods. In: Electron microscopy: methods and protocols. Springer, Microscope Laboratory, University of California, Berkeley. https://doi.org/10.1007/978-1-59745-294-6_3

Muyzer G, Stams AJM (2008) The ecology and biotechnology of sulphate-reducing bacteria. Nat Rev Microbiol. https://doi.org/10.1038/nrmicro1892

Palsdottir H, Remis JP, Schaudinn C et al (2009) Three-dimensional macromolecular organization of cryofixed *Myxococcus xanthus* biofilms as revealed by electron microscopic tomography. J Bacteriol. https://doi.org/10.1128/JB.01333-08

Ratcliffe RM (2017) Successful bioaugmentation with microbes. http://bioremediate.com/hydrocar bon.html. Accessed 25 Aug 2017

Sauer K, Camper AK, Ehrlich GD, Costerton JW, Davies DG (2002) *Pseudomonas aeruginosa*. J Bacteriol. https://doi.org/10.1128/JB.184.4.1140

Sikkema J, De Bontt J, Poolmann B (1994) Interactions of cyclic hydrocarbons with biological membranes*. J Biol Chem 269(11):8022–8028. http://www.jbc.org/content/269/11/8022.full.pdf. Accessed 22 Aug 2017

Varjani SJ (2017) Microbial degradation of petroleum hydrocarbons. Bioresour Technol 223:277–286. https://doi.org/10.1016/j.biortech.2016.10.037

Zehr JP (2010) Microbes in Earth's aqueous environments. Front Microbiol 1:4. https://doi.org/10.3389/fmicb.2010 00004

# Microbiology of Oil Fly Larvae

**28**

## K. W. Nickerson and B. Plantz

## Contents

1   Introduction .................................................................................. 420
2   What's There? ............................................................................... 420
3   Who's There? ............................................................................... 423
4   Bioremediation .............................................................................. 423
5   Tetracycline-Enhanced Solvent Tolerance ................................................. 424
6   Antibiotic Resistance ....................................................................... 425
7   Insect–Microbe Interactions ............................................................... 425
8   Global Impact .............................................................................. 426
9   Research Needs ............................................................................. 426
References ...................................................................................... 427

**Abstract**

One animal beautifully adapted to the viscous asphalt of the La Brea Tar Pits is the oil fly *Helaeomyia petrolei (Syn. Psilopa)*. As a normal part of its carnivorous existence, the oil fly larval guts are filled with tar, with no adverse effects. Surface sterilized larvae contained ca. $2 \times 10^5$ heterotrophic bacteria per larva. These bacteria have been identified as a mixture of enteric bacteria, most commonly *Providencia rettgeri* and *Acinetobacter* spp. These bacteria were clearly growing because their numbers in the larval guts were 100–1,000 times greater than in free oil/asphalt. There is no evidence yet that these bacteria can degrade the complex aromatic hydrocarbons of the tar/asphalt. However, the bacteria isolated are highly solvent tolerant, and they remain a potential source of hydrocarbon-/solvent-tolerant enzymes. Likely of greatest evolutionary interest, these bacteria were naturally resistant to 9 of 23 common antibiotics tested. This finding suggests that the oil fly bacteria have an active efflux pump for aromatic

K. W. Nickerson (✉) · B. Plantz
School of Biological Sciences, University of Nebraska, Lincoln, NE, USA
e-mail: knickerson1@unl.edu; bplantz10@gmail.com

© Springer International Publishing AG, part of Springer Nature 2018
T. Krell (ed.), *Cellular Ecophysiology of Microbe: Hydrocarbon and Lipid Interactions*,
Handbook of Hydrocarbon and Lipid Microbiology, https://doi.org/10.1007/978-3-319-50542-8_37

hydrocarbons, due to the constant selective pressure of La Brea's solvent-rich environment. We suggest that the oil fly bacteria and their genes for solvent tolerance may provide a microbial reservoir for antibiotic resistance genes.

## 1    Introduction

When the La Brea Tar Pits are mentioned in conversation, most people visualize a highly viscous petroleum quicksand, with the consistency of roofing tar on a hot day. For any animal unfortunate enough to fall into these tar pits, it is a one way trip. This view is usually based on the discovery in the tar pits of the skeletons for long extinct animals from the late Pleistocene era such as the saber-toothed tiger *Smilodon californicus*. However, one animal is beautifully adapted to the tar pits. Carnivorous larvae of the oil fly *Helaeomyia petrolei* swim happily in this viscous tar and actually prefer high viscosity oils (Chopard 1963), prompting Thorpe (1930, 1931) to refer to the oil fly as "undoubtedly one of the chief biological curiosities of the world." Oil fly larvae consume trapped carrion and the tar is ingested incidentally. We have observed tar filling the entire digestive tract with no adverse effect on the larvae. We then reasoned that any microbes in the larval gut would be of interest because they would perforce be petroleum-/solvent-resistant microbes. The novelty of this project was reinforced when a literature search revealed only one publication between the start of our own work (1994) and when the system had been described by Thorpe (1930, 1931). The present review describes our studies on the microbiology of oil fly larvae (Kadavy et al. 1999, 2000) as well as our retrospective on this work 17 years later.

The La Brea Tar Pits are the best known of a series of natural oil seeps throughout Southern California. These oil seeps emerge from the ground amidst the grass and trees of a woodland environment which is now mostly urban (Fig. 1). Oil seeps are visible around the tree in the foreground. For perspective, that's a leaf shown on the left of one of them (Fig. 2). Larvae live their entire lives submerged in pools of crude oil, but they occasionally need to surface to breathe (center of close up, Fig. 3). Larvae are typically 0.5–1.0 mm in diameter and 5–12 mm in length (Kadavy et al. 1999). Mature larvae leave the oil to pupate on nearby dry debris and vegetation (Hogue 1993), producing small black gnat-like flies which are about 1.5 mm long (Fig. 4).

## 2    What's There?

We collected larvae from three seeps within the La Brea Tar Pits (more properly called the Rancho La Brea asphalt seeps) on five occasions between 1994 and 1997 (Kadavy et al. 1999). The viscous black oil was removed, and the larvae were surface sterilized by washing in 60% linoleic acid (5×) followed by 70% ethanol

**Fig. 1** A typical oil seep at La Brea

**Fig. 2** Oil pool

**Fig. 3** Larva surfacing in pool of crude oil

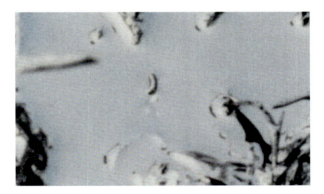

**Fig. 4** Adult oil fly

(2×), 15% hypochlorite containing 0.1% Tween 20, and phosphate-buffered saline
(PBS) containing 0.1% Tween 20 (2×). Active larvae were then allowed to crawl on
the surface of LB agar plates to prove that they had been surface sterilized. The
larvae were translucent, allowing for direct observation of the oil within the digestive
tract, and weighed ca. 3.3 mg each. Their gut system occupied ca. 50% of the total
larval volume, no ceca were detected, and the gut pH was 6.5–6.6 (Kadavy et al.
1999). Homogenized larvae from three different sites collected over a period of
4 years contained $10^5$–$10^6$ CFU when incubated aerobically on YEPM, LB, or
MacConkey agar plates. They averaged $2 \times 10^5$ heterotrophs per larva. For one
set of larvae, acridine orange direct count analysis indicated $8.6 \times 10^5$ total
microbes per larva, corresponding to $1.4 \times 10^5$ CFU per larva for that set. Thus,
16% of the total microbial population was culturable on these rich plates; a percent
culturable value comparatively high for environmental samples. Rich media were
chosen for plating because oil fly larvae are carnivorous; they do not survive on
oil/tar alone. To illustrate this point, larvae swarmed to egg meat medium (Difco) or
beef liver which were required to sustain the larvae in the lab. Without these
nutrients, the larvae would expire within 1 week of collecting.

No endospore-forming (heat-resistant) bacteria were detected under aerobic culture conditions and no fungi were found. Approximately 1% of the culturable
bacteria were nitrogen-fixing bacteria. We assumed that these nitrogen-fixing bacteria were among the enteric bacteria detected but this point was never proven. When
these isolation procedures were repeated under anaerobic conditions, very few
(≤100) CFU were found. Two of these anaerobic colonies were later identified as
*Enterococcus faecalis* and *Clostridium sporogenes* (Kadavy 2001).

The numbers of bacteria detected address the question of whether those bacteria
reside in the larval gut or are merely transient. The bacteria clearly grow in the larval
gut because, on a per weight basis, the number of bacteria in the larval guts was
always 100–1,000 times more CFU than in the oil being ingested. These calculations

are relevant because repeated cycling of oil through the larval gut results in substantially increased CFU in the oil. This point is nicely illustrated by comparing the counts found in newly emerged asphalt with no insects (30 $CFUmg^{-1}$), older asphalt containing many insects (600 $CFUmg^{-1}$), and the oil in which the larvae had been swimming just prior to surface sterilization (5,500 $CFUmg^{-1}$).

# 3    Who's There?

Which bacteria are present? Bacteria from the July 1997 collection were plated on LB, MacConkey, and blood agar plates. From 4,532 total colonies, eight different colony morphologies were recognized. After 24 h at 37 °C, roughly 70% of the colonies were 2–4 mm in diameter, whereas the remaining 30% were translucent, pinpoint colonies. All stained as Gram-negative bacteria. Fourteen representative colonies were picked for identification by the Enterotube II system (BBL), API 20E strips (bio Mērieux Vitek Inc.), and fatty acid profiles (MIDI Labs, Newark, DE). These isolates were called OF001 through OF014. Nine of the 14 isolates were identified as *Providencia rettgeri* by two or more tests, although a *Proteus* sp. often seemed like a statistically close second choice. These bacteria were not clonal in that the Enterotube and API identifications found five or six different metabolic profiles, respectively, for bacteria designated *P. rettgeri*. The *P. rettgeri/Proteus* sp. identifications were all among the larger 2–4 mm colonies. The pinpoint colonies were all identified as *Acinetobacter* sp. or as *Shewanella putrefaciens*. However, even though *P. rettgeri* and *Acinetobacter* sp. were the dominant bacteria in the 1997 samples, fatty acid profiles of the dominant colony morphologies from 1994 identified *Enterobacter* sp., *Hafnia alvei*, and *Acinetobacter radioresistens* (Kadavy et al. 1999). Moreover, the isolates we cultured from the larval guts do not align with the diversity of organisms subsequently identified in and around active tar pits by Kim and Crowley (2007). They cultured 235 bacteria and archaea from two La Brea pits being excavated for fossil remains. The lack of overlap between these two tar-/asphalt-related populations further illustrates the uniqueness of the larval gut as a selective microbial habitat.

# 4    Bioremediation

We live in a world increasingly burdened by chemical pollutants. These are often complex aromatic hydrocarbons found in solvent- or detergent-rich environments. The composition of the La Brea tar is highly weathered; few linear alkanes remain. Thin layer chromatography (TLC) analysis showed that the La Brea tar is composed of 10% branched alkanes and alkylated cyclic alkanes, 47% aromatics, 30% resins, and 13% polars, while GC/MS analysis showed significant hopanes, phenanthrene, and C1, C2 , and C3 phenanthrenes. These previously unpublished analyses were kindly provided by Roger C. Prince, ExxonMobil, Annandale, New Jersey. What would be more natural than finding polyaromatic hydrocarbon-degrading bacteria in the guts of oil fly larvae which are continuously ingesting tar/asphalt estimated to be 47% aromatic? However, for both the 1994 and 1997 collections, we did not find any bacteria able to grow on benzene, toluene, naphthalene, anthracene, phenanthrene, chrysene, benzopyrene, or camphor as the sole source of carbon and energy (Kadavy et al. 1999). However, ca. 40% of the 1994 isolates were able to grow on short-chain linear alkanes such as dodecane, and most of the 1997 isolates were able to grow on hexane. The tests for possible degradation of aromatic compounds were conducted in the basal medium of Stanier et al. (1966), so it is unlikely that the absence of growth was due to an unfulfilled metal or vitamin requirement. At that time, none of the tests were done under coculture or co-metabolism conditions or with the *Acinetobacter* sp. isolates which might have been more attractive candidates. As an example, we made several attempts to enrich for target microbes capable of mineralizing or being tolerant to aromatic solvents such as benzene or toluene, from both the insect guts and the oil seeps. Typically, the enrichment flasks became turbid, indicating growth, but that growth could not be sustained over multiple transfers into fresh, solvent-saturated media. Moreover, the isolates recovered from those initial enrichment flasks did not display increased tolerance to aromatic solvents (B. Plantz, unpublished data).

## 5     Tetracycline-Enhanced Solvent Tolerance

The efflux pumps of Gram-negative bacteria expel a remarkably broad range of antimicrobial compounds, including antibiotics, detergents, dyes, and organic solvents. The linkage between antibiotic resistance and solvent tolerance in *P. rettgeri* is nicely shown by comparing the abilities of oil fly bacteria to survive solvent overlays with and without prior exposure to tetracycline. Cell survival was dramatically enhanced by prior incubation with 20 µg of tetracycline per ml (Kadavy 2001). With tetracycline present, seven of the nine isolates tolerated 100% cyclohexane, six tolerated benzene–cyclohexane and toluene–cyclohexane (both 1:9), and three tolerated xylene–cyclohexane (1:1).

Probably of greater industrial significance would be the production of solvent-tolerant extracellular enzymes. Isolates OF007, OF010, and OF011 were selected for further study. Remember that these bacteria make their living by degrading whatever was ingested by the carnivorous oil fly larvae, usually other insects, in a high asphalt gut environment. Our initial results were promising. For each strain, at least seven extracellular proteins were detected as bands by SDS-PAGE or as peaks off a Sephadex G-150 column. We then examined enzyme activities by incorporating nine turbid polymers into agar plates and looking for zones of clearing around the respective colonies. These strains of *P. rettgeri* lysed red blood cells and produced extracellular DNAse and coagulase but, surprisingly, we did not detect any proteases (on either gelatin or casein), chitinases, keratinases, lipases, or amylases (Kadavy 2001). The extracellular proteins were not studied further.

# 6     Antibiotic Resistance

Twelve of the 14 OF cultures, including all 9 of those identified as *P. rettgeri*, were tested to determine their sensitivity or resistance to paper disks containing 23 common antibiotics (Kadavy et al. 2000). The levels and frequency of antibiotic resistance were remarkable. All nine of the *P. rettgeri* strains were highly resistant to tetracycline, vancomycin, bacitracin, erythromycin, novobiocin, polymyxin, rifampin, colistin, and nitrofurantoin, while eight of the nine were highly resistant to spectinomycin. In contrast, the *P. rettgeri* strains were highly sensitive to nalidixic acid, streptomycin, norfloxacin, aztreonam, ampicillin, ciprofloxacin, cefoxitin, cefotaxime, and piperacillin and somewhat sensitive to kanamycin and tobramycin (Kadavy et al. 2000).

Have we identified the evolutionary origin of antibiotic resistance in bacteria? The linkage between antibiotic resistance and aromatic hydrocarbons becomes obvious when one considers that tetracycline is structurally related to the polyaromatic hydrocarbons. The La Brea asphalt seeps are at least 40,000 years old, and most of the antibiotics to which our *P. rettgeri* isolates were resistant are hydrophobic molecules containing aromatic ring systems. Thus, we suggested that the oil fly bacteria and their genes for solvent tolerance provide a microbial reservoir of antibiotic resistance genes, i.e., the survival of *P. rettgeri* in the La Brea asphalt seeps depends on the ability of the cells to pump out these aromatic and polyaromatic hydrocarbons. This line of thought, which presupposes that our oil fly bacteria have never been subjected to antibiotic stress, reinforces earlier suggestions on the possible overlap between antibiotics and xenobiotics (Blasco et al. 1995).

# 7     Insect–Microbe Interactions

Is the relation between oil fly bacteria and oil fly larvae symbiotic, mutualistic, or mere coincidence? The limitation here is in the insect life cycle. Oil fly larvae (Fig. 3) are available but adult insects (Fig. 4) and eggs are not; there are no established colonies of oil flies. The eggs would, of course, be needed to obtain germ-free insects for study. At present the evidence suggests a mutual relationship whereby the bacteria gain a protected environment high in water, nitrogen, and phosphorous, while the insects gain a detoxification system and, more tentatively, the ability to extract nutrients from the oil. Chopard (1963) suggested that bacteria might assist larvae to gain energy from the paraffin fraction, even though we now know that most of the paraffins have been lost due to weathering. His view was based in part on the observation that gut petroleum appears to stay restricted within the chitinous peritrophic membrane. Similarly, the evidence for detoxification is mostly visual. As the bolus of tar passes through the larval gut, it goes from being a black and viscous tar to something having the appearance of motor oil. This change suggests that the tar has been transformed, but we have no idea into what. Clearly the insects cannot live on oil alone, and we were unable to identify any polyaromatic hydrocarbon-degrading bacteria among either the 1994 or 1997 collections. These studies should be revisited, giving particular attention to possible consortia of microorganisms that could not be sustained under the laboratory conditions originally chosen, as well as to potential resin- and asphaltene-degrading bacteria. This line of thought suggests that insect associated microbes should constitute a fertile hunting ground in the search for novel microbes. This viewpoint was a cornerstone in the rationale for establishing the Invertebrate Microbiology section of Applied and Environmental Microbiology.

## 8    Global Impact

The worldwide distribution of oil flies is not yet clear. Published reports indicate that oil flies are restricted to oil pools and natural oil seeps in Southern California, Cuba, and Trinidad (Foote 1995; Hogue 1993; Thorpe 1931), although anecdotal evidence suggests that they are far more widespread. At this point, oil flies and the microbes in their digestive tracts remain as biological curiosities and a tantalizing but untapped source of potentially useful hydrocarbon-transforming bacteria and hydrocarbon-/solvent-tolerant enzymes. But, if our antibiotic resistance hypothesis is true, the medical impact of our findings might be more significant. *Acinetobacter baumannii* is an opportunistic human pathogen and a frequent cause of nosocomial outbreaks (Richet and Fournier 2006). It is often multidrug resistant (MDR). The Centers for Disease Control and the Walter Reed Army Medical Center highlighted the enormity and gravity of MDR *A. baumannii* infections in military medical facilities treating personnel injured in Iraq and Kuwait (Hujer et al. 2006). Another study examined 20 isolates of MDR *A. baumannii* from nosocomial outbreaks in Los Angeles (Valentine et al. 2008). This juxtaposition suggests that the epidemic potential of MDR in *A. baumannii* may be related to *Acinetobacter* having first adapted to life in PAH-enriched environments such as the La Brea Tar Pits and oil-laden sands.

## 9    Research Needs

What would we have done differently? What needs to be done now? What can we do now that couldn't have been done 17 years ago?

1. We should revisit the question of which bacteria are oil fly associated using modern gene sequencing. Thus, we could find out if our previous identifications of *Providencia/Proteus* and *Acinetobacter* were an accurate sampling of the bacteria present. This knowledge might also drive improvements in the culture techniques chosen. We could also explore seasonal variations in the bacterial populations as well as variations among individual larvae. For instance, Southern California has distinct seasonal rainfall rates, a wet and dry season, during which

water content surrounding the seeps will vary and may influence the microbial species present.

2. Focus on the *acinetobacters*. Our earlier selection and growth conditions were optimized for enteric bacteria, not for hydrocarbon-metabolizing *acinetobacters*. These growth and selection procedures need to be improved. The genus *Acinetobacter* is the best known representative of the alkane-degrading α-proteobacteria, and alkane degradation seems to be a common property of most *Acinetobacter* strains (Baumann et al. 1968). Also, many bioemulsifiers are produced by the genus *Acinetobacter* (McInerney et al. 2005), while a high percentage of the oil fly bacteria we studied also produced surfactants (Kadavy 2001).

3. Heavy petroleum. We know that La Brea tar enters the larval mouth in a dark viscous state and exits as something close to motor oil. This transition should be followed by GC/MS analysis of the contents as they progress through the gut. In light of the recent economic push to recover heavy petroleum from underground sources, a consortium of larval gut microbes could be found to reduce the viscosity of the heavy petroleum and allow for easier removal.

4. Functional metagenomics. Knowledge of the genes associated with hydrocarbon mineralization has grown significantly in the years since we initiated this research, as has the ability to sequence DNA. Metagenomics of the larval gut and the surrounding environmental communities would quickly identify if the potential for hydrocarbon mineralization is present within these microbial communities. This approach would also provide a better chance of isolating novel solvent-tolerant bacteria making useful solvent-tolerant enzymes.

# References

Baumann P, Doudoroff M, Stanier RY (1968) A study of the Moraxella group. II. Oxidative-negative species (genus *Acinetobacter*). J Bacteriol 95:1520–1541

Blasco R, Wittich R-M, Mallavarapu M, Timmis KN, Pieper DH (1995) From xenobiotic to antibiotic, formation of protoanemonin from 4-chlorocatechol by enzymes of the 3-oxoadipate pathway. J Biol Chem 270:29229–29235

Chopard L (1963) La mouche du petrole et les questions qu'elle pose. La Nat (Paris) 3338:255–256

Foote BA (1995) Biology of shore flies. Annu Rev Entomol 40:417–442

Hogue CL (1993) Insects of the Los Angeles Basin, 2nd edn. Natural History Museum of Los Angeles County, Los Angeles

Hujer KM, Hujer AM, Hulten EA, Bajaksouzian S, Adams JM, Donskey CJ, Ecker DJ, Massire C, Eshoo MW, Sampath R, Thomson JM, Rather PN, Craft DW, Fishbain JT, Ewell AJ, Jacobs MR, Paterson DL, Bonomo RA (2006) Analysis of antibiotic resistance genes in multi-drug resistant *Acinetobacter* sp. Isolates from military and civilian patients treated at the Walter Reed Army Medical Center. Antimicrob Agents Chemother 50:4114–4123

Kadavy DR (2001) Characterization of microorganisms associated with *Helaeomyia petrolei* (oil fly) larvae. University of Nebraska, Ph.D. thesis, Lincoln

Kadavy DR, Plantz B, Shaw CA, Myatt J, Kokjohn TA, Nickerson KW (1999) Microbiology of the oil fly, *Helaeomyia petrolei*. Appl Environ Microbiol 65:1477–1482

Kadavy DR, Hornby JM, Haverkost T, Nickerson KW (2000) Natural antibiotic resistance of bacteria isolated from larvae of the oil fly, *Helaeomyia petrolei*. Appl Environ Microbiol 66:4615–4619

Kim J-S, Crowley DE (2007) Microbial diversity in natural asphalts of the Rancho La Brea Tar Pits. Appl Environ Microbiol 73:4579–4591

McInerney MJ, Nagle DP, Knapp RM (2005) Microbially enhanced oil recovery: past, present, and future. In: Ollivier B, Magot M (eds) Petroleum microbiology. ASM Press, Washington, pp 215–237

Richet H, Fournier PE (2006) Nosocomial infections caused by *Acinetobacter baumannii*: a major threat worldwide. Infect Control Hosp Epidemiol 27:645–646

Stanier RY, Palleroni NJ, Doudoroff M (1966) The aerobic pseudomonads: a taxonomic study. J Gen Microbiol 43:159–271

Thorpe WH (1930) The biology of the petroleum fly (*Psilopa petrolei*). Trans Entomol Soc Lond 78:331–344

Thorpe WH (1931) The biology of the petroleum fly. Science 73:101–103

Valentine SC, Contreras D, Tan S, Real LJ, Chu S, Xu HH (2008) Phenotypic and molecular characterization of *Acinetobacter baumannii* clinical isolates from nosocomial outbreaks in Los Angeles county California. J Clin Microbiol 46:2499–2507

# Part IV

# Problems of Feast or Famine

# Nitrogen Fixation and Hydrocarbon-Oxidizing Bacteria

# 29

## J. Foght

## Contents

1   Introduction ........................................................................... 432
2   Gaseous Hydrocarbons and Methanotrophs ............................................... 433
3   Liquid and Solid Hydrocarbons ......................................................... 435
    3.1   Heterotrophs .................................................................. 435
    3.2   Cyanobacteria ................................................................. 439
4   Activities in the Environment ........................................................... 441
    4.1   Sediments and Microbial Mats ................................................... 441
    4.2   Phytoremediation .............................................................. 443
5   Gaps in Knowledge and Research Needs ................................................. 444
References ................................................................................ 445

### Abstract

The ability to reduce atmospheric nitrogen to ammonia ($N_2$ fixation) is well known in diverse prokaryotes, and many microbes oxidize hydrocarbons ranging from methane to longer-chain $n$-alkanes, alkenes, and aromatics under aerobic or anaerobic conditions. Although this combination of activities should provide a selective advantage to microbes in hydrocarbon-contaminated N-limited environments, surprisingly few isolates have been reported to simultaneously, or even sequentially, fix $N_2$ while growing on hydrocarbons. Notable exceptions are methane-oxidizing bacteria that fix nitrogen in culture but whose $N_2$-fixing significance in the environment is still controversial. In axenic culture, demonstration of $N_2$ assimilation into biomass linked to hydrocarbon utilization should include confirmation of the presence and expression of appropriate genes in addition to demonstrated enzymatic activity and quantitation of hydrocarbon

J. Foght (✉)
Department of Biological Sciences, University of Alberta, Edmonton, AB, Canada
e-mail: julia.foght@ualberta.ca

© Springer International Publishing AG, part of Springer Nature 2018                                431
T. Krell (ed.), *Cellular Ecophysiology of Microbe: Hydrocarbon and Lipid Interactions*,
Handbook of Hydrocarbon and Lipid Microbiology, https://doi.org/10.1007/978-3-319-50542-8_53

removal; unfortunately, such rigorous studies are scarce. $N_2$ fixation in situ by heterotrophic and/or photosynthetic microbial communities supported by liquid hydrocarbon utilization has been inferred in recent studies, some of which are well documented but others less persuasive. An emerging model is that microbial assemblages in situ may exhibit "distributed metabolism" where one partner fixes $N_2$ and the other oxidizes hydrocarbons, with exchange of metabolic products. This model may best reflect in situ activity in photosynthetic microbial mats, but whether it also applies to anaerobic environments or aerobic habitats such as the rhizosphere is unresolved. Phytoremediation may be a valuable option for oil spill cleanup particularly in N-limited soils, but adding plants as a third partner further complicates assignation of biochemical contributions. This chapter reviews the scant literature on diazotrophic hydrocarbon degradation and highlights knowledge gaps for future research.

# 1    Introduction

In this chapter the term "hydrocarbon" includes gaseous (e.g., methane, ethane), liquid (e.g., $n$-hexadecane, toluene), and solid hydrocarbons (e.g., naphthalene and other polycyclic aromatic hydrocarbons [PAH]) as well as natural and refined hydrocarbon mixtures (e.g., crude oil, kerosene). A wide variety of prokaryotes and eukaryotes can utilize certain hydrocarbons through complete oxidation (mineralization) to $CO_2$ and water; even more species can enzymatically biotransform hydrocarbons (e.g., for detoxification purposes) merely by introducing functional groups such as hydroxyl or epoxide moities. The latter pathways generate partially oxidized end products that are more water soluble than the parent hydrocarbon and therefore deplete the transformed hydrocarbon from the nonpolar fraction that is commonly analyzed using gas chromatography (GC). Thus, "disappearance" of a hydrocarbon peak from a gas chromatogram does not indicate whether the hydrocarbon has been mineralized or only partially oxidized, an important distinction.

Nitrogen fixation refers to the exclusively prokaryotic ability to reduce $N_2$ gas to ammonia for incorporation into biomass (diazotrophy). $N_2$-fixers may be free-living or symbiotic, photosynthetic or heterotrophic, and aerobic or anaerobic. Reduction of $N_2$ is accomplished by a family of nitrogenase enzymes encoded by a large suite of structural, biosynthetic, and regulatory genes found in a wide variety of species including aerobes, facultative anaerobes, and strict anaerobes (e.g., Zehr et al. 2003). Although most of the literature cited herein infers diazotrophy from the detection of a single gene (*nifH*), Dos Santos et al. (2012) have recommended that computational prediction of nitrogen fixation potential from microbial genome sequences requires detection of at least six core genes associated with enzyme structure and biosynthesis: *nifHDK* and *nifENB*. In a survey of fully sequenced genomes, this criterion correctly identified 149 diazotrophic species plus 67 species not previously reported as $N_2$-fixers (Dos Santos et al. 2012). Such diversity suggests the existence of additional cryptic diazotrophs, some of which may oxidize hydrocarbons concurrently or sequentially with diazotrophy.

When hydrocarbons impact an environment, e.g., methane clathrate seeps, petroleum spills on soil, gasoline incursion into an aquifer or, particularly, crude oil impacting chronically N-limited systems such as warm and cold deserts or warm coastal sediments, the high carbon content of the contaminant causes an imbalance in the carbon:nitrogen ratio of the environment, and biodegradation of the hydrocarbons can be retarded or prevented by nitrogen deficiency. Therefore, crude oil spills are commonly treated by introducing fixed N sources such as nitrate or ammonium to stimulate biodegradation in situ. However, the ability to fix $N_2$ should afford a selective advantage for hydrocarbon-degrading diazotrophic bacteria in N-limited spill sites (Chu and Alvarez-Cohen 1998) and facilitate their isolation in the laboratory. Perplexingly, rarely has this combined activity been meticulously documented in axenic cultures, with the exception of methane-oxidizing bacteria (methanotrophs). Instead, the literature contains many empirical reports from laboratory and field studies that infer coincident hydrocarbon oxidation and diazotrophy but lack adequate experimental rigor to conclusively demonstrate both activities or to assign activity to individual species. Rather, it appears that the observations often are consistent either with "distributed metabolism" shared between a diazotroph and a heterotrophic hydrocarbon-oxidizing partner, or with partial oxidation of a hydrocarbon by a diazotroph, without utilization of the carbon skeleton.

## 2  Gaseous Hydrocarbons and Methanotrophs

The first intimations of $N_2$ fixation linked to hydrocarbon oxidation arose from observations that soils accidentally exposed to emissions of natural gas (predominantly methane plus ethane) became darker in color and had up to threefold higher total N content than control soil (Schollenberger 1930; Harper 1939). In moist soils exposed to methane, visible bacterial growth became apparent as masses of slime molds and myxomycetes; microbiological analysis revealed an abundance of *Clostridium* spp., leading Harper (1939) to speculate that anaerobic $N_2$ fixation was occurring in these gassed soils. Yet it was not until 25 years later that Davis et al. (1964) clearly established the ability of certain *Pseudomonas* isolates from soils to grow aerobically on methane by fixing $N_2$, measured as Kjeldahl N. This report was preceded by a patent filed in 1962 (Davis and Stanley 1965) describing the use of soil microbes to convert $N_2$ into fertilizer during growth on hydrocarbons. Subsequently, aerobic growth on gaseous hydrocarbons was used to enrich various $N_2$-fixing isolates from soils. For example, Coty (1967) isolated a strain identified at the time as *Pseudomonas methanitrificans* that grew on 30% methane and 70% air and a strain identified biochemically (and undoubtedly incorrectly) as *Mycobacterium* sp., which grew with butane and air on N-deficient agar.

To corroborate growth assays in N-free medium as evidence for diazotrophy, acetylene reduction to ethylene is commonly used to detect and quantify nitrogenase activity (ironically, the assay substrate and product are both hydrocarbons). However, de Bont and Mulder (1974) found that this method was invalid for assaying nitrogenase activity in methanotrophs, because the ethylene produced in positive cultures was co-oxidized by the bacteria, thus depleting the assay product. Hamamura et al. (1999) later examined the underlying mechanisms, discovering that acetylene can have two deleterious effects on methane oxidation. First, acetylene irreversibly inactivates methane monooxygenases (MMOs) that initiate attack on gaseous hydrocarbons. Second, the product of acetylene reduction by nitrogenase, ethylene, in turn can be oxidized (likely by the MMOs) to ethylene oxide, which is also an inactivator of MMOs. Therefore, this assay is inappropriate for measuring nitrogenase activity in methanotrophs (unless a suitable electron donor is added; Dalton and Whittenbury 1976), complicating the demonstration of $N_2$-fixing methanotrophy in the field. Buckley et al. (2008) used $^{15}N_2$-DNA-stable isotope probing to confirm incorporation of $N_2$ into methanotrophic biomass: more such rigorous studies are needed. In contrast, the alkene monooxygenase responsible for oxidation of ethene, propene, and 1-butene by *Xanthomonas* spp. is not inhibited by acetylene, enabling van Ginkel and de Bont (1986) to detect nitrogenase activity in six isolates growing on these gaseous alkenes. The isolates were unable to grow on the corresponding alkanes and required low $O_2$ levels to maintain nitrogenase activity, likely to protect the $O_2$-sensitive enzyme and ensure its expression.

In early studies Murrell and Dalton (1983) found that colonies of obligate methanotrophs could grow on N-free agar, but only some strains grew well on methane under more stringent N-free conditions in liquid medium with concomitant acetylene reduction activity. Gene and genome sequencing have since contributed to recognizing $N_2$ fixation potential in many but not all methanotrophs (Semrau et al. 2010) including Verrucomicrobia (Khadem et al. 2010). For example, Auman et al. (2001) and Boulygina et al. (2002) expanded the known diversity of $N_2$-fixing methanotrophs by screening isolates for the presence of *nifH* gene analogues that encode a nitrogenase subunit. *nifH* gene fragments were amplified from all isolates of the genera tested: *Methylococcus*, *Methylocystis*, *Methylosinus*, *Methylomonas*, and *Methylobacter*. This led to speculation that all methanotrophs have nitrogenase genes (Boulygina et al. 2002), or at least that diazotrophy is more widespread among aerobic methanotrophs than has previously been appreciated, and therefore that such organisms may be ecologically important (Auman et al. 2001). More recently, Hoefman et al. (2014) have shown that $N_2$-fixing capability is actually strain specific rather than universally present in methanotrophic bacteria, speaking to the selective role of gene acquisition and deletion in response to local environmental conditions. Although the abundance and importance of aerobic diazotrophic methanotrophs in peatlands are currently controversial (e.g., Ho and Bodelier 2015; Minamisawa et al. 2016), it is possible that such organisms are significant in the terrestrial methane budget, as recently reviewed by Bodelier and Steenbergh (2014).

Anaerobic methane oxidation also occurs globally in anoxic freshwater and marine sediments near seeps from natural methane clathrate deposits. Dekas et al. (2009, 2014) used sophisticated $^{15}N_2$-tracing analyses with fluorescent in situ hybridization coupled to nanoscale secondary ion mass spectrometry (FISH-nanoSIMS) to show that anaerobic methanotrophic archaea (ANME) are likely the prime mediators of $N_2$ fixation in symbiosis with sulfate-reducing bacteria at methane seeps. This observation has implications for global carbon budgets, but additional study is required to establish the magnitude and significance of $N_2$ fixation by ANME communities (Musat et al. 2006; Bodelier and Steenbergh 2014).

Thus, aerobic and anaerobic diazotrophic oxidation of the gaseous hydrocarbon methane, which until recently was cryptic and understudied, is now recognized as being potentially important for mitigating natural methane emissions. This contrasts with the body of literature describing diazotrophic oxidation of larger hydrocarbons, primarily associated with anthropogenic petroleum spills, reviewed below.

## 3 Liquid and Solid Hydrocarbons

### 3.1 Heterotrophs

Scott et al. (2014) noted that "How oil contamination affects nitrogen cycling processes in situ is still not well understood," and there are few convincing reports of diazotrophs simultaneously growing on liquid or solid hydrocarbons. Early accounts typically comprised simple observation of aerobic growth on aliphatic hydrocarbons in N-deficient medium, e.g., an *Azospirillum* from oil-contaminated soil growing on *n*-dodecane (Roy et al. 1988) and "*Mycococcus*" and *Arthrobacter* spp. isolated from refinery sludge growing on *n*-hexadecane (Rivière et al. 1974). The latter group demonstrated significant loss of *n*-$C_{16}$ and growth of the strains in N-deficient medium containing vitamins and cofactors, although both growth and degradation increased if fixed N was added to the medium. An estimated 9 mg N was fixed per g hexadecane consumed. Similarly, Gradova et al. (2003) found that *Azotobacter* spp. could grow on an "*n*-paraffin fraction" ($C_{14}$–$C_{18}$) in the absence of fixed N, albeit more slowly than when provided with ammonium. *Azotobacter oleovorans* grew aerobically on the aliphatic hydrocarbon tetradecane in both solid and liquid N-free medium, and another *Azotobacter* sp. grew on vapors of the monoaromatic toluene, although $N_2$-fixing liquid cultures of the latter could not be achieved (Coty 1967). Yet another species, *Azotobacter chroococcum*, grew on crude oil in N-free medium (Thavasi et al. 2006). Fries et al. (1994) isolated aromatic-degrading *Azoarcus* spp. from diverse environments by enrichment with toluene under denitrifying conditions and subsequently found that several strains also grew aerobically on benzene and ethylbenzene when supplied with nitrate as a fixed N source. Several of these isolates also demonstrated nitrogenase activity (by acetylene reduction) under microaerobic conditions with malic acid as the carbon source, and some hybridized to *nifHDK* genes encoding nitrogenase components, but Fries et al. (1994) did not test whether hydrocarbon oxidation occurred under $N_2$-fixing conditions. Martín-Moldes et al. (2015) described another *Azoarcus* sp. (strain CIB), a facultative anaerobe that is a diazotroph (inferred from the genome sequence) and that degraded toluene and *m*-xylene when provided with nitrate; however, hydrocarbon degradation under $N_2$-fixing conditions was not reported. Eckford et al. (2002) isolated $N_2$-fixing strains from fuel-contaminated Antarctic soils (but not from nearby pristine soils) and subsequently found that two isolates grew aerobically on jet fuel vapors. *Pseudomonas stutzeri* strain 5A grew on benzene, toluene, or m-xylene vapors as sole carbon source, whereas *Pseudomonas* sp. strain 5B grew on n-hexane vapors. However, both strains required the presence of fixed N to grow on these substrates, and neither demonstrated simultaneous nitrogenase activity with hydrocarbon oxidation despite extensive testing under different culture conditions including microaerobic incubation (K. Semple and J. Foght, unpublished observations). In contrast, *Polaromonas naphthalenivorans* strain CJ2 was shown, using $^{13}C$-stable isotope probing, to be a dominant aerobic naphthalene degrader in sediments contaminated with coal tar (Jeon et al. 2003). Genome sequence analysis by Hanson et al. (2012) subsequently revealed 20 *nif* genes in strain CJ2 that were absent in its closest sequenced relative, *Polaromonas* sp. strain JS666, suggesting acquisition of a ~66 kb DNA fragment encoding $N_2$ fixation in CJ2 (or its loss in strain JS666). Diazotrophy by strain CJ2 was observed as growth in semisolid N-free medium, and nitrogenase activity with and without naphthalene present was quantified using the acetylene reduction assay, confirming that the *nif* genes are functional. Nitrogenase activity increased in response to addition of naphthalene, concomitant with an increase in *P. naphthalenivorans* cell numbers, as determined using FISH-fluorescent antibody fluorescence microscopy. Thus, Hanson et al. (2012) present some of the strongest evidence to date for diazotrophic hydrocarbon oxidation by an axenic culture. They also pointed out that this strain must be able to balance the oxygen sensitivity of nitrogenase with the $O_2$ requirement of the naphthalene dioxygenase enzyme, although it is not known how this is accomplished.

Some reports of diazotrophic hydrocarbon biodegradation by aerobic heterotrophs allow for alternative interpretations. Chen et al. (1993) tested six species of diazotrophs compiled from culture collections (*Azomonas*, *Azospirillum*, *Azotobacter*, and *Beijerinckia* spp.) for growth on naphthalene and typical intermediates of aerobic aromatic hydrocarbon biodegradation (e.g., benzoate, catechol, p-toluate). All six strains demonstrated nitrogenase activity (measured as acetylene reduction) while growing on the metabolites, whereas growth on naphthalene was usually quite poor, with very low nitrogenase specific activity. This suggests that such species may contribute indirectly to the total hydrocarbon-degrading activity in an environment by removing metabolites produced by other microbes through partial oxidation, rather than having a significant direct effect on hydrocarbons. Co-metabolism (fortuitous metabolism of a nongrowth substrate by enzymes intended for growth-supporting substrates) may also occur, as suggested by an increase in naphthalene oxidation by the methanotroph *Methylosinus trichosporium* when grown under $N_2$-fixing conditions with methane (Chu and Alvarez-Cohen 1998). This activity likely resulted from induction of a soluble MMO that fortuitously oxidized naphthalene to 1- or 2-naphthol.

Unfortunately, several of the scant reports of diazotrophic hydrocarbon degradation do not provide sufficient information to validate dual activities. For example, Laguerre et al. (1987) isolated two unidentified aerobic strains from oily sludge waste containing $N_2$ fixers. Nitrogenase activity (requiring up to 2 months of incubation with acetylene, followed by $^{15}N$ incubation) was low when supported by sterilized oily sludge compared with organic acid and sugar substrates, and diazotrophic growth was not measured; therefore the report is inconclusive. Prantera et al. (2002) isolated two strains from hydrocarbon-contaminated soil, tentatively identified biochemically as *Agrobacterium* sp. and *Alcaligenes xylosoxidans*. Both were reported to fix nitrogen "in the presence of" benzene, toluene, and an undetermined xylene isomer provided in gasoline as "the main carbon source", and to degrade these substrates under $N_2$-fixing conditions. However, the mineral medium contained 0.05% yeast extract as a potential N source, and it was not determined whether the aromatic substrates were utilized for growth. A spate of recent studies of bacteria isolated from pristine and oil-polluted soils, sediments, and mudflats in the Arabian Gulf region purport to demonstrate $N_2$ fixation and hydrocarbon utilization but lack full documentation and adequate experimental controls. For example, Al-Mailem et al. (2010) studied the oil bioremediation potential of diazotrophic bacteria isolated from Arabian Gulf coastal mudflat and microbial mat communities. Hydrocarbon-degrading colonies were enumerated on nitrate-containing agar with crude oil vapors as sole carbon source, then screened for $N_2$-fixing potential by incubation with oil vapors on agar with and without nitrate as N source. However, parallel control cultures inoculated onto the same two solid media but lacking crude oil vapors were not included to account for growth on soluble organics in the agar, which was of unspecified purity and could contribute carbon and/or nitrogen unless adequately purified to remove potential nutrients. Nitrogenase activity in selected isolates was confirmed using acetylene reduction to assay cell suspensions incubated for 8 days, but the report did not disclose whether: (a) the suspending medium contained any carbon source, (b) parallel cultures containing fixed nitrogen were included as negative controls for the nitrogenase assay, or (c) the cultures were still axenic at the end of incubation. Hydrocarbon biodegradation by the isolates was tested on agar containing nitrate, pure *n*-alkanes, and PAH, scoring for relative growth but, again, without a parallel background control lacking the hydrocarbons. Crude oil degradation was assessed in liquid medium containing nitrate as a nitrogen source, and residual oil was extracted and analyzed by GC, but (a) without the addition of internal or surrogate GC standards that allow stringent comparison of chromatograms (comparisons were made only to heat-killed control cultures that also lacked internal or surrogate GC standards to account for any variation in oil addition, extraction efficiency, and analytical precision) and (b) without inclusion of live non-hydrocarbonoclastic isolates to account for any abiotic and/or non-specific biotic alteration of the crude oil such as volatilization and emulsification. Analogous methods were used to study epilithic bacteria on gravel particles from the Arabian Gulf coast (Radwan et al. 2010), to examine phytoremediation potential (Sorkhoh et al. 2010a; Dashti et al. 2009), and to study Kuwaiti desert soil bacteria (Sorkhoh et al. 2010b). In similar fashion Pérez-Vargas et al. (2000) challenged microbial "consortia" comprising three to eight bacterial strains with growth on kerosene. These mixed cultures displayed modest kerosene degradation and nitrogenase activity (determined by acetylene reduction), but the experimental design did not permit metabolic roles to be assigned to specific members of the consortia. Additionally, GC analysis was performed without internal or surrogate standards and without live and killed negative controls, as mentioned above, so that the actual amount of biodegradation is unconfirmed. (Notably, such oversights are found in numerous reports of crude oil degradation, not only in studies involving diazotrophs.)

Another case of overinterpretation is evident in a bold title declaring that "Most hydrocarbonoclastic bacteria in the total environment are diazotrophic. . ." (Dashti et al. 2015) based on analysis of 82 strains isolated from Kuwaiti sites and using the same methodology as Al-Mailem et al. (2010), a claim that cannot be justified based on the evidence presented, as the study: (a) examined only aerobic hydrocarbon degraders but did not consider anaerobes; (b) tested isolates exclusively from Arabian Gulf habitats but no other sites; (c) used denaturing gradient gel electrophoresis (DGGE) followed by amplicon sequencing to demonstrate the presence of *nif* genes in some of the isolates, but apparently without including control strains known to lack *nif* genes, nor any negative (reagent only) controls to account for possible PCR contamination; (d) apparently included neither parallel N-replete live cultures in the acetylene reduction assay nor any live non-diazotrophic cultures as negative controls.

Despite the technical errors and data overinterpretation mentioned in the paragraphs above, many or all of the isolates described in these studies may well be diazotrophic and hydrocarbonoclastic (even if not simultaneously), but stringent proof is currently lacking. If it is true that many aerobic hydrocarbon-degrading bacteria isolated from N-limited environments have the ability to fix $N_2$, it is puzzling that so few convincing reports have emerged, since such dual activity would have important ramifications for low-cost natural attenuation of remote N-limited environments (discussed below). Obviously, this is an area requiring additional rigorous research.

Another area needing more study is the role of $N_2$ fixers in anaerobic hydrocarbon degradation, where little currently is known. Whereas some methanogenic archaea can fix $N_2$, they cannot degrade solid and liquid hydrocarbons, although archaeal ANME oxidize methane and some appear to be diazotrophic, as discussed above. Some fermentative bacteria (e.g., in the order Clostridiales) as well as some respiring anaerobes (e.g., sulfate-reducing bacteria) are diazotrophic, but have not been shown to degrade nongaseous hydrocarbons under anaerobic $N_2$-fixing conditions. Conversely, the few hydrocarbonoclastic anaerobes that have been successfully isolated or whose genomes have been assembled from metagenomic sequences have not been tested for $N_2$ fixation potential. Amos et al. (2005) invoked $N_2$ fixation, either by methanogens or iron reducers, to explain reduced concentrations of $N_2$ in crude oil-contaminated subsurface sediments. More recently Collins et al. (2016) used $^{15}N_2$-incorporation into biomass to infer that anaerobes in oil sands tailings enrichment cultures were capable of fixing nitrogen under methanogenic conditions. However, although the tailings contained both bitumen (a heavy oil) and low molecular weight hydrocarbons, Collins et al. (2016) did not directly correlate diazotrophy with anaerobic hydrocarbon degradation since the enrichment cultures were amended with citrate. It is possible that $N_2$ fixation and hydrocarbon metabolism are distributed among community members, including methanogens. Metagenomic analysis may provide needed insight into this question, since cultivation of the anaerobic hydrocarbon degraders is currently difficult. An example is metabolic reconstruction of Campylobacterales metagenomic sequences from an enrichment culture derived from an anaerobic benzene-contaminated aquifer (Keller et al. 2015). The assembled genome included genes encoding nitrogenase and for sulfide assimilation, but apparently not for aromatic hydrocarbon degradation. Again, it is likely that the two activities are shared among enrichment culture members since the Campylobacterales have a gene complement that would allow mixotrophy (growth on $CO_2$ and organic compounds such as acetate) (Keller et al. 2015) and may rely on a hydrocarbon-degrading partner to provide these typical metabolites of anaerobic biodegradation.

## 3.2   Cyanobacteria

Many cyanobacteria are diazotrophic, and several species have been reported to oxidize hydrocarbons aerobically, but only when grown with a fixed N source such as nitrate. Notably though, oxidation does not necessarily indicate complete mineralization of the hydrocarbon and/or assimilation as a carbon source. As discussed above, even partial oxidation of hydrocarbons will preclude their detection using conventional solvent extraction and GC analyses, leading to the false assumption that the hydrocarbons have been completely "removed" from the system (the culture or the environment). Thus, biodegradation studies should include metabolite analysis to better understand the fate of the hydrocarbon skeleton; unfortunately, this has been done in only a minority of studies. For example, Cerniglia et al. (1980) described aromatic hydrocarbon oxidation by *Oscillatoria* sp. grown photoautotrophically (with nitrate) in the presence of biphenyl, yielding the partially oxidized metabolite 4-hydroxybiphenyl. Narro et al. (1992) similarly incubated the marine cyanobacterium *Agmenellum quadruplicatum* under photoautotrophic conditions with the PAH phenanthrene (and nitrate) and detected a suite of hydroxylated and methoxylated metabolites. Both observations suggest non-specific detoxification or co-metabolic reactions rather than hydrocarbon utilization. It is likely that other reports of cyanobacterial "removal" of PAHs from culture medium (e.g., Lei et al. 2007) also reflect partial oxidation of the hydrocarbons via cytochrome-like enzymes to form water-soluble metabolites rather than significant hydrocarbon degradation and/or utilization. Interestingly, such metabolites can be toxic to the cyanobacteria that produce them (Narro et al. 1992), which may account for reports of cyanobacterial inhibition by exposure to PAH. For example, Kumar et al. (2014) found that growth (determined as chlorophyll-a content) of three cyanobacterial species was inhibited by incubation with the PAH pyrene, with increasing sensitivity in the order *Nostoc muscorum* > *Anabaena fertillissima* > *Synechocystis* sp. The identity of any hydrocarbon metabolites was not determined, although unspecified phenols were detected in spent cultures.

Other studies of cyanobacterial oxidation of hydrocarbons have revealed further caveats for interpretation of results. For example, careful examination of some purportedly hydrocarbonoclastic cyanobacterial cultures has shown that they are not axenic; rather, they harbor heterotrophs within their mucilage sheaths (e.g., Abed et al. 2002; Abed and Köster 2005; Sánchez et al. 2005). Various combinations of roles have been ascribed to the different microbial partners. For example, Sánchez et al. (2005) discovered a consortium of heterotrophic bacteria and diazotrophic Rhizobiaceae growing within the polysaccharide sheath of a non-$N_2$-fixing cyanobacterium (*Microcoleus chthonoplastes*). In the presence of crude oil, the consortium oxidized alkylthiolanes, alkylthianes, and alkylated monoaromatics and PAH. In this case, it was conjectured that the heterotrophs and diazotrophs were exhibiting hydrocarbon oxidation and $N_2$ fixation activities, respectively, with the cyanobacterial partner providing $O_2$, organic matter, and a suitable habitat for the assemblage. Another possible scenario is partial oxidation of hydrocarbons by diazotrophic cyanobacteria to metabolites that are consumed by heterotrophic partner(s), thus relieving toxicity to the cyanobacteria and stimulating growth of both partners.

Some studies have raised awareness that the abiotic fate(s) of hydrocarbons may also be significant in cyanobacterial biomass. Chaillan et al. (2006) examined a petroleum-contaminated photosynthetic mat that demonstrated excellent hydrocarbon-degrading activity in the field, from which they isolated the dominant cyanobacterium, *Phormidium animale*. In axenic culture the strain lacked hydrocarbon oxidation activity, although it strongly sequestered oil within its biomass. This led Chaillan et al. (2006) to caution wisely that "Without a high extraction efficiency *[of oil from the culture]*, the quantity of total hydrocarbon recovered could be lower in culture than in control and interpreted as erroneous degradation by both gravimetric and GC methods." Similarly, Lei et al. (2002) found that both live and dead cells of several microalgal species were equally effective in "removing" pyrene from culture medium; however, the dominant mechanism was passive biosorption to cell walls rather than enzymatic biotransformation. Furthermore, Luo et al. (2014a) found that dead cells of some microalgal species actually were superior to live cells for transformation of several PAHs when incubated under two different light sources, suggesting that the dead cells photosensitized the PAHs and accelerated abiotic photodegradation. Thus, analytical errors may have contributed to conflicting interpretations in the literature regarding the ability of diazotrophic cyanobacteria to biodegrade oil, and claims of cyanobacterial or microalgal "degradation" of oil components must be supported by technically sound experimentation so as to discriminate among various biotic and abiotic fates of hydrocarbons.

Some researchers have constructed photosynthetic consortia to examine hydrocarbon degradation. Hamouda et al. (2016) challenged a consortium of the diazotrophic cyanobacterium *Anabaena oryzae* and the microalga *Chlorella kessleri* as well as the individual cultures with crude oil. The phototrophs were found to grow mixotrophically (measured as increase in optical density of the culture) when supplied with crude oil and nitrate as a fixed nitrogen source, but parallel incubation under diazotrophic conditions was not reported. Luo et al. (2014b) assembled a PAH-degrading consortium comprising the bacterium *Mycobacterium* sp. strain A1-PYR and the microalga *Selenastrum capricornutum*, grown with illumination and supplemented with nitrate. Pyrene metabolism by the bacterium enhanced growth of the microalga, possibly by mitigating toxicity of pyrene or its non-specific oxidation products to the alga or possibly by bacterial production of phenolic metabolites that are growth stimulants for the alga. Such interactions might explain field observations of enhanced biodegradation and phototrophy in oil-contaminated surface sediments and soils (discussed below), without invoking diazotrophy as a contributing factor.

# 4 Activities in the Environment

## 4.1 Sediments and Microbial Mats

Decades ago Knowles and Wishart (1977) noted that addition of weathered crude oil or individual liquid $n$-alkanes ($n$-$C_6$–$C_{16}$) to marine and lake sediments from northern Canada had unpredictable effects on $N_2$ fixation (measured as acetylene reduction), and the aromatic hydrocarbon trimethylbenzene actually inhibited $N_2$ fixation. A different effect was noted by Toccalino et al. (1993) who incubated sandy soil with propane or butane gas: as soil N was depleted, hydrocarbon degradation rates slowed, but bacteria in the butane-exposed soil eventually overcame N limitation, presumably by $N_2$ fixation, measured as acetylene reduction and increase in soil total N. Despite this indirect evidence of activity, no pure isolates able to grow on butane and simultaneously fix $N_2$ could be isolated from active enrichments (Toccalino et al. 1993), suggesting that the observed activity in bulk soil was due to the combined activities of indigenous heterotrophs. The best-documented field example of concomitant activities is likely that of naphthalene-degrading *P. naphthalenivorans* strain CJ2 (Hanson et al. 2012) discussed above. Screening of coal tar-contaminated sediments using a combination of nucleic acid- and antibody-based specific probes revealed that the abundance of strain CJ2 increased in sediments amended with naphthalene, as did nitrogenase activity. The inference is that diazotrophy provides a selective advantage to strain CJ2 in the contaminated sediment, especially when hydrocarbon biodegradation results in N limitation. Although the report provides only circumstantial evidence for dual activity in the field, sophisticated technical approaches such as those used by Hanson et al. (2012) should help guide future studies.

Inoculation of hydrocarbon-contaminated soils (bioaugmentation) with consortia of diazotrophs and heterotrophs has been attempted in order to provide sufficient soil N for bioremediation. Onwurah and Nwuke (2004) inoculated soil contaminated by light crude oil with differing proportions of a *Pseudomonas* sp. (previously determined to degrade hydrocarbons) and the free-living diazotroph *Azotobacter vinelandii*. They observed increased hydrocarbon removal and total soil N in soils that received both species versus heat-killed control inocula or *Pseudomonas* alone. From these data they concluded that a combined inoculum could stimulate bioremediation of oil-contaminated soil, either indirectly by providing fixed N to the *Pseudomonas* species and/or through co-metabolism of hydrocarbons by the *A. vinelandii* strain. Piehler et al. (1999) tested an unusual approach to bioremediation by adding particulate organic C (as corn slash) to coastal water samples amended with diesel fuel, with no addition of fixed N. They inferred that the observed stimulation of diesel biodegradation was attributable in part to increased $N_2$ fixation by bacterial consortia using the particulate carbon source, among other factors such as greater bioavailability of hydrocarbons and increased bacterial biomass. However, they made no attempt to isolate cultures or determine which species were responsible for the activities.

Notably, after massive contamination of the Arabian Gulf with Kuwaiti crude oil in 1990, Sorkhoh et al. (1992) observed the emergence of extensive cyanobacterial mats that were physically associated with the oil in contaminated sediments, but were absent from pristine areas. One explanation proposed for the flourishing of the mats was inhibition of eukaryotic grazers by the oil. Embedded within the mucilage of the phototrophs were heterotrophs capable of utilizing *n*-alkanes, leading Sorkhoh et al. (1992) to propose that heterotrophic aerobic oil degradation was stimulated by $O_2$ and fixed N produced by the cyanobacteria. This hypothesis was later supported by Musat et al. (2006) who detected nitrogenase gene sequences both in oil-contaminated and uncontaminated photic marine sediments, with *nifH* sequences affiliated with heterotrophs (e.g., *Azotobacter*, *Pseudomonas*, *Desulfovibrio*) being more common in the contaminated sediments. However, "expression" of *nifH* genes was almost completely attributable to cyanobacteria in the sediments, rather than to heterotrophs (Musat et al. 2006). This indicates that heterotrophic $N_2$ fixation, although genetically possible in the oil-contaminated sediments, did not contribute significantly to production of fixed N in situ. Rather, cyanobacterial $N_2$ fixation (and, presumably, phototrophic $O_2$ production) supported the aerobic oil-degrading heterotrophs even though oil addition did not stimulate $N_2$ fixation. Supporting this emerging view of distributed metabolism in phototrophic mats, Chaillan et al. (2006) concluded that hydrocarbon degradation by an oil-contaminated photosynthetic mat observed in situ was achieved exclusively by heterotrophs rather than phototrophs. Chronopoulou et al. (2013) noted an increase in abundance (but not diversity) of cyanobacteria and *nifH* genes in mudflat sediments after a simulated oil spill, as well as a decrease in dissolved inorganic nitrogen as oil degradation proceeded. Rather than attributing hydrocarbon-degrading ability to the enriched phototrophic diazotrophs, Chronopoulou et al. (2013) interpreted this community shift to result from several indirect effects of oiling, including selection for diazotrophy in the now-nitrogen-limited sediments and decreased grazing pressure from predators.

Diazotrophic Anaerobic hydrocarbon degradation is less likely to involve photosynthetic consortia; instead one or more heterotrophs may be responsible for both activities in situ. Scott et al. (2014) performed a metagenomic survey of Gulf of Mexico sediments and found that the abundance of *nifA* and *nifB* genes followed the order: pristine sites > Deepwater Horizon oil spill-impacted sediments > natural methane seep sediments. This would suggest that $N_2$ fixation is not directly coupled to anaerobic hydrocarbon utilization in these sites, but obviously additional research is needed in this and other anaerobic environments to build and test hypotheses.

## 4.2 Phytoremediation

Phytoremediation is the use of living plants and associated soil microbiota to ameliorate contamination with inorganic or organic contaminants. In this chapter, an implicit component is microbial $N_2$ fixation by plant symbionts (e.g., in nodules of legumes) or free-living $N_2$-fixers in the rhizosphere (the soil adjacent to plant roots that is directly affected by plant activity).

The potential for phytoremediation of methane emissions is currently unknown, as the scale and importance of $N_2$-fixing activity by methanotrophs in the environment are still unresolved (Bodelier and Steenbergh 2014). However, mitigating crude oil contamination of soil through plant—microbe interactions has been proposed for several decades (e.g., Radwan et al. 1995). Although nitrogen availability is obviously important to both partners in phytoremediation, the details of $N_2$ fixation (whether by symbionts or free-living diazotrophs) and hydrocarbon metabolism (whether supported directly or indirectly by $N_2$ fixation) have not been elucidated. Surveys of rhizosphere microbiota commonly report the presence of native hydrocarbon degraders. For example, do Carmo et al. (2011) isolated numerous strains from the rhizosphere of mangrove plants that could fix $N_2$ or degrade hydrocarbons aerobically when tested individually. Although concurrent activity was not assessed, the study shows the potential for distributed metabolism at the least.

Even though appropriate microbes may already exist in the rhizosphere, some studies have proposed inoculation (bioaugmentation) with free-living or symbiotic $N_2$-fixers to enhance phytoremediation. Gradova et al. (2003) assessed the potential of several free-living *Azotobacter* species for bioremediation of oil-contaminated soil. The strains apparently enhanced the hydrocarbon-degrading activity of bacteria in a commercial microbial formulation intended for bioaugmentation of oil spills in soil. However, there was little effect on wheat seedling growth after a single treatment of contaminated soil with the *Azotobacter* spp., and long-term persistence of the inoculum was not documented. *Azoarcus* sp. strain CIB, a facultative anaerobe that is a free-living diazotroph (confirmed through genome sequencing), was found to degrade toluene and *m*-xylene when provided with nitrate, although simultaneous $N_2$ fixation and hydrocarbon oxidation were not reported. Its ability to associate endophytically with rice roots led Martín-Moldes et al. (2015) to suggest that it may be able to contribute to phytoremediation. Radwan et al. (2007) tested the ability of several potential $N_2$-fixing symbionts to degrade aliphatic and aromatic hydrocarbons in vitro. Crude oil and *n*-alkanes from $C_9$ to $C_{29}$ supported growth in mineral medium containing nitrate, and biodegradation of *n*-octadecane and phenanthrene was quantified. As well, incubating nitrogen-fixing nodules of the legume *Vicia faba* (broad bean) in mineral medium with kerosene resulted in hydrocarbon depletion. The conclusion was that legume crops were good candidates for phytoremediation of oily soil by providing fixed N as well as enriching the rhizosphere with hydrocarbon-degrading bacteria (Radwan et al. 2007; Dashti et al. 2009). However, the genetic and metabolic details of diazotrophy and hydrocarbon degradation in rhizosphere communities are still lacking, as in other environments and bioremediation applications.

Thus, oxidation of liquid hydrocarbons in the field comprises a complex suite of interacting mechanisms and organisms including:

(a) Abiotic sorption of hydrocarbons onto biomass and/or sequestration in mucilage
(b) Biomass-mediated photooxidation of hydrocarbons in surface environments
(c) Partial enzymatic oxidation of hydrocarbons through co-metabolism and/or detoxification reactions, generating metabolites that may be toxic unless further metabolized by other organisms
(d) Stimulation of microbial communities through inhibition of predators by hydrocarbons
(e) Distributed metabolism, with one partner providing fixed nitrogen (and sometimes $O_2$) and the other partner(s) degrading hydrocarbons to metabolites that may or may not be utilized by the diazotrophic partner (through mixotrophy or cross-feeding)
(f) True diazotrophic hydrocarbon utilization by individual microbes

The literature currently provides the least evidence for the latter scenario, highlighting an obvious area for future investigation.

## 5 Gaps in Knowledge and Research Needs

Despite the potential importance of $N_2$ fixation in hydrocarbon-impacted environments, surprisingly few bacteria have demonstrated (or been tested for) simultaneous $N_2$ fixation and hydrocarbon oxidation abilities. It is not clear whether the paucity of well-documented cases is the result of narrow niche distribution, lack of environmental significance, or simply too little research to date. As noted by Chronopoulou et al. (2013), "there are insufficient studies in marine ecosystems to see consistent patterns between the amount of oil in an environment and rates of dinitrogen fixation." Certainly, in the case of anaerobic environments, the existence of diazotrophy coupled to hydrocarbon oxidation is an open question.

Based on the literature, one might conclude that the impact of individual hydrocarbon-degrading $N_2$-fixing species in the environment is low because of their infrequent occurrence (or detection) in environment, whereas selection for competent microbial communities or assemblages is more evident. An apparent exception is the methanotrophs, for which diazotrophy appears to be more widespread than previously appreciated (Auman et al. 2001). Even so, Bodelier and Steenbergh (2014) note that there is no consensus to date that methanotrophs fix $N_2$ in situ and that only the ANME group has been shown conclusively to assimilate $N_2$ (Dekas et al. 2009). As research into methanotrophic communities proceeds, perhaps resolution will be achieved.

Even within species known to possess both $N_2$-fixing and hydrocarbon-oxidizing potential it appears unusual for both be expressed simultaneously, for unknown reasons. As Musat et al. (2006) put it so well: "Further research is needed to understand whether there is a physiological incompatibility between active hydrocarbon degradation and nitrogen fixation under various conditions or in various microorganisms.... One may hypothesize that the co-occurrence of hydrocarbon oxidation and nitrogen fixation is rare or physiologically delicate." For example, nitrogenases are inactivated by $O_2$, whereas hydrocarbon-oxidizing mono- and dioxygenases require $O_2$. Balancing $O_2$ partial pressure to simultaneously achieve conditions suitable for both activities may be a challenge for axenic cultures but might be overcome in microniches afforded by microbial assemblages. This question might be addressed using metatranscriptome approaches. Another possible incompatibility is the requirement for different sigma factors for gene expression (e.g., $\sigma^{54}$ homologs for nitrogenase genes vs. $\sigma^{70}$ homologs for carbon metabolism genes). It may be that individual species "cycle" between activities when faced with a hydrocarbon-impacted N-deficient environment, or, more likely, that consortia offer synergistic advantages that individual species cannot achieve in natural environments. The answers to these questions are important when considering bioremediation strategies, especially phytoremediation of oil using legumes or plants that promote $N_2$-fixing rhizosphere organisms. Given the general lack of understanding of the co-occurrence of these activities within individual species and assemblages, it is not surprising that their roles in phytoremediation are currently unproven. Furthermore, the case for bioaugmentation of oil spills with photosynthetic or heterotrophic diazotrophs (individually or in mixed inocula and with or without plant partners) is a field requiring further study to determine efficacy and survival of inoculants.

The scientific community should be encouraged to address these knowledge gaps while striving for rigorously executed studies supported by appropriate technical and biological controls, as well as making use of the sophisticated methodology that is now available through 'omics and stable isotope analyses.

# References

Abed RMM, Köster J (2005) The direct role of aerobic heterotrophic bacteria associated with cyanobacteria in the degradation of oil compounds. Int Biodeter Biodegr 55:29–37

Abed RMM, Safi NMD, Köster J, De Beer D, El-Nahhal Y, Rullkötter J, Garcia-Pichel F (2002) Microbial diversity of a heavily polluted microbial mat and its community changes following degradation of petroleum compounds. Appl Environ Microbiol 68:1674–1683

Al-Mailem DM, Sorkhoh NA, Salamah S, Eliyas M, Radwan SS (2010) Oil bioremediation potential of Arabian Gulf mud flats rich in diazotrophic hydrocarbon-utilizing bacteria. Int Biodeter Biodegr 54:218–225

Amos RT, Mayer KU, Bekins BA, Delin GN, Williams RL (2005) Use of dissolved and vapor-phase gases to investigate methanogenic degradation of petroleum hydrocarbon contamination in the subsurface. Water Resour Res 41:1–15

Auman AJ, Speake CC, Lidstrom ME (2001) *nifH* sequences and nitrogen fixation in type I and type II methanotrophs. Appl Environ Microbiol 67:4009–4016

Bodelier PLE, Steenbergh AK (2014) Interactions between methane and nitrogen cycling: current metagenomic studies and future trends, Chapter 3. In: Marco D (ed) Metagenomics of the microbial nitrogen cycle: theory, methods and applications. Caister Academic Press, Norfolk, pp 33–63

Boulygina ES, Kuznetsov BB, Marusina AI, Tourova TP, Kravchenko IK, Bykova SA, Kolganova TV, Galchenko VF (2002) A study of nucleotide sequences of *nifH* genes of some methanotrophic bacteria. Microbiology 71:425–432

Buckley DH, Huangyutitham V, Hsu SF, Nelson TA (2008) [15]$N_2$-DNA-stable isotope probing of diazotrophic methanotrophs in soil. Soil Biol Biochem 40:1272–1283

Cerniglia CE, Gibson DT, Van Baalen C (1980) Oxidation of naphthalene by cyanobacteria and microalgae. J Gen Microbiol 116:495–500

Chaillan F, Gugger M, Saliot A, Couté A, Oudot J (2006) Role of cyanobacteria in the biodegradation of crude oil by a tropical cyanobacterial mat. Chemosphere 62:1574–1582

Chen YP, Lopez-de-Victoria G, Lovell CR (1993) Utilization of aromatic compounds as carbon and energy sources during growth and $N_2$-fixation by free-living nitrogen fixing bacteria. Arch Microbiol 159:207–212

Chronopoulou P-M, Fahy A, Coulon F, Païssé S, Goñi-Urriza M, Peperzak L, Acuña Alvarez L, McKew BA, Lawson T, Timmis KN, Duran R, Underwood GJC, McGenity TJ (2013) Impact of a simulated oil spill on benthic phototrophs and nitrogen-fixing bacteria in mudflat mesocosms. Environ Microbiol 15:242–252

Chu K-H, Alvarez-Cohen L (1998) Effect of nitrogen source on growth and trichloroethylene degradation by methane-oxidizing bacteria. Appl Environ Microbiol 64:3451–3457

Collins CEV, Foght JM, Siddique T (2016) Co-occurrence of methanogenesis and $N_2$ fixation in oil sands tailings. Sci Total Environ 565:306–312

Coty VF (1967) Atmospheric nitrogen fixation by hydrocarbon-oxidizing bacteria. Biotechnol Bioeng IX:25–32

Dalton H, Whittenbury R (1976) The acetylene reduction technique as an assay for nitrogenase activity in the methane oxidizing bacterium *Methylococcus capsulatus* strain Bath. Arch Microbiol 109:147–151

Dashti N, Khanafer M, El-Nemr I, Sorkhoh NA, Ali N, Radwan SS (2009) The potential of oil-utilizing consortia associated with legume root nodules for cleaning oily soils. Chemosphere 74:1354–1359

Dashti N, Ali N, Elilyas M, Khanafer M, Sorkhoh NA, Radwan SS (2015) Most hydrocarbonoclastic bacteria in the total environment are diazotrophic, which highlights their value in the bioremediation of hydrocarbon contaminants. Microbes Environ 30:70–75

Davis JB, Stanley JP (1965) Microbiological nitrogen fixation. US Patent #3,210,179

Davis JB, Coty VF, Stanley JP (1964) Atmospheric nitrogen fixation by methane-oxidizing bacteria. J Bacteriol 88:468–472

de Bont JAM, Mulder EG (1974) Nitrogen-fixation and co-oxidation of ethylene by a methane-utilizing bacterium. J Gen Microbiol 83:113–121

Dekas AE, Poretsky RS, Orphan VJ (2009) Deep-sea Archaea fix and share nitrogen in methane-consuming microbial consortia. Science 326:422–426

Dekas AE, Chadwick GL, Bowles MW, Joye JB, Orphan VJ (2014) Spatial distribution of nitrogen fixation in methane seep sediment and the role of the ANME archaea. Environ Microbiol 16:3012–3029

do Carmo FL, dos Santos HF, Martins EF, van Elsas JD, Rosado AS, Peixoto RS (2011) Bacterial structure and characterization of plant growth promoting oil and degrading bacteria from the rhizospheres of mangrove plants. J Microbiol 49:535–543

Dos Santos PC, Fang Z, Mason SW, Setubal JC, Dixon R (2012) Distribution of nitrogen fixation and nitrogenase-like sequences amongst microbial genomes. BMC Genomics 12:162

Eckford R, Cook FD, Saul D, Aislabie J, Foght J (2002) Free-living heterotrophic nitrogen-fixing bacteria isolated from fuel-contaminated Antarctic soils. Appl Environ Microbiol 68:5181–5185

Fries MR, Zhou J, Chee-Sanford J, Tiedje JM (1994) Isolation, characterization, and distribution of denitrifying toluene degraders from a variety of habitats. Appl Environ Microbiol 60:2802–2810

Gradova NB, Gornova IB, Eddaudi R, Salina RN (2003) Use of bacteria of the genus *Azotobacter* for bioremediation of oil-contaminated soils. Appl Biochem Microbiol 39:279–281

Hamamura N, Storfa RT, Semprini L, Arp DJ (1999) Diversity in butane monooxygenases among butane-grown bacteria. Appl Environ Microbiol 65:4586–4593

Hamouda RAEF, Sorour NM, Yeheia DS (2016) Biodegradation of crude oil by *Anabaena oryzae*, *Chlorella kessleri* and its consortium under mixotrophic conditions. Int Biodeter Biodegr 112:128e134

Hanson BT, Yagi JM, Jeon CO, Madsen EM (2012) Role of nitrogen fixation in the autecology of *Polaromonas naphthalenivorans* in contaminated sediments. Environ Microbiol 14: 1544–1557

Harper HJ (1939) The effect of natural gas on the growth of microorganisms and the accumulation of nitrogen and organic matter in the soil. Soil Sci 48:461–466

Ho A, Bodelier PLE (2015) Diazotrophic methanotrophs in peatlands: the missing link? Plant and Soil 389:419–423

Hoefman S, van der Ha D, Boon N, Vandamme P, De Vos P, Heylen K (2014) Niche differentiation in nitrogen metabolism among methanotrophs within an operational taxonomic unit. BMC Microbiol 14:83

Jeon CO, Park W, Padmanabhan P, DeRito C, Snape JR, Madsen EL (2003) Discovery of a bacterium, with distinctive dioxygenase, that is responsible for in situ biodegradation in contaminated sediment. Proc Natl Acad Sci U S A 100:13591–13596

Keller AH, Schleinitz KM, Starke R, Bertilsson S, Vogt C, Kleinsteuber S (2015) Metagenome-based metabolic reconstruction reveals the ecophysiological function of *Epsilonproteobacteria* in a hydrocarbon-contaminated sulfidic aquifer. Front Microbiol 6:1396

Khadem A, Pol A, Jetten MSM, Op den Camp HJM (2010) Nitrogen fixation by the verrucomicrobial methanotroph '*Methylacidiphilum fumariolicum*' SolV. Microbiology 156:1052–1059

Knowles R, Wishart C (1977) Nitrogen fixation in arctic marine sediments: effect of oil and hydrocarbon fractions. Environ Pollut 13:133–149

Kumar JIN, Patel JG, Kumar RN, Khan SR (2014) Chronic response of three different cyanobacterial species on growth, pigment, and metabolic variations to the high molecular weight polycyclic aromatic hydrocarbon – pyrene. Polycycl Aromat Compd 34:143–160

Laguerre G, Bossand B, Bardin R (1987) Free-living dinitrogen-fixing bacteria isolated from petroleum refinery oily sludge. Appl Environ Microbiol 53:1674–1678

Lei AP, Wong YS, Tam NFY (2002) Removal of pyrene by different microalgal species. Water Sci Technol 46:195–201

Lei AP, Hu ZL, Wong YS, Tam NFY (2007) Removal of fluoranthene and pyrene by different microalgal species. Bioresour Technol 98:273–280

Luo L, Wang, Lin L, Luan T, Ke L, Tam NFY (2014a) Removal and transformation of high molecular weight polycyclic aromatic hydrocarbons in water by live and dead microalgae. Process Biochem 49:1723–1732

Luo S, Chen B, Lin L, Wang X, Tam NFY, Luan T (2014b) Pyrene degradation accelerated by constructed consortium of bacterium and microalga: effects of degradation products on the microalgal growth. Environ Sci Technol 48:13917–13924

Martín-Moldes Z, Zamarro MR, del Cerro C, Valencia A, Gómez JF, Arcas A, Udaondo Z, García JL, Nogales J, Carmona M, Diaz E (2015) Whole-genome analysis of *Azoarcus* sp. strain CIB provides genetic insights to its different lifestyles and predicts novel metabolic features. Syst Appl Microbiol 38:462–471

Minamisawa K, Imaizumi-Anraku H, Bao Z, Shinoda R, Okubo T, Ikeda S (2016) Are symbiotic methanotrophs key microbes for N acquisition in rice paddy root? Microbes Environ 31:4–10

Murrell JC, Dalton H (1983) Nitrogen fixation in obligate methanotrophs. Microbiology 129:3481–3486

Musat F, Harder J, Widdel F (2006) Study of nitrogen fixation in microbial communities of oil-contaminated marine sediment microcosms. Environ Microbiol 8:1834–1843

Narro ML, Cerniglia CE, van Baalen C, Gibson DT (1992) Metabolism of phenanthrene by the marine cyanobacterium *Agmenellum quadruplicatum* PR6. Appl Environ Microbiol 58:1351–1359

Onwurah INE, Nwuke C (2004) Enhanced bioremediation of crude oil-contaminated soil by a *Pseudomonas* species and mutually associated adapted *Azotobacter vinelandii*. J Chem Technol Biotechnol 79:491–498

Pérez-Vargas J, Poggi-Varaldo HM, Calva-Calva G, Ríos-Leal E, Rodriguez-Vázquez R, Ferrera-Cerrato R, Esparza-García F (2000) Nitrogen-fixing bacteria capable of utilising kerosene hydrocarbons as a sole carbon source. Water Sci Technol 42:407–410

Piehler MF, Swistak JG, Pinckney JL, Paerl HW (1999) Stimulation of diesel fuel biodegradation by indigenous nitrogen fixing bacterial consortia. Microb Ecol 38:69–78

Prantera MT, Drozdowicz A, Gomes Leite S, Soares Rosado A (2002) Degradation of gasoline aromatic hydrocarbons by two $N_2$-fixing soil bacteria. Biotechnol Lett 24:85–89

Radwan S, Sorkhoh N, El-Nemr I (1995) Oil biodegradation around roots. Nature 376:302

Radwan SS, Dashti N, El-Nemr I, Khanafer M (2007) Hydrocarbon utilization by nodule bacteria and plant growth-promoting rhizobacteria. Int J Phytoremediation 9:475–486

Radwan SS, Mahmoud H, Khanafer M, Al-Habib A, Al-Hasan R (2010) Identities of epilithic hydrocarbon-utilizing diazotrophic bacteria from the Arabian Gulf coasts, and their potential for oil bioremediation without nitrogen supplementation. Microb Ecol 60:354–363

Rivière J, Oudot J, Jonquères J, Gatellier G (1974) Fixation d'azote atmosphérique par des bactéries utilisant l'hexadécane comme source de carbone et d'énergie. Ann Agron 25:633–644

Roy I, Shukla SK, Mishra AK (1988) *n*-Dodecane as a substrate for nitrogen fixation by an alkane-utilizing *Azospirillum* sp. Curr Microbiol 16:303–309

Sánchez O, Diestra E, Esteve I, Mas J (2005) Molecular characterization of an oil-degrading cyanobacterial consortium. Microb Ecol 50:580–588

Schollenberger CJ (1930) Effect of leaking natural gas upon the soil. Soil Sci 29:261–266

Scott NM, Hess M, Bouskill NJ, Mason OU, Jansson JK, Gilbert JA (2014) The microbial nitrogen cycling potential is impacted by polyaromatic hydrocarbon pollution of marine sediments. Front Microbiol 5:108

Semrau JD, DiSpirito AA, Yoon S (2010) Methanotrophs and copper. FEMS Microbiol Rev 34:496–531

Sorkhoh N, Al-Hasan R, Radwan S, Hopner T (1992) Self-cleaning of the gulf. Nature 359:109

Sorkhoh NA, Ali N, Salamah S, Eliyas M, Khanafer M, Radwan SS (2010a) Enrichment of rhizospheres of crop plants raised in oil sand with hydrocarbon-utilizing bacteria capable of hydrocarbon consumption in nitrogen free media. Int Biodeter Biodegr 64:659–664

Sorkhoh NA, Ali N, Dashti N, Al-Mailem DM, Al-Awadhi H, Eliyas M, Radwan SS (2010b) Soil bacteria with the combined potential for oil utilization, nitrogen fixation, and mercury resistance. Int Biodeter Biodegr 64:236–231

Thavasi R, Jayalakshmi S, Balasubramanian T, Banat IM (2006) Biodegradation of crude oil by nitrogen fixing marine bacteria *Azotobacter chroococcum*. Res J Microbiol 1:401–408

Toccalino PL, Johnson RL, Boone DR (1993) Nitrogen limitation and nitrogen fixation during alkane biodegradation in a sandy soil. Appl Environ Microbiol 59:2977–2983

van Ginkel CG, de Bont JAM (1986) Isolation and characterization of alkene-utilizing *Xanthobacter* spp. Arch Microbiol 145:403–407

Zehr JP, Jenkins BD, Short SM, Steward GF (2003) Nitrogenase gene diversity and microbial community structure: a cross-system comparison. Environ Microbiol 5:539–554

# Kinetics and Physiology at Vanishingly Small Substrate Concentrations

# 30

## D. K. Button

## Contents

1  Introduction ................................................................... 450
2  Concentration and Rate ......................................................... 450
3  Kinetic Concepts that Include Small Concentration ............................... 451
4  Rate Measurement .............................................................. 452
5  Rates in the Environment ........................................................ 453
6  Mechanisms .................................................................... 456
7  Distributions ................................................................... 458
8  Research Needs ................................................................ 459
References ........................................................................ 459

### Abstract

Hydrocarbon-using bacteria are a major portion of the aquatic microflora. They are supported in part with natural hydrocarbon sources such as terpenes along with other classes of organics. Concentrations are small, in the μg $L^{-1}$ range and generally below detection. The organisms are sometimes mischaracterized as high-affinity organisms due to small Michaelis constants, and measured affinities for these and other substrates of oligobacteria are much lower than for most organisms in laboratory culture. "Specific affinity" theory was used to predict the identity and concentrations of some naturally occurring hydrocarbons. Hydrocarbon concentrations are limited by both solubility and microbial kinetics at the steady state in an equilibrium that is analogous to chemostat culture. Both dissolved and oil phases of hydrocarbons are used. Microbial oxidation of the latter can help to cause a heavy oil to sink. Accumulation is thought to be by partitioning into the membrane lipid portion of the cell envelope for partial oxidation and trapping into the interior. One consequence of this "vectorial

D. K. Button (✉)
Institute of Marine Science, University of Alaska, Fairbanks, AK, USA
e-mail: dkbutton@ims.uaf.edu

© Springer International Publishing AG, part of Springer Nature 2018
T. Krell (ed.), *Cellular Ecophysiology of Microbe: Hydrocarbon and Lipid Interactions*,
Handbook of Hydrocarbon and Lipid Microbiology, https://doi.org/10.1007/978-3-319-50542-8_51

partitioning" is the liberation of partly oxidized hydrocarbon degradation products into the environment.

---

## 1    Introduction

Most of the vast global biomass of heterotrophic aquatic bacteria is limited by concentrations of dissolved organic carbon. All organics that dissolve, including even the heavier fraction of crude oil, are represented. The concentration of each of these solutes is affected by the kinetics of uptake. Growth rates often approach the minimum sustainable, generally generation times of days to months. Population persistence depends on the organisms' ability to effect a flow of nutrients into them. Growth rates of these oligotrophs (transthreptic or surface-feeding organisms) remain perpetually nutrient-limited by the biophysics of nutrient capture, cytoarchitecture, and bioenergetics. Increased concentrations give faster growth. At steady state, the resulting larger biomass is followed by a decrease in substrate concentration arising from increased demand. The system strives toward equilibrium where the microbial population is set by the input rate of organics, and the concentration of organics is set by the ability of the population to collect substrate. Over time frames of days or weeks concentrations remain more or less unchanged. However, the whole metabolome may downsize to match potential nutrient supply rates in the environment. This quasi equilibrium is exploited in continuous cultures or chemostats and led to early descriptions of the kinetics of microbial growth (Monod 1950). In these nutrient-limited continuous cultures, growth rates are set at some rate below the maximal, $\mu_{max}$, by the influent rate of fresh medium containing substrate. Population is set by the concentration of substrate in the influent medium. Both may be operator-controlled. Growth in the environment is basically similar except that biomass consumption is through predation rather than dilution and always nutrient-limited. Here the kinetics of hydrocarbon metabolism are explored with particular attention to the very small concentrations that persist in natural aquatic setting where a condition approaching starvation is normal. Associated kinetic parameters are defined and described.

---

## 2    Concentration and Rate

Key kinetic terms include $\mu_{max}$ and maximal utilization rate $V_{max}$, as related to $\mu_{max}$ by cell yield $Y$: $V_{max} = \mu_{max}/Y$. Substrate sequestering rate is described by the associated rate and kinetic constants. The ability to collect substrate is generally described by affinity as expressed by the Michaelis constant $K_m$ where $K_m$ is the concentration of substrate restricting uptake rate to half-maximal: $K_m = V_{max}/2$. But $\mu_{max}$ varies greatly depending on properties other than the ability to collect substrate. These include the rate of macromolecule synthesis, whether the particular substrate is a major component of the nutrient mix, and metabolome control. $V_{max}$ is a capacity factor that may depend on the surface density of transporters, the concentration of cytoplasmic enzymes, and the ability to assemble macromolecular structures. These are cytoarchitectural parameters and ones that affect terms describing the relationship between rate and concentration. They are frequently indeterminant as for both phosphate (Button et al. 1973) and toluene (Button et al. 1973; Law and Button 1986). A compilation of data shows that both cell yield (Button 1970) and the rate constant or the ability to collect substrate increases greatly with $K_m$ (Button 1998), and induction increases $V_{max}$ while leaving $K_m$ unchanged. For example $V_{max}$ for methane use by a methanotroph can increase 100-fold upon induction, but there is a concomitant increase in $K_m$ as well. And $K_m$ can increase with cell yield by over a factor of $10^7$: too much to be accounted for by kinetic theory alone. All contradict the concept that a small $K_m$ reflects high affinity, although the paradigm is supported in some 10,000 publications. Second, rate may be restricted in a concentration-dependent manner by the amount of energy available for transport. Although Michaelian concepts may be empirically accurate for single enzymes, they are incompatible with the operation of the substrate accumulation sequence within a complete organism. Further the theoretical accuracy is called into question (Kou et al. 2005), since $K_m$ can reflect the distribution of waiting times for a receptive active site rather than the probability of ineffective binding.

# 3    Kinetic Concepts that Include Small Concentration

Many pitfalls in describing microbial activity may be avoided by the use of specific affinity (Button 1983), particularly where concentrations are small as they are for environmental hydrocarbons. Specific affinity is the second-order rate constant relating the net uptake rate $vS$ of a population to the concentration of a particular substrate $S$. Here, population is given by wet weight. The maximal value is a kinetic constant and the initial slope of the kinetic ($vS$ vs $S$) curve. In enzymatic terms, it is the effective catalytic constant of the organism. As saturation ensues, specific affinity decreases from its base or maximal value $\left(a_S^o\right)$ to the rate constant $aS$ with increasing concentration of the substrate due to saturation. It is always uptake rate divided by substrate concentration, $aS = vS/S$, and is often a monotonic increasing function of S up to inhibitory concentrations for fully energized copeotrophs and "lab cultures." It is however sigmoidal in the case of positive cooperativity among substrates in energy-poor systems. To avoid association with a particular kinetic function, the point at which the rate of transport $vS$ is reduced to $V_{max}/2$ under given conditions by nutrient depravation is sometimes designated $Kt$.

For hydrocarbons, low solubility may preclude concentrations sufficient to saturate the capacity of cytoplasmic enzymes. Larger concentrations of the lighter fraction may be achieved due to increased solubility (Shaw 1989) and preference for the lipid phase, but this can lead to toxicity, particularly in the $C_6$–$C_{10}$ range. For toluene,

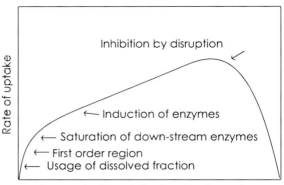

**Fig. 1** Kinetic curve for hydrocarbon uptake by aquatic microflora at steady state in the presence of cosubstrates. Initial slope and half-maximal concentrations are given in Table 1

concentrations shift from stimulatory to inhibitory at about 70 mg L$^{-1}$ (Robertson et al. 1979). Monoterpenes, plant hydrocarbons in coastal seawater (*vide post*), were surprisingly inhibitory toward toluene uptake. However, benzene added at 50 times the concentration of toluene was ineffectual as a competitive inhibitor of toluene metabolism and the induction constant for toluene was much larger (Law and Button 1986). These observations along with the selective use of alkane homologs from the mixture found in crude oil indicated metabolic pathways with unique enzymatic components (Robertson et al. 1979). No matter what the shape of the kinetic curve is, uptake rate is always given by the product of specific affinity and biomass $\nu S = aS\ SX$ where $X$ is biomass. A complete kinetic curve shape for dissolved hydrocarbons from substrate amendments to batch culture, and concentration measurements in continuous culture is suggested in Fig. 1. One application was the discovery of terpenes in seawater as previously predicted from the kinetics of toluene metabolism (*vide post*).

## 4    Rate Measurement

Rates of $^{14}$C-hydrocarbon oxidation may be determined by the rate of evolution of $^{14}$CO$_2$ from radiolabeled carbon dioxide (Button et al. 1981a). Initial concentrations are determined from specific activity along with background concentrations from gas chromatography. The hydrocarbon should first be purified to eliminate residual impurities from the chemosynthesis. Otherwise, $^{14}$CO$_2$ from isotope precursors can indicate metabolism where there is none. For $^{14}$C-toluene, the isotope is treated with NaOH to sequester $^{14}$CO$_2$ and sublimed on a cold finger. This eliminates most of the abscissa intercept in the time courses for $^{14}$CO$_2$ liberation. Where glassware has been previously exposed to $^{14}$CO$_2$, heating to 550 °C is necessary to further reduce background. Other contaminants, possibly $^{14}$C-ester formed during synthesis, are removed in a hydrophobic trap.

Microbiologically evolved $^{14}$CO$_2$ from purified $^{14}$C-toluene amendments is dried by bubbling through concentrated H$_2$SO$_4$ and remaining water removed with a cold trap of glass beads in dry-ice acetone. This is followed by the Tenax trap, cooled in the same way. Biogenic $^{14}CO_2$ in the effluent gas stream is collected in a series of small phenethylamine traps to insure complete recovery. A procedure such as this is necessary for accurate measurements in the oligotrophic environment where concentrations are small.

The ability of sediment-associated organisms to oxidize hydrocarbons can be estimated by agitating beach gavel in added seawater (Button et al. 1981a), determining their populations by flow cytometry (Robertson et al. 1998), and amending with radiolabel to give the rate of $^{14}CO_2$ evolution from that added (beach gravel, Table 1). While the absolute rate is systematically underestimated with increasing background or unlabeled substrate, the rate of isotope taken up compared with total isotope applied gives a very good estimate of the base specific affinity $(a_S^o)$ when rates are in the linear or unsaturated range with concentration (Button et al. 2004).

## 5    Rates in the Environment

Growth rate $\mu(=aS\ X\ S\ Y)$ depends on the rate of substrate uptake and cell yield $Y$. For the environmental steady state, it is also the loss rate, and for rich systems, taken as the grazing or consumption rate by bacterivores. Base values of the specific affinity $(a_s^o)$ of a particular substrate within a group $A$, $B$, $C$... are designated $(a_A^o,\ a_B^o,\ a_C^o \ldots)$ so that $(a_{\text{total}}^o = \sum a_A^o,\ a_B^o,\ a_C^o \ldots)$. For aquatic or oligobacteria the number of substrates must be very large, because the product of the concentration of any one and the associated specific affinity is small compared with measures of population growth. The small affinities may be for reasons of economy of cytoarchitecture. This is particularly true for the use of hydrocarbons by pelagic bacterial populations where the number of chemical species is large. Various hydrocarbon oxidizing isolates fed a menu of crude oil, leave unique patterns of unused components on chromatograms (Robertson et al. 1979) demonstrating that a diversity of metabolic systems is necessary for the metabolism of homologs. Many metabolic systems for hydrocabons have been described (Zylstra and Gibson 1989) including some from aquatic oligobacteria (Wang et al. 1996).

Specific affinities for toluene ranged from 84 L g-cells$^{-1}$ h in the ballast water treatment facility plume near Port Valdez down to 0.05 L g-cells$^{-1}$ h 15 km west. By comparison, the value of an isolate designated *Pseudomonas* T2 was 87 L g-cells$^{-1}$ h. $Kt$ values of toluene oxidation rates by the Port Valdez microflora were in the range of 0.5–1.5 μg L$^{-1}$ ( Table 1). Still, in the environment, specific affinities are generally comparable to oxygen-containing organics so that simultaneous use of multiple substrates is required for growth at expected rates. Microbial activity in the water is a factor; pre- *Exxon Valdez* spill oxidation rates of dodecane in Port Valdez were 0.1 μg L day$^{-1}$ and 0.001 μg L day$^{-1}$ under the Arctic Ocean ice (Arhelger et al. 1976). Values are much lower than for use of polar substrates by copeotrophs (*vide post*).

**Table 1** Kinetic parameters for hydrocarbon oxidation

| Substrate | Microflora | System | Transport constant, $Kt \times 10^6$ g L | Specific affinity, $(a_S^e)$, L (g-cells h)$^{-1}$ | Reference |
|---|---|---|---|---|---|
| Toluene | Aransas Bay Texas | Seawater | 0.26 | 1.2 | Button and Robertson (1986) |
| Toluene | Prince William Sound Alaska | Seawater | – | 0.22 | Button and Robertson (1986) |
| Toluene | Ballast water treatment pond | Ballast water | – | 244 | Button and Robertson (1988) |
| Toluene | Fresh ballast water | Oil tanker ballast water | – | 53 | Button et al. (1981b) |
| Toluene | Port Valdez near mouth | Seawater, 5 m | – | 0.054 | Button et al. (1981b) |
| Toluene | Port Valdez Near discharge | Seawater, 50 | – | 36 | Button et al. (1981b) |
| Toluene | Resurrection Bay Alaska | Treatment pond effluent | 1.9 | 390 | Button and Robertson (1986) |
| Toluene | *Pseudomonas* sp. T2 induced [a] | Batch culture | 43.9 | 126 | Button and Robertson (1986) |
| Toluene | *Pseudomonas* sp. non-induced | Batch culture | 34.5 | 4.4 | Button and Robertson (1986) |
| Toluene | *Cycloclasticus oligotrophus* | Batch culture | 60 | 52 | Button et al. 1998 |

| Toluene | PLseudomonas fluorescens | Cell suspension | 1,800 | 5.6 | Shreve and Vogel (1993) |
|---|---|---|---|---|---|
| Toluene | Cycloclasticus oligotropus | Cell suspension | 1.3 | 2,200 | Shreve and Vogel (1993) |
| Benzene | Pseudomonas sp. T2 non-induced | Amino acid grown | – | 0.03 | Law and Button (1986) |
| Benzene | Pseudomonas sp. T2 induced | Induced | – | 317 | Law and Button (1986) |
| Benzene | Pseudomonas sp. T | Continuous, amino acid limited | – | 6.8 | Law and Button (1986), Robertson et al. (1979) |
| Benzene | Mixed culture | Batch respiration | 10,800 | 56 | Alvarez et al. (1991) |
| Dodecane | 1 m seawater | Port Valdez | – | 4.1 | Robertson et al. (1979) |
| Dodecane | Under ice | Arctic Ocean | – | 0.001 | Arhelger et al. (1976) |
| Methane | Methylocystis Strain LR1 | Batch culture | 138–12,600 nM | 2–28 | Dunfield and Conrad (2000) |
| Napthalene | Gravel wash-water | Resurrection Bay, post Exxon Valdez | – | <2.1 | Button et al. 1992 |
| Napthalene | Gravel wash-water | Port Valdez, post spill. | – | 92 | Button et al. 1992 |
| Terpene mixture | Resurrection Bay, AK | Seawater, 0 and 50 m depth | Unknown ($>50$ kdpm min$^{-1}$) | 8 and 81 | Button 1984 |

[a]Half-maximum concentration $Kt$, 1.9 µg L$^{-1}$ Button and Robertson 1988

## 6    Mechanisms

Hydrocarbon use may be from either the aqueous or the nonaqueous phase. In the aqueous phase, concentration is constrained by solubility, a value that decreases with the absence of aeromaticity and the number of carbons. For alkanes, the solubility of dodecane is near 1 µg $L^{-1}$ and triples as a linear function of chain length (Button 1976), and therefore, the range is large. Absence of branching eases metabolism although simple aromatics are efficiently accumulated as well. The heavy fraction from crude oil spills floats for a time, increases in density with microbial oxidation, and may sink to the bottom. The middle oily fraction around C-10 may be penetrated, and possibly used from within. Hydrocarbon oxidizers are replete with emulsifying agents (Francy et al. 1991) and sequestered by hydrocarbon emulsions. Then organisms disappear from view through microscope but can be located as distortions of the hydrocarbon spheres. The applied oil slick in a stirred glass vessel of synthetic autoclaved seawater is stable for weeks. But addition of a little raw seawater causes the slick to disrupt and mix throughout within a day, depending on the inoculum size (Arhelger et al. 1976). Keeping track of the inoculating population and dilution allows determination of the fraction that can attack a particular substrate.

Diffusion shells can limit transport rates to high-affinity organisms. These are probably reduced in the oceans by currents that generate turbulence. But with the small affinities characteristic of hydrocarbon oxidizers, applied diffusion barriers formed by casting them in mm-sized blocks of agar made no significant difference in the kinetic constants (Button et al. 1986). That transport constants are very small for hydrocarbons (Table 1) is a likely consequence of partitioning. Membrane concentrations are increased over those external according to the size of the partition coefficient. Using octanol as a membrane surrogate, values are of the order of $10^2$–$10^5$. Membrane bound oxidases may then use the large concentration to accelerate the production of membrane-insoluble products which are trapped inside for further use (Fig. 2). Thus, unlike amino acids, hydrocarbon transport by *Pseudomonas* T2 was unaffected by high pH (small concentration of cotransporting ions) and was resistant to protonionophores such as carbonyl *m*-chlorophenylhydrazone. Moreover, liberation of metabolic products was increased by these antagonists, consistent with their production but inability to retain. Thus, the large temperature responses observed in amino acid transport, according to in situ measurements of activities and presumed to be due to cascading stimulations from energy production, would not be expected for hydrocarbons transported by vectorial partitioning (in preparation).

Hydrocarbons are high-energy substrates that require oxygen with twice the stoichiometry as carbohydrates. Cell yields therefore are large, to 80% dry weight (Johnson 1967) or 400% wet weight for oligotrophs at only 20% dry weight (Robertson et al. 1998). This, along with small size, allows aquatic bacteria to maximize surface-to-mass ratio. Most cultivars are about a third dry weight. Recent yield values for amino acids (this laboratory) are about 250% wet mass for in situ lakewater populations. Other reports are much lower but measurements of in situ values contain uncertainties. An additional factor is that at large concentrations of hydrocarbons where extensive carbon is lost to metabolic products, yields can be reduced by a factor of five (Button and Robertson 1986).

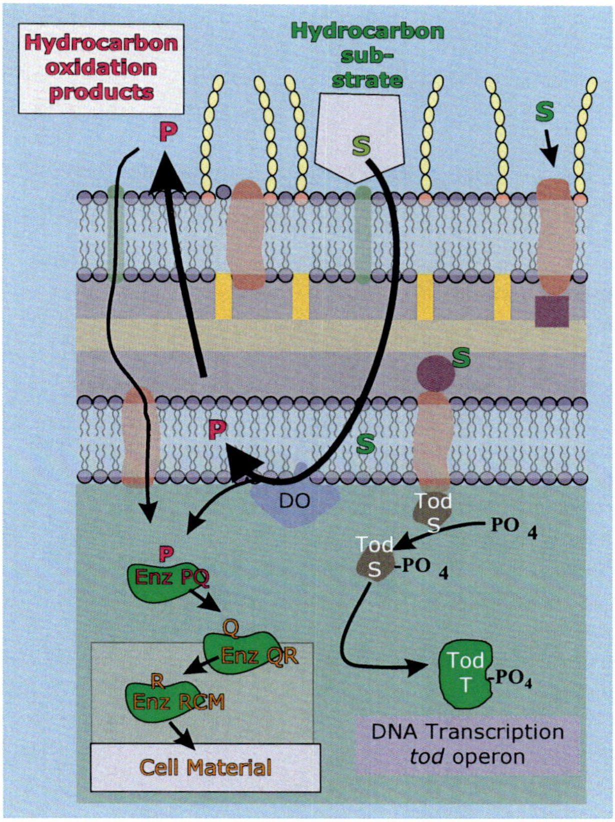

**Fig. 2** Models for use of toluene and other hydrocarbons at trace concentrations. *S* hydrocarbon, *P* oxidized organic product, *DO* dioxygenase, *TOD* toluene dioxygenase, Q, R, subsequent polar products. Hydrocarbon substrate partitions to large concentration in the oxygenase-contqining membrane where it is converted to product that is either trapped for metabolism or leaked back out to be either retransported by a permease or lost

The microbial oxidation of hydrocarbons is initially incomplete. When toluene concentrations exceeded the transport constant $K_t$, large amounts of partly oxidized products such as 2-hydroxy-6-oxohepta-2,4-dienoic acid can be released (Button and Robertson 1987). This is thought to result from a high ratio of initial pathway enzymes in the bacterial membrane to cytoplasmic catabolic enzymes and results in high specific affinities and low apparent $Kt$ values (Button 1991). According to the vectorial partitioning hypothesis (Button 1985), hydrocarbon transport depends on high hydrocarbon concentrations in the cytoplasmic membrane that result from the large partition coefficient between water and membrane lipids where the initial oxidative enzymes are located. The polar products of hydrocarbon metabolism require conventional permeases, and if the permeases are few in number, the specific affinities will be small. Values for 3-methylcatechol and toluene dihydrodiol were

$10^3$ larger (0.14 and 1.7 mg $L^{-1}$) as against about 1 μg $L^{-1}$ for the parent toluene (Button and Robertson 1988). The environmental significance of these electrophilic alkylating agents is unknown. Difficulty of concentrating a lipophilic substrate with hydrophilic permeases probably keeps potential specific affinities of hydrocarbons low. However, absence of the 5 KJ mole$^{-1}$ energy cost for permease operation makes vectorial partitioning a useful mechanism to transport substrate in extremely oligotrophic systems.

# 7    Distributions

The small concentrations of hydrocarbons in the environment are sufficient to support a large population of hydrocarbon oxidizers. Ubiquity was noted when total populations were thought to be in the hundreds rather than hundreds of thousands per mL. These organisms are mostly bacteria but taken here to include the functionally similar archaea. $^{14}$C–dodecane added to fresh seawater from the Arctic Ocean and incubated in situ produced radioactive $CO_2$ within 1.5 days (Robertson et al. 1979). Hydrocarbon-oxidizing microorganisms were more easily isolated from Cook Inlet than from Port Valdez Alaska in 1977 when oil production was limited to the inlet (Kinney et al. 1969). Later, following the operation of the ballast water treatment plant in Port Valdez, bacterial populations of 0.7 mg $L^{-1}$ appeared at a density trap-depth for the discharge of 50 m depth where the salty ballast water spread horizontally under the fresh water but above the colder saltwater in a thin layer (Button et al. 1992). Populations, located and mapped with onboard toluene oxidation rate measurements in near real time were a factor of $10^4$ larger than elsewhere. Floating slicks from spills tend to aggregate in turbulent seawater, aided by wind action, and sink. Suggested effects of clay particles on oil slicks, abundant in Alaskan glacial plumes, were not apparent in laboratory trials (Button 1976) and therefore it is the indigenous microflora that effect physical transformation of oil slicks.

Consumption rates away from the trapped discharge water from the Port Valdez ballast water treatment ponds were often no higher than those away, and consumption capacity exceeded the facility's input rate. This suggested alternative sources of hydrocarbons. Heavy loads of pollen on the seawater surface suggested that the input was terpenes from streams draining Alaska's vast conifer forests (Button 1984) suggested the presence of this plant hydrocarbon (Button 1984). Seasonal and depth dependent-concentrations of terpenes common to conifers were detected by MS/GC at the predicted μg $L^{-1}$ concentrations and brominated hydrocarbons from algae, confirmed the presence of biogenic hydrocarbons (Button and Jüttner 1989).

Terpenes are ubiquitous plant hydrocarbons with an isoprene backbone, some of which contain a heteroatom of oxygen. At least 10% of the organisms present in the Gulf of Alaska actively accumulate terpenes extracted from a $^{14}$CO$_2$-grown pine seedling as determined by autoradiography (Button 1984). Results were the same for $^3$H–toluene, while all amended E. coli cells were negative. The indigenous organisms had a relatively large specific affinity for hydrocarbons giving a turnover time of 4–19 days (Table 1): a large value for Alaskan seawater where values for toluene can approach 100 years. Sufficient quantities appear to be washed from conifers in Southeast Alaska to help base a food chain. These were discovered in a search for compounds that sustained ability of the microflora to metabolize hydrocarbons. Such plant hydrocarbons undoubtedly supported seed populations able to help dissipate oil spills.

Although a large portion of the marine microflora can metabolize petroleum hydrocarbons, specific affinities are generally only sufficient for them to supplement the many other organics used (vide supra). Enrichment cultures tend to select higher capacity but bias the selection toward copeotrophic (rich) environments (Table 1). Extinction cultures, where the population is diluted small numbers of bacteria and allowed to develop in unamended lake or sea water, produce oligobacteria more typical of the environment. One of the few cultures developed by this procedure, *Cycloclasticus oligotrophus*, was particularly good at using acetate (Button et al. 1998). Genomic analysis showed the presence of sequences consistent with the metabolism of hydrocarbons (Wang et al. 1996). Tests with toluene showed that it was in fact quite good at using hydrocarbons (Table 1), and was perhaps a rather typical marine bacterium as well. These and other data mentioned earlier underscore the ubiquity of hydrocarbons as a substrate for aquatic bacteria and their special ability to use small concentrations.

## 8    Research Needs

Three key aspects of understanding the behavior of aquatic hydrocarbons in the microbial steady state are (1) improved analyses of the dissolved organic compounds supporting growth, (2) a detailed general description of metabolic networks that integrate metabolic pathways that interface the conversion of the complex mixture of organic compounds in the environment with those used in the biosynthesis of microbial biomass, and (3) an understanding of the kinetics of hydrocarbon uptake at concentrations approaching those experienced in natural systems. The latter effort should include both the time constants for the sequence of steps between uptake and metabolism, and the loop that links energy yield from the hydrocarbon metabolized to that consumed during uptake.

## References

Alvarez PJ, Anid PJ, Vogel TM (1991) Kinetics of aerobic biodegradation of benzene and toluene in sandy aquifer material. Biodegradation 2:43–51

Arhelger SD, Robertson BR, Button DK (1976) Arctic hydrocarbon biodegradation. In: Wolf DA (ed) Fate and effects on petroleum hydrocarbons in marine ecosystems and organisms. Pergamon, London

Button DK (1970) Some factors influencing kinetic constants for microbial growth in dilute solution. In: Hood DW (ed) Organic matter in natural waters symposium. University of Alaska Press, Fairbanks

Button DK (1976) The influence of clay and bacteria on the concentration of dissolved hydrocarbon in saline solution. Geochim Cosmochim Acta 40:435–440

Button DK (1983) Differences between the kinetics of nutrient uptake by microorganisms, growth, and enzyme kinetics. Trends Biochem Sci 8:121–124

Button DK (1984) Evidence for a terpene-based food chain in the Gulf of Alaska. Appl Environ Microbiol 48:1004–1011

Button DK (1985) Kinetics of nutrient-limited transport and microbial growth. Microbiol Rev 49:270–297

Button DK (1991) Biochemical basis for whole cell uptake kinetics: specific affinity, oligotrophic capacity, and the meaning of the Michaelis constant. Appl Environ Microbiol 57:2033–2038

Button DK (1998) Nutrient uptake by microorganisms according to kinetic parameters from theory as related to cytoarchitecture. Microbiol Mol Biol Rev 62:636–645

Button DK, Jüttner F (1989) Terpenes in Alaskan waters: concentrations, sources, and the microbial kinetics used in their prediction. Mar Chem 26:57–66

Button DK, Robertson BR (1986) Dissolved hydrocarbon metabolism. Limnol Oceanogr 31:101–111

Button DK, Robertson BR (1987) Toluene induction and uptake kinetics and their inclusion into the specific affinity equation for describing rates of hydrocarbon metabolism. Appl Environ Microbiol 53:2193–2205

Button DK, Robertson BR (1988) Hydrocarbon bioconversions: sources, dynamics, products and populations. In: Shaw DG, Hameedi MJ (eds) Environmental studies in Port Valdez, Alaska: a basis for management. Springer, New York, pp 267–291

Button DK, Dunker SS, Morse ML (1973) Continuous culture of *Rhodotorula rubra*: kinetics of phosphate-arsenate uptake, inhibition, and phosphate-limited growth. J Bacteriol 113:599–611

Button DK, Schell DW, Robertson BR (1981a) Sensitive and accurate methodology for measuring the kinetics of concentration-dependent hydrocarbon metabolism rates in seawater by microbial communities. Environ Microbiol 41:936–941

Button DK, Robertson BR, Craig KS (1981b) Dissolved hydrocarbons and related microflora in a fjordal seaport: sources, sinks, concentrations, and kinetics. Appl Environ Microbiol 42:708–719

Button DK, Robertson BR, McIntosh D, Jüttner F (1992) Interactions between marine bacteria and dissolved-phase and beached hydrocarbons after the *Exxon Valdez* oil spill. Appl Environ Microbiol 58:243–251

Button DK, Robertson BR, Schmidt T, Lep P (1998) A small, dilute-cytoplasm, high-affinity, novel bacterium isolated by extinction culture that has kinetic constants compatible with growth at measured concentrations of dissolved nutrients in seawater. Appl Environ Microbiol 64:4467–4476

Button DK, Robertson B, Gustafson E, Zhao X (2004) Experimental and theoretical bases of specific affinity, a cytoarchitecture-based formulation of nutrient collection proposed to supersede the Michaelis-Menten paradigm of microbial kinetics. Appl Environ Microbiol 70:5511–5521

Dunfield PF, Conrad R (2000) Starvation alters the apparent half-saturation constant for methane in the type II methanotroph Methylocystis strain LR1. Appl Environ Microbiol 66:4136–4138

Francy DS, Thomas JM, Raymond RL, Ward CH (1991) Emulsification of hydrocarbon by surface bacteria. J Ind Microbiol 8:237–245

Johnson MJ (1967) Growth of microbial cells on hydrocarbons. Science 3769:1515–1519

Kinney PJ, Button DK, Schell DM (1969) Kinetics of dissipation and biodegradation of crude oil in Alaska's Cook Inlet. In: Proceedings of the 1969 joint conferences on prevention and control of oil spills American Petroleum Institute, Washington, DC

Kou SC, Cherayil BJ, Min W, English BP, XSB X (2005) Single-molecule Michaelis-Menten equations. J Phys Chem 109:19068–19081

Law AT, Button DK (1986) Modulation of the affinity of a marine pseudomonad for toluene and benzene by hydrocarbon exposure. Appl Environ Microbiol 49:469–476

Monod J (1950) La technique de culture continue, théory et applications. Ann Inst Pasteur Paris 79:390–410

Robertson BR, Schell DW, Button DK (1979) Dissolved hydrocarbon rates in Port Valdez, Alaska. University of Alaska Press, Fairbanks

Robertson BR, Button DK, Koch AL (1998) Determination of the biomasses of small bacteria at low concentrations in a mixture of species with forward light scatter measurements by flow cytometry. Appl Environ Microbiol 64:3900–3909

Shaw DG (1989) Part I, hydrocarbon C5 to C7. Hydrocarbons with water and seawater. Pergamon Press, Oxford

Shreve GS, Vogel TM (1993) Comparison of substrate utilization rates and growth kinetics between immobilized and suspended *Pseudomonas* cells. Biotechnol Bioeng 41:370–379

Wang Y, Lau PC, Button DK (1996) A marine oligobacterium harboring genes known to be part of aromatic hydrocarbon degradation pathways of soil pseudomonads. Appl Environ Microbiol 62:2169–2173

Zylstra GJ, Gibson DT (1989) Toluene degradation by *Pseudomonas putida* F1 Nucleotide sequence of the todC1C2BADE genes and their expression in *Escherichia coli*. J Biol Chem 264:14940–14946

# Feast: Choking on Acetyl-CoA, the Glyoxylate Shunt, and Acetyl-CoA-Driven Metabolism

# 31

M. Peña Mattozzi, Yisheng Kang, and Jay D. Keasling

## Contents

1 Introduction ................................................................................ 464
2 Acetyl-CoA Production .................................................................... 465
3 Coenzyme A (CoASH) Pools ............................................................... 467
4 Acetyl-CoA Consumption .................................................................. 467
5 Regulation of Acetate Metabolism: The Acetate Switch ................................. 470
6 Conclusions ................................................................................ 471
7 Research Needs ............................................................................ 471
References ................................................................................... 472

### Abstract

Acetyl coenzyme A (acetyl-CoA) is an essential cofactor in central metabolism: the molecule is the entry point to the tricarboxylic acid (TCA) cycle that generates biomass, energy, and intermediates for macromolecules. Its importance is not limited to biosynthetic pathways: the oxidation of carbohydrates (via pyruvate),

M. Peña Mattozzi (✉)
Department of Plant and Microbial Biology, University of California, Berkeley, CA, USA

Center for Life Sciences Boston, Harvard Wyss Institute for Biologically Inspired Engineering, Boston, MA, USA
e-mail: matt.mattozzi@gmail.com; matt.mattozzi@conagen-inc.com

Y. Kang
Department of Chemical Engineering, Washington University, St. Louis, MO, USA

J. D. Keasling
Department of Chemical Engineering, Washington University, St. Louis, MO, USA

Department of Bioengineering, University of California, Berkeley, CA, USA

Physical Biosciences Division, Lawrence Berkeley National Laboratory, Berkeley, CA, USA

Joint BioEnergy Institute, Emeryville, CA, USA
e-mail: keasling@berkeley.edu; JDKeasling@lbl.gov

© Springer International Publishing AG, part of Springer Nature 2018
T. Krell (ed.), *Cellular Ecophysiology of Microbe: Hydrocarbon and Lipid Interactions*,
Handbook of Hydrocarbon and Lipid Microbiology, https://doi.org/10.1007/978-3-319-50542-8_52

fatty acids (by the β-oxidation cycle), or aromatics (by various pathways) all produce acetyl-CoA as an end point of catabolism. Acetyl-CoA is also produced by the direct assimilation of acetate. The TCA cycle is a very efficient way to convert the acetyl-CoA pool into biomass and energy, and it results in the evolution of two $CO_2$ molecules. Growth on acetate, fatty acids, or aromatics requires the activation of the glyoxylate shunt and gluconeogenesis pathways. By converting isocitrate to malate and bypassing half the TCA cycle, these two carbons are retained at the expense of energy production (Fig. 1). During fast growth in glucose or tryptone-based medium, E. coli and several other organisms excrete acetate to regenerate $NAD^+$ and to recycle coenzyme A. The acetate acidifies the medium and can repress the production of both native and heterologous proteins. Upon depletion of other carbon sources, the cells then retool their metabolism to reactivate acetate to acetyl-CoA, the canonical "acetate switch." Finally, the excess acetyl-CoA can be harnessed for commercial interest through native or engineered pathways to produce fatty acids, bioplastics, pharmaceuticals, or biofuels.

# 1    Introduction

Acetyl-CoA is perhaps the most central molecule of metabolism: the molecule is not only the entry point to the tricarboxylic acid (TCA) cycle to generate biomass, energy, and intermediates for macromolecules but also the precursor to many commercially relevant compounds. An understanding of the pathways surrounding this central molecule is key in the study and engineering of both biodegradative and biosynthetic metabolic pathways. This chapter will discuss several aspects of acetyl-CoA metabolism, including the deleterious effects of excess acetyl-CoA ("choking" on the compound), key acetyl-CoA native and engineered pathways, and the conditions that activate the glyoxylate shunt. For a more in-depth review of some of these topics, see the review on the acetate switch (Wolfe 2005). As bacterial metabolism has been studied mostly in E. coli, we present canonical studies in that bacterium as well as relevant metabolism from other organisms.

The degradation of any hydrocarbon and subsequent use as a carbon and energy source typically involve a peripheral metabolic pathway that transforms substrate to one of a few notable intermediates (Fig. 1) (see Vol. 2, Parts 2 and 3). The cell will then typically shuttle the carbon flux from this compound into central metabolism, where the reactions of the TCA cycle or glyoxylate shunt generate energy for the cell and synthesize important compounds. As the central molecule of metabolism in most known organisms, acetyl-CoA is of utmost importance. Because of the importance of coenzyme A (CoA) to central metabolism, the production of acetyl-CoA must be carefully balanced with consumption to maintain the free CoA (CoASH) pool. Excess acetyl-CoA can also be harnessed to human benefit through native or engineered pathways, accumulating storage molecules or products of commercial interest (Vol. 4, Part 4).

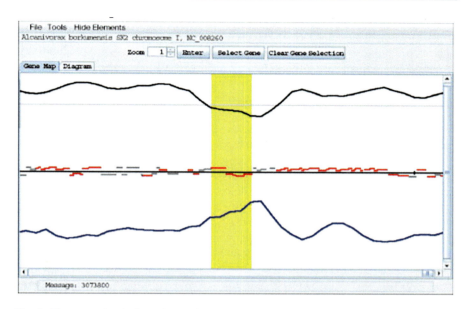

**Fig. 1** Flux must shuttle through acetyl-CoA to transfer from catabolic to anabolic pathways. A few of the important and industrially relevant pathways are shown here. *PDH* pyruvate dehydrogenase, *ADH* alcohol dehydrogenase, *ACKA* acetate kinase A, *PTA* phosphate acetyltransferase, *CS* citrate dehydrogenase, *ACN* aconitase, *ICL* isocitrate lyase, *IDH* isocitrate dehydrogenase, *KGDH* ketoglutarate dehydrogenase, *SCS* succinyl-CoA synthetase, *SDH* succinate dehydrogenase, *FUM* fumarase, *MDH* malate dehydrogenase, *ACC* acetyl-CoA carboxyltransferase, *FAB* fatty acid biosynthesis complex, *ACAT* acetyl-CoA acetyltransferase, *PHA* polyhydroxyalkanoate

## 2   Acetyl-CoA Production

Acetyl-CoA is produced in three ways: (1) activation of acetate; (2) decarboxylation of pyruvate generated primarily from oxidation of carbohydrates; and (3) oxidation of hydrocarbons such as fatty acids, alkanes, and aromatics (Vol. 2, Part 2).

*Acetate activation*: Acetate is activated (assimilated) using two routes. The first route involves phosphorylation of acetate to form the high-energy acetyl phosphate by ATP/acetate kinase A (AckA) and subsequent conversion to acetyl-CoA by acetyl-CoA (CoA)/P$_i$ acetyltransferase (Pta) (Kakuda et al. 1994a, b). The second route involves AMP-forming acetate/CoA ligase, which forms an enzyme-bound acetate-AMP that is subsequently converted to acetyl-CoA (Kumari et al. 1995).

*Production of acetyl-CoA from pyruvate*: The second major pathway for acetyl-CoA production is by decarboxylation of pyruvate, which is produced from oxidation of carbohydrates via glycolysis and the pentose phosphate pathway. Under aerobic conditions, pyruvate is decarboxylated using the pyruvate dehydrogenase complex, which also generates reducing equivalents (NADH) (Frey 1982). Under anaerobic conditions, acetyl-CoA is generated from pyruvate either via pyruvate formate lyase (PFL), which catalyzes the non-oxidative conversion of pyruvate to acetyl-CoA and formate without generation of reducing equivalents, or via pyruvate/ ferredoxin oxidoreductase (PFOR), which catalyzes the oxidative conversion for pyruvate to acetyl-CoA and $CO_2$ (Blaschkowski et al. 1982; Furdui and Ragsdale 2000; Knappe et al. 1984; Uyeda and Rabinowitz 1971).

*Fatty acid and alkane degradation*: Fatty acids are metabolized by β-oxidation, (Vol. 2, Part 2, 3) which degrades fatty acids in a cycle that removes two carbons and generates one acetyl-CoA with each turn. All fatty acids can be oxidized to acetyl-CoA, but only long-chain fatty acids (12 or longer) induce this process (Clark 1981). The β-oxidation of fatty acids is an important source of carbon and energy under what are purported to be "natural" conditions for *E. coli* in the mammalian colon. The final product of the oxidation of fatty acids is acetyl-CoA, along with a single propionyl-CoA for odd-numbered fatty acids. Growth on fatty acids is akin to growth on acetate: acetyl-CoA is the entry point into central metabolism, and the cells must activate the glyoxylate shunt and gluconeogenesis pathways in order to harness energy and building blocks for macromolecules like proteins, carbohydrates, and fats. Alkanes are similarly metabolized once the terminal carbon has been oxidized to a carboxylate via an alcohol, aldehyde, or epoxide (van Beilen and Funhoff 2007; Wentzel et al. 2007). Methanogenic communities have also been shown to degrade fatty acids.

*Degradation of aromatic compounds*: Another important source for carbon, and thus acetyl-CoA, for growth is aromatics, for which lignin is an important source. Lignins and hemicellulosic substrates are a large part of the forest detritus, and many soil-inhabiting bacteria degrade these compounds (well reviewed in Diaz et al. 2001; Harwood and Parales 1996; Masai et al. 2007 as well as in this series). The degradation of aromatic substrates by bacteria and fungi is an important part of the natural carbon cycle in the environment. The interest in the development of ligno-cellulosic biofuels (Ragauskas et al. 2006; Weng et al. 2008) has renewed study in microbial degradation of aromatics, as efficient heterologous expression of lignin-degrading enzymes could be key to economical production of petroleum alternatives (Weng et al. 2008). In the environment, peroxidases and laccases secreted from white-rot fungi break down the lignocellulosic material non-specifically and form free radical intermediates. These free radicals encourage further abiotic oxidative chemical breakdown (Masai et al. 2007). Both bacteria and fungi in those environments use the resultant lignin-derived small phenolic molecules as carbon sources. Several microorganisms, most notably *Pseudomonas* or bacteria formerly classified in that genus, catabolize these compounds into acetyl-CoA and succinyl-CoA, causing the bacteria to activate the glyoxylate shunt and gluconeogenesis. In addition, bacteria use altered versions of these pathways for the degradation of xenobiotic aromatic compounds, sometimes with the aid of laboratory evolution of these enzymes and pathways (reviewed in Parales and Ditty 2005). Even the enteric bacterium *E. coli* has been shown to degrade aromatic compounds, which, as lignin, are a major a component of plant-derived food sources (reviewed in Diaz et al. 2001).

# 3 Coenzyme A (CoASH) Pools

Free coenzyme A (CoASH), the acetyl carrier in all the reactions mentioned above, and the acyl carrier protein (ACP) are the two predominant acyl group carriers in cells. The CoA pool consists primarily of succinyl-CoA, malonyl-CoA, and acetoacetyl-CoA, which are key intermediates in TCA cycle and precursors for the biosynthesis of fatty acids and many industrially useful compounds. Microorganisms can synthesize CoASH de novo from pantothenate. The first step, catalyzed by pantothenate kinase, is the rate-limiting step in the CoA biosynthesis pathway (Alberts and Vagelos 1966; Jackowski and Rock 1981; Rock 1982). The size and composition of intracellular pools of CoASH varies depending on carbon sources and stresses (Chohnan et al. 1998). For example, at 40°C acetyl-CoA is produced, while at 50°C, no acetyl-CoA forms and the CoASH pool decreases (Chohnan et al. 1998). Compared to other carbon sources, growth on glucose generates the greatest CoA pool, predominated by acetyl-CoA. The total CoA pool fluctuates between 0.30 and 0.52 mM (Chohnan et al. 1998). Excess acetyl-CoA can result in shortages of CoASH or other esterified CoAs and repress cell growth.

El-Mansi (2004) reexamined the physiological implications of overflow metabolism. They proposed an acetate excretion mechanism to relieve the overabundance of (choking on) acetyl-CoA. When sugars are present in excess, even under fully aerobic conditions, bacterial cells can perform aerobic "fermentation," thus excreting large quantities of acetate, lactate, and other waste products. This so-called bacterial Crabtree effect is usually considered to result from an imbalance between glycolysis and the TCA cycle. The resultant excreted acetate was thus regarded as wasted carbon source. El-Mansi's experiments demonstrate that acetate excretion, in fact, facilitates faster growth and higher cell densities. Pta and acetate kinase (AckA) catalyze the conversion of acetyl-CoA to acetate and release free CoASH. The free CoASH relieves the "bottleneck" at $\alpha$-ketoglutarate dehydrogenase (in the TCA cycle) to produce sufficient succinyl-CoA for rapid cell growth.

# 4 Acetyl-CoA Consumption

Acetyl-CoA is utilized by the cell for biosynthesis of biomass components and for the generation of energy. Because cells cannot tolerate acetyl-CoA accumulation (as it would tie up CoASH), excess acetyl-CoA is either excreted as acetate (or a more reduced compound) or stored in the form of fats, oils, polyesters, or glycogen (animals).

In the presence of sufficient oxygen, acetyl-CoA is oxidized via the TCA cycle to $CO_2$ and reducing equivalents, which are then used to reduce oxygen and generate a proton motive force for synthesis of ATP. The production and activities of the enzymes of the TCA cycle are tightly regulated (this metabolism has been well reviewed in Kern et al. 2007; Kim and Copley 2007). When cells are grown in the presence of very assimilable carbon sources (such as glucose), the cell's metabolism is subject to the Crabtree effect: more glucose is consumed than the cell can completely oxidize, and the excess is excreted as ethanol or organic acids (e.g., acetate, lactate, succinate) (Ko et al. 1993).

Under excess carbon but limited electron acceptor conditions, cells will switch to fermentative metabolism in which the oxidative TCA cycle is inhibited, thus limiting carbon flux to generate energy and biomass. The reducing equivalents (in the form of NADH) generated during C-source oxidation must be consumed. Under fermentation conditions, cells can remove excess acetyl-CoA in two ways: either conversion to acetate, generating an ATP but consuming no reducing equivalents, or reduction to ethanol, which does not generate ATP but consumes the reducing equivalents. Cells will activate the appropriate pathway dependent on their need for ATP or NAD$^+$. Acetyl-CoA is converted to acetate through the same enzymes used to activate acetate, Pta and AckA. In the presence of excess carbon but limiting N, S, or P, *E. coli* recovers its NAD$^+$ with the Pta-AckA pathway by converting acetyl-CoA to acetate and excreting it. This can result in the accumulation of acetate in the growth medium, which lowers the pH and eventually poisons the cells.

*Glyoxylate shunt*: When cells are fed carbon as sugars or complex media (essentially a mix of certain amino acids), it is advantageous for the cells to shuttle their carbon flux through a complete TCA cycle. These reactions are most efficient at recapturing the energy from the carbon sources and regenerating reducing equivalents. Unfortunately, this efficient mechanism results in the evolution of two molecules of $CO_2$, which are then unavailable as biosynthetic precursors (Stryer 1995). When cells are grown on acetate as a sole carbon source (or fatty acids or aromatics, whose degradation pathways bypass glycolysis), the cells need to retain these carbons for downstream biosynthetic pathways. Under these conditions, the cell undergoes sensitive biochemical and genetic switches to activate the glyoxylate shunt, cutting out half of the TCA cycle by converting isocitrate to malate. This shunt prevents the loss of the carbons as $CO_2$, at the expense of energy production. Thus, the cells now allow the accumulation of four-carbon precursors when grown on a two-carbon substrate. When grown on acetate, the cells also activate gluconeogenesis pathways in order to provide sugar phosphates and other biomass precursors. In this case, the cells also use a branched TCA cycle in order to allow synthesis of compounds that use succinate or α-ketoglutarate as a precursor (LaPorte et al. 1984; Walsh and Koshland 1984). The work of Kornberg and Krebs (1957) answered questions as to how different cell types were able to grow without a complete TCA cycle.

The glyoxylate shunt is heavily regulated in most organisms. For the most part, the activation of the glyoxylate shunt is heralded by the expression of acetyl-CoA synthetase (ACS). In *E. coli* B strains, the genes of the glyoxylate shunt seem to be constitutively activated (Phue and Shiloach 2004) so that they are expressed alongside the enzymes of the TCA cycle. In *E. coli* K strains, however, it is not until the more labile carbon sources are used up that the cells are forced to consume acetate and activate the glyoxylate cycle. In addition to the canonical glyoxylate bypass observed in model organisms like *E. coli*, new glyoxylate pathways have been observed in *Rhodobacter sphaeroides*, which assimilates three acetate molecules and two $CO_2$ molecules through novel C4 and C5 compounds to form malate and succinate (Alber et al. 2006), and *Methylobacterium extorquens*, which assimilates two acetyl-CoA molecules as succinate through a previously unknown, complex series of anapleurotic reactions (Ensign 2006; Korotkova et al. 2005; Meister et al. 2005).

The glyoxylate shunt is conserved in organisms ranging from archaea through lower eukaryotes and plants, but not in mammals. Of particular interest, the causative agent of tuberculosis, *Mycobacterium tuberculosis*, lacks the enzyme α-ketoglutarate dehydrogenase and thus obligately expresses the glyoxylate shunt as part of its central metabolism. As a result, the enzymes exclusive to the microbial glyoxylate shunt (most notably isocitrate lyase) could be active targets for therapeutic drug design (Muñoz-Elias and McKinney 2005, 2006) (Vol. 1, Part 5).

*Fatty acid synthesis*: Acetyl-CoA is also essential in the synthesis of fatty acids. Acetyl-CoA carboxylase catalyzes an ATP-dependent condensation of bicarbonate with acetyl-CoA to form malonyl-CoA. The acyl carrier protein (ACP) replaces the CoA in malonyl-CoA. The condensation of malonyl-ACP and acetyl-CoA initiates fatty acid elongation (well reviewed in Rock and Jackowski 2002).

The limiting step in fatty acid biosynthesis is acetyl-CoA carboxylase; overproduction of this enzyme speeds synthesis of fatty acids (Davis et al. 2000) and has been shown to allow overproduction of other molecules that depend on a malonyl-CoA precursor such as flavonoids (Leonard et al. 2007). Like the microbial glyoxylate shunt enzymes, the microbial fatty acid synthesis and degradation cycles are also targets for therapeutic drug design (Freiberg et al. 2004, reviewed in Campbell and Cronan 2001).

The lipid-accumulating "oleaginous" yeasts have drawn attention as promising candidates for commercial oil production. About 30 of the 600 known yeast species are able to accumulate over 20% of their biomass by weight as oil. Some species accumulate as much as 70% of their dry biomass as oil (Angerbauer et al. 2008). Under nitrogen-limited conditions, oleaginous yeasts continue to assimilate carbon sources for lipid synthesis, while non-oleaginous microorganisms channel the extra carbohydrate substrate into polysaccharides such as glycogen, various glucans, and mannans. Oleaginous yeasts possess a unique AMP-dependent isocitrate dehydrogenase (ICDH) and ATP/citrate lyase (ACL) that can produce a continuous supply of acetyl-CoA directly in the cytosol of the cells as precursors for fatty acid synthesis. Nitrogen-limited conditions increase AMP deaminase in oleaginous cells, resulting in a lower AMP concentration. As a result, the strictly AMP-dependent ICDH stops metabolizing isocitrate, and the cells start accumulating citric acid (via aconitase) in the mitochondrion. Excess citric acid is exported to the cytosol by a malate/citrate efflux system on the mitochondrial membrane and is cleaved by ACL to generate acetyl-CoA and oxaloacetate. The resultant acetyl-CoA is then used for fatty acid biosynthesis (Ratledge 2004).

*Production of other storage compounds*: Acetyl-CoA acyltransferase (encoded by ACS) can catalyze the condensation of acetyl-CoA to acetoacetyl-CoA, which can then be converted to intracellular reserve materials in certain diverse bacteria such as *Bacillus*, *Pseudomonas*, and *Rhodopseudomonas* spp. Commonly observed reserve materials are the polyhydroxyalkanoates (PHAs): poly-3-hydroxybutyric acid (PHB), poly-3-hydroxyvaleric acid (PHBV), and poly-3-hydroxyoctanoic acids (PHOs), all of which have found applications as biodegradable plastics (Luengo et al. 2003). In fact, the specific carbon source affects the monomers incorporated into the polymers produced by the cells. In further studies cells have been engineered to excrete the PHAs, greatly aiding in their extraction for use as bioplastics (Sandoval et al. (2005), also well covered in.

*Production of secondary metabolites and other chemicals*: Just as it is the source for energy, biomass, and storage compounds in the cell, acetyl-CoA is also the precursor to many valuable secondary metabolites and other products. Acetyl-CoA is the starting point for the mevalonate-based isopentenyl pyrophosphate (IPP) biosynthetic pathway (Martin et al. 2003) that is used by many microorganisms, plants, and animals for the biosynthesis of membrane components (e.g., cholesterol), anti-pest agents, and a host of other compounds that have found use as therapeutics, flavors and fragrances, pesticides, and even fuels (Chang and Keasling 2006). Acetyl-CoA is also the starting point for polyketides, which have found wide application as antibiotics (well reviewed in Boghigian and Pfeifer 2008).

## 5    Regulation of Acetate Metabolism: The Acetate Switch

Any discussion of the metabolism surrounding acetyl-CoA would be remiss without mentioning the "acetate switch" observed in *E. coli*: cells generally excrete acetate in early growth phases as a waste product, and then as cells complete the log phase, they "switch" to consume it as a carbon source (Fig. 2). The accumulation of high concentrations of acetate both intracellularly and in the medium is toxic: acetate is an acid, which lowers the pH of the medium and represses the production of proteins (De Mey et al. 2007a, b). *E. coli* B and K strains exhibit this switch differently; K strains are more likely to excrete large amounts of acetate to the medium, whereas the more highly upregulated glyoxylate shunt in B strains consumes the acetate before it adversely affects growth (Phue et al. 2005).

**Fig. 2** A schematic showing the canonical "acetate switch" during aerobic growth in minimal medium supplemented with glucose as the sole carbon source. The *arrow* points to the physiological acetate switch, in which the cells switch from producing acetate as a waste product to using it as a carbon source (Adapted from (Wolfe 2005))

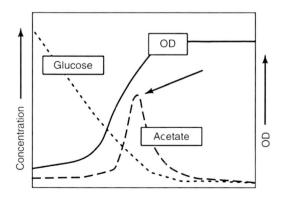

The corresponding switch from acetate excretion to acetate assimilation corresponds to a global change in gene expression, heralded by the activation of the sigma-70 dependent transcription of acetyl-CoA synthetase (encoded by ACS). Acetyl phosphate (acetyl-P) can independently function as a phosphoryl donor to a subset of response regulators to activate signaling or to ADP to generate ATP. *E. coli* assimilates the accumulated acetate primarily through the AMP-forming acetyl-CoA synthetase pathway (ACS).

Although the acetate switch has been studied most extensively in *E. coli*, there are several other organisms that also exhibit switching behavior (reviewed in Wolfe 2005). *Bacillus subtilis* has been observed to excrete acetate in late exponential growth, but avoids over-acidifying the medium by diverting some of its pyruvate flux to acetoin or 2,3-butanediol. *Bacillus* exhibits a similar switching behavior in the activation of growth on acetate, but it relies exclusively on AMP-ACS for its acetate assimilation (Grundy et al. 1993). Acetate accumulation and re-assimilation has also been observed in such diverse systems as the Gram-positive, amino acid-generating *Corynebacterium glutamicum*, halophilic archaea, and despite their lack of a glyoxylate shunt, even mammals (Crabtree et al. 1990).

# 6    Conclusions

Acetyl-CoA is arguably one of the most important metabolites in the cell. Its production and utilization are highly regulated. Understanding how to manipulate and control acetyl-CoA levels and flux is key to stimulating degradation of toxic chemicals in the environment and to the synthesis of useful chemicals, including fuels.

# 7    Research Needs

The major pathways that shuttle flux through acetyl-CoA in most organisms have been elucidated and well characterized, and the enzymes involved have been identified. Additionally, the study of central metabolic regulation, being paramount for downstream engineering purposes, is a very active area of research. Much of the regulation concerning the biochemical switches between acetate production and acetate consumption is still poorly characterized. Similar switches in other organisms have been identified, but none are as well characterized as that in *E. coli*. The glyoxylate shunt has been observed in many organisms; indeed it is the dominant biochemical cycle in several species. But why these other bacteria lack complete TCA cycles is still poorly understood. Additionally, the nascent field of synthetic biology (Vol. 3, Parts 4, 5, 8) is just beginning to realize the potential of engineered pathways that utilize acetyl-CoA to produce molecules of human interest from pharmaceutical, specialty, to commodity scales. Researchers have already attempted to manipulate CoASH pools to increase production of industrially relevant molecules. For example, the rate-limiting enzyme pantothenate kinase has been overexpressed to increase the intracellular level of CoASH (Vadali et al. 2004a, b), consequently increasing the production of compounds derived from acetyl-CoA such as isoamyl acetate. And during these processes, some old concepts of metabolism have been challenged. For example, pantothenate initially was not considered a limiting factor for CoA pools, because *E. coli* not only can synthesize pantothenate but actually excretes it as vitamin for its mammalian hosts. However, recent studies have indicated that providing external pantothenate increases CoA pools. Not only must we find the limiting factor of the CoASH pool, we must also determine its importance in the balance of other cellular cofactor pools: e.g., the NADH/NADPH pools and the ATP pool. These experiments show that although many of the pathways presented in this chapter are classical, new discoveries continually add to the canon and can shift our expectations about central metabolism.

**Financial Interest** Jay D. Keasling has a consulting relationship with and a financial interest in Amyris and a financial interest in LS9, two biofuels companies.

## References

Alber BE, Spanheimer R, Ebenau-Jehle C, Fuchs G (2006) Study of an alternate glyoxylate cycle for acetate assimilation by *Rhodobacter sphaeroides*. Mol Microbiol 61:297–309

Alberts AW, Vagelos PR (1966) Acyl carrier protein 8. Studies of acyl carrier protein and coenzyme A in *Escherichia coli* pantothenate or betaalanine auxotrophs. J Biol Chem 241:5201–5204

Angerbauer C, Siebenhofer M, Mittelbach M, Guebitz GM (2008) Conversion of sewage sludge into lipids by *Lipomyces starkeyi* for biodiesel production. Bioresour Technol 99:3051–3056

Blaschkowski HP, Neuer G, Ludwig-Festl M, Knappe J (1982) Routes of flavodoxin and ferredoxin reduction in *Escherichia coli*. CoA-acylating pyruvate: flavodoxin and NADPH: flavodoxin oxidoreductases participating in the activation of pyruvate formate-lyase. Eur J Biochem 123:563–569

Boghigian BA, Pfeifer BA (2008) Current status, strategies, and potential for the metabolic engineering of heterologous polyketides in *Escherichia coli*. Biotechnol Lett 30:1323–1330

Campbell JW, Cronan JJE (2001) Bacterial fatty acid biosynthesis: targets for antibacterial drug discovery. Annu Rev Microbiol 55:305–332

Chang M, Keasling J (2006) Production of isoprenoid pharmaceuticals by engineered microbes. Nat Chem Biol 2:674–681

Chohnan S, Izawa H, Nishihara H, Takamura Y (1998) Changes in size of intracellular pools of coenzyme A and its thioesters in *Escherichia coli* K-12 cells to various carbon sources and stresses. Biosci Biotechnol Biochem 62:1122–1128

Clark D (1981) Regulation of fatty acid degradation in *Escherichia coli*: analysis by operon fusion. J Bacteriol 148:521–526

Crabtree B, Gordon MJ, Christie SL (1990) Measurement of the rates of acetyl-CoA hydrolysis and synthesis from acetate in rat hepatocytes and the role of these fluxes in substrate cycling. Biochem J 270:219–225

Davis MS, Solbiati J, Cronan JE Jr (2000) Overproduction of acetyl-CoA carboxylase activity increases the rate of fatty acid biosynthesis in *Escherichia coli*. J Biol Chem 275:28593–28598

De Mey M, Lequeux GJ, Beauprez JJ, Maertens J, Van Horen E, Soetaert WK, Vanrolleghem PA, Vandamme EJ (2007a) Comparison of different strategies to reduce acetate formation in *Escherichia coli*. Biotechnol Prog 23:1053–1063

De Mey M, De Maeseneire S, Soetaert WK, Vandamme EJ (2007b) Minimizing acetate formation in *E. coli* fermentations. J Ind Microbiol Biotechnol 34:689

Diaz E, Ferrandez A, Prieto M, Garcia J (2001) Biodegradation of aromatic compounds by *Escherichia coli*. Microbiol Mol Biol Rev 65:523–569

El-Mansi M (2004) Flux to acetate and lactate excretions in industrial fermentations: physiological and biochemical implications. J Ind Microbiol Biotechnol 31:295–300

Ensign SA (2006) Revisiting the glyoxylate cycle: alternate pathways for microbial acetate assimilation. Mol Microbiol 61:274–276

Freiberg C, Brunner N, Schiffer G, Lampe T, Pohlmann J, Brands M, Raabe M, Habich D, Ziegelbauer K (2004) Identification and characterization of the first class of potent bacterial acetyl-CoA carboxylase inhibitors with antibacterial activity. J Biol Chem 279:26066–26073

Frey PA (1982) Mechanism of coupled electron and group transfer in *Escherichia coli* pyruvate dehydrogenase. Ann N Y Acad Sci 378:250–264

Furdui C, Ragsdale SW (2000) The role of pyruvate ferredoxin oxidoreductase in pyruvate synthesis during autotrophic growth by the Wood-Ljungdahl pathway. J Biol Chem 275:28494–28499

Grundy FJ, Waters DA, Takova TY, Henkin TM (1993) Identification of genes involved in utilization of acetate and acetoin in *Bacillus subtilis*. Mol Microbiol 10:259–271

Harwood C, Parales R (1996) The beta-ketoadipate pathway and the biology of self-identity. Annu Rev Microbiol 50:553–590

Jackowski S, Rock CO (1981) Regulation of coenzyme A biosynthesis. J Bacteriol 148:926–932

Kakuda H, Hosono K, Shiroishi K, Ichihara S (1994a) Identification and characterization of the ackA (acetate kinase A)-pta (phosphotransacetylase) operon and complementation analysis of acetate utilization by an ackA-pta deletion mutant of *Escherichia coli*. J Biochem 116: 916–922

Kakuda H, Shiroishi K, Hosono K, Ichihara S (1994b) Construction of Pta-Ack pathway deletion mutants of *Escherichia coli* and characteristic growth profiles of the mutants in a rich medium. Biosci Biotechnol Biochem 58:2232–2235

Kern A, Tilley E, Hunter IS, Legisa M, Glieder A (2007) Engineering primary metabolic pathways of industrial micro-organisms. J Biotechnol 129:6–29

Kim J, Copley SD (2007) Why metabolic enzymes are essential or nonessential for growth of *Escherichia coli* K12 on glucose. Biochemistry 46:12501–12511

Knappe J, Neugebauer FA, Blaschkowski HP, Ganzler M (1984) Post-translational activation introduces a free radical into pyruvate formate-lyase. Proc Natl Acad Sci U S A 81:1332–1335

Ko YF, Bentley WE, Weigand WA (1993) An integrated metabolic modeling approach to describe the energy efficiency of *Escherichia coli* fermentations under oxygen-limited conditions: cellular energetics, carbon flux, and acetate production. Biotechnol Bioeng 42:843–853

Kornberg H, Krebs H (1957) Synthesis of cell constituents from C2-units by a modified tricarboxylic acid cycle. Nature 179:988–991

Korotkova N, Lidstrom M, Chistoserdova L (2005) Identification of genes involved in the glyoxylate regeneration cycle in methylobacterium extorquens AM1, including two new genes, meaC and meaD. J Bacteriol 187:1523–1526

Kumari S, Tishel R, Eisenbach M, Wolfe AJ (1995) Cloning, characterization, and functional expression of acs, the gene which encodes acetyl coenzyme A synthetase in *Escherichia coli*. J Bacteriol 177:2878–2886

LaPorte DC, Walsh K, Koshland DE Jr (1984) The branch point effect. Ultrasensitivity and subsensitivity to metabolic control. J Biol Chem 259:14068–14075

Leonard E, Lim K-H, Saw P-N, Koffas M (2007) Engineering central metabolic pathways for high-level flavonoid production in *Escherichia coli*. Appl Environ Microbiol 73:3877–3886

Luengo J, García B, Sandoval A, Naharro G, Olivera E (2003) Bioplastics from microorganisms. Curr Opin Microbiol 6:251

Martin VJ, Pitera DJ, Withers ST, Newman JD, Keasling JD (2003) Engineering a mevalonate pathway in *Escherichia coli* for production of terpenoids. Nat Biotechnol 21:796–802

Masai E, Katayama Y, Fukuda M (2007) Genetic and biochemical investigations on bacterial catabolic pathways for lignin-derived aromatic compounds. Biosci Biotechnol Biochem 71:1–15

Meister M, Saum S, Alber B, Fuchs G (2005) L-Malyl-coenzyme A/{beta}-methylmalyl-coenzyme A lyase is involved in acetate assimilation of the isocitrate lyase-negative bacterium *Rhodobacter capsulatus*. J Bacteriol 187:1415–1425

Muñoz-Elias E, McKinney J (2005) Mycobacterium tuberculosis isocitrate lyases 1 and 2 are jointly required for in vivo growth and virulence. Nat Med 11:638–644

Muñoz-Elias E, McKinney J (2006) Carbon metabolism of intracellular bacteria. Cell Microbiol 8:10

Parales R, Ditty JL (2005) Laboratory evolution of catabolic enzymes and pathways. Curr Opin Biotechnol 16:315

Phue J-N, Shiloach J (2004) Transcription levels of key metabolic genes are the cause for different glucose utilization pathways in *E. coli* B (BL21) and *E. coli* K (JM109). J Biotechnol 109:21–30

Phue J-N, Noronha SB, Hattacharyya R, Wolfe AJ, Shiloach J (2005) Glucose metabolism at high density growth of *E. coli* B and *E. coli* K: differences in metabolic pathways are responsible for efficient glucose utilization in *E. coli* B as determined by microarrays and Northern blot analyses. Biotechnol Bioeng 90:805–820

Ragauskas AJ, Williams CK, Davison BH, Britovsek G, Cairney J, Eckert CA, Frederick JWJ, Hallett JP, Leak DJ, Liotta CL, Mielenz JR, Murphy R, Templer R, Tschaplinski T (2006) The path forward for biofuels and biomaterials. Science (New York, NY) 311:484–489

Ratledge C (2004) Fatty acid biosynthesis in microorganisms being used for Single Cell Oil production. Biochimie 86:807–815

Rock CO (1982) Mixed disulfides of acyl carrier protein and coenzyme A with specific soluble proteins in *Escherichia coli*. J Bacteriol 152:1298–1300

Rock CO, Jackowski S (2002) Forty years of bacterial fatty acid synthesis. Biochem Biophys Res Commun 292:1155–1166

Sandoval A, Arias-Barrau E, Bermejo F, Cañedo L, Naharro G, Olivera E, Luengo J (2005) Production of 3-hydroxy-n-phenylalkanoic acids by a genetically engineered strain of *Pseudomonas putida*. Appl Microbiol Biotechnol 67:97

Stryer L (1995) Biochemistry, 4th edn. WH Freeman, New York

Uyeda K, Rabinowitz JC (1971) Pyruvate-ferredoxin oxidoreductase. IV. Studies on the reaction mechanism. J Biol Chem 246:3120–3125

Vadali RV, Bennett GN, San KY (2004a) Applicability of CoA/acetyl-CoA manipulation system to enhance isoamyl acetate production in *Escherichia coli*. Metab Eng 6:294–299

Vadali RV, Bennett GN, San KY (2004b) Cofactor engineering of intracellular CoA/acetyl-CoA and its effect on metabolic flux redistribution in *Escherichia coli*. Metab Eng 6:133–139

van Beilen JB, Funhoff EG (2007) Alkane hydroxylases involved in microbial alkane degradation. Appl Microbiol Biotechnol 74:13–21

Walsh K, Koshland DE Jr (1984) Determination of flux through the branch point of two metabolic cycles. The tricarboxylic acid cycle and the glyoxylate shunt. J Biol Chem 259:9646–9654

Weng J-K, Li X, Bonawitz ND, Chapple C (2008) Emerging strategies of lignin engineering and degradation for cellulosic biofuel production. Curr Opin Biotechnol 19:166

Wentzel A, Ellingsen TE, Kotlar HK, Zotchev SB, Throne-Holst M (2007) Bacterial metabolism of long-chain n-alkanes. Appl Microbiol Biotechnol 76:1209–1221

Wolfe AJ (2005) The acetate switch. Microbiol Mol Biol Rev 69:12–50

# Part V

# Hydrophobic Modifications of Biomolecules

# Hydrophobic Modifications of Biomolecules: An Introduction

# 32

Álvaro Ortega

## Contents

1  Introduction ................................................................... 478
2  Hydrophobic Modifications of Nucleic Acids ............................................. 479
3  Covalent Posttranslational Modifications of Proteins ..................................... 481
4  Research Needs ................................................................... 483
References ................................................................... 484

### Abstract

Nucleic acids and proteins, the biomolecules that carry all necessary information for life in the cell, undergo very often modifications in the primary coding elements of their sequences. Some of the bases in the DNA and RNA and the majority of the amino acids in the protein can incorporate new functional groups through a covalent addition. By means of these modifications, the genetically encoded functions of active proteins or the expression patterns of the DNA are affected, leading to changes at the physiological level. These modifications are generally catalyzed by one of the most abundant enzyme families in the cell, the transferases. The importance of this enzyme family is evidenced by the fact that many of them are subject to a strict regulation since they are implicated in key cellular mechanisms. Most of these modifications cause a local increase in hydrophobicity at the biomolecule that leads to changes in protein-protein and protein-nucleic acid interactions. A relevant example for nucleic acid modification is the methylation, while alkylation, lipidation, acetylation, and ubiquitination are frequent hydrophobic modifications of proteins.

Á. Ortega (✉)
Department of Environmental Protection, Estación Experimental del Zaidín, Consejo Superior de Investigaciones Científicas, Granada, Spain
e-mail: alvaro.ortega@csic.es

© Springer International Publishing AG, part of Springer Nature 2018
T. Krell (ed.), *Cellular Ecophysiology of Microbe: Hydrocarbon and Lipid Interactions*,
Handbook of Hydrocarbon and Lipid Microbiology, https://doi.org/10.1007/978-3-319-50542-8_17

# 1     Introduction

Life is based on the fact that all information necessary for the existence of living organisms is codified in the sequence of the DNA contained inside cells. The genetic information is transcribed into individual RNA sequences that in turn, by means of the ribosome, translate into amino acid sequences that after proper folding result in active proteins. However, despite its coding capacity, the information encoded in the DNA is limited and insufficient to account for the spatial and the temporal differences of the translation into active proteins. Although each cell contains the same genetic information, some genes express preferentially in certain tissues or localizations and in determined moments and not in others. These differences are achieved through modifications of different nature, both in nucleic acids and proteins that are related to gene expression (such as histones, transcription factors, etc.). Going beyond the transmission of the genetic information to the next generation, through this type of modifications the acquired features can also be inherited. It is the field of epigenetics that studies the processes that have hereditary effects on gene expression that are not encoded in the DNA sequence. The control of the information encoded in the DNA sequence that will pass on to generate active folded proteins starts at the DNA level. Deletions, mutations, or cleavage of the backbone play an important role in how the information will be transmitted, but one of the most important processes for this control are hydrophobic modifications of the DNA, such as the methylations (Moore et al. 2013). Methylation and other modifications affect how, when, and where the transcription factors and other proteins of the transcription machinery interact with certain genes, and in nucleated cells, it controls also the chromatin states. On the next level, the first means of diversification of proteins is at the transcriptional stage, by mRNA splicing (Black 2003), where introns are removed from the nascent transcribed RNA sequence and exons are subsequently ligated to generate the final sequence to be translated. By the last mechanism, proteins are able to expand in a posttranslational way the information encoded by the limited 20 proteinogenic amino acids that are used in translation. The nascent polypeptide can undergo different modifications in reactions generally driven by specific enzymes. These modifications cover from protein cleavage or splicing, protein maturation, or proteasome automaturation to chemical modification of the side chains of their amino acids. Through the addition of external functional groups, either the protein activity or its capacity to translocate between different cellular compartments is modulated (Walsh 2005). This is of a crucial importance, and therefore very common, for proteins that are related with gene expression. Histone and transcription factors, directly linked with the transmission of the information encoded in genes, can be posttranscriptionally modified, affecting the condensation states of chromatin and the transcription of specific genes in defined locations or times. In addition, there are many proteins that change their structure and activity as a consequence of hydrophobic modification and that are not involved in gene expression. The majority of covalent modifications in proteins, like alkylation or acylation, are of hydrophobic nature although for some other modifications it is not primarily their hydrophobic nature but steric and electrostatic alterations that trigger structural and functional changes. The modification reactions are enzyme-catalysed and the corresponding enzymes represent, in higher eukaryotes for example, more than 5% of the total number of proteins encoded in the genome (Walsh 2005). Their relevance is underlined by the fact that they are under tight control and of central importance to the cell. Among these modifications, the most common are methylation and other alkylations in DNA (Liyanage et al. 2014; Chen et al. 2016), and phosphorylation, methylation, acetylation, glycosylation, and ubiquitination in proteins (Walsh 2005; Knorre et al. 2009). DNA and protein hydrophobic modifications have been found in a wide range of living organisms, from bacteria to human and play important roles in many fundamental processes as cell development, disease, and bacterial virulence (Marinus and Casadesus 2009; Korlach and Turner 2012) (Fig. 1).

## 2    Hydrophobic Modifications of Nucleic Acids

Nucleic acids are susceptible to diverse chemical modifications that influence a range of vital cellular processes. Methylation is by far the most extensively studied of these modifications and occurs both in prokaryotes and eukaryotes (Razin and Riggs 1980; Liyanage et al. 2014). Its relevance is unquestionable due to its direct influence on the complex epigenetic network of gene regulation (Hallgrimsson and Hall 2011). This modification is hereditary; the methylation pattern can be copied directly from the parent strand after each round of DNA replication (Jeltsch and Jurkowska 2014). DNA methylation is very closely related to processes that relax or condense the chromatin states, favoring or repressing as a consequence the expression of specific genes in determined cells. In most of the cases, the functional manifestations of DNA methylation are associated with gene silencing, in which DNA methylation may either lead to the recruitment of reader proteins that act as transcription repressors or prevent the binding of the transcription factors (Boyes and Bird 1991). In other cases, however, methylation favors gene expression and modulates the elongation process or regulates the RNA splicing (Jones 2012). The main biological roles induced by DNA methylation are related to gene expression (activation and repression of the transcription) through direct hindrance of transcriptional activators, recruitment of repressive protein complexes, or cross-talk with histone posttranslational modifications (Liyanage et al. 2012) but not limited to that. DNA methylation also largely affects DNA-protein interactions, influencing this way biological functions such as organization, reprogramming, and stability of the genome, along with cellular differentiation, transposon silencing, RNA splicing, and DNA repair (Ndlovu et al. 2011).

Methylation is the enzymatic transfer of a methyl group from the S-adenosylmethionine (SAM) donor molecule to any of the DNA bases, resulting in a weakening of the hydrogen bonding of the two DNA strands. The best documented case is the 5-methylcytosine (5mC) methylation (Wyatt 1951), mediated by enzymes of the family of the DNA methyltransferases (DNMTs) that act on the so-called CpG sites (Ndlovu et al. 2011; Kumar and Rao 2013). Oxidation

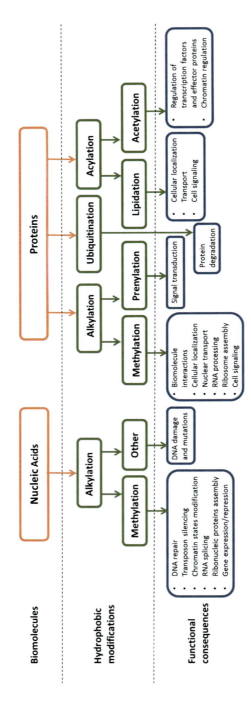

**Fig. 1** Schematic representation of the types of hydrophobic modifications and their main functional consequences that nucleic acids and proteins can suffer

derivatives of these methylated bases help modulate the effect of 5mC, working as effective demethylations (Shen et al. 2014; Chen et al. 2016). The corresponding species are 5-hydroxymethylcytosine, 5-formylcytosine, or 5-carboxylcytosine that are generated by Ten-eleven Translocation (TET) dioxygenases. Another type of methylation modification exists mainly in prokaryotes, which is $N^6$-methyladenine (Ratel et al. 2006) modification that has consequences in processes like DNA repair, replication, and cell defense. Cytosines can also suffer a methyl addition in the $N^4$ position (Ehrlich et al. 1987).

DNA can be modified by the addition of other alkyl groups, and these are among the largest factors promoting DNA damage and mutations. Some examples are $O^2$-alkylthymine, $O^4$-alkylthymine, $O^6$-methylguanine, or $O^6$-ethylguanine (Engelbergs et al. 2000). Cells, however, implement a set of mechanisms by which alkylation lesions provoked by endogenous compounds or environmental agents can be immediately repaired through the action of different alkyltransferases that protect the viability of the cell (Bouziane et al. 1998; Drablos et al. 2004).

Not only DNA but also RNA can suffer modifications. For example, tRNA modifications have a profound effect on translation, which in turn affects critical biological processes (Phizicky and Hopper 2015). MicroRNA (miRNA, a noncoding RNA that regulates gene expression at mRNA level) can be methylated and demethylated in prokaryotes and eukaryotes, thus controlling the levels of active mRNA (Chen et al. 2016). Finally, small nuclear RNA has been shown to be also posttranscriptionally modified. Methylation and pseudouridination are the most important examples of modifications in snRNA (Massenet et al. 1998) and play important roles in small ribonucleic proteins complex assembly, spliceosome formation, and splicing processes (Karijolich and Yu 2010).

# 3 Covalent Posttranslational Modifications of Proteins

The number of derivatives an active protein can generate is largely superior to the coding information received from the DNA. This is mainly due to two large groups of posttranslational modifications (PTMs): the first are covalent additions on terminal amino acids and on side chains of internal residues and the second are all the differential protease-driven cleavages of the protein backbone. The main functions of PTMs are the activation/inactivation of proteins (mainly in enzymes), protein tagging for intracellular localization or transport to the proteasome, and the influence on structure and stability of the protein (Knorre et al. 2009). Of the 20 proteinogenic amino acids, 15 are susceptible of undergoing a chemical addition. There is a great variety of possibilities in covalent posttranslational modifications, with a larger diversity than the hydrophobic modifications observed for nucleic acids (Walsh 2005). These modifications include phosphorylation (generally on hydroxyl groups of a serine residue) (Hubbard and Miller 2007), alkylation (principally methylations on carboxyl- or amino groups of arginine and lysine residues (Clarke 2013)), acylation (acetyl ($C_2$) group to lysine residues (Cole 2008), myristoyl ($C_{14}$) group to N-terminal glycine residues and palmitoyl ($C_{16}$) to the SH- group of a cysteine), prenylation (addition of isoprenoid groups, generally farnesyl and geranylgeranyl to a cysteine residue by thioether bonds (Smotrys and Linder 2004)), glycosylation (additions of a variable number of glucose or galactose molecules and can be O- on serine and threonine, widely distributed, or N- on asparagine, more complex), or ubiquitination in the C-terminal amino acid of the polypeptide sequence (Capili and Lima 2007). Other covalent additions are: carboxylation, polyglycination, sulphation, nitration, or hydroxylation (Knorre et al. 2009). The diversity of modification-mediated functional consequences cover alterations in enzymatic activities by biotinylation, lipoylation, and phosphopantetheinylation, control of the cellular localization by lipidation, or targeting of proteins for degradation by ubiquitination (Walsh et al. 2005). Among this diversity of posttranslational modifications, we wish to highlight alkylations (including methylations), acetylations (Zhang et al. 2002) and larger acylations (with lipidations as the most representative modifications) (Resh 1999; Bijlmakers and Marsh 2003).

Alkylation is the posttranslational addition of an alkyl group to the side chain of certain amino acids like lysine, arginine, cysteine, glutamic, or aspartic acid. These modifications introduce hydrophobicity on the side chains of the proteins along with steric changes, which are subtle but sufficient to trigger functional responses on the protein. The most important alkylation studied to date is methylation, which occurs mainly on the amino groups of lysine and arginine side chains (Strahl and Allis 2000) by the activity of specific arginine or lysine methyltransferases. It is also important in bacterial signaling since methylation of the cytosolic domain of chemoreceptors, catalyzed by the CheR methyltransferases (Li and Weis 2000; Hazelbauer et al. 2008), modulates the sensitivity of the chemoreceptor containing signaling complexes. Similarly to DNA methylations, chemoreceptor methylation involves the transfer of the methyl moiety from S-adenosyl methionine (SAM) to the protein. Protein methylations have a key importance in gene expression control, being histones the main targets, although many other protein families have been shown to be susceptible of methylation (Lee and Stallcup 2009; Biggar and Li 2015).

Acylation is the attachment of acyl groups, of which the best studied example is the addition of either acetyl groups, isoprenes, or fatty acids to the side chains of certain amino acid residues, in a enzymatically catalyzed reaction also known as lipidation (Nadolski and Linder 2007; Hentschel et al. 2016). N- lipidation generally occurs on glycines while C- and S- lipidation can be found on cysteines. Acylations can be classified into those occurring in the cytoplasm (or in the cytoplasmic side of the membrane) and those happening in the lumen of the secretory pathways (Knorre et al. 2009). Acetylation is one of the most important modification of proteins; it consists on the addition of an acetyl group generally to the amino group at the N-terminus of a protein chain, but it can also be observed attached to the amino-group of lysine side chains. Furthermore, proteins can be prenylated by the addition of isoprenoid moieties from isoprene residues via thioether bonds with the cysteines located at C-terminal positions of the protein with known motifs known as the CaaX (where C represents a cysteine, a is typically an aliphatic amino acid, and the identity of amino acid X determines whether the protein will be modified with the 15-carbon chain farnesyl (X = M, S, Q, A, or C) or with the 20-carbon chain geranylgeranyl

(X = L, or E) group) (Zhang and Casey 1996). On the other side, the most common modifications using fatty acids generate myristate (a 14-carbon long fatty acid) and palmitate (with 16 carbon units) adducts. Myristoylation reactions, catalyzed by a myristoyltransferase, cause stable adducts on the amino group of an N-terminal glycine (Resh 1999; Farazi et al. 2001). In contrast, S-palmitoylation generates more labile adducts forming thioesters on specific cysteine residues (Smotrys and Linder 2004). The major physiological relevance of lipidations is to facilitate the interaction and movement of protein in cell membranes (Hentschel et al. 2016). Isoprenoid and fatty acid additions are essential on signal transduction and vesicle transport, closely involved in cellular membrane anchoring and in protein-protein interactions. In particular, prenylation is important for cell growth, nuclear import, cytoskeleton organization, and apoptosis in eukaryotic cells via Ras and Rab proteins (Takai et al. 2001). Acetylation can be associated with the regulation of transcription factors and effector proteins and has a considerable impact on metabolism in prokaryotic and eukaryotic cells (Guan and Xiong 2011).

There is evidence for multiple modification as evidenced for example by mono-, di-, or even tri- modifications (as for methylation) of the same residue (Clarke 2013). Moreover, many modification events on proteins can have either additive or opposing effects (Bah and Forman-Kay 2016). There is increasing evidence for a mutual influence of different modifications; for example acetylations can block ubiquitinations (Li et al. 2002), and methylations modulate the capacity of neighboring residues to be phosphorylated (Molina-Serrano et al. 2013).

The most important effects that the proteins gain upon posttranslational modifications come from the biochemical/biophysical properties provided by the additional functional group. Due to these alterations, proteins are able to traverse membranes and relocate in different cellular compartments, are labeled for proteolysis or for binding to other molecules by ubiquitination or methylation, and change their structural or dynamic properties. PTMs transform proteins into active elements (enzymatic function generally), and they influence their spatial structure and stability. All PTMs are enzymatic processes that are strongly regulated, as they control the gene expression in transcription. Central to these reactions is the availability of the corresponding substrates, i.e., SAM for methylation, acetyl CoA for acetylation, ATP for phosphorylation, NAD-ADP for ribosylation, or UDP-glucose for glycosylation, that converts these small molecules into fundamental compounds essential for the cell survival and the transmission of hereditary features.

# 4    Research Needs

The last years have witnessed a boost in posttranslational modification research that was mainly due to the development of high resolution techniques for the discovery and definition of modified residues, namely mass spectrometry and specific antibody design. These techniques are complemented by amino acid sequencing, peptide mapping, site-directed mutagenesis, and high resolution NMR methods. However, due to the enormous variety and complexity of these modifications, a larger effort must be oriented to develop high throughput analysis of protein sequences at different states of the cell. Furthermore, the analysis of protein modifications at global scale is necessary. The diverse chemical modifications both in nucleic acids and proteins notably expand the complexity of the study needed in prokaryotic and eukaryotic organisms. For many modifications, the precise functional consequences need to be established. In addition, further research is needed to identify all cellular targets of modification. Understanding in depth all mechanisms and functions of such modifications will certainly lead to a great advance in fundamental research, but due to the implications on the control of gene transcription, among other fundamental cell processes, it will also offer possibilities to fight disease and infection.

## References

Bah A, Forman-Kay JD (2016) Modulation of intrinsically disordered protein function by post-translational modifications. J Biol Chem 291(13):6696–6705

Biggar KK, Li SS (2015) Non-histone protein methylation as a regulator of cellular signalling and function. Nat Rev Mol Cell Biol 16(1):5–17

Bijlmakers MJ, Marsh M (2003) The on-off story of protein palmitoylation. Trends Cell Biol 13(1):32–42

Black DL (2003) Mechanisms of alternative pre-messenger RNA splicing. Annu Rev Biochem 72:291–336

Bouziane M, Miao F, Ye N, Holmquist G, Chyzak G, O'Connor TR (1998) Repair of DNA alkylation damage. Acta Biochim Pol 45(1):191–202

Boyes J, Bird A (1991) DNA methylation inhibits transcription indirectly via a methyl-CpG binding protein. Cell 64(6):1123–1134

Capili AD, Lima CD (2007) Taking it step by step: mechanistic insights from structural studies of ubiquitin/ubiquitin-like protein modification pathways. Curr Opin Struct Biol 17(6):726–735

Chen K, Zhao BS, He C (2016) Nucleic acid modifications in regulation of gene expression. Cell Chem Biol 23(1):74–85

Clarke SG (2013) Protein methylation at the surface and buried deep: thinking outside the histone box. Trends Biochem Sci 38(5):243–252

Cole PA (2008) Chemical probes for histone-modifying enzymes. Nat Chem Biol 4(10):590–597

Drablos F, Feyzi E, Aas PA, Vaagbo CB, Kavli B, Bratlie MS, Pena-Diaz J, Otterlei M, Slupphaug G, Krokan HE (2004) Alkylation damage in DNA and RNA – repair mechanisms and medical significance. DNA Repair (Amst) 3(11):1389–1407

Ehrlich M, Wilson GG, Kuo KC, Gehrke CW (1987) N4-methylcytosine as a minor base in bacterial DNA. J Bacteriol 169(3):939–943

Engelbergs J, Thomale J, Rajewsky MF (2000) Role of DNA repair in carcinogen-induced ras mutation. Mutat Res 450(1–2):139–153

Farazi TA, Waksman G, Gordon JI (2001) The biology and enzymology of protein N-myristoylation. J Biol Chem 276(43):39501–39504

Guan KL, Xiong Y (2011) Regulation of intermediary metabolism by protein acetylation. Trends Biochem Sci 36(2):108–116

Hallgrimsson B, Hall BK (2011) Epigenetics: linking genotype and phenotype in development and evolution. University of California Press, Oakland

Hazelbauer GL, Falke JJ, Parkinson JS (2008) Bacterial chemoreceptors: high-performance signaling in networked arrays. Trends Biochem Sci 33(1):9–19

Hentschel A, Zahedi RP, Ahrends R (2016) Protein lipid modifications – more than just a greasy ballast. Proteomics 16(5):759–782

Hubbard SR, Miller WT (2007) Receptor tyrosine kinases: mechanisms of activation and signaling. Curr Opin Cell Biol 19(2):117–123

Jeltsch A, Jurkowska RZ (2014) New concepts in DNA methylation. Trends Biochem Sci 39(7):310–318

Jones PA (2012) Functions of DNA methylation: islands, start sites, gene bodies and beyond. Nat Rev Genet 13(7):484–492

Karijolich J, Yu YT (2010) Spliceosomal snRNA modifications and their function. RNA Biol 7(2):192–204

Knorre DG, Kudryashova NV, Godovikova TS (2009) Chemical and functional aspects of post-translational modification of proteins. Acta Nat 1(3):29–51

Korlach J, Turner SW (2012) Going beyond five bases in DNA sequencing. Curr Opin Struct Biol 22(3):251–261

Kumar R, Rao DN (2013) Role of DNA methyltransferases in epigenetic regulation in bacteria. Subcell Biochem 61:81–102

Lee YH, Stallcup MR (2009) Minireview: protein arginine methylation of nonhistone proteins in transcriptional regulation. Mol Endocrinol 23(4):425–433

Li G, Weis RM (2000) Covalent modification regulates ligand binding to receptor complexes in the chemosensory system of *Escherichia coli*. Cell 100(3):357–365

Li M, Luo J, Brooks CL, Gu W (2002) Acetylation of p53 inhibits its ubiquitination by Mdm2. J Biol Chem 277(52):50607–50611

Liyanage VR, Zachariah RM, Delcuve GP, Davie JR, Rastegar M (2012) New developments in chromatin research: an epigenetic perspective. In: Simpson NM, Stewart VJ (eds) New developments in chromatin research. Nova Science Publishers, Hauppauge, pp 29–58

Liyanage VR, Jarmasz JS, Murugeshan N, Del Bigio MR, Rastegar M, Davie JR (2014) DNA modifications: function and applications in normal and disease states. Biology (Basel) 3(4):670–723

Marinus MG, Casadesus J (2009) Roles of DNA adenine methylation in host-pathogen interactions: mismatch repair, transcriptional regulation, and more. FEMS Microbiol Rev 33(3): 488–503

Massenet S, Mougin A, Branlant C (1998) Posttranscriptional modifications in the U small nuclear RNAs. In: Grosjean H, Benne R (eds) Modification and Editing of RNA. ASM Press, Washington, DC, pp 201–227

Molina-Serrano D, Schiza V, Kirmizis A (2013) Cross-talk among epigenetic modifications: lessons from histone arginine methylation. Biochem Soc Trans 41(3):751–759

Moore LD, Le T, Fan G (2013) DNA methylation and its basic function. Neuropsychopharmacology 38(1):23–38

Nadolski MJ, Linder ME (2007) Protein lipidation. FEBS J 274(20):5202–5210

Ndlovu MN, Denis H, Fuks F (2011) Exposing the DNA methylome iceberg. Trends Biochem Sci 36(7):381–387

Phizicky EM, Hopper AK (2015) tRNA processing, modification, and subcellular dynamics: past, present, and future. RNA 21(4):483–485

Ratel D, Ravanat JL, Berger F, Wion D (2006) N6-methyladenine: the other methylated base of DNA. BioEssays 28(3):309–315

Razin A, Riggs AD (1980) DNA methylation and gene function. Science 210(4470):604–610

Resh MD (1999) Fatty acylation of proteins: new insights into membrane targeting of myristoylated and palmitoylated proteins. Biochim Biophys Acta 1451(1):1–16

Shen L, Song CX, He C, Zhang Y (2014) Mechanism and function of oxidative reversal of DNA and RNA methylation. Annu Rev Biochem 83:585–614

Smotrys JE, Linder ME (2004) Palmitoylation of intracellular signaling proteins: regulation and function. Annu Rev Biochem 73:559–587

Strahl BD, Allis CD (2000) The language of covalent histone modifications. Nature 403(6765):41–45

Takai Y, Sasaki T, Matozaki T (2001) Small GTP-binding proteins. Physiol Rev 81(1):153–208

Walsh D (2005) Posttranslational modification of proteins: expanding nature's inventory. Roberts & Company Publishers, Englewood

Walsh CT, Garneau-Tsodikova S, Gatto GJ Jr (2005) Protein posttranslational modifications: the chemistry of proteome diversifications. Angew Chem Int Ed Engl 44(45):7342–7372

Wyatt GR (1951) Recognition and estimation of 5-methylcytosine in nucleic acids. Biochem J 48(5):581–584

Zhang FL, Casey PJ (1996) Protein prenylation: molecular mechanisms and functional consequences. Annu Rev Biochem 65:241–269

Zhang K, Williams KE, Huang L, Yau P, Siino JS, Bradbury EM, Jones PR, Minch MJ, Burlingame AL (2002) Histone acetylation and deacetylation: identification of acetylation and methylation sites of HeLa histone H4 by mass spectrometry. Mol Cell Proteomics 1(7):500–508

# DNA Methylation in Prokaryotes: Regulation and Function

# 33

## Saswat S. Mohapatra and Emanuele G. Biondi

## Contents

| | | |
|---|---|---|
| 1 | Introduction | 488 |
| 2 | Bacterial Restriction-Modification Systems | 489 |
| | 2.1 Four Types of Restriction-Modification Systems | 489 |
| | 2.2 Occurrence and Prevalence of R-M Systems in the Bacterial Genome | 490 |
| | 2.3 Functions of the R-M Systems | 492 |
| 3 | Dam *Methyltransferase* | 493 |
| | 3.1 Dam-Dependent DNA Mismatch Repair | 494 |
| | 3.2 Chromosome Replication Control | 495 |
| | 3.3 Regulation of Gene Expression | 495 |
| 4 | CcrM *Methyltransferase* | 498 |
| 5 | Research Needs | 502 |
| | References | 502 |

**Abstract**

Methylation of DNA in prokaryotes is known since the 1950s, but its role is still elusive and therefore under intense investigation. Differently from eukaryotes, the most important methylation in bacteria takes place on adenines (in position N6). The enzymes responsible for DNA methylation are often associated with restriction

S. S. Mohapatra
Department of Genetic Engineering, School of Bioengineering, SRM University, Kattankulathur, TN, India
e-mail: saswatsourav.h@ktr.srmuniv.ac.in; saswatsmohapatra@gmail.com

E. G. Biondi (✉)
Aix Marseille University, CNRS, IMM, LCB, Marseille, France
e-mail: ebiondi@imm.cnrs.fr

© Springer International Publishing AG, part of Springer Nature 2018
T. Krell (ed.), *Cellular Ecophysiology of Microbe: Hydrocarbon and Lipid Interactions*,
Handbook of Hydrocarbon and Lipid Microbiology, https://doi.org/10.1007/978-3-319-50542-8_23

enzymes acting as a defense mechanism against foreign DNA (Restriction-Modification or R-M system). Other methyltransferases are solitary that function independently of the presence of a cognate restriction enzyme and are mostly involved in controlling replication of chromosome, DNA mismatch repair systems, or modulating gene expression. This is the case of the methylase Dam in gamma-proteobacteria or CcrM in alpha-proteobacteria. In this chapter, we will discuss the role of the R-M system and the activity of Dam and CcrM.

# 1    Introduction

All living organisms store the life information in a coded form in their DNA that is transmitted to the progeny. However, this inherited information can be modified by methylation of DNA, leading to an epigenetic reversible control of the genetic program. The nucleotides that can be modified by transferring a methyl group are adenosines (prevalent in prokaryotes) and cytosines (prevalent in eukaryotes). In bacteria, this DNA modification controls many important processes, including, for example, the restriction systems, regulation of gene expression, or the control of DNA replication. In this chapter we'll focus on adenosine methylation, while cytosine methylation, of which its role is less understood, is described elsewhere (Casadesús 2016; Adhikari and Curtis 2016).

The first evidence of DNA methylation in the bacteria was found studying bacterial infection by phages (Bertani and Weigle 1953). The DNA of phages and the bacterial host cell can have different methylation patterns. This mechanism encoded in many bacterial genomes is composed by a DNA methyltransferase that "marks" the DNA with specific methylation signatures, therefore protecting the DNA from the activity of the cognate restriction enzyme. This R-M (restriction-modification) system functions as a form of bacterial immune system that is able to protect the host bacterium from foreign DNA invasion.

Several DNA methyltransferases are reported which do not belong to the R-M systems. In fact, there are orphan (or solitary) DNA methyltransferases without a cognate restriction enzyme that are involved in important cellular mechanisms. Deoxyadenosine methylase or Dam is one of the most studied orphan adenine methyltransferase in the gamma-proteobacteria, being first discovered in *Escherichia coli* (Boye and Løbner-Olesen 1990). In the model alpha-proteobacterium *Caulobacter crescentus* another methyltransferase, CcrM (Cell Cycle Regulated Methylase), plays a crucial role in the regulation of cell cycle (Zweiger et al. 1994). The origin of these solitary methyltransferases could be related to a loss of the corresponding restriction endonucleases as suggested by the discovery of R-M systems in the genome of *Helicobacter pylori* in which the restriction enzyme has lost its activity while the methyltransferase is still functional (Fox et al. 2007). However, the high number of solitary methyltransferases in bacterial genomes may suggest the exact opposite hypothesis with an early evolution of methyltransferring enzymes later associated with endonucleases (Blow et al. 2016).

## 2 Bacterial Restriction-Modification Systems

DNA methylation in bacterial systems was initially discovered as a primitive defense mechanism employed by the bacteria to restrict the influx of extraneous DNA from bacteriophages (Luria and Human 1952; Bertani and Weigle 1953). This defense mechanism, known as restriction-modification (R-M) systems, consists of two enzyme components, a DNA methyltransferase (MTase) and a restriction endonuclease (REase). While the MTase methylates a specific site in the DNA sequence, the REase cleaves the DNA in a sequence specific manner. Because of the presence of both restriction and modification components of the R-M system, the bacteria are inherently capable of discriminating between the "self" and the "non-self" genetic material. As the DNA molecules originating from the bacteriophages have different methylation pattern than the host cell, the restriction enzymes produced by the host can cleave the incoming DNA and restrict the bacteriophage infection. These properties of the bacterial R-M systems are akin to an innate immune response mechanism, in contrast to the CRISPR-Cas systems that provide an adaptive form of defense against bacteriophage invasions (Vasu and Nagaraja 2013).

At the biochemical level, REases produce $5'$ or $3'$ overhangs or blunt ends at the site of their action by cleaving a phosphodiester bond. MTases modify the target nucleotides by adding a methyl group to the N6 amino group of the adenine or the C5 carbon or the N4 amino group of the cytosine residues. Both adenine and cytosine residues are modified in the bacterial and archeal genomes by methylation. At the functional level, the R-M systems exhibit considerable diversity and, therefore, can be categorized into four distinct types (Type I–IV) according to their recognition sequence, subunit composition, cofactor requirements, and site of cleavage (Roberts et al. 2003).

## 2.1 Four Types of Restriction-Modification Systems

The Type I enzymes consist of subunits for the restriction and methylation activities. The enzymes function as hetero-oligomers and cleave the DNA sequence from 100 bp to several Kbp away from their recognition sites. These enzymes are encoded by three different genes, namely *hsdR*, *hsdM*, and *hsdS* (Ershova et al. 2015). The enzyme recognition site consists of two parts separated by an intervening sequence of nucleotides, with this sequence $AAC(N)_6GTGC$ being a representative (Ershova et al. 2015). Type I MTase is a complex of two methyltransferase subunits and a specificity determining subunit ($M_2S_1$). The S subunit possesses two target recognition domains that interact with the two recognition sites, whereas the MTase dimer methylates both the DNA strands simultaneously. The restriction enzyme binds to the unmethylated target site and translocates along the DNA till it finds any other DNA binding protein, where it cleaves the DNA in an ATP-dependent manner.

Because of this activity the cleavage occurs far from the recognition site of the enzyme (Roberts et al. 2003; Ershova et al. 2015).

The Type II R-M systems consist of distinct restriction and methyltransferase enzyme components that are encoded by two different genes. The restriction enzyme functions either in homodimeric or homotetrameric form and specifically cleaves the DNA within or very close to their recognition site (Vasu and Nagaraja 2013). The recognition sequence is often a short palindrome consisting of 4–8 bp sequences (for example GAATTC). The type II MTases act as monomer and methylate both DNA strands after binding. Because of their precise site specificity and their diversity, they are very useful in genetic manipulation experiments and, therefore, most extensively studied. The type II R-M systems are further subdivided into 11 groups on the basis of their biochemical properties, though the subtyping is not mutually exclusive (Roberts et al. 2003; Ershova et al. 2015).

The type III R-M systems are encoded by the two closely located genes, *mod* and *res*, that produce the methyltransferase and the restriction components, respectively (Janscak et al. 2001; Dryden et al. 2001; Mücke et al. 2001). The Type III enzymes function as heterotrimers or heterotetramers consisting of restriction, methylation, and DNA-dependent NTPase activities. The restriction and methylation enzymes compete for the same catalytic reaction leading to incomplete reactions on most occasions. The restriction enzyme recognizes a 5–6 bp long nonpalindromic sequence and cleaves at the 3′ site at a distance of ~25–27 bp (Dryden et al. 2001; Vasu and Nagaraja 2013).

In contrast to the already described R-M systems, the enzymes belonging to the type IV systems are not R-M systems in true sense, as they do not have a methylation component. Also, the restriction enzyme only cleaves the DNA substrate that has been modified previously such as methylated, hydroxyl-methylated, and glucosyl-hydroxy-methylated, although the sequence specificity is not very well defined among these enzymes. Because of the lack or low- sequence specificity they protect the host from a broad range of extraneous DNA having different methylation patterns (Ershova et al. 2015; Loenen and Raleigh 2014). Interestingly, enzymes belonging to the type II M are also methylation directed and, therefore, have been proposed to be included in the type IV (Loenen and Raleigh 2014).

## 2.2   Occurrence and Prevalence of R-M Systems in the Bacterial Genome

With the advancement of the genome sequencing techniques and the availability of a great number of prokaryotic genomes, it is now apparent that very diverse sets of R-M systems exist in nature. The R-M systems are ubiquitous among the prokaryotes. So far approximately 4000 enzymes are known that show 300 different specificities (Roberts et al. 2010). According to the REBASE (restriction enzyme database), out of the 8500 sequenced genomes of prokaryotes only 385 have no recognizable R-M systems (Roberts et al. 2015; Ershova et al. 2015). A more recent study using SMRT (Small Molecule Real Time) sequencing methods surveying

217 bacterial and 13 archaeal species, from 19 different phyla and 37 different classes, reported an extensive dataset of novel methylated motifs in these genomes (Blow et al. 2016). This study reported a total of 858 distinct methylation motifs present in 93% of the genomes tested. Interestingly, the methyltransferases show considerable sequence conservation, while restriction enzymes are quite diverse. It could be reasoned that while sequence conservation of the MTases limits the number of recognizable extraneous sequences, at the same time broad REase specificity would provide the scope to cleave a large number of nonself DNA that the bacterial cell encounters. The genome sequences have provided information that more than 80% of the sequenced genome contains multiple R-M systems. Also, the number of R-M systems present in a genome is related to its size. For example, three R-M systems are present in organisms with a genome size between 2 and 3 Mbp, and four R-M systems in organisms with genome size between 3 and 4 Mbp (Vasu and Nagaraja 2013). However, there are notable exceptions such as *Helicobacter* and *Campylobacter* species of genome sizes between 1.5 and 2 Mbp that possess around 30 R-M systems (Vasu and Nagaraja 2013). The significance of such a large number of R-M systems in these species is not fully understood. In contrast, some obligate intracellular pathogens such as *Chlamydia*, *Coxiella*, and *Rickettsia* of genome size between 1 and 2.5 Mbp have no R-M systems. Considering the environmental niche where these organisms live, they may not be encountering any bacteriophages, therefore, the absence of the R-M systems (Vasu and Nagaraja 2013). Another interesting feature with respect to the genome size and the recognition motif of the R-M systems is that the larger proportion of the R-M systems present in organisms with a larger genome (*Bacillus*, *Pseudomonas*, *Streptomyces*, etc.) can recognize longer palindromic sequences (Vasu and Nagaraja 2013). Considering that a larger genome has a higher number of 4 bp or 6 bp recognition motifs than the 8 bp ones, it is reasonable to assume that having more number of R-M systems that prefer the longer recognition motifs would limit nonspecific cleavage of the host genome leading to random double strand breaks.

The widespread presence of the R-M systems in the prokaryotic genomes and the diversity of their recognition sequences have been attributed to their importance in their life cycle and also to their "selfish" nature by which they maintain in the genome. Previous studies have demonstrated that type II R-M systems, carried by plasmids, ensure their retention during postreplication segregation in the bacterial host, indicating the selfish nature of this R-M systems (Naito et al. 1995; Kobayashi 2001). Because of horizontal gene transfer events bacteria can also acquire new R-M systems, which is evident from the fact that different strains of the same bacterial species possess different R-M systems (Oliveira et al. 2014; Croucher et al. 2014). Specific regions in the bacterial genomes have been detected that encode several R-M systems and named as Immigration Control Regions (ICR) involved in the defense mechanisms. Investigations with the ICR flanking regions suggest that these regions are acquired from other genomes by horizontal gene transfer (Kobayashi 2001). In addition, R-M genes are colocalized with mobile genetic elements in several bacterial species such as *Staphylococcus aureus* and *Helicobacter pylori* (Alm et al. 1999; Kobayashi 2001; Lindsay 2010; Xu et al. 2011). The presence of a number of REase pseudogenes in the vicinity of the MTase genes, orphan REases that no longer function as endonuclease and nonfunctional MTases in the bacterial genomes, indicates a loss of R-M systems (Seshasayee et al. 2012; Furuta et al. 2014). Apart from the mechanisms mentioned above, prokaryotic R-M systems can alter their sequence specificity leading to the emergence of new R-M systems in the genome (Furuta et al. 2014; Sánchez-Romero et al. 2015).

## 2.3 Functions of the R-M Systems

The discovery of R-M systems was made in the context of the bacterial defense mechanisms against invading bacteriophages (Luria and Human 1952; Bertani and Weigle 1953). However, bacteriophages also employ several counterstrategies to avoid the host restriction, such as the modification of nucleotides (methylation, glucosylation, etc.). Bacterial species also possess restriction enzymes (type II M and type IV) that target such modified nucleotides from the bacteriophages leading to a "coevolutionary arms race" that has produced several diverse R-M systems. Even though it is suggested that the arms race between the host and the bacteriophages has produced so much diversity with respect to the R-M systems, however, it is not clear how a very specific restriction enzyme would provide protection against a broad range of bacteriophages. Also, it is not properly understood why certain bacterial species for example naturally competent ones like *Helicobacter pylori*, *Haemophilus influenzae*, *Streptococcus pneumoniae*, and *Neisseria gonorrohoeae* possess multiple R-M systems in their genome (Vasu and Nagaraja 2013; Ershova et al. 2015).

Discrimination between the self and nonself genetic material is the primary role of the R-M systems in the bacterial genome. However, the existence of the different subpopulations of a bacterial species is always beneficial for the species survival. In this context, R-M systems help to maintain the genetic diversity by limiting the gene transfer from other bacterial cells (Raleigh 1992; Sibley and Raleigh 2004). Horizontal gene transfer events are limited because of the presence of R-M systems in several bacterial species. The most notable examples are *Burkholderia pseudomallei*, *Streptococcus pneumoniae*, and *Neisseria meningitidis*. In the case of *B. pseudomallei*, the R-M systems have been shown to function as a barrier to genetic exchange. Three phylogenetic groups are identified among the analyzed *B. pseudomallei* genomes, showing almost no genetic exchanges between them. Representatives from the groups showed the presence of different R-M systems and further studies indicated that they are primarily responsible for restricting gene transfer among the different phylogenetic groups. In the case of *S. pneumoniae* 15 monophyletic groups has been recognized that possess distinct sets of R-M systems. In the case of *N. meningitidis*, analysis using 20 sequenced genomes indicated the presence of eight different phylogenetic groups possessing 22 different R-M systems (Ershova et al. 2015). To understand the impact of the presence of R-M systems on horizontal gene transfer events, Budroni et al. (2011) showed that the size of the fragments that were transferred among the different phylogenetic groups in

*N. meningitidis* were significantly shorter than the fragments exchanged within the group (680 bp vs. 3890 bp).

Even though the orphan methyltransferases are primarily involved in the epigenetic regulation of gene expression (see next sections), MTases that are part of the R-M systems sometimes influence the expression of genes in several bacterial species. In the case of *H. pylori,* the transcriptome analysis indicated differences in the expression of genes in the strains carrying different methylation patterns (Furuta et al. 2014). Another consequence of alternate gene expression is phase variation among the bacterial strains, which is a heritable yet on-or-off switching of transcription. This is adopted by several pathogens to increase their survival and fitness in different environmental niches (Hallet 2001; van der Woude and Bäumler 2004). Several pathogens use phase variation to generate diversity in their surface antigenic structure, such as pili, flagella, capsules, lipopolysaccharides, etc. (Weiser et al. 1990; van der Woude and Bäumler 2004). Based upon nucleotide sequence analyses and other genetic studies, it has been shown among several pathogens such as *H. influenzae, H. pylori*, and *N. meningitidis* that type I and III enzymes are involved in phase variation mechanisms (Dybvig et al. 1998; De Bolle et al. 2000; de Vries et al. 2002; Fox et al. 2007). Though the R-M systems are implicated in phase variation mechanisms in several bacterial species, their significance is not completely comprehended as yet.

The restriction enzymes cleave the nonself DNA into smaller fragments that sometimes act as substrates for homologous recombination with the host genome. Though the primary role of the R-M systems is to restrict the entry of foreign DNA, the homologous recombination that follows after an initial cleavage could be a byproduct of the process (Price and Bickle 1986; Vasu and Nagaraja 2013). In several bacterial species, nonhomologous recombination events are also a consequence of restriction mechanisms, in which a small homologous DNA sequence is used to recombine and integrate a larger nonhomologous region into the host genome. Such illegitimate recombination events, leading to genome rearrangements, have been observed with the type I R-M enzyme EcoKI (Kusano et al. 1997).

## 3   Dam *Methyltransferase*

In *gammaproteobacteria,* methylation by orphan (or solitary) methyltransferases plays several important roles, such as DNA mismatch repair, control of chromosome replication, and regulation of gene expression. Loss of function *dam* mutants in *E. coli* and *Salmonella enterica* showed an increased mutation rate (Marinus and Morris 1974; Torreblanca and Casadesús 1996) and an abnormal rate of the initiation of chromosome replication (Boye et al. 1988). The Dam methylase (Deoxyadenosine methyltransferase) is a 32 KDa monomer that uses S-adenosyl-L-methionine (AdoMet) as methyl group donor to the adenine N6 atom (Herman and Modrich 1982). Dam transfers a methyl group to 5'-GATC-3' sequences that are usually fully methylated (the adenosines on both strands are modified), as Dam is constitutively expressed, except for a short period following DNA replication

(Marinus and Casadesus 2009). Accordingly the optimal substrate of Dam is hemimethylated DNA although this enzyme can also methylate de novo unmethylated target sites (Herman and Modrich 1982). In *E. coli*, Dam methylates the ca. 20000 sites in a highly processive manner (Urig et al. 2002) ensuring a rapid and robust remethylation of the entire chromosome. But what is the function(s) of Dam methylation?

## 3.1  Dam-Dependent DNA Mismatch Repair

In *E. coli*, mismatches, due to erroneous incorporation of bases, is corrected by a mechanism called methyl-directed mismatch repair (MMR) (Fig. 1). Whereas in the base excision repair and nucleotide excision repair, the wrong nucleotide is directly detected by the repair system, in the case of a mismatch the two replicated strands are composed of unmethylated nucleotides. The presence of a methylated strand (non-mutated) dictates the correct repair (Lu et al. 1983; Pukkila et al. 1983). Mismatch regions are detected by MutS, the first DNA binding protein that then recruits MutL and MutH (Caillet-Fauquet et al. 1984; Au et al. 1992; Iyer et al. 2006). In this ternary complex, MutH has endonuclease activity on nonmethylated DNA strand near the GATC methylation site (Welsh et al. 1987), while MutL connects MutS with

**Fig. 1** Schematic view of the mismatch repair system in *E. coli*. Mismatch regions in the *E. coli* genome (*asterisk*) are detected by MutS, the DNA-binding protein that then recruits MutL and MutH. In this ternary complex, MutH has endonuclease activity on nonmethylated DNA strand near the GATC methylation site, while MutL connects MutS with MutH. The exonuclease UvrD then digests the single-strand mutated DNA and DNA polymerase III in complex with single-strand binding proteins (*dotted grey line*) and DNA ligase completely repair the mutated region. Finally Dam remethylates the repaired strand

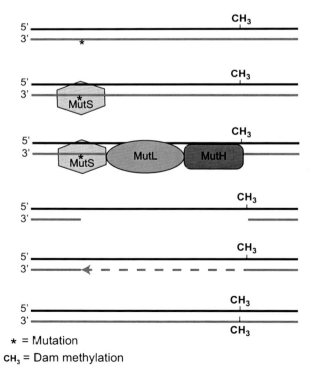

★ = Mutation

CH₃ = Dam methylation

MutH. The exonuclease UvrD then digests the single strand mutated DNA and DNA polymerase III in complex with single strand binding proteins and DNA ligase completely repairs the mutated region (Iyer et al. 2006).

Concentration of Dam methylase in the cells is tightly regulated (Løbner-Olesen et al. 1992). As mentioned earlier, the absence of Dam in *E. coli* and *Salmonella* results in an increasing rate of mutations. However, overexpression of Dam also causes an augmentation of mutations (Pukkila et al. 1983; Torreblanca and Casadesús 1996). This apparent contradiction depends on the specificity of MutH for hemimethylated DNA sequences, while the endonuclease activity of MutH is basically absent on fully methylated templates, when Dam is overexpressed (Herman and Modrich 1981). Therefore, too much or no methylation cause the same MutH inactivity and then accumulation of mutations.

## 3.2 Chromosome Replication Control

The origin of replication (*oriC*) in *E. coli* is controlled at multiple levels, including Dam methylation. As initiation of replication depends on the activity of DnaA, regulation of *dnaA* transcription and modulation of its activity are two important mechanisms at the onset of DNA replication (Mott and Berger 2007). A third mechanism directly controls the accessibility of the origin of replication (Waldminghaus and Skarstad 2009). Dam methylation is responsible for two of those mechanisms: regulation of expression of *dnaA* and *oriC* binding, both phenomena depending on SeqA (Fig. 2). Most of the 20000 GATC sites in the *E. coli* genome are remethylated in few seconds (Stancheva et al. 1999). However, *oriC* and the promoter of *dnaA*, containing respectively 11 and 8 GATC sites, remain hemimethylated for at least 10 min, although Dam methylase is present in the cells. This "protection" is performed by the dimeric protein SeqA (Campbell and Kleckner 1990), which has affinity for hemimethylated GATC sites that are also closely spaced (Kang et al. 1999, 2005). Binding of SeqA to the *oriC* and the *dnaA* promoter in turn blocks further access to those sites, stalling both the access to the origin of replication by the replisome and the transcription of *dnaA*. This time gap allows the conversion of the active DnaA-ATP to inactive DnaA-ADP and the blocking of DnaA production by SeqA repression of the promoter region (Katayama et al. 1998; Kato and Katayama 2001). In conclusion, Dam methylation and SeqA are intimately linked with respect to the initiation of DNA replication. Accordingly, *seqA* loss of function mutants show high similarity to Dam-overproducing cells (Løbner-Olesen et al. 2003).

## 3.3 Regulation of Gene Expression

Methylation sites, such as *E. coli* Dam GATC, are widespread along the genome and occur approximately every 200–300 bp; however, their presence in promoter regions, similarly to SeqA and the promoter of *dnaA*, may affect binding and activity

**G1 phase (Full methylation)**

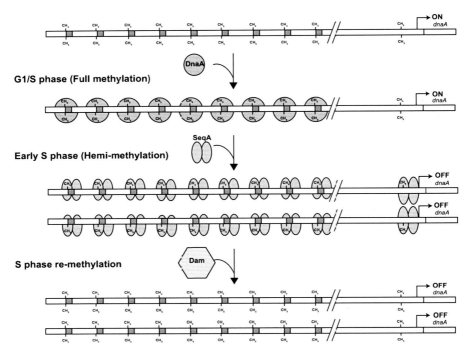

**G1/S phase (Full methylation)**

**Early S phase (Hemi-methylation)**

**S phase re-methylation**

**Fig. 2** **Mechanism of action of methylation dependent chromosome replication initiation in the *E. coli*.** In G1 cells of *E. coli*, the origin of replication and the promoter of *dnaA* are completely methylated. At the onset of DNA replication, the protein DnaA binds the *oriC* sites initiating replication. The two newly replicated *oriCs* will be then hemimethylated. In this form the protein SeqA replaces DnaA from *oriC* and blocks transcription of *dnaA*. Shortly Dam methylase remethylates all sites blocking early reinitiation of DNA replication until a new cycle is possible

of transcription factors and RNA polymerase, especially when the number of methylation sites is above average distribution. Expression profiles of *dam*-deficient cells have been recorded revealing genes whose transcription change without methylation (Torreblanca and Casadesús 1996; Oshima et al. 2002). However, not all affected genes are directly controlled by Dam methylation, such as the SOS regulon that is on the contrary activated by the abnormal activity of the MutSHL system in the absence of methylation (Marinus and Casadesus 2009). In other words, although methylation affects the expression of many genes, only a few subsets of promoters, associated with specific DNA-binding proteins, may be directly linked to a methylation control.

In *Salmonella*, among genes in the conjugative plasmid containing virulence factors, *traJ* encodes a transcription factor, whose expression is under the control of the global regulator Lrp (Camacho and Casadesús 2002, 2005). The promoter of *traJ* has two Lrp binding sites required for the activation of transcription. A fully methylated GATC site, present in one of those Lrp binding sites, is able to repress *traJ* transcription, while only hemimethylation on the adenosine of the noncoding strand is able to trigger the transcription of the gene (Camacho and Casadesús 2005). Therefore, when the plasmid is replicated, only one daughter DNA sequence is expressed although plasmids have the same genetic information. As activation of *traJ* requires DNA replication, a physiological state associated with nutrient abundance and absence of stress, this epigenetic activation ensures that energy-consuming conjugation will take place in a suitable environment. Similar mechanisms of epigenetic regulation are described in the regulation of the IS10 transposase, whose activity is kept at a low level by this kind of methylation control (Roberts et al. 1985).

Some bacterial species show the phenomenon called "phase variation" where genetically identical populations of cells express certain factors at different levels. This is a bet-hedging strategy used by the bacterial species, to prepare them to face changing environmental conditions. Although several mechanisms of phase variations require the variation of the DNA sequence, some mechanisms are epigenetically controlled by DNA methylation. For example, in the uropathogenic *E. coli*, the synthesis of Pap pili is under the control of the methylase Dam (Fig. 3). During infection, the population contains bacteria with (ON) and without Pap pili (OFF). This dual expression pattern derives from two opposite characteristics of Pap pili: on one hand those pili are highly immunogenic triggering immune responses; on the other hand Pap pili are indispensable for the colonization of the upper urinary tract. Therefore, this dual nature of the population keeps the potential ability to colonize new infection niches without triggering too much the immune response (Roberts et al. 1994; Hernday et al. 2002). This epigenetic regulation is based on the presence of two Dam methylation sites: GATC-I or distal (upstream the *papI* gene) and

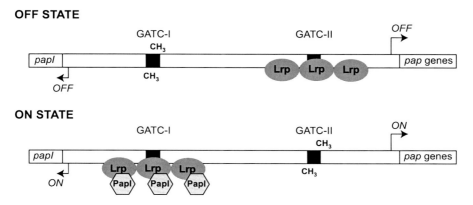

**Fig. 3 Role of DNA methylation in phase variation.** Pap pili expression is under the control of the methylase Dam. During infection, the population contains bacteria with Pap pili (ON) and cells with no pili (OFF). GATC-I and GATC-II are present (*black box*). In the OFF state, GATC-II is nonmethylated while GATC-I is methylated; in the ON state, the methylation pattern is opposite. In the OFF state, Lrp is bound to GATC-II, keeping its state non methylated and preventing Lrp to bind GATC-I, which is indeed methylated. In the OFF state, Lrp binding represses the *pap* operon transcription. In the ON-state, accumulation of PapI moves Lrp to GATC-I, freeing GATC-II allowing Dam to methylate the GATC-II site therefore stabilizing the ON state

GATC-II or proximal (upstream the *pap* operon). In the OFF state, GATC-II is nonmethylated while GATC-I is methylated; in the ON state, the methylation pattern is the opposite. Differently from the rest of the genome, GATC sites in the *pap* operon can show a nonmethylated state because of the DNA binding of the Lrp protein that prevents Dam methylation. In the OFF state, Lrp is bound to GATC-II, keeping its state nonmethylated and preventing Lrp to bind GATC-I, which is indeed methylated. In the OFF state, Lrp binding represses the *pap* operon transcription. This situation is stable unless a second factor (PapI), whose fluctuations of expression are noisy, reaches a threshold able to dislocate Lrp from GATC-II in favor of GATC-I. Once GATC-II is free from Lrp, Dam is able to methylate the GATC-II site therefore stabilizing the ON state. Several cycles of DNA replication will free the GATC-I site from methylation as Lrp and PapI stably protect this region (Casadesús and Low 2006, 2013).

Dam methylation and Lrp also control other fimbrial operons such as *foo*, *clp*, and *pef* (Casadesús and Low 2006). However, Dam can also be associated with other factors regulating phase variation. For example, the locus *agn43*, encoding a membrane protein of the outer membrane involved in biofilm formation and interaction with the infected host, is regulated by Dam methylation and the protein OxyR (Henderson and Owen 1999; Danese et al. 2000; Waldron et al. 2002; Wallecha et al. 2002, 2003; Kaminska and van der Woude 2010). Finally, OxyR and Dam are also involved in the epigenetic control of the *Salmonella opvAB* operon encoding a protein modifying the length of the O-antigen in lipopolysaccharide (Cota et al. 2012).

## 4    CcrM *Methyltransferase*

Among the cell cycle regulators in *Caulobacter crescentus*, the methyltransferase CcrM plays a crucial role in coordinating important developmental processes by transcriptional regulation (Table 1). This methylase and its functions are possibly conserved in other alpha-proteobacteria although its role has been less studied (Wright et al. 1997; Robertson et al. 2000; Kahng and Shapiro 2001). In contrast to gamma-proteobacteria in which the hemimethylation state of DNA is not stable and limited in time, in *C. crescentus* all CcrM loci remain hemimethylated during the S-phase until the replication of DNA is complete. This cell cycle-dependent methylation pattern has an important consequence on the regulation of gene expression over time in coordination with DNA replication progression (Mohapatra et al. 2014).

*C. crescentus* produces two different cell types at every division that are morphologically and functionally different, a sessile replication competent stalked cell and a vegetative and replication incompetent swarmer cell (Fig. 4a). The stalked cell, capable of replicating the circular chromosome and producing new cells (Brown et al. 2009), possesses a polar tubular appendix, the stalk, having the same composition of the cell envelope (Jenal 2000).

In *C. crescentus*, CcrM methylates adenosines of the palindromic 5′-GAnTC-3′ sites in the DNA double helix (Zweiger et al. 1994). CcrM is present and active only for a short window of time at the end of the S-phase in the late predivisional cells.

**Table 1** CcrM-dependent genes involved in cell cycle and polarity in *C. crescentus*

| Gene | Function | Regulation (hemi/full) | GcrA-dependent | Refs. |
|------|----------|------------------------|----------------|-------|
| *ctrA* | Cell cycle regulator | F– | Yes | (Reisenauer and Shapiro 2002; Holtzendorff et al. 2004; Fioravanti et al. 2013; Murray et al. 2013; Gonzalez et al. 2014) |
| *podJ* | Polarity determining protein | H+ | Yes | (Holtzendorff et al. 2004; Fioravanti et al. 2013; Murray et al. 2013) |
| *mipZ* | Cell division plane positioning ATPase | F+ | Yes | (Holtzendorff et al. 2004; Fioravanti et al. 2013; Murray et al. 2013) |
| *ftsZ* | Cell division protein | F+ | Yes | (Gonzalez and Collier 2013; Murray et al. 2013) |
| *ftsN* | Cell division protein | – | Yes | (Fioravanti et al. 2013; Murray et al. 2013; Gonzalez et al. 2014) |
| *tipF* | Polarity determining protein | – | Yes | (Fioravanti et al. 2013) |
| *pleC* | Polarity determining protein | – | Yes | (Fioravanti et al. 2013; Kozdon et al. 2013) |
| *flaY* | Motility | – | Yes | (Fioravanti et al. 2013) |
| *creS* | Cell shape | – | – | (Gonzalez et al. 2014) |
| *ftsW, ftsE* | Cell division protein | – | – | (Gonzalez et al. 2014) |
| *gyrA, gyrB* | DNA replication | – | – | (Kozdon et al. 2013; Gonzalez et al. 2014) |
| *parE* | Chromosome segregation | – | – | (Gonzalez et al. 2014) |
| *staR* | Stalk biogenesis | – | – | (Kozdon et al. 2013; Gonzalez et al. 2014) |
| *popZ* | Pole organizing protein | – | – | (Gonzalez et al. 2014) |

F+, activated when fully methylated; F–, repressed when fully methylated; H+, activated when hemimethylated, "–" no information available.

In fact, the gene encoding CcrM is under the control of the cell cycle master regulator CtrA that has its highest activity in late S-phase (Collier et al. 2007). At the same time, CcrM is degraded by the Lon protease before cell division so that swarmer cells have no CcrM (Wright et al. 1996). Although there are other methylases in *Caulobacter*, CcrM is the only one showing cell cycle-regulated activity and affecting cell cycle progression (Nierman et al. 2001; Kozdon et al. 2013). The chromosome remains fully methylated in both strands in G2 and G1 phases. The time of conversion during the S-phase into two hemimethylated copies of each GAnTC

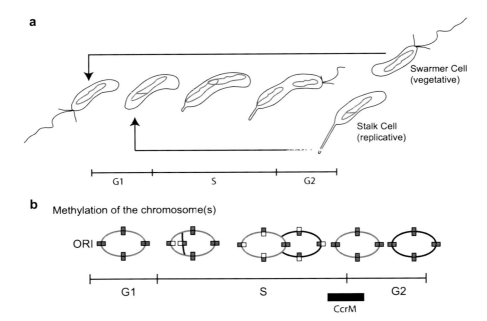

**Fig. 4 DNA Methylation and progression of cell cycle in *Caulobacter crescentus.***
(a) *C. crescentus* cell cycle progression. Every cell division *C. crescentus* produce two different
cell types: a vegetative swarmer cells, possessing a flagellum and pili, that is able to swim; a stalked
cell capable of DNA replication and attached to substrate by stalk in continuity with the cell
envelope. The swarmer cell must differentiate in a stalk cell in order to initiate a division cycle.
After cell division, the stalked cell is able to immediately initiate the DNA replication. *C. crescentus*
has a single circular chromosome, whose origin has a polar localization and it is replicated only
once per cell division. (b) The *C. crescentus* chromosome in G1 is fully methylated by CcrM (four
sites are schematically represented here in *dark grey*). As the replication initiates the sites proximal
to the origin of replication (*oriC*) become hemimethylated (*grey* and *white*). As the replication fork
proceeds, all sites become hemimethylated until the methylase CcrM (*black bar*) accumulates at the
end of S-phase

sequence depends on the chromosomal location of the individual methylation sites
(Zweiger et al. 1994; Kozdon et al. 2013) (Fig. 4b).

CcrM methylates the 4542 potential GAnTC sites in *C. crescentus* (Nierman et al.
2001) using *S*-adenosyl-L-methionine as substrate (Zweiger et al. 1994) in a distrib-
utive manner that is similar to *E. coli* Dam (Albu et al. 2012), and that is characterized by a preferential binding for a hemimethylated template (Berdis et al. 1998;
Albu et al. 2012). Approximately a quarter of these CcrM motifs are located in the
intergenic regions suggesting a potential regulation of transcription. Using single-
molecule real-time (SMRT) DNA sequencing, all sites methylated by CcrM were
identified at base-pair resolution (Kozdon et al. 2013). Accordingly to the previous
knowledge, almost all GAnTC sites progressed from full- to hemimethylation and
back again to a full methylation stage with the progression of the replication fork
during the S-phase. However, 27 GAnTC sites remained permanently unmethylated

(Kozdon et al. 2013) during the S-phase. Some of these hypomethylated sequences have been recently shown to be associated with the regulator MucR (Ardissone et al. 2016), although their functional role is still unclear.

CcrM methylation controls many important genes in *C. crescentus* (Mohapatra et al. 2014). This control is mostly positive (activation on hemi or, less frequently, on fully methylated DNAs) or negative (repression by the full methylation) (Table 1).

Under the control of CcrM-methylation, *ctrA* (Quon et al. 1996) probably is the most significant gene, as it encodes the master regulator of cell cycle in *C. crescentus*. Two promoters are responsible for *ctrA* transcription: *ctrA*P1 contains a GAnTC site at its −35 region and is activated only in the hemimethylated state, when the replication fork progresses through it, while the promoter *ctrA*P2 is auto-induced by phosphorylated CtrA (Reisenauer and Shapiro 2002).

The full methylation state keeps the promoter *ctrA*P1 in a repressed mode and only the conversion to a hemimethylated form allows transcription (Reisenauer and Shapiro 2002). The importance of temporal control of CcrM expression is evident as a strain constitutively expressing CcrM throughout the cell cycle results in aberrant cell types with multiple chromosome duplication events because of a fully methylated chromosome (Zweiger et al. 1994). The transcription from *ctrA*P1 clearly depends on the amount of time it remains fully methylated during the replication, as *ctrA*P1 when placed proximal to the origin of replication, showed highest activity, whereas a terminus proximal location of *ctrA*P1 had basically no expression (Reisenauer and Shapiro 2002). These observations prompted the question concerning the molecular mechanism connecting the methylation state to RNA polymerase activity. Studies have suggested that CcrM is associated with GcrA (Holtzendorff et al. 2004) among the members of the *Alphaproteobacteria* group (Brilli et al. 2010; Murray et al. 2013). GcrA is a DNA binding protein with sites spread all over the *Caulobacter* genome (Fioravanti et al. 2013; Haakonsen et al. 2015). Although GcrA target regions are generally associated with GAnTC sites, it also binds to regions in the chromosome having no methylation (Fioravanti et al. 2013). Coherent to previous results (Holtzendorff et al. 2004), GcrA binds the promoter P1 of *ctrA*; interestingly the CcrM full methylation, which corresponds in vivo to inhibition (Reisenauer and Shapiro 2002), shows the highest binding efficiency in comparison with the two hemimethylation arrangements, which were still more efficient than the nonmethylated sequence (Fioravanti et al. 2013). GcrA affects RNA polymerase by interacting with Sigma70 favoring the open complex formation (Haakonsen et al. 2015).

CcrM orthologs have been investigated in other alpha-proteobacteria, such as *Sinorhizobium meliloti*, *Brucella abortus*, and *Agrobacterium tumefaciens* (Wright et al. 1997; Robertson et al. 2000; Kahng and Shapiro 2001). In contrast to *Caulobacter*, in which the GAnTC methylation is dispensable (Gonzalez and Collier 2013; Fioravanti et al. 2013; Murray et al. 2013), CcrM in other alpha-proteobacteria is essential, although its exact role has not been elucidated yet (Brilli et al. 2010). Interestingly, CcrM and GcrA orthologs can complement, at least partially, functions in other alpha-proteobacteria, suggesting a similar role (Wright et al. 1997; Robertson et al. 2000; Kahng and Shapiro 2001; Fioravanti et al. 2013).

Furthermore, the cell cycle-regulated activity of CcrM, peaking at the end of S-phase, has been demonstrated in *A. tumefaciens* (Kahng and Shapiro 2001) and *S. meliloti* (De Nisco et al. 2014).

## 5    Research Needs

DNA methylation in bacteria has been described since many years, but its real importance is still not fully understood. For example, the role of the solitary methylase CcrM controlling cell cycle in *C. crescentus* appears still at the beginning of its elucidation. Multiple DNA-binding proteins controlling transcription have been associated with DNA methylation, such as SeqA, Lrp in *E. coli* or GcrA in *C. crescentus*, however, there could be many more transcription factors that could detect the methylation state and modulate global transcriptional programs in bacterial cells. It is worth noticing that many CcrM methylation sites are independent of GcrA.

New techniques, such SMRT sequencing (see previous sections), have been developed in order to understand DNA methylation at the genomic scale, revealing that multiple methylases are indeed responsible for DNA methylation in bacterial cells. Next few years will definitely open new surprising frontiers in the epigenetic regulation in bacteria. What is clear for now is that although simple, bacteria possess a complexity in regulatory mechanisms, among which DNA methylation plays a crucial role.

## References

Adhikari S, Curtis PD (2016) DNA methyltransferases and epigenetic regulation in bacteria. FEMS Microbiol Rev 40:575–591. https://doi.org/10.1093/femsre/fuw023

Albu RF, Zacharias M, Jurkowski TP, Jeltsch A (2012) DNA interaction of the CcrM DNA methyltransferase: a mutational and modeling study. Chembiochem Eur J Chem Biol 13:1304–1311. https://doi.org/10.1002/cbic.201200082

Alm RA, Ling LS, Moir DT, King BL, Brown ED, Doig PC, Smith DR, Noonan B, Guild BC, deJonge BL, Carmel G, Tummino PJ, Caruso A, Uria-Nickelsen M, Mills DM, Ives C, Gibson R, Merberg D, Mills SD, Jiang Q, Taylor DE, Vovis GF, Trust TJ (1999) Genomic-sequence comparison of two unrelated isolates of the human gastric pathogen *Helicobacter pylori*. Nature 397:176–180. https://doi.org/10.1038/16495

Ardissone S, Redder P, Russo G, Frandi A, Fumeaux C, Patrignani A, Schlapbach R, Falquet L, Viollier PH (2016) Cell cycle constraints and environmental control of local DNA hypomethylation in α-proteobacteria. PLoS Genet 12:e1006499. https://doi.org/10.1371/journal.pgen.1006499

Au KG, Welsh K, Modrich P (1992) Initiation of methyl-directed mismatch repair. J Biol Chem 267:12142–12148

Berdis AJ, Lee I, Coward JK, Stephens C, Wright R, Shapiro L, Benkovic SJ (1998) A cell cycle-regulated adenine DNA methyltransferase from *Caulobacter crescentus* processively methylates GANTC sites on hemimethylated DNA. Proc Natl Acad Sci U S A 95:2874–2879

Bertani G, Weigle JJ (1953) Host controlled variation in bacterial viruses. J Bacteriol 65:113–121

Blow MJ, Clark TA, Daum CG, Deutschbauer AM, Fomenkov A, Fries R, Froula J, Kang DD, Malmstrom RR, Morgan RD, Posfai J, Singh K, Visel A, Wetmore K, Zhao Z, Rubin EM,

Korlach J, Pennacchio LA, Roberts RJ (2016) The epigenomic landscape of prokaryotes. PLoS Genet 12:e1005854. https://doi.org/10.1371/journal.pgen.1005854

Boye E, Løbner-Olesen A (1990) The role of dam methyltransferase in the control of DNA replication in E. coli. Cell 62:981–989

Boye E, Løbner-Olesen A, Skarstad K (1988) Timing of chromosomal replication in Escherichia coli. Biochim Biophys Acta 951:359–364

Brilli M, Fondi M, Fani R, Mengoni A, Ferri L, Bazzicalupo M, Biondi EG (2010) The diversity and evolution of cell cycle regulation in alpha-proteobacteria: a comparative genomic analysis. BMC Syst Biol 4:52. https://doi.org/10.1186/1752-0509-4-52

Brown PJB, Hardy GG, Trimble MJ, Brun YV (2009) Complex regulatory pathways coordinate cell-cycle progression and development in Caulobacter crescentus. Adv Microb Physiol 54:1–101. https://doi.org/10.1016/S0065-2911(08)00001-5

Budroni S, Siena E, Dunning Hotopp JC, Seib KL, Serruto D, Nofroni C, Comanducci M, Riley DR, Daugherty SC, Angiuoli SV, Covacci A, Pizza M, Rappuoli R, Moxon ER, Tettelin H, Medini D (2011) Neisseria meningitidis is structured in clades associated with restriction modification systems that modulate homologous recombination. Proc Natl Acad Sci U S A 108:4494–4499. https://doi.org/10.1073/pnas.1019751108

Caillet-Fauquet P, Maenhaut-Michel G, Radman M (1984) SOS mutator effect in E. coli mutants deficient in mismatch correction. EMBO J 3:707–712

Camacho EM, Casadesús J (2002) Conjugal transfer of the virulence plasmid of Salmonella enterica is regulated by the leucine-responsive regulatory protein and DNA adenine methylation. Mol Microbiol 44:1589–1598

Camacho EM, Casadesús J (2005) Regulation of traJ transcription in the Salmonella virulence plasmid by strand-specific DNA adenine hemimethylation. Mol Microbiol 57:1700–1718. https://doi.org/10.1111/j.1365-2958.2005.04788.x

Campbell JL, Kleckner N (1990) E. coli oriC and the dnaA gene promoter are sequestered from dam methyltransferase following the passage of the chromosomal replication fork. Cell 62:967–979

Casadesús J (2016) Bacterial DNA methylation and methylomes. Adv Exp Med Biol 945:35–61. https://doi.org/10.1007/978-3-319-43624-1_3

Casadesús J, Low D (2006) Epigenetic gene regulation in the bacterial world. Microbiol Mol Biol Rev 70:830–856. https://doi.org/10.1128/MMBR.00016-06

Casadesús J, Low DA (2013) Programmed heterogeneity: epigenetic mechanisms in bacteria. J Biol Chem 288:13929–13935. https://doi.org/10.1074/jbc.R113.472274

Collier J, McAdams HH, Shapiro L (2007) A DNA methylation ratchet governs progression through a bacterial cell cycle. Proc Natl Acad Sci U S A 104:17111–17116. https://doi.org/10.1073/pnas.0708112104

Cota I, Blanc-Potard AB, Casadesús J (2012) STM2209-STM2208 (opvAB): a phase variation locus of Salmonella enterica involved in control of O-antigen chain length. PLoS One 7:e36863. https://doi.org/10.1371/journal.pone.0036863

Croucher NJ, Coupland PG, Stevenson AE, Callendrello A, Bentley SD, Hanage WP (2014) Diversification of bacterial genome content through distinct mechanisms over different time-scales. Nat Commun 5:5471. https://doi.org/10.1038/ncomms6471

Danese PN, Pratt LA, Dove SL, Kolter R (2000) The outer membrane protein, antigen 43, mediates cell-to-cell interactions within Escherichia coli biofilms. Mol Microbiol 37:424–432

De Bolle X, Bayliss CD, Field D, van de Ven T, Saunders NJ, Hood DW, Moxon ER (2000) The length of a tetranucleotide repeat tract in Haemophilus influenzae determines the phase variation rate of a gene with homology to type III DNA methyltransferases. Mol Microbiol 35:211–222

De Nisco NJ, Abo RP, Wu CM, Penterman J, Walker GC (2014) Global analysis of cell cycle gene expression of the legume symbiont Sinorhizobium meliloti. Proc Natl Acad Sci USA. https://doi.org/10.1073/pnas.1400421111

de Vries N, Duinsbergen D, Kuipers EJ, Pot RGJ, Wiesenekker P, Penn CW, van Vliet AHM, Vandenbroucke-Grauls CMJE, Kusters JG (2002) Transcriptional phase variation of a type III restriction-modification system in Helicobacter pylori. J Bacteriol 184:6615–6623

Dryden DT, Murray NE, Rao DN (2001) Nucleoside triphosphate-dependent restriction enzymes. Nucleic Acids Res 29:3728–3741

Dybvig K, Sitaraman R, French CT (1998) A family of phase-variable restriction enzymes with differing specificities generated by high-frequency gene rearrangements. Proc Natl Acad Sci USA 95:13923–13928

Ershova AS, Rusinov IS, Spirin SA, Karyagina AS, Alexeevski AV (2015) Role of restriction-modification systems in prokaryotic evolution and ecology. Biochemistry (Mosc) 80:1373–1386. https://doi.org/10.1134/S0006297915100193

Fioravanti A, Fumeaux C, Mohapatra SS, Bompard C, Brilli M, Frandi A, Castric V, Villeret V, Viollier PH, Biondi EG (2013) DNA binding of the cell cycle transcriptional regulator GcrA depends on N6-adenosine methylation in *Caulobacter crescentus* and other Alphaproteobacteria. PLoS Genet 9:e1003541. https://doi.org/10.1371/journal.pgen.1003541

Fox KL, Dowideit SJ, Erwin AL, Srikhanta YN, Smith AL, Jennings MP (2007) *Haemophilus influenzae* phasevarions have evolved from type III DNA restriction systems into epigenetic regulators of gene expression. Nucleic Acids Res 35:5242–5252. https://doi.org/10.1093/nar/gkm571

Furuta Y, Namba-Fukuyo H, Shibata TF, Nishiyama T, Shigenobu S, Suzuki Y, Sugano S, Hasebe M, Kobayashi I (2014) Methylome diversification through changes in DNA methyltransferase sequence specificity. PLoS Genet 10:e1004272. https://doi.org/10.1371/journal.pgen.1004272

Gonzalez D, Collier J (2013) DNA methylation by CcrM activates the transcription of two genes required for the division of *Caulobacter crescentus*. Mol Microbiol 88:203–218. https://doi.org/10.1111/mmi.12180

Gonzalez D, Kozdon JB, McAdams HH, Shapiro L, Collier J (2014) The functions of DNA methylation by CcrM in *Caulobacter crescentus*: a global approach. Nucleic Acids Res. https://doi.org/10.1093/nar/gkt1352

Haakonsen DL, Yuan AH, Laub MT (2015) The bacterial cell cycle regulator GcrA is a σ70 cofactor that drives gene expression from a subset of methylated promoters. Genes Dev 29:2272–2286. https://doi.org/10.1101/gad.270660.115

Hallet B (2001) Playing Dr Jekyll and Mr Hyde: combined mechanisms of phase variation in bacteria. Curr Opin Microbiol 4:570–581. https://doi.org/10.1016/S1369-5274(00)00253-8.

Henderson IR, Owen P (1999) The major phase-variable outer membrane protein of *Escherichia coli* structurally resembles the immunoglobulin A1 protease class of exported protein and is regulated by a novel mechanism involving dam and oxyR. J Bacteriol 181:2132–2141

Herman GE, Modrich P (1981) *Escherichia coli* K-12 clones that overproduce dam methylase are hypermutable. J Bacteriol 145:644–646

Herman GE, Modrich P (1982) *Escherichia coli* Dam methylase. Physical and catalytic properties of the homogeneous enzyme. J Biol Chem 257:2605–2612

Hernday A, Krabbe M, Braaten B, Low D (2002) Self-perpetuating epigenetic pili switches in bacteria. Proc Natl Acad Sci U S A 99(Suppl 4):16470–16476. https://doi.org/10.1073/pnas.182427199

Holtzendorff J, Hung D, Brende P, Reisenauer A, Viollier PH, McAdams HH, Shapiro L (2004) Oscillating global regulators control the genetic circuit driving a bacterial cell cycle. Science 304:983–987. https://doi.org/10.1126/science.1095191

Iyer RR, Pluciennik A, Burdett V, Modrich PL (2006) DNA mismatch repair: functions and mechanisms. Chem Rev 106:302–323. https://doi.org/10.1021/cr0404794

Janscak P, Sandmeier U, Szczelkun MD, Bickle TA (2001) Subunit assembly and mode of DNA cleavage of the type III restriction endonucleases EcoP1I and EcoP15I. J Mol Biol 306:417–431. https://doi.org/10.1006/jmbi.2000.4411

Jenal U (2000) Signal transduction mechanisms in *Caulobacter crescentus* development and cell cycle control. FEMS Microbiol Rev 24:177–191

Kahng LS, Shapiro L (2001) The CcrM DNA methyltransferase of *Agrobacterium tumefaciens* is essential, and its activity is cell cycle regulated. J Bacteriol 183:3065–3075. https://doi.org/10.1128/JB.183.10.3065-3075.2001

Kaminska R, van der Woude MW (2010) Establishing and maintaining sequestration of dam target sites for phase variation of agn43 in *Escherichia coli*. J Bacteriol 192:1937–1945. https://doi.org/10.1128/JB.01629-09

Kang S, Lee H, Han JS, Hwang DS (1999) Interaction of SeqA and dam methylase on the hemimethylated origin of *Escherichia coli* chromosomal DNA replication. J Biol Chem 274:11463–11468

Kang S, Han JS, Kim KP, Yang HY, Lee KY, Hong CB, Hwang DS (2005) Dimeric configuration of SeqA protein bound to a pair of hemi-methylated GATC sequences. Nucleic Acids Res 33:1524–1531. https://doi.org/10.1093/nar/gki289

Katayama T, Kubota T, Kurokawa K, Crooke E, Sekimizu K (1998) The initiator function of DnaA protein is negatively regulated by the sliding clamp of the *E. coli* chromosomal replicase. Cell 94:61–71

Kato J, Katayama T (2001) Hda, a novel DnaA-related protein, regulates the replication cycle in *Escherichia coli*. EMBO J 20:4253–4262. https://doi.org/10.1093/emboj/20.15.4253

Kobayashi I (2001) Behavior of restriction-modification systems as selfish mobile elements and their impact on genome evolution. Nucleic Acids Res 29:3742–3756

Kozdon JB, Melfi MD, Luong K, Clark TA, Boitano M, Wang S, Zhou B, Gonzalez D, Collier J, Turner SW, Korlach J, Shapiro L, McAdams HH (2013) Global methylation state at base-pair resolution of the *Caulobacter* genome throughout the cell cycle. Proc Natl Acad Sci U S A 110: E4658–E4667. https://doi.org/10.1073/pnas.1319315110

Kusano K, Sakagami K, Yokochi T, Naito T, Tokinaga Y, Ueda E, Kobayashi I (1997) A new type of illegitimate recombination is dependent on restriction and homologous interaction. J Bacteriol 179:5380–5390

Lindsay JA (2010) Genomic variation and evolution of *Staphylococcus aureus*. Int J Med Microbiol 300:98–103. https://doi.org/10.1016/j.ijmm.2009.08.013

Løbner-Olesen A, Boye E, Marinus MG (1992) Expression of the *Escherichia coli* dam gene. Mol Microbiol 6:1841–1851

Løbner-Olesen A, Marinus MG, Hansen FG (2003) Role of SeqA and dam in *Escherichia coli* gene expression: a global/microarray analysis. Proc Natl Acad Sci U S A 100:4672–4677. https://doi.org/10.1073/pnas.0538053100

Loenen WAM, Raleigh EA (2014) The other face of restriction: modification-dependent enzymes. Nucleic Acids Res 42:56–69. https://doi.org/10.1093/nar/gkt747

Lu AL, Clark S, Modrich P (1983) Methyl-directed repair of DNA base-pair mismatches in vitro. Proc Natl Acad Sci U S A 80:4639–4643

Luria SE, Human ML (1952) A nonhereditary, host-induced variation of bacterial viruses. J Bacteriol 64:557–569

Marinus MG, Casadesus J (2009) Roles of DNA adenine methylation in host-pathogen interactions: mismatch repair, transcriptional regulation, and more. FEMS Microbiol Rev 33:488–503. https://doi.org/10.1111/j.1574-6976.2008.00159.x

Marinus MG, Morris NR (1974) Biological function for 6-methyladenine residues in the DNA of *Escherichia coli* K12. J Mol Biol 85:309–322

Mohapatra SS, Fioravanti A, Biondi EG (2014) DNA methylation in *Caulobacter* and other Alphaproteobacteria during cell cycle progression. Trends Microbiol 22:528–535. https://doi.org/10.1016/j.tim.2014.05.003

Mott ML, Berger JM (2007) DNA replication initiation: mechanisms and regulation in bacteria. Nat Rev Microbiol 5:343–354. https://doi.org/10.1038/nrmicro1640

Mücke M, Reich S, Möncke-Buchner E, Reuter M, Krüger DH (2001) DNA cleavage by type III restriction-modification enzyme EcoP15I is independent of spacer distance between two head to head oriented recognition sites. J Mol Biol 312:687–698. https://doi.org/10.1006/jmbi.2001.4998

Murray SM, Panis G, Fumeaux C, Viollier PH, Howard M (2013) Computational and genetic reduction of a cell cycle to its simplest, primordial components. PLoS Biol 11:e1001749. https://doi.org/10.1371/journal.pbio.1001749

Naito T, Kusano K, Kobayashi I (1995) Selfish behavior of restriction-modification systems. Science 267:897–899

Nierman WC, Feldblyum TV, Laub MT, Paulsen IT, Nelson KE, Eisen JA, Heidelberg JF, Alley MR, Ohta N, Maddock JR, Potocka I, Nelson WC, Newton A, Stephens C, Phadke ND, Ely B, DeBoy RT, Dodson RJ, Durkin AS, Gwinn ML, Haft DH, Kolonay JF, Smit J, Craven MB, Khouri H, Shetty J, Berry K, Utterback T, Tran K, Wolf A, Vamathevan J, Ermolaeva M, White O, Salzberg SL, Venter JC, Shapiro L, Fraser CM, Eisen J (2001) Complete genome sequence of *Caulobacter crescentus*. Proc Natl Acad Sci USA 98:4136–4141. https://doi.org/10.1073/pnas.061029298

Oliveira PH, Touchon M, Rocha EPC (2014) The interplay of restriction-modification systems with mobile genetic elements and their prokaryotic hosts. Nucleic Acids Res 42:10618–10631. https://doi.org/10.1093/nar/gku734

Oshima T, Wada C, Kawagoe Y, Ara T, Maeda M, Masuda Y, Hiraga S, Mori H (2002) Genome-wide analysis of deoxyadenosine methyltransferase-mediated control of gene expression in *Escherichia coli*. Mol Microbiol 45:673–695

Price C, Bickle TA (1986) A possible role for DNA restriction in bacterial evolution. Microbiol Sci 3:296–299

Pukkila PJ, Peterson J, Herman G, Modrich P, Meselson M (1983) Effects of high levels of DNA adenine methylation on methyl-directed mismatch repair in *Escherichia coli*. Genetics 104:571–582

Quon KC, Marczynski GT, Shapiro L (1996) Cell cycle control by an essential bacterial two-component signal transduction protein. Cell 84:83–93

Raleigh EA (1992) Organization and function of the mcrBC genes of *Escherichia coli* K-12. Mol Microbiol 6:1079–1086

Reisenauer A, Shapiro L (2002) DNA methylation affects the cell cycle transcription of the CtrA global regulator in *Caulobacter*. EMBO J 21:4969–4977

Roberts D, Hoopes BC, McClure WR, Kleckner N (1985) IS10 transposition IS regulated by DNA adenine methylation. Cell 43:117–130

Roberts JA, Marklund BI, Ilver D, Haslam D, Kaack MB, Baskin G, Louis M, Möllby R, Winberg J, Normark S (1994) The gal(alpha 1-4)gal-specific tip adhesin of *Escherichia coli* P-fimbriae is needed for pyelonephritis to occur in the normal urinary tract. Proc Natl Acad Sci USA 91:11889–11893

Roberts RJ, Belfort M, Bestor T, Bhagwat AS, Bickle TA, Bitinaite J, Blumenthal RM, Degtyarev SK, Dryden DTF, Dybvig K, Firman K, Gromova ES, Gumport RI, Halford SE, Hattman S, Heitman J, Hornby DP, Janulaitis A, Jeltsch A, Josephsen J, Kiss A, Klaenhammer TR, Kobayashi I, Kong H, Krüger DH, Lacks S, Marinus MG, Miyahara M, Morgan RD, Murray NE, Nagaraja V, Piekarowicz A, Pingoud A, Raleigh E, Rao DN, Reich N, Repin VE, Selker EU, Shaw P-C, Stein DC, Stoddard BL, Szybalski W, Trautner TA, Van Etten JL, Vitor JMB, Wilson GG, Xu S (2003) A nomenclature for restriction enzymes, DNA methyltransferases, homing endonucleases and their genes. Nucleic Acids Res 31:1805–1812

Roberts RJ, Vincze T, Posfai J, Macelis D (2010) REBASE – a database for DNA restriction and modification: enzymes, genes and genomes. Nucleic Acids Res 38:D234–D236. https://doi.org/10.1093/nar/gkp874

Roberts RJ, Vincze T, Posfai J, Macelis D (2015) REBASE – a database for DNA restriction and modification: enzymes, genes and genomes. Nucleic Acids Res 43:D298–D299. https://doi.org/10.1093/nar/gku1046

Robertson GT, Reisenauer A, Wright R, Jensen RB, Jensen A, Shapiro L, Roop RM 2nd (2000) The *Brucella abortus* CcrM DNA methyltransferase is essential for viability, and its overexpression attenuates intracellular replication in murine macrophages. J Bacteriol 182:3482–3489

Sánchez-Romero MA, Cota I, Casadesús J (2015) DNA methylation in bacteria: from the methyl group to the methylome. Curr Opin Microbiol 25:9–16. https://doi.org/10.1016/j.mib.2015.03.004

Seshasayee ASN, Singh P, Krishna S (2012) Context-dependent conservation of DNA methyltransferases in bacteria. Nucleic Acids Res 40:7066–7073. https://doi.org/10.1093/nar/gks390

Sibley MH, Raleigh EA (2004) Cassette-like variation of restriction enzyme genes in *Escherichia coli* C and relatives. Nucleic Acids Res 32:522–534. https://doi.org/10.1093/nar/gkh194

Stancheva I, Koller T, Sogo JM (1999) Asymmetry of dam remethylation on the leading and lagging arms of plasmid replicative intermediates. EMBO J 18:6542–6551. https://doi.org/10.1093/emboj/18.22.6542

Torreblanca J, Casadesús J (1996) DNA adenine methylase mutants of *Salmonella typhimurium* and a novel dam-regulated locus. Genetics 144:15–26

Urig S, Gowher H, Hermann A, Beck C, Fatemi M, Humeny A, Jeltsch A (2002) The *Escherichia coli* dam DNA methyltransferase modifies DNA in a highly processive reaction. J Mol Biol 319:1085–1096. https://doi.org/10.1016/S0022-2836(02)00371-6

van der Woude MW, Bäumler AJ (2004) Phase and antigenic variation in bacteria. Clin Microbiol Rev 17:581–611, table of contents. https://doi.org/10.1128/CMR.17.3.581-611.2004

Vasu K, Nagaraja V (2013) Diverse functions of restriction-modification systems in addition to cellular defense. Microbiol Mol Biol Rev 77:53–72. https://doi.org/10.1128/MMBR.00044-12

Waldminghaus T, Skarstad K (2009) The *Escherichia coli* SeqA protein. Plasmid 61:141–150. https://doi.org/10.1016/j.plasmid.2009.02.004

Waldron DE, Owen P, Dorman CJ (2002) Competitive interaction of the OxyR DNA-binding protein and the dam methylase at the antigen 43 gene regulatory region in *Escherichia coli*. Mol Microbiol 44:509–520

Wallecha A, Munster V, Correnti J, Chan T, van der Woude M (2002) Dam- and OxyR-dependent phase variation of agn43: essential elements and evidence for a new role of DNA methylation. J Bacteriol 184:3338–3347

Wallecha A, Correnti J, Munster V, van der Woude M (2003) Phase variation of Ag43 is independent of the oxidation state of OxyR. J Bacteriol 185:2203–2209

Weiser JN, Williams A, Moxon ER (1990) Phase-variable lipopolysaccharide structures enhance the invasive capacity of *Haemophilus influenzae*. Infect Immun 58:3455–3457

Welsh KM, Lu AL, Clark S, Modrich P (1987) Isolation and characterization of the *Escherichia coli* mutH gene product. J Biol Chem 262:15624–15629

Wright R, Stephens C, Zweiger G, Shapiro L, Alley MR (1996) *Caulobacter* Lon protease has a critical role in cell-cycle control of DNA methylation. Genes Dev 10:1532–1542

Wright R, Stephens C, Shapiro L (1997) The CcrM DNA methyltransferase is widespread in the alpha subdivision of proteobacteria, and its essential functions are conserved in *Rhizobium meliloti* and *Caulobacter crescentus*. J Bacteriol 179:5869–5877

Xu S-Y, Corvaglia AR, Chan S-H, Zheng Y, Linder P (2011) A type IV modification-dependent restriction enzyme SauUSI from *Staphylococcus aureus* subsp. aureus USA300. Nucleic Acids Res 39:5597–5610. https://doi.org/10.1093/nar/gkr098

Zweiger G, Marczynski G, Shapiro L (1994) A *Caulobacter* DNA methyltransferase that functions only in the predivisional cell. J Mol Biol 235:472–485. https://doi.org/10.1006/jmbi.1994.1007

# DNA Methylation in Eukaryotes: Regulation and Function

# 34

Hans Helmut Niller, Anett Demcsák, and Janos Minarovits

## Contents

1   Introduction .................................................................... 510
2   DNA Methylation in Eukaryotes: Basic Facts ................................. 513
   2.1   DNMT1: The Maintenance DNA-(Cytosine-C5)-Methyltransferase in
      Human Cells .............................................................. 513
   2.2   "De Novo" Methyltransferases in Human Cells: DNMT3A and DNMT3B ........ 517
   2.3   Active DNA Demethylation and Replication-Coupled Passive DNA
      Demethylation ............................................................ 519
3   Regulation of DNA Methylation in Eukaryotes ................................. 521
   3.1   Transcriptional Regulation of Human DNA-(Cytosine-C5)-Methyltransferase
      Genes .................................................................... 522
   3.2   Expression of Human DNA-(Cytosine-C5)-Methyltransferase Genes:
      Posttranscriptional Regulation ........................................... 523
4   DNA Methylation in Mammals: Biological Functions ........................... 526
   4.1   DNA Methylation as a Regulator of Chromatin Structure and Promoter
      Activity ................................................................. 528
   4.2   The Role of DNA Methylation in Genomic Imprinting ........................ 531
   4.3   Transposon Silencing by DNA Methylation ................................. 533
   4.4   X Chromosome Inactivation ............................................... 534
   4.5   DNA Methylation as a Regulatory Mechanism of Embryogenesis in Mammals ... 534
   4.6   DNA Methylation as a Regulator of Cell Differentiation ................... 537
   4.7   The Role of DNA Methylation in the Preservation of Genome Stability ...... 539
   4.8   Regulation of Alternative Splicing by Gene Body Methylation .............. 541
   4.9   The Role of DNA Methylation in Brain Function ........................... 542

H. H. Niller
Institute of Medical Microbiology and Hygiene, University of Regensburg, Regensburg, Germany
e-mail: Hans-Helmut.Niller@ukr.de

A. Demcsák · J. Minarovits (✉)
Department of Oral Biology and Experimental Dental Research, Faculty of Dentistry, University of Szeged, Szeged, Hungary
e-mail: demi.anett@gmail.com; minimicrobi@hotmail.com

© Springer International Publishing AG, part of Springer Nature 2018                    509
T. Krell (ed.), *Cellular Ecophysiology of Microbe: Hydrocarbon and Lipid Interactions*,
Handbook of Hydrocarbon and Lipid Microbiology, https://doi.org/10.1007/978-3-319-50542-8_24

5   Research Needs ..................................................................................... 550
    5.1   Philosophical Questions ............................................................. 550
    5.2   Sequence- or Locus-Specific Regulation of Methylation ............................ 551
    5.3   Patho-Epigenetics of Disease and Perspectives for Medical Treatment ............. 552
References ............................................................................................ 552

**Abstract**

In this chapter we focus on the regulation and function of DNA methylation in mammals and especially in humans. We describe the main features of the enzymatic machinery generating 5-methylcytosine (5mC) that functions as an epigenetic mark in mammalian cells, and outline the active and passive mechanisms that can remove this reversible modification of DNA. We briefly introduce the characteristics of "maintenance" and "de novo" DNA-(cytosine-C5)-methyltransferases (DNMTs) and overview how their expression is regulated at the transcriptional, posttranscriptional, and posttranslational level. The interacting partners and chromatin marks involved in the targeting of DNMTs to the replication foci during S phase or to various chromatin domains during other phases of the cell cycle are also discussed. The enzymatic functions of DNMTs and their interactions with cellular macromolecules are involved in a series of cellular processes, some of them vital for mammals. Thus, DNA methylation has a role in the regulation of chromatin structure and promoter activity. It may silence the promoters of imprinted genes showing monoallelic expression as well as the promoters of transposons, and contributes to gene silencing on the inactive X chromosome, too. There are genome-wide demethylation and remethylation events during embryogenesis suggesting a regulatory role for DNA methylation in developmental processes, and both cytosine methylation and the active removal of 5mC from DNA is involved in the control of cell differentiation. DNA methylation plays a role in the preservation of genomic stability and gene body methylation affects the inclusion of certain exons into mature mRNA molecules by affecting – indirectly – the splicing of primary transcripts. Epigenetic regulatory mechanisms, including DNA methylation, are at the forefront of brain research these days. For this reason we outlined some of the most interesting results of this exciting new field in a separate subsection.

# 1    Introduction

DNA methyltransferase enzymes encoded by bacterial genomes catalyze the transfer of a methyl group from the cofactor S-adenosyl-L-methionine (SAM) to one of the nucleotides within the target DNA sequence, generating a methylated base and S-adenosyl-L-homocysteine. There are two major classes of DNA methyltransferases (MTases): endocyclic MTases modify the carbon 5 (C5) position of the cytosine ring, whereas exocyclic MTases methylate exocyclic nitrogens, either the N4 position of cytosine or the N6 position of adenine (Posfai et al. 1989; Bheemanaik et al. 2006; Weigele and Raleigh 2016). These covalent modifications "mark" DNA sequences

**Table 1**  Modified bases occurring in bacterial and mammalian genomes

| Domain | Oxidation | Modified bases |
|---|---|---|
| **Bacteria** | | |
| | | N6-methyladenine (6 mA) |
| | | N4-methylcytosine (4mC) |
| | | 5-methylcytosine (5mC) |
| **Mammals** | | |
| | | N6-methyladenine (6 mA) |
| | | 5-methylcytosine (5mC) |
| | Oxidation products of 5mC | |
| | | 5-hydroxymethylcytosine (5hmC) |
| | | 5-formylcytosine (5fC) |
| | | 5-carboxylcytosine (5caC) |

without altering the specificity of base pairing. Modified bases, including N4-methylcytosine, 5-methylcytosine (5mC), and N6-methyladenine, were detected in a series of bacterial DNAs (Ehrlich et al. 1987; Blow et al. 2016) (Table 1). Modifications of bacterial genomes play an important role in as diverse biological functions as methyl-dependent mismatch repair, membrane binding of the bacterial chromosome, regulation of DNA replication, regulation of gene expression, and discrimination between self and foreign DNA molecules (Kramer et al. 1984; Ogden et al. 1988; Wilson 1988; Waldminghaus et al. 2012; Makarova et al. 2013; Adhikari and Curtis 2016; Cohen et al. 2016).

The expression of bacterial DNA MTases that modify the host cell genome is frequently accompanied by the expression of a sequence-specific endo-deoxyribonuclease (restriction endonuclease) recognizing the same DNA sequence as the MTase. Such MTase/restriction endonuclease pairs may protect bacterial cells from invading foreign bacteriophage or plasmid DNA molecules that are either unmethylated or methylated at different recognition sites (Ishikawa et al. 2010). "Solitary" or "orphan" DNA methyltransferases that do not have a corresponding restriction endonuclease pair also modify bacterial genomes (reviewed by Casadesus 2016). In addition to the modified bases occurring in bacterial genomes, DNA genomes of bacteriophages may contain a series of additional modified purines and pyrimidines, possibly to avoid recognition and prevent degradation by host-encoded restriction endonucleases (Shabbir et al. 2016; Weigele and Raleigh 2016).

In contrast to bacteria, mammalian genomes typically do not contain detectable levels of the modified bases N4-methylcytosine and N6-methyladenine. N6-methyladenine was detected, however, in other organisms belonging to the domain *Eukarya*: it was enriched at active transcription start sites in the genome of *Chlamydomonas reinhardtii*, a flagellated protozoan, and it was also observed in the ciliate protozoa *Oxytricha fallax*, *Paramecium aurelia*, *Stylonichia mytilius*, and *Tetrahymena pyriformis* (Fu et al. 2015b; reviewed by Wion and Casadesus 2006). N6-methyladenine was also detected in the fungus *Penicillium chrysogenum* and in the genomes of the nematode *Caenorhabditis elegans* and the fruit fly *Drosophila melanogaster* (Rogers et al. 1986; Greer et al. 2015; Zhang et al. 2015a; reviewed by Heyn and Esteller 2015; Meyer and Jaffrey 2016). Moreover, in a recent study, Liu et al. detected N6-methyladenine, with the help of ultrahigh performance liquid chromatography coupled to triple quadrupole mass spectrometry (UHPLC-QQQ-MS/MS), during the early embryogenesis of vertebrate species, zebrafish (*Danio rerio*) and pig (*Sus domesticus*). Certain repetitive sequences of these vertebrate genomes were especially enriched, temporarily, in N6-methyladenine (Liu et al. 2016). Similar results were reported by Koziol et al. who also used UHPLC-MS/MS as well as DNA immunoprecipitation with N-6-methyl-deoxyadenosine specific antibodies followed by high throughput sequencing for the analysis of *Xenopus laevis* (African clawed frog) and *Mus musculus* (house mouse) DNA samples. Using dot blot, N-6-methyl-deoxyadenosine, the nucleoside corresponding to N6-methyladenine, was detected in the DNA of a human cell line (293T) as well (Koziol et al. 2016).

The enzymatic machinery involved in the deposition of N6-methyladenine marks on eukaryote genomes remains to be elucidated. It was observed, however, that during the embryogenesis of *Drosophila melanogaster* a DNA N6-methyladenine demethylase (DMAD) removed the methylation mark, especially from transposons, suggesting a role for DMAD in transposon suppression (Zhang et al. 2015a).

Notwithstanding the observations documenting the presence of N6-methyladenine in eukaryote genomes, the vast majority of data accumulated so far indicated that the most abundant modified base in mammalian cells is 5mC. Discovered as a component of mammalian DNA by Wyatt in 1951, 5mC occurs predominantly in the dinucleotide CpG (Wyatt 1951; reviewed by Li and Zhang 2014). Oxidation products of 5mC, including 5-hydroxymethylcytosine (5hmC), 5-formylcytosine (5fC), and 5-carboxyl-cytosine (5caC), can also be detected in mammalian genomes (reviewed by Rasmussen and Helin 2016) (Table 1). In this chapter we wish to focus primarily on the regulation and function of CpG-methylation in mammals and especially in humans. DNA methylation in other eukaryotic taxa including protists, fungi, plants, invertebrates, and vertebrates other than mammals was covered in several recent papers and reviews (Wion and Casadesus 2006; Walsh et al. 2010; He et al. 2011; Ponts et al. 2013; Vu et al. 2013; Dabe et al. 2015; Jeon et al. 2015; Huang et al. 2016; Taskin et al. 2016; Vidalis et al. 2016; Zabet et al. 2017).

Prokaryotic DNA MTases modify, with a few exceptions, all of their target sequences, and it was suggested that local hypomethylation in bacterial genomes may be due to the competition of site-specific DNA-binding proteins with DNA MTases (Ardissone et al. 2016). In contrast, in large-genome eukaryotes, including mammals, the methylation pattern of the genome is typically discontinuous and changes during developmental processes, organogenesis, and cell differentiation (Kunnath and Locker 1982; Bird 1986; Frank et al. 1990; Frank et al. 1991; Deaton and Bird 2011; Yu et al. 2011; Hansen et al. 2014; Bestor et al. 2015; Keil and Vezina 2015; Farlik et al. 2016; Li et al. 2016; Zhou et al. 2017). In mammals, a typical feature of the genome is the presence of predominantly unmethylated, short interspersed sequences called CpG-islands that are regularly associated with promoters. CpG-islands have a high GC content and they are enriched in CpG-dinucleotides compared to the average genomic pattern (Deaton and Bird 2011; Jones 2012).

## 2 DNA Methylation in Eukaryotes: Basic Facts

In eukaryotic organisms the best characterized DNA methyltransferase enzymes typically methylate the C5-position of cytosines. In mammalian and human cells, methylcytosine (5mC) can be detected predominantly within CpG-dinucleotides, and CpG-methylation in the regulatory regions of promoters frequently results in transcriptional silencing (Li and Zhang 2014). In humans, there are several DNA-(cytosine-C5)-methyltransferases that modify DNA sequences: DNMT1, a "maintenance" DNA MTase differs in protein sequence and substrate preference from the "de novo" DNA MTases DNMT3A and DNMT3B (reviewed by Jin and Robertson 2013). It was suggested that these enzymes most probably had an independent origin and their genes derived from prokaryotic DNA methyltransferase sequences (Jurkowski and Jeltsch 2011). In addition, DNMT3L, a protein related to the "de novo" DNA MTases, but lacking enzymatic activity, forms complexes with and enhances the activity of DNMT3A and DNMT3B (Jin and Robertson 2013). DNMT2, a protein with sequence similarities to DNA-(cytosine-C5)-methyltransferases, is not involved in DNA methylation in mammals: DNMT2 methylates aspartic acid transfer RNA (tRNA$^{Asp}$) (Goll et al. 2006; Jurkowski and Jeltsch 2011) (Table 2). DNA methylation is reversible: the methyl group can be removed from the C5-position of 5mC either by an active enzymatic process or by a passive mechanism, when the recruitment of DNMT1 is inefficient or its activity is inhibited during successive rounds of DNA replication and cell division (reviewed by Smith and Meissner 2013b; Wu and Zhang 2014).

## 2.1 DNMT1: The Maintenance DNA-(Cytosine-C5)-Methyltransferase in Human Cells

The C-terminal catalytic domain of DNMT1 and the corresponding domain of its mouse homolog, Dnmt1, share several conservative motifs with bacterial DNA-(cytosine-C5)-methyltransferases (reviewed by Posfai et al. 1989; Bestor 2000; Robertson 2001; Hermann et al. 2004; Zhang et al. 2015b). DNMT1 binds with high affinity to hemimethylated DNA duplexes that contain a methylated and a complementary, unmethylated DNA strand (Bestor 2000). Such hemimethylated DNA duplexes are generated during the semiconservative replication of methylated parental DNA molecules. At the replication fork, DNMT1 copies the methylation

**Table 2** DNA-(cytosine-C5)-methyltransferases and related proteins in human cells

| Protein | Main activity |
| --- | --- |
| DNMT1 | Maintenance methylation |
| DNMT3A | De novo methylation |
| DNMT3B | De novo methylation |
| DNMT3L | Regulation of DNA methylation |
| DNMT2 | Methylation of tRNA$^{Asp}$ |

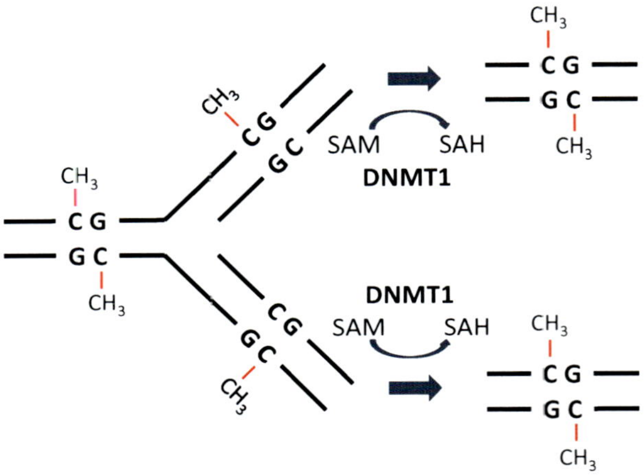

**Fig. 1** **The maintenance DNA methyltransferase function of DNMT1.** The schematic view of a replication fork is shown with two hemimethylated DNA duplexes containing a methylated CpG-dinucleotide (parental strand) and a complementary unmethylated CpG-dinucleotide (daughter strand), each. DNMT1 binds to hemimethylated DNA with high affinity and transfers a methyl group from the methyl donor S-adenosyl-L-methionine (SAM) to the C5-position of the unmethylated cytosine residue. In parallel, SAM is converted to S-adenosyl-L-homocysteine (SAH)

pattern of the parental DNA strand to the initially unmethylated daughter strand (reviewed by Robertson 2001) (Fig. 1). As a prelude to the methyl group transfer, the interaction of the mouse Dnmt1 with its double stranded DNA substrate induces "base flipping," i.e., rotation out of the target cytosine from the DNA backbone (Song et al. 2012). Base eversion permits embedding of the target cytosine into a hydrophobic pocket of the catalytic domain. This structural change is followed by the transfer of a methyl group from the methyl donor S-adenosyl-L-methionine (SAM) to the C5-position of the cytosine residue (Du et al. 2016; reviewed by Bestor 2000; Jeltsch and Jurkowska 2016). A cysteine located to a conservative Pro-Cys dipeptide plays an indispensible role in the catalytic reaction: it forms a transient covalent complex with the C6 atom of the target cytosine, resulting in the activation, i.e., an increase of the negative charge density, of the neighboring C5 atom. A nucleophilic attack of the activated C5 atom on the methyl group of cofactor SAM results in methyl group transfer to the C5-position, followed by the release of Dnmt1 (reviewed by Cheng and Blumenthal 2011; Jurkowska and Jeltsch 2016). In parallel, SAM is converted to S-adenosyl-L-homocysteine (SAH) (Fig. 1). Because CpG-methylation affects promoter activity, the "maintenance" methylase function of DNMT1 contributes to the transmission of gene expression patterns from cell generation to cell generation (epigenetic memory) (Jones and Takai 2001; Bird 2002). A model for gene activity and inactivity, based on cytosine methylation in DNA, was proposed originally by Riggs and independently by Holliday and Pugh (Holliday and Pugh 1975; Riggs 1975).

Unlike the C-terminal domain that resembles bacterial methyltransferases, the larger N-terminal domain of DNMT1 is of unknown origin and it has no significant amino acid sequence homology with the corresponding bacterial enzymes (Bestor 2000). Comparison of the mouse Dnmt1 protein with the amino acid sequences of other eukaryotic methyltransferases revealed that the N-terminal domain itself is composed of two parts, A and B (Margot et al. 2000). Thus, the mouse *Dnmt* gene, and the ancestral methyltransferase gene of *Metazoa*, was possibly generated by two fusion events: joining of the sequences coding for part A and B of the N-terminus was followed by the fusion to a putative gene coding for the C-terminus.

The N-terminal domain has multiple functions: it is necessary for the transport of DNMT1 from the cytoplasm to the nucleus, and it also plays an important role in the targeting of the enzyme to the replication foci. Interaction of the N-terminal regulatory domain with PCNA (proliferating cell nuclear antigen) and UHRF1 (ubiquitin-like protein containing PHD and RING finger domains 1) may guide DNMT1 to the replication forks. Similarly to DNMT1, its interacting partner, UHRF1, also induces base flipping: in this case a methylated cytosine is everted on the parental, methylated DNA strand (Hashimoto et al. 2008; Cheng and Blumenthal 2011). Based on molecular dynamics simulation, Bianchi and Zangi suggested that flipping out of the methylated cytosine by UHRF1, a protein which has no enzymatic activity, may facilitate the eversion of the unmethylated cytosine targeted by DNMT1 on the opposite strand (Bianchi and Zangi 2014). Thus, it is plausible that maintenance methylation at hemimethylated DNA sequences by DNMT1 proceeds via a dual base flipping mechanism (Bianchi and Zangi 2014). According to a potential scenario, UHRF1 interacts with the methylated cytosine and guides DNMT1 to the replication fork. As a next step, or simultaneously, assisted with the extra-helical conformation of the methylated cytosine, DNMT1 may induce flipping out of the unmethylated cytosine, followed by the transfer of the methyl group from SAM to the C5-position of its target (Fig. 2). In addition to PCNA and UHRF1, interaction with transcription factors may also help to deposit DNMT1 to the replication foci (Iida et al. 2002; Hervouet et al. 2010; Hervouet et al. 2012).

The N-terminal domain has a dual role in the regulation of the C-terminal catalytic domain. On the one hand, it is indispensable for the activation of the enzyme; on the other hand its interaction with the DNA binding pocket of the catalytic domain blocks substrate binding (Syeda et al. 2011; reviewed by Qin et al. 2011; Jeltsch and Jurkowska 2016). Thus, the N-terminal sequence mediating the latter autoinhibitory function should be sterically rearranged – possibly with the help of UHRF1 that interacts with DNMT1 – for the activation of the enzyme (reviewed by Qin et al. 2011; Jeltsch and Jurkowska 2016).

In addition to its enzymatic function, DNMT1 affects a series of other cellular processes by the interactions of its N-terminal regulatory domain with numerous nuclear proteins (Qin et al. 2011). Changing a critical catalytic cysteine residue to serine in the C-terminal domain of Dnmt1 abolished, however, several important biological phenomena attributed to the enzyme, suggesting that the enzymatic activity is required for in vitro differentiation of embryonic stem cells, retrotransposon suppression, and proper localization of Dnmt1 (Damelin and Bestor

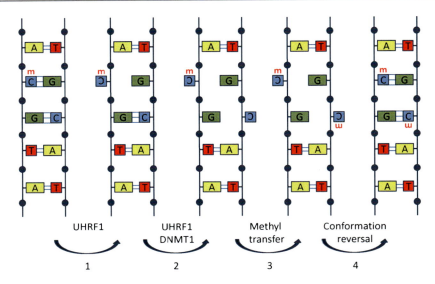

**Fig. 2 Dual base flipping: a potential mechanism for maintenance DNA methylation**. A scenario proposed by Bianchi and Zangi is illustrated here (Bianchi and Zangi 2014). The schematic view of a stretch of a hemimethylated DNA duplex is shown. As a first step (*1*), UHRF1 (Ubiquitin-like, containing PHD and RING finger domains) a multifunctional protein that specifically binds hemimethylated DNA sequences at replication forks, interacts with the methylated cytosine (shown as $^m$C on the figure), induces the breakage of the hydrogen bonds between $^m$C (located to the parental strand) and the complementary guanine (G, located to the daughter strand), and causes flipping out of $^m$C. Via its SET- and RING-associated (SRA) domain, UHRF1 associates with the replication foci targeting sequence (RFTS) of DNMT1 (Berkyurek et al. 2014) and guides DNMT1 to the replication fork, where – shown on the figure as a separate second step (*2*) – DNMT1 induces flipping out of the unmethylated cytosine (C, located to the daughter strand). This structural change is possibly facilitated by the extra-helical conformation of the methylated cytosine contacted by UHRF1 on the parental DNA strand. Next (step *3*), DNMT1 catalyzes the transfer of a methyl group from SAM (S-adenosyl-L-methionine) to the C5-position of its unmethylated, flipped-out target cytosine. Finally (step *4*), both methylated cytosines rotate back and the hydrogen bonds are reestablished between the intramolecular base pairs

2007). The very same mutation resulted in developmental arrest and embryonal lethality in *Dnmt1$^{ps/ps}$* mice carrying the catalytic-defective allele (Takebayashi et al. 2007). The embryos died shortly after gastrulation, like the Dnmt1-null mutant (*Dnmt$^{c/c}$*) embryos and compound heterozygous (*Dnmt1$^{c/ps}$*) embryos that expressed the mutant, inactive Dnmt1 protein only. In parallel, there was a significant decrease in the methyl-CpG content of repetitive DNA sequences of mutant embryos, compared to those of their wild-type counterparts (Takebayashi et al. 2007). These data suggested a vital role for the catalytic activity of the maintenance DNA-(cytosine-C5)-methyltransferase in mouse embryogenesis.

Although the purified human DNMT1 was able to bind and methylate hemimethylated oligonucleotide duplexes in vitro, it is apparent that within the cell nucleus the substrate binding of DNMT1 is influenced by a series of interacting protein partners (Pradhan et al. 1999; Qin et al. 2011). In addition to PCNA and

**Table 3**  Major functions of the DNMT1 N-terminal domain

| Function | Note |
| --- | --- |
| Transport of DNMT1 to the nucleus | NLS-mediated |
| Targeting of DNMT1 to the replication fork | Guided by PCNA and UHRF1 |
| Formation of multiprotein repressor complexes | Interactions with MBPs, de novo DNMTs, PcG proteins, chromatin remodeling ATPases, tumor suppressors |

UHRF1, methyl-CpG binding proteins, histone deacetylases, histone methyltransferases depositing heterochromatic marks, de novo DNMTs, Polycomb group (PcG) proteins, chromatin remodeling ATPases, a subset of transcription factors, tumor suppressor proteins, and other nuclear proteins may bind to DNMT1 (reviewed by Qin et al. 2011). These proteins frequently form multiprotein repressor complexes that silence promoters and maintain a compact chromatin conformation that prevents transcription factor and RNA polymerase binding (Table 3) (see Sect. 4.1).

In vitro, the purified human DNMT1 protein catalyzed the transfer of methyl groups not only to hemimethylated but also to unmethylated oligonucleotide duplexes, although it showed a strong, 7–21-fold preference for hemimethylated versus unmethylated substrates (Pradhan et al. 1999). It was also observed, that DNMT1 cooperated with the de novo DNA methyltransferases DNMT3A and DNMT3B (see Sect. 2.2) to ensure efficient methylation of unmethylated DNA duplexes, whereas DNMT3A and DNMT3B may contribute to maintenance methylation by acting on CpG-sites "missed" by DNMT1 (Fatemi et al. 2002; Liang et al. 2002; Jones and Liang 2009). Thus, DNMT1 may function both as maintenance as well as a "de novo" methyltransferase. Furthermore, DNMT1 was able to bind, in addition to CpG-dinucleotides, to non-CpG sequences as well, implicating that it may have a non-CpG methylase activity in vivo (Pradhan et al. 1999). Xu et al. speculated that induction of de novo methylation by DNMT1 by certain conditions may play a role in development or disease initiation (Xu et al. 2010).

## 2.2  "De Novo" Methyltransferases in Human Cells: DNMT3A and DNMT3B

DNMT3A and DNMT3B are expressed at high levels during embryogenesis and germ cell development, but they are typically downregulated in adult somatic cells (reviewed by Dan and Chen 2016). Unlike DNMT1, these enzymes do not discriminate between hemimethylated and unmethylated double stranded DNA substrates. They methylate, in addition to CpG-dinucleotides, non-CpG sites as well. Non-CpG-methylation was frequently observed in embryonic stem cells, and it was also detected in diverse human tissues and organ systems as well as in plants (Lister et al. 2009; Becker et al. 2011; Schultz et al. 2015). DNMT3A has two major isoforms with DNA-(cytosine-C5)-methyltransferase activity whereas there are both enzymatically active and inactive members among the more than 20 isoforms of DNMT3B (reviewed by Choi et al. 2011; Dan and Chen 2016).

DNMT3L (DNMT3-like), a member of the DNMT3 family, does not function as a DNA MTase but may act as a regulator of DNA methylation: two DNMT3L molecules form tetramers with two DNMT3A or two DNMT3B molecules, respectively. Interaction with DNMT3L within such a butterfly-like structure stimulates the DNA methylation activity of the de novo methyltransferases (Suetake et al. 2004). DNMT3A preferentially methylates CpG-dinucleotides located about one helical turn apart, and DNMT3A/DNMT3L complexes and their murine counterparts are involved in the establishment of methylation patterns during embryogenesis (reviewed by Xu et al. 2010; Chen and Chan 2014). The murine Dnmt3L has a dual function. By interacting with both Dnmt3a and the unmethylated lysine 4 of histone H3, it targets the methyltransferase enzyme to heterochromatic regions and – in parallel – it also activates the enzyme (Jia et al. 2007). In addition, the N-terminal regulatory domain of the murine Dnmt3a enzyme also binds to the histone 3 lysine 36 trimethylation mark (histone H3K36me3) and guides the enzyme to repressed nuclear regions (Dhayalan et al. 2010).

Transfection of human DNMT3 isoform constructs into HEK 293T cells revealed that DNMT3A1 preferentially targeted active genes marked with H3K4me3, whereas DNMT3B1 methylated inactive genes marked with H3K27me3 (Choi et al. 2011). In addition, the DNMT3B1, DNMT3B2, and DNMTΔ3B isoforms increased the methylation of LINE1 (Long Interspersed Nuclear Element 1) and Satellite-α (Sat-α) repeats, whereas the DNMT3A isoforms were ineffective or less effective (Choi et al. 2011). Even the DNMT3B isoforms which do not have catalytic activity may play a role in DNA methylation: Duymich et al. observed that such molecules recruited DNMT3A to gene bodies, and increased its activity in somatic cells (Duymich et al. 2016).

In murine somatic tissues, the transcriptional repressor E2F6 may recruit Dnmt3b to a set of developmental regulator genes such as *Hoxa11* and *Hoxa13* involved in the establishment of the thoracic and lumbar vertebral identity as well as germline-specific genes including *Tex11* expressed only in male germ cells, the RNA helicase *Ddx4*, and others (Velasco et al. 2010). Thus, Dnmt3b-mediated methylation at E2F6-marked promoters may ensure coordinated transcriptional silencing of a gene battery (reviewed by Walton et al. 2014). Expression of *Oct-4*, a gene coding for octamer-binding transcription factor 4 that plays an important role in the renewal and pluripotency of embryonic stem cells (ES cells), is also silenced by Dnmt3b, in collaboration with Dnmt3a during ES cell differentiation: it was observed that de novo methylation initiated at two sites upstream from the murine *Oct-4* promoter was later spread to the proximal region of the promoter (Athanasiadou et al. 2010). Lsh (lymphoid-specific helicase), a member of the SNF2 family of chromatin-remodeling ATPases, aided spreading of CpG-methylation (Athanasiadou et al. 2010). In mouse cells, Lsh collaborated with de novo DNMTs in the methylation of repetitive elements as well (reviewed by Huang et al. 2004). These included satellite repeats located to pericentromeric regions as well as retroviral LTR elements. Based on experiments with $Lsh-/-$ mouse embryo fibroblasts, Termanis et al. suggested that Lsh may interact with de novo DNMTs deposited to distinct chromatin domains and initiate local chromatin remodeling necessary for efficient de novo methylation (Termanis et al. 2016).

Zinc finger proteins containing a Krüppel-associated box (KRAB) domain may also attract de novo DNMTs to distinct DNA sequences. KRAB containing repressor proteins associate with KAP1 (KRAB-associated protein 1), a regulator of development and differentiation. Quenneville et al. observed that in embryonic stem cells the KRAB/KAP1 complex induced heterochromatin formation and de novo DNA methylation in the vicinity of KAP1 binding sites. In ES cells, CpG-methylation was spreading to the nearby CpG-islands, and the pattern of CpG-island methylation was maintained in differentiated hepatocytes (Quenneville et al. 2012).

Distinct nuclear proteins binding to characteristic sequence motifs may block de novo methylation. In the female germline, a large number of CpG-islands acquired methylation during oogenesis, whereas a set of CpG-islands containing the sequence motif CGCGC, a binding site for the transcription factors E2f1 and E2f2 involved in chromatin remodeling, resisted de novo methylation (Saadeh and Schulz 2014). Saadeh and Schulz suggested that binding of E2f1 and E2f2 caused nucleosome depletion by recruiting a nucleosome remodeling complex; such a chromatin alteration could possibly hinder binding of Dnmt3a and Dnmt3b that associate with a subset of intact nucleosomes in mammalian nuclei (Jeong et al. 2009; Saadeh and Schulz 2014).

R-loops (R standing for RNA) formed at a subset of actively transcribed promoters located in CpG-islands may also hinder de novo methylation (Ginno et al. 2012; Ginno et al. 2013). R-loops are preferably formed during the formation of primary transcripts at genes characterized by "GC skew," i.e., a significant strand asymmetry in the distribution of guanines and cytosines, immediately downstream from the transcription start site. The transcribed G-rich RNA forms a stable hetero-duplex (RNA-DNA hybrid) with the C-rich template DNA strand, while forcing the nontemplate G-rich DNA strand into a largely single-stranded "looped" conformation. Ginno et al. detected R-loop formation at a significant number of human promoters and observed a reduced efficiency of DNMT3B1-mediated methylation upon R-loop formation (Ginno et al. 2012). They also described R-loop formation at the 3′-end of human genes, a phenomenon potentially involved in transcription termination (Ginno et al. 2013). In addition to DNMT3A and DNMT3B, the maintenance methylase DNMT1 may also perform de novo methylation (see Sect. 2.1). Fatemi et al. suggested that methylated or partially methylated DNA strands generated by DNMT3A may attract and activate DNMT1 to contribute to de novo methylation (Fatemi et al. 2002; Jeltsch 2006).

## 2.3    Active DNA Demethylation and Replication-Coupled Passive DNA Demethylation

Methylated DNA sequences in mammalian genomes may undergo local demethylation initiated by pioneer transcription factors that are capable to bind heterochromatic regions (Zaret et al. 2008; Serandour et al. 2011). In addition, general demethylation events affecting most of the 5mC-residues were also observed during mammalian development (reviewed by Clark 2015). These "global" DNA demethylation events occurring in preimplantation embryos and in primordial germ cells leave, however, distinct methylated loci unaffected: certain repetitive sequences as well as a series of imprinting control regions (ICRs, see 4.2), rare intragenic regions, and CpG-islands (see Sect. 4.1) are protected from demethylation (reviewed by Wu and Zhang 2014; Clark 2015; Li et al. 2016; Rasmussen and Helin 2016). Thus, in mouse embryos, the genome is transiently hypomethylated but not completely unmethylated, even in the inner cell mass of the preimplantation blastocyst where the lowest level of CpG-methylation was detected (Smith et al. 2012). Global hypomethylation was also observed in the human preimplantation embryo (Smith et al. 2014). Imprinting control regions and a set of CpG-island promoters were protected, however, from demethylation. A variable degree of hypomethylation was observed in species-specific repetitive elements (Smith et al. 2014).

In mammals, active DNA demethylation is mediated by the TET (ten-eleven translocation) family of dioxygenases (Tahiliani et al. 2009). These enzymes mediate both locus-specific and general reversal of DNA methylation by catalyzing the successive oxidation of 5mC to 5-hydroxymethylcytosine (5hmC), 5-formylcytosine (5fC), and 5-carboxylcytosine (5caC) (reviewed by Rasmussen and Helin 2016) (Fig. 3). The activity of TET1, TET2, and TET3 depends on the binding of cofactors, Fe(II) and 2-oxoglutarate (2-OG). The DNA binding CXXC zinc finger domain of TET1 and TET3 may preferentially bind CpG-rich DNA. TET2 that lacks this domain is targeted by its partner CXXC4, a zinc finger protein (reviewed by Tsai and Tainer 2013). After iterative oxidation of the methyl group by TET dioxygenases, 5fC or 5caC can be efficiently removed by thymine-DNA glycosylase (TDG), followed by the restoration of unmethylated cytosine by base excision repair (BER) (Wu and Zhang 2014). This type of active, enzymatic DNA demethylation utilizing dioxygenases and DNA repair enzymes may proceed independently of DNA replication.

It is also possible, however, that successive oxidation of 5mC is combined with passive demethylation, because oxidized cytosine bases (5hmC, 5fC, and 5caC) may interfere with the maintenance methylation machinery, resulting in the replication-dependent passive dilution of 5mC (Wu and Zhang 2014). Such a combined mechanism, i.e., "active modification (AM) followed by passive dilution" (AM-PD) may act during the erasure of paternal genome methylation in preimplantation embryos as well as in primordial germ cells (Wu and Zhang 2014). Alternatively, passive demethylation may also occur without the involvement of 5mC-oxidation, due to the inefficient recruitment of DNMT1 to the replication foci or due to the presence of DNMT1 inhibitors during successive rounds of DNA replication and cell division (reviewed by Smith and Meissner 2013b; Wu and Zhang 2014). In the zygote, a passive dilution mechanism that follows a slow kinetics may play a role in the 5mC depletion of the maternally derived genome (Hackett and Surani 2013). This phenomenon is possibly based on the exclusion of the oocyte-derived DNMT1o and UHRF1 from the DNA replication machinery (Seisenberger et al. 2013b; Wu and Zhang 2014).

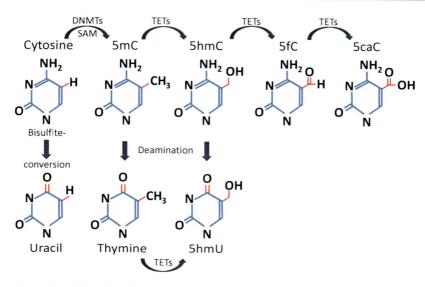

**Fig. 3 Active DNA demethylation in mammals**. In mammals, active demethylation of 5-methylcytosine (5mC) is mediated by members of the TET (ten-eleven translocation) family of dioxygenases (TET1, TET2, TET3). TET dioxygenases catalyze the successive oxidation of (5mC) to 5-hydroxymethylcytosine (5hmC), 5-formylcytosine (5fC), and 5-carboxylcytosine (5caC). 5fC or 5caC can be efficiently removed from DNA by thymine-DNA glycosylase (TDG), followed by the restoration of unmethylated cytosine by base excision repair (BER). Spontaneous deamination of cytosine yields uracil. As indicated on the figure, under experimental conditions, addition of sodium bisulfite to DNA, followed by alkaline treatment can convert cytosine, but not 5-methylcytosine (5mC) to uracil (Frommer et al. 1992). AID, activation induced deaminase, and other members of the AID/APOBEC family of cytidine deaminases, can also deaminate cytosine to uracil (reviewed by Fritz and Papavasiliou 2010; Rebhandl et al. 2015). Spontaneous deamination of 5mC yields thymine and ammonia. Deaminase enzymes can convert 5mC, but also 5hmC, to thymine and 5hmU, respectively, whereas TET enzymes can catalyze the conversion of thymine to 5hmU (Pfaffeneder et al. 2014)

## 3    Regulation of DNA Methylation in Eukaryotes

The methylation pattern of eukaryotic genomes depends on the expression levels of DNA-(cytosine-C5)-methyltransferase genes. In mammals, the transcriptional activity of these genes depends on the cell type, and the cytoplasmic level of the corresponding mRNAs is modulated by posttranscriptional regulatory mechanisms. After translation, targeting of DNA methyltransferase enzymes to various chromatin domains represents another level of DNA methylation control in mammalian cells. Targeting is affected by posttranslational modifications of DNMTs and by the interacting protein or RNA partners of these enzymes. In addition, certain activating or repressing chromatin marks recognized by DNA methyltransferases may also influence their localization to distinct chromatin domains.

## 3.1    Transcriptional Regulation of Human DNA-(Cytosine-C5)-Methyltransferase Genes

Expression of the human *DNMT1* gene or its murine counterpart is modulated by the binding of nuclear proteins to the regulatory sequences of these genes. Sp1 and Sp3, members of the specificity protein (Sp) family of transcription factors, bound to a *cis* element in the 5′-flanking region of the mouse *Dnmt1* promoter and stimulated its activity; Sp1 upregulated *DNMT1* transcription in human cell lines as well (Kishikawa et al. 2002; Lin et al. 2010; reviewed by Lin and Wang 2014). *DNMT1* transcription is also affected by BRCA1, a protein encoded by *Breast cancer-associated gene 1*. Shukla et al. observed that BRCA1, a pleiotropic regulator and breast tumor suppressor bound to the regulatory sequences of *DNMT1* at a potential OCT1 (octamer binding protein 1) site, induced activating histone modifications in the region, and upregulated *DNMT1* transcription (Shukla et al. 2010).

The activity of the human *DNMT3A* and *DNMT3B* genes coding for de novo methyltransferases is controlled by multiple promoters: there are three TATA-less promoters for *DNMT3A* and two TATA-less promoters for *DNMT3B* (Yanagisawa et al. 2002). Binding sites for the ubiquitously expressed Sp1 and Sp3 transcription factors were identified in the regulatory sequences of the *DNMT3A* third promoter and in the control regions of the first and second promoter of *DNMT3B* (Jinawath et al. 2005). Overexpression of Sp3 increased transcription of both *DNMT3A* and *DNMT3B* in a transformed human cell line, suggesting a role for Sp3 in the control of de novo MTase genes (Jinawath et al. 2005).

The regulatory region of the human *DNMT1* promoter and its murine counterpart contain binding sites for E2F1, a member of the E2F family of transcription factors involved in control of the cell cycle (McCabe et al. 2005). Binding of E2F1 to these recognition sequences activated the transcription of the maintenance methyltransferase gene both in murine and human cell lines (McCabe et al. 2005). Because E2F1 is released from its complex formed with the tumor suppressor protein RB in the late G1 phase of the cell cycle, these data indicate that the gene for maintenance DNA MTase in human and murine cells is co-regulated with a series of other genes involved in in DNA synthesis and S-phase progression.

A binding site for E2F1 was located to the regulatory region of the *DNMT3A* gene as well (Tang et al. 2012). Complexing of the tumor suppressor protein RB with E2F1 bound to this site resulted, however, in silencing of the *DNMT3A* promoter (Tang et al. 2012). In contrast, binding of WT1 (Wilms' tumor 1), a developmental master regulator, to the upstream sequences of the first and second promoter of *DNMT3A* activated transcription (Szemes et al. 2013).

In addition to RB, the tumor suppressor protein p53 also decreased *DNMT1* transcription, by binding to a region in exon 1 of the gene (Peterson et al. 2003; Lin et al. 2010). The opposite situation, repression of *p53* transcription by Dnmt1, was observed, however, in the developing pancreas of mice; Dnmt1 bound to the *p53* regulatory region in pancreatic progenitor cells (Georgia et al. 2013). Dnmt1-mediated suppression of the pro-apoptotic *p53* gene contributed to progenitor cell survival during pancreatic organogenesis.

MDM2, a human protein homologous to the murine Mdm2 (murine double minute 2) E3 ubiquitin-protein ligase, ubiquitinated p53 and also promoted proteasomal degradation of RB (Moll and Petrenko 2003; Sdek et al. 2005). Accordingly, MDM2 relieved the RB-mediated suppression of the *DNMT3A* promoter in a human lung cancer cell line (Tang et al. 2012). MDM2 also induced the expression of *DNMT3B* gene by destabilizing the repressor protein FOXO3a bound to the regulatory region of the gene (Yang et al. 2014).

## 3.2 Expression of Human DNA-(Cytosine-C5)-Methyltransferase Genes: Posttranscriptional Regulation

Processing of the primary transcripts may result in splice variant mRNAs encoding DNA MTase isoforms, whereas microRNAs attaching to mRNA molecules may facilitate the breakdown of their mRNA targets. In addition, microRNAs binding to mRNAs encoding transcription factors involved in the upregulation of MTase gene expression may affect DNMT transcript and protein levels indirectly. HuR, an RNA binding regulatory protein, may also modulate stability and translation of DNMT mRNAs.

### 3.2.1 Alternative Splicing

In human and mouse cells either the oocyte specific or the somatic cell-specific isoform of the maintenance DNA methyltransferase was detected. Due to the incorporation of a somatic cell-specific exon 1 into the mRNA, a longer protein was expressed in somatic cells, whereas the utilization of a downstream initiation codon located to the oocyte-specific exon 1 generated an N-terminal truncated enzyme (Dnmt1o in mice and DNMT1o in humans, respectively) that was expressed in oocytes and eight-cell embryos (Ratnam et al. 2002; Petrussa et al. 2014; reviewed by Dan and Chen 2016). As described above (see Sect. 2.1), both of the major isoforms of the de novo methylase DNMT3A possess enzymatic activity, whereas there are enzymatically active as well as inactive members among the more than 20 isoforms of DNMT3B (Ostler et al. 2007; reviewed by Choi et al. 2011; Dan and Chen 2016). A study by Gordon et al. revealed that inactive DNMT3B isoforms bound to the catalytically active DNMT3A and DNMT3B isoforms (Gordon et al. 2013). Similarly to the enzymatically inactive DNMT3L that interacted with and stimulated the activity of de novo MTases, binding of the inactive DNMT3B3 isoform to DNMT3A2 resulted in a modest increase in the activity of the latter. DNMT3B3 counteracted, however, the stimulatory effect exerted on DNMT3A2 activity by purified DNMT3L (Gordon et al. 2013). Interaction of DNMT3B4, another inactive isoform, with either of the active DNMT3B2, DNMT3A1, or DNMT3A2 isoforms significantly decreased de novo DNA methylation (Gordon et al. 2013). These data suggest that dysregulated expression of inactive DNMT3B variants may induce pathologic alterations by the modulation of the methylome and transcriptome in various cell types.

### 3.2.2 MicroRNAs

MicroRNAs (miRNAs) are short, nontranslated, regulatory RNA molecules modulating the level of most messenger RNAs (mRNAs) and proteins in mammalian cells (reviewed by Lin and Gregory 2015). Their precursors are either transcribed from independent transcription units or generated from the introns of primary transcripts and undergo RNase processing in the nucleus and in the cytoplasm. One strand of a mature, double stranded miRNA molecule, associated with the Argonaute protein, interacts with a short complementary sequence of a target mRNA molecule resulting in mRNA degradation or translational repression (reviewed by Lin and Gregory 2015). The mRNAs coding for DNMTs or for proteins regulating the expression of methyltransferase genes are also targeted by miRNAs in various mammalian cell types including primordial germ cells and somatic cells (reviewed by Denis et al. 2011). The expression pattern of miRNAs and the proteins involved in their biogenesis may change in various diseases including malignant tumors (reviewed by Lin and Gregory 2015). The situation is complex, however, because a miRNA may have multiple mRNA targets, and the 3′-untranslated region (UTR) of a mRNA, where most of the miRNAs attach, can interact with several miRNAs.

### 3.2.3 Stabilization of mRNAs by the RNA Binding Protein HuR

HuR (human antigen R), an RNA binding protein, is able to associate with a series of RNA molecules in the nucleus and in the cytoplasm (reviewed by Grammatikakis et al. 2017). HuR is a pleiotropic protein affecting important cellular events. One of its partners is DNMT3B mRNA: in a human carcinoma cell line, interaction with HuR stabilized and increased the expression of DNMT3B RNA (Lopez de Silanes et al. 2009; reviewed by Denis et al. 2011). López de Silanes et al. speculated that, in addition to HuR, other RNA binding proteins and miRNAs may also modulate, in concert with regulators of transcription and splicing, DNMT3B mRNA and protein levels in the cytoplasm (Lopez de Silanes et al. 2009).

### 3.2.4 Posttranslational Modifications (PTMs)

Posttranslational, reversible, covalent modifications, including acetylation, ubiquitination, phosphorylation, methylation, and sumoylation, are able to affect the stability and catalytic activity of DNMT1 (reviewed by Qin et al. 2011; Scott et al. 2014; Jeltsch and Jurkowska 2016). Although numerous PTMs were mapped on various parts of the enzyme, in most cases the functional significance of the modifications remains to be established. Acetylation of distinct lysine residues followed by UHRF1-mediated ubiquitination marked the mouse Dnmt1 for degradation whereas deubiquitination and deacetylation increased the stability of the enzyme. The outcome of phosphorylation at various serine residues was variable: depending on the targeted residue, it affected the interaction between the regulatory and catalytic domain, altered DNA binding affinity, disrupted or altered the interactions with cellular partner proteins, or stabilized the enzyme. Monomethylation at lysine 142 (K142) in late S phase promoted proteasomal degradation of the human DNMT1 whereas a lysine-specific demethylase stabilized the murine Dnmt1 enzyme (reviewed by Qin et al. 2011; Scott et al. 2014; Jeltsch and Jurkowska 2016).

Phosphorylation of the murine Dnmt3a near to the PWWP domain (see Sect. 3.2.5) increased the targeting of the enzyme to heterochromatin but reduced its activity (Deplus et al. 2014a). In contrast, peptidylarginine deiminase 4 (PADI4) citrullinated a region containing five arginines in the N-terminal domain, stabilized the enzyme, and upregulated its activity (Deplus et al. 2014b).

### 3.2.5  Targeting of DNMTs

Although DNMT1 binds with high affinity to hemimethylated DNA duplexes in vitro, during the S phase of the cell cycle it is guided to such target molecules by interacting with PCNA and UHRF1 (see 2.1). In addition to its association with the DNA replication machinery at replication foci during S phase, DNMT1 is also loaded to the chromatin during the G2 and M phases of the cell cycle (Easwaran et al. 2004). In a culture of mouse myoblast cells, Schneider et al. observed by super-resolution 3D-structured illumination microscopy and fluorescence recovery after photobleaching (FRAP) methods that in early S phase a fraction of Dnmt1 molecules binds transiently, via the PCNA-binding domain of its N-terminal regulatory region, to PCNA rings immobilized at replication foci in euchromatic regions (Schneider et al. 2013). This interaction may facilitate complex formation with hemimethylated CpG-sites located nearby. In late S phase, however, a stronger interaction, mediated by TS (targeting sequence, also called RFTS, replication foci targeting sequence, located to the N-terminal part of Dnmt1), directed the enzyme to replication foci as well as to the pericentromeric heterochromatin (Schneider et al. 2013). Schneider et al. suggested that maintenance methylation in densely methylated regions may continue even during the G2 phase (Schneider et al. 2013). DNMT1 is expressed in nonproliferating, postmitotic (G1 or G0) cells as well. Using in situ hybridization, Dnmt1 mRNA was detected in mature neurons of adult mice, and Dnmt1 protein was present in the nuclear and cytoplasmic fraction of mouse brain and spinal cord (Goto et al. 1994; Chestnut et al. 2011). These data suggested that in neurons Dnmt1 may fulfill a unique biological function unrelated to maintenance DNA methylation (see Sect. 4.9).

The nucleosomal structure of chromatin may limit the accessibility of DNA for DNMT1. In vitro model experiments with reconstituted nucleosomes revealed that Dnmt1 preferentially bound to and methylated linker DNA sequences at the entry and exit sites of the nucleosome (Schrader et al. 2015). In contrast, the nucleosomal core particle served as a substrate for Dnmt1 only in the presence of the chromatin remodeling ATPases Brg1 (Brahma-related gene 1) and ACF (ATP-dependent chromatin assembly factor) that regulate nucleosome movement and spacing (Schrader et al. 2015).

DNMT1 interacts with methyl-CpG binding proteins which bind to methylated CpG-sites with high affinity. These binding partners may target DNMT1 to highly methylated heterochromatic regions and may ensure promoter silencing in such regions by complexing with histone deacetylases (HDACs) (Qin et al. 2011). HDACs interact with DNMT1 as well, and by the deacetylation of lysine or arginine residues of histones, they facilitate the establishment of a more condensed chromatin structure. Thus, hypermethylated DNA sequences regularly associate with hypoacetylated histones in transcriptionally inactive chromatin domains (Qin et al. 2011). In addition, histone methyltransferases that deploy chromatin marks associated with transcriptional repression may also recruit DNMT1 to target genes (reviewed by Qin et al. 2011).

Unmethylated histone H3 lysine 4 (H3K4me0), a histone mark characteristic for heterochromatic regions, may affect target selection of the *de novo* DNA-(cytosine-C5)-methyltransferases DNMT3A and DNMT3B by binding to the ADD domain located to the N-terminal regulatory region of these enzymes (reviewed by Denis et al. 2011). The ADD (ATRX–DNMT3–DNMT3L) domain of ATRX, a chromatin remodeling protein mutated in the alpha-thalassemia and mental retardation X-linked syndrome (ATR-X), contains a H3K4me0-binding pocket, similarly to DNMT3 and DNMT3L (Dhayalan et al. 2011). Binding of the unmodified histone H3 tail to the DNMT3A ADD domain stimulates the activity of the enzyme by disrupting the interaction of the ADD domain with the catalytic domain (Guo et al. 2015c).

DNMT3L, the catalytically inactive regulatory factor of de novo DNMTs, also possesses an ADD domain in the N-terminal regulatory region. Introduction of a point mutation (*D124A*) into this domain resulted in the defect of spermatogenesis in mice homozygous for the mutant *Dnmt3L* gene (Vlachogiannis et al. 2015). The mutation disrupted histone H3 binding by DNMT3L and caused a reduction of both CpG- and non-CpG methylation in prospermatogonia. In these cells, hypo-methylation of retrotransposons and endogenous retroviruses and alteration of gene expression pattern was also observed. In addition, defective spermatogenesis was characteristic for male mice (Vlachogiannis et al. 2015). Thus, DNMT3L, the adaptor protein of de novo DNMTs, plays an important role in the reprogramming of DNA methylation during spermiogenesis in mice (see Sect. 4.5).

Via its PWWP domain, characterized by a proline-tryptophan-tryptophan-proline motif, the N-terminal regulatory region of DNMT3A interacts with H3K36me3, another repressive histone mark. As outlined in Sect. 2.2, this interaction facilitates the association of DNMT3A with heterochromatin (Dhayalan et al. 2010). The PWWP domain of DNMT3B is also involved in directing the enzyme to repetitive sequences of pericentric heterochromatin (Chen et al. 2004).

DNMT3A and DNMT3B may directly interact with a series of sequence-specific transcription factors that target them to distinct sequences causing aberrant de novo DNA methylation during tumorigenesis (Hervouet et al. 2009; Blattler and Farnham 2013). Such interactions may also occur in normal cells. The exact mechanism for such targeted methylation events remains to be elucidated (Ficz 2015).

## 4 DNA Methylation in Mammals: Biological Functions

In bacterial cells, DNA methylation fulfills vital functions such as DNA repair, attachment of the bacterial chromosome to the cell membrane, regulation of DNA replication, and resistance to invading phage or plasmid DNA molecules (Kramer et al. 1984; Ogden et al. 1988; Wilson 1988; Waldminghaus et al. 2012; Makarova et al. 2013; Adhikari and Curtis 2016; Cohen et al. 2016). The latter function depends on the joint action of DNA MTase and restriction endonuclease pairs recognizing the same DNA sequence. To protect the bacterial genome from degradation by the corresponding endonucleases, prokaryotic methyltransferases modify essentially all of their recognition sequences. Unmethylated DNAs or DNAs methylated at other recognition sites are recognized as "foreign" by restriction endonucleases. This "protective" function of DNA methylation results in a monotonous, continuous, relatively stable modification of bacterial genomes. In contrast, in large-genome eukaryotes, including mammals, there are alternating methylated and unmethylated regions and certain sequences may gain or lose methyl groups, even within the same cell. Thus, mammalian and human genomes are characterized by discontinuous and changing patterns of DNA methylation (Bird 1986, 1987).

The genes coding for DNA-(cytosine-C5)-methyltransferase enzymes of eukaryotes and related genes encoding proteins with an altered enzymatic function or without enzymatic activity originated, most likely, from bacterial methyltransferases of restriction-modification systems (Rodriguez-Osorio et al. 2010; Iyer et al. 2011; Jurkowski and Jeltsch 2011). DNMT1 and the related eukaryotic enzymes had an independent origin from the family of enzymes where DNMT3A and DNMT3B belong. DNMT2, an RNA-(cytosine-C5)-methyltransferase with sequence homology to the DNA-(cytosine-C5)-methyltransferases apparently changed its substrate preference from DNA to RNA during evolution (Goll et al. 2006; Jurkowski and Jeltsch 2011). The catalytically inactive DNMT3L has sequence similarities to DNMT3A and DNMT3B but it is characterized by a shorter N-terminal and a truncated C-terminal domain. Comparison of DNMT1, DNMT3A, and DNMT3B with bacterial DNA-(cytosine-C5)-methyltransferases revealed that the eukaryotic enzymes acquired distinct N-terminal regulatory sequences that are absent from their bacterial counterparts. Based on the analysis of the mouse Dnmt1, Bestor proposed that acquisition of the N-terminal regulatory domain was associated with a novel function for DNA methylation: in large-genome eukaryotes cytosine methylation may play a role in the compartmentalization of the genome (Bestor 1990). This idea fits well to the presence of alternating domains of methylated and unmethylated DNA regions in mammalian genomes. Compartmentalization may facilitate the interaction of diffusible regulatory factors to the unmethylated domains, whereas the methylated regions would be inaccessible for such regulators (Bestor 1990).

Several lines of evidence support a role for cytosine methylation in the regulation of chromatin structure by facilitating the establishment of closed, heterochromatic domains (reviewed by Jin et al. 2011; Moore et al. 2013). Thus, DNA methylation may indeed affect, in concert with other epigenetic regulatory mechanisms, the accessibility of promoters and regulatory sequences for transcription factors and RNA polymerases (reviewed by Minarovits et al. 2016). Interactions of methylated DNA sequences with methylcytosine-binding proteins and other components of the epigenetic regulatory machinery, including DNMTs, form the basis for the documented functions of cytosine methylation in mammals: (1) promoter silencing, transcriptional regulation; (2) imprinting, allele-specific gene expression; (3) transposon silencing; (4) X chromosome inactivation; (5) regulation of embryogenesis; (6) regulation of cell differentiation; (7) preservation of genome stability; (8) splicing; (9) brain function (reviewed by Smith and Meissner 2013a; Meng et al. 2015) (Table 4).

**Table 4** Biological functions of DNA-(cytosine-C5)-methyltransferases in human and mouse cells

| Function | Enzyme involved |
|---|---|
| Maintenance CpG-methylation | DNMT1, aided by DNMT3A and DNMT3B; During preimplantation development: DNMT1s (somatic form); 8-cell stage: DNMT1o (oocyte-specific isoform) |
| De novo DNA methylation | DNMT3A and DNMT3B aided by DNMT1 |
| Promoter silencing | All DNMTs |
| Imprinting | De novo MTases; DNMT1o: maintenance of genomic imprinting in preimplantation embryos |
| Transposon silencing | DNMT1, DNMT3A, DNMT3B |
| X chromosome inactivation | De novo MTases are involved in stably silencing *Xist* on the active X chromosome (Xa); Dnmt1 is involved in gene silencing on the inactive X chromosome (Xi) |
| Regulation of embryogenesis | DNMT1o, DNMT1s, DNMT3A, DNMT3B |
| Regulation of cell differentiation | DNMT3A, DNMT3B, DNMT1 |
| Preservation of genome stability | DNMT1, DNMT3A, DNMT3B |
| Regulation of splicing | Methylation of weak exons by de novo DNMTs possibly excludes their transcripts from mRNAs |
| Brain function | DNMT1, DNMT3A, DNMT3B |

## 4.1   DNA Methylation as a Regulator of Chromatin Structure and Promoter Activity

In eukaryotic cells, epigenetic regulatory mechanisms ensure the heritable alterations of cellular states and the inheritance of gene expression patterns from cell generation to cell generation. The best studied epigenetic regulatory mechanisms include DNA methylation, histone modifications, and the Polycomb and Trithorax protein complexes that also modify histone molecules. Other epigenetic regulators that do not necessarily rely on reversible covalent modifications include variant histones, pioneer transcription factors, long noncoding RNAs, and nuclear proteins mediating long-distance chromatin interactions (reviewed by Minarovits et al. 2016). Furthermore, the localization of distinct promoters to open or closed chromatin domains (see below) within the nucleus also influences transcriptional activity, and a change in nuclear position may activate or silence promoters (reviewed by Gyory and Minarovits 2005). Thus, there is a complex, multilayered epigenetic regulatory system in mammalian cells.

DNA methylation, histone modifications, as well as Polycomb and Trithorax protein complexes modulate the structure of chromatin, i.e., the architecture of the DNA-protein complex characteristic for eukaryotes. Chromatin is built up from repeating units, called nucleosomes, that contain eight histone proteins (histone octamer) and a stretch of 147 bp long DNA wrapped around the histone octamer

(Kornberg 1974). The nucleosomes are connected by linker DNA of variable length and stabilized by histone H1. Two molecules of each histone H2A, H2B, H3, and H4 form the octamer, which is organized as a heterotypic tetramer $(H3–H4)_2$ with two associated dimers (H2A–H2B) (Kelley 1973; reviewed by Marino-Ramirez et al. 2005). Epigenetic modifications usually affect the long tails of histone H3 and H4 although modifications of H2A and H2B occur as well (reviewed by Minarovits et al. 2016).

In vertebrates, the typical epigenetic mark of DNA is cytosine methylation (see Sect. 2). In mammalian cells, methylated cytosines are frequently present in the control regions of inactive promoters and methylation of such sequences in reporter constructs suppresses transcription (reviewed by Robertson 2001). In contrast, a series of active promoters are typically located to unmethylated chromatin domains, called CpG-islands characterized by an elevated G + C content and an increased density of CpG-dinucleotides compared to the rest of the genome (Deaton and Bird 2011). DNA sequences within CpG-islands are kept methylation-free by the presence of histone H3 trimethylated at lysine 4 (H3K4me3), an activating chromatin mark associated with an "open" chromatin architecture. H3K4me3 may block de novo methylation by interfering with the association of the murine Dnmt3a to the histone H3 tail (Zhang et al. 2010). Active DNA demethylation is mediated by Tet1 in mouse cells (see Sect. 2.3). Tet1 may protect unmethylated CpG-rich sequences from stochastic, aberrant methylation by converting the newly methylated 5mC to its oxidation products that block Dnmt1 (Williams et al. 2011; Ji et al. 2014). The enrichment of histone H3K4me3 at CpG-islands is due to the preferential binding of CFP1 (CxxC finger protein 1), a component of the mammalian SETD1 (SET domain 1) complex involved in histone H3 K4 methylation, to unmethylated DNA sequences (Thomson et al. 2010). Tet2 and Tet3 also facilitate SET1 binding and transcriptional activation (Deplus et al. 2013; reviewed by Delatte et al. 2014).

Unmethylated CpG-islands and activating histone modifications mark euchromatic domains that are typically associated with active promoters. In contrast, highly methylated DNA sequences and repressive histone modifications are characteristic for heterochromatic regions where silent promoters are located. Promoter inactivation is frequently achieved by multiple epigenetic regulators that may form repressor complexes. Methyl-CpG binding domain proteins (MBDs) preferentially bind to DNA sequences containing one or more symmetrically methylated CpG-dinucleotides (reviewed by Bogdanovic and Veenstra 2009). MBDs interact with histone deacetylases, nucleosome remodeling complexes, histone methyltransferases, and Polycomb group complexes involved in heterochromatin formation and promoter silencing (Jones et al. 1998; Nan et al. 1998; Fujita et al. 2003; Sakamoto et al. 2007; Gopalakrishnan et al. 2008).

Recent studies established that the effect of DNA methylation on promoter activity depends on the density of CpG-dinucleotides in the promoter region. High CpG-density promoters are frequently located to unmethylated CpG-islands in various cell types and can be efficiently inactivated by DNA methylation (reviewed by Messerschmidt et al. 2014). We would like to note, however, that methylation-mediated silencing of CpG island-associated promoters is a frequent event during carcinogenesis. CpG-methylation can suppress the activity of intermediate CpG-density promoters as well. Such events may occur at the promoters of pluripotency genes or germ cell-specific genes in differentiated cells. CpG-methylation may not necessarily influence, however, the activity of low CpG-density promoters (reviewed by Messerschmidt et al. 2014). Jiang et al. argued that in vertebrate species high CpG-density promoters are typically associated with housekeeping genes that are active in a wide spectrum of tissues, whereas tissue-specific functions are characteristic for genes with low CpG-density promoters (Jiang et al. 2014).

In addition to the methylation pattern of promoters, CpG-methylation at enhancer sequences also affects gene transcription. Aran et al. observed that hypomethylated enhancers were associated with genes upregulated in cancer whereas enhancer hypermethylation correlated with the downregulation of gene activity (Aran et al. 2013). Aran et al. suggested that enhancers do not necessarily act in a cell-type-specific manner; instead, an unmethylated enhancer may regulate transcription levels in more than one cell type provided that the relevant promoter is unmethylated (the "dimmer switch" model for enhancer action) (Aran et al. 2013).

Although methylation of mammalian promoters is usually associated with gene silencing, there are examples for CpG methylation-mediated activation of gene expression, too. Bahar Halpern et al. observed that a core CpG-island (CpG2a) and a neighboring CpG-rich region located upstream of the *FoxA2* promoter were highly methylated in the FoxA2 expressing human pancreatic islet cells and early human endoderm stage cells (CXCR4+ cells) derived from human embryonic stem cells (hES cells) (Bahar Halpern et al. 2014). In contrast, the very same sequences were hypomethylated in undifferentiated hES cells that did not express FoxA2. Two other CpG-islands located upstream or downstream from CpG2a were hypo-methylated independently of *FoxA2* promoter activity. Bahar Halpern et al. specu-lated that the paradoxically high level of CpG-methylation at CpG2a may prevent the binding of a repressor protein to *FoxA2* regulatory sequences (Bahar Halpern et al. 2014). Thus, methylation-dependent gene activation – possibly mediated by DNMT3B – activated by an unknown signal – may switch on the expression of *FoxA2*, a master regulator of early endoderm formation.

In human osteoblast-like MG63 cells, CpG-methylation in the distal promoter region of the *PDPN* (*podoplanin*) gene also stimulated promoter activity (Hantusch et al. 2007). In breast carcinoma cell lines, high levels of *IL-8* (*interleukin 8*) gene transcription correlated with methylation of two isolated CpG-sequences that were located outside CpG-islands, far upstream of the *IL-8* promoter (De Larco et al. 2003). De Larco et al. suggested that methylation at these CpG-sites possibly prevented the binding of a repressor or affected chromatin remodeling (De Larco et al. 2003). Methylation of sequences located far downstream from the promoter may also upregulate transcription. Unoki et al. found that the methylated intron 1 of *EGR2* (*early growth response 2*), a gene coding for a putative tumor suppressor protein, enhanced the activity of the *EGR2* promoter (Unoki and Nakamura 2003).

We would like to add that BZLF1, an immediate early protein of Epstein-Barr virus (EBV, a human gamma herpesvirus), preferentially associated to its methylated recognition sequences and activated Rp, one of its target promoters (Bhende et al. 2004). High affinity of BZLF1 to its methylated responsive elements permitted the induction of lytic, productive replication of highly methylated, latent EBV genomes. Similarly to BZLF1, RFX, a cellular regulatory protein, was also able to bind to and activate a methylated promoter – without inducing local demethylation – at the major histocompatibility complex (MHC) (Niesen et al. 2005).

## 4.2    The Role of DNA Methylation in Genomic Imprinting

Genomic imprinting is a complex and fast evolving area of research. Here we outline only some of the most essential points related to the topic and refer only to a selected number of papers and reviews covering the field. In mammals, a set of "imprinted" genes shows monoallelic expression, i.e., they are transcribed either from the maternal or the paternal allele in diploid cells. Imprinted genes play an important role in the control of embryogenesis and in the formation of the placenta that regulates nutrient transfer between mother and fetus. Various genes located to different chromosomes in mammals became independently imprinted during evolution (Edwards et al. 2007). DNA methylation plays an important role in the regulation of imprinted genes, as suggested by the data of Li et al., who observed activation or repression of various imprinted alleles in Dnmt1 deficient mice (Li et al. 1993). Imprinted genes frequently form gene clusters and their expression is controlled by regulatory sequences called imprinting control regions (ICRs) or germline differentially methylated regions (gDMRs). The latter name refers to the fact that the parental alleles are distinguished from each other by epigenetic marks deposited to the ICRs in gametes and indicates that DNA methylation is the key epigenetic mechanism involved in the establishment and maintenance of the imprinting signal (Tucker et al. 1996; Reik and Walter 1998). Thus, the methylation pattern of the ICR influences whether the maternal or paternal allele will be silenced in somatic cells (reviewed by Colot and Rossignol 1999; Bartolomei and Ferguson-Smith 2011; Renfree et al. 2013; Barlow and Bartolomei 2014). Maternally methylated gDMRs are typically located at CpG-rich promoters, whereas paternally marked gDMRs are situated in intergenic regions (Renfree et al. 2013).

A methylated gDMR may act as a selector: in the case of the reciprocally regulated mouse *Igf2* and *H19* genes, the methylated paternal gDMR inactivated the adjacent *H19*, but permitted activation of *Igf2* by preventing the buildup of an insulator complex between downstream enhancers and *Igf2*. In contrast, the unmethylated gDMR on the maternal chromosome permitted binding of CTCF, a regulator of nuclear architecture, that insulated *Igf2* from the enhancers and prevented its expression. In parallel, the unmethylated *H19* was transcribed from the maternal chromosome (Kurukuti et al. 2006; reviewed by Sasaki et al. 2000; Renfree et al. 2013) (Fig. 4).

A methylated gDMR may also hinder the expression of a regulatory RNA molecule involved in gene silencing: in the placenta the methylated gDMR directly blocked the activity of a promoter where the transcripts of the *Airn* lncRNA (long

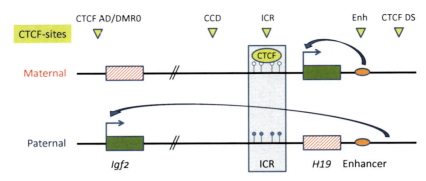

**Fig. 4** **Monoallelic expression of *Igf2* and *H19* genes regulated by the methylation of the imprinting control region**. A schematic view of the mouse *Igf2/H19* locus is shown to demonstrate the insulator model of imprinting (Bell and Felsenfeld 2000). On the maternal allele, binding of CTCF to the unmethylated imprinting control region (ICR) insulates *Igf2* from the enhancer that activates *H19*. On the paternal allele, methylation of the ICR prevents CTCF binding and inactivates the *H19* promoter; in the absence of the insulator, *Igf2* is activated by the enhancer. Striped boxes: inactive genes; filled boxes: actively transcribed genes; straight arrows: active promoters; curved arrows: enhancer-promoter interactions; open lollipops: unmethylated CpGs; closed lollipops: methylated CpGs. Triangles stand for CTCF binding sites mapped in the orthologous human *IGF2/H19* locus that permit the formation of alternative chromatin loops by CTCF-CTCF and CTCF-cohesin interactions positioning the enhancer either in the vicinity or far from the *IGF2* promoter (Nativio et al. 2009). Abbreviations: CTCF AD/DMR0: a CTCF site adjacent to DMR0 (a differentially methylated region at the 5'end of *IGF2*); CCD: Centrally Conserved DNase I hypersensitive domain; ICR: Imprinting Control Region; Enh: Enhancer; CTCF DS: CTCF binding Downstream Site

noncoding RNA) are initiated on the maternal chromosome (reviewed by Barlow and Bartolomei 2014). Switching off the *Airn* promoter permitted the expression of neighboring genes of the *Igf2r* cluster on the maternal chromosome, whereas on the paternal chromosome the unmethylated *Airn* promoter was active and the *Airn* lncRNA "coated" and suppressed – in *cis* – the genes of the *Igf2r* cluster (reviewed by Braidotti et al. 2004; Barlow and Bartolomei 2014). The *Airn* lncRNA may inactivate genes by helping to establish a repressive chromatin structure, similarly to the *Xist* lncRNA which mediates X-chromosome inactivation (Brockdorff and Turner 2015) (see Sect. 4.4).

Recent studies indicated that in the gametes the number of differentially methylated genes was much higher than the number of imprinted genes (reviewed by Kelsey and Feil 2013). These data suggested that DNA methylation at ICRs is perhaps not a specifically targeted process, but the DNA methylation marks deposited during gametogenesis are preserved after fertilization at the imprinted loci, even during a wave of genome-wide DNA demethylation occurring in the preimplantation embryo (reviewed by Kelsey and Feil 2013; Barlow and Bartolomei 2014) (see Sect. 4.5). DNMT1o, the maternal isoform of DNMT1, may play a role in the maintenance of genomic imprinting in preimplantation embryos (Howell et al. 2001).

## 4.3   Transposon Silencing by DNA Methylation

A predominant fraction of mammalian genomes consists of repetitive sequences including satellite repeats, tandem repeats, and mobile genetic elements. Mobile genetic elements or transposable elements are capable to change their genomic position that may cause genomic instability due to the insertions, deletions, and genomic rearrangements associated with transposition events. Transposable elements or transposons may relocate within the genome either via an RNA intermediate that is reverse transcribed to DNA, or with the help of transposase enzymes that cut out and reinsert them into a new target site. Retrotransposons or retrotransposable elements comprise around 50% of mammalian genomes (Munoz-Lopez and Garcia-Perez 2010; Zamudio and Bourc'his 2010). They are relocating via an RNA intermediate and include the class of LTR retrotransposons, such as the proviruses of endogenous retroviruses (ERVs) that are flanked with long terminal repeats (LTRs). The other class of retrotransposons called non-LTR retrotransposons lack LTRs and comprise long interspersed nuclear elements (LINEs) and short interspersed nuclear elements (SINEs). DNA transposons transpose via a nonreplicative "cut and paste" mechanism. They make up around 3% of the human genome (Zamudio and Bourc'his 2010). Because transposons impose a potential threat to genome integrity and most of the 5mC in mammalian genomes resides in transposons, Yoder et al. proposed that a major function of DNA methylation is the suppression of transposons which they regarded as "parasitic" sequence elements (Yoder et al. 1997). The idea of methylation-mediated transposon silencing was supported by experiments demonstrating the reactivation of retroelements in mice with various defects of the DNA methylation machinery. Inactivation of Dnmt1, Dnmt3A and Dnmt3B, as well as deficiency in Dnmt3L, the de novo MTase cofactor "reanimated," i.e., increased the expression of LINE and IAP (intracysternal A particle, an endogenous retrovirus) retroelements (reviewed by Zamudio and Bourc'his 2010). Global hypomethylation in cancer cells may also induce aberrant expression of LINE-1 (Miousse and Koturbash 2015).

Apparently, there is more than one layer of protection against transposon reactivation in mammalian cells. Induction of DNA hypomethylation in embryonic stem cells activated transposon transcription but also resulted in the accumulation of the repressive histone mark H3K27me3 at various transposon families that became re-silenced by the Polycomb-mediated reconfiguration of the chromatin (Walter et al. 2016). In addition, RISC (RNA-induced silencing complex) and piRNA (piwi-interacting RNA) pathway proteins may also suppress transposon activity by the degradation of retroelement RNAs (reviewed by Goodier 2016). Although transposable elements may cause genetic instability, their expression was regularly detected in mammalian oocytes and early embryos (reviewed by Evsikov and Marin de Evsikova 2016). These observations suggested that transposons – regulated by the cellular epigenetic machinery – may have a physiological function in mammals, especially during oogenesis and early embryogenesis. In these developmental processes or in stem cells of adult organisms, transposons may serve as a tool for the control of co-expressed gene batteries (Peaston et al. 2004; Macfarlan et al. 2012; Gifford et al. 2013; Evsikov and Marin de Evsikova 2016).

## 4.4    X Chromosome Inactivation

In somatic cells of female mammals, random inactivation of one X chromosome ensures dosage compensation for the products of X-linked genes between males and females. The majority of genes carried by Xi, the inactive X chromosome, are silent due to the concerted action of several epigenetic regulatory mechanisms (Csankovszki et al. 2001; Chaligne and Heard 2014). X chromosome inactivation (XCI) results in heterochromatinization of Xi which is relocated to the nuclear periphery in the G1 phase of the cell cycle as a Barr body and replicates late during S phase (Chen et al. 2016a). During the initiation of XCI, transcription of *XIST* (*X-inactive specific transcript*) is switched on (Gilbert et al. 2000). *XIST* is active on Xi but silent on the active X chromosome, Xa. On Xa, de novo DNMTs are involved in stably silencing *Xist* (Kaneda et al. 2004). The *XIST* transcript, a long noncoding RNA (lncRNA), spreads in *cis* with the aid of L1 repetitive elements, coats the entire X chromosome to be inactivated, and induces the deposition of heterochromatic marks on the future Xi (Engreitz et al. 2013; Bala Tannan et al. 2014; Cerase et al. 2015). Transcriptional silencing of numerous genes located on Xi is possibly due to the eviction of RNA polymerase II from the chromatin or blocking its access to the chromatin due to the collapse of interchromatin channels (Smeets et al. 2014; Cerase et al. 2015). Cerase et al. suggested that silencing of transcription may actually initiate removal of euchromatic histone marks, deposition of heterochromatic histone marks, and DNA methylation (Cerase et al. 2015). Xi is characterized by a unique three-dimensional structure which is smoother and more spherical than its active counterpart, Xa (Pandya-Jones and Plath 2016). Silenced promoters are associated with deacetylated histones and PRC2, the Polycomb repressor complex depositing a repressive chromatin mark by methylating histone H3 lysine 27 (Gilbert and Sharp 1999; Brockdorff 2013). Although the *XIST* transcript initiates facultative heterochromatinization and gene silencing during random XCI, maintenance of the silenced state is due to the concerted action of several regulatory mechanisms, including *XIST* lncRNA, histone hypoacetylation, and DNA methylation (Csankovszki et al. 2001). Methylation profiling of human Xi and Xa revealed an increased methylation of CpG-islands associated with most of the silenced genes on the Xi. A subset of CpG-islands were less methylated on the Xi. Accordingly, some of the associated genes escaped XCI and showed biallelic expression (Sharp et al. 2011).

## 4.5    DNA Methylation as a Regulatory Mechanism of Embryogenesis in Mammals

The catalytic activity of the maintenance DNA-(cytosine-C5)-methyltransferase plays a vital role in mouse embryogenesis (Takebayashi et al. 2007) (see Sect. 2.1).

The activity of *Dnmt3a* and *Dnmt3b* genes coding for de novo DNA methyltransferases (see Sect. 2.2) was also detected in mouse embryos using an expression reporter cassette targeted to these genes: during the early phase of embryogenesis, *Dnmt3a* was expressed at a moderate level in embryonic ectoderm and at a low level in mesodermal cells, whereas there was a high level of *Dnmt3b* expression in the embryonic ectoderm, accompanied with a weak expression in mesodermal and endodermal cells (Okano et al. 1999). At a later stage there was a ubiquitous expression of *Dnmt3a* in the embryos and a predominant expression of *Dnmt3b* in the forebrain and in the eyes (Okano et al. 1999). Inactivation of both genes in *Dnmt3a*$^{-/-}$, *Dnmt3b*$^{-/-}$ mice caused embryonic lethality due to an arrest of growth and morphogenesis shortly after gastrulation (Okano et al. 1999).

Whereas somatic cells maintain high 5mC-levels and lineage-specific methylomes, in murine primordial germ cells (PGCs) there is a genome-wide demethylation during gametogenesis. This demethylation causes an almost complete erasure of methylation marks and is followed by the establishment of gamete-specific DNA methylation patterns (Hackett and Surani 2013). During mouse development, PGCs are derived from somatic precursors located initially posterior to the primitive streak and carry methylated, silenced pluripotency genes such as *Oct4* and *Nanog* and repressed germline-specific genes (Seisenberger et al. 2013a; Dean 2014; Messerschmidt et al. 2014). The demethylation of the genome in PGCs starts as a passive process during the migration to the genital ridges, because the cells also proliferate in parallel, resulting in the dilution of 5mC in the daughter cells (Messerschmidt et al. 2014). Active demethylation is characteristic, however, for postmigratory PGCs: at this stage, a transient increase of 5hmC accompanies the accelerated loss of 5mC (Messerschmidt et al. 2014) (Fig. 5).

In parallel with genome-wide demethylation, the pluripotent genes as well as a set of "germline defence" genes implicated in the regulation of transposon activity are reactivated, and – in female mouse embryos – Xi, the inactive X chromosome is activated (Hackett et al. 2012; reviewed by Saitou and Yamaji 2012; Messerschmidt et al. 2014). After colonization of the genital ridges the PGCs continue to proliferate, until they enter meiotic prophase in females and mitotic arrest in males (Hackett and Surani 2013; Dean 2014).

In PGCs, 5mC-erasure extends to the genomic imprints. It was suggested that this may permit the reestablishment of methylation marks at ICRs during oogenesis and spermatogenesis according to the sex of the cell. Transposons are also demethylated in murine PGCs, with the exception of certain IAPs (Hackett and Surani 2013). The process of "epigenetic reprogramming" was defined as the erasure of DNA methylation followed by the establishment of a new methylome mediated by de novo DNMTs (Morgan et al. 2005). Morgan et al. suggested that during gametogenesis such a reprogramming may build up sex-specific epigenotypes in mature gametes (Morgan et al. 2005). Thus, after the erasure of the paternal imprints the newly established imprints would correspond to the sex of the individual.

A global demethylation was also observed in PGCs isolated from human embryos, although the relative timing of distinct demethylation events such as the erasure of imprints differed between humans and mice (Guo et al. 2015a; reviewed

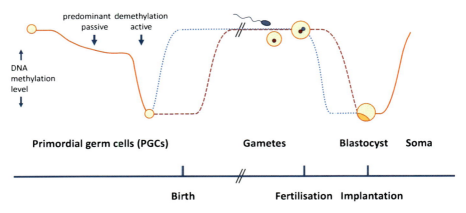

**Fig. 5 Dynamic changes of DNA methylation levels during mouse gametogenesis and embryogenesis**. A continuous line indicates the relative 5mC-content beginning with primordial germ cells (PGCs). There is a genome-wide demethylation during gametogenesis, followed by the establishment of gamete-specific DNA methylation patterns within the fetal gonads of male mice (dotted line) and in the female germline (dashed line). After fertilization, during early embryogenesis, both the paternal genome (dotted line) and the maternal genome (dashed line) undergo almost complete demethylation. The lowest level of DNA methylation is reached in the inner cell mass (shown as a filled elliptoid) of the blastocyst. After implantation, de novo methylation reestablishes a high 5mC-level during the later stages of embryogenesis both in somatic cell precursors and in early PGCs

by von Meyenn and Reik 2015). In vitro studies of PGC-like cells also indicated species-specific differences in reprogramming (von Meyenn et al. 2016). Demethylation of PGC methylomes in mouse and human embryos did not necessarily correlate, however, with gene activation: loss of 5mC did not result in a significant increase in the transcription of transposable elements (reviewed by von Meyenn and Reik 2015).

After genome-wide demethylation in PGCs, de novo methylation starts at embryonic day E13.5 in prospermatogonia or gonocytes within the fetal gonads of male mice, and the buildup of the male methylome is completed before birth (Stewart et al. 2016). In the female germline of mice, de novo methylation starts somewhat later, after birth, and the oocyte methylome is established by day 21 of postnatal development (Stewart et al. 2016). The intricate details of the consecutive epigenetic events shaping the sperm and oocyte methylomes in mice and humans are discussed in the recent review by Stewart et al. 2016.

Although the pattern of DNA methylation is usually stable in adult somatic cells, there is a spectacular change in the 5mC-content of mammalian cells during development: the genome undergoes almost complete demethylation during early embryogenesis (Guibert et al. 2012; reviewed by Hackett and Surani 2013). Between fertilization and implantation, the lowest level of DNA methylation was detected in the inner cell mass of the blastocyst. It is important to note, however, that although the erasure of the methylation mark begins already in the zygote, distinct DNA sequences in the maternal genome, marked by H3K9me2 and by binding of the maternal factor STELLA, stay methylated even in preimplantation embryos (Nakamura et al. 2012; reviewed by Kang et al. 2013). Paternal imprinted regions bound by the maternal factor STELLA also keep their methylation marks, similarly to IAP retrotransposons and DNA sequences located to centromeric heterochromatin (reviewed by Seisenberger et al. 2013a; Messerschmidt et al. 2014). In contrast, 5mC in the bulk of the paternal genome is accessible for TET3-mediated oxidation and the products of this process are removed by passive replication-coupled dilution. Alternatively, the activation of the base excision repair (BER) pathway may contribute to the demethylation of the paternal genome, independently of TET3 activity (Amouroux et al. 2016). Although the maternal pronucleus is not subjected to active demethylation, it is undergoing passive demethylation during subsequent cell divisions (reviewed by Seisenberger et al. 2013a; Messerschmidt et al. 2014) (Fig. 5).

Erasure of DNA methylation during early embryogenesis may play a role in the acquisition of pluripotency by the cells of the inner cell mass and by embryonic stem cells. Demethylation of the genes encoding pluripotency transcription factors such as *Nanog* that is methylated in the sperm facilitated this process (reviewed by Feng et al. 2010). After implantation, de novo methylation reestablishes distinct methylomes in somatic cell precursors and in early primordial germ cells (PGCs) and the methylomes are further modified during lineage-specific differentiation (Ji et al. 2010; Hackett and Surani 2013).

In summary, embryogenesis in mammals is a complex process, viewed as alternating phases of cellular differentiation that usually restricts developmental potential, and reprogramming events that allow the reconfiguration of the epigenome in the zygote and in developing germ cells, permitting the reconstruction of pluripotent or totipotent states (Reik 2007; Hackett and Surani 2013). Hackett and Surani speculated that epigenetic mechanisms, including DNA methylation may prevent reversion to preceding cellular states during mammalian development and may contribute to the maintenance of cell identity (Hackett and Surani 2013). In addition, cellular methylomes regularly undergo dynamic changes that create specialized, hypomethylated epigenotypes to ensure the realization of complex developmental programs and the successful reproduction of mammalian species (Reik 2007; Dean 2014; Messerschmidt et al. 2014).

## 4.6   DNA Methylation as a Regulator of Cell Differentiation

Although inactivating the maintenance DNA-(cytosine-C5)-methyltransferase Dnmt1 or knocking out of both *de novo* methyltransferase genes, *Dnmt3a* and *Dnmt3b*, disturbed embryogenesis and caused embryonic lethality in mice, undifferentiated mouse embryonic stem cells (ES cells) lacking all three of these enzymes survived and proliferated in tissue culture (Li et al. 1992; Okano et al. 1999; Tsumura et al. 2006; Takebayashi et al. 2007). These in vitro growing "triple knockout" mammalian cells lacking CpG-methylation kept their stem cell properties, expressed typical markers of undifferentiated cells, and their growth characteristics were comparable to that of their wild-type counterparts in a competition assay.

Upon induction of cell differentiation by embryoid body formation, however, the growth of the "triple knockout" cells was delayed compared to the wild-type embryonic stem cells (Tsumura et al. 2006). These data suggested that CpG-methylation is dispensable for the maintenance of stem cell properties and proliferative capacity of ES cells. The above mentioned in vivo and in vitro observations also indicated, however, that DNA methylation plays an important role in the establishment and maintenance of cell differentiation.

Comparison of promoter methylation patterns ("the promoter epigenome") of pluripotent mouse ES cells and embryonic germ (EG) cells with the promoter epigenome of differentiated primary embryonic fibroblasts revealed that a set of pluripotency–related genes including *Nanog*, *Lefty1*, and *Tdgf1* were hypomethylated in ES cells and EG cells, but hypermethylated in fibroblasts (Farthing et al. 2008). In ES cells, promoter hypomethylation correlated with an increased gene expression in most cases. In contrast, targeted de novo promoter methylation silenced *Nanog*, a gene actively transcribed in ES cells under normal circumstances. Analysis of the promoter methylome in trophoblast stem cells (TS cells) representing the extraembryonic lineage showed a distinct pattern that differed from that of the other cell types (Farthing et al. 2008). Fouse et al. also observed that the housekeeping and pluripotency-related genes were unmethylated in mouse ES cells whereas most of the differentiation associated genes were repressed and methylated (Fouse et al. 2008). Comparison of wild-type ES cells with "triple knockout "ES cells lacking CpG-methylation revealed that demethylation was able to upregulate the expression of certain X-linked genes and genes implicated in cell differentiation. A series of genes, however, were not upregulated in the absence of CpG-methylation, possibly due to the action of other epigenetic regulators (Fouse et al. 2008).

Analysis of CpG-island methylation showed that ES cells, embryoid bodies, and teratomas had a characteristic, complex methylation pattern and indicated that cellular differentiation was associated with both de novo methylation and demethylation processes (Kremenskoy et al. 2003). In line with this observation, Dawlaty et al. studied the importance of cytosine demethylation in cellular differentiation (Dawlaty et al. 2014). They observed that knocking out of all three genes, *Tet1*, *Tet2*, and *Tet3* that code for dioxygenase enzymes converting 5mC to 5hmC resulted in the complete loss of 5hmC in mouse ES cells and impaired the capacity of the "triple knockout (TKO)" cells to differentiate. The impaired potential for embryoid body and teratoma formation correlated with the hypermethylation and dysregulation of promoters driving the expression of genes involved in developmental processes and cell differentiation (Dawlaty et al. 2014). Most of the analyzed genes that were silenced in TKO cells by hypermethylation were active in wild-type ES cells. Thus, Tet-mediated DNA demethylation is indispensable for the establishment of the proper epigenotype and gene expression pattern involved in ES cell differentiation.

Tet enzymes play a role in the maintenance of chromosomal stability as well by regulating sub-telomeric methylation levels (Yang et al. 2016b). In addition, Wiehle et al. observed that in mouse fibroblasts Tet dioxygenases prevented the methylation of distinct genomic domains called DNA methylation canyons or DNA methylation valleys (Wiehle et al. 2015). Because genes involved in cellular differentiation are frequently located to such genomic areas, Wiehle et al. suggested that Tet activity may ensure the undisturbed access of chromatin regulators to key topological domains controlling developmental processes (Wiehle et al. 2015).

## 4.7   The Role of DNA Methylation in the Preservation of Genome Stability

In addition to transposon silencing (see Sect. 4.3), DNA methylation may contribute to the maintenance of genomic integrity by the stabilization of microsatellite repeats and via the recruitment of the mismatch repair surveillance complex to hemimethylated DNA sequences at the replication fork (reviewed by Putiri and Robertson 2011; Wang et al. 2016). Furthermore, establishment of an appropriate methylation pattern of centromeric and pericentromeric DNA sequences decreases the probability of chromosomal segregation defects and chromosome rearrangements. Finally, DNA methylation-mediated transcriptional silencing may preserve genome stability by decreasing the frequency of mitotic recombination. We would like to mention, however, that local CpG-hypermethylation may promote genome instability by silencing the genes encoding DNA repair enzymes, and 5mC is a potential source of mutations in itself because of its propensity to spontaneous deamination that results in thymine (5mC to T transition) which also contributes to human inherited disease (Cooper et al. 2010; reviewed by Putiri and Robertson 2011; Meng et al. 2015).

### 4.7.1   Stabilization of Microsatellite Repeats by Cytosine Methylation

Microsatellites are genomic elements composed of short tandem repeats of simple DNA motifs that are one to nine nucleotides in length (Sawaya et al. 2013). Formation of single-stranded DNA during the processes of DNA replication, transcription, or repair facilitates the formation of various secondary DNA structures including hairpins, Z-DNA, H-DNA (a DNA triplex), and G-quadruplex (G4) (reviewed by Putiri and Robertson 2011; Sawaya et al. 2013; Sawaya et al. 2015). During transcription, distinct trinucleotide repeats may transiently form R-loops containing an 8 bp long RNA:DNA hybrid (Reddy et al. 2011). The presence of these secondary structures may facilitate genomic rearrangements due to faulty repair or to replication slippage by DNA polymerase. Changes in repeat number within protein coding sequences cause neurological diseases in humans, and repeat number variations in protein-noncoding regulatory regions and promoter elements were also implicated in pathogenesis (Pearson et al. 2005; Sawaya et al. 2013). DNA methylation may stabilize repeat length by transcriptional repression of genes carrying either CpG-containing or non-CpG-containing microsatellites (reviewed by Putiri and Robertson 2011). Association of 5mC-rich sequences with distinct methyl-CpG binding proteins and repair proteins may also contribute to microsatellite stabilization, although in malignant cells heterochromatic histone modifications and CpG-island hypermethylation correlated with a high level of microsatellite instability (reviewed by Putiri and Robertson 2011).

### 4.7.2 Recruitment of the Mismatch Repair Surveillance Complex to Hemimethylated DNA Sequences

Wang et al. observed that in murine cells the postreplicative DNA mismatch repair (MMR) machinery was recruited to a complex of Dnmt1 and Uhrf1, bound to hemimethylated CpG-sites in newly synthesized DNA. The authors argued that the collaboration of the maintenance MTase and the MutSα MMR complex probably preceded chromatin assembly on the newly synthesized DNA duplex, because nucleosome formation could potentially block mismatch recognition by the DNA mismatch repair system (Wang et al. 2016). Such a coordinated action may ensure the correct transmission of both epigenetic and genetic information from cell generation to cell generation in mammals.

### 4.7.3 Methylation Patterns of Centromeric and Pericentromeric DNA Sequences: Functional Consequences

Yamagata et al. observed that the centromeric and pericentromeric DNA was hypomethylated in male and female germline cells of mice but hypermethylated in adult tissues with the exception of testis and epididymal sperm. In contrast, the IAP1 and LINE1 endogenous retroviral repetitive elements were highly methylated in the testis and sperm, like in other adult tissues. The authors speculated that the germline-specific hypomethylation of satellite sequences from centric and pericentric regions was associated with the expression of germ cell-specific genes transcribed in the testis and ovary (Yamagata et al. 2007). As a matter of fact, hypomethylated "pockets" were identified within hypermethylated sequences of a functional centromere (Wong et al. 2006). The kinetochore protein CENP-B (centromere protein B) preferentially attached to nonmethylated recognition sequences located within extensively methylated regions (Tanaka et al. 2005).

CENP-C, another constitutive centromere protein, interacted with DNMT3B and facilitated methylation of centromeric and pericentromeric satellite sequences (Gopalakrishnan et al. 2009). Both CENP-B and CENP-C enhanced the formation of heterochromatin and interacted with the centromere-specific histone variant CENP-A and possibly with a network of centromere-associated proteins to maintain centromere integrity (Okada et al. 2007; Gopalakrishnan et al. 2009; Putiri and Robertson 2011; Giunta and Funabiki 2017). Local or general hypomethylation of satellite CpG-dinucleotides caused, however, either chromosome rearrangements or chromosomal segregation defects in a subset of patients with hepatocellular carcinoma or in patients with ICF syndrome (immunodeficiency, centromeric instability, facial anomalies syndrome), respectively (reviewed by Putiri and Robertson 2011).

### 4.7.4 Inhibition of Homologous Recombination by DNA Methylation

Dnmt1 deficiency in mice enhanced the rate of mitotic recombination, activated the transcription and transposition of endogenous retroviral elements, and facilitated tumorigenesis (Eden et al. 2003; Gaudet et al. 2003; Howard et al. 2008). DNA methylation also suppressed meiotic recombination in several organisms (reviewed by Termolino et al. 2016). In contrast, Sigurdsson et al. observed, based on the analysis of data sets from various sources, that in human male germ cells there was an elevated level of DNA methylation at recombinational hot spots (Sigurdsson et al. 2009). Sigurdsson et al. speculated that methylated regions could be preferential sites of meiotic recombination or recombined sequences could perhaps be targeted by DNMTs (Sigurdsson et al. 2009).

## 4.8   Regulation of Alternative Splicing by Gene Body Methylation

In eukaryotes, most of the genes are composed of exons and introns and their primary transcripts are processed by spliceosomes that excise the protein-non coding introns and unite the protein-coding exons to form mRNA molecules. Spliceosomes are large ribonucleoprotein complexes assembled co-transcriptionally (Will and Luhrmann 2011). In higher eukaryotes, alternative splicing of the same primary transcript frequently generates mRNAs with different combinations of exons: the so-called constitutive exons are regularly included into the mRNA whereas – depending on the rate of transcription elongation – "alternative" or "weak" exons are either included or skipped. Pausing of RNA polymerase II favors inclusion of "alternative" exons into the mRNA (reviewed by Shukla and Oberdoerffer 2012). The choice of splice sites is also affected by the structure of the chromatin and by RNA binding proteins. Recently, DNA methylation emerged as an important regulator of splicing. Shukla et al. observed that binding of the zinc finger protein CTCF to the unmethylated exon 5 of the *CD45* gene elicited RNA polymerase II pausing that promoted the inclusion of this "weak" exon into the mRNA (Shukla et al. 2011). CTCF does not bind DNA, however, when its recognition sequence is methylated (Hark et al. 2000). Accordingly, Shukla et al. found that CpG-methylation at the alternative exon 5 of *CD45* resulted, during the processing of the primary transcript, in the exclusion of exon 5 from the mRNA (Shukla et al. 2011). A possible explanation for this phenomenon was that cytosine methylation blocked the binding of CTCF to its recognition sequence, and in the absence of a local barrier created for RNA polymerase II by CTCF, the rapid rate of elongation quickly resulted in the synthesis of competing "strong" downstream splice sites, before spliceosome assembly at the "weak" upstream splice site. Thus, spliceosome assembly occurred only at these downstream sites, skipping the weak upstream site (Shukla et al. 2011; Shukla and Oberdoerffer 2012). Using CTCF depleted cells and a combination of ChIP-seq and RNA-seq methods, Shukla et al. demonstrated that in addition to *CD45*, processing of a series of other cellular transcripts is also regulated by CTCF-mediated RNA polymerase II pausing (Shukla et al. 2011). In contrast, Maunakea et al. found that DNA methylation and binding of the methyl-CpG-binding protein MeCP2 to alternatively spliced exons (ASEs) facilitated the inclusion of ASEs into mRNAs during the processing of the primary transcripts (Maunakea et al. 2013). Further studies may illuminate how CpG-methylation in gene bodies, especially in exons, may affect the complex process of RNA splicing in *trans*.

In a recent work, Marina et al. described that 5hmC and 5caC – the TET generated oxidation products of 5mC – facilitated CTCF binding to the *CD45* "model gene" in T lymphocytes and promoted the inclusion of the alternative exon 5 into mRNA. They also observed that in vitro activation of naïve CD4$^+$ T cells induced exclusion of exon 5, in parallel with a decrease in 5hmC and increase in 5mC at exon 5. The authors argued that 5hmC and 5caC promoted CTCF association and exon inclusion whereas 5mC blocked CTCF binding and facilitated exon exclusion (Marina et al. 2016). Oxidation products of 5mC bind a set of distinct nuclear proteins (Spruijt et al. 2013). This observation suggests that they may act as signals in important biological processes including splicing (see Sect. 4.9.2).

## 4.9 The Role of DNA Methylation in Brain Function

Epigenetic regulatory mechanisms, among them DNA methylation, are at the fore-front of brain research these days. For this reason, we wished to outline some of the most interesting results of this exciting new field in a separate subsection. Oxygen is required for demethylation processes, and thus it is believed that first the increase in atmospheric oxygen allowed the appearance of reversible methylation and therefore regulable enzymatic methylation systems for DNA and proteins. Regulated methyl-ation in turn was required for the emergence of multicellular animals and for organ development. Therefore, the Cambrian radiation was probably dependent on the parallel increase in atmospheric oxygen (Jeltsch 2013). Regulable DNA methylation at CpG-dinucleotides plays also an important role in the delineation of the human phenotype in comparison with other primate lineages. There is an estimated number of ~1.19 million nonpolymorphic species-specific CpG-dinucleotides, termed CpG-beacons, part of them extremely densely clustered into 21 genomic locations which are functionally associated with CpG-island evolution, human traits and diseases. Beacon clusters have been predicted to colocalize with accessible chromatin (Bell et al. 2012). Comparative primate epigenomic data from prefrontal cortex neurons showed that beacon clusters are indeed enriched for human-specific H3K4me3 peaks. Both beacon clusters and permissive chromatin signatures were enriched at telomeric chromosomal regions supporting a predisposition for recombination. In summary, epigenomic analysis was able to pinpoint chromatin structures that con-tribute to the human-specific phenotype (Bell et al. 2014). Regulated methylation may be of even higher importance for highly adaptive organ systems, like the immune system and the brain, than for the function of other organs. Regulated methylation within the brain may, of all biological properties, even be the most discerning between species. Comparative methylome analysis of human and chim-panzee brain tissue uncovered 85 human-specific and 102 chimp-specific differen-tially methylated regions (DMRs), approximately half of them located in intergenic regions or gene bodies. DMRs were enriched in active chromatin loops suggesting a human-specific organization of higher-order chromatin structure (Mendizabal et al. 2016).

### 4.9.1 Dynamic Methylation Within the Brain During Development and Disease

Methylation-profiling of CpG-dinucleotides in fetal frontal cortices using 450 K methylation arrays showed that methylation was continuously increasing at 1767 positions and decreasing at 1149 positions with gestational age. These altogether 2916 developmentally regulated differentially methylated positions (dDMPs) were enriched in gene bodies but underrepresented in CpG-islands. Overall, during the first trimester the human fetal brain is globally hypomethylated with methylation increasing in the second and third trimester. Because genes associated with dDMPs are assumed to be important for brain function, it is remarkable that dDMPs were also enriched for regions that have been associated with schizophrenia and autism. The authors concluded that this gestational age-matching set of dDMPs may have been adopted for both brain evolution and ontogeny (Schneider et al. 2016).

Organoids derived from pluripotent human cells recapitulated the very early three-dimensional organization of the fetal brain and were used to compare epigenomic and transcriptional programs and to establish suitability as an in vitro model for gene expression dynamics in the early to mid-term fetal human brain. Non-CpG methylation was enriched at super-enhancers in cerebral organoids and fetal brains, and non-CpG methylation in vitro was a predictor of impending super-enhancer repression in vivo. Overall, cerebral organoids recapitulated the large-scale epigenomic remodeling during the early fetal brain development. However, generally organoids tend to a higher methylation level at CpG-positions, and wide-spread pericentromeric demethylation was observed in the in vitro model but not in the fetal brain (Luo et al. 2016). The brain methylome undergoes a widespread reconfiguration from fetal to young adult age corresponding to synaptogenesis. Concomitantly, non-CpG methylation becomes the more prevalent form of methylation in neurons, but not glia. Interestingly, non-CpG methylation identified genes escaping X-chromosome inactivation (Lister et al. 2013). While above studies did not differentiate methylation states between gray and white matter, a first 450 K chip- and pyrosequencing-analysis of Brodmann area 9 which is located within the dorsolateral prefrontal cortex and is important for higher cognitive skills and affected in diverse neurological disorders demonstrated robust gray-white methylation differences. Gray matter corresponds predominantly to neuronal cells whereas white matter is mainly built up from glial cells. Especially cell type-specific markers were enriched among differentially methylated genes. A large number of the identified DMRs had previously been associated with degenerative neurological diseases, like Alzheimer's (AD), Parkinson's (PD), Huntington's diseases (HD), amyotrophic lateral sclerosis (ALS), and multiple sclerosis (MS) (Sanchez-Mut et al. 2017).

It is commonplace that genetic diseases associated with aberrant methylation, e.g., fragile X syndrome, alpha-thalassemia X-linked mental retardation syndrome, Angelman syndrome, Beckwith-Wiedemann syndrome, Prader-Willi syndrome, ICF (immunodeficiency, centromere instability, facial abnormalities) syndrome, Rett syndrome, Rubinstein-Taybi syndrome, and others, are also associated with cognitive impairment (reviewed by Calfa et al. 2012; Eggermann 2012; Gatto et al. 2012). It is also to be expected that inherited dysregulation of methylation interferes with the differentiation of brain cells and therefore brain function. However, the role of Epigenetics in normal brain function and of Patho-Epigenetics of the previously functional diseased brain is only beginning to be uncovered (reviewed by Klausz et al. 2012; Zelena 2012).

### 4.9.2 Normal Brain Function and Memory

Evidence is mounting that epigenetic processes are involved in the everyday operation of the central nervous system, and that long-term memory formation and synaptic plasticity needs accompanying epigenetic modifications of DNA and histones. The role of epigenetic regulation in memory formation has recently been reviewed (Kim and Kaang 2017). Therefore, only a selection of illustrative examples with a focus on CpG-methylation is mentioned here. In order to understand the epigenetic changes associated with learning, genome-wide histone and methylation profiles from neuronal and nonneuronal cells of the CA1 field of the hippocampus and the anterior cingulate cortex of adult mice at three time points before and 1 h and 4 weeks after learning were established. This provided a physiological gene regulatory network of learning as a background for the analysis of neurological and psychiatric diseases and potential development of medical treatment (Centeno et al. 2016). In contrast to short-term memory, long-term memory formation requires de novo protein synthesis (Davis and Squire 1984; Igaz et al. 2002). Early seminal experiments on individual *Aplysia californica* (California sea hare, a mollusk) neurons demonstrated that facilitating signals through serotonin (5-HT, 5-hydroxytryptamin), which are needed for new synapse formation, led to acetylation at the promoter of the C/EBP transcription factor gene, while inhibitory input signals through FMRF amide caused deacetylation at the C/EBP promoter. If both signals were simultaneously applied, the inhibitory signal dominated and yielded histone deacetylation and long-term synaptic depression (Guan et al. 2002). Recent experimentation on *Aplysia* sensory neurons uncovered a mechanism for sequence-specific methylation in the serotonin-dependent synaptic facilitating signal pathway. The CREB2 protein is the major inhibitory constraint on memory formation in *Aplysia*. A Piwi/piRNA (piwi-interacting RNA) complex conveyed the local methylation of a conserved CpG-island at the *CREB2* promoter (Rajasethupathy et al. 2012). However, it is still unclear whether piRNAs or other microRNAs are playing a general major role in mammalian sequence-specific methylation in the brain and interconnected memory formation (reviewed by Sweatt 2013).

After fear conditioning, i.e., learning, Dnmt gene expression was increased in the adult rat hippocampus leading to CpG-methylation at the memory suppressor gene *PP1* (protein phosphatase 1), while the promoter of the synaptic plasticity gene *RELN* (Reelin) was demethylated and expressed. These methylation changes were highly dynamic and back to normal within 24 hours. Accordingly, the gene for *BDNF* (brain-derived neurotrophic factor) responded with a quick change of exon-specific methylation, histone modifications and mRNA expression, which went back to normal within 24 hours. In addition, NMDA receptor blockade prevented these methylation changes and inhibited memory formation in rats (Lubin et al. 2008). Earlier yet, a hypomethylation at the *Bdnf* promoter region IV upon depolarization, the dissociation of the MeCP2-HDAC1-mSin3A repressive complex from the promoter, and a corresponding transcriptional increase of exon IV mRNA were observed in cortical mouse neurons cultured in vitro, 3 days after stimulation (Martinowich et al. 2003; reviewed by Grigorenko et al. 2016). Furthermore, Dnmt inhibition led to both a disturbed regulation of methylation and functional impairment of memory formation in the treated animals (Miller and Sweatt 2007). Transcriptional repression by methylation of another memory suppressor gene, *PPP3CA* (Calcineurin), in cortical neurons persisted up to 30 days after the learning experiments. Again, treatment with Dnmt inhibitors disrupted long-term fear memory (Miller et al. 2010).

Neurons are strongly expressing DNMT1, DNMT3A, and DNMT3B, and are highly enriched in 5hmC (reviewed by Delgado-Morales and Esteller 2017). The abundance of 5hmC in the brain suggested a functional role for 5hmC in the adult brain (Kriaucionis and Heintz 2009). It has been further suggested that 5hmC in the brain is not just an intermediate stage of the demethylation process operated by TET proteins, but also a key epigenetic mark for neurological disorders (Chen et al. 2014; Cheng et al. 2015). Whole genome analyses of 5hmC distribution with single-nucleotide resolution showed that 5hmC was even more enriched at constitutive exons of genes with synapse-related function, both in the human and the mouse brain (Khare et al. 2012). 5hmC in fetal brains was a marker for regulatory regions, which were poised for demethylation and activation in the adult brain. Demethylation was dependent on Tet2 (Lister et al. 2013). The regulation of methylation and demethylation relying on the interplay between DNMTs and TET-dioxygenases has been shown to play a major functional role in memory formation. Neuronal activity upon electro-convulsive treatment modified the neuronal DNA methylation landscape. About 1.4% of 219,991 CpG-sites of the adult mouse hippocampus showed a rapid active demethylation or de novo methylation. Regulated CpG-sites were significantly enriched in brain-specific genes related to neuronal plasticity (Guo et al. 2011; reviewed by Sweatt 2013; Kim-Ha and Kim 2016). Tet1 expression levels in the mouse hippocampus were decreased 3 to 4 h after neuronal activity, induced by either KCl depolarization, flurothyl-induced seizure, or fear conditioning. Following neuronal activation by flurothyl-induced seizure, 5mC and 5hmC levels underwent a decrease at 24 hours postseizure. Adeno-associated virus (AAV)-mediated overexpression of Tet1 in the dorsal hippocampus led to decreased 5mC and increased levels of 5hmC and of unmodified cytosines, 2 weeks after AAV injection, accompanied by an altered transcription of genes involved in memory formation and synaptic plasticity, and an impaired contextual fear memory formation at 24 hours after training (Kaas et al. 2013). Furthermore, *Tet1*-knockout mice suffer dysregulated hippocampal gene expression programs, impaired synaptic plasticity, and impaired memory extinction in learning experiments (Rudenko et al. 2013). Contrary to Tet1 expression, Tet3 expression was significantly increased shortly after neuronal stimulation and in learning experiments. Fear-extinction learning led to a Tet3-mediated accumulation of 5hmC, which was accompanied by changes in gene expression and rapid behavioral adaptation (Li et al. 2014). In mice, the cognitive decline associated with aging is accompanied by diminished expression of hippocampal Dnmt3a2 levels. After restoring Dnmt3a2 expression in old mice by injecting recombinant AAV carrying the Dnmt3a2 open reading frame, their cognitive ability was improved back to normal levels in fear conditioning experiments (Oliveira et al. 2012). Uhrf2 is a *bona-fide* 5hmC reader protein. *Uhrf2*-knockout mice were viable and exhibited no gross defects, but showed a reduced level of 5hmC and altered neuronal gene expression pattern and a partial impairment in spatial memory acquisition and retention. Therefore, Uhrf2 plays a functional role in long-term spatial memory (Chen et al. 2017).

Experience-dependent epigenomic changes in the offspring remind us of Jean-Baptiste de Lamarck's theory which is therefore reconsidered again. One aspect of Lamarck's theory, frequently exemplified by the famous giraffe stretching its neck to reach leaves high up in a tree, postulates that exercise will modify and strengthen organs whose enhanced function, in this case a longer and stronger neck, will in turn be inherited by the offspring. Also Charles Darwin's Pangenesis theory, developed to accommodate his predecessor Lamarck, and bashfully ignored for decades by evolutionary biologists, is being rescued from oblivion, due to the resemblance of Darwin's "gemmules" with extracellular vesicles, exosomes, or small RNAs (reviewed by Liu and Li 2012; Chen et al. 2016c; Minarovits and Niller 2017). There are two types of transgenerationally transmitted epigenomic changes in the central nervous system, which are produced by experience, one of them culturally transmitted, but not biologically inherited, while the other one is apparently inherited via the germline (reviewed by Sweatt 2013; Bohacek and Mansuy 2015). Maternal nurturing behavior regarding newborn pups triggers durable methylation changes in the glucocorticoid receptor (GR) signaling pathway in the central nervous system which has profound and durable effects on stress response behavior into the adult life. "Stressed" mice transmit their disturbed stress response behavior to their offspring in a cultural way by lack of appropriate grooming behavior, thereby creating the next generation of stressed or depressed animals with the same methylation changes in the GR promoter within the hippocampus (Weaver et al. 2004; reviewed by Champagne and Curley 2009; Anacker et al. 2014). Interestingly however, learning effects can be passed on to the offspring of mice not only through early-life experience but also through durable epigenetic changes which are inherited via the germline (reviewed by Bohacek and Mansuy 2015). For example, in the case of parental traumatic olfactory exposure with acetophenone, the olfactory receptor gene *Olfr151* became hypomethylated. Hypomethylation and enhanced neuroanatomical representation of the Olfr151 pathway persisted to the F2 generation. The transgenerational effects were inherited through the parental gametes (Dias and Ressler 2014). Early life traumatic experience led to an altered microRNA expression. The behavioral and epigenetic characteristics of the affected mice were transmitted through injection of sperm RNA into fertilized wild-type oocytes (Gapp et al. 2014).

An important mechanism of genetic neuronal plasticity interconnected with epigenetics is L1-retrotransposition which leads to genomic plasticity of high numbers of individual brain cells. This phenomenon resembles somewhat the somatic diversity of the adaptive immune system (reviewed by Sweatt 2013). Long interspersed nuclear elements-1 (LINE-1 or L1s) are abundant and comprise roughly 20% of the human genome. A massive somatic L1 insertional activity was observed in the normal adult human brain, but not other adult tissues, which leads to a physiological level of genetic mosaicism of individual neuronal genomes. Insertional activity is down-modulated by methyl-DNA binding protein MeCP2 which is mutated and therefore dysfunctional in patients with Rett syndrome who have a higher retrotranspositional activity in their brain cells (Muotri et al. 2010; reviewed by Erwin et al. 2014). Somatic L1-associated variants depending on either retrotransposition or deletion events occur in crucial neural genes and are hotspots for somatic copy number variation in the healthy human brain. An estimated 50% of all brain cells, including progenitor, glia, and neuronal cells, are affected by such somatic events (Erwin et al. 2016).

Besides and interconnected with CpG-methylation, a large number of posttranslational histone modifications and an abundance of specific microRNAs in the brain have been described in association with memory regulation (reviewed by Kim-Ha and Kim 2016; Kim and Kaang 2017). Furthermore, the regulation of the three-dimensional conformation and reconfiguration of neuronal chromatin in association with normal brain function and disease is just beginning to be explored and is probably most important (Juraeva et al. 2014; Flavahan et al. 2016; Mendizabal et al. 2016).

The rather new discipline of "neuroimaging epigenetics" tries to correlate epigenetic alterations of transcriptional promoters which are important for brain function, e.g., for neurotransmitter or brain receptor genes, with imaging techniques. The methylation status of promoters is correlated with visible changes in structural or functional magnetic resonance tomography (fMRT) or positron emission tomography (PET) images (reviewed by Nikolova and Hariri 2015). The exact mapping of the methylomes of different brain areas showed greater between-tissue variability than interindividual methylation variability. Nevertheless, some interindividual variability of brain methylation was reflected by the methylation status of peripheral DNA. Therefore, for practicability, neuroimaging epigenetics uses the peripheral methylation status, which is measured from blood or saliva as a proxy for brain methylation (Davies et al. 2012; Smith et al. 2015; reviewed by Nikolova and Hariri 2015).

### 4.9.3  Epigenetics of Psychiatric Disorders and Dementia

There is a broad overlap between epigenetic alterations associated with normal aging and dementia-associated epigenetic alterations. This fact is not surprising, because the most important single risk factor for dementia is aging (Heyn et al. 2012; Horvath et al. 2012). However, the distinction of the epigenomic dysregulation coming along with aging, mild cognitive impairment (MCI), or with Alzheimer dementia (AD) has been made. Paradoxically, in MCI patients, gene expression programs associated with synaptic plasticity and facilitation were connected with a lower score in mental performance testing (Berchtold et al. 2014). The overall methylation status seems to change bidirectionally with aging. While repetitive elements become hypomethylated in many tissues of mice and humans, many developmental genes are being hypermethylated with aging (Maegawa et al. 2010). Interestingly, the interindividual epigenomic profiles of the human brain become more similar in the later stages of life and also more similar between cerebral cortex and cerebellum. Furthermore, a loss of boundaries between transcriptional domains was observed, altogether resembling cell dedifferentiation (Oh et al. 2016; reviewed by Kim-Ha and Kim 2016).

Dementia is a disease trait common to various neurodegenerative disorders, like AD, Parkinson's disease (PD), Lewy body dementia (LBD), and frontotemporal dementia (FTD, also named Pick's disease). It has been postulated that the accumulation of epigenetic alterations and as a consequence alterations of gene expression during the lifetime might seed and sustain dementia-associated disorders which has been termed the "Latent Early-Life Associated Regulation" (LEARn) hypothesis. The LEARn model implies that incipient dementia may be diagnosed before the onset of symptoms and ideally, as a future perspective, be halted in time or even reverted (Lahiri et al. 2009; Maloney and Lahiri 2016). Interestingly, AD, PD, LBD, and also Down syndrome share a common set of genes which undergo aberrant methylation shifts (Sanchez-Mut et al. 2016). Future work on epigenomics of dementia must distinguish methylation patterns of different cell types and brain regions and also differentiate between 5mC, 5hmC, and all other forms of modified nucleotide bases. Because both 5mC and 5hmC protect cytosines from conversion to uracil, this cannot be accomplished through mere bisulfite sequencing. Interestingly, the epigenetic age of different brain regions differs. This applies strikingly to the cerebellum (Horvath et al. 2015). However, also the prefrontal cortex seems to age slower than other parts of the human brain and body (Klein and De Jager 2016). The current knowledge of the role of methylation and hydroxymethylation in the major dementia-related disorders AD, PD, LBD, and FTD has recently been comprehensively reviewed. Causality has still to be established for most of the alterations (Klein and De Jager 2016; Delgado-Morales and Esteller 2017). A genome-wide profiling for 5hmC in the AD brain identified 517 differentially hydroxymethylated regions (DhMRs) associated with neuritic plaques and 60 DhMRs associated with neurofibrillary tangles. Hydroxymethylated gene loci were significantly enriched for functioning in neurobiological processes (Zhao et al. 2017). Repeated gas-chromatographic measurements of overall 5mC and 5hmC levels from four different brain regions of different types and stages of dementia (AD, LBD, FTD) showed significant alterations especially in the preclinical stages of AD. This suggested that epigenetic alterations may play an early role in the progression of AD and other forms of dementia (Ellison et al. 2017). A transgenic mouse expressing the AD-associated protein p25 in dependence of doxycyclin regulation, exhibited memory deficits which were rescued by environmental enrichment. Memory restoration and synaptic function was accompanied by increased histone acetylation and hippocampal and cortical chromatin remodeling. Furthermore, HDAC inhibitor treatment of those mice induced neurite sprouting, increased synapse numbers and restored learning behavior and access to long-term memory as well (Fischer et al. 2007). In the inducible mouse model for AD, transcriptomic and epigenomic profiling of the hippocampus demonstrated a coordinated downregulation of synaptic plasticity and upregulation of immune response genes and regulatory regions over time. The situation was reflected in human brain tissue making the inducible AD mouse a useful model (Gjoneska et al. 2015).

Epigenetic mechanisms may be of particular importance for multifactorial diseases with low genetic penetrance (reviewed by Sweatt 2013). Epigenetic alterations are certainly associated with the development of psychosis, i.e., schizophrenia, bipolar disorder, and major depression which may also open up new avenues of future treatment (Oh et al. 2015; reviewed by Labrie et al. 2012). A recent genomewide association study (GWAS) on schizophrenia (SCZ) described 108 distinct SCZ-associated loci, 83 of which had been newly detected (Schizophrenia Working Group of the Psychiatric Genomics Consortium 2014). Another genome-wide approach is the definition of methylation quantitative trait loci (meQTLs) by charting methylation levels at specific loci which are dependent on individual genetic differences, i.e., mostly SNPs, at other defined loci. Cis-acting and trans-acting meQTLs are distinguished and are enriched at regulatory sites. Several meQTL studies have been conducted on psychiatric diseases (reviewed by Hoffmann et al. 2016). A landmark study on the genomic landscapes of the brains of 335 nonaffected control individuals across all ages from the 14th gestational week fetal stage to 80 years of age, and of 191 patients with schizophrenia identified 2104 CpG-sites that differed between SCZ-patients and controls. SCZ-GWAS-risk loci were slightly enriched among those sites. The SCZ-related CpG-sites were strongly enriched for neurodevelopmental and differentiation genes. Furthermore, they strongly correlated with changes at the prenatal-to-postnatal transition, but not with later changes at the transition from adolescence to later adult life (Jaffe et al. 2016). Similarly, more than 16,000 meQTLs from 166 fetal brains were found to be fourfold enriched in SCZ-GWAS-risk loci (Hannon et al. 2016). While the effects of individual meQTLs are small, the location of meQTLs to genomic regions important for methylome reconfiguration during early brain development might hint at early vulnerable periods for the development of SCZ (reviewed by Hoffmann et al. 2016; Bale 2015). A study examining the epigenome of different postmortem brain regions of schizophrenic patients using the 450 K methylation array identified 139 differentially methylated CpG-sites and yielded a complex picture. Differences were located at known and novel candidate gene sequences, included gene bodies and CpG-islands, shores, and shelves, and were dependent on the brain area analyzed. Furthermore, the methylation states in brain tissue were largely not reflected by blood cell sampling underscoring the need to analyze the appropriate tissues for meaningful conclusions (Alelu-Paz et al. 2016). Valproic acid (VPA) has long been in use and still is for the treatment of epilepsy and bipolar disorder, depression and schizophrenia and more recently off-label for migraine-associated cluster-headaches. The mechanisms of action are assumed to be the blockade of neuronal ion channels, the increased synthesis of the inhibitory neurotransmitter GABA (gamma-aminobutyric acid), and HDAC inhibition. VPAs general long-term neuroprotective effect is ascribed to its epigenetic activity which prevents seizure-dependent aberrant neurogenesis in the adult hippocampus and the seizure-associated cognitive decline (Jessberger et al. 2007; reviewed by Monti et al. 2009). The implications of epigenetics in the workings of the central nervous system, especially of the above-mentioned "Lamarckian" experiments, of the wide-spread genetic mosaicism within the brain, and of the apparently very-early-in-life risk for schizophrenia are somewhat mind-boggling.

## 5     Research Needs

Research on epigenetics within many subdisciplines of biology has – figuratively speaking – taken off after the year 2000 (Sweatt 2013). Due to the development of powerful high-resolution and high-throughput methods, and due to the involvement of epigenetic regulation in all sorts of biological mechanisms (see Sect. 4), this great interest in epigenetics has not come by chance.

### 5.1     Philosophical Questions

From nineteenth-century philosophy, a notorious sentence by the then world-renowned physician and physiologist Jakob Moleschott was passed down to us: "Ohne Phosphor kein Gedanke" (no thinking without phosphorus). This sentence highlighting the philosophical direction of scientific materialism of the times was ridiculed by Arthur Schopenhauer as "Barbiergesellen-Philosophie" (barber-shop philosophy). Varying this phosphorus-sentence, today we might say "no thinking without methylation" or more precisely and less catchpenny-like "there may be no memory formation or thinking without the dynamic regulation of methylation states in the central nervous system." Perhaps, Schopenhauer would not criticize us anymore for that latter phrase, as long as we would not imply any philosophical overstatement, because latter sentence represents a differentiated and experiment-based view on the functioning of the brain, as far as we know today. However, some of the fundamentals, especially the dichotomy between brain and mind, and the question of how both relate to each other, are still an unsolved riddle, just like 165 years ago. Beyond the fundamental mind-brain-problem, there is the not nearly as fundamental "nature versus nurture"-problem which has found the entrance to its solution (Sweatt 2009). Clearly, epigenetic regulation is the biological interface for the dynamic interplay between genes and environmental exposure or experience (Sweatt 2013).

Epigenetic mechanisms are at the center of cell differentiation and organ development, but – as we have seen above – are also involved in everyday function of the brain and, of course, also of other organs (Sweatt 2013). It will be interesting to continue to comprehensively map the epigenome of the developing brain and of the normally operating young and adult brain. This map should sort out the epigenetic marks and genes involved in cell fate determination and the epigenetic marks associated with organ function. The same should be tried for, let's say, the liver and all other organs. An associated experimental question might be whether the epigenomes of specific single cells or structured cell groups in the liver are the same or very similar to the epigenomes of brain cells, neuronal or glial. To answer those questions, single cell methylomes of myriad cells would need to be established, and large amounts of bioinformatic power need to be tapped (Guo et al. 2013; Moroz and Kohn 2013; Smallwood et al. 2014; Farlik et al. 2015; Guo et al. 2015b; Angermueller et al. 2016; Gravina et al. 2016; Hou et al. 2016; Hu et al. 2016). After all, we might be in for a surprise about some large overlaps between the epigenetics of development and the epigenetics of normal function and, possibly, about some striking epigenetic similarities between different organs in their proper mode of functioning. Catchily speaking: why does the brain do the thinking and not the liver? Not a wee bit?

Furthermore, the act of thinking certainly relies in part on prior memory formation. The question then is whether quick thinking within seconds or decision-making within split-seconds may also rely on the (quick) dynamic regulation of CpG-methylation, of posttranslational modification of histones, or of three-dimensional chromatin structures in the brain. However, it may be hard to design experimental approaches to address such a question. For a further reading on neuroepigenetics and the associated major biological questions we refer to a most excellent perspective by J. David Sweatt (Sweatt 2013).

## 5.2   Sequence- or Locus-Specific Regulation of Methylation

A major question of epigenetic regulation is how CpG-methylation and other epigenetic marks may be targeted in a sequence-specific or locus-specific manner (see Sect. 3.2.5). While RNA-directed localization of epigenetic marks is an important mechanism in protozoa, plants, flies, and nematodes, mammalian cells in general seem to lack the respective biochemical machineries (Zhang and Zhu 2011; Yang et al. 2016a; reviewed by Joh et al. 2014). An important highly specific example of locus-specific methylation has been found by Rajasethupathy et al. who described a Piwi/piRNA (piwi-interacting RNA) complex which imparted the local methylation of a conserved CpG-island at the *CREB2* promoter of *Aplysia californica* (Rajasethupathy et al. 2012). A mammalian example showing the requirement for components of the PIWI-interacting RNA (piRNA) pathway in local DNA methylation has been described for a differentially methylated region (DMR) of the paternally imprinted mouse Rasgrf1-locus (Watanabe et al. 2011). Nevertheless, it is still unclear whether piRNAs or other microRNAs are playing a general dominant role in mammalian sequence-specific methylation and what the exact mechanisms are (Stewart et al. 2016). However, since piRNAs have been described to play a role in reshaping the epigenome of cancer cells, and the epigenomes of mice were shaped by the injection of sperm RNAs or tRNA-derived small sperm RNAs (tsRNAs) into fertilized oozytes, a prominent role for small RNAs in the directed placement of epigenetic marks is to be expected in all likelihood (Gapp et al. 2014; Fu et al. 2015a; Chen et al. 2016b). Remarkably, the relative spatiotemporal distribution of small RNAs changes considerably during the development of primordial spermatogonia to mature spermatozoa (reviewed by Chen et al. 2016c). May the different amounts of piRNAs, miRNAs, and tsRNAs in different developmental stages of spermatogenesis reflect on a changing biochemical RNA-processing machinery available to the cell in the respective phase? Furthermore, the question may be asked whether sperm RNAs are under certain physiological circumstances able to influence the epigenomes of somatic cells of adult females.

## 5.3    Patho-Epigenetics of Disease and Perspectives for Medical Treatment

Not only has research addressing the basic mechanisms of methylation taken off in recent years, but also research on epigenetics in the context of diseases and their potential treatment. The description of epigenetic alterations and the accumulation of data on methylomes of diseases as diverse as cancer of many subtypes, autoimmune, infectious, genetic, imprinting, genetic, cardiovascular, psychiatric and neurologic disease, and as a subspecialty of malignant disease, the epigenetics of virus- or microbe-associated cancer has progressed quite far. This is exemplified by a large body of literature and a series of textbooks on the Patho-Epigenetics of Human Disease (Minarovits and Niller 2012; Tollefsbol 2012; Minarovits and Niller 2016). In some cases, especially in connection with infectious disease, e.g., HIV-infection, epigenetic alterations may imply the use of miRNAs and epigenetic drugs, like HDAC-inhibitors, as promising therapeutic approaches (Takacs et al. 2009; Ay et al. 2013; Szenthe et al. 2013; Minarovits and Niller 2017). A very interesting and promising gene-therapeutic approach to rewriting the epigenome may be the employment of CRISPR/Cas or other sequence-specific binding proteins to target localized methylation and demethylation enzymes at will (Xu et al. 2016; reviewed by Cano-Rodriguez and Rots 2016). Epigenetic approaches may be considered even for genetic syndromes, like trisomy 21 (Dekker et al. 2014).

**Acknowledgments** This work was supported by the grant GINOP-2.3.2-15-2016-00011 to a consortium led by the University of Szeged, Szeged, Hungary (participants: the University of Debrecen, Debrecen, and the Biological Research Centre, Hungarian Academy of Sciences, Szeged, Hungary), project leader Janos Minarovits. The grant was funded by the European Regional Development Fund of the European Union and managed in the framework of Economic Development and Innovation Operational Programme by the Ministry of National Economy, National Research, Development and Innovation Office, Budapest, Hungary.

## References

Adhikari S, Curtis PD (2016) DNA methyltransferases and epigenetic regulation in bacteria. FEMS Microbiol Rev 40:575–591
Alelu-Paz R, Carmona FJ, Sanchez-Mut JV, Cariaga-Martinez A, Gonzalez-Corpas A, Ashour N, Orea MJ, Escanilla A, Monje A, Guerrero Marquez C, Saiz-Ruiz J, Esteller M, Ropero S (2016) Epigenetics in schizophrenia: a pilot study of global DNA methylation in different brain regions associated with higher cognitive functions. Front Psychol 7:1496

Amouroux R, Nashun B, Shirane K, Nakagawa S, Hill PW, D'souza Z, Nakayama M, Matsuda M, Turp A, Ndjetehe E, Encheva V, Kudo NR, Koseki H, Sasaki H, Hajkova P (2016) De novo DNA methylation drives 5hmC accumulation in mouse zygotes. Nat Cell Biol 18:225–233

Anacker C, O'Donnell KJ, Meaney MJ (2014) Early life adversity and the epigenetic programming of hypothalamic-pituitary-adrenal function. Dialogues Clin Neurosci 16:321–333

Angermueller C, Clark SJ, Lee HJ, Macaulay IC, Teng MJ, Hu TX, Krueger F, Smallwood SA, Ponting CP, Voet T, Kelsey G, Stegle O, Reik W (2016) Parallel single-cell sequencing links transcriptional and epigenetic heterogeneity. Nat Methods 13:229–232

Aran D, Sabato S, Hellman A (2013) DNA methylation of distal regulatory sites characterizes dysregulation of cancer genes. Genome Biol 14:R21

Ardissone S, Redder P, Russo G, Frandi A, Fumeaux C, Patrignani A, Schlapbach R, Falquet L, Viollier PH (2016) Cell cycle constraints and environmental control of local DNA Hypo-methylation in alpha-Proteobacteria. PLoS Genet 12:e1006499

Athanasiadou R, De Sousa D, Myant K, Merusi C, Stancheva I, Bird A (2010) Targeting of de novo DNA methylation throughout the Oct-4 gene regulatory region in differentiating embryonic stem cells. PLoS One 5:e9937

Ay E, Banati F, Mezei M, Bakos A, Niller HH, Buzas K, Minarovits J (2013) Epigenetics of HIV infection: promising research areas and implications for therapy. AIDS Rev 15:181–188

Bahar Halpern K, Vana T, Walker MD (2014) Paradoxical role of DNA methylation in activation of FoxA2 gene expression during endoderm development. J Biol Chem 289:23882–23892

Bala Tannan N, Brahmachary M, Garg P, Borel C, Alnefaie R, Watson CT, Thomas NS, Sharp AJ (2014) DNA methylation profiling in X;autosome translocations supports a role for L1 repeats in the spread of X chromosome inactivation. Hum Mol Genet 23:1224–1236

Bale TL (2015) Epigenetic and transgenerational reprogramming of brain development. Nat Rev Neurosci 16:332–344

Barlow DP, Bartolomei MS (2014) Genomic imprinting in mammals. Cold Spring Harb Perspect Biol 6:a018382

Bartolomei MS, Ferguson-Smith AC (2011) Mammalian genomic imprinting. Cold Spring Harb Perspect Biol 3:a002592

Becker C, Hagmann J, Muller J, Koenig D, Stegle O, Borgwardt K, Weigel D (2011) Spontaneous epigenetic variation in the Arabidopsis thaliana methylome. Nature 480:245–249

Bell AC, Felsenfeld G (2000) Methylation of a CTCF-dependent boundary controls imprinted expression of the Igf2 gene. Nature 405:482–485

Bell CG, Wilson GA, Beck S (2014) Human-specific CpG 'beacons' identify human-specific prefrontal cortex H3K4me3 chromatin peaks. Epigenomics 6:21–31

Bell CG, Wilson GA, Butcher LM, Roos C, Walter L, Beck S (2012) Human-specific CpG 'beacons' identify loci associated with human-specific traits and disease. Epigenetics 7:1188–1199

Berchtold NC, Sabbagh MN, Beach TG, Kim RC, Cribbs DH, Cotman CW (2014) Brain gene expression patterns differentiate mild cognitive impairment from normal aged and Alzheimer's disease. Neurobiol Aging 35:1961–1972

Berkyurek AC, Suetake I, Arita K, Takeshita K, Nakagawa A, Shirakawa M, Tajima S (2014) The DNA methyltransferase Dnmt1 directly interacts with the SET and RING finger-associated (SRA) domain of the multifunctional protein Uhrf1 to facilitate accession of the catalytic center to hemi-methylated DNA. J Biol Chem 289:379–386

Bestor TH (1990) DNA methylation: evolution of a bacterial immune function into a regulator of gene expression and genome structure in higher eukaryotes. Philos Trans R Soc Lond Ser B Biol Sci 326:179–187

Bestor TH (2000) The DNA methyltransferases of mammals. Hum Mol Genet 9:2395–2402

Bestor TH, Edwards JR, Boulard M (2015) Notes on the role of dynamic DNA methylation in mammalian development. Proc Natl Acad Sci U S A 112:6796–6799

Bheemanaik S, Reddy YV, Rao DN (2006) Structure, function and mechanism of exocyclic DNA methyltransferases. Biochem J 399:177–190

Bhende PM, Seaman WT, Delecluse HJ, Kenney SC (2004) The EBV lytic switch protein, Z, preferentially binds to and activates the methylated viral genome. Nat Genet 36:1099–1104

Bianchi C, Zangi R (2014) Dual base-flipping of cytosines in a CpG dinucleotide sequence. Biophys Chem 187-188:14–22

Bird A (2002) DNA methylation patterns and epigenetic memory. Genes Dev 16:6–21

Bird AP (1986) CpG-rich islands and the function of DNA methylation. Nature 321:209–213

Bird AP (1987) CpG islands as gene markers in the vertebrate nucleus. Trends Genet 3:342–347

Blattler A, Farnham PJ (2013) Cross-talk between site-specific transcription factors and DNA methylation states. J Biol Chem 288:34287–34294

Blow MJ, Clark TA, Daum CG, Deutschbauer AM, Fomenkov A, Fries R, Froula J, Kang DD, Malmstrom RR, Morgan RD, Posfai J, Singh K, Visel A, Wetmore K, Zhao Z, Rubin EM, Korlach J, Pennacchio LA, Roberts RJ (2016) The epigenomic landscape of prokaryotes. PLoS Genet 12:e1005854

Bogdanovic O, Veenstra GJ (2009) DNA methylation and methyl-CpG binding proteins: developmental requirements and function. Chromosoma 118:549–565

Bohacek J, Mansuy IM (2015) Molecular insights into transgenerational non-genetic inheritance of acquired behaviours. Nat Rev Genet 16:641–652

Braidotti G, Baubec T, Pauler F, Seidl C, Smrzka O, Stricker S, Yotova I, Barlow DP (2004) The air noncoding RNA: an imprinted cis-silencing transcript. Cold Spring Harb Symp Quant Biol 69:55–66

Brockdorff N (2013) Noncoding RNA and Polycomb recruitment. RNA 19:429–442

Brockdorff N, Turner BM (2015) Dosage compensation in mammals. Cold Spring Harb Perspect Biol 7:a019406

Calfa G, Percy AK, Pozzo-Miller L (2012) Dysfunction of methyl-CpG-binding protein MeCP2 in Rett syndrome. In: Minarovits J, Niller HH (eds) Patho-epigenetics of disease. Springer, New York, pp 43–69

Cano-Rodriguez D, Rots MG (2016) Epigenetic editing: on the verge of reprogramming gene expression at will. Curr Genet Med Rep 4:170–179

Casadesus J (2016) Bacterial DNA methylation and Methylomes. Adv Exp Med Biol 945:35–61

Centeno TP, Shomroni O, Hennion M, Halder R, Vidal R, Rahman RU, Bonn S (2016) Genome-wide chromatin and gene expression profiling during memory formation and maintenance in adult mice. Sci Data 3:160090

Cerase A, Pintacuda G, Tattermusch A, Avner P (2015) Xist localization and function: new insights from multiple levels. Genome Biol 16:166

Chaligne R, Heard E (2014) X-chromosome inactivation in development and cancer. FEBS Lett 588:2514–2522

Champagne FA, Curley JP (2009) Epigenetic mechanisms mediating the long-term effects of maternal care on development. Neurosci Biobehav Rev 33:593–600

Chen BF, Chan WY (2014) The de novo DNA methyltransferase DNMT3A in development and cancer. Epigenetics 9:669–677

Chen CK, Blanco M, Jackson C, Aznauryan E, Ollikainen N, Surka C, Chow A, Cerase A, McDonel P, Guttman M (2016a) Xist recruits the X chromosome to the nuclear lamina to enable chromosome-wide silencing. Science 354:468–472

Chen Q, Yan M, Cao Z, Li X, Zhang Y, Shi J, Feng GH, Peng H, Zhang X, Zhang Y, Qian J, Duan E, Zhai Q, Zhou Q (2016b) Sperm tsRNAs contribute to intergenerational inheritance of an acquired metabolic disorder. Science 351:397–400

Chen Q, Yan W, Duan E (2016c) Epigenetic inheritance of acquired traits through sperm RNAs and sperm RNA modifications. Nat Rev Genet 17:733–743

Chen, R., Zhang, Q., Duan, X., York, P., Chen, G.D., Yin, P., Zhu, H., Xu, M., Chen, P., Wu, Q., Li, D., Samarut, J., Xu, G., Zhang, P., Cao, X., Li, J., and Wong, J. (2017). The 5-hydroxymethylcytosine (5hmC) reader Uhrf2 is required for normal levels of 5hmC in mouse adult brain and spatial learning and memory. J Biol Chem.

Chen T, Tsujimoto N, Li E (2004) The PWWP domain of Dnmt3a and Dnmt3b is required for directing DNA methylation to the major satellite repeats at pericentric heterochromatin. Mol Cell Biol 24:9048–9058

Chen Y, Damayanti NP, Irudayaraj J, Dunn K, Zhou FC (2014) Diversity of two forms of DNA methylation in the brain. Front Genet 5:46

Cheng X, Blumenthal RM (2011) Introduction–epiphanies in epigenetics. Prog Mol Biol Transl Sci 101:1–21

Cheng Y, Bernstein A, Chen D, Jin P (2015) 5-Hydroxymethylcytosine: a new player in brain disorders? Exp Neurol 268:3–9

Chestnut BA, Chang Q, Price A, Lesuisse C, Wong M, Martin LJ (2011) Epigenetic regulation of motor neuron cell death through DNA methylation. J Neurosci 31:16619–16636

Choi SH, Heo K, Byun HM, An W, Lu W, Yang AS (2011) Identification of preferential target sites for human DNA methyltransferases. Nucleic Acids Res 39:104–118

Clark AT (2015) DNA methylation remodeling in vitro and in vivo. Curr Opin Genet Dev 34:82–87

Cohen NR, Ross CA, Jain S, Shapiro RS, Gutierrez A, Belenky P, Li H, Collins JJ (2016) A role for the bacterial GATC methylome in antibiotic stress survival. Nat Genet 48:581–586

Colot V, Rossignol JL (1999) Eukaryotic DNA methylation as an evolutionary device. BioEssays 21:402–411

Cooper DN, Mort M, Stenson PD, Ball EV, Chuzhanova NA (2010) Methylation-mediated deamination of 5-methylcytosine appears to give rise to mutations causing human inherited disease in CpNpG trinucleotides, as well as in CpG dinucleotides. Hum Genomics 4:406–410

Csankovszki G, Nagy A, Jaenisch R (2001) Synergism of Xist RNA, DNA methylation, and histone hypoacetylation in maintaining X chromosome inactivation. J Cell Biol 153:773–784

Dabe EC, Sanford RS, Kohn AB, Bobkova Y, Moroz LL (2015) DNA methylation in basal metazoans: insights from ctenophores. Integr Comp Biol 55:1096–1110

Damelin M, Bestor TH (2007) Biological functions of DNA methyltransferase 1 require its methyltransferase activity. Mol Cell Biol 27:3891–3899

Dan J, Chen T (2016) Genetic studies on mammalian DNA methyltransferases. Adv Exp Med Biol 945:123–150

Davies MN, Volta M, Pidsley R, Lunnon K, Dixit A, Lovestone S, Coarfa C, Harris RA, Milosavljevic A, Troakes C, Al-Sarraj S, Dobson R, Schalkwyk LC, Mill J (2012) Functional annotation of the human brain methylome identifies tissue-specific epigenetic variation across brain and blood. Genome Biol 13:R43

Davis HP, Squire LR (1984) Protein synthesis and memory: a review. Psychol Bull 96:518–559

Dawlaty MM, Breiling A, Le T, Barrasa MI, Raddatz G, Gao Q, Powell BE, Cheng AW, Faull KF, Lyko F, Jaenisch R (2014) Loss of Tet enzymes compromises proper differentiation of embryonic stem cells. Dev Cell 29:102–111

De Larco JE, Wuertz BR, Yee D, Rickert BL, Furcht LT (2003) Atypical methylation of the interleukin-8 gene correlates strongly with the metastatic potential of breast carcinoma cells. Proc Natl Acad Sci U S A 100:13988–13993

Dean W (2014) DNA methylation and demethylation: a pathway to gametogenesis and development. Mol Reprod Dev 81:113–125

Deaton AM, Bird A (2011) CpG islands and the regulation of transcription. Genes Dev 25:1010–1022

Dekker AD, De Deyn PP, Rots MG (2014) Epigenetics: the neglected key to minimize learning and memory deficits in down syndrome. Neurosci Biobehav Rev 45:72–84

Delatte B, Deplus R, Fuks F (2014) Playing TETris with DNA modifications. EMBO J 33:1198–1211

Delgado-Morales R, Esteller M (2017) Opening up the DNA methylome of dementia. Mol Psychiatry 22:485–496. https://doi.org/10.1038/mp.2016.242

Denis H, Ndlovu MN, Fuks F (2011) Regulation of mammalian DNA methyltransferases: a route to new mechanisms. EMBO Rep 12:647–656

Deplus R, Blanchon L, Rajavelu A, Boukaba A, Defrance M, Luciani J, Rothe F, Dedeurwaerder S, Denis H, Brinkman AB, Simmer F, Muller F, Bertin B, Berdasco M, Putmans P, Calonne E, Litchfield DW, De Launoit Y, Jurkowski TP, Stunnenberg HG, Bock C, Sotiriou C, Fraga MF, Esteller M, Jeltsch A, Fuks F (2014a) Regulation of DNA methylation patterns by CK2-mediated phosphorylation of Dnmt3a. Cell Rep 8:743–753

Deplus R, Delatte B, Schwinn MK, Defrance M, Mendez J, Murphy N, Dawson MA, Volkmar M, Putmans P, Calonne E, Shih AH, Levine RL, Bernard O, Mercher T, Solary E, Urh M, Daniels DL, Fuks F (2013) TET2 and TET3 regulate GlcNAcylation and H3K4 methylation through OGT and SET1/COMPASS. EMBO J 32:645–655

Deplus R, Denis H, Putmans P, Calonne E, Fourrez M, Yamamoto K, Suzuki A, Fuks F (2014b) Citrullination of DNMT3A by PADI4 regulates its stability and controls DNA methylation. Nucleic Acids Res 42:8285–8296

Dhayalan A, Rajavelu A, Rathert P, Tamas R, Jurkowska RZ, Ragozin S, Jeltsch A (2010) The Dnmt3a PWWP domain reads histone 3 lysine 36 trimethylation and guides DNA methylation. J Biol Chem 285:26114–26120

Dhayalan A, Tamas R, Bock I, Tattermusch A, Dimitrova E, Kudithipudi S, Ragozin S, Jeltsch A (2011) The ATRX-ADD domain binds to H3 tail peptides and reads the combined methylation state of K4 and K9. Hum Mol Genet 20:2195–2203

Dias BG, Ressler KJ (2014) Parental olfactory experience influences behavior and neural structure in subsequent generations. Nat Neurosci 17:89–96

Du Q, Wang Z, Schramm VL (2016) Human DNMT1 transition state structure. Proc Natl Acad Sci U S A 113:2916–2921

Duymich CE, Charlet J, Yang X, Jones PA, Liang G (2016) DNMT3B isoforms without catalytic activity stimulate gene body methylation as accessory proteins in somatic cells. Nat Commun 7:11453

Easwaran HP, Schermelleh L, Leonhardt H, Cardoso MC (2004) Replication-independent chromatin loading of Dnmt1 during G2 and M phases. EMBO Rep 5:1181–1186

Eden A, Gaudet F, Waghmare A, Jaenisch R (2003) Chromosomal instability and tumors promoted by DNA hypomethylation. Science 300:455

Edwards CA, Rens W, Clarke O, Mungall AJ, Hore T, Graves JA, Dunham I, Ferguson-Smith AC, Ferguson-Smith MA (2007) The evolution of imprinting: chromosomal mapping of orthologues of mammalian imprinted domains in monotreme and marsupial mammals. BMC Evol Biol 7:157

Eggermann T (2012) Imprinting disorders. In: Minarovits J, Niller HH (eds) Patho-epigenetics of disease. Springer, New York, pp 379–395

Ehrlich M, Wilson GG, Kuo KC, Gehrke CW (1987) N4-methylcytosine as a minor base in bacterial DNA. J Bacteriol 169:939–943

Ellison EM, Abner EL, Lovell MA (2017) Multiregional analysis of global 5-methylcytosine and 5-hydroxymethylcytosine throughout the progression of Alzheimer's disease. J Neurochem 140:383–394

Engreitz JM, Pandya-Jones A, McDonel P, Shishkin A, Sirokman K, Surka C, Kadri S, Xing J, Goren A, Lander ES, Plath K, Guttman M (2013) The Xist lncRNA exploits three-dimensional genome architecture to spread across the X chromosome. Science 341:1237973

Erwin JA, Marchetto MC, Gage FH (2014) Mobile DNA elements in the generation of diversity and complexity in the brain. Nat Rev Neurosci 15:497–506

Erwin JA, Paquola AC, Singer T, Gallina I, Novotny M, Quayle C, Bedrosian TA, Alves FI, Butcher CR, Herdy JR, Sarkar A, Lasken RS, Muotri AR, Gage FH (2016) L1-associated genomic regions are deleted in somatic cells of the healthy human brain. Nat Neurosci 19:1583–1591

Evsikov AV, Marin De Evsikova C (2016) Friend or foe: epigenetic regulation of retrotransposons in mammalian oogenesis and early development. Yale J Biol Med 89:487–497

Farlik M, Halbritter F, Muller F, Choudry FA, Ebert P, Klughammer J, Farrow S, Santoro A, Ciaurro V, Mathur A, Uppal R, Stunnenberg HG, Ouwehand WH, Laurenti E, Lengauer T,

Frontini M, Bock C (2016) DNA methylation dynamics of human hematopoietic stem cell differentiation. Cell Stem Cell 19:808–822

Farlik M, Sheffield NC, Nuzzo A, Datlinger P, Schonegger A, Klughammer J, Bock C (2015) Single-cell DNA methylome sequencing and bioinformatic inference of epigenomic cell-state dynamics. Cell Rep 10:1386–1397

Farthing CR, Ficz G, Ng RK, Chan CF, Andrews S, Dean W, Hemberger M, Reik W (2008) Global mapping of DNA methylation in mouse promoters reveals epigenetic reprogramming of pluripotency genes. PLoS Genet 4:e1000116

Fatemi M, Hermann A, Gowher H, Jeltsch A (2002) Dnmt3a and Dnmt1 functionally cooperate during de novo methylation of DNA. Eur J Biochem 269:4981–4984

Feng S, Jacobsen SE, Reik W (2010) Epigenetic reprogramming in plant and animal development. Science 330:622–627

Ficz G (2015) New insights into mechanisms that regulate DNA methylation patterning. J Exp Biol 218:14–20

Fischer A, Sananbenesi F, Wang X, Dobbin M, Tsai LH (2007) Recovery of learning and memory is associated with chromatin remodelling. Nature 447:178–182

Flavahan WA, Drier Y, Liau BB, Gillespie SM, Venteicher AS, Stemmer-Rachamimov AO, Suva ML, Bernstein BE (2016) Insulator dysfunction and oncogene activation in IDH mutant gliomas. Nature 529:110–114

Fouse SD, Shen Y, Pellegrini M, Cole S, Meissner A, Van Neste L, Jaenisch R, Fan G (2008) Promoter CpG methylation contributes to ES cell gene regulation in parallel with Oct4/Nanog, PcG complex, and histone H3 K4/K27 trimethylation. Cell Stem Cell 2:160–169

Frank D, Keshet I, Shani M, Levine A, Razin A, Cedar H (1991) Demethylation of CpG islands in embryonic cells. Nature 351:239–241

Frank D, Lichtenstein M, Paroush Z, Bergman Y, Shani M, Razin A, Cedar H (1990) Demethylation of genes in animal cells. Philos Trans R Soc Lond Ser B Biol Sci 326:241–251

Fritz EL, Papavasiliou FN (2010) Cytidine deaminases: AIDing DNA demethylation? Genes Dev 24:2107–2114

Frommer M, McDonald LE, Millar DS, Collis CM, Watt F, Grigg GW, Molloy PL, Paul CL (1992) A genomic sequencing protocol that yields a positive display of 5-methylcytosine residues in individual DNA strands. Proc Natl Acad Sci U S A 89:1827–1831

Fu A, Jacobs DI, Hoffman AE, Zheng T, Zhu Y (2015a) PIWI-interacting RNA 021285 is involved in breast tumorigenesis possibly by remodeling the cancer epigenome. Carcinogenesis 36:1094–1102

Fu Y, Luo GZ, Chen K, Deng X, Yu M, Han D, Hao Z, Liu J, Lu X, Dore LC, Weng X, Ji Q, Mets L, He C (2015b) N6-methyldeoxyadenosine marks active transcription start sites in Chlamydomonas. Cell 161:879–892

Fujita N, Watanabe S, Ichimura T, Tsuruzoe S, Shinkai Y, Tachibana M, Chiba T, Nakao M (2003) Methyl-CpG binding domain 1 (MBD1) interacts with the Suv39h1-HP1 heterochromatic complex for DNA methylation-based transcriptional repression. J Biol Chem 278:24132–24138

Gapp K, Jawaid A, Sarkies P, Bohacek J, Pelczar P, Prados J, Farinelli L, Miska E, Mansuy IM (2014) Implication of sperm RNAs in transgenerational inheritance of the effects of early trauma in mice. Nat Neurosci 17:667–669

Gatto S, D'Esposito M, Matarazzo MR (2012) The role of *DNMT3B* mutations in the pathogenesis of ICF syndrome. In: Minarovits J, Niller HH (eds) Patho-Epigenetics of disease. Springer, New York, pp 15–41

Gaudet F, Hodgson JG, Eden A, Jackson-Grusby L, Dausman J, Gray JW, Leonhardt H, Jaenisch R (2003) Induction of tumors in mice by genomic hypomethylation. Science 300:489–492

Georgia S, Kanji M, Bhushan A (2013) DNMT1 represses p53 to maintain progenitor cell survival during pancreatic organogenesis. Genes Dev 27:372–377

Gifford WD, Pfaff SL, Macfarlan TS (2013) Transposable elements as genetic regulatory substrates in early development. Trends Cell Biol 23:218–226

Gilbert SL, Pehrson JR, Sharp PA (2000) XIST RNA associates with specific regions of the inactive X chromatin. J Biol Chem 275:36491–36494

Gilbert SL, Sharp PA (1999) Promoter-specific hypoacetylation of X-inactivated genes. Proc Natl Acad Sci U S A 96:13825–13830

Ginno PA, Lim YW, Lott PL, Korf I, Chedin F (2013) GC skew at the 5′ and 3′ ends of human genes links R-loop formation to epigenetic regulation and transcription termination. Genome Res 23:1590–1600

Ginno PA, Lott PL, Christensen HC, Korf I, Chedin F (2012) R-loop formation is a distinctive characteristic of unmethylated human CpG island promoters. Mol Cell 45:814–825

Giunta S, Funabiki H (2017) Integrity of the human centromere DNA repeats is protected by CENP-A, CENP-C, and CENP-T. Proc Natl Acad Sci U S A 114:1928–1933

Gjoneska E, Pfenning AR, Mathys H, Quon G, Kundaje A, Tsai LH, Kellis M (2015) Conserved epigenomic signals in mice and humans reveal immune basis of Alzheimer's disease. Nature 518:365–369

Goll MG, Kirpekar F, Maggert KA, Yoder JA, Hsieh CL, Zhang X, Golic KG, Jacobsen SE, Bestor TH (2006) Methylation of tRNAAsp by the DNA methyltransferase homolog Dnmt2. Science 311:395–398

Goodier JL (2016) Restricting retrotransposons: a review. Mob DNA 7:16

Gopalakrishnan S, Sullivan BA, Trazzi S, Della Valle G, Robertson KD (2009) DNMT3B interacts with constitutive centromere protein CENP-C to modulate DNA methylation and the histone code at centromeric regions. Hum Mol Genet 18:3178–3193

Gopalakrishnan S, Van Emburgh BO, Robertson KD (2008) DNA methylation in development and human disease. Mutat Res 647:30–38

Gordon CA, Hartono SR, Chedin F (2013) Inactive DNMT3B splice variants modulate de novo DNA methylation. PLoS One 8:e69486

Goto K, Numata M, Komura JI, Ono T, Bestor TH, Kondo H (1994) Expression of DNA methyltransferase gene in mature and immature neurons as well as proliferating cells in mice. Differentiation 56:39–44

Grammatikakis I, Abdelmohsen K, Gorospe M (2017) Posttranslational control of HuR function. Wiley Interdiscip Rev RNA 8:e1372

Gravina S, Dong X, Yu B, Vijg J (2016) Single-cell genome-wide bisulfite sequencing uncovers extensive heterogeneity in the mouse liver methylome. Genome Biol 17:150

Greer EL, Blanco MA, Gu L, Sendinc E, Liu J, Aristizabal-Corrales D, Hsu CH, Aravind L, He C, Shi Y (2015) DNA methylation on N6-adenine in C. elegans. Cell 161:868–878

Grigorenko EL, Kornilov SA, Naumova OY (2016) Epigenetic regulation of cognition: a circumscribed review of the field. Dev Psychopathol 28:1285–1304

Guan Z, Giustetto M, Lomvardas S, Kim JH, Miniaci MC, Schwartz JH, Thanos D, Kandel ER (2002) Integration of long-term-memory-related synaptic plasticity involves bidirectional regulation of gene expression and chromatin structure. Cell 111:483–493

Guibert S, Forne T, Weber M (2012) Global profiling of DNA methylation erasure in mouse primordial germ cells. Genome Res 22:633–641

Guo F, Yan L, Guo H, Li L, Hu B, Zhao Y, Yong J, Hu Y, Wang X, Wei Y, Wang W, Li R, Yan J, Zhi X, Zhang Y, Jin H, Zhang W, Hou Y, Zhu P, Li J, Zhang L, Liu S, Ren Y, Zhu X, Wen L, Gao YQ, Tang F, Qiao J (2015b) The transcriptome and DNA methylome landscapes of human primordial germ cells. Cell 161:1437–1452

Guo H, Zhu P, Guo F, Li X, Wu X, Fan X, Wen L, Tang F (2015c) Profiling DNA methylome landscapes of mammalian cells with single-cell reduced-representation bisulfite sequencing. Nat Protoc 10:645–659

Guo H, Zhu P, Wu X, Li X, Wen L, Tang F (2013) Single-cell methylome landscapes of mouse embryonic stem cells and early embryos analyzed using reduced representation bisulfite sequencing. Genome Res 23:2126–2135

Guo JU, Ma DK, Mo H, Ball MP, Jang MH, Bonaguidi MA, Balazer JA, Eaves HL, Xie B, Ford E, Zhang K, Ming GL, Gao Y, Song H (2011) Neuronal activity modifies the DNA methylation landscape in the adult brain. Nat Neurosci 14:1345–1351

Guo X, Wang L, Li J, Ding Z, Xiao J, Yin X, He S, Shi P, Dong L, Li G, Tian C, Wang J, Cong Y, Xu Y (2015a) Structural insight into autoinhibition and histone H3-induced activation of DNMT3A. Nature 517:640–644

Gyory I, Minarovits J (2005) Epigenetic regulation of lymphoid specific gene sets. Biochem Cell Biol 83:286–295

Hackett JA, Reddington JP, Nestor CE, Dunican DS, Branco MR, Reichmann J, Reik W, Surani MA, Adams IR, Meehan RR (2012) Promoter DNA methylation couples genome-defence mechanisms to epigenetic reprogramming in the mouse germline. Development 139:3623–3632

Hackett JA, Surani MA (2013) DNA methylation dynamics during the mammalian life cycle. Philos Trans R Soc Lond Ser B Biol Sci 368:20110328

Hannon E, Spiers H, Viana J, Pidsley R, Burrage J, Murphy TM, Troakes C, Turecki G, O'Donovan MC, Schalkwyk LC, Bray NJ, Mill J (2016) Methylation QTLs in the developing brain and their enrichment in schizophrenia risk loci. Nat Neurosci 19:48–54

Hansen KD, Sabunciyan S, Langmead B, Nagy N, Curley R, Klein G, Klein E, Salamon D, Feinberg AP (2014) Large-scale hypomethylated blocks associated with Epstein-Barr virus-induced B-cell immortalization. Genome Res 24:177–184

Hantusch B, Kalt R, Krieger S, Puri C, Kerjaschki D (2007) Sp1/Sp3 and DNA-methylation contribute to basal transcriptional activation of human podoplanin in MG63 versus Saos-2 osteoblastic cells. BMC Mol Biol 8:20

Hark AT, Schoenherr CJ, Katz DJ, Ingram RS, Levorse JM, Tilghman SM (2000) CTCF mediates methylation-sensitive enhancer-blocking activity at the H19/Igf2 locus. Nature 405:486–489

Hashimoto H, Horton JR, Zhang X, Bostick M, Jacobsen SE, Cheng X (2008) The SRA domain of UHRF1 flips 5-methylcytosine out of the DNA helix. Nature 455:826–829

He XJ, Chen T, Zhu JK (2011) Regulation and function of DNA methylation in plants and animals. Cell Res 21:442–465

Hermann A, Goyal R, Jeltsch A (2004) The Dnmt1 DNA-(cytosine-C5)-methyltransferase methylates DNA processively with high preference for hemimethylated target sites. J Biol Chem 279:48350–48359

Hervouet E, Nadaradjane A, Gueguen M, Vallette FM, Cartron PF (2012) Kinetics of DNA methylation inheritance by the Dnmt1-including complexes during the cell cycle. Cell Div 7:5

Hervouet E, Vallette FM, Cartron PF (2009) Dnmt3/transcription factor interactions as crucial players in targeted DNA methylation. Epigenetics 4:487–499

Hervouet E, Vallette FM, Cartron PF (2010) Dnmt1/transcription factor interactions: an alternative mechanism of DNA methylation inheritance. Genes Cancer 1:434–443

Heyn H, Esteller M (2015) An adenine code for DNA: a second life for N6-methyladenine. Cell 161:710–713

Heyn H, Li N, Ferreira HJ, Moran S, Pisano DG, Gomez A, Diez J, Sanchez-Mut JV, Setien F, Carmona FJ, Puca AA, Sayols S, Pujana MA, Serra-Musach J, Iglesias-Platas I, Formiga F, Fernandez AF, Fraga MF, Heath SC, Valencia A, Gut IG, Wang J, Esteller M (2012) Distinct DNA methylomes of newborns and centenarians. Proc Natl Acad Sci U S A 109: 10522–10527

Hoffmann A, Ziller M, Spengler D (2016) The future is the past: methylation QTLs in schizophrenia. Genes (Basel) 7:E104

Holliday R, Pugh JE (1975) DNA modification mechanisms and gene activity during development. Science 187:226–232

Horvath S, Mah V, Lu AT, Woo JS, Choi OW, Jasinska AJ, Riancho JA, Tung S, Coles NS, Braun J, Vinters HV, Coles LS (2015) The cerebellum ages slowly according to the epigenetic clock. Aging (Albany NY) 7:294–306

Horvath S, Zhang Y, Langfelder P, Kahn RS, Boks MP, Van Eijk K, Van Den Berg LH, Ophoff RA (2012) Aging effects on DNA methylation modules in human brain and blood tissue. Genome Biol 13:R97

Hou Y, Guo H, Cao C, Li X, Hu B, Zhu P, Wu X, Wen L, Tang F, Huang Y, Peng J (2016) Single-cell triple omics sequencing reveals genetic, epigenetic, and transcriptomic heterogeneity in hepatocellular carcinomas. Cell Res 26:304–319

Howard G, Eiges R, Gaudet F, Jaenisch R, Eden A (2008) Activation and transposition of endogenous retroviral elements in hypomethylation induced tumors in mice. Oncogene 27:404–408

Howell CY, Bestor TH, Ding F, Latham KE, Mertineit C, Trasler JM, Chaillet JR (2001) Genomic imprinting disrupted by a maternal effect mutation in the Dnmt1 gene. Cell 104:829–838

Hu Y, Huang K, An Q, Du G, Hu G, Xue J, Zhu X, Wang CY, Xue Z, Fan G (2016) Simultaneous profiling of transcriptome and DNA methylome from a single cell. Genome Biol 17:88

Huang J, Fan T, Yan Q, Zhu H, Fox S, Issaq HJ, Best L, Gangi L, Munroe D, Muegge K (2004) Lsh, an epigenetic guardian of repetitive elements. Nucleic Acids Res 32:5019–5028

Huang R, Ding Q, Xiang Y, Gu T, Li Y (2016) Comparative analysis of DNA methyltransferase gene family in fungi: a focus on Basidiomycota. Front Plant Sci 7:1556

Igaz LM, Vianna MR, Medina JH, Izquierdo I (2002) Two time periods of hippocampal mRNA synthesis are required for memory consolidation of fear-motivated learning. J Neurosci 22:6781–6789

Iida T, Suetake I, Tajima S, Morioka H, Ohta S, Obuse C, Tsurimoto T (2002) PCNA clamp facilitates action of DNA cytosine methyltransferase 1 on hemimethylated DNA. Genes Cells 7:997–1007

Ishikawa K, Fukuda E, Kobayashi I (2010) Conflicts targeting epigenetic systems and their resolution by cell death: novel concepts for methyl-specific and other restriction systems. DNA Res 17:325–342

Iyer LM, Abhiman S, Aravind L (2011) Natural history of eukaryotic DNA methylation systems. Prog Mol Biol Transl Sci 101:25–104

Jaffe AE, Gao Y, Deep-Soboslay A, Tao R, Hyde TM, Weinberger DR, Kleinman JE (2016) Mapping DNA methylation across development, genotype and schizophrenia in the human frontal cortex. Nat Neurosci 19:40–47

Jeltsch A (2006) On the enzymatic properties of Dnmt1: specificity, processivity, mechanism of linear diffusion and allosteric regulation of the enzyme. Epigenetics 1:63–66

Jeltsch A (2013) Oxygen, epigenetic signaling, and the evolution of early life. Trends BiochemSci 38:172–176

Jeltsch A, Jurkowska RZ (2016) Allosteric control of mammalian DNA methyltransferases – a new regulatory paradigm. Nucleic Acids Res 44:8556–8575

Jeon J, Choi J, Lee GW, Park SY, Huh A, Dean RA, Lee YH (2015) Genome-wide profiling of DNA methylation provides insights into epigenetic regulation of fungal development in a plant pathogenic fungus, Magnaporthe oryzae. Sci Rep 5:8567

Jeong S, Liang G, Sharma S, Lin JC, Choi SH, Han H, Yoo CB, Egger G, Yang AS, Jones PA (2009) Selective anchoring of DNA methyltransferases 3A and 3B to nucleosomes containing methylated DNA. MolCell Biol 29:5366–5376

Jessberger S, Nakashima K, Clemenson GD Jr, Mejia E, Mathews E, Ure K, Ogawa S, Sinton CM, Gage FH, Hsieh J (2007) Epigenetic modulation of seizure-induced neurogenesis and cognitive decline. J Neurosci 27:5967–5975

Ji D, Lin K, Song J, Wang Y (2014) Effects of Tet-induced oxidation products of 5-methylcytosine on Dnmt1- and DNMT3a-mediated cytosine methylation. Mol BioSyst 10:1749–1752

Ji H, Ehrlich LI, Seita J, Murakami P, Doi A, Lindau P, Lee H, Aryee MJ, Irizarry RA, Kim K, Rossi DJ, Inlay MA, Serwold T, Karsunky H, Ho L, Daley GQ, Weissman IL, Feinberg AP (2010) Comprehensive methylome map of lineage commitment from haematopoietic progenitors. Nature 467:338–342

Jia D, Jurkowska RZ, Zhang X, Jeltsch A, Cheng X (2007) Structure of Dnmt3a bound to Dnmt3L suggests a model for de novo DNA methylation. Nature 449:248–251

Jiang N, Wang L, Chen J, Wang L, Leach L, Luo Z (2014) Conserved and divergent patterns of DNA methylation in higher vertebrates. Genome Biol Evol 6:2998–3014

Jin B, Li Y, Robertson KD (2011) DNA methylation: superior or subordinate in the epigenetic hierarchy? Genes Cancer 2:607–617

Jin B, Robertson KD (2013) DNA methyltransferases, DNA damage repair, and cancer. Adv Exp Med Biol 754:3–29

Jinawath A, Miyake S, Yanagisawa Y, Akiyama Y, Yuasa Y (2005) Transcriptional regulation of the human DNA methyltransferase 3A and 3B genes by Sp3 and Sp1 zinc finger proteins. Biochem J 385:557–564

Joh RI, Palmieri CM, Hill IT, Motamedi M (2014) Regulation of histone methylation by noncoding RNAs. Biochim Biophys Acta 1839:1385–1394

Jones PA (2012) Functions of DNA methylation: islands, start sites, gene bodies and beyond. Nat Rev Genet 13:484–492

Jones PA, Liang G (2009) Rethinking how DNA methylation patterns are maintained. NatRevGenet 10:805–811

Jones PA, Takai D (2001) The role of DNA methylation in mammalian epigenetics. Science 293:1068–1070

Jones PL, Veenstra GJ, Wade PA, Vermaak D, Kass SU, Landsberger N, Strouboulis J, Wolffe AP (1998) Methylated DNA and MeCP2 recruit histone deacetylase to repress transcription. Nat Genet 19:187–191

Juraeva D, Haenisch B, Zapatka M, Frank J, Investigators G, Group, P.-G.S.W, Witt SH, Muhleisen TW, Treutlein J, Strohmaier J, Meier S, Degenhardt F, Giegling I, Ripke S, Leber M, Lange C, Schulze TG, Mossner R, Nenadic I, Sauer H, Rujescu D, Maier W, Borglum A, Ophoff R, Cichon S, Nothen MM, Rietschel M, Mattheisen M, Brors B (2014) Integrated pathway-based approach identifies association between genomic regions at CTCF and CACNB2 and schizophrenia. PLoS Genet 10:e1004345

Jurkowska RZ, Jeltsch A (2016) Enzymology of mammalian DNA methyltransferases. Adv Exp Med Biol 945:87–122

Jurkowski TP, Jeltsch A (2011) On the evolutionary origin of eukaryotic DNA methyltransferases and Dnmt2. PLoSOne 6:e28104

Kaas GA, Zhong C, Eason DE, Ross DL, Vachhani RV, Ming GL, King JR, Song H, Sweatt JD (2013) TET1 controls CNS 5-methylcytosine hydroxylation, active DNA demethylation, gene transcription, and memory formation. Neuron 79:1086–1093

Kaneda M, Sado T, Hata K, Okano M, Tsujimoto N, Li E, Sasaki H (2004) Role of de novo DNA methyltransferases in initiation of genomic imprinting and X-chromosome inactivation. Cold Spring Harb Symp Quant Biol 69:125–129

Kang J, Kalantry S, Rao A (2013) PGC7, H3K9me2 and Tet3: regulators of DNA methylation in zygotes. Cell Res 23:6–9

Keil KP, Vezina CM (2015) DNA methylation as a dynamic regulator of development and disease processes: spotlight on the prostate. Epigenomics 7:413–425

Kelley RI (1973) Isolation of a histone IIb1-IIb2 complex. Biochem Biophys Res Commun 54:1588–1594

Kelsey G, Feil R (2013) New insights into establishment and maintenance of DNA methylation imprints in mammals. Philos Trans R Soc Lond Ser B Biol Sci 368:20110336

Khare T, Pai S, Koncevicius K, Pal M, Kriukiene E, Liutkeviciute Z, Irimia M, Jia P, Ptak C, Xia M, Tice R, Tochigi M, Morera S, Nazarians A, Belsham D, Wong AH, Blencowe BJ, Wang SC, Kapranov P, Kustra R, Labrie V, Klimasauskas S, Petronis A (2012) 5-hmC in the brain is abundant in synaptic genes and shows differences at the exon-intron boundary. Nat Struct Mol Biol 19:1037–1043

Kim-Ha J, Kim YJ (2016) Age-related epigenetic regulation in the brain and its role in neuronal diseases. BMB Rep 49:671–680

Kim S, Kaang BK (2017) Epigenetic regulation and chromatin remodeling in learning and memory. Exp Mol Med 49:e281

Kishikawa S, Murata T, Kimura H, Shiota K, Yokoyama KK (2002) Regulation of transcription of the Dnmt1 gene by Sp1 and Sp3 zinc finger proteins. Eur J Biochem 269:2961–2970

Klausz B, Haller J, Tulogdi A, Zelena D (2012) Genetic and epigenetic determinants of aggression. In: Minarovits J, Niller HH (eds) *Patho-Epigenetics of Disease*. Springer, New York, pp 227–280

Klein HU, De Jager PL (2016) Uncovering the role of the Methylome in dementia and neurodegeneration. Trends Mol Med 22:687–700

Kornberg RD (1974) Chromatin structure: a repeating unit of histones and DNA. Science 184:868–871

Koziol MJ, Bradshaw CR, Allen GE, Costa AS, Frezza C, Gurdon JB (2016) Identification of methylated deoxyadenosines in vertebrates reveals diversity in DNA modifications. Nat Struct Mol Biol 23:24–30

Kramer B, Kramer W, Fritz HJ (1984) Different base/base mismatches are corrected with different efficiencies by the methyl-directed DNA mismatch-repair system of E. coli. Cell 38:879–887

Kremenskoy M, Kremenska Y, Ohgane J, Hattori N, Tanaka S, Hashizume K, Shiota K (2003) Genome-wide analysis of DNA methylation status of CpG islands in embryoid bodies, teratomas, and fetuses. Biochem Biophys Res Commun 311:884–890

Kriaucionis S, Heintz N (2009) The nuclear DNA base 5-hydroxymethylcytosine is present in Purkinje neurons and the brain. Science 324:929–930

Kunnath L, Locker J (1982) Variable methylation of the ribosomal RNA genes of the rat. Nucleic Acids Res 10:3877–3892

Kurukuti S, Tiwari VK, Tavoosidana G, Pugacheva E, Murrell A, Zhao Z, Lobanenkov V, Reik W, Ohlsson R (2006) CTCF binding at the H19 imprinting control region mediates maternally inherited higher-order chromatin conformation to restrict enhancer access to Igf2. Proc Natl Acad Sci U S A 103:10684–10689

Labrie V, Pai S, Petronis A (2012) Epigenetics of major psychosis: progress, problems and perspectives. Trends Genet 28:427–435

Lahiri DK, Maloney B, Zawia NH (2009) The LEARn model: an epigenetic explanation for idiopathic neurobiological diseases. Mol Psychiatry 14:992–1003

Li E, Beard C, Jaenisch R (1993) Role for DNA methylation in genomic imprinting. Nature 366:362–365

Li E, Bestor TH, Jaenisch R (1992) Targeted mutation of the DNA methyltransferase gene results in embryonic lethality. Cell 69:915–926

Li E, Zhang Y (2014) DNA methylation in mammals. Cold Spring Harb Perspect Biol 6:a019133

Li N, Shen Q, Hua J (2016) Epigenetic Remodeling in Male Germline Development. Stem Cells Int 2016:3152173

Li X, Wei W, Zhao QY, Widagdo J, Baker-Andresen D, Flavell CR, D'alessio A, Zhang Y, Bredy TW (2014) Neocortical Tet3-mediated accumulation of 5-hydroxymethylcytosine promotes rapid behavioral adaptation. Proc Natl Acad Sci U S A 111:7120–7125

Liang G, Chan MF, Tomigahara Y, Tsai YC, Gonzales FA, Li E, Laird PW, Jones PA (2002) Cooperativity between DNA methyltransferases in the maintenance methylation of repetitive elements. Mol Cell Biol 22:480–491

Lin RK, Wang YC (2014) Dysregulated transcriptional and post-translational control of DNA methyltransferases in cancer. Cell Biosci 4:46

Lin RK, Wu CY, Chang JW, Juan LJ, Hsu HS, Chen CY, Lu YY, Tang YA, Yang YC, Yang PC, Wang YC (2010) Dysregulation of p53/Sp1 control leads to DNA methyltransferase-1 overexpression in lung cancer. Cancer Res 70:5807–5817

Lin S, Gregory RI (2015) MicroRNA biogenesis pathways in cancer. Nat Rev Cancer 15:321–333

Lister R, Mukamel EA, Nery JR, Urich M, Puddifoot CA, Johnson ND, Lucero J, Huang Y, Dwork AJ, Schultz MD, Yu M, Tonti-Filippini J, Heyn H, Hu S, Wu JC, Rao A, Esteller M, He C,

Haghighi FG, Sejnowski TJ, Behrens MM, Ecker JR (2013) Global epigenomic reconfiguration during mammalian brain development. Science 341:1237905

Lister R, Pelizzola M, Dowen RH, Hawkins RD, Hon G, Tonti-Filippini J, Nery JR, Lee L, Ye Z, Ngo QM, Edsall L, Antosiewicz-Bourget J, Stewart R, Ruotti V, Millar AH, Thomson JA, Ren B, Ecker JR (2009) Human DNA methylomes at base resolution show widespread epigenomic differences. Nature 462:315–322

Liu J, Zhu Y, Luo GZ, Wang X, Yue Y, Wang X, Zong X, Chen K, Yin H, Fu Y, Han D, Wang Y, Chen D, He C (2016) Abundant DNA 6mA methylation during early embryogenesis of zebrafish and pig. Nat Commun 7:13052

Liu Y, Li X (2012) Darwin's pangenesis and molecular medicine. Trends Mol Med 18:506–508

Lopez De Silanes I, Gorospe M, Taniguchi H, Abdelmohsen K, Srikantan S, Alaminos M, Berdasco M, Urdinguio RG, Fraga MF, Jacinto FV, Esteller M (2009) The RNA-binding protein HuR regulates DNA methylation through stabilization of DNMT3b mRNA. Nucleic Acids Res 37:2658–2671

Lubin FD, Roth TL, Sweatt JD (2008) Epigenetic regulation of BDNF gene transcription in the consolidation of fear memory. J Neurosci 28:10576–10586

Luo C, Lancaster MA, Castanon R, Nery JR, Knoblich JA, Ecker JR (2016) Cerebral organoids recapitulate epigenomic signatures of the human fetal brain. Cell Rep 17:3369–3384

Macfarlan TS, Gifford WD, Driscoll S, Lettieri K, Rowe HM, Bonanomi D, Firth A, Singer O, Trono D, Pfaff SL (2012) Embryonic stem cell potency fluctuates with endogenous retrovirus activity. Nature 487:57–63

Maegawa S, Hinkal G, Kim HS, Shen L, Zhang L, Zhang J, Zhang N, Liang S, Donehower LA, Issa JP (2010) Widespread and tissue specific age-related DNA methylation changes in mice. Genome Res 20:332–340

Makarova KS, Wolf YI, Koonin EV (2013) Comparative genomics of defense systems in archaea and bacteria. Nucleic Acids Res 41:4360–4377

Maloney B, Lahiri DK (2016) Epigenetics of dementia: understanding the disease as a transformation rather than a state. Lancet Neurol 15:760–774

Margot JB, Aguirre-Arteta AM, Di Giacco BV, Pradhan S, Roberts RJ, Cardoso MC, Leonhardt H (2000) Structure and function of the mouse DNA methyltransferase gene: Dnmt1 shows a tripartite structure. J Mol Biol 297:293–300

Marina RJ, Sturgill D, Bailly MA, Thenoz M, Varma G, Prigge MF, Nanan KK, Shukla S, Haque N, Oberdoerffer S (2016) TET-catalyzed oxidation of intragenic 5-methylcytosine regulates CTCF-dependent alternative splicing. EMBO J 35:335–355

Marino-Ramirez L, Kann MG, Shoemaker BA, Landsman D (2005) Histone structure and nucleosome stability. Expert Rev Proteomics 2:719–729

Martinowich K, Hattori D, Wu H, Fouse S, He F, Hu Y, Fan G, Sun YE (2003) DNA methylation-related chromatin remodeling in activity-dependent BDNF gene regulation. Science 302:890–893

Maunakea AK, Chepelev I, Cui K, Zhao K (2013) Intragenic DNA methylation modulates alternative splicing by recruiting MeCP2 to promote exon recognition. Cell Res 23:1256–1269

McCabe MT, Davis JN, Day ML (2005) Regulation of DNA methyltransferase 1 by the pRb/E2F1 pathway. Cancer Res 65:3624–3632

Mendizabal I, Shi L, Keller TE, Konopka G, Preuss TM, Hsieh TF, Hu E, Zhang Z, Su B, Yi SV (2016) Comparative methylome analyses identify epigenetic regulatory loci of human brain evolution. Mol Biol Evol 33:2947–2959

Meng H, Cao Y, Qin J, Song X, Zhang Q, Shi Y, Cao L (2015) DNA methylation, its mediators and genome integrity. Int J Biol Sci 11:604–617

Messerschmidt DM, Knowles BB, Solter D (2014) DNA methylation dynamics during epigenetic reprogramming in the germline and preimplantation embryos. Genes Dev 28:812–828

Meyer KD, Jaffrey SR (2016) Expanding the diversity of DNA base modifications with N(6)-methyldeoxyadenosine. Genome Biol 17:5

Miller CA, Gavin CF, White JA, Parrish RR, Honasoge A, Yancey CR, Rivera IM, Rubio MD, Rumbaugh G, Sweatt JD (2010) Cortical DNA methylation maintains remote memory. Nat Neurosci 13:664–666

Miller CA, Sweatt JD (2007) Covalent modification of DNA regulates memory formation. Neuron 53:857–869

Minarovits J, Banati F, Szenthe K, Niller HH (2016) Epigenetic regulation. Adv Exp Med Biol 879:1–25

Minarovits J, Niller HH (2012) Patho-epigenetics of disease. Springer, New York

Minarovits J, Niller HH (2016) Patho-epigenetics of infectious disease. Springer, New York

Minarovits J, Niller HH (2017) Current Trends and alternative scenarios in EBV research. Methods Mol Biol 1532:1–32

Miousse IR, Koturbash I (2015) The fine LINE: methylation drawing the cancer landscape. Biomed Res Int 2015:131547

Moll UM, Petrenko O (2003) The MDM2-p53 interaction. Mol Cancer Res 1:1001–1008

Monti B, Polazzi E, Contestabile A (2009) Biochemical, molecular and epigenetic mechanisms of valproic acid neuroprotection. Curr Mol Pharmacol 2:95–109

Moore LD, Le T, Fan G (2013) DNA methylation and its basic function. Neuropsychopharmacology 38:23–38

Morgan HD, Santos F, Green K, Dean W, Reik W (2005) Epigenetic reprogramming in mammals. Hum Mol Genet 14(1):R47–R58

Moroz LL, Kohn AB (2013) Single-neuron transcriptome and methylome sequencing for epigenomic analysis of aging. Methods Mol Biol 1048:323–352

Munoz-Lopez M, Garcia-Perez JL (2010) DNA transposons: nature and applications in genomics. Curr Genomics 11:115–128

Muotri AR, Marchetto MC, Coufal NG, Oefner R, Yeo G, Nakashima K, Gage FH (2010) L1 retrotransposition in neurons is modulated by MeCP2. Nature 468:443–446

Nakamura T, Liu YJ, Nakashima H, Umehara H, Inoue K, Matoba S, Tachibana M, Ogura A, Shinkai Y, Nakano T (2012) PGC7 binds histone H3K9me2 to protect against conversion of 5mC to 5hmC in early embryos. Nature 486:415–419

Nan X, Ng HH, Johnson CA, Laherty CD, Turner BM, Eisenman RN, Bird A (1998) Transcriptional repression by the methyl-CpG-binding protein MeCP2 involves a histone deacetylase complex. Nature 393:386–389

Nativio R, Wendt KS, Ito Y, Huddleston JE, Uribe-Lewis S, Woodfine K, Krueger C, Reik W, Peters JM, Murrell A (2009) Cohesin is required for higher-order chromatin conformation at the imprinted IGF2-H19 locus. PLoS Genet 5:e1000739

Niesen MI, Osborne AR, Yang H, Rastogi S, Chellappan S, Cheng JQ, Boss JM, Blanck G (2005) Activation of a methylated promoter mediated by a sequence-specific DNA-binding protein, RFX. J Biol Chem 280:38914–38922

Nikolova YS, Hariri AR (2015) Can we observe epigenetic effects on human brain function? Trends Cogn Sci 19:366–373

Ogden GB, Pratt MJ, Schaechter M (1988) The replicative origin of the E. coli chromosome binds to cell membranes only when hemimethylated. Cell 54:127–135

Oh G, Ebrahimi S, Wang SC, Cortese R, Kaminsky ZA, Gottesman I, Burke JR, Plassman BL, Petronis A (2016) Epigenetic assimilation in the aging human brain. Genome Biol 17:76

Oh G, Wang SC, Pal M, Chen ZF, Khare T, Tochigi M, Ng C, Yang YA, Kwan A, Kaminsky ZA, Mill J, Gunasinghe C, Tackett JL, Gottesman I, Willemsen G, De Geus EJ, Vink JM, Slagboom PE, Wray NR, Heath AC, Montgomery GW, Turecki G, Martin NG, Boomsma DI, McGuffin P, Kustra R, Petronis A (2015) DNA modification study of major depressive disorder: beyond locus-by-locus comparisons. Biol Psychiatry 77:246–255

Okada T, Ohzeki J, Nakano M, Yoda K, Brinkley WR, Larionov V, Masumoto H (2007) CENP-B controls centromere formation depending on the chromatin context. Cell 131:1287–1300

Okano M, Bell DW, Haber DA, Li E (1999) DNA methyltransferases Dnmt3a and Dnmt3b are essential for de novo methylation and mammalian development. Cell 99:247–257

Oliveira AM, Hemstedt TJ, Bading H (2012) Rescue of aging-associated decline in Dnmt3a2 expression restores cognitive abilities. Nat Neurosci 15:1111–1113

Ostler KR, Davis EM, Payne SL, Gosalia BB, Exposito-Cespedes J, Le Beau MM, Godley LA (2007) Cancer cells express aberrant DNMT3B transcripts encoding truncated proteins. Oncogene 26:5553–5563

Pandya-Jones A, Plath K (2016) The "lnc" between 3D chromatin structure and X chromosome inactivation. Semin Cell Dev Biol 56:35–47

Pearson CE, Nichol Edamura K, Cleary JD (2005) Repeat instability: mechanisms of dynamic mutations. Nat Rev Genet 6:729–742

Peaston AE, Evsikov AV, Graber JH, De Vries WN, Holbrook AE, Solter D, Knowles BB (2004) Retrotransposons regulate host genes in mouse oocytes and preimplantation embryos. Dev Cell 7:597–606

Peterson EJ, Bogler O, Taylor SM (2003) p53-mediated repression of DNA methyltransferase 1 expression by specific DNA binding. Cancer Res 63:6579–6582

Petrussa L, Van De Velde H, De Rycke M (2014) Dynamic regulation of DNA methyltransferases in human oocytes and preimplantation embryos after assisted reproductive technologies. Mol Hum Reprod 20:861–874

Pfaffeneder T, Spada F, Wagner M, Brandmayr C, Laube SK, Eisen D, Truss M, Steinbacher J, Hackner B, Kotljarova O, Schuermann D, Michalakis S, Kosmatchev O, Schiesser S, Steigenberger B, Raddaoui N, Kashiwazaki G, Muller U, Spruijt CG, Vermeulen M, Leonhardt H, Schar P, Muller M, Carell T (2014) Tet oxidizes thymine to 5-hydroxymethyluracil in mouse embryonic stem cell DNA. Nat Chem Biol 10:574–581

Ponts N, Fu L, Harris EY, Zhang J, Chung DW, Cervantes MC, Prudhomme J, Atanasova-Penichon V, Zehraoui E, Bunnik EM, Rodrigues EM, Lonardi S, Hicks GR, Wang Y, Le Roch KG (2013) Genome-wide mapping of DNA methylation in the human malaria parasite plasmodium falciparum. Cell Host Microbe 14:696–706

Posfai J, Bhagwat AS, Posfai G, Roberts RJ (1989) Predictive motifs derived from cytosine methyltransferases. Nucleic Acids Res 17:2421–2435

Pradhan S, Bacolla A, Wells RD, Roberts RJ (1999) Recombinant human DNA (cytosine-5) methyltransferase. I. Expression, purification, and comparison of de novo and maintenance methylation. J Biol Chem 274:33002–33010

Putiri EL, Robertson KD (2011) Epigenetic mechanisms and genome stability. Clin Epigenetics 2:299–314

Qin W, Leonhardt H, Pichler G (2011) Regulation of DNA methyltransferase 1 by interactions and modifications. Nucleus 2:392–402

Quenneville S, Turelli P, Bojkowska K, Raclot C, Offner S, Kapopoulou A, Trono D (2012) The KRAB-ZFP/KAP1 system contributes to the early embryonic establishment of site-specific DNA methylation patterns maintained during development. Cell Rep 2:766–773

Rajasethupathy P, Antonov I, Sheridan R, Frey S, Sander C, Tuschl T, Kandel ER (2012) A role for neuronal piRNAs in the epigenetic control of memory-related synaptic plasticity. Cell 149:693–707

Rasmussen KD, Helin K (2016) Role of TET enzymes in DNA methylation, development, and cancer. Genes Dev 30:733–750

Ratnam S, Mertineit C, Ding F, Howell CY, Clarke HJ, Bestor TH, Chaillet JR, Trasler JM (2002) Dynamics of Dnmt1 methyltransferase expression and intracellular localization during oogenesis and preimplantation development. Dev Biol 245:304–314

Rebhandl S, Huemer M, Greil R, Geisberger R (2015) AID/APOBEC deaminases and cancer. Oncoscience 2:320–333

Reddy K, Tam M, Bowater RP, Barber M, Tomlinson M, Nichol Edamura K, Wang YH, Pearson CE (2011) Determinants of R-loop formation at convergent bidirectionally transcribed trinucleotide repeats. Nucleic Acids Res 39:1749–1762

Reik W (2007) Stability and flexibility of epigenetic gene regulation in mammalian development. Nature 447:425–432

Reik W, Walter J (1998) Imprinting mechanisms in mammals. Curr Opin Genet Dev 8:154–164

Renfree MB, Suzuki S, Kaneko-Ishino T (2013) The origin and evolution of genomic imprinting and viviparity in mammals. Philos Trans R Soc Lond Ser B Biol Sci 368:20120151

Riggs AD (1975) X inactivation, differentiation, and DNA methylation. Cytogenet Cell Genet 14:9–25

Robertson KD (2001) DNA methylation, methyltransferases, and cancer. Oncogene 20:3139–3155

Rodriguez-Osorio N, Wang H, Rupinski J, Bridges SM, Memili E (2010) Comparative functional genomics of mammalian DNA methyltransferases. Reprod Biomed Online 20:243–255

Rogers SD, Rogers ME, Saunders G, Holt G (1986) Isolation of mutants sensitive to 2-aminopurine and alkylating agents and evidence for the role of DNA methylation in Penicillium chrysogenum. Curr Genet 10:557–560

Rudenko A, Dawlaty MM, Seo J, Cheng AW, Meng J, Le T, Faull KF, Jaenisch R, Tsai LH (2013) Tet1 is critical for neuronal activity-regulated gene expression and memory extinction. Neuron 79:1109–1122

Saadeh H, Schulz R (2014) Protection of CpG islands against de novo DNA methylation during oogenesis is associated with the recognition site of E2f1 and E2f2. Epigenetics Chromatin 7:26

Saitou M, Yamaji M (2012) Primordial germ cells in mice. Cold Spring Harb Perspect Biol 4:a008375

Sakamoto Y, Watanabe S, Ichimura T, Kawasuji M, Koseki H, Baba H, Nakao M (2007) Overlapping roles of the methylated DNA-binding protein MBD1 and polycomb group proteins in transcriptional repression of HOXA genes and heterochromatin foci formation. J Biol Chem 282:16391–16400

Sanchez-Mut JV, Heyn H, Vidal E, Delgado-Morales R, Moran S, Sayols S, Sandoval J, Ferrer I, Esteller M, Graff J (2017) Whole genome grey and white matter DNA methylation profiles in dorsolateral prefrontal cortex. Synapse 71:e21959. https://doi.org/10.1002/syn.21959

Sanchez-Mut JV, Heyn H, Vidal E, Moran S, Sayols S, Delgado-Morales R, Schultz MD, Ansoleaga B, Garcia-Esparcia P, Pons-Espinal M, De Lagran MM, Dopazo J, Rabano A, Avila J, Dierssen M, Lott I, Ferrer I, Ecker JR, Esteller M (2016) Human DNA methylomes of neurodegenerative diseases show common epigenomic patterns. Transl Psychiatry 6:e718

Sasaki H, Ishihara K, Kato R (2000) Mechanisms of Igf2/H19 imprinting: DNA methylation, chromatin and long-distance gene regulation. J Biochem 127:711–715

Sawaya S, Bagshaw A, Buschiazzo E, Kumar P, Chowdhury S, Black MA, Gemmell N (2013) Microsatellite tandem repeats are abundant in human promoters and are associated with regulatory elements. PLoS One 8:e54710

Sawaya S, Boocock J, Black MA, Gemmell NJ (2015) Exploring possible DNA structures in real-time polymerase kinetics using Pacific biosciences sequencer data. BMC Bioinformatics 16:21

Schizophrenia Working Group of the Psychiatric Genomics Consortium (2014) Biological insights from 108 schizophrenia-associated genetic loci. Nature 511:421–427

Schneider E, Dittrich M, Bock J, Nanda I, Muller T, Seidmann L, Tralau T, Galetzka D, El Hajj N, Haaf T (2016) CpG sites with continuously increasing or decreasing methylation from early to late human fetal brain development. Gene 592:110–118

Schneider K, Fuchs C, Dobay A, Rottach A, Qin W, Wolf P, Alvarez-Castro JM, Nalaskowski MM, Kremmer E, Schmid V, Leonhardt H, Schermelleh L (2013) Dissection of cell cycle-dependent dynamics of Dnmt1 by FRAP and diffusion-coupled modeling. Nucleic Acids Res 41:4860–4876

Schrader A, Gross T, Thalhammer V, Langst G (2015) Characterization of Dnmt1 binding and DNA methylation on nucleosomes and nucleosomal arrays. PLoS One 10:e0140076

Schultz MD, He Y, Whitaker JW, Hariharan M, Mukamel EA, Leung D, Rajagopal N, Nery JR, Urich MA, Chen H, Lin S, Lin Y, Jung I, Schmitt AD, Selvaraj S, Ren B, Sejnowski TJ, Wang W, Ecker JR (2015) Human body epigenome maps reveal noncanonical DNA methylation variation. Nature 523:212–216

Scott A, Song J, Ewing R, Wang Z (2014) Regulation of protein stability of DNA methyltransferase 1 by post-translational modifications. Acta Biochim Biophys Sin Shanghai 46:199–203

Sdek P, Ying H, Chang DL, Qiu W, Zheng H, Touitou R, Allday MJ, Xiao ZX (2005) MDM2 promotes proteasome-dependent ubiquitin-independent degradation of retinoblastoma protein. Mol Cell 20:699–708

Seisenberger S, Peat JR, Hore TA, Santos F, Dean W, Reik W (2013a) Reprogramming DNA methylation in the mammalian life cycle: building and breaking epigenetic barriers. Philos Trans R Soc Lond Ser B Biol Sci 368:20110330

Seisenberger S, Peat JR, Reik W (2013b) Conceptual links between DNA methylation reprogramming in the early embryo and primordial germ cells. Curr Opin Cell Biol 25:281–288

Serandour AA, Avner S, Percevault F, Demay F, Bizot M, Lucchetti-Miganeh C, Barloy-Hubler F, Brown M, Lupien M, Metivier R, Salbert G, Eeckhoute J (2011) Epigenetic switch involved in activation of pioneer factor FOXA1-dependent enhancers. Genome Res 21:555–565

Shabbir MA, Hao H, Shabbir MZ, Wu Q, Sattar A, Yuan Z (2016) Bacteria vs. bacteriophages: parallel evolution of immune arsenals. Front Microbiol 7:1292

Sharp AJ, Stathaki E, Migliavacca E, Brahmachary M, Montgomery SB, Dupre Y, Antonarakis SE (2011) DNA methylation profiles of human active and inactive X chromosomes. Genome Res 21:1592–1600

Shukla S, Kavak E, Gregory M, Imashimizu M, Shutinoski B, Kashlev M, Oberdoerffer P, Sandberg R, Oberdoerffer S (2011) CTCF-promoted RNA polymerase II pausing links DNA methylation to splicing. Nature 479:74–79

Shukla S, Oberdoerffer S (2012) Co-transcriptional regulation of alternative pre-mRNA splicing. Biochim Biophys Acta 1819:673–683

Shukla V, Coumoul X, Lahusen T, Wang RH, Xu X, Vassilopoulos A, Xiao C, Lee MH, Man YG, Ouchi M, Ouchi T, Deng CX (2010) BRCA1 affects global DNA methylation through regulation of DNMT1. Cell Res 20:1201–1215

Sigurdsson MI, Smith AV, Bjornsson HT, Jonsson JJ (2009) HapMap methylation-associated SNPs, markers of germline DNA methylation, positively correlate with regional levels of human meiotic recombination. Genome Res 19:581–589

Smallwood SA, Lee HJ, Angermueller C, Krueger F, Saadeh H, Peat J, Andrews SR, Stegle O, Reik W, Kelsey G (2014) Single-cell genome-wide bisulfite sequencing for assessing epigenetic heterogeneity. Nat Methods 11:817–820

Smeets D, Markaki Y, Schmid VJ, Kraus F, Tattermusch A, Cerase A, Sterr M, Fiedler S, Demmerle J, Popken J, Leonhardt H, Brockdorff N, Cremer T, Schermelleh L, Cremer M (2014) Three-dimensional super-resolution microscopy of the inactive X chromosome territory reveals a collapse of its active nuclear compartment harboring distinct Xist RNA foci. Epigenetics Chromatin 7:8

Smith AK, Kilaru V, Klengel T, Mercer KB, Bradley B, Conneely KN, Ressler KJ, Binder EB (2015) DNA extracted from saliva for methylation studies of psychiatric traits: evidence tissue specificity and relatedness to brain. Am J Med Genet B Neuropsychiatr Genet 168B:36–44

Smith ZD, Chan MM, Humm KC, Karnik R, Mekhoubad S, Regev A, Eggan K, Meissner A (2014) DNA methylation dynamics of the human preimplantation embryo. Nature 511:611–615

Smith ZD, Chan MM, Mikkelsen TS, Gu H, Gnirke A, Regev A, Meissner A (2012) A unique regulatory phase of DNA methylation in the early mammalian embryo. Nature 484:339–344

Smith ZD, Meissner A (2013a) DNA methylation: roles in mammalian development. Nat Rev Genet 14:204–220

Smith ZD, Meissner A (2013b) The simplest explanation: passive DNA demethylation in PGCs. EMBO J 32:318–321

Song J, Teplova M, Ishibe-Murakami S, Patel DJ (2012) Structure-based mechanistic insights into DNMT1-mediated maintenance DNA methylation. Science 335:709–712

Spruijt CG, Gnerlich F, Smits AH, Pfaffeneder T, Jansen PW, Bauer C, Munzel M, Wagner M, Muller M, Khan F, Eberl HC, Mensinga A, Brinkman AB, Lephikov K, Muller U, Walter J, Boelens R, Van Ingen H, Leonhardt H, Carell T, Vermeulen M (2013) Dynamic readers for 5-(hydroxy)methylcytosine and its oxidized derivatives. Cell 152:1146–1159

Stewart KR, Veselovska L, Kelsey G (2016) Establishment and functions of DNA methylation in the germline. Epigenomics 8:1399–1413

Suetake I, Shinozaki F, Miyagawa J, Takeshima H, Tajima S (2004) DNMT3L stimulates the DNA methylation activity of Dnmt3a and Dnmt3b through a direct interaction. J Biol Chem 279:27816–27823

Sweatt JD (2009) Experience-dependent epigenetic modifications in the central nervous system. Biol Psychiatry 65:191–197

Sweatt JD (2013) The emerging field of neuroepigenetics. Neuron 80:624–632

Syeda F, Fagan RL, Wean M, Avvakumov GV, Walker JR, Xue S, Dhe-Paganon S, Brenner C (2011) The replication focus targeting sequence (RFTS) domain is a DNA-competitive inhibitor of Dnmt1. J Biol Chem 286:15344–15351

Szemes M, Dallosso AR, Melegh Z, Curry T, Li Y, Rivers C, Uney J, Magdefrau AS, Schwiderski K, Park JH, Brown KW, Shandilya J, Roberts SG, Malik K (2013) Control of epigenetic states by WT1 via regulation of de novo DNA methyltransferase 3A. Hum Mol Genet 22:74–83

Szenthe K, Nagy K, Buzas K, Niller HH, Minarovits J (2013) MicroRNAs as targets and tools in B-cell lymphoma therapy. J Canc Ther 4:466–474

Tahiliani M, Koh KP, Shen Y, Pastor WA, Bandukwala H, Brudno Y, Agarwal S, Iyer LM, Liu DR, Aravind L, Rao A (2009) Conversion of 5-methylcytosine to 5-hydroxymethylcytosine in mammalian DNA by MLL partner TET1. Science 324:930–935

Takacs M, Segesdi J, Banati F, Koroknai A, Wolf H, Niller HH, Minarovits J (2009) The importance of epigenetic alterations in the development of Epstein-Barr virus-related lymphomas. Mediterr J Hematol Infect Dis 1:e2009012

Takebayashi S, Tamura T, Matsuoka C, Okano M (2007) Major and essential role for the DNA methylation mark in mouse embryogenesis and stable association of DNMT1 with newly replicated regions. Mol Cell Biol 27:8243–8258

Tanaka Y, Kurumizaka H, Yokoyama S (2005) CpG methylation of the CENP-B box reduces human CENP-B binding. FEBS J 272:282–289

Tang YA, Lin RK, Tsai YT, Hsu HS, Yang YC, Chen CY, Wang YC (2012) MDM2 overexpression deregulates the transcriptional control of RB/E2F leading to DNA methyltransferase 3A over-expression in lung cancer. Clin Cancer Res 18:4325–4333

Taskin KM, Ozbilen A, Sezer F, Hurkan K, Gunes S (2016) Structure and expression of DNA methyltransferase genes from apomictic and sexual Boechera species. Comput Biol Chem 67:15–21

Termanis A, Torrea N, Culley J, Kerr A, Ramsahoye B, Stancheva I (2016) The SNF2 family ATPase LSH promotes cell-autonomous de novo DNA methylation in somatic cells. Nucleic Acids Res 44:7592–7604

Termolino P, Cremona G, Consiglio MF, Conicella C (2016) Insights into epigenetic landscape of recombination-free regions. Chromosoma 125:301–308

Thomson JP, Skene PJ, Selfridge J, Clouaire T, Guy J, Webb S, Kerr AR, Deaton A, Andrews R, James KD, Turner DJ, Illingworth R, Bird A (2010) CpG islands influence chromatin structure via the CpG-binding protein Cfp1. Nature 464:1082–1086

Tollefsbol TO (2012) Epigenetics in human disease. Academic, Waltham

Tsai CL, Tainer JA (2013) Probing DNA by 2-OG-dependent dioxygenase. Cell 155:1448–1450

Tsumura A, Hayakawa T, Kumaki Y, Takebayashi S, Sakaue M, Matsuoka C, Shimotohno K, Ishikawa F, Li E, Ueda HR, Nakayama J, Okano M (2006) Maintenance of self-renewal ability of mouse embryonic stem cells in the absence of DNA methyltransferases Dnmt1, Dnmt3a and Dnmt3b. Genes Cells 11:805–814

Tucker KL, Beard C, Dausmann J, Jackson-Grusby L, Laird PW, Lei H, Li E, Jaenisch R (1996) Germ-line passage is required for establishment of methylation and expression patterns of imprinted but not of nonimprinted genes. Genes Dev 10:1008–1020

Unoki M, Nakamura Y (2003) Methylation at CpG islands in intron 1 of EGR2 confers enhancer-like activity. FEBS Lett 554:67–72

Velasco G, Hube F, Rollin J, Neuillet D, Philippe C, Bouzinba-Segard H, Galvani A, Viegas-Pequignot E, Francastel C (2010) Dnmt3b recruitment through E2F6 transcriptional repressor mediates germ-line gene silencing in murine somatic tissues. Proc Natl Acad Sci U S A 107:9281–9286

Vidalis A, Zivkovic D, Wardenaar R, Roquis D, Tellier A, Johannes F (2016) Methylome evolution in plants. Genome Biol 17:264

Vlachogiannis G, Niederhuth CE, Tuna S, Stathopoulou A, Viiri K, De Rooij DG, Jenner RG, Schmitz RJ, Ooi SK (2015) The Dnmt3L ADD domain controls cytosine methylation establishment during spermatogenesis. Cell Rep 10:944–956. https://doi.org/10.1016/j.celrep.2015.01.021

Von Meyenn F, Berrens RV, Andrews S, Santos F, Collier AJ, Krueger F, Osorno R, Dean W, Rugg-Gunn PJ, Reik W (2016) Comparative principles of DNA methylation reprogramming during human and mouse in vitro primordial germ cell specification. Dev Cell 39:104–115

Von Meyenn F, Reik W (2015) Forget the parents: epigenetic reprogramming in human germ cells. Cell 161:1248–1251

Vu TM, Nakamura M, Calarco JP, Susaki D, Lim PQ, Kinoshita T, Higashiyama T, Martienssen RA, Berger F (2013) RNA-directed DNA methylation regulates parental genomic imprinting at several loci in Arabidopsis. Development 140:2953–2960

Waldminghaus T, Weigel C, Skarstad K (2012) Replication fork movement and methylation govern SeqA binding to the Escherichia coli chromosome. Nucleic Acids Res 40:5465–5476

Walsh TK, Brisson JA, Robertson HM, Gordon K, Jaubert-Possamai S, Tagu D, Edwards OR (2010) A functional DNA methylation system in the pea aphid, Acyrthosiphon pisum. Insect Mol Biol 19(Suppl 2):215–228

Walter M, Teissandier A, Perez-Palacios R, Bourc'his D (2016) An epigenetic switch ensures transposon repression upon dynamic loss of DNA methylation in embryonic stem cells. elife 5:e11418

Walton EL, Francastel C, Velasco G (2014) Dnmt3b prefers germ line genes and centromeric regions: lessons from the ICF syndrome and cancer and implications for diseases. Biology (Basel) 3:578–605

Wang KY, Chen CC, Tsai SF, Shen CJ (2016) Epigenetic enhancement of the post-replicative DNA mismatch repair of mammalian genomes by a hemi-mCpG-Np95-dnmt1 axis. Sci Rep 6:37490

Watanabe T, Tomizawa S, Mitsuya K, Totoki Y, Yamamoto Y, Kuramochi-Miyagawa S, Iida N, Hoki Y, Murphy PJ, Toyoda A, Gotoh K, Hiura H, Arima T, Fujiyama A, Sado T, Shibata T, Nakano T, Lin H, Ichiyanagi K, Soloway PD, Sasaki H (2011) Role for piRNAs and noncoding RNA in de novo DNA methylation of the imprinted mouse Rasgrf1 locus. Science 332:848–852

Weaver IC, Cervoni N, Champagne FA, D'alessio AC, Sharma S, Seckl JR, Dymov S, Szyf M, Meaney MJ (2004) Epigenetic programming by maternal behavior. Nat Neurosci 7:847–854

Weigele P, Raleigh EA (2016) Biosynthesis and function of modified bases in bacteria and their viruses. Chem Rev 116:12655–12687

Wiehle L, Raddatz G, Musch T, Dawlaty MM, Jaenisch R, Lyko F, Breiling A (2015) Tet1 and Tet2 protect DNA methylation canyons against Hypermethylation. Mol Cell Biol 36:452–461

Will CL, Luhrmann R (2011) Spliceosome structure and function. Cold Spring Harb Perspect Biol 3:a003707

Williams K, Christensen J, Pedersen MT, Johansen JV, Cloos PA, Rappsilber J, Helin K (2011) TET1 and hydroxymethylcytosine in transcription and DNA methylation fidelity. Nature 473:343–348

Wilson GG (1988) Type II restriction–modification systems. Trends Genet 4:314–318

Wion D, Casadesus J (2006) N6-methyl-adenine: an epigenetic signal for DNA-protein interactions. Nat Rev Microbiol 4:183–192

Wong NC, Wong LH, Quach JM, Canham P, Craig JM, Song JZ, Clark SJ, Choo KH (2006) Permissive transcriptional activity at the centromere through pockets of DNA hypomethylation. PLoS Genet 2:e17

Wu H, Zhang Y (2014) Reversing DNA methylation: mechanisms, genomics, and biological functions. Cell 156:45–68

Wyatt GR (1951) Recognition and estimation of 5-methylcytosine in nucleic acids. Biochem J 48:581–584

Xu F, Mao C, Ding Y, Rui C, Wu L, Shi A, Zhang H, Zhang L, Xu Z (2010) Molecular and enzymatic profiles of mammalian DNA methyltransferases: structures and targets for drugs. Curr Med Chem 17:4052–4071

Xu X, Tao Y, Gao X, Zhang L, Li X, Zou W, Ruan K, Wang F, Xu GL, Hu R (2016) A CRISPR-based approach for targeted DNA demethylation. Cell Discov 2:16009

Yamagata K, Yamazaki T, Miki H, Ogonuki N, Inoue K, Ogura A, Baba T (2007) Centromeric DNA hypomethylation as an epigenetic signature discriminates between germ and somatic cell lineages. Dev Biol 312:419–426

Yanagisawa Y, Ito E, Yuasa Y, Maruyama K (2002) The human DNA methyltransferases DNMT3A and DNMT3B have two types of promoters with different CpG contents. Biochim Biophys Acta 1577:457–465

Yang DL, Zhang G, Tang K, Li J, Yang L, Huang H, Zhang H, Zhu JK (2016a) Dicer-independent RNA-directed DNA methylation in Arabidopsis. Cell Res 26:66–82

Yang J, Guo R, Wang H, Ye X, Zhou Z, Dan J, Wang H, Gong P, Deng W, Yin Y, Mao S, Wang L, Ding J, Li J, Keefe DL, Dawlaty MM, Wang J, Xu G, Liu L (2016b) Tet enzymes regulate telomere maintenance and chromosomal stability of mouse ESCs. Cell Rep 15:1809–1821

Yang YC, Tang YA, Shieh JM, Lin RK, Hsu HS, Wang YC (2014) DNMT3B overexpression by deregulation of FOXO3a-mediated transcription repression and MDM2 overexpression in lung cancer. J Thorac Oncol 9:1305–1315

Yoder JA, Walsh CP, Bestor TH (1997) Cytosine methylation and the ecology of intragenomic parasites. Trends Genet 13:335–340

Yu NK, Baek SH, Kaang BK (2011) DNA methylation-mediated control of learning and memory. Mol Brain 4:5

Zabet NR, Catoni M, Prischi F, Paszkowski J (2017) Cytosine methylation at CpCpG sites triggers accumulation of non-CpG methylation in gene bodies. Nucleic Acids Res 45:3777–3784. https://doi.org/10.1093/nar/gkw1330

Zamudio N, Bourc'his D (2010) Transposable elements in the mammalian germline: a comfortable niche or a deadly trap? Heredity (Edinb) 105:92–104

Zaret KS, Watts J, Xu J, Wandzioch E, Smale ST, Sekiya T (2008) Pioneer factors, genetic competence, and inductive signaling: programming liver and pancreas progenitors from the endoderm. Cold Spring Harb Symp Quant Biol 73:119–126

Zelena D (2012) Co-regulation and epigenetic dysregulation in schizophrenia and bipolar disorder. In: Minarovits J, Niller HH (eds) Patho-epigenetics of disease. Springer, New York, pp 281–347

Zhang G, Huang H, Liu D, Cheng Y, Liu X, Zhang W, Yin R, Zhang D, Zhang P, Liu J, Li C, Liu B, Luo Y, Zhu Y, Zhang N, He S, He C, Wang H, Chen D (2015a) N6-methyladenine DNA modification in drosophila. Cell 161:893–906

Zhang H, Zhu JK (2011) RNA-directed DNA methylation. Curr Opin Plant Biol 14:142–147

Zhang Y, Jurkowska R, Soeroes S, Rajavelu A, Dhayalan A, Bock I, Rathert P, Brandt O, Reinhardt R, Fischle W, Jeltsch A (2010) Chromatin methylation activity of Dnmt3a and Dnmt3a/3L is guided by interaction of the ADD domain with the histone H3 tail. Nucleic Acids Res 38:4246–4253

Zhang ZM, Liu S, Lin K, Luo Y, Perry JJ, Wang Y, Song J (2015b) Crystal structure of human DNA methyltransferase 1. J Mol Biol 427:2520–2531

Zhao J, Zhu Y, Yang J, Li L, Wu H, De Jager PL, Jin P, Bennett DA (2017) A genome-wide profiling of brain DNA hydroxymethylation in Alzheimer's disease. Alzheimers Dement 13:674–688. https://doi.org/10.1016/j.jalz.2016.10.004

Zhou Y, Song N, Li X, Han Y, Ren Z, Xu JX, Han YC, Li F, Jia X (2017) Changes in the methylation status of the Oct3/4, Nanog, and Sox2 promoters in stem cells during regeneration of rat tracheal epithelium after injury. Oncotarget 8:2984–2994

# Methylation of Proteins: Biochemistry and Functional Consequences

# 35

Álvaro Ortega

## Contents

1 Introduction ................................................................. 572
2 Arginine Methylation ......................................................... 573
3 Lysine Methylation .......................................................... 577
4 Other Methylations .......................................................... 580
5 Research Needs .............................................................. 581
References ..................................................................... 582

**Abstract**

Methylation is one of the most abundant modifications that the proteome of living cells undergo. Catalyzed by enzymes of the methyltransferase family, it occurs in many biological processes of prokaryotes and eukaryotes. The most common methylations occur on the amino groups of lysine and arginine side chains providing them with hydrophobic and steric properties that affect the way they behave and recognize other proteins and nucleic acids. Methylation of proteins occurs at a posttranslational level, and its main function is the effective control of the gene expression by histones and transcription factors. Other functions are protein labeling for cellular localization, RNA processing, ribosome assembly, or cell signaling. Methylations also occur at the N- and C-termini of proteins or on carboxyl and thiol groups of histidine, cysteine, proline, or glutamate side chains.

Á. Ortega (✉)
Department of Environmental Protection, Estación Experimental del Zaidín, Consejo Superior de Investigaciones Científicas, Granada, Spain
e-mail: alvaro.ortega@csic.es

© Springer International Publishing AG, part of Springer Nature 2018
T. Krell (ed.), *Cellular Ecophysiology of Microbe: Hydrocarbon and Lipid Interactions*,
Handbook of Hydrocarbon and Lipid Microbiology, https://doi.org/10.1007/978-3-319-50542-8_25

## 1    Introduction

Post-translational modifications (PTMs) are, along with mRNA splicing, the main mechanisms cells implement to expand the information encoded in the primary sequence of their genomes. The most common type of PTM is the covalent addition of a functional group to the side chain of specific residues in the protein. The same translated protein may be thus transformed into thousands of variants considering all possible combinations of enzyme-catalyzed PTMs. When these PTMs are reversible, they represent an additional control and signaling of the modified proteins.

Along with phosphorylation, acylation, or glycosylation, alkylation brings in the most common of the covalent additions to the peptide. One of the best studied alkylations is methylation: the addition of a methyl group. Since its discovery more than half a century ago (Murray 1964), the scientific interest in methylation has grown and is now known to participate in a great number of cellular processes. Based on the number of putative sites, methylation is ranked as the fourth more common modification type (Khoury et al. 2011; Low and Wilkins 2012). Although phosphorylation or ubiquitination was discovered at approximately the same time and were the object of numerous investigations, the prevalence and importance of methylation in pro- and eukaryotes has mainly been elucidated in the last 15 years. This has been primarily motivated by the rapid expansion and growing interest of the scientific community in the field of epigenetics (methylation plays here an important role), but also by the advances in the technology used to identify methylation sites and states on proteins. It has been estimated that 12 ATP molecules are necessary for a single methyl group addition (Lake and Bedford 2007). This energetic cost endured by the cell gives us an idea of the essentiality of this modification. The fundamental contribution of methylation to cellular biology and as a regulator of the physiology and pathology of the cell starts now to become more obvious.

Methylation is the posttranslational addition of the methyl moiety of the S-adenosylmethionine (SAM, also called AdoMet) donor molecule either to the amino- or to the carboxyl-terminal group of a functional polypeptide chain or to the nitrogen or carbon atoms of the side chains of certain amino acids (Fig. 1). N-methylation has been observed mainly on lysine, arginine, glutamine, and histidine, while aspartic, glutamic, cysteine, and leucine showed methylation on their carboxyl groups (Polevoda and Sherman 2007). Lysine and arginine methylations (both N-methylations on the side chain) are the most common among these modifications in eukaryotic cells.

The enzymes catalyzing these reactions, methyltransferases, comprise about 1–2% of genes in a great number of prokaryotic and eukaryotic organisms (Clarke 2013) and have been classified into five (I–V) different classes according to their structure (Grillo and Colombatto 2005; Schubert et al. 2006). Although the majority of methyltransferase genes in a genome can be identified by sequence analyses, we are not remotely close to identify their targets.

The relevance of protein methylation in the cell is unquestionable if we take into account the number of proteins susceptible of being methylated that have already

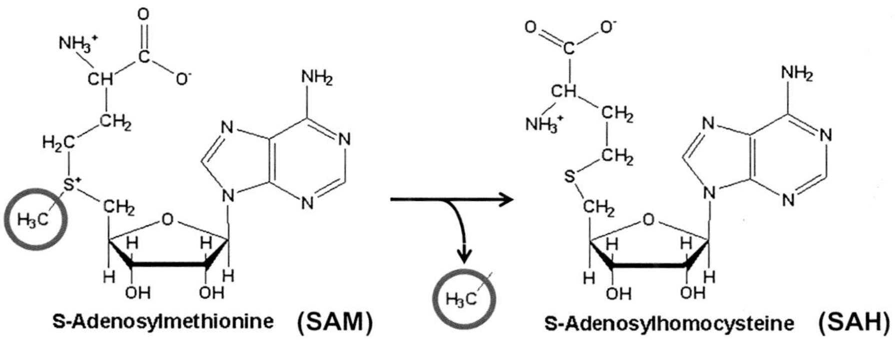

**Fig. 1** Conversion of S-adenosyl methionine (SAM) into S-adenosyl homocysteine (SAH) as a result of the donation of the methyl group to the substrate to be methylated

been described (mainly histones, but also riboproteins, transcription factors, chemo-receptors, or nuclear receptors) and the cellular processes for which this PTM is essential, such as protein-protein and protein-nucleic acid interactions, cellular localization, nuclear transport, RNA processing, ribosome assembly, or cell signaling (Walsh 2006; Polevoda and Sherman 2007).

This chapter primarily summarizes knowledge on the main methylation types, namely that involving arginine and lysine residues. The last section deals with other methylation targets, giving particular emphasis on the biological functions and the chemistry involved in the methylation process.

## 2    Arginine Methylation

Along with lysine, arginine methylation is probably the methylation that draws most scientific attention, mainly in eukaryotes. An increasing volume of research is being dedicated to its study and description. This is mainly due to the direct linkage that has been observed to many human diseases and in particular to cancer. It is an extraordinarily common event, as illustrated by the fact that more than 2% of the arginine residues in rat liver nuclei are methylated (Lee and Stallcup 2009). Methylation on arginines has not been observed to date in prokaryotes (Fisk and Read 2011).

Enzymatic methylation of arginine residues is always performed by members of the protein arginine methyltransferases (PRMT) family (Bedford and Clarke 2009; Blanc and Richard 2017) that catalyze the transfer of a methyl group from SAM to the nitrogen atoms of the arginine guanidino group giving rise to methylarginine and S-adenosyl homocysteine (SAH) (Fig. 1). There are currently four different classes of methylarginines identified (Blanc and Richard 2017), according to the terminal arginine modification produced by each type of PRMT: in $\omega$-N$^G$-mono-methylarginine (MMA), it is the amino group that undergoes methylation (Fig. 2a). Subsequently, dimethylarginine is generated either symmetrically by the

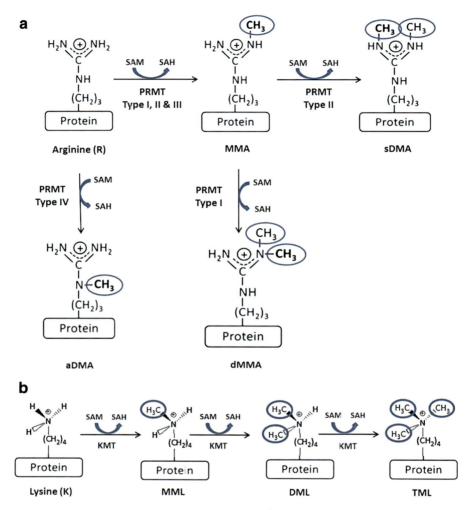

**Fig. 2** (a) Methylation of arginine residues (MMA: ω-N$^G$-monomethylarginine, sDMA: ω-N$^G$,N$^G$-symmetric dimethylarginine, aDMA: ω-N$^G$,N$^G$-asymmetric dimethylarginine, dMMA: δ-N-mono-methylarginine). The type of the methyltransferases (PRMT) implicated in each of the reactions is indicated. (**b**) Methylation of lysine residues (MML: monomethyllysine, DML: dimethyllysine, TML: trimethyllysine)

modification of the second nitrogen of the guanidino group, resulting in ω- N$^G$,N$^G$-symmetric dimethylarginine (sDMA), or asymmetrically adding the methyl group to the same nitrogen that was methylated in the first event, ω- N$^G$,N$^G$-asymmetric dimethylarginine (aDMA) (Fig. 2a). The fourth methylarginine has been observed only in yeast (possibly also in plants) and is characterized by a modification of the amino group at the delta position: δ-N-monomethylarginine (dMMA) (Low and Wilkins 2012).

Based on the chemistry of their reactions, PRMTs are classified into four categories (I–IV). The great majority of these enzymes belong to the class I, which are in charge of generating MMAs, as a first step to the asymmetric dimethylation, producing aDMAs. PRMTs of the class II dimethylate symmetrically, again following the initial step of monomethylation, generating sDMAs. Class III of PRMTs are only capable of catalyzing the formation of MMA, and PRMTs of class IV, which have been observed only on *Saccharomyces cerevisiae* and *Candida albicans*, catalyze the dMMA formation (Bedford and Clarke 2009; Low and Wilkins 2012). Ten different PRMTs have been identified in mammals (Krause et al. 2007): PRMT1, PRMT2, PRMT3, PRMT4, PRMT6, and PRMT 8 belong to the type I. PRMT1 was the first human PRMT identified, and it has a broad substrate range and performs 80% of PRMT activity in cells. In contrast, PRMT4 and PRMT6 show restricted substrate profiles (PRMT4 is also called CARM1, as co-activator-associated methyltransferase). PRMT3 and PRMT8 are also characterized by an elevated substrate specificity and, as PRMT2, occur only in the nucleus (the others have activity also in the cytoplasm). To the type II family belongs PRMT5, one of the best characterized methyltransferases, as well as PRMT9 and PRMT7. The latter posses, along with PRMT10, two methyltransferase domains (Di Lorenzo and Bedford 2011; Biggar and Li 2015). In yeast, there is only one PRMT per class (Hmt1/Rmt1, Rmt2, Hsl7, and Sfm1, respectively) (Low and Wilkins 2012).

PRMTs are highly conserved in their quite unique structure, and they are composed of one methyltransferase domain (except PRMT7 and PRMT10 that have two) and a dimerization domain, which are both generally conserved in other methyltransferases. These domains are fused to a 7 beta strand (7BS) barrel domain (containing the so-called Rossmann fold), a feature that is unique to PRMTs (Cheng et al. 2005). PRMTs are primarily monomeric, but dimeric, tetrameric, or even higher oligomeric forms have been observed (Weiss et al. 2000). In yeast, three of the PRMTs have this Rossmann fold, but the fourth (belonging to the type IV of PRMTs) differs in that it conserves a SPOUT domain. The PRMT target sites are arginine and glycine rich motifs, termed RGG/RG motifs, where the presence of glycine enhances the flexibility of the protein facilitating the access of the PRMT to the target arginine (Gary and Clarke 1998). Other examples of methylations sites include a proline, glycine, methionine motif (PGM) typical of PRMT4, and a motif surrounded by lysine residues used by PRMT7 (Wei et al. 2014).

Unlike other PTMs like phosphorylation or ubiquitination, arginines do not alter their net charge upon methylation, indicating that their biological effect is unrelated to charge modifications (Low and Wilkins 2012; Blanc and Richard 2017). Instead, the functional consequences of methylations are frequently related to changes in DNA or protein affinity caused by subtle increases in hydrophobicity and steric hindrance of the arginine side chain. Arginine is the only residue that has five potential hydrogen bond donors in its guanidino group, which are frequently involved in hydrogen bonding with DNA, RNA loops, etc. Each methyl group addition not only changes the conformation of the side chain, but also prevents a hydrogen bond from being formed. An additional effect is the attenuation of the capacity for hydrophobic ring stacking between the arginine guanidium ion and the nucleic bases (Allers and Shamoo 2001). At least 40% of the arginines form ionic salt bridges and 74% of interactions between cations and aromatic rings involve arginines (Gallivan and Dougherty 1999; Gowri Shankar et al. 2007). The alteration of protein-protein, protein-DNA, and protein-RNA interactions is the primary functional consequence of the methylation reaction. However, other effects have been observed in which these interactions are not affected, one example is their function in protein labeling, which is a necessary reprequisite for the efficient protein import into the nucleus or RNA binding proteins export from the nucleus to the cytoplasm.

Owing to these effects, arginine methylation is important for gene regulation, principally by activating or repressing the histones but also in a more direct manner in transcription, cellular signaling, mRNA translation, DNA damage signaling and repair, receptor trafficking, and pre-mRNA cutting (examples of nonhistone protein methylations) (Biggar and Li 2015). The main biological functions of arginine methylation are the coactivation of transcription by histone methylation (it can modify the histone code) or methylation of proteins that modify histones, transcription factors, or transcription coactivators and also indirectly by affecting methylation of adjacent lysines in transcription factors. Histone methylation occurs mostly in the exposed tails of the histones H3 and H4 and a crosstalk between methylation in the same histone (*cis*) or among different histones (*trans*) can occur, as well as between histone and nonhistone proteins, leading to a fine tuning of the transcription control (Binda 2013). Chromatin must be in an open conformation (euchromatin switched to heterochromatin) and transcription factors, RNA polymerases, and other regulator proteins must be able to bind or release the DNA, all these mechanisms are affected by arginine methylation. Other functions are co-repression of transcription, nuclear/cytoplasmic shuttling of different proteins (first observed in yeast), DNA repair, and signal transduction (Lee and Stallcup 2009). DNA repair needs similar conditions of the chromatin as transcription activations. Another property of arginine methylation is that the signal generated by the methylations can have agonistic or antagonistic effects with other PTMs, such as phosphorylation (Molina-Serrano et al. 2013). Alternatively, the main group of proteins susceptible to arginine methylation are RNA binding proteins. Methylation of this protein family directly affects mRNA splicing, RNA stability, translation (through riboprotein methylation), nucleocytoplasmic RNA transport, and small RNA pathways.

Methylation events can be detected and interpreted in the cell by the so-called "reader" proteins, which interpret and transmit the information represented by an arginine methylation and serve as the output agents. These proteins contain a specific methylation-binding domain called Tudor (Lu and Wang 2013), best described on SMN, TRDR, and SPF30 proteins (Tripsianes et al. 2011; Blanc and Richard 2017). Methyl groups produce a gain in hydrophobicity and therefore an increase in the affinity of their target proteins with the Tudor domains. However, in spite of their significance, only a few proteins with Tudor domains could be described, and their number is well below that of the methylation susceptible domains.

There are many examples of cellular pathways regulated in this manner by arginine methylation. For example, receptor tyrosine kinase signaling can be controlled through the lysine and arginine methylation of many of the protein kinases (Biggar and Li 2015).

The effects of arginine methylation can be externally controlled by different means. Although methylation is a highly stable modification, demethylation reactions may occur. However, their existence remains controversial since evidence for methyl esterase activities could not be obtained so far (Low and Wilkins 2012; Biggar and Li 2015; Blanc and Richard 2017). S-Adenosyl homocysteine (SAH) is the methylation product released (Fig. 1), and it acts as a potent inhibitor of PRMTs, since its affinity for the PRMT is superior to that of SAM. Other control mechanisms are the regulation of the cellular localization of the PRMTs, their oligomerization, and the interplay of arginine methylation with other PTMs such as acetylation or phosphorylation of the arginine residues themselves or neighboring residues (Molina-Serrano et al. 2013).

# 3    Lysine Methylation

Lysine methylation was the first methylation to be discovered. The first protein methylation ever described was in the flagella protein of *Salmonella typhimurium* (Murray 1964; Paik et al. 2007). It shows many parallelisms with the above-commented arginine methylation, although their differences are important in the chemistry of reaction, functional consequences, methyltransferases involved, and their mode of action. Although lysine modification was discovered in the 1960s, it was not until the 2000s when T. Jenuwein and colleagues identified the first lysine methyltransferase (KMT) (Rea et al. 2000; Black et al. 2012). This enzyme (KMT1A) was demonstrated to be conserved from yeast to human. Many KMTs have been discovered since and research has largely been facilitated by the availability of specific antibodies for methylated lysine in its different methylation modes and the use of advanced genomics technologies.

Lysine methylation is widely distributed, and it can be found both in eukaryotes and in prokaryotes (unlike arginine methylation). Currently more than 2000 lysine methyl modifications have been reported in human, with more than 1200 proteins affected (Falnes et al. 2016). All additions of methyl groups to a lysine are catalyzed by the lysine methyltransferases (KMT) that modify specifically the lysine ε-amine group. This amine group can be mono-, di-, or even tri-methylated (Fig. 2b), and the degree of methylation presents another layer of information that goes beyond the simple fact that the lysine is methylated. As for the arginine methylation, SAM is also the substrate for lysine methylations.

The majority of the methyltransferases belong to the group of enzymes that contain the SET domain (from SU(var), Enhancer of Zeste and Trithorax) (Rea et al. 2000), which carried out the methyltransferase reaction. SET domains are characterized by three regions folded into a β-sheet structure that forms the active site. The channel formed by this domain allows the protein and SAM to enter the active site but at the same time controls the number of methylation events that occur. This SET domain containing family can be further subdivided into seven subfamilies based on sequence alignments. However, there are other KMTs that have different types of seven β-strand (7BS) domain-containing Rossman fold structures, similar to the PRMTs (Class I) described above. These enzymes show a broad phylogenetic distribution and can be found in eukaryotes, prokaryotes, and archaea (Falnes et al. 2016).

The chemistry and biophysics of the modification reaction is very similar to that of the arginine methylation, i.e., methylation on lysine residues does not change the net nor local charges and causes only minor alteration to the mass, steric volume, and hydrophobicity. Its mechanism of action is therefore based on the interaction of the methylated proteins with other partners, either proteins, DNA, or RNA. Methylation can be considered as a regulator of ubiquitination or other PTMs, since methyltransferases compete with the ubiquitin-conjugating set of enzymes (E1, E2, E3) for the same lysine residues or act as "methyl switches" where one methylation stimulates or inhibits the action of another PTM in neighboring residues (Moore and Gozani 2014). Methylation can also promote the binding of other proteins, known as the "reader" proteins that have a higher affinity for the substrate protein when in its methylated state, due to a better binding surface exposure, a structural rearrangement, or a methylation induced increase in stability. "Reader" proteins can distinguish between different methylation states and react accordingly. "Reader" proteins can be compared to an aromatic cage with hydrophobic contacts and cation-pi interactions towards the region involving the methylated residue (Wozniak and Strahl 2014). Examples of "reader" protein domains are the Royal domain family, Plant HomeoDomain (PHM), or the Tudor domains, recognizing both histone and nonhistone lysine methylated proteins (Black et al. 2012; Moore and Gozani 2014). Other adaptations induced by lysine methylation are increases in the binding affinity for DNA and increased resistance to proteases or heat denaturation of the methylated protein.

The existence of lysine demethylases (KDM) has been unambiguously demonstrated, in contrast to arginine demethylases (Shi et al. 2004). KDMs discovered to date are able to demethylate mono-, di-, and tri-methylated lysines on histones, but also nonhistone proteins such as transcription factors and DNA methyltransferases (Lanouette et al. 2014). Demethylation has been shown to be a key component of the signalization and modulation dynamics of the proteome. In addition, KMTs and KDMs are also highly regulated through complex mechanisms of posttranslational control. For example, ubiquitination of these enzymes favors their degradation by the proteasome, thereby regulating KMT and KDM function (Black et al. 2012) by controlling their abundance. Other KMTs and KDMs can be phosphorylated. Additional regulation mechanisms observed are through miRNAs, which can be also regulated, and through metabolites acting as cofactors. Examples of regulator metabolites are oxygen, FAD, alfa-ketoglutarate, succinate, 2-OH glutarate, or SAM and SAH (Shi and Whetstine 2007).

The functional roles of lysine methylation are comparable to those of arginine methylation, although histone methylations at lysines are more abundant as compared to those at arginines. These modifications are a major determinant for the formation of active and inactive regions in the genome, depending on the sites that are methylated and also on the degree of methylation (Black et al. 2012).

Histone methylation is a fundamental process affecting transcription and epigenetics. The degree of specificity of the complex that associates to bind the DNA sequence in transcription cannot be explained solely by the recognition of the DNA sequence. To add complexity, different cell types express different genes. This specificity is achieved through a codification of the chromatin. Chromatin states are defined by their histone methylation degree and pattern. These states are defined by the subclassification of promoter states, transcribed states, active and repressed intergenic regions, and repetitive elements. Histones are the actors that carry out this definition through an array of PTMs that serve, among other functions, as recognition points for the transcriptional factors. Lysine methylation is one of the most abundant PTM on histones. Of the 4 histone classes that constitute the nucleosome: H2A, H2B, H3, and H4, lysine methylation occurs mainly on the flexible tails of histones H3 and H4 in a process that is strongly regulated by KMT and KDMs, linking lysine methylation with euchromatin and heterochromatin dynamics and structure to maintain the genomic stability and the cell development. KMTs and KDMs are recruited during transcription, modifying protein domains that recognize chromatin states or interacting in a direct manner with transcription factors, which have "reader" domains that are capable of interpreting the methylation, lack of methylation, or PTM combinations found on histones (Black et al. 2012). Lysine methylation on histones is thus associated with the activity at the transcription start site, heterochromatin formation, chromosome silencing and transcriptional repression, transcriptional elongation, histone exchange in chromatin, and DNA damage responses.

In addition to histone methylation, lysine methylation has other functions, all derived from its activity as regulator of protein-protein interactions. It is also widespread in nonhistone proteins, controlling downstream effects such as protein stability, subcellular localization, or DNA binding (Huang and Berger 2008; Lanouette et al. 2014; Biggar and Li 2015). An important number of proteins involved in transcription and translation are affected; however, there is not a general rule for the mechanism, it has become clear that the methylation of the same lysine on the same protein promotes different outcomes depending on the cellular localization (Lanouette et al. 2014). There is less information on methylation affecting translation, although methylation of ribosomes has been shown in mammals, plants, yeast, or bacteria (Moore and Gozani 2014). Other examples of proteins regulated by lysine methylation are chaperones, kinetochores, DNA methyltransferases, or the Rubisco in eukaryotes, pili and GTPases in bacteria, and even in viral proteins lysine methylation has been reported (Huang and Berger 2008).

# 4    Other Methylations

Of no less physiological relevance are the methylations of residues other than arginine or lysine, which include methylation of the side chains of glutamic acid, aspartic acid, asparagine, glutamine, and cysteine residues as well as carboxyl- and amino- termini of proteins.

Glutamic acid methylation occurs in prokaryotes (not found in eukaryotes thus far) where it modulates the chemotactic response in bacteria such as *E. coli*, *B. subtilis*, or *Pseudomonas* (Li and Weis 2000; Hazelbauer et al. 2008). The methylation of the cytosolic part of chemoreceptors by CheR methyltransferases adjusts the sensitivity of the chemoreceptor/CheA/CheW complex to the local abundance of the chemoreceptor signals. (Antommattei and Weis 2006). This adjustment implies that further chemoreceptor stimulation is only achieved by higher signal concentrations, which in turn permits chemosensory pathways to respond to signal gradients and thus permitting chemotaxis. *Pseudomonas* species were found to possess multiple CheR paralogues, of which each reacts with a defined set of chemoreceptors (Garcia-Fontana et al. 2013; Garcia-Fontana et al. 2014). CheB methylesterases reverse the CheR-mediated methylation. CheR methylates specific sequences in the cytosolic domain of chemoreceptors using SAM as the methyl donor (Jurica and Stoddard 1998). Methylation of aspartate is known both in pro-karyotes and eukaryotes and is mainly involved in protein repair mechanisms (Sprung et al. 2008). In contrast to arginine and lysine residues, the methylation at aspartate and glutamate cancel the net charges of their side chains, and these electrostatic alterations account largely for the functional output of methylation. Other non arginine or lysine methylations have been observed in the carboxyl groups of C-terminal residues. Leucine methylation has been shown to occur in the mammalian protein phosphatase 2A on its C-terminal carboxyl group (Lee and Stock 1993; Wu et al. 2000). The methyl group can be removed by a specific methylesterase. It is in general accepted that an inhibition of this leucine methylation reduces the phosphatase activity of the protein. There are other examples of C-terminal methylation, on isoaspartylated (spontaneous alterations of aspartyl residues due to age damage) or isoprenylated (farnesyl modified-) cysteine residues (Grillo and Colombatto 2005).

Cysteine methylation was observed in the yeast ribosome, where it may be involved in DNA repair (as a potential acceptor for alkylated DNA). In addition, it participates in intermediate steps of catalysis and in automethylation reactions of Dnmt3 methyltransferase leading to autoregulation (Siddique et al. 2011; Clarke 2013). Cysteine methylations also occur in zinc-cysteine clusters with diverse functions (Young et al. 2012). Regarding histidine methylation, the methylation of the N-1($\pi$) or N-3($\tau$) atoms of the imidazole ring has been established in prokaryotic and eukaryotic cells (Clarke 2013). The best characterized process is the histidine methylation of actin (Nyman et al. 2002), but it has also been shown to act on a yeast ribosomal protein (Lee et al. 2002), as is also the case of lysine, arginine, and cysteine methylations.

Although the most extensive modification in the N-terminal amino group of proteins group corresponds to acetylation, the relevance of N-terminal methylation has also been shown in prokaryotes and eukaryotes (Stock et al. 1987). It occurs in yeast, plants, mammals, and humans. Its functional output can be associated with a net charge modulation, similarly to methylations of glutamate and aspartate residues. It is frequent in ribosomes, but also in enzymes like Rubisco or structural proteins like myosin (Webb et al. 2010). The majority of N-terminal methylated residues form part of a conserved sequence containing the N-methylated methionine, which will be cleaved, followed by prolines or alanines that are subject to methylation at their amino groups. In yeast ribosomes, histidine methylation has been documented (Webb et al. 2010); it is specifically a 3-methylhistidine, potentially responsible for the modification of Rpl3, a protein of the large ribosomal subunit on yeast.

## 5     Research Needs

While sequence analysis permits the identification of methyltransferases genomes, their physiological roles remain frequently unknown. Future research will initially need to identify the methylation targets which then serve as a basis to decipher the functional consequences of methylation reactions. There is also the need for further technological advances. Research would largely benefit from techniques that allow high throughput detection of protein methylation beyond the limits of mass spectrometry and antibody-based approaches that are currently employed. Analytical techniques should be able to discriminate between mono-, di-, or tri-methylation events.

Another research need is to cast further light into possible crosstalk between different PTMs and to decipher the relationship between methylation and protein-RNA interactions. Methylation plays a central role in systems biology and its physiological roles are crucial in the understanding of interactome dynamics and structure. Although great advances have been achieved over the last 15 years in the comprehension of the mechanisms and functional effects of methylation, a central remaining question is that of the possible existence of demethylases, primarily those for methylated arginines. As methylation, a stable modification, is expected to be a dynamic process, demethylation appears to be a plausible control mechanism. Future research will focus on unveiling demethylation processes.

Further functions for methylation will be discovered, which may not be related to histone coding and transcription, currently the primary cellular functions of methylation. Research will also lead to the identification of other "reader" proteins that execute the information generated by methyltransferases, and deciphering their mechanism of action will be among the major research topics in the field. Finally, the involvement of protein methylation in human health and disease is well known. Several pathologies are related to methylation, such as cancer, and therefore the complete understanding of the methylation process is fundamental for developing diagnostic biomarkers and to understand the functional link between methylation and pathogenicity. The knowledge on intracellular regulatory networks is essential for the development of therapeutic strategies. Methyltransferase inhibitors are among the options to fight methylation-related pathologies.

# References

Allers J, Shamoo Y (2001) Structure-based analysis of protein-RNA interactions using the program ENTANGLE. J Mol Biol 311(1):75–86

Antommattei FM, Weis RM (2006) 12 reversible methylation of glutamate residues in the receptor proteins of bacterial sensory systems. Enzyme 24:325–382

Bedford MT, Clarke SG (2009) Protein arginine methylation in mammals: who, what, and why. Mol Cell 33(1):1–13

Biggar KK, Li SS (2015) Non-histone protein methylation as a regulator of cellular signalling and function. Nat Rev Mol Cell Biol 16(1):5–17

Binda O (2013) On your histone mark, SET, methylate! Epigenetics 8(5):457–463

Black JC, Van Rechem C, Whetstine JR (2012) Histone lysine methylation dynamics: establishment, regulation, and biological impact. Mol Cell 48(4):491–507

Blanc RS, Richard S (2017) Arginine methylation: the coming of age. Mol Cell 65(1):8–24

Cheng X, Collins RE, Zhang X (2005) Structural and sequence motifs of protein (histone) methylation enzymes. Annu Rev Biophys Biomol Struct 34:267–294

Clarke SG (2013) Protein methylation at the surface and buried deep: thinking outside the histone box. Trends Biochem Sci 38(5):243–252

Di Lorenzo A, Bedford MT (2011) Histone arginine methylation. FEBS Lett 585(13):2024–2031

Falnes PO, Jakobsson ME, Davydova E, Ho A, Malecki J (2016) Protein lysine methylation by seven-beta-strand methyltransferases. Biochem J 473(14):1995–2009

Fisk JC, Read LK (2011) Protein arginine methylation in parasitic protozoa. Eukaryot Cell 10(8):1013–1022

Gallivan JP, Dougherty DA (1999) Cation-pi interactions in structural biology. Proc Natl Acad Sci USA 96(17):9459–9464

Garcia-Fontana C, Reyes-Darias JA, Munoz-Martinez F, Alfonso C, Morel B, Ramos JL, Krell T (2013) High specificity in CheR methyltransferase function: CheR2 of *Pseudomonas putida* is essential for chemotaxis, whereas CheR1 is involved in biofilm formation. J Biol Chem 288(26):18987–18999

Garcia-Fontana C, Corral Lugo A, Krell T (2014) Specificity of the CheR2 methyltransferase in *Pseudomonas aeruginosa* is directed by a C-terminal pentapeptide in the McpB chemoreceptor. Sci Signal 7(320):ra34

Gary JD, Clarke S (1998) RNA and protein interactions modulated by protein arginine methylation. Prog Nucleic Acid Res Mol Biol 61:65–131

Gowri Shankar BA, Sarani R, Michael D, Mridula P, Ranjani CV, Sowmiya G, Vasundhar B, Sudha P, Jeyakanthan J, Velmurugan D, Sekar K (2007) Ion pairs in non-redundant protein structures. J Biosci 32(4):693–704

Grillo MA, Colombatto S (2005) S-adenosylmethionine and protein methylation. Amino Acids 28(4):357–362

Hazelbauer GL, Falke JJ, Parkinson JS (2008) Bacterial chemoreceptors: high-performance signaling in networked arrays. Trends Biochem Sci 33(1):9–19

Huang J, Berger SL (2008) The emerging field of dynamic lysine methylation of non-histone proteins. Curr Opin Genet Dev 18(2):152–158

Jurica MS, Stoddard BL (1998) Mind your B's and R's: bacterial chemotaxis, signal transduction and protein recognition. Structure 6(7):809–813

Khoury GA, Baliban RC, Floudas CA (2011) Proteome-wide post-translational modification statistics: frequency analysis and curation of the swiss-prot database. Sci Rep 1:pii: srep00090

Krause CD, Yang ZH, Kim YS, Lee JH, Cook JR, Pestka S (2007) Protein arginine methyltransferases: evolution and assessment of their pharmacological and therapeutic potential. Pharmacol Ther 113(1):50–87

Lake AN, Bedford MT (2007) Protein methylation and DNA repair. Mutat Res 618(1–2):91–101

Lanouette S, Mongeon V, Figeys D, Couture JF (2014) The functional diversity of protein lysine methylation. Mol Syst Biol 10:724

Lee YH, Stallcup MR (2009) Minireview: protein arginine methylation of nonhistone proteins in transcriptional regulation. Mol Endocrinol 23(4):425–433

Lee J, Stock J (1993) Protein phosphatase 2A catalytic subunit is methyl-esterified at its carboxyl terminus by a novel methyltransferase. J Biol Chem 268(26):19192–19195

Lee SW, Berger SJ, Martinovic S, Pasa-Tolic L, Anderson GA, Shen Y, Zhao R, Smith RD (2002) Direct mass spectrometric analysis of intact proteins of the yeast large ribosomal subunit using capillary LC/FTICR. Proc Natl Acad Sci USA 99(9):5942–5947

Li G, Weis RM (2000) Covalent modification regulates ligand binding to receptor complexes in the chemosensory system of *Escherichia coli*. Cell 100(3):357–365

Low JK, Wilkins MR (2012) Protein arginine methylation in *Saccharomyces cerevisiae*. FEBS J 279(24):4423–4443

Lu R, Wang GG (2013) Tudor: a versatile family of histone methylation 'readers'. Trends Biochem Sci 38(11):546–555

Molina-Serrano D, Schiza V, Kirmizis A (2013) Cross-talk among epigenetic modifications: lessons from histone arginine methylation. Biochem Soc Trans 41(3):751–759

Moore KE, Gozani O (2014) An unexpected journey: lysine methylation across the proteome. Biochim Biophys Acta 1839(12):1395–1403

Murray K (1964) The occurrence of epsilon-N-methyl lysine in histones. Biochemistry 3:10–15

Nyman T, Schuler H, Korenbaum E, Schutt CE, Karlsson R, Lindberg U (2002) The role of MeH73 in actin polymerization and ATP hydrolysis. J Mol Biol 317(4):577–589

Paik WK, Paik DC, Kim S (2007) Historical review: the field of protein methylation. Trends Biochem Sci 32(3):146–152

Polevoda B, Sherman F (2007) Methylation of proteins involved in translation. Mol Microbiol 65(3):590–606

Rea S, Eisenhaber F, O'Carroll D, Strahl BD, Sun ZW, Schmid M, Opravil S, Mechtler K, Ponting CP, Allis CD, Jenuwein T (2000) Regulation of chromatin structure by site-specific histone H3 methyltransferases. Nature 406(6796):593–599

Schubert HL, Blumenthal RM, Cheng X (2006) 1 Protein methyltransferases: their distribution among the five structural classes of AdoMet-dependent methyltransferases. Enzyme 24:3–28

Shi Y, Whetstine JR (2007) Dynamic regulation of histone lysine methylation by demethylases. Mol Cell 25(1):1–14

Shi Y, Lan F, Matson C, Mulligan P, Whetstine JR, Cole PA, Casero RA, Shi Y (2004) Histone demethylation mediated by the nuclear amine oxidase homolog LSD1. Cell 119(7):941–953

Siddique AN, Jurkowska RZ, Jurkowski TP, Jeltsch A (2011) Auto-methylation of the mouse DNA-(cytosine C5)-methyltransferase Dnmt3a at its active site cysteine residue. FEBS J 278(12):2055–2063

Sprung R, Chen Y, Zhang K, Cheng D, Zhang T, Peng J, Zhao Y (2008) Identification and validation of eukaryotic aspartate and glutamate methylation in proteins. J Proteome Res 7(3):1001–1006

Stock A, Clarke S, Clarke C, Stock J (1987) N-terminal methylation of proteins: structure, function and specificity. FEBS Lett 220(1):8–14

Tripsianes K, Madl T, Machyna M, Fessas D, Englbrecht C, Fischer U, Neugebauer KM, Sattler M (2011) Structural basis for dimethylarginine recognition by the Tudor domains of human SMN and SPF30 proteins. Nat Struct Mol Biol 18(12):1414–1420

Walsh CT (2006) Posttranslational modification of proteins: expanding Nature's invention. Roberst Publishers, Greenwood Village

Webb KJ, Zurita-Lopez CI, Al-Hadid Q, Laganowsky A, Young BD, Lipson RS, Souda P, Faull KF, Whitelegge JP, Clarke SG (2010) A novel 3-methylhistidine modification of yeast ribosomal protein Rpl3 is dependent upon the YIL110W methyltransferase. J Biol Chem 285(48):37598–37606

Wei H, Mundade R, Lange KC, Lu T (2014) Protein arginine methylation of non-histone proteins and its role in diseases. Cell Cycle 13(1):32–41

Weiss VH, McBride AE, Soriano MA, Filman DJ, Silver PA, Hogle JM (2000) The structure and oligomerization of the yeast arginine methyltransferase, Hmt1. Nat Struct Biol 7(12):1165–1171

Wozniak GG, Strahl BD (2014) Hitting the 'mark': interpreting lysine methylation in the context of active transcription. Biochim Biophys Acta 1839(12):1353–1361

Wu J, Tolstykh T, Lee J, Boyd K, Stock JB, Broach JR (2000) Carboxyl methylation of the phosphoprotein phosphatase 2A catalytic subunit promotes its functional association with regulatory subunits in vivo. EMBO J 19(21):5672–5681

Young BD, Weiss DI, Zurita-Lopez CI, Webb KJ, Clarke SG, McBride AE (2012) Identification of methylated proteins in the yeast small ribosomal subunit: a role for SPOUT methyltransferases in protein arginine methylation. Biochemistry 51(25):5091–5104

# Index

**A**

AadR, 155
ABC transport, 292
ABC transporters, 294
*ABC1* gene, 67
ABS, *see* Activation binding
    site (ABS)
Absorption, 35
Accessibility, 249, 319
    of promoters, 527
Acetylation, 479, 482, 483, 524, 581
Acetyl-coenzyme A
    consumption, 467–470
    metabolism, 470–471
    pools, 467
    production, 465–466
Acetylene reduction, 433
*Acinetobacter* sp, 423
    *A. baumannii*, 426
    *A. baylyi* ADP1, 156
    *A. calcoaceticus*, 399
    *A. radioresistens*, 401
    *A. venetianus*, 400, 401
AcpR, 155
Activation binding site (ABS), 149
Activating histone modifications, 529
Activator, 140
Activator protein, 181
Active demethylation, 519–520, 535
Active efflux system, 292
Active transport of hydrocarbons,
    *see* Hydrocarbons
Acyl-CoA oxidases (Aox), 70
Acyl-homoserine lactones, 121
*N*-Acyl-homoserine lactones (AHLs), 274
Adaptation, 403
Adenylate energy charge (E.C.), 404
Adhesins, 109
Adhesion, 21

to hydrocarbons, 401
to the substrate, 399
Aerobic degradation pathways, 142
A-factor, 265
Aging, 41, 547
Agonists, 123, 140
*Airn* long non-coding RNA (lncRNA), 532
Air-water interface, 11
*Alcanivorax borkumensis*, 393
Alkane(s), 61, 66, 290
    degradation, 128
Alkane monooxygenase system (AMOS), 67
*n*-Alkanes, 400
*n*-Alkanols, 342
AlkL, 291
Alkylation(s), 479, 481, 482, 572
2-Alkyl-4-quinolones (AQs), 274
Alpha-thalassemia and mental retardation
    X-linked syndrome (ATR-X), 526
Altered microRNA expression, 546
Alternatively spliced exons (ASEs), 541
Alzheimer dementia (AD), 547, 548
Amphipathic molecules, 339
Amphiphilic lipids, signaling molecules
    and quorum sensing, *see* Quorum
    sensing (QS)
Anaerobic degradation pathways, 142
Anaerobic hydrocarbon degradation, 438
Analytical and microscopic techniques, 28
*ANT1* gene, 67
Antagonists, 123, 140
Anthracene, 295
Anthranilate, 154
Antibiotic resistance, 363, 424–425
Antimicrobials, 109
*Aplysia californica* (California sea hare, a
    mollusk), 544
AraC/XylS, 152–154
AreR, 152

Arginine residues, 573
Argonaute, 524
Aromatic acid:H$^+$ symporter (AAHS), 294
Aromatic compounds, 138
Aromatic degradation, 466
Aromatic hydrocarbons, 107, 178, 244, 288,
    290, 292, 296, 379–382
Atmospheric oxygen, 542
ATP, 398, 404, 405
ATP-dependent chromatin assembly factor
    (ACF), 525
ATRX–DNMT3–DNMT3L (ADD)
    domain, 526
AtzR, 149
Autoinducers, 276
Autophosphorylation, 202
*Azoarcus* sp. 155

**B**

Bacterial adhesion, 21
Bacterial cells
    adaptations, changing environment, 379
    biosynthesis, phospholipids, 376
    branched chain fatty acids, 381
    cardiolipin, 376, 381
    *cis-trans* isomerase, 380
    cyclopropane fatty acids, 381
    cytoplasmic membrane (*see* Cytoplasmic
        membrane)
    fatty acids, 376, 381
    lipid domains, 378
    structure, 374–375
Bacterial cell-to-cell signals, 256
Bacterial Crabtree effect, 467
Bacterial floc groups, 414
Bacterial plasticity, 311
Bacterial restriction-modification (R-M system)
    functions, 492–493
    occurrence and prevalence, 490–492
    types, 489–490
Bacterial-fungal interactions (BFI), 320
BadM, 164
BadR, 163
BadR protein, 155
Barr body, 534
Base excision repair (BER), 537
Base flipping, 514, 516
Bench-scale experiments, 229
BenK, 294
BenM, 150
Benzoate, 143, 294

Benzoate dioxygenase, 153
Benzoyl-CoA, 159
Bet-hedging strategy, 497
Biallelic expression, 534
Bifunctional crotonase, 263
Bioaccessibility, 10, 42, 304
Bioaccessible, 247
    fraction, 11
Bioaccumulation, 6, 18
Bioaugmentation, 246, 441
Bioavailability, 6, 8, 19, 28, 36, 40, 242, 244,
    246, 247, 250, 295, 304, 319, 398
Bioavailability of hydrophobic substrates,
    biosurfactants, *see* Biosurfactants (BSs)
Biodegradability of hydrocarbons, 412
Biodegradation, 18, 36, 40, 107, 242, 243,
    247, 249–250, 328, 332, 352, 356
Biodegraders, 332
Bio-emulsans, 54
Biofilm(s), 21, 106, 276, 295
Biofilm formation, 402
    as adaptive response, 53
    cell adhesion to hydrophobic compounds,
        50–51
    multidisciplinary approaches, 55
    multispecies biofilms on hydrophobic
        interface, 49–50
    regulation and determinism of, 51
Biological capacity, 8
Biological interface, 550
Biological membranes, 328
Biological treatment, 328
Bioreactors, 111
Bioremediation, 7, 77, 78, 91–92, 242, 243,
    245, 247, 250
Bioremediation strategies, 216
Biosensors, 141, 151
    protein-based biosensor, 216
    styrene biosensor, 216
Biosolvents, 22
Biosurfactants (BSs), 295, 331, 400
    actinobacteria, 88–89
    biodiversity of, 77–78
    conversion of renewable resources, 92–93
    environmental bioremediation, 91
    glycolipids, eukaryotes, 89–91
    lipopeptides, 87
    microemulsion-based technologies, 92
    rhamnolipids, *Pseudomonas* spp. 78–86
Biotechnology, 112
Biotransformation, 106, 242, 249, 432
Biphenyl, 291

Biphenyl catabolic genes, 160
Bipolar disorder, 549
Black carbon, 308
Bottom-up interactions, 183
Boundary layers, 18
  diffusion, 18
BoxR, 160
Brahma-related gene, 1 (Brg1), 525
Brain-derived neurotrophic factor (BDNF), 544
Brain methylome, 543
Branched chain fatty acids (BCFAs), 381
BRENDA, 128
BrlR-SagS two-component systems, 182
BTEX, 203
Bulk phase, 6
*Burkholderia* signal (BDSF), 262
BzdR protein, 159
BZLF1, 530

**C**
Cambrian radiation, 542
cAMP, 155
Capillary assay, 224
Capsular polysaccharides, 51
Carbazole, 154, 161
5-Carboxylcytosine (5caC), 512, 520, 521, 542
Cardiolipin (CL), 375–376, 378, 381
Carrier-sorbed hydrocarbons, 28
Carrying capacity of microbial habitats, 10
Catabolic genes, 108
Catabolic pathways, 332
Catabolic repressor control protein (Crc), 151
Catabolism, 331–332
  of aromatic compounds, 143
Catechol, 143
CatM, 150
CbnR, 150
CcrM *methyl-transferase*, 498–502
Cell envelopes, 398
Cell membrane, 305
Cells lacking CpG-methylation, 537
Cellular signaling, 576
Centromere, 540
Centromere-associated proteins, 540
Chaperones, 108
CheB methylesterases, 580
Chemical activity, 37, 304
Chemical microscopy, 321
Chemoattraction, 243, 247, 250
Chemoreceptor based signaling, 124
Chemostat, 8

Chemotactic responses, 309
Chemotaxis, 50, 120, 124, 222, 242, 243, 247,
  250, 293, 329, 348
  to alkanes and alkenes, 228–229
  to aromatic hydrocarbons, 226–228
  assays, 223–226
  and biodegradation, 249
  of bioremediation bacteria, 248
  to chlorinated hydrocarbons, 231–232
  implication of, 250
  to (methyl)phenol, 232–233
  negative, 246
  to nitroaromatic compounds and
    explosives, 229–231
  parameters, 248
  positive, 243, 247, 250
  of *Pseudomonas putida*, 246
  repellent responses, 233
CheR methyltransferases, 580
CheY-P, 223
Chimeric regulator, 160
ChIP-chip, 131
Chlorobenzoates, 162, 231
Chlorocatechols, 150
ChIP-seq, 131
Chromatin, 528
  loops, 542
  marks, 521, 526, 529, 534
  remodeling, 519, 526, 530, 548
  remodeling ATPases, 517, 518, 525
  states, 478, 579
*Cis*-2-decenoic acid, 263, 280
*Cis*-11-methyl-2-dodecenoic acid, 279
*Cis/trans* fatty acids, 295
*Cis–trans*-isomerase (Cti), 380
  activity regulation, 389–392
  bacterial genomes, 389
  molecular biology and biochemistry,
    388–389
  physiological function, 386–388
Co-inducer-responsive transcriptional
  regulators, 143
Colloidal interactions, 21
Column systems, 309
Combustion residues, 35
Co-metabolism, 436
Composition of the La Brea tar, 423
Concentration gradient, 19
Concentration-gradient driven diffusion, 18
Confocal scanning laser microscopy (CSLM),
  49, 52
Contact angle, 34

Contaminants, 316
Control of co-expressed gene batteries, 534
Coordinated transcriptional silencing, 518
Co-transport of hydrocarbons, 28
*p*-Coumaroyl-CoA molecules, 163
*p*-Coumaroyl-HSL, 163
CouR, 163
Covalent modifications, 510, 524, 528
CpG-beacons, 542
CpG-dinucleotides, 512, 514, 517, 518, 529,
     540, 542, 543
CpG-island methylation, 538
CpG-islands, 512, 519, 520, 529, 530, 534,
     543, 549
CpG methylation-mediated activation of
     gene expression, 530
CprK proteins, 156
Creosote, 307
Critical micelle concentration (CMC), 53, 54
Cross-regulation, 142
Cross-talk, 182, 277, 581
CRP/FNR proteins, 154–156
CTCF, 531, 541, 542
CTCF-mediated RNA polymerase II
     pausing, 541
C-terminal catalytic domain, 513, 515
CXXC zink finger domain, 520
Cyclohexanecarboxylate, 155, 163
Cyclopropanation, 381
Cyclopropane fatty acids (CFAs), 381
*p*-Cymene, 157
CymR, 157
Cysteine methylation, 580
Cytoplasmic membrane
     acid resistance, 381
     adaptations, changing environment,
         379–380
     branched chain fatty acids, 381
     cardiolipin, 376–378, 381
     cellular division, 379
     *cis-trans* isomerase, 380
     composition, 375–376
     cyclopropanation, 381
     cyclopropane fatty acids, 381
     energy functions, 378
     fluidity changes, 382
     gram-negative bacteria, 375
     gram-positive bacteria, 374
     lipid domains, 378
     modifications, 382
     phospholipid biosynthesis, 376
     protein translocation, 379

     signaling functions, 379
     structure, 375
     transport functions, 378–379
Cytoplasmic streaming, 317
Cytoplasmic transport, 28

**D**
2,4-D, *see* 2,4-Dichlorophenoxy
     acetate (2,4-D)
Dam, 488
     chromosome gene expression, 495–498
     chromosome replication control, 495
     methyl-directed mismatch repair, 494–495
Decarboxylation of pyruvate, 465
Degradation of *N*-AHL signals, 259–260
Degradation pathways, 121
Degrading microorganisms, 28
Dehalorespiration, 156
Dementia, 548
Dementia-related disorders, 548
*De novo* DNA methylation, 519, 523, 526
*De novo* methylation, 536
*De novo* synthesis, 70
Derjaguin-Landau-Verwey-Overbeek (DLVO)
     theories, 50
Desorption extraction, 304
*Desulfovibrio vulgaris*, 411–412, 415
Developing brain, 550
Developmental arrest, 516
Developmental programs, 537
Diazotrophy, 432
Dichloroethylene, 231
2,4-Dichlorophenoxyacetate (2,4-D),
     294, 337
Differentially hydroxymethylated regions
     (DhMRs), 548
Differentially methylated genes, 532
Differentially methylated
     positions (dDMPs), 543
Differentiation, 515, 518, 519, 537–539, 550
Differentiation associated genes, 538
Diffusible signal factor (DSF), 279
     BDSF, 262
     RpfB and signal degradation, 264
     RpfF and signal synthesis, 263
     Rpf proteins and DSF signaling in
         xanthomonads, 261–262
Diffusion rates, 275
3,5-Dihydroxybenzoate, 149
1,6-Diphenyl-1,3,5 hexatriene, 340
Direct interfacial transport, 61

Dispersal of microorganisms, 309
Dissolved organic carbon (DOC), 13, 24
Dissolved organic matter (DOM), 308
Distributing contaminants, 246
DmpR protein, 151
DNA-binding motif, 142
DNA looping, 150
DNA methylation, 488
    active DNA demethylation and passive
        DNA demethylation, 519
    alternative splicing, 523, 541–542
    in bacterial systems (*see* Bacterial
        restriction-modification (R-M system))
    canyons/DNA methylation valleys, 538
    in *Caulobacter crescentus*, 500
    cell differentiation, 537
    chromatin structure and promoter activity,
        528–531
    DNMT1, 513, 525
    DNMT3A and DNMT3B, 517, 526
    embryogenesis, 535
    functional consequences, 540
    genomic imprinting, 531–532
    homologous recombination, inhibition of,
        540–541
    HuR, 524
    in phase variation, 497
    microRNAs, 523–524
    microsatellite stabilization, 539
    mismatch repair, 540
    normal brain function and memory,
        544–547
    psychiatric disorders and dementia,
        547–550
    PTMs, 524–525
    transcriptional regulation, 522–523
    transposon silencing, 532–534
    X chromosome inactivation, 534
DNA methyltransferase enzymes, 510, 513,
    521
DNA methyltransferases (DNMTs), 479
DNA motifs, 539
DNA N6-methyladenine demethylase
    (DMAD), 512
DNA-protein recognition, 190
DNA repair, 576
DNMT1, 513–517, 513, 519, 520, 524, 525,
    527
DNMT1o, 532
DNMT2, 527
DNMT3A, 513, 517, 527, 517–519
DNMT3B, 513, 517, 527, 530

DNMT3L, 513, 518, 526, 527
DntR, 150
DOC-mediated transport, 26
DOT-T1E, 109
Drop assay, 224
DSF, *see* Diffusible signal factor (DSF)
Dual base flipping, 515, 516
Dynamic regulation of methylation
    states, 550
Dysregulated expression of inactive
    DNMT3B variants, 523


E
Early brain development, 549
Early S phase, 525
Ecosystem property, 14
E2F1, 522
E2F6, 518
Effector modulation, 190
Effector molecules, 190
Effector-specific transcriptional
    regulation, 138
Efflux pumps, 107, 296, 347, 403
Efflux systems, 330
Eight-cell embryos, 523
Electrokinetics, 307
Electron microscopy, 414, 415
Embryogenesis, 534–537
Embryonal lethality, 516
Embryonic stem cells (ES cells), 537
    differentiation, 538
Emulsan, 400
Emulsifiers, 34
Emulsions, 88, 92
Endocytotic uptake, 23
Energetics, hydrocarbons, *see* Hydrocarbons
Energy-dependent processes, 178
Enhanced diffusion, 24
Environmental hazard, 244
Enzymatic processes, 483
Enzyme kinetics, 9
Enzymes, 318
Epigenetic(s), 478, 479, 572
    mark, 545
    memory, 514
    regulatory mechanisms, 528
    reprogramming, 535
Epigenome, 550
Epigenotype, 538
Epstein-Barr virus (EBV), 530
Erasure of DNA methylation, 537

Erasure of imprints, 535
ES cell, *see* Embryonic stem cells (ES cells)
*Escherichia coli*, 339, 376, 468, 470
*p*-Ethylphenol, 152
Eukaryote genomes, 512
Evolution, 112, 141
Evolutionary origin of antibiotic
    resistance, 425
Exo-enzymes, 319
Exon inclusion, 542
Exons, 541
Exon-specific methylation, 544
Exopolysaccharides (EPS) , 111,
    295, 415
Experience-dependent epigenomic
    changes, 546
Expression, tolerance mechanisms, 111
Extracellular polymeric substances
    (EPS), 54
Extractability, 42
Extraembryonic lineage, 538
Extrusion mechanisms, 107


F
Facilitated diffusion, 24
FadL family, 289
Farnesol, 281
Fatty acid methyl esters, 274
Fatty acid transporter, 277
Fatty acids (FA), 121, 308
    degradation of, 70–71, 466
    synthesis, 66
Fick's law, 18
5caC, *see* 5-Carboxylcytosine (5caC)
5fC, *see* 5-Formylcytosine (5fC)
5hmC, *see* 5-Hydroxymethylcytosine (5hmC)
5mC, *see* 5-Methylcytosine (5mC)
*Flavimonas oryzihabitans*, 228
Flow cytometry, 453
Fluidity, 329
Fluoranthene, 292
5-Formylcytosine (5fC), 512, 520, 521
*FoxA2* promoter, 530
FOXO3a, 523
Freely dissolved pool, 34
Frontotemporal dementia (FTD), 548
Fruit fly, 511
Functional groups, 478, 483
Functional magnetic resonance tomography
    (fMRT), 547

Fungal highways, 316
Fungal mycelia, 28
Fungal pipelines, 316
Fungus (fungi), 28, 316, 511


G
Gametogenesis, 535
Gamma-butyrolactones (GBLs), 264–266
Gas chromatography (GC) content, 512
Gene(s)
    bodies, 541, 549
    chips, 130
    clusters, 531
    expression, 274, 348
    mutations, 60
    with synapse-related function, 545
General demethylation, 520
Genetic circuits, 191
Genetic instability, 533
Genetic mosaicism, 550
Genetic traps, 166
GenK, 294
Genome wide association study
    (GWAS), 549
Genome-wide demethylation, 535
Genome-wide responses, hydrocarbons
    ChIP-seq, 131
    metabolome, 130
    microbial communities, 132–133
    proteome, 130
    RNA-seq, 130
    in silico analyses of genes and
        oligonucleotide signatures, 128–129
    transcriptome, 130
Genomic Islands Database, 128
Gentisate, 155, 294
Geosorbents, 35
Germ cell-specific genes, 540
Germline defence genes, 535
Germline differentially methylated regions
    (gDMRs), 531
Glial cells, 543
Global demethylation, 535
Global regulators, 141, 181
Glutamic acid methylation, 580
GlxR, 155
Glycolipids, 79, 88, 89
Glyoxylate shunt, 468
GntR protein, 160–162
G2 phase, 525

G-quadruplex (G4), 539
Gradient plate assay, 224
Gradients, 250
Gram-negative bacteria, 179, 180, 182, 387
  cell envelope structure, 374, 375
  fatty acids, 376
  phospholipids, 375–376
Gram-positive bacteria, 374
Gratuitous inducer, 140
Gratuitous induction, 149
Growth-phase state, 183

**H**
Habitat, 8
Hard natural organic matter, 35
HbaR, 155
HcaR, 162
HcaT, 294
HDAC-inhibitors, 552
H-DNA, 539
Health hazard, 244
$\sigma^{32}$ Heat-shock sigma factor, 153
*Helaeomyia petrolei* (Syn. Psilopa), *see* Oil
  fly larvae
Hemi-methylated CpG-sites, 540
Hemi-methylated DNA, 513, 515, 525,
  539–540
Heterochromatin, 519, 540
Heterochromatinization, 534
*n*-Hexadecane, 292
Hfq, 151
High CpG-density promoters, 530
High-pressure freezing/freeze-substitution
  (HPF/FS), 414, 415
High-resolution, 550
High-throughput methods, 550
Hippocampus, 544
Histone acetylation, 548
Histone H1, 529
Histone modifications, 528
Histone octamer, 528
HmgR protein, 157
Homeoviscous adaptation, 379
Homoprotocatechuate, 162
Horizontal gene transfer (HGT), 320
HpaR, 162
HppK, 294
Human antigen R (HuR), 524
Human embryos, 535
Human fetal brain, 543

HuR, *see* Human antigen R (HuR)
Hydrocarbon/aqueous environment
  interface, 410
  biofuel production, 412–413
  fossil fuels production and spills,
    410–411
  microbe-hydrocarbon research, imaging in,
    413–416
  oil well souring, 411
  sulfate reduction, *Desulfovibrio*
    *vulgaris*, 411
Hydrocarbon compounds, *see* Hydrocarbons
Hydrocarbon efflux pump, 296
  conceptual features, 191
  coordinated expression, 188
  inducible resistance, 189
  intrinsic resistance, 189
  regulatory networks, 180–183
Hydrocarbon-oxidizing bacteria and nitrogen
  fixation
  cyanobacteria, 439–441
  gaseous hydrocarbons and
    methanotrophs, 433
  heterotrophs, 435–439
  phytoremediation, 442–444
  sediments and microbial mats, 441–442
Hydrocarbons, 328, 346
  aliphatic, 346
  aromatic, 346
  biodegradation, 249
  bioremediation, 242
  chemotactic bacteria, 246–249
  chemotaxis, 348–349
  contaminants, 244–246
  cytoplasmic membrane, 291
  gram negative bacteria, outer
    membrane of, 289
  hydrocarbon acquisition, 295
  physical and chemical remediation
    techniques, 242
  physico-chemical surface properties,
    398–403
  sensing mechanisms, 120
  TOD pathway for catabolism, 350–352
  TOL pathway for catabolism, 352
  toxic hydrocarbons, effect of, 403–405
  uptake, 124
Hydrocarbon-utilizing bacteria,
  *see* Hydrocarbons
Hydrodynamic mixing, 39
Hydrophilic interaction, 4

Hydrophobicity, 4, 111, 328, 398, 575
Hydrophobic organic compounds (HOCs), 36,
    39–41, 55
  insoluble HOCs, 53–55
  regulation and determinism of biofilm
    formation, 51–53
Hydrophobic substrates (HS), 60
  binding of, 62
  degradation of, 61
  modification of, 68–70
  solubilization, 62–64
  transport/export, 64–68
3-Hydroxybenzoate, 155, 294
4-Hydroxybenzoate, 293
Hydroxycinnamoyl-CoA thioesters, 162
Hydroxylated fatty acid esters, 266
5-Hydroxymethylcytosine (5hmC), 512, 520,
    535, 538, 542, 545, 548
6-Hydroxynicotinic acid, 159
4-Hydroxyphenylacetate, 294
4-Hydroxyphenylacetic acid, 154
3-Hydroxyphenylpropionic acid, 157
3-(3-Hydroxyphenyl) propionic acid, 294
(S)-3-Hydroxytridecan-4-one, 267–268
Hyperchemotaxis, 226
Hyphosphere, 319


I
IclR proteins, 156–157
IHF, see Integration host factor (IHF)
Immunodeficiency, centromeric instability,
    facial anomalies syndrome (ICF
    syndrome), 540
Impermeable surface skins, 38
Imprinted genes, 531
Imprinting control regions (ICRs), 520, 531,
    532, 535
Inactive DNMT3B isoforms, 523
Indole, 154
Inducer molecule, 140
Induction of cell differentiation, 538
Infectious disease, 552
Inner cell mass, 536
Integration host factor (IHF), 151
Interface fertilizer, 310
Interfaces, 21
Interfacial area, 21
Internalization systems, 277
Interphase, 18
Interspecies and inter-kingdom signaling,
    268–269

Interspecific signaling, 280
Intramolecular phosphotransfer
    mechanism, 206
Intrinsic logic, 191
Introns, 541
In vivo expression technology
    (IVET), 132
Irreversibly-bound pool, 34
Isoforms, 518, 523, 532
Isophthalate metabolism, 157
Isothermal titration calorimetry
    (ITC), 205


K
β-Ketoadipate pathway, 143, 293
Krüppel-associated box (KRAB), 519


L
Labile contaminant pools, 42
La Brea tar pits, 420, 426
Larvae, surface sterilized, 420
Latent EBV genomes, 531
Lateral diffusion, 291
Lateral transport, 290
Late S phase, 525
Learning, 544, 545
Lewy body dementia (LBD), 548
Lineage-specific differentiation, 537
Lineage-specific methylomes, 535
Linker DNA, 525, 529
Lipidation, 482, 483
Lipids
  biosynthesis, 377
  branched chain fatty acids, 381
  cardiolipin, 376, 381
  composition, bacterial cytoplasmic
    membrane, 376
  cyclopropane fatty acids, 381
  domains, bacterial membrane, 378
  fatty acids, 381
  modification, 382
  phospholipid, 376
Lipopeptides (LPs), 86–88
Lipopolysaccharides (LPS), 22, 51, 375,
    400, 415
Local demethylation, 519, 531
Local regulators, 181
Locus-specific methylation, 551
Logicome, 165
$logP_{ow}$, 178

Long interspersed nuclear elements-1
    (LINE-1/L1s), 547
Long non-coding RNA (lncRNA), 534
Long-term memory, 544
Long-term spatial memory, 546
Loss of 5hmC, 538
Low CpG-density promoters, 530
*LuxI*, 275
LuxI family proteins, 258
LuxMN, 259
*LuxR*, 275
LuxR family, 258
Lymphoid-specific helicase (Lsh), 518
Lysine demethylases (KDM), 578
Lysine methylation, 577–579
Lysine methyltransferase (LMT), 577
LysR family of transcriptional regulators,
    143–150


**M**
Maintenance rate coefficient, 9
Major depression, 549
Major facilitatory superfamily (MFS), 293
Malignant disease, 552
Mammalian genomes, 511, 512, 519,
    527, 533
Manual temporal assays, 226
*Marinobacter hydrocarbonoclasticus* SP17
    biofilms, 48
MarR-type regulators, 162–164
Mass flux, 24, 37
Mass spectrometry (MS), 131, 581
Mass transfer, 18
Matrix, biofilm, 106
Maximum membrane concentration (MMC),
    340–343
MbdR protein, 159
McpT, 348, 349
MDM2, *see* Murine double minute, 2 (MDM2)
MeCP2, 541, 547
Meiotic recombination, 541
Membrane active compounds, 340
Membrane fatty acid, 22
Membrane lipids, 22
Membrane proteins, 367–368
Membrane vesicles, 403
MEME Suite, 128
Memory formation, 545, 550, 551
Memory regulation, 547
Memory restoration, 548

Memory suppressor gene *PP1* (protein
    phosphatase, 1), 544
*Meta* pathway, 154
Metabolic diversification of single cells, 165
Metabolic interactions, 108
Metabolome, 130
Metagenome, 132
Metagenomic libraries, 166
Metal toxicity, 106
Metatranscriptome, 132
Methane monooxygenases (MMOs),
    434, 436
Methanotrophs, 433–435
Methyl-accepting chemotaxis
    proteins (MCPs), 222, 293
$N^6$-Methyladenine, 511, 512
Methylated intron, 530
Methylated promoter, 531
Methylation(s), 478, 479, 481, 483, 524, 572
    arginine, 573–574
    crosstalk, 581
    cysteine, 580
    glutamic acid, 580
    lysine, 577
    mass spectrometry, 581
    physiological roles, 581
Methylation-mediated transposon silencing,
    533
Methylation quantitative trait loci (meQTLs),
    549
3-Methylbenzoate, 153
3-Methylbenzoyl-CoA, 159
Methyl-CpG binding domain proteins
    (MBDs), 529
Methyl-CpG binding proteins, 525
5-Methylcytosine (5mC), 511, 513, 520, 529,
    533, 535, 539, 542, 545, 548
Methylcytosine-binding proteins, 527
Methyl-directed mismatch repair (MMR), 494
Methyl group transfer, 514
Methyl switches, 578
12-Methyl-tetradecanoic acid, 280
Methyltransferases, 482, 572
MhbT, 294
MhpR protein, 157
MhpT, 294
Michaelian concepts, 451
Michaelis-Menten curve, 6
Microbial attachment, 310–311
Microbial degradation, 6
Microbial ecology, 36
Microbial logistics, 316

Microbial oxidation of hydrocarbons, 457
Microbial reservoir of antibiotic resistance
    genes, 425
Microbial surfactants, 308
Microbial uptake, 23
Microfluidic devices, 226
Microhabitat, 316, 320
Micropollutants, 321
Microporous minerals, 35
MicroRNAs (miRNAs), 524, 551
Microsatellite stabilization, 539
Microsatellites, 539
Microvoids, 35
Mild cognitive impairment (MCI), 547
Mineral surfaces, 35
Mineralization, 432
miRNAs, see MicroRNAs (miRNAs)
Mismatch repair (MMR), 540
Mitotic recombination, 540
Mixotrophy, 439
Mobile sorbents, 24
Mobilizing factor, 244
Modified bases, 511
Modified strains, 112
Modulators, 140
Molecular diffusion, 37
Molecular mechanisms, 250
Monod kinetics, 8
Mono-disperse dissolution, 18, 23
Monoterpenes, 452
Motile organisms, 26
Multicomponent NAPL mixtures, 38
Multiple drug efflux pumps, 179
Murine double minute, 2 (MDM2), 523
Mutualistic interactions, 28
Mycelial pathways, 310
Mycelium, 316
Mycolic acids, 21, 399
Mycophagy, 320
Mycorrhiza, 318
Mycosphere, 316
Myristoylation, 483

N
Nanoparticles, 309
Nanopod, 296
Naphthalene, 291
Naphthalene chemotaxis, 227
Narcotic effect, 337
Natural attenuation, 328
Nematode, 511, 551
Network organisation, 317

Neuroepigenetics, 551
Neuroimaging epigenetics, 547
Neuronal cells, 543
Neuronal plasticity, 546
Neurons, 545
Nicotiana benthaminia, 69
Nicotine, 158
Nicotinic acid, 159
NicR repressor, 164
NicS, 159
nifH, 432
Nitrogenase, 432
Nitrogen fixation
    definition, 432
    and hydrocarbon-oxidizing bacteria
        (see Hydrocarbon-oxidizing bacteria
        and nitrogen fixation)
4-Nitrotoluene (4NT) degrading strains, 230
NodV/NodW TCS, 214
Non-aqueous phase liquids (NAPLs), 13, 34,
    310
Non-CpG methylation, 517, 543
Non-CpG sequences, 517
N-terminal domain, 515, 517, 525
NtrC, 150
Nucleic acids, 479–481, 484
Nucleophilic attack, 514
Nucleosomal core particle, 525
Nucleosome(s), 528
    formation, 540


O
Octanol-water, 311
Octanol-water partition coefficient, 4
Oil fly larvae
    antibiotic resistance, 425
    bioremediation, 423–424
    global impact, 426
    insect-microbe interactions, 425–426
    tetracycline enhanced solvent
        tolerance, 424
Oil hydrocarbons, 36, 41
Oil spill, 410, 411
Oil well souring, 410, 411
Oleaginous yeast, 469
Oligotrophs, 450, 456
Omics, 321
OmpW/AlkL family, 291
One-component systems, 121–123, 139
Oocyte-derived DNMT1o, 520
Oocytes, 523, 533, 536
Oocyte-specific exon, 1, 523

Operator regions, 140
Operon, 181
OphD, 294
Organic-carbon based distribution
    coefficients, 311
Organic matter, 13
Organic solvents, 328
Organoids, 543
Orthogonalization, 154
*Ortho* pathway, 154
Outer membrane porins, 403
Outer membrane protein, 289
Outer membrane vesicles, 23
Oxidative stress, 108
Oxygen demands, 307
OxyR, 108

**P**
p53, 522
PaaR protein, 158
PaaX repressor, 161
PaaY protein, 161
PadR-type regulators, 164
PAH, *see* Polycyclic aromatic
    hydrocarbons (PAH)
Parasitic sequence elements, 533
Parkinson's disease (PD), 548
Partition coefficients, 4
Passive demethylation, 520
Passive diffusion, 124, 291
Passive replication-coupled dilution, 537
Passive sampling, 304
Patho-Epigenetics of Human Disease, 552
Pathogenicity and methylation, 582
PcaK, 293
*pcaK* gene, 153
PcaR, 156
PcaU protein, 156
PcaY, 293
PCNA, *see* Proliferating cell nuclear
    antigen (PCNA)
Penicillins, 161
Peptidoglycan, 374–375
Perchloroethylene, 231
Permeability, 329
Persistence, 109
Petroleum, 244
Petroleum hydrocarbons, 459
Petroleum/solvent resistant microbes, 420
PfmR, 158
Phase exchange, 307–309
Phenanthrene, 292

Phenol, 151, 233
3-Phenoxybenzoate, 157
Phenylacetate, 158
Phenylacetyl-CoA, 158
Phenylacetyl-CoA catabolon, 161
Phenylpropanoids, 294
3-Phenylpropionic acid, 143, 294
Phosphatidylcholine (PC), 376
Phosphatidylethanolamine (PE), 375, 378
Phospholipids
    adaptations, environment, 381–382
    biosynthesis, 376
    composition, bacterial cytoplasmic
        membrane, 375–376
Phosphorylation, 524
    of acetate, 465
Photoconversion, 416
Phthalate, 157, 294
Physiological roles, 581
Phytoremediation, 318
Pick's disease, *see* Frontotemporal
    dementia (FTD)
Pioneer transcription factors, 519, 528
Piwi-interacting RNA (piRNA) pathway
    proteins, 533
Plants, 306
Pluripotency, 537
Pluripotency genes, 535
*Pm* promoter, 153
*Po* promoter, 151
PobR protein, 154, 156
Polycomb and Trithorax protein
    complexes, 528
Polycyclic aromatic hydrocarbons (PAHs), 4,
    48, 50, 52, 227, 243, 244, 250, 295,
    304, 432, 437, 439, 440
Polyparameter linear free energy
    relationships, 22
Pores, 124
Positron emission tomography (PET), 547
Posttranslational modifications (PTMs), 481,
    483, 521, 572
Post-translational regulation, 183
*POX* genes, 70
Prenatal-to-postnatal transition, 549
Prenylation, 482, 483
Primordial germ cells (PGCs), 535
Proliferating cell nuclear antigen (PCNA),
    515, 516, 525
Promiscuous regulators, 182
Promoter epigenome, 538
Promoters, 140
Prospective risk assessment approach, 14

Protective function of DNA methylation, 527
Protein arginine methyltransferases (PRMT), 573, 575
Protein(s), 478, 481–483
  BLAST-analysis, 390
  stability, 579
γ-Proteobacteria, 214, 215
Proteome, 130
Protocatechuate, 156, 294
Protomers, 150
Proton antiporters, 368
Protozoa, 511, 551
*Providencia rettgeri*, 423
*Pseudomonas*, 466
  *P. aeruginosa*, 402
  *P. mendocina*, 204, 211
  *P. putida*, 204, 216, 226, 231, 338, 347, 348, 354, 355, 362, 365, 366, 376, 381–382, 386, 402, 403
Pseudomonas putida KT2440, 109
Pseudomonas putida mt-2, 109
Pump mechanism, 367
*Pu* promoter, 151
PWWP domain, 525, 526

**Q**
Quinaldine, 161
Quinoline, 154
Quorum sensing (QS), 121, 256, 274
  degradation of *N*-AHL signals, 259
  DSF-mediated signaling (*see* Diffusible signal factor (DSF))
  hydroxylated fatty acid esters, 266
  *(S)*-3-hydroxytridecan-4-one, 267
  interspecies and inter-kingdom signaling, 268
  LuxIR, 258–259
  LuxMN, 259
  molecules, 183

**R**
Ralsotonia sp. SJ98, 229
RB, 522
RBS, *see* Recognition binding site (RBS)
Reader proteins, 479, 576
Recognition binding site (RBS), 149
Reconfiguration of neuronal chromatin, 547
Regulable DNA methylation at CpG-dinucleotides, 542
Regulation of tolerance mechanism, 107

Regulator receiver domain (REC), 204
Regulatory cascades, 141
Regulatory networks, 142
Regulatory noise, 141
Regulatory RNA molecules, 524, 531
Regulatory sequences, 522, 527, 530, 531
Remediation, 106
Repellent responses, 124
Repetitive sequences, 533
Replication fork, 513, 515, 539
Repressed nuclear regions, 518
Repressive chromatin structure, 532
Repressive histone mark H3K27me3, 533
Repressive histone modifications, 529
Repressor, 140
Repressor complexes, 529
Resistance, 111
Resistance-nodulation-cell division (RND) family, 296
Resorcinol, 158
Response regulator (RR) output domain, 202, 207
Restriction endonuclease, 511, 526
Retrospective risk assessment approach, 14
Retrotransposons, 533
  suppression, 515
Reversibly-bound pool, 34
Rhamnolipids, 278–279, 295, 308
Rhizoremediation, 250
Rhizosphere, 443
RhlR, 278
*Rhodococcus*, 399
*Rhodopseudomonas palustris*, 155
*Rhodotorula glutinis*, 71
RNA-(cytosine-C5)-methyltransferase, 527
RNA-DNA hybrid, 519
RNA-induced silencing complex (RISC), 533
RNA polymerase II, 541
RNA polymerases, 527
RNA-processing machinery, 552
RNase processing, 524
RNA-seq, *see* RNA sequencing (RNA-seq)
RNA sequencing (RNA-seq), 130
RND superfamily, 179
RND type efflux pumps, 363–365
Root-nodulation-cell division (RND) efflux pumps, 121
Rosmarinic acid, 277
Rossmann fold, 575
RpfB, 264
RpfF, 261, 263–264
*rpf* gene cluster, 261
Rrf2 family, 164

**S**

*Saccharomyces cerevisiae*, 71
S-adenosyl-L-homocysteine (SAH), 514
S-adenosyl-L-methionine (SAM), 510,
    514, 515
S-adenosylmethionine (SAM), 572
Saturated aliphatics, 4
Schizophrenia (SCZ), 549
Secondary DNA structures, 539
Selective capture of transcribed sequences
    (SCOTS), 132
Self-transmissible plasmid, 189
Sensor histidine kinase (SHK) input domains,
    207, 217
Sequence-specific methylation, 544
SeqWord Genomic Island Sniffer, 128
Sessile lifestyle, 110
SET domain, 577
7BS, *see* 7 beta strand (7BS)
Shelves, 549
Shikimate kinases, 159
Shores, 549
Short interspersed sequences, 512
Signal molecule, 275
Signature-tagged mutagenesis (STM), 132
Single-cell assays, 133
Single cell methylomes, 551
Sliding dimer model, 149
Slow desorption, 307
Small and mobile sorbing phases, 40
Small-RNAs, 183
Soft natural organic matter, 35
Soil, 307
Solid phase microextraction (SPME), 7
Solubility, 311
Solvent contaminants, 328
Solvent resistance, 362
Solvent tolerance, 329–331, 403
Solvent tolerance phenomena, 189
Solvent tolerant extracellular enzymes, 424
Solvents, 106, 404
Somatic cell-specific exon, 1, 523
Sorption, 309
Sorption strengths, 11
Sp1, 522
Sp3, 522
Spatial distribution, 8, 42
Spatio-temporal ecosystem, 28
Species-specific CpG-dinucleotides, 542
Specific affinity, 6, 451
Spliceosomes, 541
Spontaneous deamination, 539
SPOUT domain, 575

Spreading of CpG-methylation, 518
Steady-state mass transfer rate, 6
Steric hindrance, 575
Steryl esters (SE), 71
Stress, 106
Stressors, 183
Styrene, 161
StyR, phosphorylated (StyR-P), 212
StyS-StyR TCS
    function, 211
    styrene, 213
    StyR binding, 212
    StyS approaches, 213
Subcellular localization, 579
Subpopulations, 142
Substrate preference, 513, 527
Substrate promiscuity, 366–367
Substrate-induced gene expression
    (SIGEX), 165
Sumoylation, 524
Super-enhancers, 543
Surface charge, 22, 399
Surface hydrophobicity, 21, 399
Surface-mediated transport, 61
Surface properties, hydrocarbons,
    *see* Hydrocarbons
Surfactant(s), 24
    action-process, 306
    micelles, 13
Surrogate GC standards, 437
Synaptic plasticity, 544
Synaptic plasticity gene *RELN* (Reelin), 544
Synaptogenesis, 543
Synthetic biology, 191

**T**

Targeted methylation events, 526
Targeting of DNA methyltransferase
    enzymes, 521
TATA-less promoters, 522
TbmR, 152
TbuT, 152
Temporal assays, 226
Tenax, 305
Ten-eleven translocation (TET) family of
    dioxygenases, 520
Terephthalate, 157, 295
Terpenes, 458
Tet enzymes, 538
Tetracycline enhanced solvent tolerance, 424
*n*-Tetradecane, 292
Tetralin, 149

TetR protein, 157–159
TfdK, 294
*Thauera aromatica*, 214, 338
ThnY, 149
TmoS/TmoT TCS, 211
*Tod* operon, 204
TodS homologues, 214
TodS-TodT system
    agonists, 206, 207
    antagonists, 206, 207
    autophosphorylation assays, 205
    intramolecular phosphotransfer
        mechanism, 206
    ITC analysis, 205
    oxidative stress agents, 207
    PAS binding domain, 205
    phylogenetic distribution, 215
    REC, 204
    RR output domain, 207
    TOD pathway, 204, 209
    $P_{todX}$ promotor activation, 209, 210
TodS/TodT two-component regulatory
    system (TCS), 350, 356
TodX, 290
TOL catabolic plasmid, 153
TOL pathway, 352–354
TOL plasmid, 108
Tolerance mechanisms, 107
Toluene, 107, 151, 290, 338, 350, 352, 353,
    355, 452
    antagonists, 206
    binding and PAS1 ligands, 208
    degradation pathways, 123
    monooxygenase, 152
    PAS binding, 205
Toluene dioxygenase (TOD) pathway, 205, 350
*p*-Toluenesulfonate, 143
TouR, 152
Toxicity, 36, 328
Transcription, 576
    factors, 478, 479, 483, 515, 517, 519,
        522, 526, 528, 537, 544
    repressors, 479
Transcriptional initiation, 190
Transcriptional regulators, 179, 203
Transcriptional repressors, 188
Transcriptome, 130
Transfer-limited conditions, 21
Transgenerational effects, 546
Transmission electron microscopy (TEM),
    414, 416

Transmission of gene expression
    patterns, 514
Transphosphorylation, 205
Transposable elements/transposons, 533
Transposon suppression, 512
Tricarboxylic acid (TCA) cycle, 467, 468
Trichloroethylene, 231
Triglycerides (TAG), 71–72
2,4,6-Trinitrophenol, 157
Tripartite aromatic acid transporter, 295
Tripartite efflux pump genes, 181
tRNA-derived small sperm RNAs
    (tsRNAs), 551
Trophoblast stem cells (TS cells), 538
TutC/TutB TCS, 214
Two-component systems (TCS), 123
    miniaturizing biosensors, 216
    NodV/NodW TCS, 214
    phosphorelay mechanism, 217
    phylogenetic distribution, 215
    StyS-StyR TCS, 211
    TmoS/TmoT TCS, 211
    TodS homologues, 214
    TodS-TodT system (*see* TodS-TodT
        system)

**U**
Ubiquitination, 524
Ubiquitin-like protein containing PHD and
    RING finger domains, 1 (UHRF1),
    515, 517, 520, 524, 525
Uhrf2, 546
Unmethylated histone H3 lysine
    4 (H3K4me0), 526
Unmethylated substrates, 517
Unsaturated aliphatics, 4
Unstirred boundary layers, 37
Upstream activator sequences (UASs), 150

**V**
Valproic acid (VPA), 549
Vanillate demethylase, 160
VanK, 294
VanR protein, 160
Variant histones, 528
Vectorial partitioning hypothesis, 457
*Vibrio* sp. 386
Virus- /microbe-associated cancer, 552
Viscosity, 34

**W**
Wastewater treatment plants, 215
Water contact angles, 399
Weathering, 38, 41
Wilms' tumour, 1 (WT1), 522

**X**
X chromosome inactivation (XCI), 534
Xenobiotics, 141
X-inactive specific transcript (XIST), 534
*m*-Xylene, 151, 290
XylN, 290
XylR protein, 151, 353, 356

XylS, 353, 354, 356
XylS protein, 153

**Y**
*Yarrowia lipolytica*, 60, 72
Yeast ribosomes, 581

**Z**
Z-DNA, 539
Zeta potentials, 22, 399
Zoospores, 310